The karoo
Ecological patterns and processes

The succulent and Nama-karoo form part of the arid south-western zone of Africa, a vast region of rugged landscapes and low treeless vegetation. Studies of this unique biome have yielded fascinating insights into the ecology of its flora and fauna. This book is the first to synthesise these studies, presenting information on biogeographic patterns and life processes, form and function of animals and plants, foraging ecology, landscape-level dynamics and anthropogenic influences. Novel analyses of the factors distinguishing the biota of the Karoo from that of other temperate deserts are given and generalizations about semi-arid ecosystems challenged. The ideas expounded, the ecological principles reviewed and the results presented are relevant to all those working in the extensive arid and semi-arid regions of the world.

RICHARD DEAN and SUZANNE MILTON are researchers at the Percy FitzPatrick Institute of African Ornithology, University of Cape Town.

The karoo
Ecological patterns and processes

Edited by W. Richard J. Dean and Suzanne J. Milton

CAMBRIDGE UNIVERSITY PRESS
Cambridge, New York, Melbourne, Madrid, Cape Town, Singapore,
São Paulo, Delhi, Dubai, Tokyo

Cambridge University Press
The Edinburgh Building, Cambridge CB2 8RU, UK

Published in the United States of America by Cambridge University Press, New York

www.cambridge.org
Information on this title: www.cambridge.org/9780521126878

First published 1999
This digitally printed version 2009

A catalogue record for this publication is available from the British Library

Library of Congress Cataloguing in Publication data

The karoo: ecological patterns and processes/edited by W. Richard J. Dean and
  Suzanne J. Milton.
    p. cm.
  Includes bibliographical references and index.
  ISBN 0 521 55450 0 (hardback)
   1. Ecology – South Africa – karoo. I. Dean, W. Richard J.
(William Richard John), 1940-. II. Milton, Suzanne J. (Suzanne Jane), 1952–.
QH195.S6K36 1998
577'.09687'15–dc21   98-25771 CIP

ISBN 978-0-521-55450-3 Hardback
ISBN 978-0-521-12687-8 Paperback

Contents

Part two Form and function   87
B. G. Lovegrove

6.   Form and function in perennial plants G. F. Midgley and F. van der Heyden   91

7.   Functional aspects of short-lived plants M. W. van Rooyen   107

9. Animal form and function   B. G. Lovegrove   145

10. Animal foraging and food   W. R. J. Dean and S. J. Milton   164

Part five Comparisons

18.   Comparison of ecosystem processes in the Nama-karoo and other deserts W. G. Whitford

19.   The succulent karoo in a global context: plant structural and functional comparison with North American winter-rainfall deserts K. J. Esler, P. W. Rundel and R. M. Cowling

20.   The karoo: past and future S. J. Milton and W. R. J. Dean

Contributors

Johan Booysen Research Institute for Reclamation Ecology, Potchefstroom University for CHE, Potchefstroom, 2520 South Africa

Ben Cousins Programme for Land and Agrarian Studies (PLAAS), School of Government, University of the Western Cape, Bellville, 7535 South Africa

Richard M. Cowling Institute for Plant Conservation, Botany Department, University of Cape Town, Rondebosch, 7701 South Africa

Robert A. G. Davies PO Box 1390, Halfway House, 1685 South Africa

W. Richard J. Dean Percy FitzPatrick Institute of African Ornithology, University of Cape Town, Rondebosch, 7700 South Africa

Phil G. Desmet Institute for Plant Conservation, Botany Department, University of Cape Town, Rondebosch, 7701 South Africa

Karen J. Esler Botany Dept, University of Stellenbosch, Private Bag X1, Matieland, 7602 South Africa

Howard Hendricks South African National Parks, PO Box 110040, Hadison Park, Kimberley, 8306, South Africa

Craig Hilton-Taylor Ecology and Conservation, National Botanical Institute, Private Bag X7, Claremont, 7735 South Africa

John H. Hoffmann Dept of Zoology, University of Cape Town, Rondebosch, 7700 South Africa

M. Timm Hoffman Ecology and Conservation, National Botanical Institute, Private Bag X7, Claremont, 7735, South Africa

Florian Jeltsch UFZ – Centre for Environmental Research, Postfach 2, D-04301, Leipzig, Germany

Klaus Kellner Dept of Plant and Soil Sciences, Potchefstroom University for CHE, Potchefstroom, 2520, South Africa

Graham I. H. Kerley Dept of Zoology, University of Port Elizabeth, PO Box 1600, Port Elizabeth, 6000 South Africa

J. Wendy Lloyd ARC-Institute for Soil, Climate & Water, PO Box 12455, Die Boord, 7613 South Africa

Barry G. Lovegrove Department of Zoology and Entomology, University of Natal, Private Bag X01, Scottsville, 3209 South Africa

Mike E. Meadows Dept of Environmental and Geographical Science, University of Cape Town, Rondebosch, 7701 South Africa

Theunis Meyer Northwest Agricultural Development Institute, Private Bag X804, Potchefstroom, 2520, South Africa

Guy F. Midgley Ecology and Conservation, National Botanical Institute, Private Bag X7, Claremont, 7735 South Africa

Suzanne J. Milton Percy FitzPatrick Institute of African Ornithology, University of Cape Town, Rondebosch, 7700 South Africa

Peter A. Novellie South African National Parks, PO Box 787, Pretoria, 0001 South Africa

Anthony R. Palmer ARC – Range & Forage Institute, PO Box 101, Grahamstown, 6140, South Africa

Ashia Petersen Ecology and Conservation, National Botanical Institute, Private Bag X7, Claremont, 7735 South Africa

Piet Roux PO Box 377, Middelburg, 5900 South Africa

Philip W. Rundel Department of Biology, UCLA, Los Angeles, CA 90024–1786, USA

W. Roy Siegfried PO Box 395, Constantia, 7848 South Africa

Andrew B. Smith Department of Archaeology, University of Cape Town, Rondebosch, 7701, South Africa

François van der Heyden ENVIRONMENTEK, PO Box 320, Stellenbosch, 7599 South Africa

Margaretha W. van Rooyen Dept of Botany, University of Pretoria, Pretoria, 0002 South Africa

Carl J. Vernon PO Box 19592, Tecoma, 5214 South Africa

Mike K. Watkeys Department of Geology and Applied Geology, University of Natal, Durban, 4041 South Africa

Walter G. Whitford US Environmental Protection Agency, National Exposure Research Laboratory, Characterization Research Division, Las Vegas, Nevada, USA (Postal address: USDA-ARS Jornada Experimental Range, Dept 3JER, New Mexico State University, Las Cruces, N. M. 88003, USA)

Thorsten Wiegand UFZ – Centre for Environmental Research, Postfach 2, D-04301 Leipzig, Germany

Christian Wissel UFZ – Centre for Environmental Research, Postfach 2, D-04301, Leipzig, Germany

Helmuth G. Zimmerman Plant Protection Research Institute, Private Bag X134, Pretoria, 0001 South Africa

Foreword

Since the turn of the twentieth century, and in particular since the 1920s, grave concern has mounted over the evident degradation of the semi-arid and arid rangelands of South and southern Africa. Climate change, desert encroachment, soil erosion, salinization of the few irrigated lands and, above all, overstocking by domestic livestock, were thought to be responsible for changes in the apparent production potential of the land. This led to a number of research programmes being set up by the Department of Agriculture, all designed to obtain some predictive understanding of the biology of organisms in the southern African arid and semi-arid rangelands, particularly the karoo. The karoo covers 35% of South Africa, and extends into neighbouring Namibia, and, as such, represents a significant proportion of southern Africa. Despite the accumulation of a large amount of knowledge on vegetation, grazing effects and the management of rangelands in the karoo over several decades, there still remained a serious gap in understanding the dynamics of vegetation and plant–animal interactions in this region.

The various 'Biome Projects' (see, for example, Scholes and Walker, 1993), set up by the South African National Programme for Ecosystem Research from the early 1970s, were designed to develop some understanding of ecosystem functioning in the major biomes and inland waters in South Africa. Research within the biome projects focussed on climate and soils, adaptive physiology and behaviour, reproductive biology, population dynamics, species interactions and community processes, and the protection of biodiversity and ecosystem modelling. Projects were funded mostly by the National Programme for Ecosystem Research, the Department of Environment Affairs and Tourism and the Department of Agriculture through the universities, nature conservation organizations and NGOs.

Despite the concern over the 'degradation' of the karoo, and the perception that agriculture (*sensu lato*, including ranching) in the karoo was less productive than it could have been, the Karoo Biome Project was only set up in 1985 (Cowling, 1986) and had been largely dismantled by the early 1990s. It was thus the youngest and most short-lived of the biome projects, but it produced exciting new findings and provided training for many biologists at a relatively low cost. Goegap Nature Reserve and the Richtersveld National Park in Namaqualand, where studies on plant and animal population dynamics were carried out, the Worcester Veld Reserve in the south-western succulent karoo, where numerous studies of vegetation and soils were done, Grootfontein Agricultural College at Middelburg, where research on management of karoo rangelands was carried out, and the Tierberg Karoo Research Centre at Prince Albert, where more academic research on plant physiology, vegetation dynamics, plant–animal interactions and animal population dynamics was carried out, are among the places in the karoo where many studies were successfully completed. All these sites, with the exception of Tierberg, were in use by researchers before the Karoo Biome Project was initiated. Recently, research in the karoo has focussed on natural resources in communal rangelands. Research in the karoo is never easy, with a harsh and unpredictable climate, rough terrain and long distances to contend with, but the rewards were often great, and results of experiments often totally unexpected.

The information on patterns and processes in the karoo is fairly widely scattered and has never been synthesized or made available in a single volume. The objectives of this book are to succinctly review the state of knowledge of patterns and processes in the karoo. The book is primarily aimed at researchers, lecturers, graduate students, conservationists and other land managers in southern Africa and elsewhere. This review differs from others in that, in addition to information on biogeographic patterns and life forms, a great deal of the book focusses on form and function of plants and animals, foraging ecology, landscape-level dynamics of plants and animals, models that simulate the dynamics of various organisms, and Man's past, present and perhaps future place in the karoo. The ideas expounded here, the ecological principles reviewed and the results that are presented here are applicable and relevant to a large number of more or less similar sites and conditions elsewhere in the world.

The authors of the various chapters can all be regarded as experts and experienced in their fields of research. It stands to reason that all the aspects pertaining to the karoo could not be treated in the space of a single book. Of special interest to me, as these fit in more with my field of interest and experience, are the chapters on the driving variables, plant biogeographic patterns, animal foraging and food, modelling karoo populations and dynamics, human impacts, historical and contemporary land use, and comparisons of ecosystem processes in the karoo and other deserts. From a South African point of view, this book can be regarded as a significant advance in the understanding of the karoo.

This book, and the reports and papers published by the various contributing authors, are prime examples of what can be achieved through a small amount of funding in the hands of enthusiastic and dedicated researchers. Many projects in the karoo, have been extremely cost-effective, and the scientific outputs from such projects have been of a high standard. As such, results have been incorporated into the curricula of universities and other institutions of learning. From a practical point of view, results from research in the karoo have contributed significantly to the interpretation of past land use and to policy for land use in the present and future.

As a member of the 'old school' of biologists, to whom field work was the most important part of the project, I am appalled at the fact that field studies have been relegated to subordinate positions by today's young biologists. Reworking old databases has become a popular pastime (and in some cases with exciting new interpretations), but there is the danger that no new databases are being assembled at the same time. Breaks in historic databases are disastrous and irretrievable, and cannot be contemplated in ecosystems, such as the karoo, where turnover is slow and projects on population dynamics almost worthless unless funded over long periods. It is essential, for the sake of further advancement of ecosystem research, environmental conservation and management in the karoo, that adequate funding and support for research is at least maintained, if not increased. Funding for research in this region has always been a problem and the Karoo Biome Project was always the 'Cinderella' of the South African Biome Projects. This book is an example of the results that can be obtained through personal motivation, initiatives, a positive approach and a burning curiosity about South African arid and semi-arid ecosystems. There never was much money, but it was money that was very well spent.

P. W. Roux,
Former Chairman of the Karoo Biome Project Steering Committee

Preface

The arid south-western zone of Africa, which includes the Namib Desert, the arid savanna of the Kalahari, and the succulent and Nama-karoo (Fig. 0.1), is a vast region of rugged landscapes and low treeless vegetation, bounded on the west by the cold Atlantic coastline, in the south by the winter rainfall fynbos and evergreen forest biomes, and in the north and east by arid and mesic savannas. The fauna and flora of the karoo, *sensu lato*, combine elements from the desert, arid and moist savannas, grasslands and, in sheltered sites, from the forest. There is a gradient from succulent dwarf shrublands to woody dwarf shrublands and to grasslands. The animals of the karoo have been drawn from the surrounding biomes, and the level of endemism among the best-studied groups is not high. Conversely, although the vegetation has been similarly drawn from surrounding biomes, there is a high percentage of endemics, particularly in the succulent karoo.

The karoo is an ancient landscape. The varied rocks and sediments that underlie it span 500 million years of geological time and range from glacial moraines and lacus-

tral deposits to recent aeolian sands. Fossil-rich sediments bear testimony to the changing environments through which the karoo has passed. Within the archaeological record, recent climate change has modified plant distributions, animal assemblages and human behaviour. The karoo has a long history of utilization by hunter-gatherers and herders whose populations remained low, whose shifting settlements were unstable, and whose impact on the landscape was localized in space and time, like those of the indigenous and naturalized plants and animals on which these peoples depended.

Two hundred years ago, the karoo was colonized by peoples of European origins, who brought with them agricultural traditions, livestock and crops more appropriate for a less stochastic mesic climate. Settled agriculture dependent on underground water combined with ploughing of alluvial soils for dryland crops has since changed the structure and composition of karoo habitats and biota (Roux and Vorster, 1983; Macdonald, 1989; Hoffman and Cowling, 1990b; Milton and Hoffman, 1994; Dean and Macdonald, 1994; Dean and Milton, 1995; Steinschen et al., 1996). Ploughing lands to plant crops was a novelty in the karoo ecosystem, and a largely unsuccessful experiment during the first 150 years of occupation (Macdonald, 1989; Dean and Milton, 1995). Grazing by domestic livestock was thought to be sustainable in the karoo and to be the best agricultural use for this arid region, but it, too, has associated problems.

Research in the karoo was motivated by the need to develop a predictive understanding of ecosystem functioning so that this knowledge could be applied to grazing management systems and thus increase, through sound management, the proportion of the gross national product that came from the karoo. Research in the karoo was in four phases:

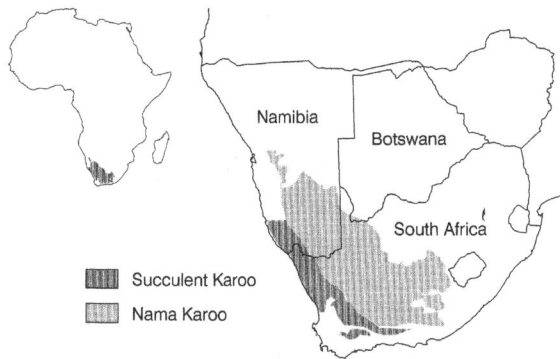

Figure 0.1 **Southern Africa, showing the succulent and Nama-karoo biomes.**

- Up to the early twentieth century, most biologists simply collected organisms in the karoo.
- From about 1920 to the early 1950s, applied, rather than basic research was encouraged because of the perceived need to solve problems associated with grazing or aridification (Anon, 1923, 1951; Schumann and Thompson, 1934; Wallis, 1935; Kokot, 1948; Tidmarsh, 1948).
- From the 1950s to the 1980s, research focussed on succulent plants, centres of endemism and the biogeography of plants in the karoo. This was really phytosociology on a broad scale.
- Since the 1980s, research has been directed towards gaining an understanding of ecosystem processes in the karoo (Cowling, 1986).

Conceptual models of ecosystem function are urgently needed for conservation and land use planning in the karoo, and for addressing the question of how the vast and biologically diverse, but unproductive karoo region should be used in a country with a growing land-hungry population.

In this book, we have attempted both to bring together the findings of basic and applied ecological research in the succulent and Nama-karoo, and to highlight fields that are still poorly known. Subject reviews have been grouped into those dealing with: broad-scale geographical patterns (that set the biotic and physical stage for the book); the links between form and function in living organisms characteristic of the region; population and community dynamics; and brief reviews of the past and present influences of humans on the karoo ecosystems. The final section presents comparisons between the karoo and other similar arid regions.

W. R. J. Dean and S. J. Milton

Acknowledgements

Financial support for this research was provided by the Department of Environmental Affairs and Tourism, the Foundation for Research Development, WWF SOUTH AFRICA (formerly the Southern African Nature Foundation) and the University of Cape Town. The Foundation for Research Development also funded the participation of the authors at a workshop. We thank the following for advice, refereeing chapters, providing additional information, and for providing technical support: William Bond, Mike Cameron, Richard Cowling, Morné du Plessis, Danelle du Toit, Eryn Griffin, Phil Hockey, Timm Hoffman, Graham Kerley, Sue Jackson, Steve Johnson, Norbert Jürgens, Peter Linder, Jeremy Midgley, Norman Pammenter, Dave Richardson, Piet Roux, Roy Siegfried, Vivienne Stiemens, Willy Stock, Chris Tobler, Gretel van Rooyen, Jan Vlok and Mike Watkeys.

We thank Phil Hockey for facilitating the authors workshop, Timm Hoffman for arranging the venue for the workshop and Caz Thomas for acting as the scribe at the workshop.

Biogeographic patterns and the driving variables

S. J. Milton and W. R. J. Dean

More species of plants are packed into the small succulent karoo landscape than in any other desert on earth, yet the vegetation is strangely homogeneous in appearance. This pattern is not repeated in the Nama-karoo, or reflected in the vertebrates of either karoo biomes.

The geological history of the two biomes is similar, being sedimentary, but influenced by folding along the southern and western edges of the karoo, and by igneous intrusions in the northern and eastern parts. The soils of both biomes are typical of arid regions, and are poorly developed with little organic matter. Folding, endorrheic drainage and wind-blown movement of sand has resulted in localized patchiness in substrata which influences the biota.

Anomalous soils developed along the present inland boundary between the winter and summer rainfall regions indicate past shifts in rainfall seasonality (Watkeys, this volume). Pollen core data for the past 10 000 years that indicate fluctuations between grass and shrub dominance have occurred repeatedly throughout the karoo (Meadows and Watkeys, this volume). The grass : shrub ratio still follows patterns in present-day rainfall seasonality (Palmer et al., this volume). There can, however, be little doubt that the climate of the western succulent karoo has fluctuated less in the recent past and is currently more predictable than that of the Nama-karoo (Desmet and Cowling, this volume).

The biota of the Succulent and Nama-karoo appear to have different geographical origins (Cowling and Hilton-Taylor, this volume; Vernon, this volume). Desert and fynbos genera are prevalent among succulent karoo plants, arachnids, insects and reptiles, whereas Nama-karoo genera have been drawn from savanna, grassland and forest, as well as from the adjacent succulent karoo and desert biomes.

Climatic patterns offer some insight into reasons for the explosive speciation of succulent plants and geophytes in the western parts of the succulent karoo. In their chapter on plant biogeograhy, Cowling and Hilton-Taylor point out that the predictability of the winter rainfall in the succulent karoo is unique among desert regions. This peculiar climate has favoured short-lived shrubs that store winter moisture in their leaves but die when drought-stressed. Limited lifespan and continuous turnover in populations result in lottery replacement and minimize competitive interactions resulting in structural similarity among species. Short seed dispersal distances and dependence on insect pollinators has favoured speciation. Succulent karoo plant communities are thus species-rich, but uniform in structure. Insects and the invertebrates and reptiles that prey on them show centres of species-richness in the succulent karoo that may reflect patterns of plant speciation. For example, Vernon (this volume) shows that indigenous solitary bee assemblages are most species-rich in Namaqualand, and species forage on a limited array of plants. Similarly, the masarid wasps and vespid wasps include many geographically restricted species that forage on the pollen of Mesembryanthema.

Some of the ideas discussed in the next five chapters are of necessity speculative, because the disciplines of soil science, palynology and metereology are undergoing rapid development in the karoo. Analyses of biogeographic patterns are severely hampered by systematic problems in the most speciose groups, and by many undescribed taxa among the invertebrates. Nevertheless, as the following five chapters show, advances in knowledge and techniques over the past few years have changed our perception of the ways in which climates and soils are influencing the evolution and distribution of karoo biota.

1 The climate of the karoo – a functional approach

P. G. Desmet and R. M. Cowling

1.1. Introduction

The arid lands of southern Africa occupy the area west of approximately 27 °E and north of 34 °S. North of approximately 27 °S, the arid zone becomes confined to coastal belt and plateau of southern Namibia. North of 22 °S, this zone is confined to the hyper-arid coastal belt of the Namib Desert and stretches as far north as 12 °S in southern Angola. The climate is dominated by, and indeed, the aridity is largely caused by, the southern subtropical high-pressure (anticyclone) belt. To the south, the region is influenced by the circumpolar westerly airstream (Schulze, 1965). Only the southern and south-western arid regions are influenced by this belt of temperate cyclones. Local modifications occur as a result of the Cape Fold Mountains in the south, the mountains of the Great Escarpment, the raised interior plateau and the cold, north-flowing Benguela current that washes the west coast of the subcontinent. Incursions of moisture into the region are associated with the advection of air across the warm Indian Ocean (maximum precipitation from this source is largely confined to the eastern parts of the sub-continent) and the circumpolar westerlies to the south. Southern African arid lands are geographically marginal to these rain-producing systems.

The first part of this chapter provides an overview of the contemporary climate of the region. The second part provides a description of the weather systems that influence the karoo. This approach is taken to highlight the great diversity of systems that are responsible for the varied karoo climates, something which has not been appreciated by earlier reviews. This functional understanding of the weather patterns in the karoo is essential for understanding the landscape–vegetation patterns (i.e. at the level of biome and veld type). In the final part of this chapter, we present a new analysis of the climate of the karoo, which comprises a multivariate model that illustrates vegetation–climate relationships quantitatively. We do not, however, discuss long- and medium-term changes in climate, nor predictions of future climate change.

1.2. A general overview of the climate of the karoo

The climate of the karoo is summarized in the form of climate diagrams (Figs. 1.1(a) and 1.1(b)). The focus of this section is a discussion of the three primary limiting climatic factors which influence plant growth in arid lands, namely precipitation, temperature and light (Schulze and McGee, 1978).

1.2.1. Precipitation

Rainfall
The overall feature apparent from the distribution of mean annual rainfall in the karoo is that south of the Tropic of Capricorn, precipitation decreases uniformly westwards from the eastern escarpment across the plateau. Only in the extreme south do the isohyets follow an east-west trend. This is due to the topographic irregularities of the Cape Fold Belt and associated orographic rain linked to westerly frontal and post-frontal systems. In the west, north of the tropic, the isohyets follow a similar north–south trend, although steeper; thus, in the Namib Desert of southern Angola, coastal stations receive <100 mm rainfall, whereas stations 200 km to the east on top of the escarpment receive in excess of 800 mm

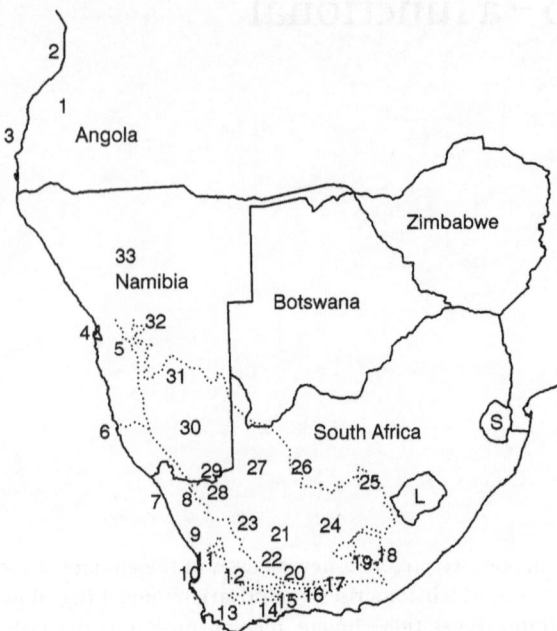

Figure 1.1(a) **Map of southern Africa showing the localities of numbered weather stations from which climate data were obtained to generate the climate diagrams shown in Fig. 1.1(b) on p. 6. The boundary of the karoo sensu lato is shown as a dotted line. L = Lesotho, S = Swaziland**

- a change in amplitude, thus higher or lower peaks in monthly rainfall, or more pronounced seasonality;
- a shift in phase (winter- or summer-rainfall maxima or minima), which is essentially a change in seasonality;
- and, alternatively, a change of curve shape from the general parabolic curve to sinusoidal or aseasonal.

Thus, within the karoo, for the same given mean annual rainfall there are a number of different possible combinations of rainfall distribution (Figs. 1.1(a) and 1.1(b)). Coupled to this variation are differences in reliability and intensity of rainfall events. This diversity in the occurrence of rainfall regimes arises as a result of the location of the southern African arid zone between two weather systems. The regional weather patterns that are responsible for this variation are discussed in more detail in the following sections of the chapter. The biogeographical implications of this diversity in rainfall regimes is discussed in section 1.7.

annually (Figs. 1.1(a) and 1.1(b)). The desert biome, namely the Namib Desert, occupies a narrow range of mean annual rainfall (<100 mm) whereas the succulent and Nama-karoo experience a broader range in rainfall, between 50 and 500 mm and 50 and 600 mm respectively (Fig. 1.2).

At a finer scale, the distribution of mean monthly rainfall across the subcontinent highlights two import trends in the annual march of rainfall, namely, summer and winter maxima regimes. Over most of the eastern parts of the country, summer rainfall regimes dominate. Over the western interior, to the east of the escarpment, and north of the Cape Fold Belt, this summer maximum is less marked and grades into a predominately winter-rainfall regime, or Mediterranean climate, in the south and south-west mountains and coastal belt.

At an even finer scale, considered much more meaningful in terms of the distribution and characteristics of plant cover than the general patterns already discussed, are the duration, time of occurrence, and the degree of intensity of the rainy season and dry season; particularly when viewed from the perspective of plant available soil moisture being able to meet evaporative demand. The annual march of mean monthly rainfall across the karoo shows a number of distinct patterns (Figs. 1.1(a) and 1.1(b)). These patterns can be summarized in terms of three distinct parameters, namely amplitude, phase and shape:

Fog

Fog is an important alternative source of moisture for plants. Although there is generally no direct precipitation of water on the soil surface, interception by vegetation may lead to significant amounts of water entering the soil below the plant. Fog 'precipitation', although recognized as significant (Schulze and McGee, 1978; Walter, 1986; Werger, 1986; Lancaster et al., 1984; Pietruszka and Seely, 1985; Olivier, 1995), is not measured by standard rain gauges, and thus its importance in ecological studies has been very difficult to estimate. The measurement of moisture derived from fog is dependent on the type of obstacle used to catch fog moisture (Walter, 1986). Finely branched structures that transmit wind, such as a fine mesh or the canopy of a shrub, are much more efficient at combing out fog moisture than a smooth, solid structure.

Advective sea fog is characteristic of the entire west coast of the subcontinent, essentially the coastal Namib Desert (including both winter-rainfall strandveld, lowland succulent karoo (Low and Rebelo, 1996) and summer-rainfall portions of the desert) (Schulze and McGee, 1978; Olivier, 1995). Locally, fog is referred to by a number of different names: *Cacimbo* in Angola (Jackson, 1951), *Nieselregen* in Namibia (Walter, 1986) and *Mal-mokkie* in Namaqualand (A. Kotze, personal communication). We do not know of any studies on the occurrence or significance of fog (radiation fog) elsewhere in the karoo.

Clouds form when air is supersaturated with respect to water or ice (Preston-Whyte and Tyson, 1988). One manner in which this can occur is by the mixing of air. Advection fog occurs when warm air with high relative humidity is advected over a cool surface. The temperature differential between air and surface must be sufficiently large to enable the air to reach saturation after a small amount of cooling. Medium velocity winds are also necessary in the advection process, since strong winds would cause too much turbulence and vertical mixing to maintain the fog, whereas low wind speeds would provide too little advection and mixing. When air over the Atlantic Ocean moves across the leading edge of the cold Benguela current, temperature is depressed to dew point and fog forms. The coastline constitutes another leading edge with air moving over a hot, arid desert. Inland movement of fog is therefore limited by the arid nature of the new surface conditions, and the fog thins and evaporates downwind. By day, this process is hastened by surface heating.

The predominance of colder coastal ocean surface temperatures during summer, as a result of the seasonal intensification of the mid-Atlantic Ocean high, creates conditions more favourable for fog formation. The dominant flow of air during this period is westerly, thus warm moist air from the mid-Atlantic is cooled near the coast and fog forms. As midsummer wind velocities are too high to maintain the integrity of the fog bank, the coast experiences fog predominantly during spring and autumn, when the wind velocities are lower, but the flow of air is still predominately onshore.

The frequency of fog occurrence along and perpendicular to the coast varies considerably (Olivier, 1995). Using Meteosat images, Olivier (1995) estimated the highest occurrence to be between Sandwich Bay and Cape Cross in the central Namib with an excess of 100 days per year. South of the Orange River, the value is less than 75 days and in southern Angola less than 50 days. Fog also penetrates as far inland as the foothills of the escarpment and beyond where less than 10 fog days may be expected. Major river courses, such as that of the Orange River allow fog to penetrate deeper into the valleys and foothills of the escarpment mountain ranges than elsewhere. There is, however, little quantitative understanding of how fog is distributed in the landscape south of the Orange River.

Fog also plays an important amelioratory role in the local climate. From South African Weather Bureau data, the average total number of days per annum during which fog is recorded at Port Nolloth is 148, or 41% of the total days. As a result, the sunshine duration averages less than 70% of the possible total and this has a significant ameliorating effect (Burns, 1994).

The potential amount of water that can be derived from a fog event, relative to the mean annual rainfall, is substantial. For Swakopmund, with 121 fog days per annum, the amount of water intercepted in 1958 was equivalent to 130 mm of rainfall. More than seven times the mean annual rainfall (Schulze and McGee, 1978), but this amounts to an average of <1.0 mm (average of 0.2 mm) per fog event (Walter, 1986). Minimum and maximum annual fog-water totals along a latitudinal transect from Walvis Bay to Gobabeb were 49–158 mm (Rooibank, 20 km inland); 88–271 mm (Swartbank 40 km inland); and 8–48 mm (Gobabeb 60 km inland). The annual coefficient of variation for fog at the three same stations was 29%, 29%, and 36%, respectively, whereas that for rainfall was 123% at Gobabeb and 106% at Walvis Bay (Pietruszka and Seely, 1985). These coefficients for rainfall and fog are similar to those for the southern Namib (Desmet, 1996). Fog is a potentially significant source of water in the desert environment, and also a far more predictable source of moisture than rainfall (Pietruszka and Seely, 1985).

If and how desert plants derive any benefit from fog moisture is unclear. There is, however, no evidence for direct uptake of fog condensation on leaves by plants (Danin, 1991). A notable exception is *Trianthema hereroensis* from the sand erg of the central Namib Desert (Louw and Seely, 1982). Von Willert et al. (1990, 1992) argue that any leaf structure capable of absorbing water on the leaves is also a potential route via which water can evaporate. Thus, there would be little benefit for plants in a hyper-arid environment to absorb fog moisture directly from the leaves. A more likely route whereby plants could benefit from fog moisture would be by absorbing condensation on the sand surface (Danin, 1991) and as a result of stem flow. This route would facilitate the uptake of both fog and dew condensation on the soil surface. Louw and Seely (1982) sprayed tritiated water on the top 1 cm of soil near *Salsola subulicola* growing in the Namib, and found efficient water absorption by the plant. Certain plants growing in the fog zone of the Namib have well-developed superficial root networks (Danin, 1991) or efficient mycorrhizal relationships to be able to benefit from alternative moisture sources such as fog and dew (see below).

Dew

In the absence of coastal advection fog, the potential still exists for plants to obtain moisture from heavy dews. Within the karoo, the occurrence of dew is a more widespread phenomenon than fog (Werger, 1986). Although it is a parameter that is difficult to quantify, moisture derived from dew condensation on plants and the ground is probably significant and worthy of some investigation.

Dew-point temperature is that to which air at a constant pressure and water vapour content must be cooled in order to become saturated and for dew to precipitate (Preston-Whyte and Tyson, 1988). At night, radiative cooling of the air to below dew-point temperature causes dew to form on the ground. The extraction of water vapour from the overlying air causes an inversion to form in the water vapour profile. The depth and strength of this inversion is determined by the downward flux of water vapour

1. LUBANGO	2. LOBITO	3. NAMIBE	4. WALVIS-BAY
14°56' 13°34' 1761m	12°22' 13°32' 1m	15°12' 12°09' 45m	22°57' 14°30' 2m
904.0mm 18.4°	239.0mm 24.1°	43mm 20.7°	15.2mm 17.3°

5. GOBABEB	6. DIAZ-POINT	7. PORT NOLLOTH	8. OKIEP
23°34' 15°03' 407m	26°38' 15°06' 16m	29°15' 16°52' 5m	29°36' 17°53' 915m
27.2 21.1°	17.3mm 15.9°	66.1mm 14.9°	159.6mm 17.9°

9. GARIES	10. LAMBERTS BAY	11. VREDENDAL	12. ELANDSVLEI
30°34' 18°00' 240m	32°02' 18°20' 92m	31°40' 18°30' 30m	32°19' 19°34' 121m
141.4mm 18.7°	222.3 mm 16.3°	145.5mm 18.8°	94.7mm 20.5°

13. AAN-DE-DOORNS	14. RIVERSDALE	15. OUDTSHOORN	16. WILLOWMORE
33°42' 19°29' 220m	34°06' 21°16' 128m	33°35' 22°12' 315m	33°18' 23°29' 831m
247mm 17.6°	433.6mm 17.2°	248.8 mm 17.5°	272.7mm 16.1°

17. JANSENVILLE	18. QUEENSTOWN	19. CRADOCK	20. BEAUFORT WEST
32°56' 24°40' 427m	31°54' 26°52' 1067m	32°10' 25°37' 875m	32°21' 22°35' 899m
264.8mm 18.7°	496.5 mm 16.4°	315.7mm 16.8°	240.8 mm 17.5°

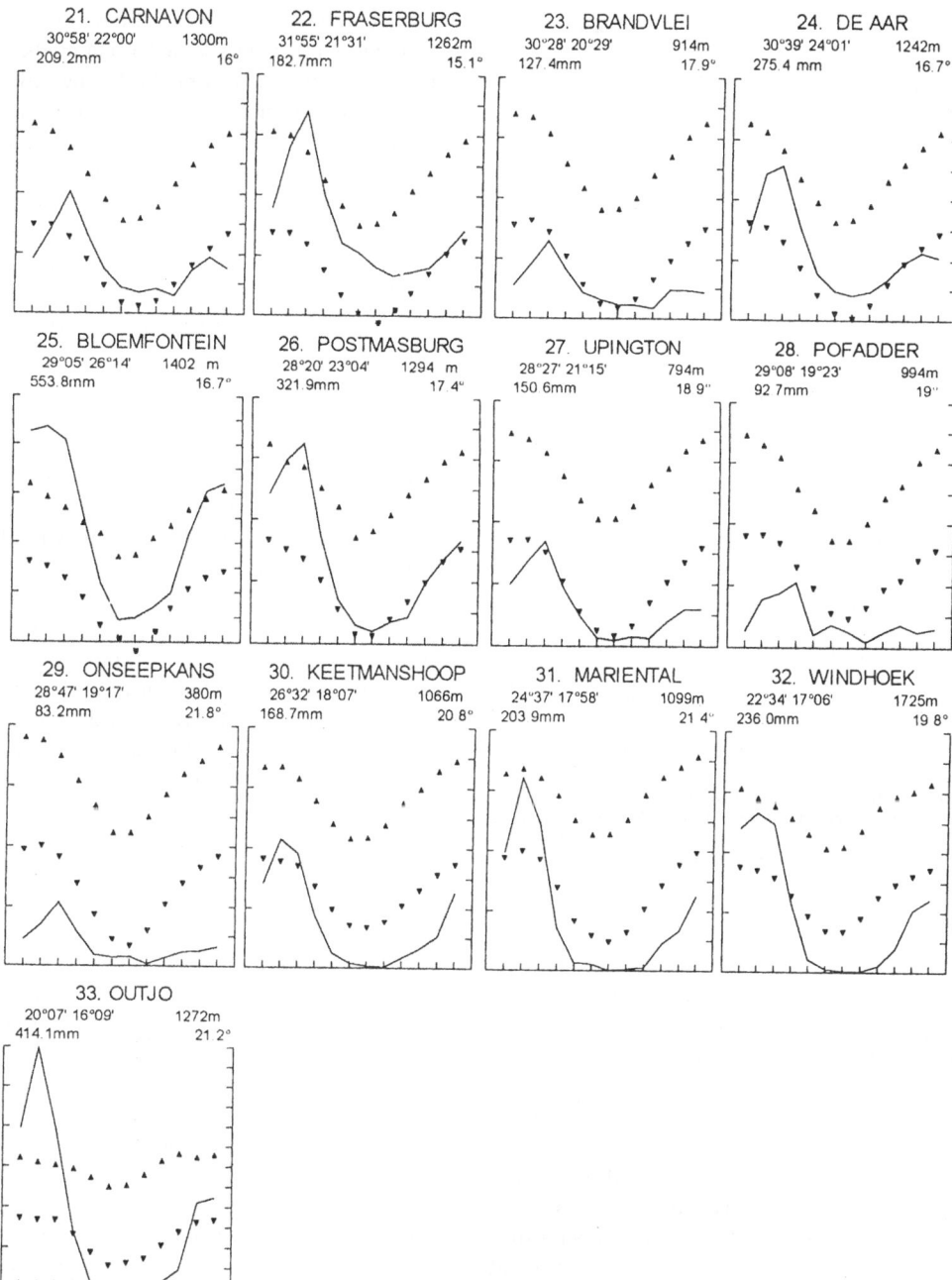

Figure 1.1(b) **Climate diagrams for selected weather stations in the karoo and Namib Desert, and surrounding non-arid zone vegetation types. The title for each diagram contains the town name, co-ordinates, altitude (m), mean annual rainfall (mm) and mean annual temperature (°C). Points on each graph represent mean monthly maximum and minimum temperature, and the curve mean monthly rainfall. In all cases, the rainfall scale (mm), in increments of 10 units, equals 2 x that of temperature (°C). Months on the horizontal axis are from January to December**

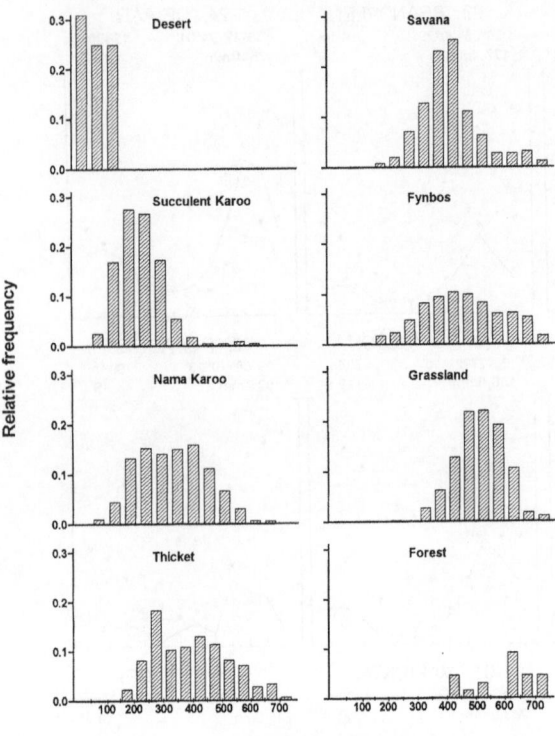

Figure 1.2 **The relative frequency distribution of mean annual rainfall for rainfall stations in the arid zone of southern Africa in relation to surrounding biomes. The data used in this figure were obtained from the Computing Centre for Water Research (University of Natal) database for South Africa weather stations. Additional data for Angola and Namibia were obtained from Lebedev (1970) and the South African Weather Bureau, respectively. Stations were classified according to biome and Veld Type**

in a suitable turbulent environment. Thus, the level of turbulence is critical. If it is too low (i.e. calm conditions), dew ceases to form since the ground cannot be replenished by water vapour from above. If the turbulence is too high, mixing inhibits surface radiative cooling to below dew-point temperature.

The probability of occurrence of heavy dews would be highest when the difference between mean minimum monthly temperature and dew-point temperature is smallest combined with the highest mean monthly relative humidity values and lowest mean monthly night-time wind speeds. These conditions are most frequent during the autumn (April–May) in summer-rainfall areas, and during mid winter (July–August) in winter-rainfall areas. Overall, dew as a phenomenon, and especially as a potential source of water not only for higher plants, but more so for the lower plant components of soil crusts (e.g. algae), has received little investigation.

The amount of moisture delivered by dew and its utilization is uncertain, but dew differs measurably from rain in terms of its predictability. Long-lived perennial plants

would only be able to survive in a desert receiving less than 50 mm yr^{-1} if there was some form of predictability in the moisture regime. Low rainfall is highly variable. Dew is a common occurrence, but how much water this makes available to plants is unknown. Fog is potentially a substantial water source and its predictability is far greater than that of rainfall. Plants inhabiting the fog zone of the southern Namib Desert should possess a unique suite of ecological characteristics of morphological/physiological features which enable them to utilize these alternative sources of moisture. This is a ripe area for further research.

1.2.2.   Reliability of sources of moisture

The reliability of different sources of moisture across the karoo has important ecological implications. In 1.2.1. (Fog), it was shown that, for the west coast, fog as a source of moisture for plant growth is far more reliable in terms of frequency or predictability of occurrence than rainfall. It is not known if the same holds true for dew.

Rainfall across the karoo decreases from east to west and from south to north. Similarily, rainfall variability, expressed as co-efficient of variation (cv), follows a similar trend. This is to be expected, since cv is log-linearly related to mean annual rainfall (Fisher, 1994). What is more interesting to compare is cv for different stations with the same mean annual rainfall. In Fig. 1.3, the cv of mean annual rainfall is compared between stations in the Nama- (summer rain) and succulent (winter rain) karoo. On average, the rainfall in the winter rainfall karoo is 1.15 times more reliable than corresponding rainfall in the summer-rainfall karoo. This difference has important implications for the type of plant life-history strategies and plant community structure and dynamics prevalent in the different regions of the karoo (e.g. Hoffman and Cowling, 1987; Cowling and Hilton-Taylor, this volume).

1.2.3.   El Niño in southern Africa

El Niño is a phenomenon that usually begins with the relaxation of the normally intense easterly trade winds that drive the westward equatorial surface currents and expose cold waters to the eastern Pacific surface (Philander, 1992; Preston-Whyte and Tyson, 1988). When these winds relax, they allow the warm surface waters, that have piled up in the western Pacific, to surge eastwards taking with them the region of heavy rainfall. Thus, the central Pacific, usually an arid zone, receives abnormally high torrential rains. In contrast, eastern Australia and the western Pacific islands, the usual recipients of these rains, experience drought. However, this effect is not restricted to the Pacific, but is linked to similar phenomena in both the Indian and Atlantic Oceans by what has been termed the Southern Oscillation (Philander, 1992),

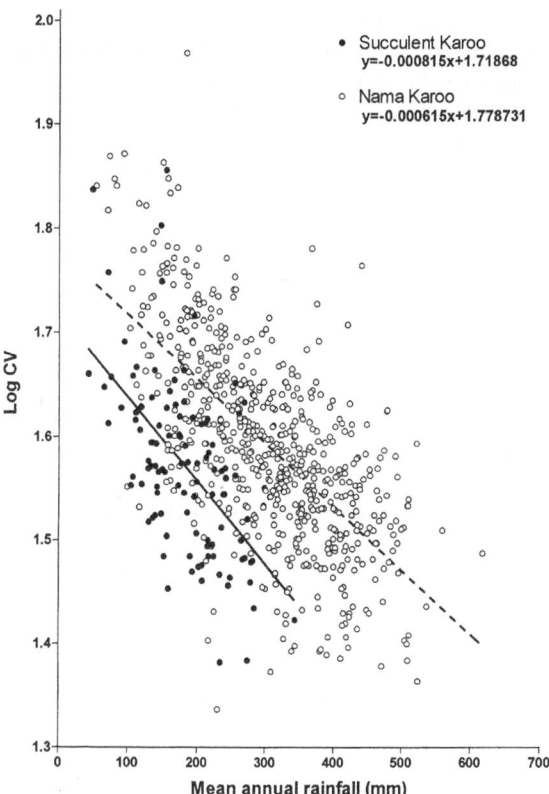

Figure 1.3 **Co-efficient of variation (CV) of mean annual rainfall for the succulent karoo vs. Nama-karoo. Equations for the curves are** $y = 0.0008148x + 1.71868$ **and** $y = 0.0006154x + 1.778731$**, respectively. The slopes are not significantly different (**$p = 0.1004$**) but the intercepts are significantly different (**$p = <0.0001$**)**

an irregular, interannual fluctuation in ocean air pressure.

The implications for southern Africa are profound. With the relaxation of the easterly Trades, the inflow of moist tropical air over the subcontinent subsides. Thus, regions of the karoo that rely on this tropical source of moisture, essentially the summer-rainfall karoo, experience abnormally low rainfalls. By contrast, the belt of westerly cyclones that brings winter rain to the region remains unaffected by the Southern Oscillation. This, in part, could explain the differences in reliability of annual rainfall in the Nama- and succulent karoo discussed in the previous section.

1.3.   Temperature

There are five major characteristics in the distributions of mean annual temperature across the subcontinent (Schulze and McGee 1978):

- an expected overall temperature increase towards the equator;
- isotherms parallel to the coast over most of the area, which exhibit decreasing values with distance inland, reflecting the effects of continentality;
- the effects of the cold Benguela and warm Agulhas currents moving northwards and southwards on the west and east coast, respectively;
- the temperature irregularities induced by topographic variation on the subcontinent, for instance the lower temperatures along the escarpments on the perimeter of the subcontinent (<14 °C) or the higher temperatures along the Orange River valley (>22 °C);
- highest mean annual temperatures which occur in areas with highest continentality, namely the Orange River trough. These areas also experience the greatest range in mean annual temperature (Werger, 1986);

The annual range of temperature shows a matching characteristic with the smallest ranges (<6 °C) along the west coast; and the greatest values (>16 °C) over the southern Kalahari and northern karoo, where the ameliorating effect of cloud is generally absent. High temperatures, low relative humidity and little to no cloud cover is characteristic of the karoo, especially the central areas. This results in large annual and daily ranges in temperature. This is a characteristic of arid climates generally (McGinnis, 1979). The exception is the west coast where there is an abundance of moisture in the air due to the predominately onshore sea breeze, relative humidity is high, and temperature is regulated by the cold Benguela current

The annual march of temperature in the region reflects both the coastal and continental patterns characteristic of the subcontinent. Coastal stations along the west coast generally show a lag of one month in maximum and minimum temperatures; thus, February and August are the two extremes. This is due to the lag in the heating and cooling of the ocean current, which exerts a strong regulating effect on these coastal climates. These coastal stations also show the temperature anomaly of recording their highest maximum temperatures in midwinter due to the sudden heating effect of warm berg winds blowing off the plateau this time of year (Schulze, 1965).

Mean daily minimum temperatures in the region are highest along the west coast and increase equatorwards with again the escarpment and high-lying areas of the plateau showing the coldest minimum temperatures (Schulze and McGee, 1978). In the karoo, the lowest mean minimum temperatures are found towards the centre of the subcontinent, and show the effect of both continentality and altitude (Werger, 1986). With the exception of the coastal and northern subtropical Namib Desert, the entire karoo area falls within the line of 50% probability of receiving frosts during winter (Schulze, 1965). High-lying areas of the escarpment and central plateau are especially frost-prone.

1.4.   Cloud and light

At the regional scale, light is not considered limiting to plant growth in the karoo. Skies are normally clear and sunshine is abundant. Cloudiness is normally at a maximum during the morning hours in the winter rainfall areas, including parts of the central Namib (Schulze, 1965), but an afternoon maximum is normal for most of the summer rainfall area. During winter, cloudless conditions in the central karoo may persist for weeks on end.

The average annual duration of bright sunshine is more than 80% of the daylight hours in most of the region. Along the coast of Namibia, the average duration may drop below 50% in some places owing to fog and low cloud (see 1.2.1. *Fog*). Despite the abundance of light, energy for growth is limiting for many organisms growing during the cool winter months of the succulent karoo. Plant traits such as the northward curvature of the stem in *Pachypodium namaquanum* (Rundel et al., 1995), psammophily across a range of genera (Jürgens, 1996) and spiral-surfaced leaves in bulbous monocots, have been suggested as being adaptations to maximizing energy absorption by these winter growing plants (Midgley and Van Der Heyden, this volume).

1.5.   Wind

The prevailing wind direction along the west coast is parallel to the coast, predominately from the southern quarter. Ecologically, these southerly sea breezes and the frequent warm, dry offshore berg winds play an important driving role in this arid coastal system (Desmet, 1996; Louw and Seely, 1982; Lancaster, 1989). In the western interior, winds in summer are mainly from the south-west and in winter from the north; in the eastern interior they are south-easterly and north-westerly, respectively. Mostly, these winds are dry. Dust devils and small whirlwinds are frequent in the interior in summer, but dust-storms are uncommon except in the coastal belt in winter due to berg winds. In the interior, winds are, for the most part, local in nature, such as valley winds due to local topographic relief (Schulze and McGee, 1978; Werger, 1986).

1.5.1.   Berg winds

Berg winds are important features of coastal climates and are associated with large-scale pre-frontal divergence and dynamic warming of subsiding air moving offshore from the plateau (Preston-Whyte and Tyson, 1988). Berg winds may blow for several days or only a few hours, and are most common in late winter and early spring. They result in the anomaly of highest maximum temperatures being recorded in winter at many coastal stations. The strong offshore effect of berg winds on the west coast may produce significant dust plumes blowing over the coastal plain, and across the ocean on the west coast. The impact of berg winds on plants is discussed in Von Willert et al. (1992)

1.6.   The weather systems of the karoo

Reference has already been made to the average atmospheric circulation patterns that determine the climate of southern Africa. In this section we explore some of the major deviations from these average conditions – events that influence the weather of the karoo. This account draws mainly on Preston-Whyte and Tyson (1988), while other sources include Heydorn and Tinley (1980), Schulze (1965) and Tyson (1987).

Throughout the year, the average circulation of the atmosphere over southern Africa is anticyclonic. This subtropical control is effected through the South Indian anticyclone, the continental high and the South Atlantic anticyclone. During winter, this continuous band of high pressure intensifies and moves northwards, while the upper level westerlies expand and displace the tropical easterlies equatorward. During summer, the continental high is displaced by low pressure conditions that arise in response to surface heating, and the oceanic anticyclones move southwards (about 6°), displacing the westerly airstream. Following Preston-Whyte and Tyson (1988), we recognize three categories of circulation patterns that influence the weather of the karoo: (1) fine-weather conditions, (2) tropical disturbances, and (3) temperate disturbances.

1.6.1.   Fine-weather conditions

Subtropical anticyclones
These conditions, associated with a strongly subsiding air mass, fine and clear conditions, and no rainfall, prevail over most of the interior plateau (including the bulk of the karoo) during the winter months. Anticyclonic conditions during summer are less common; however, when they prevail for extended periods they result in severe heat waves and desiccation. In the already arid karoo, ecosystems are

subject to severe stress when summer heat waves persist for more than a week.

Coastal lows

Coastal lows, which have their highest frequency during the winter months, are associated with the generation of localized cyclonic vorticity as a result of the westward movement of air off the high plateau. They are initiated on the west coast, move southward to Cape Town, and thence eastward and north-eastward along the coast. Like cold fronts, they produce a substantial drop in temperature. However, they seldom result in precipitation other than orographic mist and fine drizzle, usually confined to the coastal margin. Thus, their influence on the climate of the karoo is largely restricted to the arid and semi-arid west and south coast regions (Namib Desert, Namaqualand and Little Karoo).

1.6.2.   Tropical disturbances

Easterly waves and lows

Easterly waves and lows result from disturbances in the tropical easterly flow, at the junction of the inter-tropical convergence zone and the subtropical high-pressure belt. Moist air, sucked in from the north, is carried upwards by the diverging air mass, resulting in widespread and pro-longed rains behind (to the east) the trough. These rains, whose regularity distinguishes abnormally wet years in the summer-rainfall region, have their highest frequency in mid-summer. Ahead of, and to the west of the trough (the region which includes much of the karoo) subsiding air masses ensure no rainfall, clear skies and hot conditions.

Subtropical lows

During summer, when the upper westerly waves are remote, low-pressure cells may develop in the upper troposphere. These conditions are usually associated with heavy rainfall in the central and eastern parts of the subcontinent.

1.6.3.   Temperate disturbances

Westerly waves

Westerly waves are associated with disturbances in the westerly airstream. To the rear of the surface trough, cloud and precipitation occur in unstable air; ahead of the trough, stable air ensures clear, fine weather. These disturbances, which are rarely observed during winter, have their highest frequency in the spring and autumn months. Rainfall seldom extends inland of the Great Escarpment.

Cut-off lows

Cut-off lows, which have a profound influence on the climate of southern karroid regions, are a more intense form of the westerly trough. The depression starts as a trough in the upper westerlies and deepens, extending downwards to the surface. In doing so, the low is displaced northwards and 'cut-off' from the westerly current. These deep lows are a source of major divergence and account for many flood-producing rains in the southern karroid regions (e.g. the Laingsburg floods of 1981). Cut-off lows have their highest frequency during spring and autumn, but their overall occurrence is highly unpredictable.

Southerly meridional flow

When a deeply penetrating cold front (see *Cold fronts*) is followed by a well-developed high-pressure cell, a strong zonal pressure gradient develops between the two systems. These condition produce a trough in the upper atmosphere which overlies the convergence zone west of the surface trough or cold front. The resulting vertical motion produces rain, usually confined to the coastal seaboard west of Cape Agulhas. Thus, only parts of the Little Karoo are affected by these conditions, which have their highest frequency in the spring months. The southerly meridional flow is also associated with a sharp drop in temperatures over most of the subcontinent, as cold Antarctic air is advected inland.

Ridging anticyclones

When the South Atlantic High ridges around the subcontinent after the passage of a westerly wave in the upper atmosphere, widespread rains often fall in the eastern parts of the subcontinent. This rain results form the steep pressure gradients which advect moist, unstable air over the land. Orographic rain may be particularly intense. In the south-west, subsiding air associated with anticyclonic conditions brings clear, fine and hot weather, often accompanied by strong south-easters. This circulation type brings rainfall to the eastern part of the country throughout the summer months but with a slight tendency for maximum frequencies of occurrence in October and February.

West-coast troughs

Widespread rains over western South Africa occur with the coincident appearance of a surface trough on the west coast and an upper tropospheric westerly wave to the west of the continent. These conditions most frequently (albeit rarely) occur in early summer and autumn.

Cold fronts

Cold fronts are major disturbances in the westerly air stream that produce characteristic cold snaps. They occur together with westerly waves, depressions or cut-off lows, and should not be considered in isolation from these systems. However, they occur most frequently in winter, when the westerly belt penetrates furthest northwards. Ahead of the front, northerly winds associated with divergence and subsidence produce cloud-free conditions. At the rear of the front, conditions favourable for convection result in widespread rain, especially along the west and south-western coasts. Depending on the strength of the front, rain may be very widespread. Snow may fall on high-lying ground. Deep fronts penetrate well beyond the Cape Fold Belt and even the Great Escarpment. Post-frontal conditions are invariably cold and sometimes wet (in the east), especially if the front is followed by a well-developed high pressure cell (see *Ridging anticyclones*).

1.6.4.   Other rain-producing systems

Thunderstorms

Much of the rainfall in the eastern, summer rainfall region is of convective origin. Thunder-storm activity is a complex phenomenon, being dependent on the diurnal heating cycle, synoptic conditions and regional and local effects. However, karroid regions, especially along the west coast, experience very few thunderstorms: fewer than 20 thunder days per year are experienced in the western Upper karoo (as opposed to 80 days on the eastern highveld).

Development of the continuous high-pressure cell

At the end of the summer rainfall season, towards the end of March, a single high-pressure cell (linking the South Atlantic and Indian anticyclones) develops over the subcontinent. This results in a northerly flow of moist air from the tropics over the western parts of southern Africa, including much of the karoo. The influx of moist air is largely responsible for the autumn rainfall maximum for these arid, western regions.

1.7.   **An ecological interpretation of the weather patterns of the karoo: the role of climate in understanding vegetation patterns**

Previous analyses of the climate of southern Africa have failed to provide a convincing classification of the climate, i.e. one that provides an adequate and meaningful biogeographical subdivision of the subcontinent. In this section, we do not attempt a comprehensive reclassification of the subcontinents' climate, but instead present a new analysis of the climate of the karoo to stress the driving role that climate plays in determining landscape-scale vegetation patterns.

The availability of water is generally considered as the greatest limitation of plant growth and distribution (Woodward, 1987). There have been a number of classifications of southern Africa's climate, the most well known are probably those of Köppen, Holdridge and Thornthwaite. No attempt is made here to discuss these classifications further, as these have been adequately reviewed elsewhere (Schulze, 1947; Preston-Whyte, 1974; Schulze and McGee, 1978).

Other biogeographic climatic classifications of southern Africa include those of Jackson (1951), Preston-Whyte (1974) and Rutherford and Westfall (1986). In all cases, rainfall emerges as the primary driving variable. More importantly, though, the distinction between summer and winter rainfall maxima has emerged as the primary explanatory variable (Preston-Whyte, 1974; Rutherford and Westfall, 1986). The bulk of this chapter has been aimed at providing ecologist with a functional understanding of rainfall patterns in the karoo.

Consequently, the models produced attempt to summarize the range and variation in rainfall in a few meaningful indices. Such indices include the usual descriptive statistics of climate (e.g. mean annual rainfall, percentage winter rainfall, mean monthly temperature, etc.); climatic indices such as the summer aridity index (SAI) (Rutherford and Westfall, 1986) or Thornthwaite's climatic indices (Schulze, 1947; Schulze and McGee, 1978); and CV (Jackson 1951). These indices fail to integrate a number of important features of the regions' climate. These features are firstly, the lower, more variable rainfall expected in an arid zone (e.g. summer aridity index). Secondly, the three dimensions of the annual march of rainfall discussed earlier that arise as a result of the different weather systems influencing the regions' climate. Thirdly, the variability in occurrence and intensity of individual rainfall events. Consequently, the analyses fail to produce a climatic map of southern Africa that adequately explain plant biogeographic patterns. In the following sections, we have attempted to address these problems.

In addition to moisture, temperature needs to be considered. Temperature alone is not a significant factor in determining major regional vegetation patterns, although its indirect influence on water availability through its effects on, for instance, evapotranspiration rates is of primary importance (Schulze and McGee, 1978). On a meso- or micro-scale it does play a major part in determining plant patterns; this scale of variation will not be addressed here. Critical temperature indices therefore, like summer and winter maxima and minima (and associated frosts) or ranges are of more significance to plant distribution. The important distinction between temperature and rainfall patterns is that the annual march of temperature follows a relatively simple curve, readily tractable with these summary statistics.

1.7.1. The model

In our analyses, we have not attempted to provide a detailed classification of karoo climates, but have rather emphasized the intimate link between landscape vegetation patterns and climate. This classification differs significantly from previous efforts principally in the manner in which rainfall is incorporated into the analysis. Instead of using solely descriptive statistics (means or CV) or climatic indices (SAI), we use a novel approach of fitting the monthly rainfall data to a mathematical model to approximate the actual shape of the annual march of rainfall. Thus, it is possible to explicitly incorporate, in a relatively few values (model parameters), all three dimensions of the annual march of rainfall. What we fail to incorporate, however, is the variability between individual rainfall events, as this would require a considerably greater amount of time for collation of the raw data.

$$Y = k + (c_1.\cos q_i + s_1.\sin q_i) + (c_2.\cos 2q_i + s_2.\sin 2q_i) + (c_3.\cos 3q_i + s_3.\sin 3q_i)$$

Monthly rainfall and temperature data for all karoo weather stations (with both rainfall and temperature data) were extracted from the Computational Center for Water Research (CCWR). Additional data for Namibia and Angola were obtained from the South African Weather Bureau and Lebedev (1970). The monthly rainfall data for each station was fitted to Equation 1, where, y represents mean monthly rainfall; and, q_i the month expressed in degrees, such that January equals 15°, February 45°, March 75°, etc. Thus, for the ordination, rainfall was represented by the above seven parameters (k, c_1, s_1, c_2, s_2, c_3, s_3); plus, the amount of summer (September to March, **summer**) and winter (March to September, **winter**) rainfall; percentage winter rainfall (% **winter**); and, total annual rainfall (**avg rain**).

Temperature, on the other hand, is comparatively simple to model. Temperature does not show the same degree of plasticity as rainfall. Generally, it is easily represented by a simple sinusoidal curve. Thus, descriptive variables such as mean annual maximum (**avg max**) and minimum (**avg min**); and highest maximum (**max**) and lowest minimum (**min**) temperature, adequately describe the annual march of temperature. All these variables, except average annual temperature were used in the analysis.

The final data set with 100 weather stations and 15 climatic variables were subjected to correspondence analysis (CA). The eigenvalues and percentage variance explained for the first four axes of the ordination are presented in Table 1.1. The ordination of the first and second

Table 1.1. *Eigenvalues and cumulative percentage variance explained for the first four axes of the correspondence analysis of the climate data for the karoo*

Axes	1	2	3	4	Total inertia
Eigenvalues	0.094	0.038	0.013	0.009	0.177
Cumulative percentage variance of climate data	53.4	75.2	82.7	88.0	

axis and first and third are presented in Figs. 1.4(a) and 1.4(b), respectively.

The first axis of the ordination is representative of seasonality of rainfall and explains most of the variance in the data. Percentage winter rainfall clearly separates the Succulent from the Nama-karoo. This axis also correlates well with the type of rainfall curve, where c_1, s_1 and s_2 are indicative of summer maxima regimes; and, c_3 and s_3 with winter maxima. c_2 represents bimodal rainfall curves and lies near the origin of the axis. Stations lying far from the first axis have less distinct rainfall curves tending towards aseasonal rainfall types. The second and third axes separate stations based on temperature. The results are discussed with regard to the vegetation types in the following section.

1.7.2. Discussion: vegetation–climate relationships

This analysis has shown a clear and effective separation of karoo climate stations that is consistent with a biome-level, and, to a lesser degree, vegetation type-level classification. In this section we discuss the relationships between vegetation and climate, with special emphasis on the weather systems presented in section 12.

Succulent karoo

At the biome scale, succulent karoo sites separate from the rest on the basis of low annual rainfall (Fig. 1.2), high percentage winter rain, high absolute and average minimum temperatures, and parameters from the non-linear regression model associated with strong winter peaks (c_3 and s_3) in the annual march of rainfall (Figs. 1.4(a) and 1.4(b)). Some of these associations have been described by many authors in the past (Werger, 1986; Rutherford and Westfall, 1986; see also Cowling and Hilton Taylor, this volume). However, little attempt has been made to explain these patterns in terms of the frequency and reliability of occurrence of the prevailing weather systems. This we do below.

The entire succulent karoo receives its rainfall from weather systems associated with disturbances in the westerly stream. The three western vegetation types of the Namaqualand–Namib Domain (Cowling and Hilton Taylor, this volume), namely strandveld, lowland and upland succulent karoo, receive the bulk of their rain from cold fronts during the winter months. Peak occur-

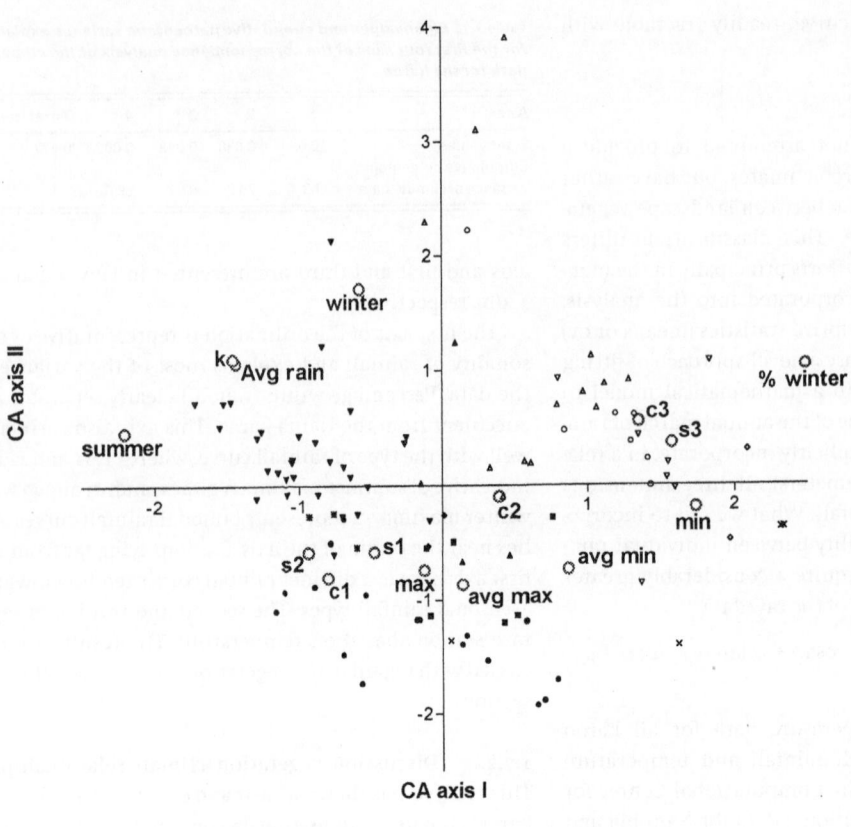

Figure 1.4(a) **Ordination axes I and II of weather stations in the karoo. See text for explanation of climatic parameters**

- ■ Bushmanland Nama Karoo
- ▲ Central Nama Karoo
- ▼ Eastern Mixed Nama Karoo
- ♦ Great Nama Karoo
- ● Orange River Nama Karoo
- ■ Upper Nama Karoo

- ▲ Little Succulent Karoo
- ▼ Lowland Succulent Karoo
- ○ Strandveld Succulent Karoo
- ○ Upland Succulent Karoo

- × Desert

rence of fronts, and hence rainfall, is during the mid-winter months (Fig. 1.1(b)). In southern Africa, the equatorwards penetration of the westerly airstream is greatest among all continents (Preston-Whyte and Tyson, 1988); hence the high frequency and reliability of winter rainfall events in both a regional (Fig. 1.3) and global context (Esler et al., this volume).

The two strandveld succulent karoo stations were separated on the basis of highest percentage winter rain and highest minimum temperatures. These coastal sites receive almost no summer rain and are under the ameliorating influence of the Atlantic Ocean. The remaining sites show a clear trajectory in the multivariate graph, associated with increasing annual, summer, and winter rainfall. In geographical space this gradient moves in an easterly (lowland to upland succulent karoo) and south-easterly (Little Succulent Karoo) direction. The former areas receive more, albeit unpredictable warm season (mainly February-April) rainfall associated with west-coast troughs, thunderstorms

and the autumnal northerly flow of moist, tropical air (see 1.6.4. *Development of the continuous high-pressure cell*). Higher altitudes result in a more pronounced continentality.

The Little Succulent Karoo covers a large tract of multivariate space. This is consistent with its location as transitional between winter and summer rainfall conditions. Some sites cluster near upland succulent karoo, others near central and eastern Nama-karoo in the south and south-eastern karoo regions, respectively (Figs. 1.4(a) and 1.4(b)). While the Little Karoo does receive a substantial proportion of its rain from winter, westerly fronts (Fig. 1.1(b)), most frontal rains fail to penetrate the barriers afforded by the Cape Fold Belt. The largest rainfall events in the Little Karoo are invariably associated with the less predictable cut-off lows, westerly waves, southern meridional flows, and ridging anticyclones. These systems generally have their highest frequency of occurrence in spring and autumn, thus explaining the bimodal peaks in the annual march of rainfall (Fig. 1.1(b)). This is also consistent

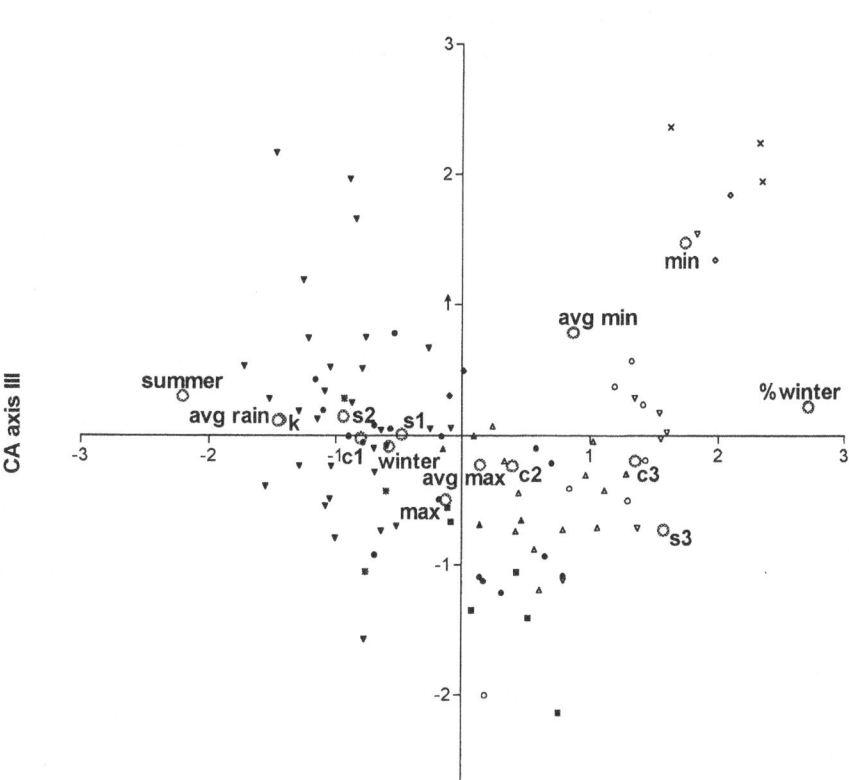

Figure 1.4(b) **Ordination axes I and III of weather stations in the karoo. See text for explanation of climatic parameters. Symbols for stations correspond to those used in Fig. 1.4(a)**

with the non-linear regression parameter which describes bimodal rainfall curves, c_2, located in the centre of the ordination diagrams (Figs. 1.4(a) and 1.4(b)). The relatively strong association between Little Karoo sites and the amount of winter rainfall stems from our delineation of winter to include March and September, prime months for equinoctial rains associated with the weather systems mentioned above. Indeed, the proximity of total winter and summer rainfall in the ordination space (Fig. 1.4(a)) is a result of the inclusion of these months in delineating both seasons, and the fact that many karoo sites receive substantial equinoctial rain (Fig. 1.1(b)).

Nama-karoo

Nama-karoo sites are associated with higher maximum temperatures, bimodal (c_2) or strongly seasonal (c_1, s_1) rains, and, for some areas at least, relatively high rainfall, especially during summer (see also Rutherford and Westfall, 1986).

The eastern mixed Nama-karoo, which grades into the grassland biome of the east-central plateau of South Africa, is largely distinguished on the basis of higher rainfall, especially during summer (Figs. 1.4(a) and 1.4(b)). This vegetation receives a great deal of its rainfall from tropical disturbances during the summer months. However, equinoctial rains associated with cut-off lows, etc. are also important, as are winter rains (and snowfalls) derived from occasional, deep cold fronts. Being centrally located, this vegetation type receives the fringe of all major weather systems in southern Africa.

The remaining Nama-karoo vegetation types are separated from eastern mixed Nama-karoo by stronger bimodality (central Nama-karoo), stronger seasonality (upper Nama-karoo), higher maximum temperatures (Orange River Nama-karoo), and higher maximum and higher minimum temperatures (Bushmanland Nama-karoo). With the exception of the central Nama-karoo sites which receive their rainfall from the same weather systems as the eastern mixed karoo, these more western Nama-karoo areas receive their largely autumnal rains (Figs. 1.4(a) and 1.4(b)) from west-coast troughs, thunderstorms and late-season influx of tropical air. The occurrence of these rainfall events is highly unpredictable. The combination of low and unpredictable rainfall, and extremely high summer temperatures, makes these some of the harshest environments in the karoo.

Desert

Rather than a biogeographically and climatically delineated region, the Namib Desert is a loosely defined geographical area (Jürgens, 1991; Cowling and Hilton Taylor, this volume). This is evident from the large range of multivariate space occupied by the four desert sites used in this analysis (Figs. 1.4(a) and 1.4(b)). Low-rainfall coastal sites in the central Namib (Luderitz and Walvis Bay) are closely clustered with strandveld succulent karoo, regarded by many (e.g. Jürgens 1991; Desmet, 1996) as a southern extension of the Namib Desert. Closely related to these two sites is Namibe in southern Angola. The fourth site, Lobito, at the most northern extremity of the Namib cannot be separated from Orange River Nama-karoo and can be regarded as having a similar climate (Fig. 1.1(b)).

In the context of the arid zone of southern Africa, the Namib is a special case, as the hyper-arid conditions that prevail are regarded as a palaeo feature (Ward et al., 1983). This is due to a permanent temperature inversion over cold Benguela current and adjacent landmass. As a result, warm air-currents from the east are blocked and the daily south-westerly sea breeze dominates, bringing cool, humid air (Walter, 1986). Thus, the core area of this climatic zone has remained stable in the face of the palaeo climatic fluctuations that affected other parts of the karoo. The consequences for the evolution of life forms unique to this system are discussed elsewhere (Cowling and Hilton-Taylor, this volume; Vernon, this volume).

1.8.   Conclusions

Two-thirds of southern Africa have an arid to semi-arid climate, but the causes of this aridity are varied. Generally, aridity of the subcontinent is due to the presence of subtropical descending air (high-pressure cells), although the Namib Desert is a special exception. Higher rainfall regions in the karoo are due to the penetration of tropical systems (north-east) and regular penetration of westerly fronts and associated weather systems (south-west). The remainder of the region is located in a position that is marginal to these systems. Aridity is most pronounced along the west coast, but the succulent karoo has most reliable rainfall. Aridity is least pronounced to the north-east where the karoo grades into grasslands, and central areas have the least reliable rainfall and most extreme energy conditions. The great diversity of climatic determinants and associated patterns must play a pivotal role in the extremely varied patterns and processes associated with the biota of the karoo.

2   Soils of the arid south-western zone of Africa

M. K. Watkeys

2.1.   Introduction

Soils represent the interface between the lithosphere and the atmosphere where the elements in inorganic minerals are released for utilization in the organic biosphere. In arid to semi-arid regions, soils form slowly (Buol, 1965; Dan, 1973; Dunne et al., 1978) and are an important control of the abundance and distribution of plants and animals (Leonard et al., 1988), as has been demonstrated in the karoo (Van Rooyen and Burger, 1974; Vorster, 1986; Vorster et al., 1987; Palmer et al., 1988; Lloyd, 1989a). The objective of this chapter is to give an overview of the soils across the arid south-west of Africa and to outline some general controls due to geology, geomorphology and climate. The summary will be confined to the karoo south of the Orange River because that is the only portion with a reliable soils map (Ellis and Lamprechts, 1986; Ellis, 1988).

For the purposes of this chapter, the karoo has been subdivided into 6 regions on geological, geomorphological and pedological factors: (1) west coast, (2) Namaqualand, Bushmanland and Korannaland, (3) Great Karoo (north), (4) Great Karoo (south), (5) Little Karoo and (6) Great Escarpment (Fig. 2.1). These subdivisions do not quite correspond to the biogeographic regions of the karoo (Hilton-Taylor and Le Roux, 1989). The succulent karoo biome occurs in Region 1, the western parts of Regions 4 and 6, and Region 5, while the Nama-karoo biome is found over the rest of the area.

2.2.   Controlling factors on soil variability

2.2.1.   General factors
Soil is a mixture of inorganic and organic particles with variable amounts of water and air. Its formation is initi-

ated by the breakdown of rock to form a regolith which may ultimately develop into a soil (Rowell, 1994). This breakdown is through a combination of physical processes

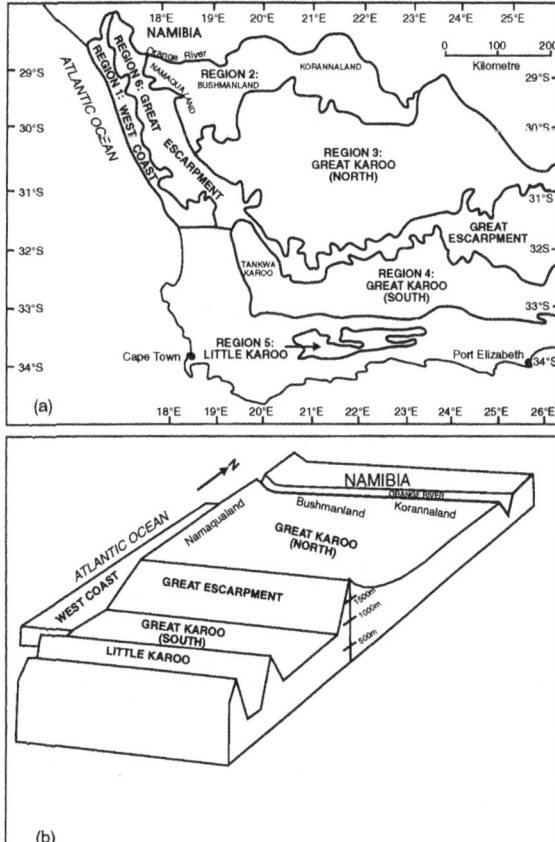

Figure 2.1 **(a) Subdivisions of the karoo based on geological geomorphological and pedological criteria; (b) a schematic perspective showing the topographic relationships between the subdivisions**

(mechanical weathering, thermal expansion and contrac-tion, etc.,) chemical processes (involving water) and biolog-ical processes (which move particles to create spaces or yields metabolic products utilized in chemical weather-ing). Consequently, the two most important controls on soil formation are the climatic conditions of the region and the nature of the parent material.

The lack of moisture in arid regions results in less weathering and leaching when compared to humid regions, so that large areas of bare rock may be present and very old soils are preserved. The slow soil formation results in shallow coarse soils with sharp soil boundaries due to the abrupt thresholds in soil behaviour, causing the soils to be very sensitive to degradation. The amount of rain is important in determining the movement of chlorides, car-bonates and sulphates to the subsoils. In increasingly wet-ter climates, the sequence of decreasing solubility is sodium, potassium, sulphates, magnesium and calcium (Drever and Smith, 1978) so that sodium salts are retained in only very arid topsoils, gypsum in fairly arid soils and carbonates in slightly wetter topsoils. A mean annual rain-fall of even 50 mm may wash sodium salts out of the sur-face (Blume et al., 1985). Nutrients in the soils concentrate at the wetting front and may become stranded by evapora-tion resulting in a patchy distribution (Hunter et al., 1982).

The breakdown of the parent rock results in two impor-tant characteristics in soils: texture and chemical compo-sition. Texture is a physical characteristic depending on the nature of the mineral particles in the soil (size, shape) and their spacing. The particles may be partly primary (inherited) minerals from the parent rock, but with a smaller grain size, or pedogenic (secondary) clay minerals developed in the soil (Schwertmann, 1971; Dixon and Weed, 1989). With regards to chemical composition, the most important elements in soils from an ecological standpoint are potassium, calcium, magnesium, phospho-rus, sulphur and nitrogen. All but the last two occur as major element oxides in most common rocks, whilst sul-phur is a minor element and nitrogen is derived from the atmosphere. Other elements such as copper, lead, zinc, nickel and chromium are only found in trace amounts in most common rocks, but may be concentrated locally to have a profound affect on the biota as documented in southern Africa (Soane and Saunder, 1959; Jacobsen, 1967a and b; Shewry and Petersen, 1976; Wild, 1978; Cole and Le Roex, 1978; Cole, 1980).

The pore spaces between particles allow penetration of air and water, providing a link between the parent mater-ial and the climate. The presence of water affects both the physical properties of soil (Du Plessis and Shainberg, 1985) as well as the chemical nature, especially alkalinity, acid-ity and salinity. In general, alkali soils occur in drier regions, whereas acidity is associated with leached soils of

wetter areas, although the amount of leaching is depen-dent on the relationship between rainfall, temperature and organic material. Salinity is a problem encountered with irrigation of drier regions where rainfall is too low to wash away soluble salts (Van Schilfgaarde, 1974). Such chemical processes may redistribute elements such as iron to form ferricrete (FitzPatrick, 1987) or silicon to form silcretes (Flach et al., 1969; Smale, 1973) which substan-tially alter the physical nature of the soil.

The main constraint on organic matter in arid soils is the low productivity of the ecosystem and the low shoot : root ratios of the plants (Orians and Solbrig, 1977). As a consequence, the nitrogen content of arid soils is low, but both nitrogen and phosphorus tend to accumulate close to the top of the soil profiles so that there is a fast turnover when plants and soil micro-organisms are briefly active (Charley and Cowling, 1968). Biological processes may also reconstitute the soil not only by the addition of organic material, but also by physical reworking. The effect of ter-mites and ants on soil formation is generally low in arid conditions (Chhotani, 1988), becoming increasingly important under more humid conditions (Pullen, 1979), and has been investigated in the karoo (Midgley and Musil, 1990; Dean and Yeaton, 1993a).

2.2.2.   Geology and geomorphology of the karoo

The geology of the karoo is described by Meadows and Watkeys (this volume). From a pedological perspective, it can be broadly subdivided into two informal units (Table 2.1). The western and northern areas are underlain by a 'basement complex' of Precambrian metamorphic rocks which, in the central and southern areas, are covered by Phanerozoic sediments intruded by dolerites. From Table 2.1, it can been seen that there is a wide variety of parent material in the karoo ranging from ultramafic igneous rocks through to highly siliceous quartzites.

Superimposed on the geology are a number of land sur-faces, the numbers, correlations and ages of which have been the subject of various conflicting interpretations (Moon and Dardis, 1988; Gilchrist et al., 1994) (Fig. 2.2). In their overview, Partridge and Maud (1987) adopted King's (1967) terminology but not his denudaation chronology, and this system will be retained here (Table 2.2).

The oldest geomorphological feature is the Great Escarpment which is approximately parallel to the coast. Ollier and Marker (1985) consider that it was initiated by erosion on down-warped margins to the base level of the new coastline that was formed when Africa and South America separated about 138 million years ago. Since then it has retreated inland about 100–200 km and is still being eroded. The African land surface, found at different eleva-tions inland and seawards of the Great Escarpment, repre-sents the product of a long period of erosion backwearing

Table 2.1. *Summary of the geology of the karoo*

Geographical Area	Geology			
	Age	Informal Name	Unit	Rock Types
West Coast Namaqualand			Tertiary	sediments
Little Karoo	Phanerozoic	Cover	Jurassic	sediments
Great Karoo			Karoo Supergroup	sediments (glacial, shale, sandstone) with dolerite intrusions
southern and western Cape mountains			Cape Supergroup	folded sediments (quartzites and shales)
West Coast, Great Escarpment	Proterozoic	Basement Complex	Pan-African belts	deformed and metamorphosed volcanaic and sedimentary rocks with some granitic intrusions
West Coast, Great Escarpment, Namaqualand, Bushmanland, Korannaland			Namaqualand Metamorphic Complex	highly deformed and metamorphosed sedimentary, volcanic and intrusive rocks with granitic gneisses dominant
North-eastern Karoo	Archaean		Kaapvaal Craton	Metasedimentary and metavolcanic rocks

After Tankard et al. (1982)

Table 2.2. *Geomorphological land surfaces of the karoo*

Geographical Area	Geomorphological land-surfaces		
	Age	Name	Main Features
Great Karoo (south)	late Pliocene	post-African II	major valley incision
Great Karoo, Namaqualand, Bushmanland, Korannaland	mid-Miocene to late Pliocene	post-African I	partly planed in interior; dissected tableland below Great Escarpment
West Coast, Namaqualand, Bushmanland, Great Karoo	early Cretaceous to early Miocene	African	represents an erosion period of >100 Ma; characterized by kaolinitized weathering profiles and duricrusts
separates West Coast and Great Karoo (south) from the Great Karoo (north) and Namaqualand	initiated in early Cretaceous; still actively retreating	Great Escarpment	contains mountainous areas above the African surface

After Partridge and Maud (1987).

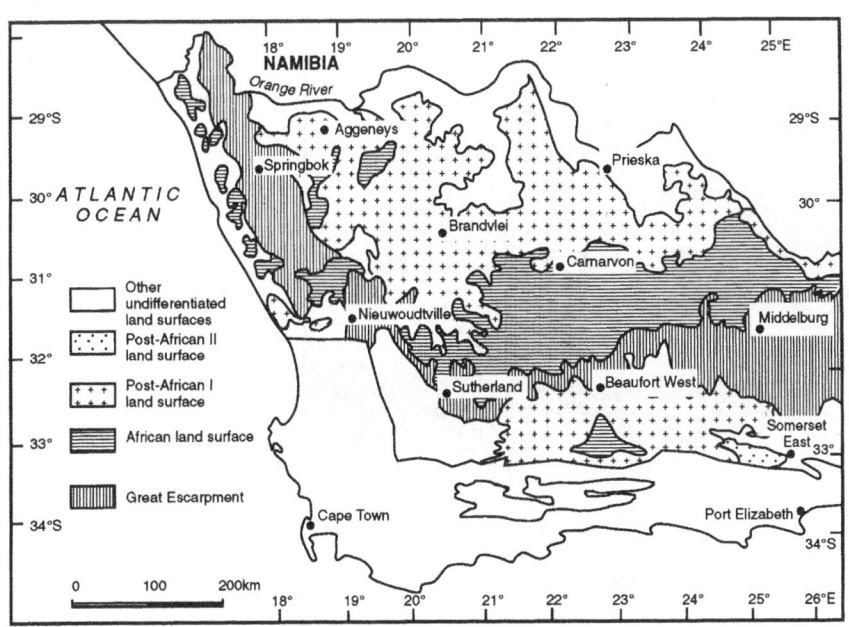

Figure 2.2 **Geomorphological map of the karoo (modified after Partridge and Maud, 1987)**

and pediplanation from the Cretaceous to the early Miocene. An important aspect of this gentle pediplain from a pedological viewpoint is the deep kaolinization of the bedrock beneath a laterite or silcrete duricrust which developed during a warm and humid but seasonally dry climate.

The mid Miocene to late Pliocene post-African I surface is planed some 100–300 m below the African surface. In the interior, behind the Great Escarpment, it is generally partly planed whereas seaward of the Escarpment it is usually a dissected tableland. The late Pliocene post-African II surface is mostly expressed in regional incision resulting in dissection of the coastal plains.

2.2.3.   Climate of the karoo

The climate of the karoo is described by Desmet and Cowling (this volume). At any locality, it depends on the latitude, longitude and topography, but essentially it is semi-arid with temperatures that in winter fall to freezing and in summer rise above 40 °C. Such ranges in temperature are conducive to mechanical weathering, whilst the low precipitation does not favour chemical weathering. The west receives reliable winter rainfall, whereas summer rains dominate the east. In between these two areas, rainfall is low, with Bushmanland having an annual mean precipitation of less than 100 mm.

2.3.   Soils of the karoo

The early work on the soils of the arid and semi-arid regions of South Africa was by Van der Merwe (1955, 1962).

However, studies of the soils of the karoo remained largely neglected until investigations summarized by Ellis (1988) (Fig. 2.3) which involved mapping units with uniform terrain form and soil pattern known as pedosystems (Land Type Survey Staff, 1984).

Soil classification in South Africa has recently been revised and now revolves around two levels (Soil Classification Working Group, 1991). A general level of soil *forms* is based on unique vertical sequences of diagnostic horizons and materials. All forms are divided at a more specific level into two or more soil *families* on the basis of 19 diagnostic properties. Soil texture is used in this system in conjunction with forms and families rather than as a differentiating criterion. In an overview such as this, it is necessary to make generalizations and, consequently, the general level of soil forms has to be used although soil families can clearly play an important ecological role on a smaller scale.

Region 1: West Coast

This coastal region (Fig. 2.1), approximately 50 km wide and 350 km long, is generally flat, rising up to about 300 m above mean sea-level (amsl) at the break with the Great Escarpment. The plain is broken by isolated ridges and low rocky hills (*koppies*) reaching up to over 700 m amsl. It is underlain by Proterozoic igneous and metamorphic rocks of the Namaqualand Metamorphic Complex for most of its length, with rocks of the Pan-African belts occurring at the northern and southern ends (Table 2.1), covered in places by a thin veneer of Tertiary and unconsolidated Quaternary sediments. Remnants of the dissected African land surface (Fig. 2.2 and Table 2.2) are preserved along the interfluves between rivers flowing west from the

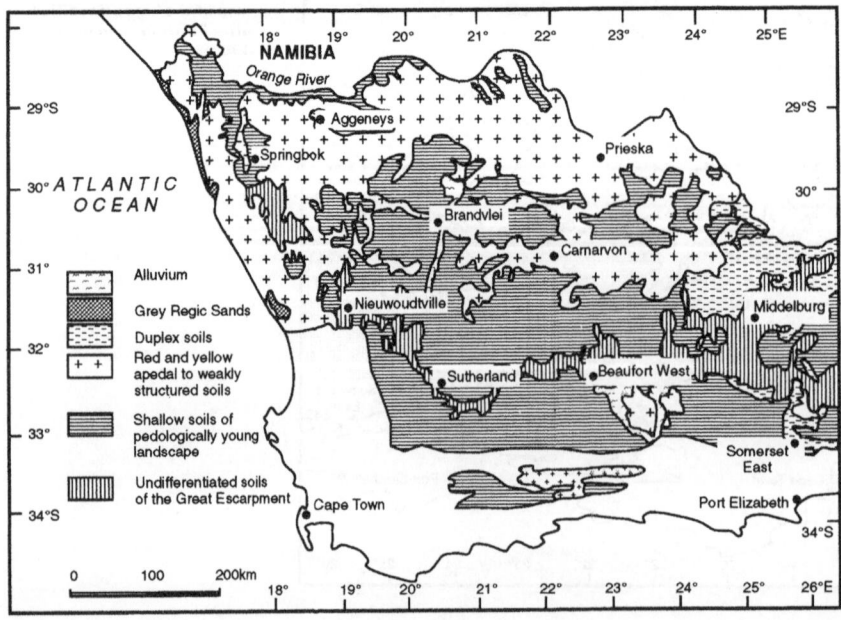

Figure 2.3 **Soils map of the karoo (modified after Ellis and Lamprecht, 1986)**

Escarpment to the Atlantic Ocean. This region receives precipitation from winter rain as well as coastal fog due to the cold, northwards flowing Benguela current.

The grey regic sands bordering the ocean are recent aeolian deposits that show little evidence of pedogenesis. At the southern end of the plain, these sands are associated with remnants of the Post-African I surface. Inland of these, the soils are derived from the underlying Proterozoic rocks with variable amounts of aeolian contribution decreasing towards the Escarpment (Fig. 2.3). Yellow, high-base soils (Clovelly Form) (Table 2.3) occur nearest the coast, consisting of moderately deep uniform coarse-textured sand, usually underlain by a more clayey neocutanic horizon with high pH (6.5–9.5). A transition zone of yellow-red soils separates these from the most common soil of the West Coast, a red, high-base status soil (usually Hutton Form) which is apedal to weakly structured, freely drained with a medium to coarse texture and variable thickness. The soils at the northern and southern ends of the West Coast associated with the Pan-African lithologies tend to be richer in plant nutrient than those overlying the granitic gneisses in the central portion.

The northernmost parts of the coastal plain contain remnants of reddish-coloured hardpan or duribanks which are fairly common throughout the arid south-west (Ellis and Schloms, 1982). Areas underlain by reddish-coloured hardpan cemented by silica (known locally as 'dorbank') are frequently covered by shallow (less than 400 mm), usually coarse, sand to loamy sand soils with high pH values (pH$_{H_2O}$ 8.0).

Region 2: Namaqualand, Bushmanland and Korannaland

This region extends eastwards from the Great Escarpment and is over 500 km long and up to 250 km wide, consisting of part of Namaqualand, Bushmanland and Korannaland (Fig. 2.1). It is underlain by highly deformed and metamorphosed rocks of the Namaqualand Metamorphic Province which is dominated by granitic gneisses (Table 2.1). It represents an interior plateau 900–1000 m amsl sloping very gently northwards, with rocky hills rising about 200 m above the surrounding plains. The rivers, which are usually dry, flow downslope to the Orange River which has incised a valley up to 500 m below the plateau. The west receives winter rains, while the east receives summer rains, with Bushmanland in the middle having the worst of both regimes.

The flatness of the region is a combination of exhumation of the pre-Karoo glaciated surface (and possibly the older pre-Nama surface) in combination with the Mesozoic and Tertiary land surfaces. Most of the region consists of the partly planed post-African I surface which has cut down into the African surface (Fig. 2.2 and Table 2.2). The only pristine remnants of this latter surface are adjacent to the Great Escarpment and in Bushmanland where the deep weathering and kaolinization is exposed (McCarthy et al., 1985). This is indicative of weathering in a warm and humid, but seasonally dry climate during the Tertiary.

The most common soils of this region are red and yellow soils reflecting weathering of the often granitic parent material in a well-drained, oxidizing environment (Fig. 2.3 and Table 2.3). This has given rise to coatings of iron oxides on the soils particles (hence the reddish colour) and the development of non-swelling clays. As is typical in the more arid areas of South Africa, the A horizon is orthic (i.e. there is a lack of an organic, humic, vertic or melanic topsoil). It is free from waterlogging and, in the west, is typically very sandy (<6% clay), whereas in the east it is sandy (6–15% clay).

The low hills that rise above the surrounding plain often consist of a capping of quartzite, with the slopes consisting of schists and amphibolites, while the base of the slope is granitic gneiss. The slopes are often steep and covered with a coarse mantle of cobbles and boulders. Where they are shallow, lithosols are developed on these pedologically young features and the change in lithologies with topography results in a variation in soil fertility. The highly siliceous quartzites are infertile, whereas the schists and amphibolite contain important major elements such as potassium, calcium and magnesium with variable amounts of copper and zinc. The granitic gneisses have lower concentrations of these elements, and comparisons of the fresh and weathered gneisses reveals remarkably little redistribution of elements during the process of soil formation (Duncan et al., 1985). In the vicinity of Aggeneys, major base-metal sulphide ore bodies occur (Moore et al., 1990), but the environmental response to the elevated copper, lead, zinc and silver in the soils and effects of mining have yet to be investigated.

In dune areas, such as the Koa River Valley of Bushmanland between Springbok and Aggeneys (Dean et al., 1991), the soil has little profile and is deep, uniform and coarse-textured sand poor in plant nutrients. Duribank areas occur throughout the region, and the coarse soils associated with them may be high in most plant nutrients, particularly potassium. Both the K- and P-status are high in shallow, coarse, sand to sandy loam soils which overlie hardpan calcrete (Netterberg, 1980). These soils, which also have a high pH$_{H_2O}$, are often found in the interdune areas of Bushmanland. In detail, the soils associated with dorbanks and hardpan calcretes may vary as shown in Table 2.4.

The Orange River Valley, which forms the northern border to Region 1, is a pedologically young landscape containing structureless to weakly structured soils, developed

Table 2.3. *Summary of the main soil forms found along the west coast, in Namaqualand, Bushmanland and Korannaland*

Soil Form	Mispah Form	Glenrosa Form	Oakleaf Form	Hutton Form	Clovelly Form	Augrabies Form
Top Soil	orthic A	orthic A	orthic A	orthic A	orthic A	orthic A
Subsoil	hardrock	lithocutanic B	neocutanic B	red apedal B	yellow-brown apedal B	neocarbonate B
				unspecified		

After Soil Classification Working Group (1991).

Table 2.4. *Summary of the main soil forms found associated with dorbanks and hardpan calcretes (the latter in italics)*

Soil Form	Knersvlakte Form (Coega Form)	Oudsthoorn Form (Gamoep Form)	Garies Form (Plooysburg Form)	Trawal Form (Prieska Form)
Top Soil	orthic A	orthic A	orthic A	orthic A
Subsoil		neocutanic B	red apedal B	neocarbonate B
			dorbank (hardpan carbonate horizon)	

After Soil Classification Working Group (1991).

mainly from *in situ* weathering. These tend to be Mispah form, where the orthic A horizon is immediately underlain by hard rock, or Glenrosa form, where the orthic horizon overlies a zone of minimal *in situ* weathering, resulting in the development of a lithocutanic illuvial B horizon (Table 2.3). The parent rock types in this area are all rich in plant nutrients, particularly in the west, but the poorly developed nature of the soil, together with the low rainfall and extremely high summer temperatures, inhibits growth. Along the rivercourse, the alluvial terraces contain deep (>1000 mm), stratified and weakly structured soils, with sandy loam to sandy-clay loam A horizons. Common forms are Dundee, where the orthic A horizon is underlain by stratified alluvium, or Oakleaf (Table 2.3).

Region 3: The Great Karoo (north)

This region lies between Region 2 and the Great Escarpment, and is over 500 km long and up to 300 km wide (Fig. 2.1). The northern part is the southward extension of the post-African I land surface (Fig. 2.2 and Table 2.2), being flat and about 900–1000 m amsl; however, it is distinguished by major differences in geology and consequent differences in geomorphology, particularly the presence of numerous pans. The southern part is the African land surface which gradually increases in elevation towards the Great Escarpment.

The whole region is underlain by flat-lying, sedimentary rocks of the Karoo Supergroup with the Dwyka and Ecca Groups in the north and the Beaufort Group in the south (Table 2.1). These have been intruded by dolerite dykes and sills. The latter are responsible for the flat-topped low hills often depicted as being typical of the karoo landscape. In the eastern parts of the region the sills have saucer shapes, resulting in the circular basins of differing sizes which subdivide the landscape. They provide a vertical variation in what otherwise would have been a flat, monotonous landscape produced by the African and post-African I land surfaces.

The majority of soils on the African land surface are shallow (less than 300 mm), structureless to weakly structured soils, mainly derived from *in situ* weathering on a pedologically young landscape (Fig. 2.3). Mispah and Glenrosa Forms (Table 2.3) dominate, while shallow Oakleaf Form transported soils are generally found on low-lying flat pediment slopes or in pans. The north-west, south-west and east areas of Region 2 have loamy topsoil (15–35% clay), while elsewhere the topsoil is sandy (6–15% soil). Soils overlying the karoo sediments tend to have lime in upland and bottomland positions; this feature is associated with low rainfall and decreases eastwards as the rainfall increases.

In the Carnarvon–Brandvlei area of the Great Karoo, red and yellow apedal to weakly structured, freely drained soils are present, characterized by an abundance of $CaCO_3$. These are associated with the Dwyka Group glacial sediments at the base of the Karoo Supergroup, and with pans and palaeodrainage lines. The soils are usually medium textured (sandy loam to sandy-clay loam) and calcareous throughout their profile. Their surface is often covered by a prominent layer of stones (desert pavement) which are clasts weathered out of the Dwyka diamictite (tillite).

The dolerites develop shallow to moderately deep calcareous, medium textured (sandy-clay loam) soils which contain calcrete and calcareous horizons. This calcareous nature is also present in the deep (>1000 mm), stratified and weakly structured soils (Dundee and Oakleaf forms) of the alluvial terraces of the Sak River transecting this area. Within the pan of Brandvlei, the orthic A horizon is underlain by a soft carbonate subsoil (Table 2.5).

Duplex soils dominate the eastern parts of Region 3. In such soils, the clay percentage of the B horizon is at least twice that of the A horizon and the A/B horizon boundaries

Table 2.5. *Summary of the main soil forms found associated with soft carbonate horizons*

Soil Form	Brandvlei Form	Etosha Form	Kimberley Form	Addo Form
Top Soil	orthic A	orthic A	orthic A	orthic A
Subsoil	neocutanic B	red apedal B	neocarbonate B	
		soft carbonate horizon		

After Soil Classification Working Group (1991).

are abrupt. The most common type have red B horizons. This change in soils is associated not with the stratigraphic difference of rock-types across the karoo, but rather with the increase in rainfall eastwards.

Region 4: Great Karoo (south)

This region lies between the Great Escarpment and the Cape mountains (Fig. 2.1). It is 60–100 km wide by over 600 km long, and is about 900–700 m amsl, generally sloping gently to the south. It largely lies in the winter rain shadow of the Cape mountains, and this effect is taken to the extreme in the west where the Tankwa Karoo is one of the driest places in the whole arid south-west. The eastern parts of this region, however, receive summer rains.

Apart from the Tankwa Karoo, where the drainage is north-west, rivers generally flow southwards from the Escarpment through gaps in the Cape mountains. Their tributaries tend to be aligned east–west, picking out a tectonic fabric in the underlying rocks. These rocks are mainly Beaufort Group sediments of the karoo Supergroup (Table 2.1), which become deformed in the Cape Fold Belt in the southern parts of the region. Here and in the Tankwa karoo, the underlying Ecca and Dwyka Groups and the uppermost parts of the Cape Supergroup are exposed. The western half of this region is devoid of dolerite, while the eastern half contains dykes and sills, but far fewer than Region 3. Quaternary sediments form a large alluvial fan appropriately called *Die Vlakte* (The Flats), up to 100 km long and wide, south-east of Beaufort West in the vicinity of the Sout and Kariega rivers.

Most of this region consists of the post-African I surface (Fig. 2.2 and Table 2.2). Due south of Beaufort West there are remnants of the African land surface, while part of the post-African II surface occurs in the east, south-west of Somerset East. The presence of dolerite dykes and sills in the eastern half of this region has geomorphological and hydrological implications which, in turn, influence the soils. Dolerite not only shapes the landscape and provides locally distinct soils, but also affects groundwater flow and aquifers.

As with Region 3, the shallow soils of Region 4 are developed on a pedologically young landscape. They are structureless to weakly structured, resulting from *in situ* weath-

ering, and their composition reflects the underlying parent material with lime generally present over the entire landscape. In the vicinity of *Die Vlakte*, the water-courses contain unconsolidated deposits while, on the Quaternary sediments on either side, red, high-base, apedal to weakly structured, freely drained soils have developed. The lack of lithification of these sediments and their ease of erosion have led to areas of shallow, weakly structured soils (mainly Oakleaf Form) (Table 2.3) formed from pedisediment.

In the eastern parts of this region, Oakleaf Form soils occur on the post-African II land surface while, around Somerset East, duplex soils have developed. In the latter case, non-red B horizons are more common, unlike those of Region 3 above the Great Escarpment in which red B horizons dominate.

Region 5: Little Karoo

This region essentially consists of valleys, up to 50 km wide and 200 km long, lying 400–600 m amsl between the east–west mountain range of the southern Cape (Fig. 2.1). The mountains consist of the Cape and Karoo Supergroups (Table 2.1) which were folded during the Permo-Triassic to form the Cape Fold Belt. The valleys, on the other hand, owe their origin in part to Jurassic rifting of Gondwana, and are one of the few areas in South Africa to contain onshore Jurassic sediments.

Shallow (<300 mm) soils of pedologically young landscapes dominate this topographically variable region. Red, high-base status apedal to weakly structured soils (Hutton Form) with a medium texture (sandy loams to sandy-clay loams) overly the Jurassic sediments. The relationship of geology to soils in this region is well established and partly determines the distribution of fynbos vegetation in this region (Deacon et al., 1992).

Region 6: The Great Escarpment

The Great Escarpment is a geomorphologically dominant feature that separates the interior plateau of the arid south-west from seaward areas to the west and south (Fig. 2.1). It runs southwards from the Orange River defining the east side of the West Coast plain and then turns eastwards, separating the Great Karoo (south) from the Great Karoo (north). It varies in width from being virtually a single cliff to a wide dissected region (Fig. 2.2). In the west, it rises from about 300 m amsl to peaks of over 1700 m amsl in the Kamiesberg south of Springbok while, in the east, it rises from about 900 m to peaks of over 2000 m in the Sneeuberge south-west of Middelburg. These high regions may be snow-covered during winter, providing moisture conditions and mechanical weathering (such as frost action) generally absent throughout the rest of the karoo.

Soils of adjacent areas are known to penetrate this

Table 2.6. *Summary of the main soil forms found on dolerites south-west of Middelburg*

Soil Form	Shortlands Form	Inhoek Form	Mayo Form	Bonheim Form
Top Soil	orthic A	melanic A	melanic A	melanic A
Subsoil	red structured B	unspecified	lithocutanic A	pedocutanic B
		unspecified		

After Soil Classification Working Group (1991).

mountainous area, with shallow soils of the pedologically young landscape being most common (Fig. 2.3). The Escarpment also contains soils that are difficult to classify into the current classification scheme (Soil Classification Working Group, 1991). This is partly because of the lack of data and partly because the terrain contains large areas of exposed rocks. In such areas, podzolization is the main soil-forming process. In the initial stages of soil formation, the aspect of the locality is important, as well as are joints and other fractures in the rocks which concentrate water in sites where chemical weathering is increased. This early stage of soil formation is occurring in Namaqualand along the western part of the Escarpment where large areas of granitic gneisses are exposed. The lithosols that eventually develop are highly variable in depth. This process eventually leads to podzolic B horizons and leaching with eluviation of possible clays and soluble salts from the overlying A horizon. This yields a soil low in plant nutrients, being deficient in P and bases.

Despite problems with classification, the Great

Escarpment contains some interesting soils. At the eastern end of the Great Escarpment, the soils developed on dolerites south-west of Middelburg (Table 2.6) are markedly different from the soils on dolerites in the more arid Carnarvon–Brandvlei area and reflect a transition towards wetter conditions. The melanic horizons, more typical of sub-humid climates, do not have sufficient organic carbon to qualify as diagnostic organic and are distinguished from humic A horizons by a higher-base status derived by weathering of the parent dolerite.

Adjacent to the western sections of the Great Escarpment, plinthic soils (Table 2.7) and hydromorphic duplex soils overlie ferricrete west of the Nieuwoudtville. The orthic A horizon may be underlain by an E horizon of quartz or other resistant fine to medium-grained minerals. Plinthic soils are indicative of localization and accumulation of iron and/or manganese under conditions of a fluctuating water table. This suggests that these particular soils developed under wetter conditions and are, effectively, palaeosols reflecting an older, different climatic regime during the Tertiary.

The Okiep Copper District of Namaqualand is one of the few areas of mining in the arid south-west, producing mainly copper, with some nickel and lesser gold and silver. Although the stream sediments of the region display elevated copper values due to mining, no obvious effects have been observed in the ecosystem (Wulfse and Holdsworth, 1994). The effects of radioactive waste from a monazite mine in southern Namaqualand (Andreoli et al., 1994) have yet to be investigated.

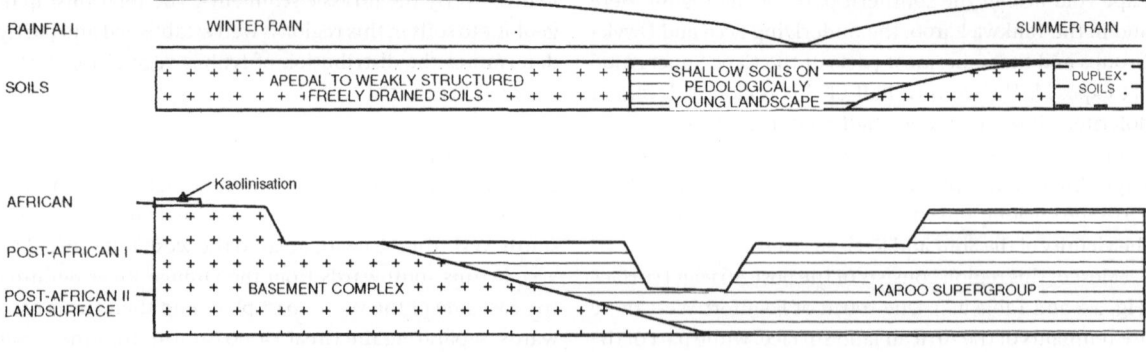

Figure 2.4 **A schematic E–W cross-section showing the relationships between geology, climate and soils**

Table 2.7. *Summary of the main plinthic soil forms found west of Nieuwoudtville*

Soil Form	Westleigh Form	Bainsvlei Form	Avalon Form	Longlands Form	Wasbank Form	Glencoe Form
Top Soil	orthic A	orthic A	orthic A	orthic A	orthic A	orthic A
Subsoil	red apedal B	yellow-brown apedal B		E horizon	E horizon	yellow-brown apedal B
		soft plinthic			hard plinthic	

After Soil Classification Working Group (1991).

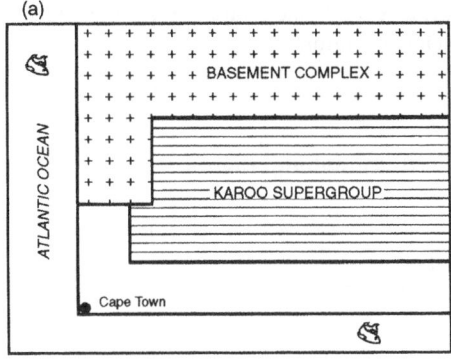

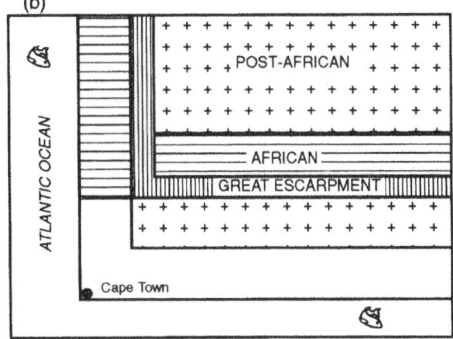

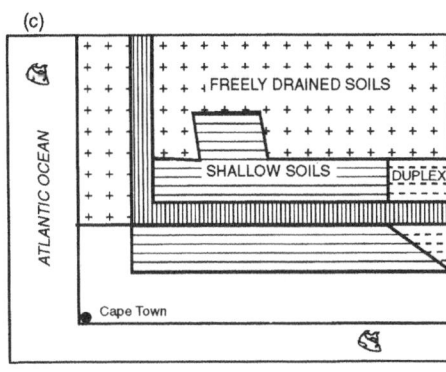

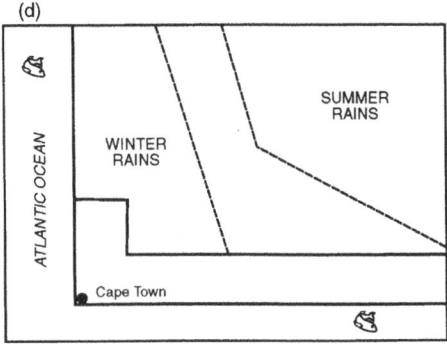

Figure 2.5 **A model for the broad soil pattern: (a) simplified geology; (b) simplified land surfaces; (c) simplified soils; (d) simplified rainfall**

2.4.   Discussion of the soil pattern

The soils of the karoo are typical of an arid to semi-arid region, displaying the full spectrum of soil development, including the preservation of palaeosols. In places, bare rock is being subjected to weathering resulting in loose, coarse weathered mantles that eventually form shallow soils on a pedologically young landscape. Further soil formation yields free-draining soils, while depressions (pans) have developed their own soil types. Throughout most of the karoo, there is little organic material to provide organic topsoil or to assist with biological and chemical breakdown of parent material. Duplex soils occur towards the east where rainfall increases, or locally on the Great Escarpment in areas of higher precipitation.

It is clear from the geological map (Meadows and Watkeys, this volume) and the above description that there are broad correlations between rock types and soil types. In general, the freely drained soils are associated with the Basement Complex. This is because granitic gneisses, which dominate the region, consist mainly of feldspar and quartz. In the absence of chemical weathering, these minerals form medium to coarse primary particles in the soil which, together with the lack of formation of secondary clays, yields a free-draining soil. On the other hand, in the west, the Karoo Supergroup is associated with shallow soils of a pedologically young landscape. This difference is lithological leading to differences in both the rock composition and the geomorphology. In the latter case, the horizontal bedding and closely spaced vertical jointing of the sediments is conducive to formation of small scarps which, in a predominantly mechanical weathering environment, remain vertical during retreat. This effect is enhanced by the presence of dolerite sills parallel to bedding.

Such broad generalities allow for simplified diagrams and models of the geology, geomorphological land surfaces, soils and rainfall to be constructed (Figs. 2.4, 2.5). When this is undertaken, it is apparent that the northern part of Region 3 is anomalous: the same lithologies on the same land surface are overlain by different soils. West of Carnarvon, the Karoo Supergroup is overlain by freely drained soils, whilst between Carnarvon and Brandvlei shallow soils of a pedologically young landscape dominate.

The solution to this enigma probably lies in the climate. The eastern half of this region falls within the summer rainfall regime, whereas the western half has a very low rainfall as it lies in the zone between the summer and winter rainfall areas. Therefore, soil-forming processes in the east are occurring more rapidly and are tending towards the duplex soils seen in the easternmost part of the region. However, the Brandvlei–Carnarvon area does contain freely drained soils, and one possible explanation is that they are relics of a soil type that once dominated the area but which has eroded and transported downstream to expose a pedologically young landscape.

If this explanation is valid, the climatic changes during the Quaternary would have seen the migration of the winter:summer rainfall interface back and forth across the karoo as alternate glacial and interglacial periods occurred. In turn, this would have affected the soil-forming processes in different regions, and, in the light of the present slow rate of formation, the soils observed today have characteristics derived from past climates. Furthermore, it implies that there has been a net soil loss during the general amelioration of the climate since the last glacial maximum.

2.5.   Acknowledgements

This overview would not have been possible without the mapping undertaken by the staff of the Soil and Irrigation Research Institute, Stellenbosch and without the compilation of the soils of the karoo undertaken by Dr F. Ellis.

3 Palaeoenvironments

M. E. Meadows and M. K. Watkeys

3.1. Introduction

The contemporary biogeography of any region such as the karoo is the culmination of a complex of processes operating over geological, evolutionary and ecological time. The Victorian geologist Charles Lyell put forward the principle of uniformitarianism – that the present is a key to the past – but to explain the biogeography of the karoo is to accept that it is the *past* that represents the true key to unlocking the observable patterns of the *present* (George, 1976). It is vital to know something of the history of a biological community in order that a reasoned explanation can be offered for its present distribution and ecology; in this way, the fourth dimension, time, is an important component of understanding the present and future distribution of that community. Different timescales are applicable to different kinds of geological and biological changes, for example broadscale geological changes are apparent only over longer time periods compared with, say, changes in the characteristic geomorphological or pedological processes of a region. This chapter offers a narrative history of the development of the phytogeography and zoogeography of the karoo seen against the backdrop of geological, geomorphological and climatic changes and, in so doing, emphasizes environmental change as a significant and consistent feature of the karoo. Such an understanding is crucial to the more appropriate management of karoo communities in a future which, like the past before it, will offer its unique array of environmental changes and challenges. One of the most potent of these challenges faced by biological communities is climate change.

Since the dawn of time, climate conditions on earth have fluctuated markedly in all directions, so that at various times the climate of the karoo, as elsewhere, has been warmer or cooler, wetter or drier than today. The variability of climate under the direction of both macro- and micro-scale climatological, geological and, most recently, anthropogenically induced processes has had major implications for the prevailing biota. Evolutionary adaptations provide for a degree of flexibility and survival of relatively small-scale or slow climate change, but any shift of greater magnitude or faster rate is a factor both in promoting extinction and in providing opportunities for the evolution of novel species. Indeed, Stebbins (1952) has even gone so far as to argue that aridity acts as a stimulus to speciation; this fails, however, to account for the apparent contradiction that arid and semi-arid regions like the karoo are often interpreted to be impoverished with respect to species numbers in comparison with more mesic environments (although Cowling and Hilton-Taylor, this volume, demonstrate otherwise). We are still left, however, with the inescapable conclusion that climate exerts a controlling influence on global patterns of biological diversity (Tallis, 1991) and that changes in climate impact on these patterns. An accurate reconstruction of climate change and its impacts in the past is, accordingly, an important goal in biogeography.

Two major methods exist which facilitate the reconstruction and interpretation of the biogeography of the past. *Phytogeographical and zoogeographical methods* rely on identification and analysis of contemporary biogeographical patterns and on an accurate systematic classification of the taxa concerned. This has several constraints, not least of which is the general unreliability of most taxonomic distribution maps, particularly so in those far remote areas such as the karoo which are probably underdescribed in this respect. An additional problem is the controversial issue of interpreting the mechanisms which produce the characteristic distribution, i.e. the problem as to whether the patterns were produced by dispersal from a centre of origin, or by vicariance through geological or climatic change (see Poynton, 1983, for a review of this debate in the South African context). The controversy is further intensified because of contrasting interpretations offered by the proponents of either punctuated evolution or phylogenetic gradualism (Vrba, 1985; Hilton-Taylor, 1987). In some respects, the second major tool of biogeographical reconstruction, namely, *palaeontology and palynology*, which involve the more direct analysis of biological and other evidence actually preserved from the past, is the 'final arbiter' (Cowling, 1985) but is also subject to some important interpretive constraints. The fossil record is incomplete both spatially and temporally, and is biased in so far as only certain types of organisms have the physical and chemical characteristics promoting their preservation over time. It is further skewed by differential diagenesis of the fossils which form. These difficulties are compounded in the karoo, where the semi-arid and arid environments have provided relatively rare opportunities for the accumulation of reliable fossil assemblages, a problem which is especially marked in respect of fossil pollen (see Horowitz, 1992). Thankfully, there are some important exceptions described in this chapter which focusses mainly on the palaeontological and palynological record of the karoo.

Before examining in detail the evolutionary history and development of the karoo, it is apposite to consider briefly the issue of the origin of its biota. The problem has been addressed by Hilton-Taylor (1987) who reviews the various hypotheses put forward to account for the mixture of plant taxa that constitute the contemporary karoo, and by Vernon (this volume) who addresses the origins of the animals. Essentially, there are three major groups of interpretations, namely, those that focus on the links between the karoo flora and that of the subtropical and tropical flora to the north and east; those that focus on the associations between the flora of the karoo and that of the fynbos to the south and west; and, finally, those that recognize significant disjunctions between the karoo and the arid regions of North Africa which are, of course, non-contiguous in the present day but which suggest ancient linkages through an arid or semi-arid corridor. Hilton-Taylor (1987) concludes that, because the floristic data for the region are neither totally reliable nor completely analysed, it is not yet possible to eliminate any or all of the hypotheses. However, the strong associations between the western (winter rainfall) karoo and the fynbos, and the northern (summer rainfall) karoo and the flora of regions beyond its northern limits, suggest that, phytogeographically at least, the karoo is not an entirely coherent entity (Gibbs Russell, 1987; Jürgens, 1991; Hartmann, 1991; Bruyns and Linder, 1991; Cowling and Hilton-Taylor, this volume; Vernon, this volume).

The following narrative is based largely on a review of actual fossil evidence for previous plant and animal distributions and palaeoenvironmental conditions.

3.2.   Long-term geological and biological history

The karoo has always been subject to environmental changes, many of which have been reflections of major global geological events associated with continental drift and polar wandering, although intrinsic changes in climate are also apparent in the fossil record. The long-term geological development of the arid south-west may be regarded in four main stages: pre-Gondwana, Gondwana assembling, Gondwana assembled and Gondwana breaking up (Table 3.1). This has resulted in six main geological entities: the Kaapvaal craton, the Namaqualand Metamorphic Province, the Pan-African belts, the karoo Supergroup and the Kalahari sands (Tankard et al., 1982) (Table 3.1 and Fig. 3.1).

3.2.1.   Pre-Gondwana (3400–1000 Ma)

This stage produced the Kaapvaal craton and the Namaqualand Metamorphic Province in two periods of major crustal growth. The Kaapvaal craton impinges on the eastern sections of the arid south-west and contains granite–greenstone terrains which developed between 3400 Ma and 3100 Ma (De Wit et al., 1992). The greenstones are deformed volcano-sedimentary sequences in which unicellular and filamentous structures of biogenic origin have been identified (Engel et al., 1968). By analogy with modern examples, the occurrence of stromatolites (layers of calcium carbonate or silica, often with a hemispherical shape) indicates that this earliest life form on Earth comprised prokaryotes, probably blue-green algae (cyanobacteria) living in shallow water. As the Kaapvaal craton represents the early stages of growth of the Earth's crust, life in the karoo is, indeed, as old as its rocks.

The granite–greenstone terrains are overlain by a number of younger sedimentary and volcanic sequences, representing a range of environments between 3000 Ma and about 2000 Ma. Stromatolites have been found in the 2750 Ma Ventersdorp Supergroup (Grobler and Emslie, 1986) and in the Griqualand West Supergroup where they contain thick, extensive mats which grew in shallow water and tidal flat conditions about 2550 Ma ago (Beukes, 1977). The biomass of stromatolites releasing oxygen into the atmosphere became so great that it eventually resulted in a shift from anoxygenic to oxygenic atmospheric conditions at about 1800 Ma. This change is reflected in the first appearance of so-called *red beds* in the overlying

Table 3.1. *The geological timescale in relation to fossil evidence for the arid south-west of Africa*

RADIOMETRIC AGE (my)	GEOLOGICAL AGE	STRATIGRAPHY	FOSSIL EVIDENCE: PLANTS	FOSSIL EVIDENCE: ANIMALS	MAJOR EXTINCTION EVENTS
	QUARTERNARY	Quarternary successions	Angiosperms: the modern flora	Mammals etc	
1.8	TERTIARY	Tertiary successions	Gymnosperms and angiosperms	Mammals etc	[K/T Extinction]
65	CRETACEOUS	Uitenhage Group	Ferns, cycads and gymnosperms		
141	JURASSIC	Stromberg Basalts	*Dicroidium* flora including ferns, seed ferns, cycads, *Ginkgo* and gymnosperms	Dinosaurs	
195	TRIASSIC			Mammal-like reptiles	[P/T Extinction]
230	PERMIAN	Beaufort Group	Pure *Glossopteris* flora		
		Ecca Group		Reptiles	
280		Dwyka Group	Mixed *Glossopteris* flora	Amphibians	
345	CARBONIFEROUS		Lycopods	Arthropods	[Late Devonian Extinction]
		Witteberg Group		Fishes	
	DEVONIAN	Bokkeveld Group	Pterophytes	Brachiopods, trilobites and other invertebrates	
395			First land plants: psilophytes		
435	SILURIAN	Table Mountain Group		Trilobites etc	
500	ORDOVICIAN			Trace fossils (Arthropods)	[O/S Extinction]
570	CAMBRIAN	Nama Group		Metazoan fauna	
2000	PRECAMBRIAN	Namaqualand Province	Algal stromatolites		
		Griqualand West Supergroup			
3400		Kaapvaal craton	Algal stromatolites		

(Karoo Sequence spans Beaufort Group through Dwyka Group; Cape Supergroup spans Witteberg Group through Table Mountain Group)

Oliphantshoek Supergroup, the red colouration being due to iron oxide minerals (Botha et al., 1976). It was followed closely by the appearance of eukaryotes.

The Namaqualand Metamorphic Province lies to the west of the Kaapvaal craton and developed in the Proterozoic between about 2000 Ma until about 1000 Ma (Stowe et al., 1984). It represent the roots of a mountain belt that formed between 1200 Ma and 1000 Ma when the region became accreted to the Kaapvaal craton. Unequivocal establishment of palaeoenvironments and

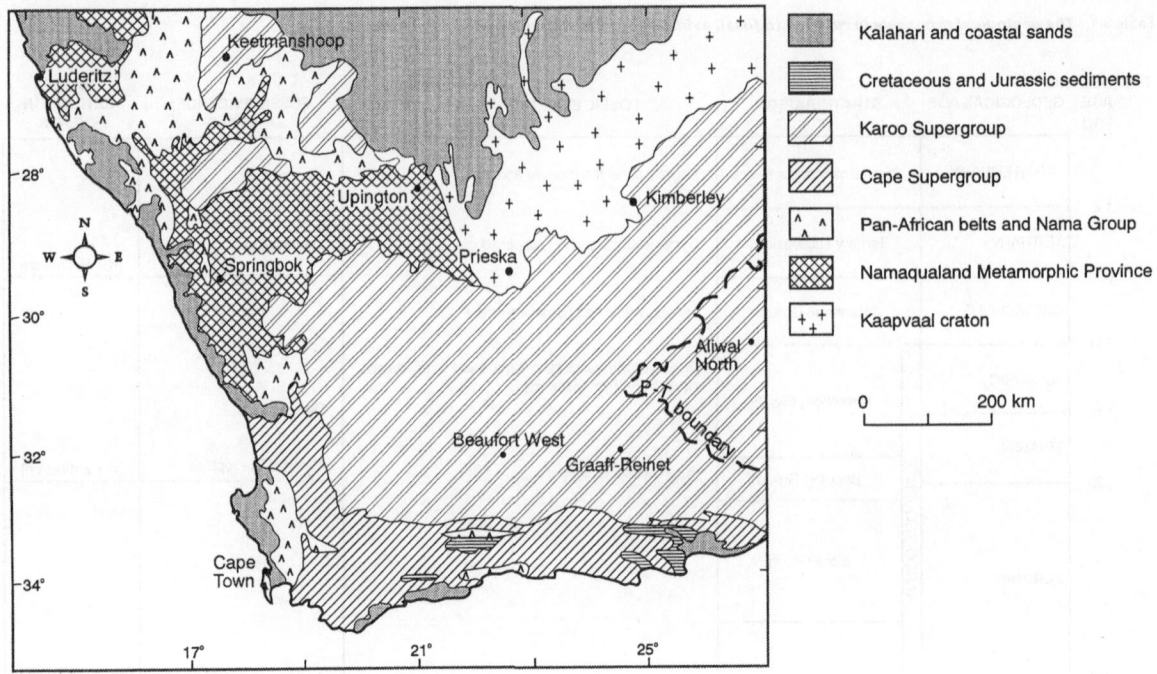

Figure 3.1 **Geological map of the karoo**

any associated life forms that might have existed then is impossible because of the complexity of the history of this phase.

3.2.2.   Gondwana assembling (800–250 Ma)

This stage involves the development of the Pan-African belts, the Cape Supergroup and the lower part of the Karoo Supergroup (Table 3.1 and Fig. 3.1). The Pan-African event commenced at about 800 Ma when the Proterozoic crust of the Namaqualand Metamorphic Province rifted and then drifted apart, forming a sedimentary basin and an intervening ocean (Stowe et al., 1984). This ocean then closed and a collision took place between Proterozoic continents, resulting in a mountain chain, an event that was completed by about 500 Ma.

While the Pan-African event was taking place, the first appearance of body fossils occurred in the geological record. These were first discovered at Edicara, Australia and have been dated at 640 Ma (Clarkson, 1993). A similar assemblage occurs in the Nama Group of southern Africa which, as with the Edicara situation, is associated with late Precambrian glaciation (Germs, 1974). The metazoan fauna, including coelenterates and medusoids, inhabited shallow seas and tidal flats in a foreland basin adjacent to the Pan-African mountain chain.

Erosion and peneplanation of the Pan-African mountains was followed by deposition of the Cape Supergroup across the southern parts of the region. The Ordovician

Table Mountain Group sediments were initially deposited on tidal flats but, as the region subsided, this changed to shallow marine and continental shelf-environments inhabited by marine brachiopods and the oldest known arthropods, trilobites. A glacial interlude, responsible for the Ordovician–Silurian extinction event (Rong and Harper, 1988), was followed by a post-glacial transgression and more tidal flats with marine invertebrates (Cramer et al., 1974).

These conditions became deltaic and moderately to high-energy marine in the Devonian, as recorded by the Bokkeveld Group, which contains a diverse suite of invertebrates, fish and plants (Boucot et al., 1969; Chaloner et al., 1980; Plumstead, 1967). The plants represent some of the first terrestrial communities, with the earliest in the Bokkeveld being semi-aquatic psilophytes. Despite their primitive status, they possessed the necessary support, conducting and reproductive systems to enable them to cope with the challenge of life on land (Ingrouille, 1992). This terrestrial invasion by plants is one of the most significant events in evolution because the invertebrates and vertebrates eventually followed them out onto land.

In the upper Bokkeveld, the psilophytes were replaced by more advanced pterophytes (ferns) as deltaic conditions gave way to the tidal, lacustrine and finally glaciogenic environments of the Witteberg Group (Hiller and Dunlevey, 1978). These sediments were deposited during the time of the world-wide Devonian extinction zone. In this region, however, marine bivalves, brachiopods and trilobites were replaced by freshwater plants and palaeoniscoid fish, reflecting the change in environment rather than the extinction.

The top of the Witteberg Group is probably Carboniferous in age (Gardiner, 1969) and represents the onset of a major period of glaciation as southern Africa drifted towards the South Pole. After a hiatus (Dunlevey and Hiller, 1979), the succeeding Karoo Supergroup was deposited virtually uninterrupted from about 300 to 180 Ma as southern Africa migrated away from the South Pole. Glacial environments were succeeded progressively by tundra, temperate and then high latitude desert conditions, until the eruption of karoo basalts (Smith, 1990). During this time, an amazing radiation of the amniotes took place, resulting in the first terrestrial ecosystems comparable to modern situations and having profound implications for subsequent life on Earth.

The roots of this diversification lie in the Carboniferous period which contains the oldest tetrapod fossil (an amphibian) and the oldest reptile. After evolving from the amphibians, the reptiles quickly split into three main lineages (Benton, 1990), based on the presence, absence and number of fenestra (skull openings behind the eye sockets). The anapsids have no opening, and are ancestral to turtles and tortoises; the synapsids have a single lower opening and ultimately became the mammals, via the mammal-like reptiles; the diapsids have two openings as seen in birds, lizards, snakes and crocodiles, as well as in dinosaurs. This tripartition occurred while southern Africa was positioned over the South Pole and largely covered by ice sheets. The deposits of this time are represented by the early stages of the Permo-Carboniferous Dwyka Group rocks of the Karoo Supergroup. As the region moved away from the South Pole, the continental ice sheet retreated to the highlands lying to the north-east, and a shallow epicontinental sea widened between this sheet and the semi-grounded and rafted ice shelf to the south and west. Opening up of the sea facilitated the influx of marine flora and fauna such as radiolaria, foraminifera, crinoids, echinoids, asteroids, brachiopods, gastropods, the bivalve *Eurydesma*, cephalopods *Orthoceras* and *Eosianites*, and palaeoniscid fish (McLachlan and Anderson, 1973).

The amelioration of climate as the region moved out of polar conditions saw the further retreat of the ice sheets and deposition of the Ecca Group (Smith, 1990). Around the continental highlands in the east and west, deltaic and lacustrine environments represent tundra-like conditions, akin to the present situation of the MacKenzie delta on the Canadian–Alaskan border. Further south, submarine fans spilled northwards onto a continental slope. Molluscan fossils are absent, probably due to dissolution of the shells but, most notably, the lowermost early Permian Ecca contains *Mesosaurus* (McLachlan and Anderson, 1973). This is an anapsid reptile, which is the first known reptile to return to an aquatic environment, probably to catch fish such as *Palaeoniscus capensis*, which is also found in the Ecca.

The characteristic Gondwana fossil plant *Glossopteris* makes its first appearance during this period. Consisting mainly of trees (up to 6 m tall) and shrubs, these seed-bearing ferns had a deciduous habit which suggests that the climate at the time was seasonally dry and at least locally semi-arid, as evidenced by stunted forms and small-leaved structures of some fossils (Oelofsen and Loock, 1987). Still others have argued that the climate of the time must have remained cool (Plumstead, 1967) owing to the fact that the karoo lay some 65 degrees south (Smith et al., 1973). Rayner (1995) has, however, analysed the flora at three fossil localities and concluded, on the basis of dominant large-leaved forms of *Glossopteris*, that the growing conditions, at least during summer months at this time, were in fact highly favourable; daily temperatures exceeding 30 °C and annual rainfall values of around 3000 mm or more are interpolated. The large forests on the eastern Ecca deltas, which must have been composed of a closed canopy mixed woodland (Plumstead, 1969), eventually formed the coal deposits of that region.

The succeeding Beaufort Group was deposited in a foreland basin during early stages of Cape Fold Belt development along the southern parts of the region (Tankard et al., 1982). This period of mountain building was caused by the collision of Patagonia with Africa and the rest of South America (Lock, 1980); it represents the final assembling of Gondwana. Rivers flowing northwards off the Cape Fold Belt and south-westwards from the continental interior formed fine-grained alluvial floodplains with low to high sinuosity channels (Smith, 1990). By these lower Beaufort times the climate had become more equitable and the vegetation on these floodplains included trees such as *Dadoxylon*, early conifers, club mosses, horsetails, tree ferns, seed ferns and a fossil species of *Ginkgo*. These terrestrial niches became exploited by terrestrial vertebrates (Rubidge, 1987), giving rise to a splendid array of species (Kitching, 1977; Keyser and Smith, 1979).

The synapsid lineage was represented by the therapsids (Parrish et al., 1986) which displayed advances over their early Permian sail-backed pelycosaur ancestors with respect to their jaws and limbs – enabling a wider range of life styles. In the late Permian, the floodplains were inhabited by the suborder Dinocephalia, which comprised both giant herbivores, such as *Moschops* (>5 m long), as well as smaller carnivores like *Titanosuchus*. The dominant carnivores of the time, however, belong to the suborder Gorgonopsia (Benton, 1990), with *Lyacaeops* displaying the prominent canines and large gape necessary for killing prey and rapidly consuming meat.

Included in its prey would have been members of the suborder Dicynodontia, of which Endiothioda is a primitive example (King, 1990). This suborder ranged in size from a few centimetres to 3 m long, with only some species, such as *Dicynodon*, retaining the two canines that give the suborder its name. Rather suprisingly for an entirely herbivorous suborder, most species lacked any teeth; instead, a horny beak was used for tackling the tough vegetation of the period. Fully articulated, curled-up specimens of *Diictodon* have been discovered in the terminal chambers of underground burrows (Smith, 1987). Burrows in the river-banks may have been susceptible to flooding, but they provided protection against the cold and defence against such carnivores as the suborder Cynodontia. *Procynosuchus* is a good example of these small to medium-sized animals which had differentiated teeth and the makings of a secondary palate, characteristics which were to be ultimately inherited by their mammalian descendants (Kemp, 1982).

The therapsids were not the only terrestrial inhabitants of the region in the late Permian. The squat, bulky anapsids, and pareiasaurs, such as *Scutosaurus*, held stage for a while and probably helped marginalize the diapsids. A particular diapsid of the Beaufort, *Youngina*, is ancestral to the lizards and snakes (Carroll, 1977), whereas most diapsids were thecondonts which gave rise to the archosaurs.

3.2.3.   Gondwana assembled (250–155 Ma)

After the accretion of Patagonia, the fully assembled supercontinent of Gondwana existed as a single entity for just under 100 Ma. It comprised South America, Africa, Arabia, Madagascar, India, Antarctica and Australia, plus various fragments that are now attached to Europe and Asia. In the northern hemisphere, North America, Greenland, Europe and much of Asia formed another supercontinent called Laurasia. As this abutted against north-west Gondwana, during this period the terrestrial world essentially comprised a single landmass, known as Pangaea, enclosed by a single ocean (Scotese and Sager, 1989).

At 248 Ma, the most devastating extinction event recorded in geological history took place (Maxwell, 1989). It is marked by the change from the Palaeozoic ('old life') to the Mezozoic ('middle life') on the boundary between the Permian and Triassic. The cause of this event is uncertain but it coincides with extrusion of the continental flood basalts of Siberia. It occurs right in the middle of the Beaufort Group, affording an excellent opportunity to study the effect of the extinction on terrestrial vertebrates in the karoo region.

The upper Beaufort Group, younger than 248 million years, records the recovery and brief flourishing of the therapsids after the extinction event. The dicynodont *Lystrosaurus* was a true world citizen, taking advantage of the single landmass and exploiting its semi-aquatic existence to the fullest (King, 1990). It was by far the dominant animal in every early Permian terrestrial community from Antarctica to Russia. A second radiation gave rise to more massive varieties such as *Kannemeyeria*, but eventually the dicynodonts went extinct before the Jurassic.

The Permo-Triassic extinction decimated most of the therapsid carnivores with the notable exception of the cynodonts. This group displayed significant advances in the Triassic (Kemp, 1982). The dog-like *Thrinaxodon* displays a well-developed hearing mechanism, while *Cynognathus* may have had whiskers. It is speculated that both may have acquired yet another mammalian characteristic, namely fur. This lineage survived across the Triassic–Jurassic boundary to give rise to the shrew-like *Megazostrodon*, a typical early mammal.

In the early Triassic, the descendants of the surviving thecondonts, the archosaurs, made use of the niche left vacant by the demise of most of the therapsid carnivores (Benton, 1990). One group, the proterosuchids, retained the thecondont sprawling gait, whereas another, the crocodylotarsi, developed a semi-erect gait, giving rise to *Orthosuchus*, *Lesothosuchus* and, eventually, crocodiles. A further group, the erythrosuchids, developed a new stance with legs directly under the body. This gave them a locomotive edge which was used for carnivorous advantage, as shown by the impressive dental system of *Erythrosuchus*, a giant carnivore (up to 5 m long), and *Euparkeria*, the World's first biped (Ewer, 1965). These are the early Triassic forerunners of the dinosaurs.

The Molteno Formation overlying the Beaufort Group represents fluvial environments and is poorly fossiliferous as far as vertebrates are concerned, although it is rich in plants and insects (Cairncross et al., 1995). It was during this time that the dinosaurs divided into the entirely herbivorous order Ornithischia ('bird-hipped') and the order Saurischia ('lizard-hipped') consisting of the carnosaurs and sauropodmorphs (Benton, 1990). The succeeding lacustrine and desert environments of the Elliot and Clarens Formations (Smith, 1990) contain some of the earliest dinosaurs of both orders. *Hetereodontosaurus* was an early ornithischian (Charig and Crompton, 1974) while *Massospondylus*, was the first prosauropod. This flourishing fauna perished at about 180 Ma when the continental flood basalts of the karoo Supergroup erupted and covered southern Africa (Erlank, 1984), to be followed by an hiatus until about 155 Ma.

3.2.4.   Gondwana breaking up (155 Ma to present)

At about 155 million years ago, Gondwana began to break-up, initially through West Gondwana (South America,

Africa and Arabia) sliding past East Gondwana (Madagascar, India, Antarctica and Australia). Then, about 138 Ma, the present continents drifted apart and the intervening oceans formed (Scotese and Sager, 1989). This resulted in isolation of biological communities and duplication of environments on separate continents. The development of new oceans with fast spreading mid-ocean ridges reduced the volume of the ocean and produced a world-wide rise in sea-level to heights of up to 400 m above the present level (Haq et al., 1987), a situation which prevailed until around 30 Ma when levels returned to approximately those of today. These changing levels and the dynamics of continental geography were instrumental in the cooling of the southern oceans and the subsequent aridification of the west coast and adjacent interior of southern Africa, a circumstance with dramatic implications for the biogeography of the karoo throughout the Tertiary and Quaternary, discussed below.

The arid south-west of Africa has retained a rather poor record of the events of the break-up of Gondwana as most of the fossil evidence probably lies offshore on the continental shelf (Dingle et al., 1983). Noteworthy assemblages are found at Kangnas, 100 km south of the Orange River, where

dinosaur bones and teeth of Cretaceous age are preserved (Haughton, 1915), and at Vaalputs (Fig. 3.2), where there are fossils preserved in the Dasdap alluvial fan, of late Cretaceous to early Tertiary age (Levin et al., 1986).

This review of its longer-term geological and biological development reveals that the flora and fauna of the modern karoo environment are hardly *direct* descendants of the forms of life which inhabited its environs for much of the geological record. Rather, they are forms derived or inherited from assemblages which emerge mainly in the upper part of the record, identifiable only from the Tertiary and Quaternary sediments of the past 65 million years, to which we now turn.

3.3.   Tertiary historical biogeography

3.3.1.   Palaeoenvironmental indications during the Tertiary

The last 65 million years of Earth history witnessed arguably its most dramatic physical and biological changes. The continents, having been subject to major

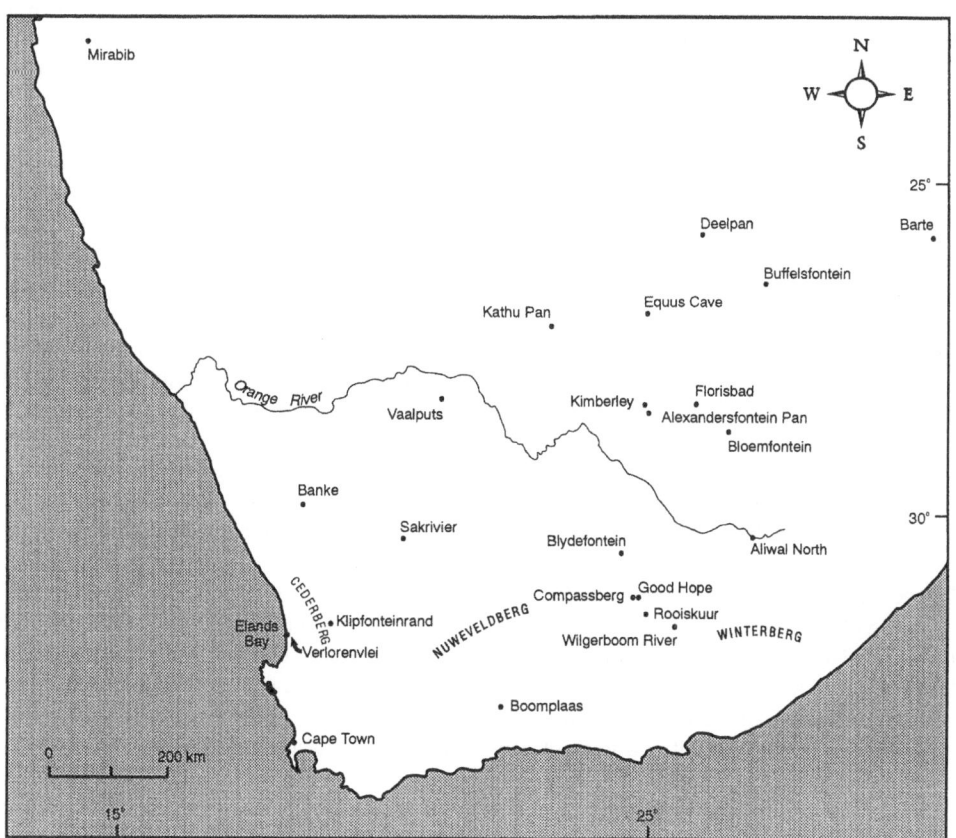

Figure 3.2 **Location map for places mentioned in the text.**

fragmentation since the onset of the Mesozoic, gradually took up their contemporary geographical configuration and, in so doing, appear both to have pushed some pieces of the terrestrial jigsaw puzzle into line while at the same time once again separating oceans and isolating certain ocean currents. The biogeographical impact of this alone was enormous (Tallis, 1991), but was also accompanied by widespread climatic change which further compounded the evolutionary effects. Earlier in the Tertiary, global climates seem to have been relatively benign, with most regions, particularly those at lower latitudes, experiencing both temperature and moisture values more favourable to mesic plant communities than today, such that there appear to have been predominantly woody floras at all latitudes (Axelrod and Raven, 1978; Wolfe, 1985; Tallis, 1991). But global temperatures, now known from detailed oxygen-isotope analyses on ocean sediments (Shackleton, 1986), declined steadily throughout the early Tertiary and reached minimum values, appreciably lower than those of today, in the late Oligocene, perhaps 25 Ma (Tallis, 1991). After recovering in the early Miocene, temperatures declined again and there was undoubtedly glacial ice present in Antarctica by the mid Miocene, a situation which would have had important implications for the generation of the Benguela current offshore south-western Africa (Linder et al., 1992). Falling sea levels (Haq et al., 1987) intensified continental aridity, as indeed did the geological uplift simultaneously occurring in many regions, and the overriding climatic signal, particularly for the later Tertiary of the continental interiors, is one of deteriorating temperature and moisture conditions. All this adds up to an expectation that the subtropical conditions of the karoo should be replaced by cooler and semi-arid climates later in the Tertiary; this hypothesis can thus be tested against the fossil record for that period.

The palaeogeography of the continents is particularly important at 65 Ma, when a major extinction event occurred at the Cretaceous–Tertiary (KT) boundary. This event, usually ascribed to a meteorite impact on the Yucatan peninsular of Central America, had a significant impact on biological evolution. The extinction of the dominant terrestrial group, the dinosaurs, was followed by rapid diversification of mammals and grasses in the early Tertiary. The geological separation of the continents, combined with world-wide high sea levels, enabled the marsupial mammals to evolve and dominate the fauna of South America, Madagascar, Antarctica and Australia. In Africa, however, which was loosely connected to Europe and Asia, the marsupials lost the competition with invading placental mammals.

The arid south-west of Africa has little record of the Tertiary period until the late Miocene when many of the recorded species originated elsewhere in the world. Their evolution during this time is related to world-wide climatic changes which were, in turn, related to geological events. During the Palaeocene, the climate world-wide was warmer than the present. These conditions assisted the spectacular adaptive radiation of the placental mammals which gave rise to many of the modern orders, as well as numerous extinct orders. The condylarths represent the archaic ungulate root-stock from which the perissodactyls and the artiodactyls evolved. During the same period, the carnivores separated into the creodonts and the miacids, the ancestors of the extant carnivores. In the Old World, the miacids gave rise to the viverravine branch which yielded cats and hyaenas, whereas, in the New World, the vulpavine branch eventually produced dogs and bears.

The Eocene saw the diversification of the perissodactyls, which became the main browsers, and then the artiodactyls. While both these orders were herbivorous, some condylarths became hoofed-carnivores, competing successfully with the creodonts, whereas others returned to the sea, evolving into whales and their relatives. During the early to mid Eocene, the climate was warmer than in the Palaeocene, but the onset of break-up between Australia and Antarctica at about 50 Ma was to have far-reaching consequences for world climate. It allowed the development of the circum-Antarctic current which chilled the Southern Ocean and commenced a world-wide cooling from the late Eocene.

The Eocene–Oligocene boundary (38 Ma) is known as the 'Great Divide' because of the change in fauna brought about by the drying of the Turgai Straits west of the Urals which separated Asia from Europe. The condylarths became extinct and the perissodactyls were particularly badly affected, with horses only surviving in North America. With the development of the Antarctic ice-cap at 32 Ma and accompanying sea-level fall, another land bridge emerged, this time in the Bering Sea between Asia and North America. It afforded the opportunity of expansion for the ungulates as well as interchange between the Old World and New World carnivores as they pursued their prey.

The Miocene is recorded in fossil materials found at Sakrivier (Fig. 3.2) between Brandvlei and Sakrivier station (Bamford and De Wit, 1993; De Wit and Bamford, 1993). The older terraces are mid Miocene, with extant angiospermous families (Dipterocarpaceae, Fagaceae, Myrtaceae, Oleaceae, Rutaceae) and indicate a moist subtropical climate. At the end of the Miocene, a significant shift to more arid conditions is indicated by the development of calcretes. The younger terraces in this area document periods of greater moisture availability, also known as pluvials.

3.3.2.   The Banke flora

The most important Tertiary fossil site in southern Africa hails from the Arnot kimberlite pipe on the farm Banke, some 80 km south of the Orange River in Namaqualand (Fig. 3.2). The Arnot Pipe is a sediment-filled volcanic pipe which was formed during crustal activity around 70 Ma. Several types of fossils have been identified in the sediment, including petrified wood (Adamson, 1931), leaves (Rennie, 1931) and even frogs (Haughton, 1931), although it is the palynomorphs, particularly of the upper 20 m of sediment that have produced the most valuable palaeoenvironmental information (Kirchheimer, 1934; Scholtz, 1985). Scholtz (1985) provides a detailed account of the palynology of seven levels within the sediments, which appear to have accumulated during the Palaeocene and offer a fascinating insight as to the nature of the vegetation which occurred in what is today the karoo–Savanna ecotone. Scholtz describes 72 forms of spores, gymnosperm and angiosperm pollen with affinities to some 28 plant families including, most significantly, members of the extant Proteaceae, Restionaceae and Ericaceae families, all of which are prominent members of the present-day fynbos. By this time, there appear to have been marked differences between the climate and vegetation of the Namaqualand region and that of subtropical Africa to the north and east. Climate was warmer than today and the vegetation is suggested to have formed a type of dry forest with fynbos shrubs forming an understorey (Scholtz, 1985, p. 100). There are no indications of the asteraceous shrub flora that dominates the region in the present day. The development of karroid plant and animal communities in this part of the present-day karoo biome is, therefore, seen to post-date the early Tertiary and is more likely to have occurred in response to the increase in aridity associated with the growth of the Antarctic ice sheet during the Miocene.

Hilton-Taylor (1987) agrees that the dry forests seem to have persisted into the Miocene, at least along river courses, although progressively more open habitats appear to have developed in response to dryness. Unfortunately, the replacement of these woodlands by karoo vegetation proper can only be indirectly inferred, because there is no fossil plant record for the region spanning the late Tertiary. The woody tree type of vegetation was certainly absent by the time we are next able to witness any direct evidence for environmental conditions, which is for deposits dating only to the latter part of the Quaternary period. Perhaps the most intriguing component of the biological history of the karoo, i.e. the arrival and dispersal of xeric shrubland and its associated animal communities, has remained elusive.

3.3.3.   Geomorphological development

Partridge and Maud (1987), following King's (1955) philosophy of land surfaces, propose that the geomorphological development of the karoo region since the Mesozoic consists of a number of uplifts with intervening periods of pediplanation. The oldest land surface is the African land surface with its diagnostic deep weathering profiles and duricrust capping. Its inception was in the early Cretaceous and, by the early Miocene, a gentle pediplain had been produced at an elevation of 500–600 m. Another uplift of 150–300 m in the Miocene tilted the continent slightly west, producing the post-African I surface. This event terminated at the end of the Pliocene when further uplift raised the eastern interior by as much as 900 m to commence the post-African II cycle. This is manifest in downcutting of the rivers in the interior and deep incision of the coastal hinterland. It is proposed that the Great Escarpment developed at the time of continental rifting. Its survival as a topographic feature throughout these pediplain events is considered to be a consequence of the high elevation of southern Africa prior to Gondwana break-up.

This scheme indicates that most of the denudation of southern Africa took place during the Cenozoic. It has been challenged by numerical experiments which indicate that the majority of the denudation took place before the beginning of the Cenozoic (Gilchrist et al., 1994). It is a problem that has yet to be resolved satisfactorily because the dating of landscapes is notoriously difficult.

3.4.   The Quaternary period including the Holocene

3.4.1.   Fluctuating climates of the Quaternary

The final two-million-year period of Earth history is generally known in greater detail in respect of its historical biogeography, not least because the relatively recent geological sediments have been better preserved. It is a period characterized by widespread and repeated fluctuations in climate and, in its upper part, the establishment of human populations. Both climate change and human activity have been shown to have significantly impacted plant and animal communities (Goudie, 1991), and the karoo is no exception. The most obvious environmental signal of the period is indicated in the successive expansion and contractions of the major global ice sheets.

While southern Africa was not directly influenced by these fluctuations in a physical sense, the variations in global temperature and moisture conditions were reflected here in important ways. Later Quaternary climate changes in southern Africa have recently been reviewed by Partridge et al. (1990) and Partridge (1993), although the data prior to 40 000 y BP are very sparse. The last full glacial–interglacial cycle is therefore poorly represented, although it is possible to glean some idea of the magnitude of change in the subcontinent from the more complete record of the last 20 000 years, which encompasses the Last Glacial Maximum and the Holocene so-called 'climatic optimum' or hypsithermal. Based on a wide range of palaeoenvironmental data, Partridge (1993) has suggested that much of the interior of southern Africa, including the karoo, would have been five to six degrees Celsius cooler on average around the time of the maximum global extent of ice sheets between 16 000 and 21 000 y BP. These cooler climates were almost certainly associated with intensified aridity in the summer rainfall region. The Holocene hypsithermal occurred in southern Africa between 7000 and 6500 y BP when temperatures may have been as much as two degrees warmer than today (Partridge, 1993). The magnitude of such changes is likely to have had a marked impact on the plant and animal communities of the karoo which, it can be hypothesized, would have responded by adjustments in range as well as through speciation and extinction events.

Much of what is known of the fluctuating environmental conditions in the karoo during the late Quaternary is based on pollen analysis and related palaeoecological techniques, despite the obvious problems faced by the technique in arid and semi-arid circumstances (Horowitz, 1992). A variety of karoo environments have yielded data on this issue, and the more important sites are now reviewed.

3.4.2.   Florisbad

The Florisbad site is located 45 km north of Bloemfontein in what is today the ecotone between the karoo and grassland biomes. It consists of a mound of sand, silt and organic clay sediments which have accumulated at a spring on the edge of a large salt-pan depression (Van Zinderen Bakker, 1989). These sediments have yielded well-preserved fossil pollen which, although difficult to date very precisely, has resulted in a detailed palaeoecological reconstruction (Van Zinderen Bakker, 1989).

The sediments probably accumulated at the spring site, over an extended period of the later Quaternary, intermittently from the time of the penultimate glacial period, perhaps 200 000 years ago (Van Zinderen Bakker, 1989) during the so-called Middle Stone Age. At this time, the pollen spectra are dominated by grasses and suggest extensive grassland environments in the vicinity of Florisbad. A complex sequence of fluctuations follows, in which, at various stages through to what is probably the Holocene, grasses (indicative of greater moisture) alternate in dominance with pollen types more indicative of aridity (in particular the Chenopodiaceae, presumed to be the halophytic *Salsola* which is associated with brackish habitats often following rivers or around pans and floodplains).

The alternation of grassland and arid karoo vegetation throughout the sequence is seen as a response to fluctuating dry–warm/cold–moist environments, an interpretation which is based partly on the correspondence of other pollen taxa and partly on associated animal fauna (Van Zinderen Bakker, 1989; Brink, 1988). The association of cool climates at Florisbad with conditions in which moisture is more freely available is not in agreement with most reconstructions of cooler Quaternary climates for the interior of southern Africa (see Partridge et al., 1990) which are usually thought to have been more arid. Van Zinderen Bakker's (1989) reconstruction, however, does not rule out the possibility that temperature reduction produced a depression of the altitudinal vegetation zones around Florisbad and introduced upland elements to the vegetation; in this sense greater moisture does not, therefore, necessarily mean greater precipitation. The difficulty of establishing an adequate chronology for the Florisbad sequence, however, reduces the palaeoecological value of what is otherwise an important Quaternary fossil site at the northeastern margins of the contemporary karoo. An additional problem is highlighted by the sedimentological work of Visser and Joubert (1991), who note that the interpreted arid phases based on the palynological work of Van Zinderen Bakker (1989) paradoxically appear to coincide with higher lake stands. Scott and Brink (1992) suggest that there may be post-depositional movements or differential corrosion of pollen resulting from surficial cracks in the sediment surface or root penetration.

3.4.3.   Aliwal North

A site analogous to the Florisbad spring is found at Aliwal North (Coetzee, 1967), a few kilometres south of the Orange River (Fig. 3.2), again in the present-day karoo–grassland transition. A sequence of organic peat deposits accumulating at the Aliwal spring has been radiocarbon dated and represents accumulation of pollen-rich deposits around the Pleistocene–Holocene transition, between approximately 12 600 and 5000 y BP. Regrettably, the last few thousand years of deposit contain little or no pollen so that the palaeoecological picture is restricted to the immediate post-glacial period. Here, as at Florisbad, the vegetation around the site is shown to be in a state of dynamic flux throughout the period of sedimentation, with grass-dominated communities alternating frequently with those dominated by karroid vegetation. Coetzee (1967) interprets the periods of prominent grassland as cooler and moister than the present day, while the periods characterized by Chenopodiaceae and Asteraceae pollen are thought to have been warmer and drier (cf. Van Zinderen Bakker, 1989). Again, there exists the possibility that this situation reflects depression of altitudinal belts of vegetation in the region. The more recent record for this region is provided by pollen analysis of sediments at nearby Badsfontein by Scott and Cooremans (1990) which points to prominent aridity in the earlier Holocene followed by a period of greater moisture availability after 4000 y BP.

Most interestingly, Coetzee (1967) records a marked shift in prevailing vegetation at Aliwal between around the time of the so-called Younger Dryas/Allerød transition well documented in northern hemisphere sites (Williams et al., 1993). Since this event has been clearly identified in other southern hemisphere localities, particularly in South America, it is not surprising that sediments dated to an equivalent age should illustrate a marked contemporaneous temperature oscillation in southern Africa. The Aliwal sequence strongly suggests that environments such as the karoo–grassland ecotone are highly sensitive to environmental changes.

3.4.4.   Pan sites of the karoo

Until relatively recently, it was believed that only the scarce sedimentary accumulations of organic matter could provide sufficient palynomorph fossils to produce reliable reconstructions of Quaternary palaeoecology. Numerous studies of the salt-pans of semi-arid southern Africa, however, using such evidence as animal bones (see references in Scott and Brink, 1992), molluscs and ostracods (Partridge and Dalbey, 1986), ostrich eggshell (Horowitz et al., 1978) and, most particularly, pollen (Scott and Brink, 1992) have revealed a rich palaeoenvironmental record preserved in their sediments. The Alexandersfontein pan (Fig. 3.2), east of Kimberley, is one of the largest in the region, and was occupied by a lake from around 16 000 to 13 000 y BP (Butzer et al., 1973) suggesting that, even in the event of significantly cooler climates, precipitation was somewhat higher immediately following the Last Glacial Maximum. Pollen from pan and spring sediments at Deelpan (Fig. 3.2) has been interpreted by Scott (1988; Scott and Brink, 1992) as indicating fluctuating environments during the Holocene. Changes in the ratio of xeric indicators, principally pollen of Chenopodiaceae, are used by Scott and Brink (1992) to infer moisture fluctuations in the second half of the Holocene. Moister conditions appear to have characterized much of the later Holocene, in contrast to the Kathu pan (Fig. 3.2) reconstruction of Beaumont et al. (1984), which points to swampy vegetation, and thereby moist local conditions, associated with the first half of the Holocene. Whether the difference suggests regional contrasts in climatic response, or interpretive difficulties associated with the over-representation of local pollen in such environments remains an unsolved problem. In any event, it is clear that more palaeoenvironmental studies on sediments of this type are required before their significance can properly be assessed.

3.4.5.   Valley deposits of the karoo

Meadows (1988) has presented argument alluding to the potential of several types of valley and valley-head deposits in the search for pattern in the Quaternary of southern Africa. The recent environmental history of the karoo is clearly preserved in such sediments, and a range of types of palaeoenvironmental evidence has been forthcoming. In sediments which are in any way organic, pollen is generally preserved in adequate quantities, notwithstanding the normally high rates of decomposition associated with arid and semi-arid conditions. Palynological studies on sites in the southern central karoo have been conducted by Bousman et al. (1988) in the Blydefontein basin, by Sugden and Meadows (1989) in the Nuweveldberg mountains near Beaufort West and at Compassberg, near Graaff-Reinet by Meadows and Sugden (1988).

An insight into the Holocene environmental history of the southern central karoo is provided by Bousman et al. (1988) employing evidence from a sequence of valley cut-and-fill deposits at Blydefontein, near Hanover in the Eastern Cape (Fig. 3.2). Pollen and sedimentary evidence indicates fluctuations in environmental conditions, with greater moisture apparent in the later part of the Holocene. Grassier fynbos prevails between around 5000 y BP and 1000 y BP, while the shrub component dominates in a poorly dated early Holocene fill. The sediments appear to record a recurrent sequence of cut-and-fill cycles, which Bousman et al. (1988) attest may be a function of adjustments of the local sedimentary system to small-scale climatic changes. Clearly, even in these central karoo environments, the prevailing vegetation is sensitive to relatively minor changes in precipitation quantity or seasonality.

The studies of Meadows and Sugden (1988) on a similar valley fill sequence at Compassberg (Fig. 3.2) also suggest subtle, but ecologically significant changes in climate during the Holocene. Pollen studies on an organic deposit in an upland valley head record significant variations in the ratio of types indicative of greater abundances of grass in the region to those types suggesting more xeric conditions.

A picture of environmental changes which correspond to the last few hundred years only in the karoo is given by palaeoecological reconstructions at Nuweveldberg (Sugden and Meadows, 1989), in the mountains north of Beaufort West. The record is characterized by relatively minor shifts in vegetation patterns, most likely associated with small-scale climate change, although the influence of human activity, in the form of adjustments to the grazing regime, may explain some of the variations. Grass and asteraceous (karroid shrub) pollen frequencies, as in the other central karoo sites at Blydefontein and Compassberg, seem to fluctuate out of phase and again suggest the periodic expansion and subsequent contraction of grassland elements. Paradoxically, however, the near-surface (i.e. recent) spectra at Nuweveldberg are characterized by increased pollen frequencies in both grasses and arid-indicators, a phenomenon which Sugden and Meadows (1989) suggest may be a function of increased grazing pressure reducing the abundance of palatable grasses but favouring the expansion of the non-palatable grass *Merxmuellera*.

The palynology of all the valley sites in the central and eastern karoo reveals what may be described as the ebb and flow of grassland at relatively frequent intervals in these landscapes at various stages during the late Pleistocene and Holocene. Fig. 3.3 represents an attempt

to collate the pollen data from Blydefontein, Compassberg and Nuweveldberg valley sites supplemented by the important late Pleistocene data from Aliwal North (Florisbad was omitted due to the dating problems), and reveals regularly oscillating grass and karroid-indicator (Chenopodiaceae plus Asteraceae) pollen frequencies suggesting that grassiness in the karoo is a consistent, if fluctuating, element of the vegetation.

Some additional sites central to and on the periphery of the contemporary karoo are worthy of mention. Holmes and Marker (1995) have examined exposures of palaeosols and other deposits at Good Hope, Rooiskuur and Wilgerboom River, all in the Eastern Cape (Fig. 3.2) and conclude that a marked environmental change took place

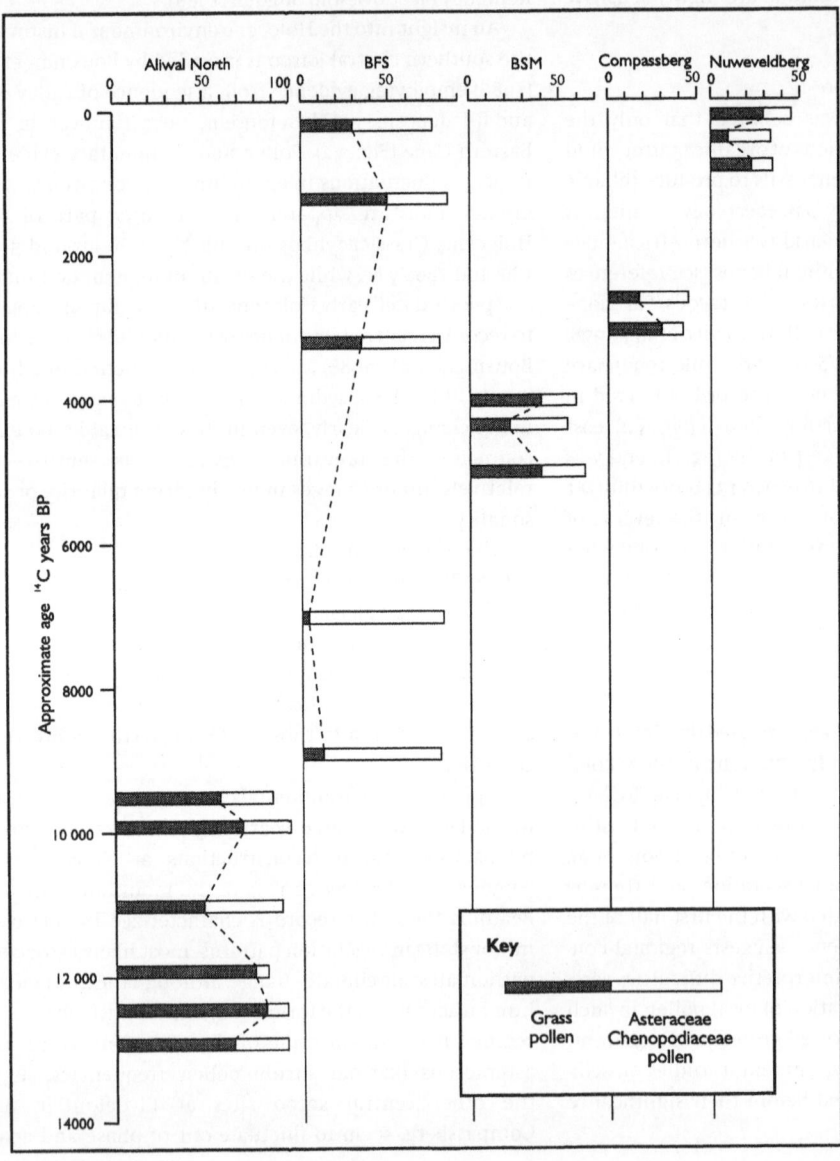

Figure 3.3 **Summary pollen diagrams for sites at Aliwal North (after Coetzee, 1967), Blydefontein (BFS = Blydefontein section, BSM = Blydefontein stream mouth, after Bousman et al., 1988), Compassberg (after Meadows and Sugden, 1988) and Nuweveldberg (after Sugden and Meadows, 1989)**

in the mid-Holocene of the karoo. The implication is that this corresponds to a shift in moisture conditions associated with a largely more arid earlier Holocene and greater moisture availability in the later Holocene. Meadows and Meadows (1988) examined late Pleistocene and Holocene polleniferous sediments in the Winterberg mountains (Fig. 3.2), rather to the east of the contemporary karoo boundary. Karroid elements were apparently more abundant towards the end of the Pleistocene, and appear to have invaded these mountain sites to a degree, perhaps indicating the coincidence of cooler with drier conditions in this region at that time. The waxing and waning of karroid vegetation communities in the eastern and central karoo is, however, not so apparent at its western margins; Meadows and Sugden (1991) could find no evidence for the invasion of fynbos by karoo vegetation at any time during the last 14 000 years at higher altitudes in the Cederberg mountains (Fig. 3.2), although the possibility that this occurred at lower altitudes during glacial times cannot be ruled out, as suggested by Scott (1994). At the western margins of the karoo, where karroid vegetation merges into fynbos, Meadows et al. (1994) and Baxter and Meadows (1994) have examined the later Holocene vegetation history of the Verlorenvlei (Fig. 3.2). The pollen preserved in a complex series of lacustrine and estuarine sediments provides an insight into the dynamics of vegetation in the catchment over a period of approximately 5000 years. Many of the changes are indicative of fluctuations in sea level, for example, the lower portions of this valley effectively formed a wide marine embayment during the mid Holocene high sea-level stands and freshwater conditions only became established within the last few hundred years. From the perspective of karoo vegetation, the most interesting observations (Baxter and Meadows, 1994) are those which suggest widespread physical and biological disturbance of the catchment following occupation by European settler farmers. Natural communities were markedly reduced in extent through the introduction of wheat farming and greater numbers of stock (Baxter and Meadows, 1994) and through the extermination of large indigenous herbivorous mammals (Baxter and Davies, 1994).

3.4.6.   Other deposits

The relative scarcity of accumulations of organic material usually considered most advantageous to palaeoecological reconstructions has led to the search for alternative types of sediment. Scott (1990, 1994), Scott and Bousman (1990), Scott and Cooremans (1992) and Scott and Vogel (1992) have pioneered the technique of interpreting palaeoecological spectra which are preserved in the dung middens of the common rabbit-sized herbivore *Procavia capensis* (rock hyrax). The reliability of such sequences was

tested by Scott and Vogel (1992) and indicates that even short-term variations in vegetation may be accurately recorded in such sequences. The palaeoenvironmental reconstruction for Blydefontein (see section 3.4.5) is further elucidated by palynological analysis of hyrax middens in the area which reveals further fluctuations in the grass/karroid vegetation over the past 1000 years or so. The potential for this kind of sediment to produce reliable and detailed palaeoecological information is further indicated by Scott (1987) in his examination of pollen in hyena coprolites in the southern Kalahari and again in his documentation of fluctuations in the late Pleistocene and Holocene vegetation pattern north of the contemporary karoo margin.

Still other Quaternary palaeoecological analyses for the karoo region have been conducted involving elements of the animal, rather than plant, communities, as the micromammal fossil work of Avery (1988, 1990a, 1990b and 1993) testifies, although there appear to be no suitable sites within the central area of the karoo proper (see Avery, 1993). Using this form of proxy palaeoenvironmental evidence emanating from cave deposits generally produces support for the palynological interpretations where appropriate. The micromammal evidence is concordant with the idea that the first part of the Holocene was more arid than the second across most of the region (Avery, 1993). At several sites around the margins of the present-day karoo, Boomplaas, Klipfonteinrand and Mirabib (Fig. 3.2), Avery (1993) has used the evidence from changing frequencies of micromammalian fossils to infer changes in the seasonality of rainfall in these areas. The evidence for the mid Holocene appears to indicate lower annual rainfall in the summer rainfall region in contrast to greater moisture availability in the contemporary winter rainfall region. This points to a difference in the direction of response of climate to globally induced changes between those areas of southern Africa which receive their rainfall predominantly in the winter, as opposed to summer, months and suggests, perhaps, that the winter rainfall-dominated areas of the western karoo and succulent karoo may be characterized by greater moisture availability during periods of reduced temperature and vice versa.

Details as to the nature of karoo environments around the time of the Last Glacial Maximum have indeed proved difficult to glean from the plethora of palaeoecological evidence presented. The occurrence of an outlier of red sand near Buffelsfontein (Fig. 3.2) east of the contemporary karoo margins, and yielding a thermoluminescence date of around 20 500 y BP (Marker and Holmes, 1993), is strongly suggestive of considerably cooler, windier and more arid climates in the area at that time. Lee-Thorpe and Beaumont (1990) examined carbon isotope ratios of enamel apatite in fossil grazing bovids from Equus Cave

(Fig. 3.2) in the Northern Cape Province. Increased proportions of C_3 over C_4 grasses are evident around the time of the Last Glacial Maximum, suggesting that winter rainfall was possibly relatively more important in this area around that time. In contrast to the situation evident elsewhere within the central parts of the karoo biome, the Last Glacial Maximum in the west coast region appears to have been associated with greater moisture availability, as indicated by fossil pollen preserved in cave sediments at Eland's Bay (Baxter, 1996).

3.4.7.   Climate change or human impact?

The record of environmental change in the karoo and the responses of the biota to those changes implicate fluctuating climate, more particularly precipitation, as the forcing factor in most instances. In the later part of the Holocene, an additional ecological factor – human population – must have added to the rigours of biological survival in these environments. It is frequently difficult to isolate the influence of human activity, usually associated with changes in the intensity and type of grazing and with cultivation, from those 'external' forcing factors involving climate. The matter is dealt with more fully by Hoffman et al. (this volume); suffice to say that there is palaeoecological evidence for the increasing impact of people on the karoo landscape through time (see Sugden and Meadows, 1989, for example). Not all the most recent vegetation changes, however, can be attributed to human management (or mismanagement) in the karoo and, as Bousman and Scott (1994) have noted, overgrazing alone does not account for the shift from grassland to karroid vegetation that appears to have begun in the eastern karoo in the early 1700s.

3.5.   The karoo as a dynamic landscape

The environmental history of the karoo is a long and distinguished one. Almost since the dawn of time, it seems, the karoo has been subject to climatic fluctuation, with aridity consistently following more mesic conditions which, in turn, appear to be replaced once again by aridity. The contemporary karoo flora and fauna may be seen as a product of environmental circumstances which only came about, at least in a geological sense, relatively recently. It is only since the late Tertiary and throughout the Quaternary that the essentially modern karoo biota has flourished in the region. During that time, the climatic conditions appear to have continually fluctuated from arid and semi-arid to sub-humid and back again.

Fig. 3.4 offers a simplified diagrammatic summary of the shifting moisture and temperature conditions during the later Quaternary for which the record is more complete and, therefore, more reliable. The interpretation is supported by the recent review work of Partridge et al. (1990) and suggests that conditions during the Last Glacial Maximum were certainly cooler than at present by up to five degrees Celsius and that precipitation values where somewhat lower than today.

The late Pleistocene (late glacial) appears to have been a time of considerable climatic dynamism, although greater moisture availability is a common conclusion of analyses of sediments and fossils dating to this period. The Holocene was certainly warmer than today, perhaps by two degrees Celsius in the earlier part, a situation which seems to have been coincident with more intensive aridity. The final part of the Holocene has, in general, been characterized by a return to more mesic climates,

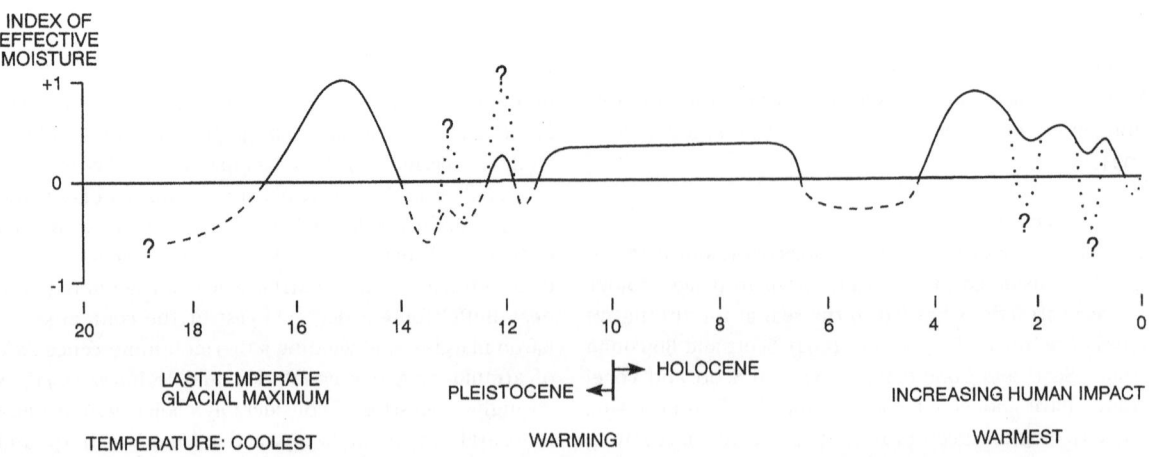

Figure 3.4 **Summary of moisture changes in the karoo during the late Pleistocene and Holocene, in 1000s y BP (modified after Meadows and Sugden, 1988)**

although at several intervals there are indications of short periods of drier climates. The last 2000 years and, especially, the colonial and post-colonial period of European settlement have featured the increased imprint of human activity upon the karoo landscape.

The arid south-west of Africa displays unusually high biological diversity and abuts directly against the Cape Floristic Region which is another region of elevated biological richness. The richness can be seen to be a consequence of underlying geological variations coupled with the environmental changes that have characterized the history of the region. In essence, the geographical position of the karoo has resulted in species being trapped during times of climatic change, being unable to move southwards or westwards and only a relatively short distance eastwards (Cowling and Hilton-Taylor, this volume; Vernon, this volume). Movement towards the north either involves adaptation to thick sandy substratals or movement along narrow and restricted corridors. Consequently, the biogeography of the karoo has been largely a situation of adapt or die. The present system can be seen to be a mixture of its indigenous Gondwanan inheritance and Laurasian invaders which successfully penetrated southwards into Africa.

The overwhelming message emanating from the numerous studies of the historical biogeography and palaeoenvironment of the karoo is one of *change*. This is neither a new idea, nor is it a situation unique to the karoo. Change in the characteristic climatic conditions is recorded in the fossil record, and inferred from contemporary biogeographical distributions, for the entire documented geological history of the proto-karoo and karoo. These changes have had major impacts on the biological communities throughout the passage of time, and a picture of the karoo as a fundamentally dynamic entity emerges. Aridity in central southern Africa is no recent phenomenon, and those who argue that the most crucial environmental problem facing the region at present is desertification would do well to look at the situation through a lens which provides a longer time perspective. The record remains inadequate in the sense in which it is spatially and temporally incomplete, but the key lesson it teaches us remains that to expect the karoo to be 'stable', 'static' or even 'predictable' over time (many of the landowners and politicians responsible for its ecological management have done so) is to expect the impossible. Although the environmental changes of the past are difficult to utilize directly to *predict* future karoo environments, they do indicate that continual change is the most likely scenario.

4 Plant biogeography, endemism and diversity

R. M. Cowling and C. Hilton-Taylor

4.1. Introduction

The extraordinary plant diversity and floristic peculiarity of southern Africa's arid lands are often overshadowed by the better studied and more widely publicized Cape Floristic Region. While Cape fynbos has its analogue in the Kwongan of Mediterranean south-western Australia (Cowling et al., 1996), the flora of southern Africa's winter-rainfall karoo, with its vast number of succulent species, is in a league of its own. A major objective of this chapter is to describe this diversity and seek explanations for its origins. We do this by analysing patterns and correlates of restricted range size (or endemism), and patterns and determinants of diversity at different spatial scales.

This chapter also provides the phytogeographical context for many of the other chapters in this book. We skim through this section, reluctant to become embroiled in the typological controversies that have characterized karoo phytogeography. Readers are referred to Werger (1978a, 1978b) and Jürgens (1991) for details.

Finally, we wish to clarify the use of some terms in this chapter. First, we use the term 'Karoo–Namib Region' to describe, in a phytogeographical context, the semi-arid shrubland and desert areas of southern Africa (Werger, 1978a). Although Jürgens (1991) provides justification for dividing this region into two phytochoria of equal rank, we retain the term as a matter of convenience. Second, our species counts always refer to vascular plants only. Third, we use the terms 'species' instead of the more long-winded 'species and infraspecific taxa'. Fourth, as a result of ongoing taxonomic debate, we have (reluctantly) included the Mesembryanthemaceae, southern Africa's most speciose and interesting arid-land taxon, in the Aizoaceae (Bittrich and Hartmann, 1988; Hartmann, 1991). Aizoaceae *sensu stricto* is not a big family in the Karoo–Namib Region.

Wherever possible, we refer to the 'mesems' as the Mesembryanthema (*sensu* Hartmann, 1991).

4.2. Biogeographical background

Since Bolus (1875) first recognized the karoo as a phytogeographical unit distinct from the Kalahari and other arid lands of sub-Saharan Africa, there has been much debate on the affinities and origins of its flora. Werger (1978b) provides a historical overview of the phytogeographic delimitation for the period 1875–c. 1975. In this section we briefly review the recent literature on this topic and discuss the floristic composition and climatic regimes of the major phytochoria.

4.2.1. Biogeographical units and delimitation

Werger (1978a, 1978b) followed White (1976, 1983) in recognizing a Karoo–Namib phytochorion (Fig. 4.1), and agreed with earlier workers that this unit forms part of the Palaeotropical Kingdom as opposed to the Cape Kingdom.

He did, however, emphasize the transitional nature of the region's flora, with strong Sudano-Zambezian affinities towards its northern and eastern boundaries, and strong Cape affinities towards the south-west. Werger (1978a, 1978b) delimited four domains within the Karoo–Namib Region (Namib, Namaqualand, western Cape and Karoo) as well as the Southern Kalahari Subdomain (Fig. 4.1(a)). There is still no agreement as to whether the arid savannas of the driest, southern part of the Kalahari belong in the Karoo–Namib or Sudano-Zambezian Regions. For example, Jürgens (1991) excludes most of the southern Kalahari from his Nama-karoo Region (Fig. 4.1(b)).

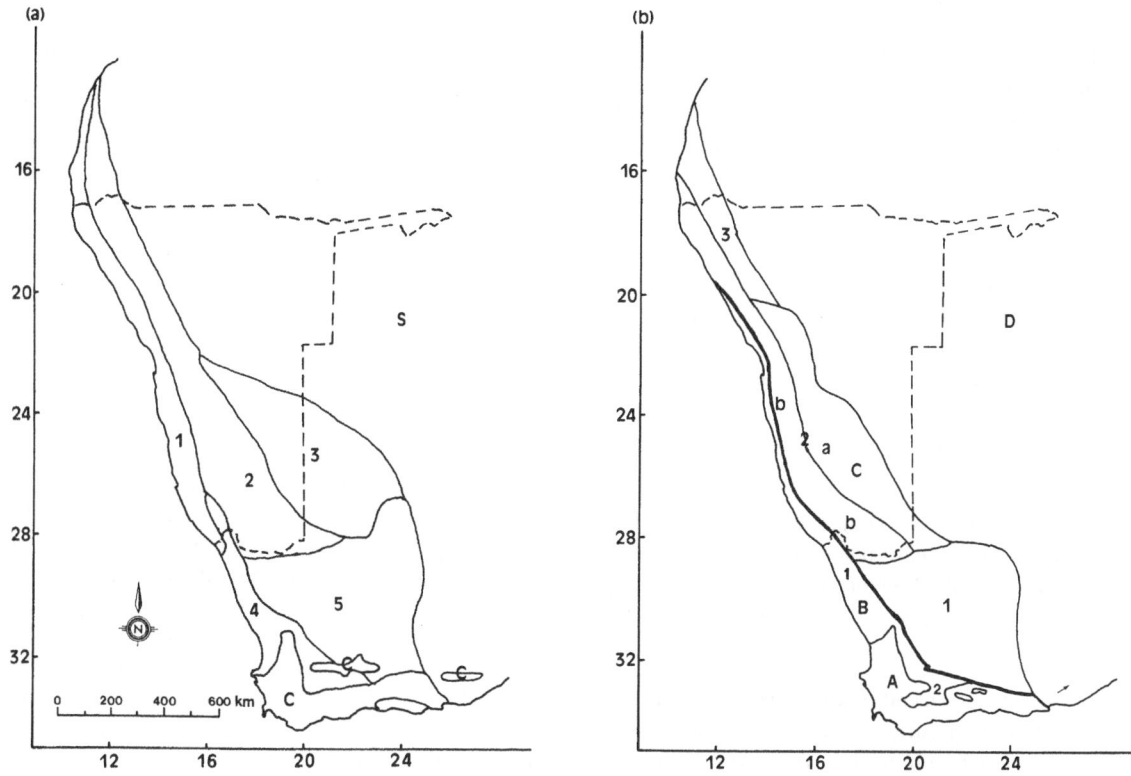

Figure 4.1 **Phytogeographic division of the Karoo–Namib Region. (a) Werger's (1978b) system: 1 = Namib Domain; 2 = Namaland Domain, 3 = Southern Kalahari Subdomain; 4 = Western Cape Domain: 5 = Karoo Domain; C = Cape Region; S = Sudano-Zambezian Region, and other phytochoria to the east. (b) Jürgens' (1991) system: the Greater Cape Floral Kingdom comprises the Cape Floristic Region (A) and afromontane regions further east as indicated by the arrow, and the Succulent Karoo Region (B). The Palaeotropical Kingdom comprises the Nama-karoo Region (C), the Sudano-Zambezian Region (D), and other phytochoria to the north and east. The Nama-karoo Region is subdivided into: 1 = Eastern Karoo Domain; 2 = Namaland Domain (subdivided into a = Namaland Subdomain; b= Namib Subdomain; 3 = Damaraland– Kaokoland Domain. The Succulent Karoo Region is divided into: 1 = Namaqualand–Namib Domain; 2 = Southern Karoo Domain (redrawn from Werger, 1978b and Jürgens, 1991)**

Werger (1978a, 1986) provides a detailed characterization of the various domains in terms of environment, flora and vegetation. We provide only a brief summary below. The Namib Domain includes the hyper-arid and arid Namib Desert, stretching from Alexander Bay at the mouth of the Orange River to 150 km south of Lobito in Angola (Fig. 4.1(a)). The vegetation is sparse and much influenced by fog, especially close to the cold Atlantic Ocean (see Walter (1986) and Jürgens et al. (1997) for comprehensive reviews). The Namaland Domain is centred on the shrub-covered elevated plateau of Namaland in Namibia, but also includes the narrow escarpment zone inland of the Namib Domain, and the sandy flats of the southern Kalahari. The Western Cape Domain corresponds to the predominantly winter-rainfall part of the Karoo–Namib Region in which leaf succulent shrubs predominate. The Karoo Domain comprises the extensive area of dwarf, grassy shrubland on the summer rainfall, inland plateau of South Africa.

Using distribution data for 1700 species, Jürgens (1991)

developed a new phytogeographical subdivision of southern Africa's arid lands (Fig. 4.1(b)). Rather than tinkering with boundaries, his proposals have far-reaching implications for African biogeography. The fundamentals of his scheme are summarized below:

- The division of the Karoo–Namib Region into two phytochoria of equal rank, namely the *Succulent Karoo Region* (more-or-less coincident with Werger's (1978a) Western Cape Domain) and the *Nama-Karoo Region*.

- The recognition that the Succulent Karoo Region forms part of a greater Cape Flora rather than part of the Paleotropical Kingdom (see Hilton-Taylor, 1987).

- The recognition of a new system of domains in both regions, delimited on the basis of the distribution of species from zonal vegetation complexes. Thus, the Namaqualand–Namib Domain (strongly winter rainfall) and the Southern Karoo Domain (non-seasonal rainfall) are recognized for the Succulent Karoo Region; and the Namaland Domain (including most of Werger's (1978a) Namaqualand and Namib Domains), Damaraland–Kaokoland Domain (see also Nordenstam, 1974; Hilton-Taylor, 1994a), and the Eastern Karoo Domain (corresponding to Werger's (1978a) Karoo Domain) are recognized for the Nama-karoo Region.

- The division of the Namib Desert (as a geographic entity) into two phytochoria: a south-central coastal portion as part of the succulent karoo (Namaqualand–Namib Domain); and a northern and inland zone as part of the Nama-karoo (Namib Subdomain of the Namaland Domain) (see also Robinson, 1978).

Recently, Hilton-Taylor (1994b, 1996) has delimited a series of 'centres of endemism' for the succulent karoo (see also Nordenstam, 1967; Werger, 1978a). In this chapter, we refer exclusively to Jürgens' (1991) scheme of regions and domains.

4.2.2.   Composition and characteristics of floras
The flora of southern Africa's arid lands is unusually rich and compositionally interesting. The entire flora of the Karoo–Namib Region (sensu Werger, 1978a) probably exceeds 7000 species of which up to 50% are endemic (Hilton-Taylor and Le Roux, 1989). There are 4849 species in the succulent karoo (40.3% endemic) (Hilton-Taylor, 1996) and 1427 species (34.5% endemic) in the Namibian part of the Namib Desert (partly Namaqualand–Namib Domain and partly Namib Subdomain of the Namaland Domain) (Robinson, 1978). There are no reliable estimates of the size of the Nama-karoo flora: Gibbs Russell (1987) provides a value of 2147 species in a 'core area' of 198 500 km², about 30% of the region.

Floras from the succulent karoo differ compositionally from Nama-karoo floras and those from other arid lands (Table 4.1).

They are uniquely characterised by high numbers of Aizoaceae (especially Mesembryanthema), and relatively high numbers of Iridaceae and Geraniceae. The last two-mentioned families are well represented in all fynbos floras of the Cape Region, and Aizoaceae rank highly in lowland fynbos and renosterveld floras (Cowling and Holmes, 1992). Succulents are extremely well represented in succulent karoo floras and are associated with many of the larger families (Aizoaceae, Asteraceae, Liliaceae, Crassulaceae, Geraniaceae, Euphorbiaceae and Asclepiadaceae). Indeed, the succulent karoo probably harbours about one third of the world's approximately 10 000 succulent species (Van Jaarsveld, 1987; Smith et al., 1993).

The succulent karoo flora includes 730 genera of which

67 (9.2%) are endemic (Hilton-Taylor, 1996). This is more than three times the number of endemic genera in the Sahara–Arabian and North American arid lands (Shmida, 1985), which are orders of magnitude larger than the succulent karoo. Another unusual feature of the flora is the high species to genus ratio of 6.9. The region includes several unusually large genera for an arid-land flora, including Ruschia (Aizoaceae: 136 spp.), Conophytum (Aizoaceae: 85 spp.), Oxalis (Oxalidaceae: 114 spp.), Euphorbia (Euphorbiaceae: 77 spp.), Pelargonium (Geraniaceae: 72 spp.), Senecio (Asteraceae: 72 spp.), Eriospermum (Liliaceae: 65 spp.), Othonna (Asteraceae: 61 spp.) and Drosanthemum (Aizoaceae: 55 spp.) (Hilton-Taylor, 1996). Most of these large genera are comprised entirely of either succulent shrubs or geophytes.

The Nama-karoo flora is not as well known as the flora of the succulent karoo. Generally, regional floras (see Table 4.1) are more typical of those from other arid lands where Asteraceae, Poaceae and Fabaceae are the most frequent top-ranking families (Shmida, 1985). As with the succulent karoo, Nama-karoo floras are unusual among arid-land floras in the low importance of Chenopodiaceae. Interestingly, the Damaraland–Kaokoland Domain flora (Robinson, 1978; Nordenstam, 1974; Hilton-Taylor, 1994a) bears a strong resemblance to Thar (Indian) Desert floras in the relative importance of Acanthaceae, Capparaceae and Cucurbitaceae (Shmida, 1985). Although no data have been published, there are certainly fewer genera endemic to the Nama-karoo than the succulent karoo. However, the former region is home to the only exclusively desert family – the Welwitschiaceae (Shmida, 1985).

At the generic and lower taxon levels, the floras of both the succulent karoo and Nama-karoo show very weak affinities with other arid-land floras of the world (Shmida, 1985). Much has been made of the disjunctions among plants between the arid lands of southern and northern Africa (see Werger, 1978a for a review of the literature). However, these amount to a handful of shared genera, about a dozen shared species (mostly well-dispersed grasses) and a few vicariant species pairs. Similarities are much stronger between the floras of North and South America's arid regions (Raven, 1963; Armesto and Vidiella, 1993) than between the disjunct arid lands of Africa. As is the case for other continents (Shmida, 1985), the similarity between karroid floras and those of adjacent Mediterranean shrubland (fynbos) and grassland/savanna regions (Gibbs Russell, 1987), is much stronger than with arid areas elsewhere in the world. There is general consensus that the succulent karoo flora is largely derived from an ancient Cape stock, and the Nama-karoo flora from both this and the adjacent Sudano-Zambezian flora (Acocks, 1953; Raven and Axelrod, 1978; Goldblatt, 1978; Werger, 1978a).

Table 4.1. *Composition of floras (ten largest families) from the Karoo–Namib Region. Values in brackets are the percentage contribution of each family to the local flora*

	Succulent karoo		Nama-karoo	
Namaqualand–Namib Domain, semi-arid[a]	Southern Karoo Domain, semi-arid[a]	Namaqualand–Namib Domain, arid[b]	Eastern Karoo Domain, semi-arid[a]	Damaraland–Kaokoveld Domain, arid[c]
n = 748	n = 1195	n = 300	n = 715	n = 337
Asteraceae (20.2)	Asteraceae (17.9)	Aizoaceae (20.1)	Asteraceae (20.7)	Asteraceae (16.0)
Aizoaceae (15.0)	Aizoaceae (11.1)	Asteraceae (17.7)	Poaceae (10.8)	Poaceae (9.6)
Liliaceae (8.3)	Poaceae (8.4)	Crassulaceae (9.0)	Liliaceae (9.1)	Fabaceae (5.0)
Poaceae (6.4)	Liliaceae (8.0)	Scrophulariaceae (7.7)	Aizoaceae (6.9)	Scrophulariaceae (4.2)
Crassulaceae (5.6)	Scrophulariaceae (4.8)	Liliaceae (7.3)	Scrophulariaceae (5.3)	Acanthaceae (3.7)
Scrophulariaceae (5.2)	Crassulaceae (4.4)	Poaceae (5.7)	Fabaceae (2.9)	Aizoaceae (3.7)
Fabaceae (4.0)	Fabaceae (3.6)	Geraniaceae (3.7)	Chenopodiaceae (2.7)	Sterculiaceae (3.7)
Iridaceae (3.3)	Asclepiadaceae (2.6)	Fabaceae (3.3)	Crassulaceae (2.5)	Euphorbiaceae (2.9)
Geraniaceae (2.9)	Euphorbiaceae (2.4)	Euphorbiaceae (3.0)	Sterculiaceae (2.2)	Capparaceae (2.4)
Chenopodiaceae (2.7)	Iridaceae (2.0)	Iridaceae (2.3)	Iridaceae (2.1)	Liliaceae (2.4)

[a] Data from Werger (1986)
[b] Data from Desmet (1996)
[c] Data from Nordenstam (1974)

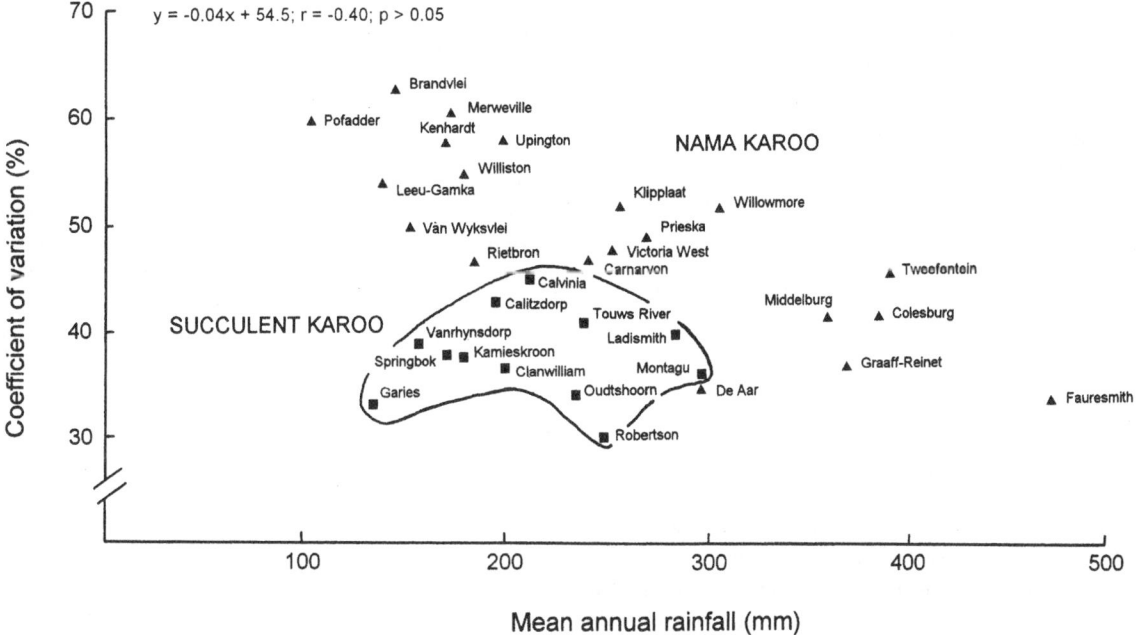

Figure 4.2 **Relationship between mean annual rainfall and coefficient of variation of mean annual rainfall for succulent karoo (square) and Nama-karoo (triangle) sites (Redrawn from Hoffman and Cowling, 1987)**

4.2.3.   Selective regimes

The succulent karoo and Nama-karoo differ fundamentally in their climatic, and hence, selective regimes. These differences, which have profound implications for endemism, diversity and rates of diversification of floras, are summarized as follows:

- The succulent karoo derives most of its rainfall from depressions associated with the circumpolar

westerly belt (Schulze and McGee, 1978; Desmet and Cowling, this volume). Most rain falls during winter (May–September) (Werger, 1978a, 1986; Rutherford and Westfall, 1986) and, even at low annual totals, rainfall is remarkably predictable (Hoffman and Cowling, 1987; Esler et al., this volume; Fig. 4.2).

- Nama-karoo rainfall is derived from a number of sources to the east and north of the region. Most rain falls during the late summer months (or with strong autumn and weak spring peaks in the eastern karoo) (Hoffman and Cowling, 1987) and, for a given annual total, rainfall is much less predictable than in the succulent karoo (Fig. 4.2).

- In the succulent karoo, rainfall events are widespread, of long duration and are mostly gentle showers; rainfall events in the Nama-karoo are invariably highly localized, short, intense thundershowers (Desmet and Cowling, this volume).

- Temperature extremes in the succulent karoo are ameliorated by proximity to the ocean and the high incidence, in the Namaqualand–Namib Domain, of fog during the summer months (Schulze and McGee, 1978; Olivier, 1995; Desmet and Cowling, this volume). Frosts are rarely, if ever, experienced over most of the area (Rutherford and Westfall, 1986; Werger, 1986). In the Nama-karoo, continentality is more pronounced and temperature extremes (including freezing conditions in winter) prevail.

- In the succulent karoo, high humidity (especially at night and in the early morning) and frequent fog (at least in the Namaqualand–Namib Domain) provide additional sources of moisture during the dry summer months (Walter, 1986; Von Willert et al., 1992). In the Nama-karoo, rainfall events are the principal source of precipitation (Werger, 1986; Desmet and Cowling this volume).

Thus, the succulent karoo has a warm temperate and oceanic climate characterized by relatively mild winters with low but reliable winter rainfall, and relatively mild summers where drought is ameliorated by heavy dew and frequent fog. The Nama-karoo has a more continental climate, characterized by hot summers with low and unreliable rainfall, and very dry, mild to cool winters with frequent night frosts. It is important to emphasize that the environmental conditions in the Southern Karoo Domain of the succulent karoo are, in many respects, transitional between those in the Namaqualand–Namib Domain and the Nama-karoo.

4.3.   Endemism

The patterns, determinants and correlates of plant species endemism have been poorly studied in the Karoo–Namib Region. The only detailed research that has been carried out in the Namaqualand–Namib Domain of the succulent karoo (Cowling and Hilton-Taylor, 1994; Hilton-Taylor, 1996). In this section, we attempt to provide a meaningful profile of the region's endemic flora. We focus on extent of occurrence as a measure of endemism or range size, since

few data exist on area of occupancy (i.e. the number of sites within a particular range occupied by a species) (Gaston, 1994).

4.3.1.   Patterns

Generally, for equal sized areas, endemism in the succulent karoo is considerably higher than in the Nama-karoo (Table 4.2, Fig. 4.3), but lower than in Cape fynbos landscapes (Cowling and McDonald, 1997).

The only data on endemism in other warm deserts and semi-arid areas are for the Sonoran Desert (26.6% endemism or 650 endemics in 310 000 km^2) and the Saharan Desert (25% endemism or 162 endemics in 9 million km^2) (Major, 1988). By comparison, the succulent karoo has three times as many endemics as the Sonoran Desert in a third of its area, and the Namib Desert has three times more endemics than the Saharan Desert in an area a fraction (0.015) of its size. Although more data are required to establish clearer patterns, it would appear that levels of endemism in the Karoo–Namib Region (especially the succulent karoo) are extraordinarily high.

4.3.2.   Taxonomic aspects

Endemics in the Karoo–Namib Region are not a random assemblage phylogenetically. In the western and eastern Gariep, an area encompassing parts of both the succulent karoo and Nama-karoo regions, local endemics were significantly overrepresented in three families, namely the Aizoaceae (Mesembryanthema), Crassulaceae and Asclepiadaceae (Cowling and Hilton-Taylor, 1994), all families of predominantly succulent species (Van Jaarsveld, 1987). A similar pattern was recorded for the flora of the coastal belt between Alexander Bay and Port Nolloth, in the Namaqualand–Namib Domain (Desmet, 1996); here, Mesembryanthema were massively overrepresented among regional endemics. In an analysis of the entire flora of the succulent karoo, significantly higher than average endemism was recorded for the above-mentioned families as well as the entirely succulent Aloaceae (Liliaceae *sensu lato*), and four predominantly geophytic families (Amaryllidaceae, Eriospermaceae (Liliaceae *sensu lato*), Iridaceae and Oxalidaceae) (Hilton-Taylor, 1996). Overrepresentation of endemics in the Aizoaceae (Mesembryanthema), Crassulaceae and Asclepiadaceae, as well as succulent Euphorbiaceae, is also evident for the succulent thorn thickets on the south-eastern margin of the Eastern Karoo Domain (Cowling and Holmes, 1991). In all floras studied so far, families with lower than expected levels of endemism include Asteraceae, Poaceae and Scrophulariaceae.

Detailed data on the taxonomic correlates are lacking for elsewhere in the Karoo–Namib Region. According to Robinson (1978), 85% of the 492 Namib Desert endemics

Table 4.2. **Patterns of endemism in the Karoo–Namib Region**

Region	Area (10^3km^2)	No. spp	No. endemics (%)	Phytochorion[a]
Succulent karoo[b]	112.2	4849	1954 (40.3)	Succulent Karoo Region
Succulent karoo	50.6	2125	616 (29.0)	Succulent Karoo Region (core biome)[c]
W & E Gariep[d]	40.9	2010	397 (19.8)	Namaqualand–Namib/Namaland domains
Nama-karoo (core biome)	198.5	2147	377 (17.6)	Nama-karoo Region
Namib Desert[e]	134.4	1427	492 (34.5)	Namaqualand–Namib/Namaland/Damaraland–Kaokoland domains
Kaokoveld[f]	70.0	952	116 (12.2)	Damaraland–Kaokoland Domain
Brandberg[g]	0.75	337	11 (3.3)	Damaraland–Kaokoland Domain

[a] According to Jürgens (1991) (see text and Fig. 4.1(b))
[b] Data from Hilton-Taylor (1996)
[c] Data from Gibbs Russell (1987)
[d] Data from Cowling and Hilton-Taylor (1994)
[e] Data from Robinson (1978)
[f] Data from Hilton-Taylor (1994a)
[g] Data from Nordenstam (1974)

are confined to the southern, winter-rainfall area, and the vast majority of these were succulent shrubs (Aizoaceae, Crassulaceae) or geophytes (Liliaceae, Oxalidaceae). Northern desert (Nama-karoo) endemics are associated with a wider range of growth forms (trees, shrubs, grasses, geophytes) and families (Burseraceae, Capparidaceae, Acanthaceae, Cucurbitaceae, Fabaceae), suggesting taxonomic and growth form harmony (*sensu* Auerbach and Shmida, 1985) between the endemic and non-endemic floras.

4.3.3.   Biological aspects

In all the floras studied thus far (western and eastern Gariep, Alexander Bay–Port Nolloth area, the succulent karoo flora, and the succulent thorn thickets), endemics are significantly overrepresented among succulent (dwarf and low) shrubs and, to a lesser extent, geophytes, and underrepresented among annuals, forbs, trees and non-succulent

shrubs (Cowling and Holmes, 1991: Cowling and Hilton-Taylor, 1994; Desmet, 1996; Hilton-Taylor, 1996; Fig. 4.4). Of particular interest is the relatively high number of locally endemic dwarf succulent shrubs in the succulent karoo (Fig. 4.4(b)), including many species of *Anacampseros* (Portulacaceae), *Crassula* (Crassulaceae), *Cheirodopsis*, *Conophytum*, *Gibbaeum* and *Glottiphylum* (Mesembryanthema) (Van Jaarsveld, 1987; Hammer, 1993; Desmet, 1996).

Data from the Namaland Domain suggest that endemics are distributed in the same frequency among growth forms as non-endemics. The Damaraland–Kaokoveld Domain is the only part of the Karoo–Namib Region with a disproportionately high number of non-succulent woody endemics, many of which are taxonomically isolated (Nordenstam, 1974; Hilton-Taylor, 1994a).

4.3.4.   Habitat aspects

At a very broad scale, high levels of local endemism in the

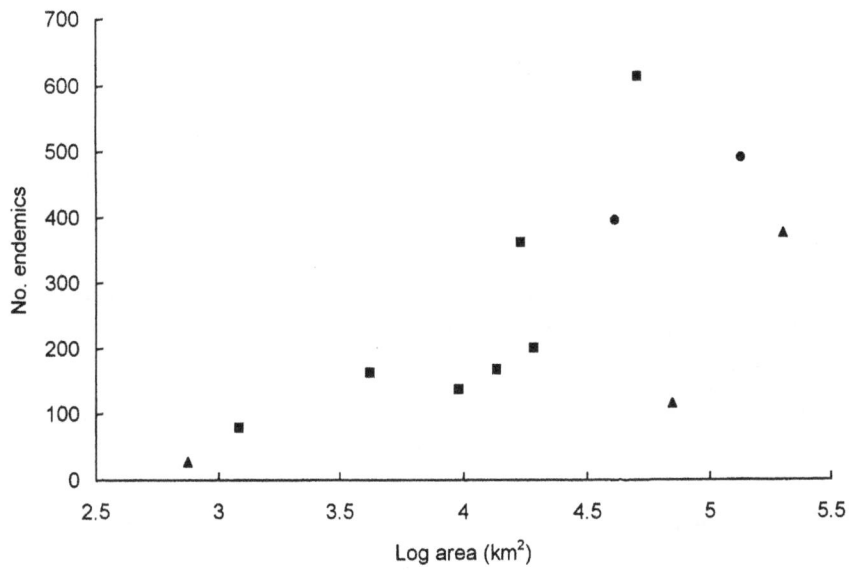

Figure 4.3 **Relationship between numbers of endemic species and area in the succulent karoo (square), Nama-karoo (triangle), and areas including parts of both regions (dot) (data from Table 4.2 and Hilton-Taylor, 1996)**

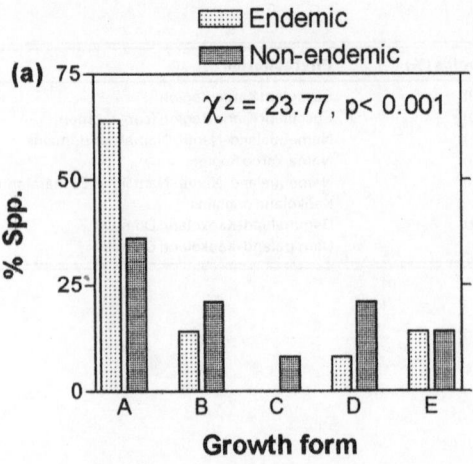

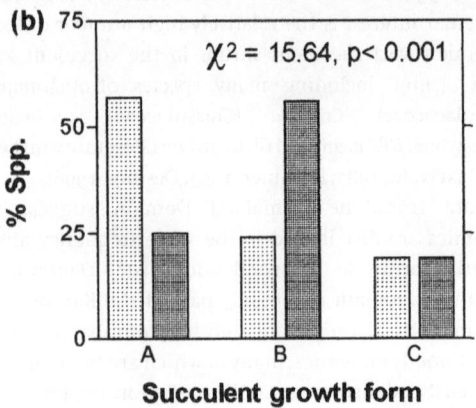

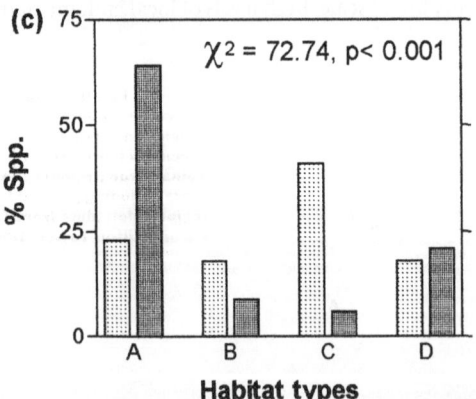

Figure 4.4 **Association in the flora (*n* = 300 spp.) of the arid coastal belt between Port Nolloth and Alexander Bay (Namaqualand–Namib Domain) between endemism and (a) growth form (A = succulent shrub, B = non-succulent shrub, C = graminoid, D = forb/annual, E = geophyte); (b) succulent growth form (A = dwarf leaf-succulent shrub, B = other leaf succulent shrub, C = stem succulent shrub); and (c) habitat type (A = dunes, B = rocky hills, C = pebble plains, D = <1 habitat type, salt-pans and disturbed sites). The chi-square tests the null hypothesis that the frequency of species in the endemic categories are similarly distributed with regard to growth form and habitat categories. From Desmet (1996)**

Karoo–Namib Region are overwhelmingly associated with winter rainfall conditions, especially the foggy, coastal areas of the Namaqualand–Namib Domain. At a finer scale, endemics (especially dwarf succulent shrubs) are clustered in broken, rocky habitats rather than sandy or loamy flats (Fig 4.4(b)) (Jürgens, 1986; Van Jaarsveld, 1987; Hammer, 1993; Desmet, 1996; personal observations), or on regionally unusual substrata such as quartzites and weathered quartz veins (so called *knersvlakte*) (Jürgens, 1986). However, much more research is required to characterize endemic-rich habitats in the Karoo–Namib Region.

So what is a reasonably comprehensive profile of a local endemic in the Karoo–Namib Region? The odds are strongly in favour that it will be a dwarf succulent shrub in the Aizoaceae, Crassulaceae or Asclepiadaceae, and associated with rocky hills and outcrops in the winter-rainfall succulent karoo. The remaining endemics are likely to be geophytic members of the Iridaceae, Amaryllidaceae and Geraniaceae, also confined to winter-rainfall areas. The vast majority of these species are neoendemics: they are often members of large genera and have many close relatives, often species in different habitats within the same landscape or geographic vicariants (Hammer, 1993; Cowling and Hilton-Taylor, 1994; Ihlenfeldt, 1994). Taxonomically isolated endemics (paleoendemics) are few and are mainly non-succulent shrubs of the Nama-karoo Region, especially Damaraland–Koakoland Domain (Nordenstam, 1974).

4.4. Local diversity

By local diversity, we mean within-community or alpha diversity (Whittaker, 1977). Given the low moisture availability, many arid and semi-arid areas have surprisingly high local diversities (Whittaker and Niering, 1975; Cowling et al., 1989; Aronson and Shmida, 1992). In this section, we analyse the patterns and determinants of local diversity in the Karoo–Namib Region. First, we provide a brief overview of the theory invoked to explain plant diversity in arid lands.

4.4.1.   Theory

Competition for limited soil moisture is widely recognized as the overriding determinant of plant community structure in arid and semi-arid regions, and there is much inferential evidence of its effects on plant spacing and performance (e.g. Phillips and MacMahon, 1981; Yeaton and Esler, 1990). It is not surprising, therefore, that the predictions of classical niche theory are most commonly invoked to explain patterns of diversity in these communities. We detail some of these below:

- Diversity increases with increasing rainfall (or resource availability) up to a maximum of about 400 mm yr^{-1}, after which it declines (Whittaker and Niering, 1975; Whittaker, 1977; Noy-Meir, 1985; Shmida, 1985). The argument is that conditions of extreme moisture stress provide limited opportunities for specialization, resulting in depauperate communities; conditions of intermediate moisture stress provide the greatest array of opportunities for niche differentiation, thereby reducing competition and increasing diversity; under conditions of low moisture stress, competition from the dominant growth form (usually trees or tall shrubs) reduces diversity (Aronson and Shmida, 1992).

- Cody (1989) has taken the climatic argument further by suggesting that conditions that provide the widest variety of growth opportunities throughout the year would enable the coexistence of the widest range of growth forms, and hence, the greatest number of species. Thus, under conditions of a strong limiting resource (e.g. soil moisture in arid and semi-arid environments), each growth form represents a unique solution, and community level growth form diversity parallels species diversity (see also Rundel and Mahu, 1976). Regions with similar annual rainfall but differences in rainfall seasonality and thermal regimes, may have very different growth form mixes and, consequently, species diversities.

- Within a particular climatic regime, conditions

which provide a high heterogeneity of soil moisture conditions (e.g. run-on sites and rocky substrata) will have longer structural niche axes, thereby supporting more growth forms and species than less heterogeneous habitats (e.g. Barbour and Diaz, 1973; Olsvig-Whittaker et al., 1983; Noy-Meir, 1985; Shmida, 1985; Montana, 1990).

4.4.2.   Patterns

At the 0.1 ha scale, succulent karoo communities have the third highest richness (mean = 74; range = 32–115) of all the biomes in southern Africa, and are significantly richer than those from the Nama-karoo (mean = 47; range = 22–76) (Fig. 4.5). The latter have values slightly higher than those recorded for Sonoran Desert communities, regarded by Whittaker and Niering (1975) as some of the most species-rich vegetation in North America.

Diversity in the succulent karoo is comparable to values from the winter-rainfall, semi-arid steppes and deserts of the Middle East (Aronson and Shmida, 1992), where more than 100 species per site is not uncommon. However, a major difference between the two areas is that most species in the succulent karoo communities are dwarf, leaf succulent shrubs (largely Mesmbryanthema) (Jürgens, 1986; Cowling et al., 1994; Desmet, 1996; Midgley and Van Der Heyden, this volume), whereas annuals predominate in the Middle East (Aronson and Shmida, 1992).

Cowling et al. (1994) sampled 9 sites (25 m² plots) in the

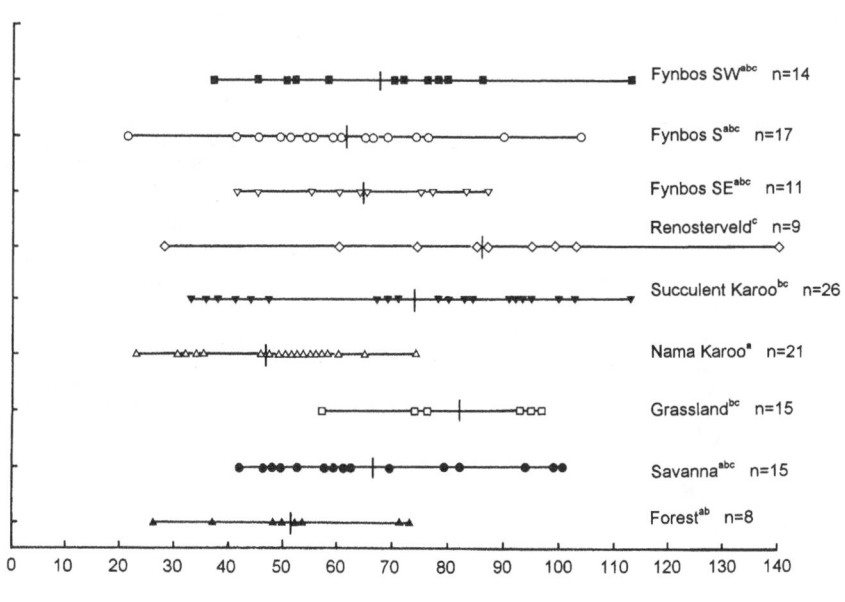

Figure 4.5 **Patterns of local diversity (0.1 ha) in the biomes of southern Africa. Biomes with different superscripts (or groups of superscripts) have significantly different richness at this scale ($p < 0.05$, Tukey's test) (redrawn from Cowling et al. (1989)**

Nama-karoo and 10 in the succulent karoo, ranging in both regions from extreme desert conditions to the semi-arid margin (250–350 mm yr⁻¹). Succulent karoo communities were about 1.6 times richer than those in the Nama-karoo. On average, succulent karoo sites had more than 4 times the numbers of succulent species and about 7 times the number of Mesembryanthema (Aizoaceae) than Nama-karoo sites.

4.4.3.   Explanations

Hoffman (1989a) documented patterns of local diversity at four sites in both well-managed and overgrazed vegetation, spanning the transition from dwarf karroid shrublands (c. 200 mm yr⁻¹) to succulent thorn thickets (c. 450 mm yr⁻¹) in the south-eastern zone of the Eastern Karoo Domain. Diversity was not significantly related to either rainfall or growth form diversity, but overgrazed vegetation was significantly poorer in species than well-managed rangeland.

Data from the southern boundary of the Nama-karoo supports the notion that run-on habitats with heterogeneous soil moisture conditions have higher local diversity and numbers of growth forms than less heterogeneous plains habitats (Milton, 1990). At this site, plots (25 m²) in drainage lines supported twice as many species (20–24) and a more diverse array of growth forms than plots on adjacent, loamy plains (10–12 spp.).

Cowling et al. (1994) investigated the climatic correlates of species and growth form diversity in rocky slope and flats habitats at 9 Nama-karoo sites and 10 succulent karoo sites located throughout southern Africa (see above). Contrary to the predictions of niche theory, the more heterogeneous rocky habitats did not have

significantly more species or higher growth form diversity than adjacent loamy flats at the community level (25 m² plot). However, significantly more species were recorded in rocky than flats habitats at the site scale (10 000 m²), probably reflecting the longer resource axes and higher internal beta diversity (Whittaker, 1977) in the former habitat. Species-richness showed a significant positive relationship with growth form diversity (Fig. 4.6), a pattern also evident in South American (Rundel and Mahu, 1976) and North American (Cody, 1989) arid lands.

However, this relationship was much stronger for Nama-karoo than succulent karoo sites, where many different species of dwarf, leaf succulent shrubs (mainly Mesembryanthema) coexist (see also Jürgens, 1986; Desmet, 1996). An index of climatic heterogeneity (reflecting the diversity of conditions suitable for plant growth) and measures of rainfall evenness throughout the year, emerged as the strongest predictors of growth form diversity, and, hence, species-richness. These results are consistent with Cody's (1989) structural niche hypothesis. However, significant deviations from this model were recorded for sites in the strongly winter-rainfall Namaqualand–Namib Domain with their large numbers of coexisting leaf succulents.

Cowling et al.'s (1994) study suggests that Nama-karoo and succulent karoo communities, especially those in the Namaqualand–Namib Domain, are structured in fundamentally different ways. The former show similarities to New World regions where climatic factors determine the length of structural niche axes and, hence, local diversity. The shrub matrix in these communities comprises long-lived individuals which, over time, have been organized by competitive interactions (Yeaton and Esler, 1990) to parti-

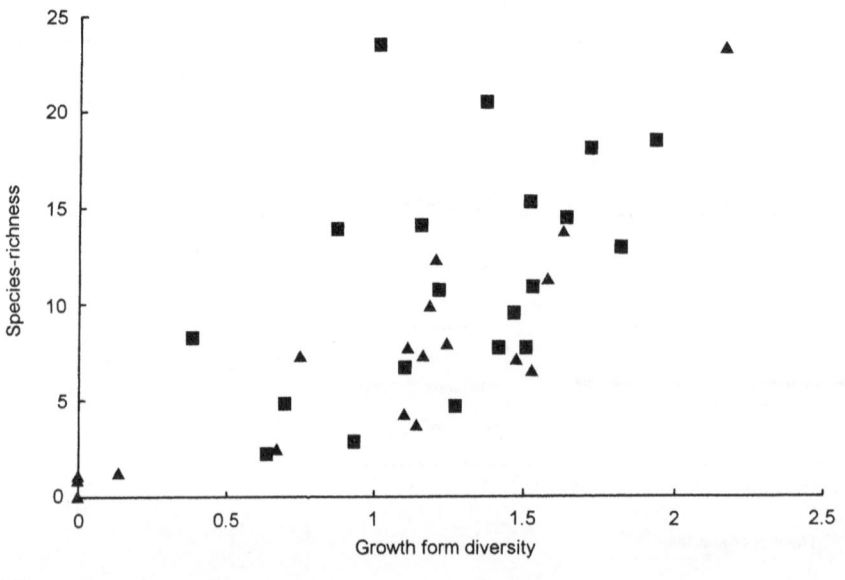

Figure 4.6 **Relationship between species-richness and growth form diversity in 25 m² plots located in two habitats (sand/loam flats and rocky slopes) in, 19 sites throughout the Karoo–Namib Region. Results of a simple linear regression of the full data set: $y = 0.294 + 7.88x$, $r^2 = 46.80$, $p < 0.001$; succulent karoo (square): $y = 3.091 + 6.50x$, $r^2 = 20.41$, $p < 0.05$; Nama-karoo (including central Namib Desert) (triangle): $y = -0.743 + 7.73x$, $r^2 = 71.84$, $p < 0.001$ (redrawn from Cowling et al. (1994)**

tion resources via assembly rules operating at the growth form or functional type level. Community structure, other than the short-lived grass component (Hoffman and Cowling, 1990b), is relatively stable in Nama-karoo communities where competition from long-lived shrubs limits recruitment (Milton, 1995a; Wiegand et al., 1995).

The succulent karoo is unique amongst the world's arid lands, since the vast majority of its perennial component comprises relatively short-lived shrubs, most of which are succulents (Cowling et al., 1994). Unlike perennial shrubs in other arid lands, where recruitment events are highly episodic and populations are largely even-aged (Zedler, 1981; Jordan and Nobel, 1982), the age structure of these succulents is uneven: plants die and are replaced continuously, often resulting in significant compositional change (Von Willert et al., 1985; Jürgens et al., 1997; R. M. Cowling and J. J. Midgley, unpublished data). As expected, there is little inferential evidence for both inter- and intraspecific competition between succulent shrubs in these communities (Werger, 1986; R. M. Cowling and J. J. Midgley, unpublished data). It appears, therefore, that community membership in the Namaqualand–Namib Domain of the succulent karoo is determined by a lottery process whereby functionally equivalent shrubs coexist in highly dynamic communities. Predictable winter rain and fog-ameliorated summers provide conditions for continuous recruitment, and occasional droughts rearrange emerging competitive hierarchies through the mass death of shallow-rooted and drought intolerant succulent shrubs (Von Willert et al., 1985; Jürgens et al., 1997). The result is species-rich communities dominated by functionally equivalent shrubs (Cowling et al., 1994).

In conclusion, Nama-karoo communities are similar in terms of patterns and determinants of local diversity to other arid lands, particularly those of the New World. Succulent karoo communities have similarly high species-richness and low growth form diversity to communities in the Old World (Shmida and Whittaker, 1979; Shmida, 1985). However, in the succulent karoo, most of this diversity is associated with functionally similar succulent shrubs; in the latter, it is largely associated with annuals. In both regions, high population turnover and associated lottery processes may be responsible for the maintenance of diversity (Aronson and Shmida, 1992; Cowling et al., 1994).

4.5.   Differentiation diversity

Differentiation diversity refers to compositional change along habitat gradients (beta diversity) and along geographical gradients (gamma diversity) (Cowling et al., 1992). High differentiation diversity is largely the product of the evolution of habitat specialists and geographic vicariants (Cody, 1986; Cowling et al., 1992). This aspect of diversity has been very poorly studied in the arid lands of the world.

Despite the paucity of data, it is clear that differentiation diversity is much higher in the succulent karoo than the Nama-karoo. Using Jürgens (1986) data, Cowling et al. (1989) computed a compositional shift of 1.5 (Wilson and Shmida's (1984) B-value) for four sites along a gradient of increasing soil depth, spanning approximately 100 m horizontal distance, in the Namaqualand–Namib Domain (*knersvlakte*) of the succulent karoo. Ihlenfeldt (1994) describes the extraordinary differentiation diversity within the genus *Argyroderma* (Mesembryanthema) in the same area. What is especially remarkable is the fine-scale habitat (mainly edaphic) differentiation in an otherwise relatively homogeneous environment. Similar patterns are evident for many other genera of Mesembryanthema (e.g Hammer, 1993), as well as geophytes (Goldblatt and Manning, 1996).

By comparison, beta diversity in the Nama-karoo is low. Hoffman (1989) recorded a compositional shift (Wilson and Shmida's (1984) B-value) of 1.9 for four sites spanning about 250 km along a gradient of increasing aridity from succulent thorn thicket (450 mm yr^{-1}) to karroid shrubland (200 mm yr^{-1}) in the eastern karoo. This gradient crosses a major phytochorological boundary. Also in the eastern karoo, Palmer and Cowling (1994) recorded B-values of 1.1 and 1.5 for five sites spanning topo-moisture gradients of 500 m and 300 mm yr^{-1} on dolerite and sandstone substrata, respectively.

These patterns suggest that diversification has been more pronounced in the succulent karoo than the Nama-karoo. In the former area, explosive diversification within certain lineages has resulted in the fine-scale discrimination of habitats, and the existence of many related species occurring in similar habitats separated by a few to tens of kilometres (Hammer, 1993; Ihlenfeldt, 1994; Goldblatt and Manning, 1996). Interestingly, many of these localized specialists occur in very small populations (Ihlenfeldt, 1994: personal observations). These aspects of rarity in the succulent karoo are in urgent need of research.

4.6.   Regional diversity

Regional diversity refers to the richness of areas that encompass more than one community (10^1–10^6 km^2). At this scale, diversity is the product of the number of species within communities, the compositional change along environmental gradients (beta diversity), and the compositional change between equivalent environments along geographical gradients (Cowling et al., 1992). Any conditions that enhance the value of these diversity components will increase regional diversity. In recent years, there has been much interest in describing and explaining patterns of regional diversity (Ricklefs and Schluter, 1993; Rosenzweig, 1995). In this section, we briefly review the theory invoked to explain regional diversity; present patterns in the southern African karroid phytochoria; compare these patterns with those in other arid lands of the world; and discuss the determinants of regional richness in the Karoo–Namib Region.

4.6.1.   Theory

A number of hypotheses have been invoked to explain patterns of diversity at the regional scale. We summarize these below.

- *Area*: Species number increases in samples of larger area, and that rate of increase decreases with progressively larger area, so that, over most area sizes, the double logarithmic form of the species–area curve provides the best fit (Williamson, 1988). For the area sizes of the scale we are considering, many have argued that area is merely a surrogate for environmental heterogeneity (Williamson, 1988; Rosenzweig, 1992). However, area is also a surrogate measure of resource availability: larger areas support larger total resources, and, hence, larger population sizes, thus facilitating the persistence of more species, especially naturally rare ones (Diamond, 1988).

- *Heterogeneity*: Environmentally heterogeneous landscapes support more habitats and, hence, more species than more uniform landscapes (Williamson, 1988).

- *Favourableness*: The notion that climatically benign or favourable conditions promote regional richness has a long history in ecology (e.g. Dobzhansky, 1950). A problem with this hypothesis is that conditions are defined relative to species' preferences within the regional pool, rendering it tautologous (Terborgh, 1973).

- *Energy*: Simply stated, 'energy-diversity' or 'species-energy' theory argues that the number of organisms and, hence, the number of species a region can support, increases monotonically with environmentally available energy (Wright, 1983). The energy hypothesis, however, fails to explain why energy should be partitioned among increasing numbers of species rather than increasing number of individuals of a few species within a regional biota (Currie, 1991).

- *Seasonality and irregularity*: The hypothesis that large variations in resource abundance within and between years (seasonality and irregularity, respectively) lead to reduced regional richness, is linked to the notion that stability augments diversity by permitting specialization (e.g. Sanders, 1968).

- *Speciation history*: Speciation augments regional richness whereas extinction reduces it (Rosenzweig, 1995). In areas where ecological factors that promote diversification of resident lineage have persisted for a long time (so-called time hypothesis) (Whittaker, 1977), or where diversification-prone lineages have experienced explosive speciation (Vrba, 1980; Cowling et al., 1992), species will accumulate without apparent limit.

4.6.2.   Patterns

Cowling et al. (1998) described and compared patterns of regional diversity within the Karoo–Namib Region, as well as for arid lands globally. They found significant log-log species–area curves for both the succulent karoo ($p < 0.001$) and the Nama-karoo ($p < 0.05$) (Fig. 4.7). Since the slopes of these two curves were homogeneous, and the intercepts significantly different, they computed the k- or intercept-ratios (Gould, 1979). This value is 2.65, indicating that, for any given region within the range of areas in Fig. 4.7, succulent karoo has at least 2.5 times as many species as the Nama-karoo.

Comparisons at a global scale were limited by the lack of data from other arid lands of the world (Cowling et al., (1998). Regional diversity in the Nama-karoo does not appear to be substantially higher than that in other summer rainfall, semi-arid areas (Fig. 4.8(a)), although more data from New World sites would enable more meaningful comparisons. Certainly, Australian regions are considerably poorer than those from southern Africa, but the Sonoran Desert (the only North American site) has similar richness to an equivalent area of the Nama-karoo. When compared to other winter rainfall, semi-arid zones of the world, the succulent karoo emerges as having the highest regional diversity over the full range of areas (Fig. 4.8(b)). For example, sites in the succulent karoo are on average 3.8 times richer than similar-sized regions in the winter rainfall arid lands of North America (Cowling et al., 1998).

Species-richness in desert regions, defined here as both summer- and winter-rainfall areas with an annual rainfall of less than 100 mm, showed a non-significant relationship with area (Cowling et al., 1998) (Fig. 4.8(c)). None the less, desert regions in southern Africa generally have more species when compared to similar-sized desert regions in North Africa, including the topographically complex

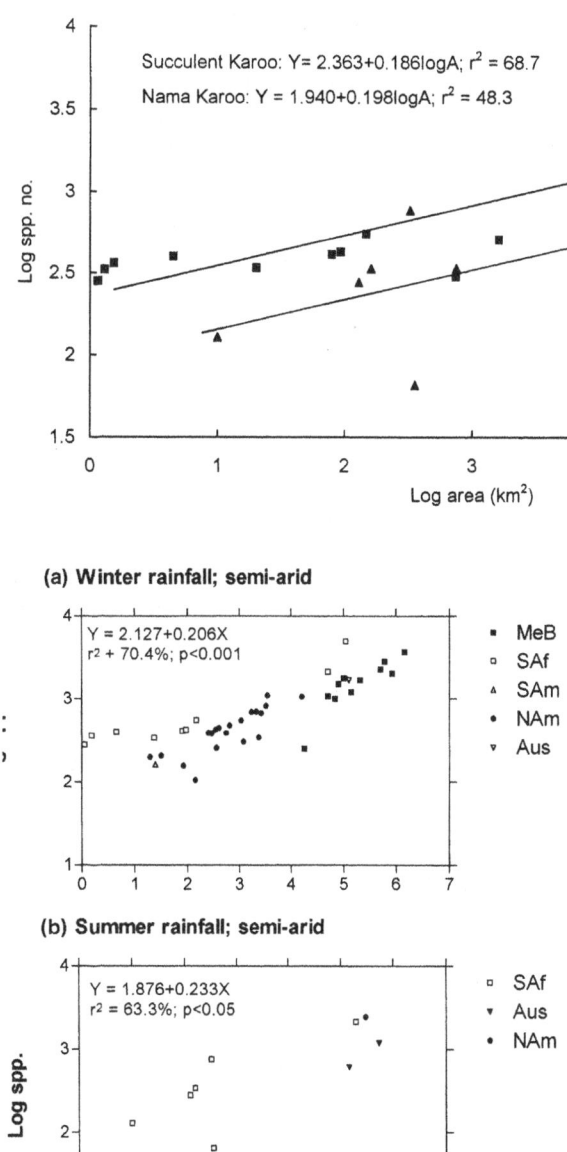

Figure 4.7 **Species–area relations for succulent karoo (square) and Nama-karoo (triangle) sites in southern Africa (reproduced with permission from Cowling et al. (1998)**

uplands of the central Sahara. For example, the Namib Desert has between two and four times as many species as equivalent-sized areas of the Sahara.

Interestingly, the trend for winter-rainfall arid lands to be richer than those from summer rainfall regions holds at a global scale (Cowling et al., 1998). A comparison of the two species–area curves (Figs. 4.8(a) and (b)) revealed that the slopes were not significantly different. The corresponding *k*-ratio is 1.78, indicating that winter-rainfall regions are nearly twice as rich as equivalent sized summer-rainfall regions. The analysis is constrained by the lack of data from summer-rainfall areas outside southern Africa.

4.6.3.   Explanations

Hoffman et al. (1994) investigated the correlates of species-richness in 34 arid and semi-arid Veld Types (homogeneous agro-ecological regions) (Acocks, 1953) of South Africa. Their sample included areas beyond the boundary of the Karoo–Namib Region. They found a non-significant log area–log species relationship, but did record a weak positive correlation between species number and mean annual rainfall (*r* = 0.44), and a stronger, negative correlation between richness and potential evapotranspiration (*r* = 0.64). Combining both explanatory variables and area (i.e. to provide an estimate of total resources in a region (Wright, 1983)) in a multiple regression, they derived the following relationship:

$$\text{Log}S = 16.5 - 1.74\log\text{PET} + 0.18\log A + 0.33\log\text{MAR}; \ r^2 = 0.57$$

Figure 4.8 (left) **Species–area relations for warm-temperate and subtropical semi-arid shrublands and deserts of the world. (a) Winter rainfall, semi-arid (100–400); (b) summer rainfall, semi-arid; (c) desert (100 mm yr⁻¹) regions. Aus = Australia, ME = Middle East, MeB = Mediterranean Basin, NAm = North America; SAf = southern Africa, SAm = South America. Reproduced with permission from Cowling et al. (1998)**

where S = species number, PET = potential evapotranspiration, A = area, and MAR = mean annual rainfall. They interpreted the relationship with PET as the negative arm of a more general hump-backed curve between energy availability and species-richness. Their argument is that in arid areas, high energy availability, as estimated by PET (which is a measure of the integrated, crude ambient energy to evaporate water and is independent of water availability), has an antagonistic effect on plant performance, resulting in reduced diversity. However, Cowling et al. (1997) found no evidence of a curvilinear relationship between regional richness and three measures of energy availability for sites distributed across South Africa, including many low-energy regions. Besides, this curvilinear relationship holds at the local rather than regional scale (Rosenzweig and Abramsky, 1993). Hoffman et al.'s (1994) results suggest that energy-rich, dry areas support fewer species than wetter areas with less available energy, such as the Veld Types within the species-rich fynbos region which were included in their sample. Moreover, these areas also have significantly higher environmental heterogeneity than the Karoo-Namib Veld Types (Cowling et al., 1997). The study by Hoffman et al. (1994) did not include measures of edaphic or topographical heterogeneity as explanatory variables.

Cowling et al. (1997) investigated the relationships between the richness of 15 variously sized regions in the Karoo-Namib Region of South Africa (10 in the succulent karoo and 5 in the Nama-karoo), and 13 explanatory variables. The latter were chosen to test hypotheses on the role of area, heterogeneity, favourableness, energy, seasonality, and irregularity in determining patterns of regional richness. In bivariate analyses, the strongest relationship was between richness and the range in mean annual rainfall (difference between highest and lowest value of mean annual rainfall) in a region – a measure of heterogeneity (Fig. 4.9). A stepwise multiple regression, employing the explanatory variable for each hypothesis that provided the best fit in bivariate analyses, gave the following result:

$$\text{Log}S = 1.51 + 0.35\text{logRAR} - 0.303\text{logRCV}; r^2 = 0.59$$

where S = species number, RAR = rainfall range, and RCV = coefficient of variation of monthly rainfall averaged for the wettest three consecutive months. This model suggests that the heterogeneity in moisture conditions is the major predictor of regional diversity in the Karoo-Namib Region, but that irregularity (as measured by RCV) also plays a role. Thus, the most species-rich regions are those with long moisture gradients and predictable seasonal rainfall, such as would be found in the escarpment zone of the Namaqualand-Namib Domain of the succulent karoo.

4.7.   General discussion

4.7.1.   Phytochoria compared

There are profound differences in patterns of endemism and diversity between the succulent karoo (especially the Namaqualand-Namib Domain) and Nama-karoo. The former region has higher levels of local endemism as well as higher local and differential diversity. Therefore, it is not surprising that considerably more species are packed into succulent karoo landscapes (Cowling et al., 1998). What explains these differences?

The answer lies in the unique climatic regime of the succulent karoo. Mild, oceanic and fog-ameliorated conditions, combined with predictable (albeit low) winter showers have selected for relatively short-lived shrub life styles. The overwhelming majority of these species are leaf succulents (largely Mesembryanthema) with limited water storage capacity and shallow root systems (Von Willert et al., 1992). This relatively benign environment enables regular germination. Occasional rainfall failure that causes widespread mortality of these drought-intolerant plants (Von Willert et al., 1985; Jürgens et al., 1997), ensures that there is usually sufficient space for seedling recruitment. The result is weakly structured communities of high diversity (Cowling et al., 1994). But what is the link between endemism and regional diversity, and these local-scale processes?

The answer lies in differential rates of diversification (Cowling et al., 1998). Lineages with the traits described above (the majority of succulent karoo species, especially local endemics), have high numbers of sexually produced generations – attributes that would invariably lead to high rates of genetic recombination and rapid speciation (Rosenzweig, 1995). Moreover, for the Mesembryanthema at least, limited gene flow (in terms of seed and pollen dispersal), peculiar genetic controls on morphology (Ihlenfeldt, 1994), and weakly persistent seed banks (Esler, this volume), would enhance diversification. The rapid accumulation of taxa would result in a very fine-scale discrimination of habitats, such as Ihlenfeldt (1994) describes for *Argyroderma* (Mesembryanthema) in the Namaqualand-Namib Domain. The result is very species-rich landscapes with numerous highly localized habitat specialists, often capable of existing in small and isolated populations. Clearly, landscapes with greater heterogeneity will support more species (Cowling et al., *1997*), but the capacity for fine-scale habitat discrimination of explosively diversifying lineages lends support to Rosenzweig's (1992, 1995) argument that habitat diversity is a co-evolved property of species diversity.

Two outstanding issues in the succulent karoo remain unexplained. Why are there so few annuals, especially endemics; and why are there so many geophytes, includ-

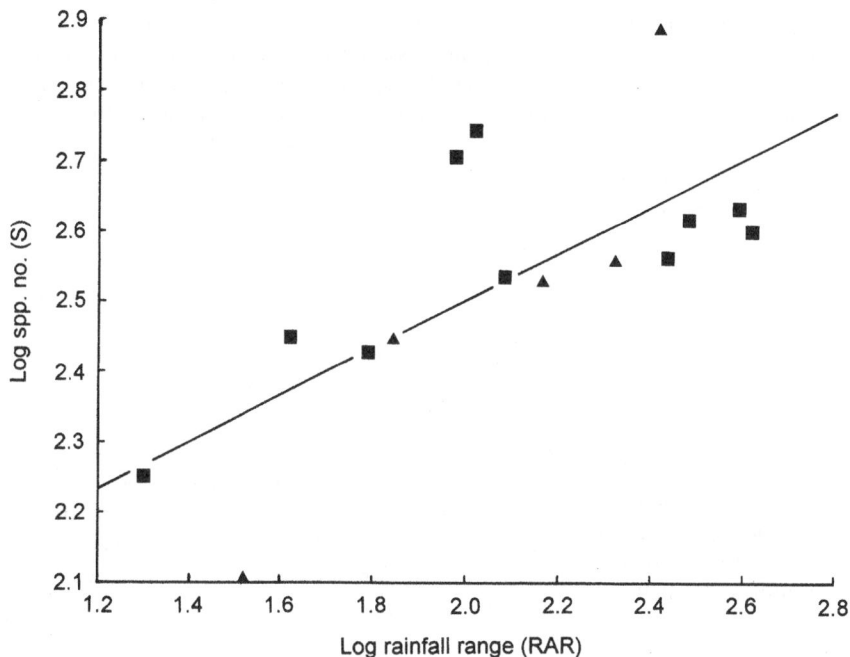

Figure 4.9 **Relationship between regional richness and rainfall range (difference between highest and lowest value of mean annual rainfall in a region) in the Karoo–Namib region of South Africa (succulent karoo (square), Nama-karoo (triangle)) (redrawn with permission from Cowling et al., 1997).**

ing many highly localized species? Given their short generation times, and the fact that environmental conditions in the succulent karoo favour this life form (Van Rooyen, this volume), the low diversity and endemism of annuals is puzzling. Diversification of geophytes may have been promoted by pollinator–flower co-evolution, as has been demonstrated in the geophyte-rich fynbos landscapes (Johnson, 1992). However, despite a very rich assemblage of insect pollinators in the succulent karoo (Struck, 1994a; Vernon, this volume), the pollinator fauna is dominated by generalist foragers (Struck, 1994b). None the less, there is evidence for specialist pollinator interactions between certain geophytes groups, for example Iridaceae and Geraniaceae with long-tongued flies (Nemestrinidae) (Goldblatt et al., 1995; Manning and Goldblatt, 1996), and Iridaceae, Liliaceae (*sensu lato*) and Orchidaceae with monkey beetles (Scarabaeidae: Hopliini) (Picker and Midgley, 1996) (see also Esler, this volume). Goldblatt and Manning (1996) suggest that edaphic specialization provided the initial impetus for genetic differentiation between founder and parent populations of the predominantly succulent karoo irid genus *Lapeirousia*. Specialization to different pollinators (long-tongued flies) is invariably associated with parapatric species pairs. The causes of diversification within geophytic lineages, especially in the Namaqualand–Namib Domain, remains an intriguing research question.

In the Nama-karoo, community structure, diversity patterns and rates of diversification are not unusual for an arid land (Cowling et al., 1998). Shrubs are mainly long-

lived and well dispersed, and the short-lived component is comprised almost entirely of widespread, wind-pollinated, and wind- or animal-dispersed grasses with persistent seed banks (Esler, this volume). Shrub community structure is relatively stable, and competition from this long-lived component (Yeaton and Esler, 1990) limits recruitment (Milton, 1995a; Wiegand et al., 1995). These factors would militate against rapid diversification and habitat specialization. Even in the environmentally heterogeneous escarpment zones, beta diversity is very low.

Areas of higher than average diversity and endemism occur in the south-eastern zone of Eastern Karoo Domain (Hoffman and Cowling, 1991; Cowling et al., 1994), where karroid vegetation abuts on succulent thorn thickets; and in the Damaraland–Kaokoland Domain, a complex phytochorological transition area of steep environmental gradients (Huntley and Matos, 1994). In the former zone, community structure and patterns of endemism show strong similarities to the succulent karoo (Cowling and Holmes, 1991), whereas the Damaraland–Kaokoland Domain appears to have provided refuge for many taxonomically isolated endemics (Nordenstam, 1974; Hilton-Taylor, 1994a).

4.7.2.   Global comparisons
Here we ask the question: is the succulent karoo unique when compared to other winter-rainfall arid lands of the world? This question is addressed in more detail by Esler et al. (this volume). Our focus is exclusively on patterns of diversity and endemism. We summarize some points

below, bearing in mind that the lack of comparative data from other arid lands is a major constraint.

- In the succulent karoo, total diversity at the regional scale (Cowling et al., 1998), and the diversity of leaf succulent shrubs at the local and regional scales, is without parallel (Cowling et al., 1994; 1998). The only other part of the world with an even moderate concentration of non-halophytic leaf succulents is the Canary Islands (Jürgens, 1986). Interestingly, these species are concentrated in foggy, coastal environments which receive low but predictable rainfall (Shmida and Werger, 1992).

- The low diversity of annuals and high numbers of geophytes at all spatial scales, is a distinctive feature of the succulent karoo (Hilton-Taylor, 1996). Other winter-rainfall arid lands, such as the Negev (Shmida and Whittaker, 1979), Mohave (Rundel and Gibson, 1996) and northern Chilean (Armesto and Vidiella, 1993) deserts, are rich in winter annuals, and endemics are well represented in this life form. It is of interest that annuals are also poorly represented in the arid zone of the Canary Islands (Shmida and Werger, 1992). Geophytes are reasonably common in the semi-arid regions of the Middle East (Danin, 1983) and Chile (Armesto and Vidiella, 1993) but numbers in both relative and absolute terms are much lower than in the succulent karoo.

- The explosive diversification of certain succulent karoo genera, resulting in large species numbers, and the high level of generic endemism, is unique among the world's arid lands (Cowling et al., 1998). We elaborate on this in the next section.

4.7.3.   Diversification in an arid world

There are two extra-tropical areas that are recognized as extraordinary hot-spots of plant diversity and endemism: the Mediterranean-climate regions of South Africa and south-western Australia (Cowling et al., 1998). Clearly, the succulent karoo is a third such region (Cowling et al., 1998). The patterns and determinants of diversity in the succulent karoo bear a striking resemblance to those in these regions, even though the predominant vegetation (fynbos in South Africa and kwongan in Australia) is a dense, sclerophyllous and fire-prone shrubland. Both regions have very species-rich landscapes with high differ-

entiation diversity and large numbers of habitat-specific, local endemics associated with a limited number of genera (Cowling et al., 1996). Diversification is largely associated with fire-sensitive lineages (short generation times) that have limited gene flow. Thus, fire plays a key role in promoting diversification and the fine-scale partitioning of habitat space. We suggest that occasional severe drought in the succulent karoo, by promoting generation turnover and population fragmentation, is analogous to fire in these Mediterranean-climate shrublands (Cowling et al., 1998).

There is no evidence to suggest that the massive diversification of the succulent karoo flora is the result of an unusually long and stable history. Like the Mediterranean shrublands, karroid vegetation became a widespread and zonal formation in south-western Africa fairly recently, probably in the Pliocene (Deacon et al., 1983; Scott et al., 1997; Meadows and Watkeys, this volume). As was the case globally, arid-adapted lineages have occupied azonally dry sites since the late Cretaceous, especially after the Oligocene (Axelrod, 1972; Shmida, 1985), and many have persisted as paleoendemics (e.g. *Welwitschia mirabilis*). However, the explosive speciation of the Mesembryanthema in southern Africa (2000 species, 116 genera), largely concentrated in the succulent karoo, is a relatively recent phenomenon, and is probably unrivalled in angiosperms (Ihlenfeldt, 1994).

4.8.   Acknowledgements

This chapter is dedicated to H-D Ihlenfeldt at the Institut fur Allgemeine Botanik und Botanischer Garten, University of Hamburg. Professor Ihlenfeldt fostered an enormous amount of research on the Mesembryanthema, and, through his guidance and insight, the mysteries of this group are beginning to unravel. We also acknowledge the role of other German researchers, many students or associates of Ihlenfeldt, who have done so much to understand the plant life of southern Africa's arid lands. In particular, we express our appreciation to: V. Bittrich, M. Dehn, M. Gerbaulet, H. Hartmann, S. Liede, M. Struck, D. J. von Willert and, especially, N. Jürgens. Others who have shared with us their valuable insights on the plants of arid lands are M. B. Bayer, W. R. J. Dean, P. Desmet, K. J. Esler, S. Hammer, M. T. Hoffman, S. J. Milton, S. M. Pierce, P. W. Rundel, E. van Jaarsveld and G. Williamson.

5 Biogeography, endemism and diversity of animals in the karoo

C. J. Vernon

5.1. Introduction

The south-western arid zone of Africa, which includes the Namib and Kalahari Deserts, and the succulent and Nama-karoo biomes, is rich in plant and animal species, and in endemism. The distribution of many animals, however, in this region is poorly known, and is further complicated by the presence of many undescribed species, particularly among the invertebrates.

In this chapter, a species is defined as a discrete entity which remains in equilibrium during phases of environmental stability. It is assumed that there is only one type of species. A species is a set of populations of individuals spatio-temporally bounded by birth, death and finite geographical boundaries. The individuals of a species have phenotypic resemblance and shared reproductive systems. They are capable of recognizing and responding to each other, breeding together and producing fertile offspring. There are no adaptations which are the property of a species and not of individuals, and it is the individuals which are the units of natural selection. The attributes ascribed to species such as habits, niche, breeding system, are in fact the properties of the individuals. A relict species is one which clearly is isolated from conspecifics, and here considered to be a faunal remnant from a prior, different epoch. An isolated species is one with a restricted distribution which is disjunct from congeners.

Biogeography takes cognizance of evolutionary and speciation events and aims to explain the distribution patterns of all organisms. An endemic species or higher taxon is one whose distribution is geographically limited, but this endemicity provides no information about the place where that organism evolved. However, historical biogeographic patterns caused by physical factors induced by the alternation of climatic epochs are a basis for understanding the present-day fauna of the karoo (Van Zinderen Bakker, 1978; Werger, 1978a; Vrba, 1985; Deacon and Lancaster, 1988; Bond and Richardson, 1990; Avery, 1992; Marker, 1995). The animals and the resources they exploit are discussed here in three sections, (1) the origins of the karoo fauna, (2) endemicity and distribution patterns in the Nama- and succulent karoo, and (3) anthropogenic factors influencing the present-day distribution of animals. The treatment is uneven because of the differential availability of information.

5.1.1. Environmental influences

In order to maximize their use of both abiotic and biotic resources, animals respond to environmental cues and have interactions with other organisms. In the karoo biomes, rainfall is the most important environmental factor. In the Nama-karoo, the rainfall is low and erratic, with long periods of drought punctuated by brief falls of rain (Venter et al., 1986). The succulent karoo is also arid, but rain falls regularly in the winter (Hoffman and Cowling, 1987).

Seasonal movements

All organisms are persistent in a suite of environmental conditions to which they adapted during speciation. Thus their present-day habitat preferences are similar to those in which the organisms evolved. The resource phenology determines the survival tactics of individuals (Dean and Milton, this volume). A stable environment permits the individual to be sedentary and resident. The dispersion of the resources will determine whether they are worth defending and then individuals may be territorial. A seasonal environment has regular seasonal fluctuations which compel the individual to react to those changes, either through migration or dormancy. Alternatively, individuals may have strategies such as seasonal switching in diet or foraging site, or local itinerancy, that permit them to reside in seasonal environments. Individuals in such seasonal environments can be seasonally territorial.

An episodic environment is where there are short periods of resource renewal, followed by gradual decreases and long periods of depleted resources. The episodes occur randomly in time and space. The individuals in such environments tend to be nomadic, and, although their wandering may appear random, it may depend upon local knowledge and environmental clues of how to locate resources.

5.1.2.   Evolution

The process of evolution is neither pre-ordained, purposeful, directional nor regular (Gould, 1989). Although the finer details of evolutionary theory are a matter of continuing debate, the concept of punctuated equilibrium provides a basis for what is written here (Vrba, 1980, 1985; Endrödy-Younga, 1988).

Punctuated equilibrium describes the sequence of speciation in which there are long periods of equilibrium punctuated by episodic bursts of speciation. A speciation event may give rise to many new species contrasting with the bimodal branching of gradual phyletic evolution. The apparent barriers to species distribution and the accidental formation of fragmented populations are essential preconditions for speciation.

An example of speciation is provided by the 14 species of flightless stag-beetles in the fynbos endemic genus *Colophon* (subfamily Lucaninae). The genus is a relic of Gondwana, with its nearest relatives in Brazil and Australia. Isolated populations of *Colophon* are found in montane grasslands at altitudes over 1000 m with only one species at each mountain.

It is postulated that the speciation of *Colophon* was consequent of a climatic change. In a climatic epoch of increasing cold, the habitat, currently only on mountain tops, shifted down the mountains and onto the surrounding plains. Much of those plains are today part of the karoo biome. The ancestral *Colophon* population shifted with the habitat and colonized the newly available habitat. As the climatic epoch ended and temperatures became warmer, both the habitat and the beetles withdrew to isolated mountains. Speciation followed and the extant array and dispersion of *Colophon* species are the survivors of that event (Endrödy-Younga, 1988).

5.2.   **Origins of the karoo fauna**

5.2.1.   Gondwana

The breakup of Gondwana about 160 Myr ago is a biological benchmark. It is the last time when organisms could disperse freely across a landscape that is now divided into separate continents. Gondwanan fauna has two characteristics; groups which had evolved into recognizable entities by the Mesozoic, and a common distribution pattern in the southern continents (Mackerras, 1970).

There are families whose cosmopolitan distribution patterns indicate that they were widespread before Gondwana broke up. Subsequently there has been speciation and there may be a different suite of genera and species in each continent. There are also families with few archaic species which are interpreted as being relicts of Gondwana. These relicts, such as the Onychophora, have remained in equilibrium, unchanged for at least 160 Myr (Newlands and Ruhberg, 1978; Endrödy-Younga and Peck, 1983).

The nucleus of the biota of southern Africa has its origins in Gondwana (Van Zinderen Bakker, 1978), as does the insect fauna in Australia (Mackerras, 1970).

The termites (Isoptera) are an example of a Gondwana relic fauna (Table 5.1). Termites are cosmopolitan, suggesting ancestral origins in Pangea.

There are six families of termites with about 2000 species throughout the world. Three families have representatives in Africa, America, Asia, Australia and Madagascar, and no family is endemic to any one continent. Within those families there are at least two genera, *Glyptotermes* and *Nasutitermes* that have species in all four continents and Madagascar. There are another 24 genera which have species in Africa and in at least one of the other continents (Gray, 1970; Scholtz and Holm, 1985; Snyder, 1961, 1968). There are no termite species that are common to these continents. There are two relict families, Mastotermitidae and Termopsidae, each with only a few genera and species. The former is monotypic with a species of *Mastotermes* found in Australia and the Oriental region. In the Termopsidae there are only two small genera *Porotermes* and *Stolotermes*. *Porotermes* has one species in Australia, another in Chile and a third in South Africa.

5.2.2.   Palaearctic links and the mussel fauna

The breakup of Gondwana was probably the last time when freshwater organisms could move between the Ethiopian and Palaearctic regions. Thereafter the oreogenesis of mountains created physical barriers disrupting old waterways and creating new watersheds and drainage basins (Meadows and Watkeys, this volume).

The freshwater bivalves are relicts of a prior southern African fauna related to that in the Palaearctic. Fossil mus-

Table 5.1. **Gondwana termites showing the number of species on each continent**

Family and genus	No. of species in genus				
	Africa	Madagascar	America	Asia	Australia
Mastotermitidae					
Mastotermes				1	1
Miotermes			1		
Kalotermitidae					
Bicornitermes	2				
Bifiditermes	2	1		2	
Comatermes			1		
Cryptotermes	3	3	12	11	4
Epicalotermes	3		1	3	
Eucryptotermes			1		
Glyptotermes	6	1	9	16	5
Incisitermes			15	1	
Kalotermes	2		6	9	7
Lobitermes	1				
Neotermes	6	6	6	28	1
Paraneotermes			1		
Postelectrotermes	1	1		3	
Procryptotermes	1	2	3	5	
Proneotermes			2		
Pterotermes			1		
Rugitermes			12		
Tauritermes			2		
Termopsidae					
Termopsinae					
Archotermopsis				1	
Hodotermopsis				2	
Zootermopsis			3		
Porotermitinae					
Porotermes	1		1		1
Stolotermitinae					
Stolotermes	1				6
Hodotermitidae					
Hodotermes	2				
Microhodotermes	3				
Anacanthotermes	1			13	
Rhinotermitidae					
Coptotermitinae					
Coptotermes	12	1	4	25	6
Heterotermitinae					
Heterotermes	2	1	11	8	8
Reticulitermes	3	3	6	8	
Psammotermitinae					
Glossotermes			1		
Psammotermes	3	1		3	
Termitogetoninae					
Termitogeton				2	
Stylotermitinae					
Stylotermes				4	
Rhinotermitinae					
Acorhinotermes			1		
Dolichorhinotermes			4		
Operculitermes			1	1	
Parrhinotermes			5		1
Rhinotermes					1
Sarvaritermes				1	
Prorhinotermitinae					
Schedorhinotermes	2			21	4
Prorhinotermes	3	1	1	8	
Serritermidae					
Serritermes			1		
Termitidae					
Macrotermitinae					
Acanthotermes	1				
Allodontermes	3				
Ancistrotermes	12				
Bellicositermes	5				
Euscaiotermes				1	
Hypotermes				6	
Macrotermes	34			11	
Microtermes	43	2		13	
Odontotermes	90			79	
Protermes	1				
Pseudacanthotermes	3				
Sphaerotermes	1				
Synacanthotermes	2				
Apicotermitinae					
Acholotermes	1				
Acutidentitermes	1				
Adaiphrotermes	1				
Aganotermes	1				
Allognathitermes	4				
Alyscotermes	1				
Anoplotermes	10		36	1	
Apicotermes	14				
Astalotermes	1				
Coxotermes	1				
Duplidentitermes	1				
Eburnitermes	1				
Firmitermes	1				
Heimitermes	2				
Hoplognathitermes	1				
Jugositermes	1				
Rostrotermes	1				
Skatitermes	2				
Speculitermes					
Trichotermes	1				
Termitinae					
Ahamitermes					3
Amitermes	17		14	7	50
Amphidotermes	1				
Angulitermes	7				
Apilitermes	1				
Basidentitermes	3				
Capritermes		1		6	
Cavitermes			2		
Cephalotermes	1				
Cornicapritermes			3		
Crepititermes			1		
Crenetermes	3				
Cubitermes	19				
Cylindrotermes			4		
Dentispicotermes			1		
Dicuspiditermes				13	
Dihoplotermes			1		
Drepanotermes					2
Eremotermes	2				
Euchilotermes	3				
Fastigitermes	1				
Foraminitermes	6				
Forficulitermes	1				
Furculitermes	6				
Genuotermes			1		
Gnathitermes			4		
Homallotermes				2	
Incolitermes					1
Lepidotermes	2				
Megagnathotermes	1				
Microcerotermes	51	15	6	23	11
Miricapritermes				1	
Mirocapritermes				6	
Mucrotermes	1				
Neocapritermes			14		
Nitiditermes	1				
Noditermes	5	5			
Okavangotermes	1				
Ophiotermes	1				
Orthognathotermes			5		
Orthotermes	1				
Ovambotermes	1				
Paracapritermes					2
Pericapritermes				6	
Pilotermes	1				
Planicapritermes			1		
Proboscitermes	1				
Procapritermes				6	

Table 5.1. *cont*

Family and genus	No. of species in genus				
	Africa	Madagascar	America	Asia	Australia
Procubitermes	2				
Profastigitermes	1				
Promirotermes	3				
Protocapritermes					2
Pseudocapritermes				3	
Pseudomicrotermes	1				
Quasitermes		1		1	
Spicotermes			1		
Spinitermes			2		
Synhamitermes			1	1	
Termes	5	1	11	19	19
Thoracotermes	2				
Unguitermes	3				
Unicornitermes	1				
Nasutitermitinae					
Aciculitermes				2	
Afrosubulitermes	1				
Ampoulitermes				1	
Angularitermes			1		
Australitermes				1	
Baucaliotermes	1				
Bulbitermes				23	
Ceylonitermellus				1	
Ceylonitermes				1	
Coarctotermes		3		3	
Constrictotermes			3		
Convexitermes			8		
Cornitermes			15		
Curvitermes			2		
Diversitermes			5		
Emersonitermes				1	
Eutermellus	4				
Eutermes	2			2	
Fulleritermes	4				
Grallatotermes	1				
Havilanditermes				1	
Hirtitermes				3	
Hospitalitermes				26	
Labiotermes			5		
Lacessititermes				15	
Leptomyxotermes	1				
Leucopitermes				2	
Longipeditermes				2	
Macrosubulitermes				1	
Malagasiotermes		1			
Mimeutermes	5				
Mycterotermes	1				
Nasutitermes	47	10	68	66	19
Obtusitermes			2		
Occasitermes					2
Occultitermes					1
Orientsubulitermes				1	
Paracornitermes			4		
Parvitermes			6		
Postsubulitermes	1				
Procornitermes			6		
Rhadinotermes	1	1			
Rhynchotermes			2		
Rotunditermes			1		
Spatullitermes	2				
Subulitermes			11		
Syntermes			20		
Tarditermes	1	1			
Tenuirostritermes			4		
Terrentitermes			1		
Trinervitermes	57			7	
Tumultitermes					25
Velocitermes			6		
Verrucositermes	1				

Sources: Coaton (1963); Snyder (1961, 1968); Gray (1970); Coaton and Sheasby (1973–1978); Ruelle et al. (1975); Mitchell (1980); Pearce (1997).

sels (*Unio* spp.) are recorded from Ecca beds in South Africa (Rilett, 1951) as well as from beds of Triassic age in East Africa and the Holarctic. *Unio cafer*, that has congenerics in Eurasia, is endemic to southern Africa.

There are populations of *U. cafer* in at least three rivers in the Nama-karoo. There are also middens and excavations of mussel shells in many other rivers (Fig. 5.1). Those along the Fish River near Cradock are *c.* 1000 years old (Jubb, 1976). These mussel shells are indicators that *U. cafer* was formerly found in rivers across what is now the karoo, linking the extant *U. cafer* populations in the fynbos, grassland and savanna biomes. *Unio cafer* is a fluviatile species that lives in permanent pools. It cannot cope with drought, and dies soon after a pool dries up, so its presence in the karoo is an indicator of a prior wetter epoch.

The *U. cafer* populations in the karoo have become increasing isolated and relict. Residual populations are found around mountain outliers of the Drakensberg abutting the karoo and the mountains between Cradock and Murraysberg which are surrounded by the karoo. The ability of *U. cafer* to resist the increasingly aridity is partly related to the rate of erosion. The Sneeuberge range in the Graaff Reinet area is drained to the south by the Sundays River and to the north by the Seacow River (Seekoeirivier), a tributary of the Orange River. The Sundays River descends rapidly, about 800 m in 50 km, while the Seacow River descends more gradually and takes 200 km to drop to the same elevation. *Unio cafer* is extant along the Seacow River, but only shells in middens are found along the Sundays River.

Unio cafer disperses in the larval stage as an endoparasite of fish. The fish host of *U. cafer* is unknown but by comparison of distribution patterns, *Barbus* spp. fish are the most probable hosts. Thus, what follows about *U. cafer* is also applicable, in the main, to *Barbus*. The initial colonization of southern Africa by *U. cafer* occurred when there was a single drainage system. This was fragmented so that *U. cafer* populations were isolated in the Orange, Limpopo and Zambesi river systems as well as the rivers draining east of the Drakensberg and south of the Cape Fold mountains. In the Orange River system, *U. cafer* has become fragmented to small populations. Most of these are in tributary rivers in the grassland biome. The Aughrabies falls is a barrier which separates those small populations from another population in the lower reaches of the Orange River. Throughout time, *U. cafer* has remained a single species. This contrasts with *Barbus* which is now an array of species (Skelton, 1993).

The cosmopolitan freshwater bivalve genera *Corbicula*, *Eupera*, *Pisidium* and *Sphaerium* have isolated species or populations in southern Africa, then reappear in Africa north of the equator and again in the Palaearctic. *Eupera ferruginea* is found in the ephemeral pans in the Nama-karoo. The

Figure 5.1 **Past (—) and present
(■) distribution of the
freshwater mussel *Unio cafer*
in Western South Africa**

animal is inconspicuous and most likely to be detected when pans dry up leaving myriads of minute shells on the shore (personal observations). How the life cycle of *E. ferruginea* has adjusted to cope with the episodic and erratic water regime of those pans is not known. Presumably *Eupera* has a rapid life cycle and lays eggs which can withstand long periods of desiccation.

5.2.3.   Drainage basins and the fish fauna

The southern African continental raft warped and fractured to create an inland basin with a surrounding perimeter of mountains. The seaward rivers draining the mountains cut back and eventually captured inland waters, diverting them to the coast, giving rise to the extant drainage basins (Dingle and Hendy, 1984). At one stage, the Orange River flowed into the sea at the present Olifants River mouth. Relict beds of the Orange River in the karoo indicate that the river moved gradually northwards, until it was captured by the Fish River and cut its present channel over the past 10 Myr.

The fish fauna of southern Africa can be divided geographically into regions (Skelton, 1993), which can be equated with biomes. A temperate Southern Faunal Area (SFA) includes ecoregions that are equivalent to the forest, fynbos, and grassland biomes. The desert, Nama-karoo,

and succulent karoo biomes have been alternately part of the SFA or a southern extension of a tropical faunal area. At some time in the past, the SFA may have covered all of South Africa and the southern half of Botswana. Conversely, at times the tropical fauna extended south to encompass three biomes (Skelton, 1993). There are no extant fish species of the karoo biome that speciated following the southern extension of the tropical fish fauna.

The fauna of the SFA is characterized by a low diversity of species with a high degree of endemism. Within the SFA, the fishes of the karoo biomes are mostly hardy species or relicts (Skelton, 1993), and the endemic fish species of the karoo biome are found along the Orange River. There are no fish in the karoo which are specific to ephemeral pans or subterranean caverns.

In southern Africa, there are 217 freshwater fish species of which about 82% are part of a tropical fish fauna whose southern limits are the Orange river (Skelton, 1986). In past epochs, this fauna may have extended south to encompass the karoo (Skelton, 1993). The tropical fish *Mesobola brevianalis*, that has an isolated population below the Aughrabies Falls on the Orange River and is also found in the Limpopo River system, may be a relic of this past epoch.

As a consequence of the aridity, the Nama- and succulent karoo biomes now have a poor fish fauna (Appendix 5.6). Endemic karoo fish species are members of only three genera, *Austroglanis*, *Barbus* and *Labeo*. All three genera also have representative endemic species in the fynbos biome, and indicate a prior link between the rivers of the fynbos and karoo biomes. The klipbaber, *Austroglanis sclateri*, is endemic to the Orange River. The other two members of the genus are endemic to the fynbos.

Of 53 *Barbus* species in southern Africa, only one, *Barbus hospes*, is confined to the karoo. One or more species of the genus *Barbus* may be host to the larvae of the mussel *Unio cafer*. This suggests that the link of the mussel with the Palaearctic may be mirrored between the genus *Barbus* and the Palaearctic. There is no single *Barbus* species whose distribution encompasses that of the mussel, suggesting that the southern African speciation of *Barbus* took place subsequent to the co-evolution between the fish and mussel.

5.2.4   Rivers as dispersal routes

Rivers which flow through arid regions form ribbons of mesic habitat. These provide pathways for organisms both to pass through and to gain access to the arid hinterland (see Clancey, 1994 for birds). Such habitats and passage of organisms would be greatest in pluvial epochs. For example, it is suggested that the division between the northern and southern Namib Desert is the consequence of a wetter epoch when riparian woodland extended along watercourses forming a barrier between north and south (Endrödy-Younga, 1986).

The Orange River flows through the karoo biomes, but the birds associated with the river are not part of the karoo biome avifauna. As a waterway, the river provides a safe passage for dispersing and migrating birds. It also provides a corridor of woodland in which some birds are resident. There are relict populations of birds such as Cape Francolin *Francolinus capensis*, goldentailed woodpecker *Campethera abingoni*, Cape robin *Cossypha caffra*, olive thrush *Turdus olivaceus* and Cape white-eye *Zosterops pallidus* which are dependent upon the riverine vegetation. These species have used the river both as a refuge and as a basis from which to expand their populations into new habitats created around human habitation.

Within the karoo, and excluding the Orange River, most rivers are episodic water-courses which soon reduce to residual pools. Perennial water is scarce, but it did exist as four bird species, black stork *Ciconia nigra*, Hamerkop *Scopus umbretta*, African black duck *Anas sparsa* and giant kingfisher *Megaceryle maxima*, that inhabit rivers and streams, are commonly found in the karoo. The decline in tenancy of black stork eyries and the presence of old, abandoned Hamerkop nests indicate that populations of these species have declined, and may be associated with a general trend towards increasing aridity accelerated by anthropogenic factors.

5.2.5   Afro-montane zone

The Afro-montane zone is composed of those areas of Africa that have a temperate climate. In the tropics, this can only occur at high altitudes, while in the temperate latitudes of southern Africa the Afro-montane zone approaches sea level. The Afro-montane zone extends along the mountains running the length of Africa from Ethiopia to the southern Cape Province. In places it is disrupted by intervening lowland. The Afro-montane zone suggests a temperate corridor along which there could have been floral and faunal exchange between the Ethiopian and Palaearctic regions, so that in every glacial epoch waves of Eurasian animals could emigrate south into Africa (Vrba, 1985).

In warmer epochs than the present, the Afro-montane zone became even more fragmented, whereas in cooler epochs the boundary would have descended to lower altitudes and greatly increased the area of the zone. The distribution pattern of Afro-montane organisms is 'J'-shaped and illustrated by that of the black eagle *Aquila verreauxi* (Fig. 5.2).

Examples of species with an Afro-montane origin are the rupicolous lizards of the genus *Afroedura*. These lizards are endemic to the mountainous parts of Africa south of 11 °S, and occur in the karoo *sensu lato*, in isolated populations on dolerite koppies (Onderstall, 1984). The polytypic

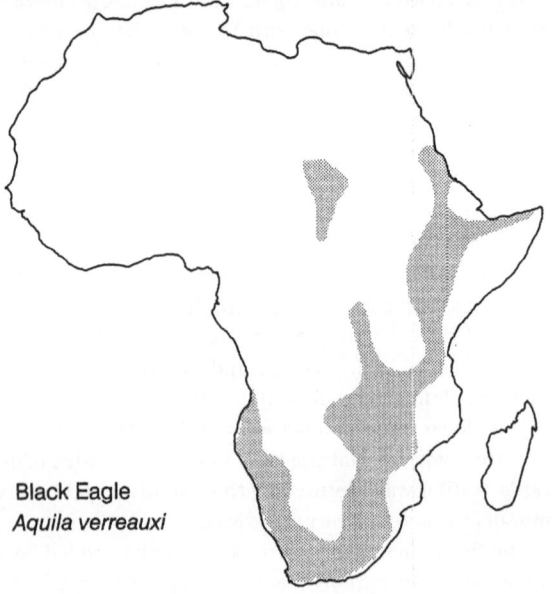

Black Eagle
Aquila verreauxi

Figure 5.2 **The distribution of the black eagle *Aquila verreauxi* in Africa, showing a typical 'J'-shaped Afro-montane distribution pattern**

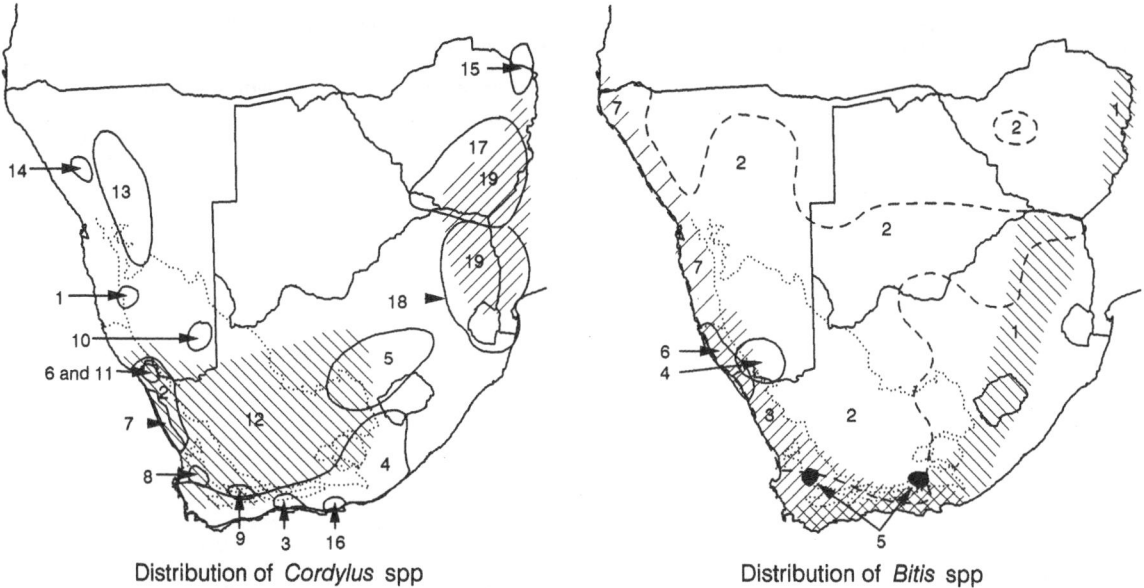

Figure 5.3 **The radiation of species from Afro-montane ancestors. (Left), the distribution of *Cordylus* spp. lizards in southern Africa:**
1 = *Cordylus campbelli*; 2 = *C. cataphractus*; 3 = *C. coreruleopunctatus*; 4 = *C. cordylus*; 5 = *C. giganteus*; 6 = *C. lawrencei*; 7 = *C. macropholis*;
8 = *C. mclachlani*; 9 = *C. minor*; 10 = *C. namaquensis*; 11 = *C. peersi*; 12 = *C. polyzonus*; 13 = *C. jordani*; 14 = *C. pustulatus*; 15 = *C. rhodesianus*;
16 = *C. tasmani*; 17 = *C. tropidosternum*; 18 = *C. vittifer*; 19 = *C. warreni*

(Right), the distribution of *Bitis* spp. snakes in southern Africa: 1 = *Bitis atropos*; 2 = *B. caudalis* (distribution enclosed within the dotted line);
3 = *B. cornuta*; 4 = *B. xeropaga*; 5 = *B. inornata*; 6 = *B. schneideri*; 7 = *B. peringueyi*. The approximate boundaries of the Nama- and succulent
karoo are shown as a fine dotted line

lizard genera *Cordylus* (Fig. 5.3) and *Pachydactylus* have an array of afro-montane species that were in place prior to their profuse speciation and their spread into the karoo biomes. Snakes of the genus *Lamprophis* probably radiated from afro-montane stock, as the spotted house snake *Lamprophis guttatus* has a 'J'-shaped distribution pattern. The genus *Bitis* has a group of species (*atropos*, *cornuta*, *schneideri* and *xeropygia*) which collectively show an Afro-montane distribution pattern (Fig. 5.3). The rodent genus *Otomys* forms a complex of species which have diffused from an original 'J'-shaped Afro-montane distribution pattern, and moles (*Amblysomus*, *Chlorotalpa* and *Chrysochloris* spp.) collectively show an Afro-montane distribution pattern.

5.2.6.   The arid corridor and other biogeographic patterns

In the past, there were arid epochs which favoured the increase of the desert and savanna biomes. Those biomes then were able to expand and link with a contemporary expansion of similar biomes in north-east Africa (Balinsky, 1962; Werger, 1978b). The link between the two regions has been termed an arid corridor providing a pathway for flora and fauna to move between the two regions. The disjunct populations of the same species, of genera and families which have representative species in each region, indicate that the corridor has been formed many times. Some

examples of this distribution are shown in Fig. 5.4, and some species are given in Table 5.2. The most recent corridor occurred about 18 000 years ago (Hilton-Taylor, 1987, citing Van Zinderen Bakker, 1982). The present distribution of forms of the Sabota lark *Mirafra sabota* and related species suggests that the arid corridor repeatedly opened and closed (Fig. 5.4).

The Cradock corridor

The southward extension of savanna would have gone round the grassland biome. On the west, the spread would have been south from Prieska through Cradock to Port Elizabeth, forming the Cradock corridor of savanna between the Nama-karoo and grassland biomes.

The bullfrog *Pyxicephalus adspersus*, a savanna species found in the eastern Nama-karoo (Passmore and Carruthers, 1979), has a distribution pattern suggesting a savanna corridor linking Port Elizabeth with the Orange River. The leopard tortoise *Geochelone pardalis* is a savanna species which has an arid corridor distribution pattern and extends into the fynbos and karoo biomes. *Geochelone pardalis* is absent from the central parts of the Nama-karoo, and occurs only on the eastern and southern margins of the karoo biomes (Greig and Burdett, 1976; Branch, 1988a). Its range surrounds but does not encompass the grassland, and shows a corridor from the Orange River

Table 5.2(a). *The animals of the arid corridor.*
Species found in southern Africa and in northeast Africa

Frogs
Bufo garmani
Tomopterna cryptotis

Birds
Ardeotis kori
Calandrella cinerea
Chersomanes albofasciata
Eremopterix leucotis
Eupodotis ruficrista
Eupodotis senegalensis
Mirafra africanoides
Oena capensis
Plocepasser mahali
Poliohierax semitorquatus
Rhinoptilus africanus
Struthio camelus

Mammals
Acomys subspinosus
Canis mesomelas
Lepus capensis
Madoqua kirkii
Manis temminckii
Oryxa gazella
Orycteropus afer
Otocyon megalotis
Otomys saundersiae
Pedetes cafer
Proteles cristatus
Raphicerus campestris
Redunca fulvorufula
Tadarida aegyptica

Table 5.2(b). *Pairs of super-species or closely related ecologically equivalent species*

Southern Africa	Northeast Africa
Invertebrates (Coleoptera)	
Acmaeodera spp.	*Acmaeodera* spp. (Holm, 1990;)
Julodella bicolor.	*Julodella* sp. (Holm, 1990)
Vertebrates	
Frogs	
Bufo vertebralis group	*B. parkeri, B. lughensis* (Poynton, 1995)
Snakes	
Bitis caudalis	*B. nasicornis*
Birds (all from Hall and Moreau, 1970 and Snow, 1978)	
Ammomanes grayi	*A. deserti*
Bubalornis niger	*B. albirostris*
Bubalornis niger niger	*B. niger intermedius*
Eremomela gregalis	*E. canescens?*
Eurocephalus anguitimens	*E. rüppellii*
Galerida magnirostris	*G. fremantlii*
Meleirax metabates	*M. sp.*
Mirafra apiata	*M. collaris*
Mirafra cheniana	*M. cantillans*
Mirafra passerina	*M. albicauda*
Mirafra ruddi	*M. archeri*
Myrmecocichla formicivora	*M. aethiops*
Oenanthe monticola	*O. leucopyga*
Parisoma layardi	*P. lugens*
Parisoma subcaeruleum	*P. boehmi*
Sproropipes squamifrons	*S. frontalis*
Uraeginthus granatina	*U. ianthinogaster*
Vidua regia	*V. fischeri*
Mammals (Smithers, 1983 and Kingdon, 1974)	
Chlorotalpa sclateri	*C. tytonis*
Cryptochloris spp	*C. stuhlmanni*
Cryptomys hottentotus	*C. ochraceocinereus*
Desmodillus auricularis	*D. braueri*
Gerbillurus spp.	*Gerbillus* spp.
Hyaena brunnea	*H. hyaena*
Otomys spp	*O. denti, O. typus*
Xerus inauris	*X. rutilans*

south through Cradock to Port Elizabeth. Similarly, the South African hedgehog *Erinaceus frontalis* (Fig. 5.5) and the white-bellied korhaan *Eupodotis cafra* have distribution patterns in the eastern karoo following the Cradock corridor. In general, both species have a mesic savanna distribution in southern Africa.

The springbok *Antidorcas marsupialis* formerly had a distribution which extended from the savanna biome into the desert, grassland and Nama-karoo biomes. The distribution extended down the Cradock corridor and excluded the Transkei (Skinner and Louw, 1996). A similar distribution pattern is shown by the elephant shrew *Elephantulus rupestris*, and there is a parallel in the discontinuous distribution of *Otomys saundersiae* which has populations in the south-west Cape and from the Albany Centre through the Cradock corridor to Lesotho.

The Transkei gap
In the east, the spread of savanna would have extended along the coast and down to Durban. This would have left an area, approximately that of the former Transkei, which remained grassland or forest.

Marine regression
Changes in sea level have altered the amount of coastal lowlands. In past epochs, marine regression caused the

appearance of a more extensive coastal plain, so that the western coastal succulent karoo may have been connected with the southern Agulhas coastal fynbos, with consequent expansion of the distribution ranges of some organisms.

5.2.7.   Savanna–fynbos links
There have been prior epochs when the savanna biome encompassed the area of the existing karoo biomes. This is inferred from the mammals of the Tertiary recorded at Langebaanweg that included bears (*Agriotherium*), giraffe, okapi, rhino, elephant, peccaries, wolverine and musk oxen, sabre-toothed cats and five species of hyaena. The closest modern relative of *Agriotherium* is the giant panda of China (Hendey, 1982). In periods when the lowland areas had a savanna fauna including a now extinct springbok *Antidorcas australis*, the fynbos biome would have been restricted to montane areas. Past links between savanna and fynbos can be inferred from closely related bird and reptile species that occur in karoo, fynbos and savanna.

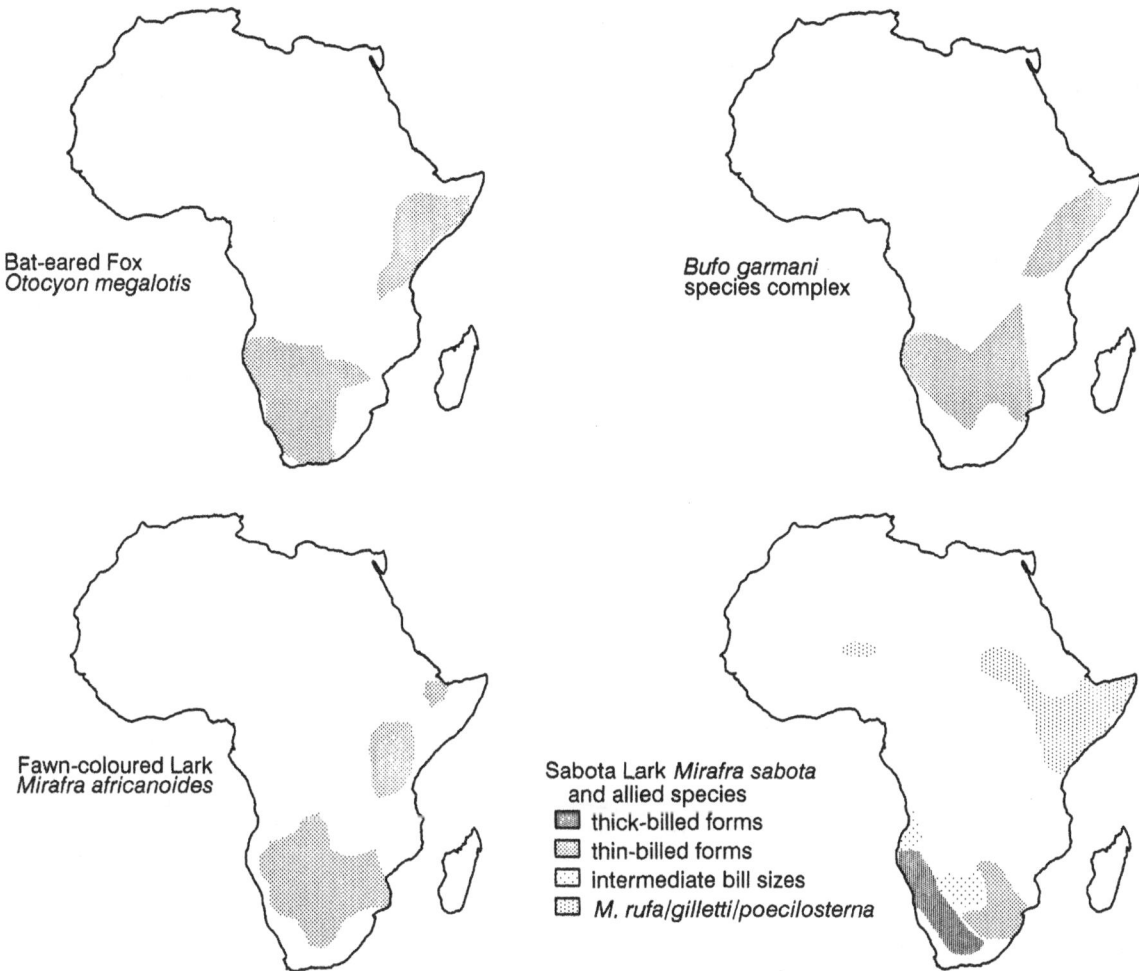

Figure 5.4 **Examples of arid corridor distributions in Africa. The distribution of two wide-ranging species, the bat-eared fox *Otocyon megalotis* and the olive toad *Bufo garmani*, the distribution of the fawn-coloured lark *Mirafra africanoides*, showing the extension of range into the karoo of a savanna species with an arid corridor distribution, and the distribution of adjacent morphs of the Sabota lark *Mirafra sabota*, suggesting that the arid corridor repeatedly opened and closed**

For example, the distribution of the spotted sand lizard *Pedioplanis lineocellata* encompasses the karoo biomes. There are two subspecies, *P. l. lineocellata* in the savanna and *P. l. pulchella* in the fynbos, whose distribution ranges meet in the central Nama-karoo. Within the bustards, the fynbos and succulent karoo black korhaan *Eupodotis afra* forms a species pair with the savanna white-quilled korhaan *E. afraoides* (considered by some taxonomists to be a single species). In terms of the degree of difference between the *afra* and *afraoides*, it suggests a sequence for environmental events: (i) the division of the *E. afra* complex from the karoo korhaan *E. vigorsii* complex by separating the fynbos from the grassland, (ii) speciation of the *E. vigorsi* complex into the white-bellied Korhaan *E. cafra*, the blue korhaan *E. caerulescens*, Rüppells Korhaan *E. rueppellii* and karoo korhaan by breaking the grassland into four

refugia with expansion of the fynbos biome, (iii) simultaneous expansion of the range of *E. afra* over the fynbos from the then adjacent savanna and, (iv) the recent speciation of the *E. afra* complex with the retreat of the fynbos, resulting in the isolation of *E. afra* in the fynbos and *E. afraoides* in the savanna (Fig. 5.6, Fig. 5.7). Many of the bird species in the karoo that are members of polytypic genera may have evolved in a similar manner.

5.3.   Endemicity and distribution patterns – Nama-karoo

The Nama-karoo has a relatively species-poor fauna. The majority of animals have extended their ranges from adja-

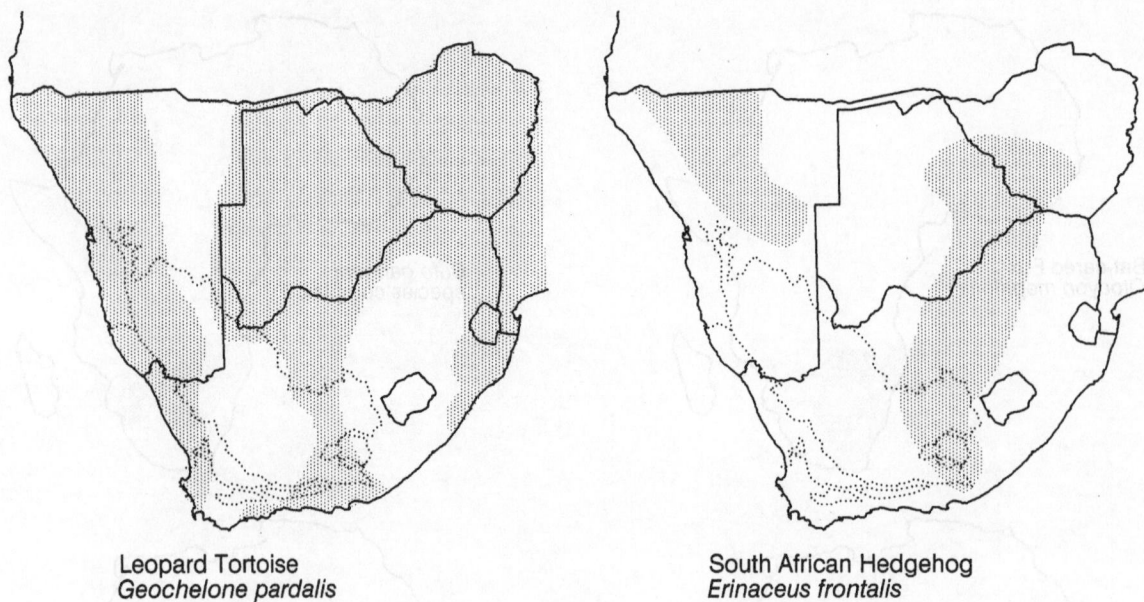

Leopard Tortoise
Geochelone pardalis

South African Hedgehog
Erinaceus frontalis

Figure 5.5 **The distribution of the leopard tortoise *Geochelone pardalis* and the South African hedgehog *Erinaceus frontalis* in southern Africa, showing the Cradock corridor along the eastern Nama-karoo**

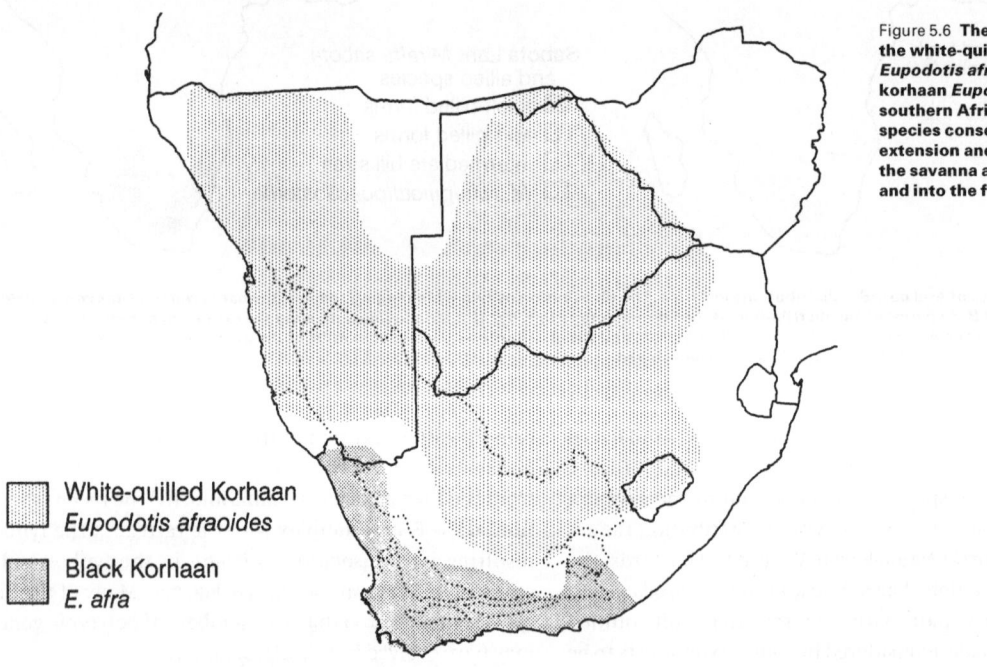

Figure 5.6 **The distribution of the white-quilled korhaan *Eupodotis afraoides* and black korhaan *Eupodotis afra* in southern Africa: a pair of species consequent of the extension and withdrawal of the savanna across the karoo and into the fynbos**

White-quilled Korhaan
Eupodotis afraoides

Black Korhaan
E. afra

cent biomes into the Nama-karoo. There are few endemic species and many of these are relicts of epochs when other biomes extended over the Nama-karoo.

5.3.1.   Invertebrates
The invertebrates are probably the most important, but also the poorest-known faunal group in the Nama-karoo.

This generalization is valid in terms of both pattern and process, the number of species, the number of individuals and ecological adaptations. Some examples of endemic families and genera are:

Arachnida: Of a total of 124 species of scorpion (Scorpionida) in southern Africa (Prendini, 1995), 79

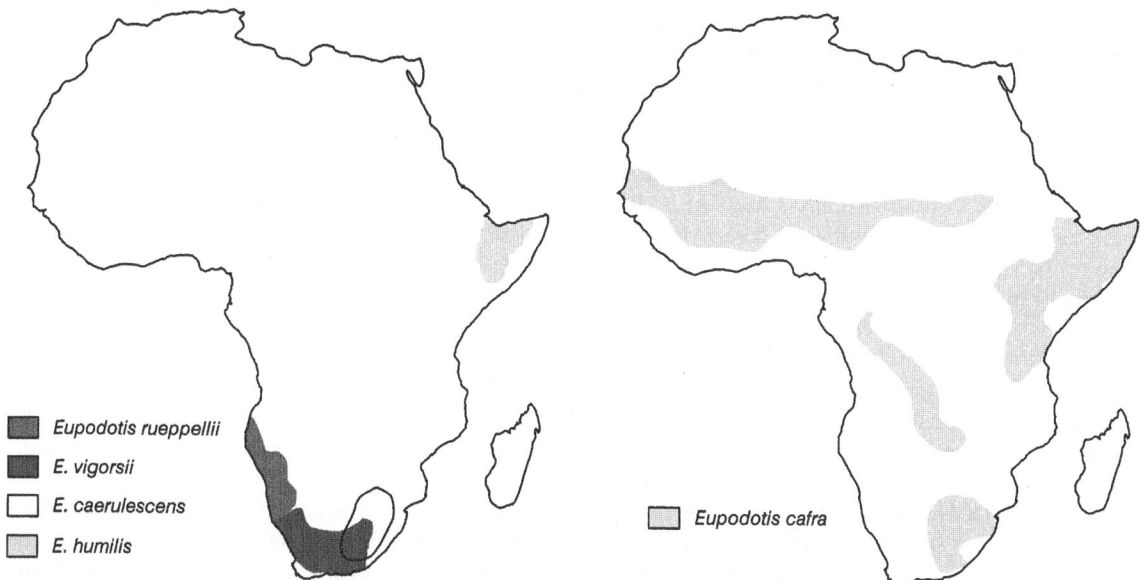

Figure 5.7 **The distribution of Rueppell's korhaan *Eupodotis rueppellii*, karoo korhaan *E. vigorsii* and blue korhaan *E. caerulescens* in southern Africa, the little brown bustard *E. humilis* in Somalia, and the distribution in Africa of the white-bellied korhaan *E. cafra***

species occur in the south-western arid zone of Africa. Compared with the Namib Desert and the succulent karoo, the Nama-karoo is not particularly rich in scorpions (Appendix 5.1), but is species-rich compared with the arid savanna, where only 14 species occur. Six species of scorpion in 4 genera are endemic to the Nama-karoo, of which *Opistophthalmus* (Scorpionidae) with 3 species is best represented. The Namib Desert has 34 species in total, with 14 endemic species. The genera *Opistophthalmus* (6 spp.) and *Parabuthus* (Buthidae: 3 spp.) are most species-rich.

The centre of solifuge (Solpugida) distribution is the Namib Desert where there are about 32 species in 16 genera, of which 22 species are endemic (Wharton, 1981; Eryn Griffin, personal communication). The solifuge fauna of the karoo (Appendix 5.3) is part of that of the Namib Desert, and all genera are common to the karoo and Namib Desert. In the desert/savanna transition zone within the Nama-karoo, there are 17 genera and 38 species (Eryn Griffin, personal communication). In Namaqualand, there are 27 species in 14 genera (Wharton, 1981), while on the southern Nama-karoo/succulent karoo interface there are 9 species in 7 genera (Dean and Griffin, 1993).

A number of species of ticks (Acari: Mestigmata) in the genera *Amblyomma*, *Argas*, *Dermacentor*, *Haemaphysalis*, *Hyalomma*, *Ixodes*, *Margaropus*, *Ornithodoros*, *Rhipicentor* and *Rhipicephalus* occur in the Nama-karoo, but only *Hyalomma glabrum* is endemic (Theiler, 1965; Meyer and Loots, 1978).

One family of termite-eating spiders (Araneae: Ammoxenidae) is endemic to the arid and semi-arid

regions of southern Africa. There are two genera, *Ammoxenus* and *Rastellus*, with six described species in each genus. Three *Ammoxenus* species can occur sympatrically in the Nama-karoo, whereas elsewhere only two, or more often only one species occurs in any locality. *Ammoxenus* spp. live in sandy soils and can dig rapidly to escape or bury their prey. A species characteristic is that they bury themselves upside down, holding prey above them and feeding from below. Their main prey are harvester termites *Hodotermes mossambicus* in the savanna (Wilson and Clark, 1977) and *Microhodotermes viator* in the southern Nama-karoo (Dean, 1988) and, in general, their distribution follows that of the hodotermitid termites in the karoo. *Rastellus* spp. are very small spiders, and their biology is poorly known.

Chilopoda: There are 26 species of centipedes in the karoo which are part of a Namibian fauna and not related to that of the savanna (Lawrence, 1975).

Isoptera: There are about 208 termite species, in 52 genera in southern Africa (Scholtz and Holm, 1985). Most southern African termites are members of the Termitidae and species of tropical savanna. There are 12 species in the karoo, *sensu lato*, all of which occur in the Nama-karoo (Appendix 5.4). The most abundant termite in the biome is *Hodotermes mossambicus* (Hodotermitidae) of the savanna biome (Coaton, 1958). There are two species of the relict family Termopsidae that are endemic to the fynbos. In terms of biomass, the termites are probably the most important faunal component of the energy cycle and an important component of the food of a number of animals (Kok and Hewitt, 1990; Dean and Milton, this volume).

Termites of the genera *Allodontermes*, *Lepidotermes*, *Odontotermes*, *Promirotermes* and *Unicornitermes* occur on the northern periphery of the Nama-karoo (Coaton and Sheasby, 1973–8). *Lepidotermes* has a trail of isolated populations which extend south-west across the Nama-karoo to the succulent karoo. *Promirotermes* has isolated populations in the succulent karoo. As such they reflect a prior history of Savanna extending over the Nama-karoo. The distribution patterns of *Microcerotermes* and *Odonto-termes* reflects the savanna corridor on the eastern side of the karoo between Port Elizabeth and the Orange River.

Rainfall is the principal factor determining the distribution of the harvester termite *Hodotermes mossambicus*. Following years of low rainfall, there is an increase in density and termites extend their foraging range (Coaton, 1958). The distribution and density of termites should vary with fluctuations in rainfall, with fluctuations large enough to affect the populations of termitivores (Macdonald, 1982). *Hodotermes mossambicus* and *Micro-hodotermes viator* are both active during the day in winter and during the night in summer (Lourens and Nel, 1990) It is possible that their ecology and population dynamics are also similar. Their distributions overlap in the Nama-karoo, but in general *M. viator* is a succulent karoo and fynbos species, and *H. mossambicus* is a savanna species. The Nama-karoo may be where both species are episodic. *Hodotermes* may advance in years of low rainfall and *Microhodotermes* may advance in years of winter rainfall.

Orthoptera: The endemic Nama-karoo grasshopper *Karruhippus albicornis* is a relict in a subfamily, Gomphocerinae. It is 'yet another component of the highly autochtonous acridid fauna that has evolved in the Nama-karoo' (Brown, 1989).

The brown locust *Locustana pardalina* (Acrididae) is one of the most important and characteristic animals of the Nama-karoo and savanna biome (Lea, 1968, 1969). During episodic irruptions, the locust is the most numerous animal in the karoo. The locust irruptions are a response which synchronizes the animal with its environment. Its life-history mirrors the patchily dispersed and temporally erratic occurrence of resources in the karoo environment.

Diptera: Two species of *Nemopalpus* (Psychodidae) occur in the Nama-karoo in Namibia (Stuckenberg, 1978). These are typically forest flies, and their presence in Namibia could be a relict of a prior temperate, wet epoch.

Lepidoptera: Caterpillars of the moth *Loxostege frustalis*

(Pyralidae) are an important faunal component of the karoo (Annecke and Moran, 1977) (in terms of biomass), but in general the Nama-karoo has a poor butterfly fauna. There are 786 species of butterflies in southern Africa (Cottrell, 1985). Only 33 of 232 common species, of which three are endemic, occur in the Nama-karoo (Migdoll, 1987). Most butterfly species are found on the periphery of the karoo as extensions of range from adjacent biomes and may be isolated populations associated with particular food plants. For example, the fig tree blue *Myrina silenus* (Lycaenidae) occurs in the succulent karoo only in the Gariep area where the required fig trees are found.

Hymenoptera: The Apocrita, comprising bees (Apoidea), wasps (Vespoidea) and ants (Formicoidea), are well represented in the karoo.

There are 593 species and 69 genera of ants (Formicidae) in southern Africa (Scholtz and Holm, 1985) of which more than 100 species occur in the Nama-karoo and Namib Desert (Marsh, 1985a; Dean and Bond, 1990; Milton et al., 1992a; Willis et al., 1992). The Nama-karoo ant fauna, like that of the desert, may derive from the savanna. Ants are one of the dominant insect taxa in the Namib Desert, and there are 37 species, probably derived from the savanna (Marsh, 1990). The dominant genus is *Monomorium* with 13 species. The gravel plains have 33 species and the species-richness and abundance of ant species on the plains are correlated with rainfall (Marsh, 1985a).

5.3.2.   Vertebrates

Amphibia: There are 11 species of frog in the Nama-karoo, out of about 84 species in southern Africa (Appendix 5.5). The Platanna *Xenopus laevis* (Pipidae) is widespread in southern Africa, including the Nama-karoo. It is related to the fossil frog *Eoxenopoides reuningi* from Cretaceous deposits near Banke in Namaqualand, which has a common ancestor with modern *Xenopus* spp. (Estes, 1977). *Bufo gariepensis* (Bufonidae) is a fynbos species that also occurs in the succulent karoo. *Tomopterna cryptotis* (Ranidae) is widespread in the northeastern Nama-karoo and savanna biomes, and is widespread in Somalia (Lanza, 1981, cited by Poynton, 1995). It occurs throughout the arid and moist savanna, and in the north-eastern Nama-karoo (Passmore and Carruthers, 1979). The remaining species, *Phrynomerus annectans* (Microhylidae), *Bufo gutturalis*, *B. garmani*, *B. vertebralis*, *Pyxicephalus adspersus*, *Rana angolensis*, *R. fuscigula* and *Cacosternum boettgeri* (all Ranidae) peripherally enter the Nama-karoo from the savanna biome. *Bufo vertebralis* and *B. garmani* (Fig. 5.4) are also members of groups of species that have arid corridor distributions (Poynton, 1995).

Reptilia: Six species of tortoise (Chelonii) occur in the Nama-karoo, of which three *Homopus* spp. are largely confined to the Nama-karoo (Fig. 5.8, Appendix 5.7). They are part of a species-complex with other species in the succulent karoo and adjacent coastal fynbos. The tent tortoise *Psammobates tentorius* occurs throughout the Nama- and succulent karoo, and in the past has been subdivided into 6 different species and 30 different subspecies (Greig and Burdett, 1976; Branch, 1988a). Current opinion is that there are three polymorphic species of *Psammobates* (Fig. 5.8). The polymorphism of *P. tentorius* is a consequence of a mosaic of habitats in the karoo and tortoise dispersal inhibited by mountain barriers.

There are 100 species of snakes (Squamata: Serpentes) in southern Africa of which 50 are endemic to the subregion (Branch, 1988a). Three species are endemic to the karoo biomes, *Prosymna frontalis*, *Psammophis notostictus* and *Telescopus beetzii* (all Colubridae). The Nama-karoo has a low diversity of snakes with 38 species. There are no snakes endemic to the Nama-karoo, and, with the exception of 4 species (see Appendix 5.8), all species found in the Nama-karoo are also found in the succulent karoo. Most of the snakes are from other biomes, and more than half of the species only occur on the periphery of the Nama-karoo (Appendix 5.8).

With the exception of the genus *Bitis* (Viperidae), there are few desert snakes amongst the Nama-karoo fauna. Speciation of *Bitis* in the arid areas of southern Africa may have occurred with repeated alternation of tropical and temperate epochs. Seven of 11 species in the genus *Bitis*

occur in the karoo *sensu lato*. The genus divides into two groups. There are the savanna species (*caudalis*, *peringueyi* and *schneideri*) that live on sand where they bury themselves and ambush their prey. The horned adder *B. caudalis* may be a recent colonist of the savanna, while Peringuey's adder *B. peringueyi* and the Namaqua dwarf adder *B. schneideri* may be relics of a prior arid epoch. The second group (*atropos*, *cornuta* and *xeropaga*) are grassland species that live in rocky mountainous places. The seventh species, the puff adder *B. arietans* is a tropical savanna snake widespread in southern Africa.

The snake fauna partially derives from that of the savanna biome, from where 26 of the 38 species originate (Appendix 5.8). Seven savanna species occur only on the periphery of the Nama-karoo. One savanna species, the coral snake *Aspidelaps lubricus*, has three populations, one in the karoo biomes, and two in Namibia. Presumably this snake is part of the arid corridor fauna and its nearest relative is to be found in East Africa. The division of the species into three populations is similar to the *Eupodotis* bird species and *Pachydactylus* lizard species.

There are about 193 species of lizards (Squamata: Sauria) in southern Africa of which 75% are endemic to the region. The Nama-karoo has a low diversity of lizards and only 47 species, in 23 genera, are found there. The degree of endemism to the Nama-karoo is low and most species derive from the succulent karoo or the savanna (Appendix 5.9).

There are three lizards that are endemic to the karoo biomes *sensu lato*, *Nucras tessellata*, *Pedioplanis laticeps* (Lacertidae) and *Pachydactylus mariquensis* (Gekkonidae),

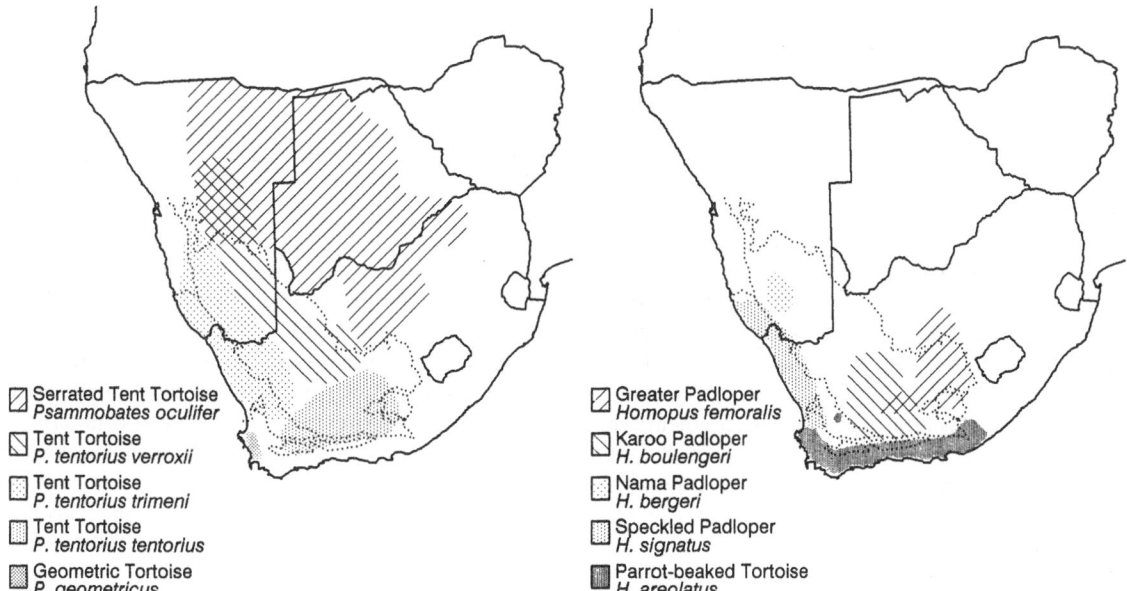

Serrated Tent Tortoise
Psammobates oculifer

Tent Tortoise
P. tentorius verroxii

Tent Tortoise
P. tentorius trimeni

Tent Tortoise
P. tentorius tentorius

Geometric Tortoise
P. geometricus

Greater Padloper
Homopus femoralis

Karoo Padloper
H. boulengeri

Nama Padloper
H. bergeri

Speckled Padloper
H. signatus

Parrot-beaked Tortoise
H. areolatus

Figure 5.8 **Speciation in tortoises within the karoo *sensu lato*. The distribution of species and subspecies in *Psammobates*, and species in the endemic genus *Homopus***

and there are three lizards, *Pachydactylus oculatus*, *P. serval* and *Typhlosaurus gariepensis* (Scincidae) that are endemic to the Nama-karoo. *Typhlosaurus gariepensis* is restricted to the Pofadder district and is the only animal known to be limited to what is botanically termed the Pofadder Centre (Hilton-Taylor and le Roux, 1989). Of the 47 species in the Nama-karoo, 18 species originate from the Savanna biome, although 14 species only occur on the periphery of the Nama-karoo (Appendix 5.9). There are at least three grassland species, *Agama atra* (Agamidae), *Afroedura karroica* (Gekkonidae) and *Mabuya capensis* (Scincidae), whose distribution extends into the Nama-karoo.

Aves: There are c. 650 species of birds that breed in southern Africa and 170 (26%) are endemic to the region (Clancey, 1986). Over 400 species have been recorded in the karoo biomes (Dean, 1995) and, of these, 78 species are southern African endemics. The majority of karoo species are found in two or more biomes (Appendix 5.10).

On the basis of records by observers contributing to the Southern African Bird Atlas Project (Harrison, 1992), Dean (1995) considered that species that were recorded 10% more frequently in either of the two karoo biomes showed a preference for that biome. Only eight species, *sensu stricto*, are endemic to the karoo (Appendix 5.10), and none are restricted to the Nama-karoo or the succulent karoo. Three species, the karoo korhaan, Sclater's lark *Spizocorys sclateri* and black-eared finchlark *Eremopterix australis* (both Alaudidae) show a preference for the Nama-karoo. *Certhilauda*, *Spizocorys* and *Eremopterix* are polytypic genera of larks, which, with the polytypic bustards *Eupodotis* (Otididae), have several species endemic to southern Africa. These karoo species fit in a spatial mosaic with congeneric species as ecological counterparts in other biomes.

The karoo korhaan *Eupodotis vigorsii* is a member of a species complex which includes Rüppell's korhaan *E. rueppellii* in the desert, white-bellied korhaan *E. cafra* in the savanna and the blue korhaan *E. caerulescens* in the grassland. All look alike, have similar calls and are distinct from the other southern African *Eupodotis* species. Presumably the *E. vigorsii* species complex radiated after the closure of the last arid corridor, and their nearest relative occurs or occurred in East Africa (Fig. 5.7) (Snow, 1978). The karoo korhaan favours habitats with a diversity of perennial shrubs where there is little grass and few succulents and feeds on both plants and animals. It utilizes a wide range of patchily distributed food resources, and capitalizes upon episodic ephemeral abundances (Boobyer and Hockey, 1994).

It is postulated that the karoo korhaan evolved in the karoo biome where an omnivore could thrive on a perennial diversity of patchy resources and exploit ephemeral resources in dry seasons (presumably lean periods). In

areas degraded by anthropogenic factors and a consequent reduction of resource diversity, 'what evolved as a means of overcoming lean periods in the past is now a daily survival strategy' (Boobyer, 1989).

The karoo korhaan lives in large permanent territories which are occupied by a pair of birds and their offspring who delay their dispersal from the natal territory. Group size is small and correlated with territory area and inversely with annual rainfall. The other members of the species complex all have a larger mean group size (Hockey and Boobyer, 1994). If it is assumed that delayed dispersal is the norm amongst korhaans then the karoo korhaan has a lower fecundity than the other members of the species complex. This suggests that the karoo *sensu lato* is a less favourable environment for korhaans than is either the desert, savanna, or grassland.

The nomadic black-eared finchlark *Eremopterix australis* is a member of a species complex which occupies the southern savannas and extends up the arid corridor. It is the only finchlark that is endemic to the karoo and occurs most commonly on grassy plains on sands in the Nama-karoo. Of the other *Eremopterix* spp., the distribution of the grey-backed finchlark *E. verticalis* encompasses the karoo while the chestnut-backed finchlark *E. leucotis* occurs on the periphery of the Nama-karoo.

Sclater's lark is nomadic and restricted to gravel plains (Steyn and Myburgh, 1991). These plains are in the more xeric parts of the karoo biomes and encompass the Gariep and Pofadder centres as well as the fossil courses of the Orange River.

There are seven desert bird species, all members of polytypic genera, that are found in both the karoo biomes. Some of these also extend into the fynbos and the savanna. The chat flycatcher *Melaenornis infuscatus* (Muscicapidae) and lark-like bunting *Emberiza impetuani* (Fringillidae) are found more widely in the Nama-karoo than the succulent karoo. The flycatcher is confined to plains with some bushes and appears sedentary. In contrast, the bunting is a nomad that is sometimes the most abundant species in stony habitats. The tractrac chat *Cercomela tractrac*, which has a distinct association with calcrete outcrops, and dusky sunbird *Nectarinia fusca*, associated with drainage lines and rocky hillsides, are found equally in both biomes. Three species endemic to the desert which have no apparent southern African counterparts are: Gray's lark *Ammomanes grayi*, the rockrunner *Achaetops pycnopygius* (Sylviidae), and the Herero chat *Namibornis herero* (Turdidae). The lark is associated with the arid corridor as other *Ammomanes* species are found in North Africa. *Achaetops* and *Namibornis* are monotypic genera.

In the karoo and savanna biomes there are 23 species endemic to southern Africa whose distributions extend peripherally into the desert and fynbos biomes. The Long-billed crombec *Sylvietta rufescens* (Sylviidae) is commonest in the succulent karoo; 6 species are found equally in both karoo biomes, and 16 species are found more widely in the Nama-karoo. The overlap in species between the Nama-karoo and succulent karoo is because these biomes grade into each other.

The habitat preferences of these savanna species range between two extremes. Some are specific to, and resident in, a habitat and extend into the karoo only where their habitat occurs. Examples include fawn-coloured lark *Mirafra africanoides*, that occurs on patches of Kalahari sand from Somalia to the karoo (Fig. 5.4), and sociable weaver *Philetarius socius* (Ploceidae) which is confined to kalahari sand patches and associated vegetation.

Some of the savanna species are members of the same polytypic genera that occur in the karoo while others are southern representatives of tropical genera. For example, the pale chanting goshawk *Melierax canorus* (Accipitridae) is a bird of the savanna biome whose distribution encompasses the karoo biomes, and is a member of a species complex which involves the arid corridor. There is a species in the northern savanna and another in the central African woodlands.

There are fynbos bird species, endemic to southern Africa, whose distribution encompass the Nama-karoo. These are found more widely in the succulent karoo than the Nama-karoo. Only five species that commonly occur in fynbos (whitebacked mousebird *Colius colius* (Coliidae), karoo robin *Erythropygia coryphaeus*, fiscal flycatcher *Sigelus silens*, fairy flycatcher *Stenostira scita* and Cape sparrow *Passer melanurus*) are found equally in both karoo biomes.

Seven endemic grassland species may be found on the periphery of the Nama-karoo, and there are 24 grassland species that extend their distribution further into the karoo. Four species (blue korhaan, melodious lark *Mirafra cheniana*, pink-billed lark *Spizocorys conirostris* and black-chested Prinia *Prinia flavicans* (Sylviidae)) are found only in the Nama-karoo and not in the succulent karoo. The blue korhaan occurs only on the eastern periphery of the karoo, while the melodious lark and pink-billed lark enter the karoo only in favourable years.

Mammalia: There are 284 mammal species in southern Africa of which 23% are endemic to the region (Coe and Skinner, 1992). The majority of mammals in the karoo are species with widespread distributions that originate from the savanna and grassland biomes. Nine species are endemic to the karoo *sensu lato* (Appendix 5.11).

The Nama-karoo has a mammal fauna of 83 species and a low degree of endemism, with only 3 endemics. Grant's rock mouse *Aethomys granti* (Muridae) is a member of a polytypic genus with other species in the savanna and grassland biomes. The riverine rabbit *Bunolagus monticu-*

laris (Leporidae) is a relict, with no close relatives in the Ethiopian region. Visagie's golden mole *Chrysochloris visagiei* (Chrysochloridae), occurs in the Nama-karoo west of Williston (Smithers, 1983) and may be a relict population of a species closely related to the endemic *Cryptochloris wintoni* of the succulent karoo.

The yellow mongoose *Cynictis penicillata* (Viveridae) is widespread in the Nama- and succulent karoo, and is a communal breeding species, with one or two successive litters of only one or two young per year. Parents and helpers provision the young in the natal den until they are two months old (Rasa et al., 1992). This contrasts with other social viverrids that live in larger groups and have larger litters, and where the young go out from the den with the adults at about four weeks of age. Presumably these differences between the yellow mongoose and other viverrids represent adaptations by a tropical savanna species to the temperate savanna and karoo environments.

The grassland mammals form the second largest component of the Nama-karoo fauna, and include Sclater's golden mole *Chlorotalpa sclateri*, mountain zebra *Equus zebra* (Equidae), Smith's red rock rabbit *Pronolagus rupestris* and several rodents. The mountain zebra is found on the periphery of the karoo and only persists there in protected areas. The rock rabbit is widespread in upland areas of the karoo and prefers places where there are dolerite outcrops. The rodent genera *Otomys* and *Parotomys* (Otomyinae) form a complex of species which have diffused from an original 'J'-shaped Afro-montane grassland distribution pattern. The two *Parotomys* species, and the bush karoo rat *Otomys unisulcatus* are karoo endemics which have diverged most radically from ancestral species. The closest *Otomys* relative of *Parotomys* remains to be determined. *Otomys unisulcatus* may be most closely related to *O. irroratus* as both build lodges. Three *Otomys* species (*irroratus*, *saundersiae* and *sloggetti*) are found only on the periphery of the karoo.

The whistling rat *Parotomys brantsii* is a herbivore, and includes toxic succulent and woody plants in its diet (see Dean and Milton, this volume). This adaptation may have allowed the whistling rat to invade the karoo from the grassland and to survive in a harsh environment.

Independence from free water is a prerequisite for desert rodents, but the ability to survive indefinitely on a diet of air-dried seed is characteristic of only some desert species (see Lovegrove, this volume). The degree of physiological adaptation to desert conditions is associated with kidney morphology (Schmidt-Nielsen, 1983). *Aethomys* is probably the most recent invader of the desert, as it has the least specialized kidney. All desert rodents minimize their water and energy costs of reproductive effort and maximize the survival of the adult individuals. Adaptive extremes are shown by *Petromus*, which is a highly specialized diurnal herbivore with high survival rate and low reproductive potential, and *Aethomys*, which is unspecialized, but has low survival and high fecundity (Withers et al., 1980).

The bush karoo rat builds a lodge of sticks up to 3.9 m³ in size, inhabited by a few individuals, and usually contains two nests and two latrines (Vermeulen and Nel, 1988). The physiology of the bush karoo rat indicates that it has colonized the karoo from adjacent more temperate biomes, and has adjusted to the warmer and drier karoo by building permanent lodges, so creating micro-climatic conditions more appropriate to its physiological abilities (Kerley and Erasmus, 1992a, 1992b).

5.4.   Endemicity and distribution patterns – succulent karoo

In contrast to the Nama-karoo, the succulent karoo has its own characteristic fauna. The dominant animals are invertebrates, more especially hymenoptera that have a special interrelationship with the flora. There is also a large suite of endemic vertebrates with the reptiles particularly well represented.

The succulent karoo has a strong faunal relationship with, and is intermediate between, the fynbos and the desert. This relationship is also illustrated by the grasses. There are 101 grass species in the succulent karoo, of which 19% are endemic. There are many malate species, indicating that the main floral relationship is with the fynbos (Cowling et al., 1989). Grasses originating from the desert biome mainly occur as range extensions down the coast. Species that are endemic to the Gariep Centre include *Chaetobromus involucaris*, *Dregeochloa pumila* and *Stipagrostis brevifolia*.

5.4.1.   Invertebrates

Gastropoda: The karoo biomes have a poor snail fauna. The terrestrial snail *Tulbaghinia isomeroides* (Dorcasiidae) of the succulent karoo and fynbos is a relict species found in mountains above 1000 m. (Van Bruggen, 1978; Grindley, 1987).

Arachnida: The succulent karoo is particularly rich in scorpions (Scorpionida), with centres of high species-richness and endemism in the Namaqualand highlands and in the southwest at Worcester. Fifty species occur in the biome, of which 22 are endemic (Prendini, 1995). Genera that are particularly species-rich are *Opistophthalmus* (Scorpionidae), with 24 spp. (15 endemic), *Parabuthus*

(Buthidae), with 10 spp. (3 endemic) and *Uroplectes* (Buthidae) with 6 spp. (2 endemic) (Appendix 5.1). The scorpion fauna of the succulent karoo shows fairly strong links with that of the Nama-karoo, with 19 shared species, and to the Namib Desert, where there are 17 species common to both regions.

There are five species of whip-scorpions (Amblypygi) in southern Africa, mostly confined to the eastern and northern parts of the subcontinent. However, one species, *Myodalis scullyi* is endemic to the northwestern succulent karoo (Newlands, 1978).

There are several genera of spiders (Araneae) that are known from the arid areas of southern Africa (Lawrence, 1964) (Appendix 5.2). These are interpreted to represent a link between the succulent karoo and the desert biomes. The genus *Cangoderces* (Telemidae), with one species (*C. lewisi*) endemic to the Cango Caves at Outdshoorn (Lawrence, 1964), also occurs in Cameroon (Eryn Griffin, personal communication) suggesting that it is a relic of a former mesic epoch.

Amblysoma sylvaticum (Acari: Mestigmata) is the only tick confined to the succulent karoo. All ticks of this genus are parasitic on tortoises (Theiler, 1965).

Thysanura: The psammophilous Thysanuran fauna of the succulent karoo is thought to be more recent than the fauna of other substrata (Irish, 1990). The former is thought to have originated from a northern expansion of the Cape fauna, whereas the latter derives from a westward expansion of the Savanna fauna. The Kalahari sand of the savanna biome has a distinct Thysanuran fauna. This fauna has intermingled with that of the Namib in the Bushmanland region of the Nama-karoo along the Orange River (Irish, 1990).

Isoptera: Most of the termite species found in the succulent karoo are also found in the Nama-karoo and derive from the savanna fauna (Appendix 5.4). Three species (*Baucaliotermes hainesi*, *Fulleritermes mallyi* and *Microhodotermes viator*) are endemic to the succulent karoo (Coaton and Sheasby, 1973–1978). *Baucaliotermes hainesi* is a Namibian species whose range extends south into Namaqualand. It is a monotypic genus in the subfamily Nasutitermitinae which has few species in South Africa. Its closest relatives may be *Trinervitermes*, which are widespread outside of the range of *B. hainesi*.

Fulleritermes mallyi is another nasutitermite and the southern representative of an African genus of four species. It is confined to the succulent karoo and separated from the other two southern African species (*F. contractus* and *F. coatoni*) by the Kalahari Desert. Presumably the genus is not psammophilous.

The small harvester termite *Microhodotermes viator* is the

most abundant isopteran, and may be a dominant element of the fauna of the succulent karoo. *Microhodotermes* colonies, locally termed *heuweltjies*, are conspicuous, regularly spaced in the landscape and used by a variety of other animals (Lovegrove and Siegfried, 1989; Milton and Dean, 1990a), including secondary colonization by *Amitermes* spp. (Dean, 1991).

Dermaptera: *Esphalmenus peringueyi* (Pygidicranidae) is found in montane grassland from Lesotho south to Caledon and extends into the succulent karoo at Nieuwoudtville (Hincks, 1955).

Hemiptera (Homoptera): At least five species of cicada in three genera, *Quintillia*, *Henicotettix* and *Platypleura* (Cicadidae) occur at Tierberg in the Nama-karoo/succulent karoo interface (Milton et al., 1992a), but species-richness of cicadas generally in the succulent karoo is largely unknown.

Coleoptera: The genus *Onymacris* (Tenebrionidae), a dominant genus in the desert biome, also has representatives in the karoo (Penrith, 1984a). Six of 14 species (*boschimana, hottentota, lobicollis, multistriata, plana, unguicularis*) extend their distribution into the Namaqualand sectors of the Nama-karoo and the succulent karoo. It is suggested that ancestral *Onymacris* stock developed in the Namib, extended south to Namaqualand with the accumulation of sand in an arid epoch, then spread northwards into the Kalahari with the northward spread of sand (Penrith, 1984a, 1984b, 1984c).

Psammophilous tenebrionid beetles *Calaharena dutoiti, C. irishi, Syntyplus subterraneus* and *S. namaquensis* are members of genera which were previously considered monotypic endemics of the Namib. They provide evidence of faunal links between the Kalahari sands in the savanna biome and the Namib Desert in the desert biome (Penrith, 1983b, 1983c).

Tenebrionid (Cryptochilini) beetles occur from the south-western succulent karoo to Angola. There are at least 30 species in the succulent karoo, 10 in the Desert, 8 in the fynbos and 3 in the Nama-karoo. All species are opportunistic omnivorous feeders who appear seasonally and are capable of surviving without a period of adult emergence for several seasons (Penrith and Endrödy-Younga, 1994).

The Julodinae (Buprestidae) are a subfamily of the southern arid region, that have root-feeding larvae. There are 27 species in the genus *Julodis*, the majority of which occur in the succulent karoo (Gussmann, 1995).

In the genus *Acmaeodera* (Cetoninae) there are 42 savanna species and 29 species of a Cape (succulent karoo and fynbos) fauna, whose individual ranges extend vari-

ously into the desert (Holm, 1990). Most of the Cape beetles occur in the desert as isolated populations on inselbergs, even though related populations are found further south on coastal plains. This is interpreted to indicate that the desert fauna is a relict from a recent epoch when the succulent karoo extended further north into the Namib Desert.

The monkey beetles (Scarabaeidae; Rutelinae, Hopliini) are largely endemic to South Africa, with the most species-rich area in the south-western succulent karoo (Peringuey, 1902; Picker and Midgley, 1996). Monkey beetles are visitors to annual Asteraceae in the succulent karoo (Giliomee, 1986). Different genera of monkey beetles are associated with particular flower colours and species of Asteraceae, with clear differences in methods of feeding on flower parts and differential roles as pollinators (Picker and Midgley, 1996).

Diptera: Bombyliid flies (Syrphidae) and kelp flies (Coelopidae) appear to have filled some of the vacant butterfly niches, and the succulent karoo, particularly in the southwest, is a centre of fly speciation (Whitehead et al., 1987).

Lepidoptera: Despite the abundance of flowering plants, the succulent karoo has a poor butterfly fauna, owing to the lack of water and the lack of larval foods during the dry summers (Whitehead et al., 1987). Of 232 common butterfly species in the karoo, only 21 occur in the succulent karoo (Migdoll, 1987).

Within the Satyridae, the monotypic genus *Melampias* is essentially a succulent karoo endemic, and 5 (of 7) *Tarsocera* species are largely endemic to the succulent karoo, with ranges extending from the Gariep Centre to south-western fynbos (Vari, 1971; Dickson and Kroon, 1978).

Hymenoptera: This is one of the major invertebrate groups in the succulent karoo, and the species-richness of hymenopterans there apparently parallels the vast array of flowering annuals.

Anthophorid bees (Anthophoridae) are represented by 11 genera in southern Africa and are relatively specialized in their choice of plants. Their distribution patterns are correlated with rainfall as the number of species declines with latitude and decreasing longitude (Eardley, 1983, 1988, 1989, 1991a, 1991b, 1991c, 1993, 1994; Eardley and Brooks, 1989; Eardley and Schwarz, 1991). The abundance of species in Namaqualand suggests that the distribution of these bees may be determined by the local abundance and diversity of floral resources (Struck, 1994a).

Colletid bees (Colletidae) have both specialist (oligolectic) and generalist (polylectic) species. The species of the genera *Colletia*, *Scrapter* and *Polyglossa* appear to prefer flowers of the Mesembryanthema (Struck, 1994b). There are 30 species of *Scrapter* in southern Africa, and of these 22 occur in the succulent karoo, mainly in the Gariep Centre (Eardley, 1996).

Fideliid bees (Fideliidae) have a centre of distribution near the north-western coast of the succulent karoo, close to the Gariep Centre, where seven species can occur sympatrically. From there they extend into the Nama-karoo and into the Kalahari sector of the savanna biome. *Fidelia* species specialize on the pollen and nectar of flowers of the Mesembryanthema, and *Parafidelia* collect pollen from *Tribulus, Tribulocarpus, Sisyndite* and *Sesamum* (Whitehead, 1984). The fideliids are solitary bees which nest in sandy soils, and their distribution may be limited to particular soil types.

Rediviva bees (Melittidae) collect oil from the floral spurs of *Diascia, Hemimeris* and *Bowkeria* flowers, and are essential for pollination of *Diascia* in montane grassland. Oil-collecting bees occur in the savanna, grassland, fynbos and western succulent karoo. At least 12 species of *Rediviva* bees occur in south-western South Africa. *Rediviva* bees have both oligolectic and polylectic species, and there appears to be some concordance between the species-richness of *Diascia* and *Hemimeris* species and the species-richness of *Rediviva* bees in the Western Cape Province (Whitehead and Steiner, 1985).

Ants (Formicidae), together with termites, are the most abundant animals resident in the succulent karoo. Their numbers are only exceeded temporarily by episodes of population increases by other invertebrates. There are at least 25 ant species at one site in the succulent karoo (Milton et al., 1992a), and 34 species have been collected at another (Arnold, 1915–26). The number of ant species found in five localities throughout the karoo *sensu lato* show a correlation with rainfall (Fig. 5.9) in an exponential relationship that suggests that no more than 40 species of ants will be found at any one locality. The Namib Desert is exceptional.

Masarid wasps (Masarinidae) are characterized by their habits of generally feeding on specific flower taxa and provisioning their young with pollen and nectar. The collection of resources from a narrow range of plants is a feature of semi-arid biomes where the climatic conditions lead to simultaneous flowering of many kinds of plants (Gess, 1992). The evolution of the flowers has provided a fulcrum for the radiation of masarine wasps which overrides basic predictions of species-richness based on climatic or latitudinal correlates. The distribution maps of Mesembryanthema and of masarine wasps show a remarkable degree of concordance, with similar centres of species-richness (Gess, 1992).

There are 28 species in the genus *Ceramias* (Vespidae) of which 18 are found in South Africa. All *Ceramias* spp. are specific to semi-arid winter or between-seasons rainfall

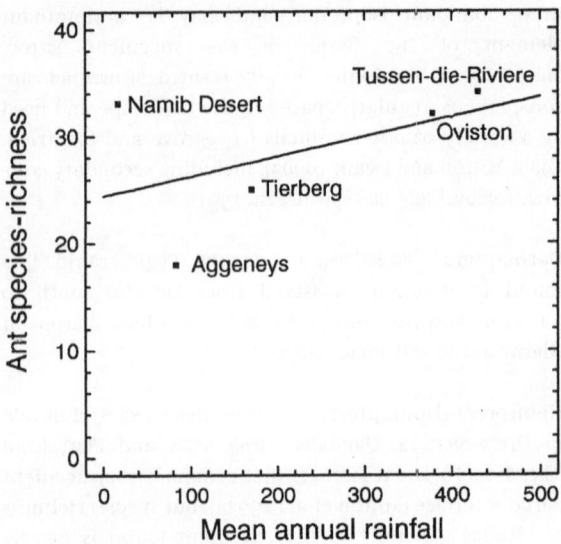

Figure 5.9 **The relationship between the species-richness of ants (Formicidae) and rainfall in the karoo and Namib Desert**

biomes. *Ceramias* spp. wasps are pollen-feeders, that make subterranean burrow nests, typically with mud-lined entrances and mud turrets of various forms (Gess and Gess, 1988). Many species feed on pollen of Mesembryanthema and all *Ceramias* species are dependent on the availability of particular soil types and water. The South African species are only found in the succulent karoo and species-richness tends to be highest along the succulent karoo/fynbos interface (Gess and Gess, 1988).

5.4.2.   Vertebrates

Amphibia: Only 10 species of frogs occur in the succulent karoo (Drinkrow and Cherry, 1995). The frog fauna shows links with the adjacent fynbos frog fauna (Appendix 5.5). Three genera, *Breviceps, Cacosternum* and *Rana* have 13 species in the fynbos and succulent karoo, of which three species are found in both biomes. The distribution of *Rana grayi* extends into Namibia where it is restricted to isolated populations on inselbergs (Poynton, 1962).

The six *Breviceps* species each have restricted populations extending from the fynbos northwards along the coast through the succulent karoo to the Orange River mouth. The coastal facies of the succulent karoo is a region of high diversity of animals, and besides frogs there are 2 snakes and 10 endemic lizards there (Branch, 1988b).

Reptilia: Eight species of tortoise (Chelonii: Testudinae) are found in the karoo *sensu lato*, and except for *Geochelone pardalis* are all endemic to the karoo biomes (Appendix

5.7). This is relative to a total of 12 species in South Africa and 40 world wide.

The succulent karoo has six species of tortoise. Three of these are shared with the fynbos biome. One of these, the Geometric Tortoise *Psammobates geometricus*, is classed as vulnerable because its major habitat, *Elytropappus rhinocerotis* shrubland (renosterveld), is threatened, and the tortoise population estimated at *c.* 2000 animals (Branch, 1988b). As there are no species confined to fynbos it is presumed that the karoo tortoises evolved in the succulent karoo and dispersed from there into the adjacent biomes.

There are 38 species of snakes (Squamata: Serpentes) found in the succulent karoo, four of which (*Bitis schneideri*, *B. xeropaga*, *Lamprophis fiskii* and *Leptotyphlops gracilior*) are endemic. The most speciose genus (*Bitis*) with four succulent karoo species is discussed under 5.3.2. The majority of the snakes of the succulent karoo originate from the savanna and the grassland biome (Appendix 5.8).

There are 71 species of lizards (Squamata: Sauria) in the succulent karoo and 33 of them are endemic (Appendix 5.9). A further 13 species are shared with the Nama-karoo. Most of the endemic species are found along the coast between the Cape peninsula and Angola. The majority of these, 18 species, have ranges that overlap with the Gariep Centre. It would appear that the expansion and contraction of the fynbos, succulent karoo and desert biome boundaries along the coast has led to the isolation of populations of many species. This trend has been enhanced by the rugged mountainous terrain of the Gariep Centre.

The chameleon *Brachypodion karroicum* (Chamaeleonidae) and the gecko *Phyllodactylus peringueyi* (Gekkonidae) have distribution patterns which link the Gariep and the Albany Centres. The skink *Acontias gracilicauda* (Scincidae) has populations in the Gariep Centre and an isolated population along the Cradock Corridor to the Albany Centre (Branch, 1988a).

Aves: Of the eight endemic bird species in the karoo *sensu lato*, the karoo lark *Certhilauda albescens* (Alaudidae), karoo eremomela *Eremomela gregalis* and Namaqua warbler *Phragmacia substriata* (Sylviidae) are more frequently recorded in the succulent karoo, but the red lark *Certhilauda burra* and cinnamon-breasted warbler *Euryptila subcinnamomea* show no biome preference.

The karoo lark is regionally polymorphic and several subspecies have been described. This suggests that the species has persisted through epochs when the succulent karoo has expanded and contracted. The red lark is part of a species complex that includes the karoo lark, Barlow's lark *C. barlowi* and the dune lark *C. erythrochlamys* (Ryan, 1996; Ryan and Bloomer, 1997). The number of species in the complex is a matter for continuing debate. The red

lark is a scarce resident which has a very restricted distribution and is confined to sand dunes in the Pofadder centre (Dean et al., 1991). These dunes are relics of a prior xeric, psammophilous epoch. A possible biogeographic interpretation is that the red lark and dune lark are relict species and, while they have remained in refugia, the karoo lark has expanded its range to abut on the other two species (Fig. 5.10).

The karoo eremomela occurs in succulent and woody dwarf shrubland on plains. The species is a member of a polytypic genus whose closest relatives, in terms of their behaviour, are the dusky eremomela *E. scotops* and the black-collared eremomela *E. atricollis* from the woodlands of central Africa (Hall and Moreau, 1970). Those species are separated from the karoo eremomela by the savanna biome, which is occupied by *Eremomela* species that have different behaviour patterns and that are probably not closely related to the karoo eremomela.

The cinnamon-breasted warbler *Euryptila subcinnamomea* is a scarce resident on boulder-strewn hills in both karoo biomes. These hills tend to be isolated and cinnamon-breasted warblers only occur on some of them, so that the bird has a discontinuous and patchy distribution. *Euryptila* is a monotypic genus whose closest relatives are not known. As no races of *Euryptila* have been described, it is concluded that it is a relict species persisting in refugia. Its habitat preference is similar to that occupied by such other endemic southern African monotypic genera as *Achaetops*, *Chaetops*, *Namibornis* and *Sphenoeacus*. While these birds may not be related, their common status suggests that they all may be relics.

The Namaqua warbler *Phragmacia substriata* is found in rank vegetation along water-courses. It is widespread in both karoo biomes and common in some places. It occurs along dry water-courses and so is independent of permanent surface water. *Phragmacia* is now considered a monotypic genus (Brooke and Dean, 1990) although the species was long considered a member of the polytypic genus *Prinia*.

Five of seven species endemic to the desert biome extend their distribution into the succulent karoo. The species are tractrac chat *Cercomela tractrac*, karoo chat *C. schlegelii* (Turdidae), pale-winged starling *Onychognathus nabouroup* (Sturnidae), dusky sunbird *Nectarinia fusca* (Nectarinidae) and white-throated canary *Serinus albogularis* (Fringillidae). The seven desert endemics are all members of polytypic genera. The pale-winged starling is the exception amongst them as it has only one local, and rather distant relative, the red-winged starling *O. morio*, with more closely related congeners in East Africa.

The karoo chat and pale-winged starling, together with the white-throated canary, are found more widely in the succulent karoo than in the Nama-karoo. They are resident

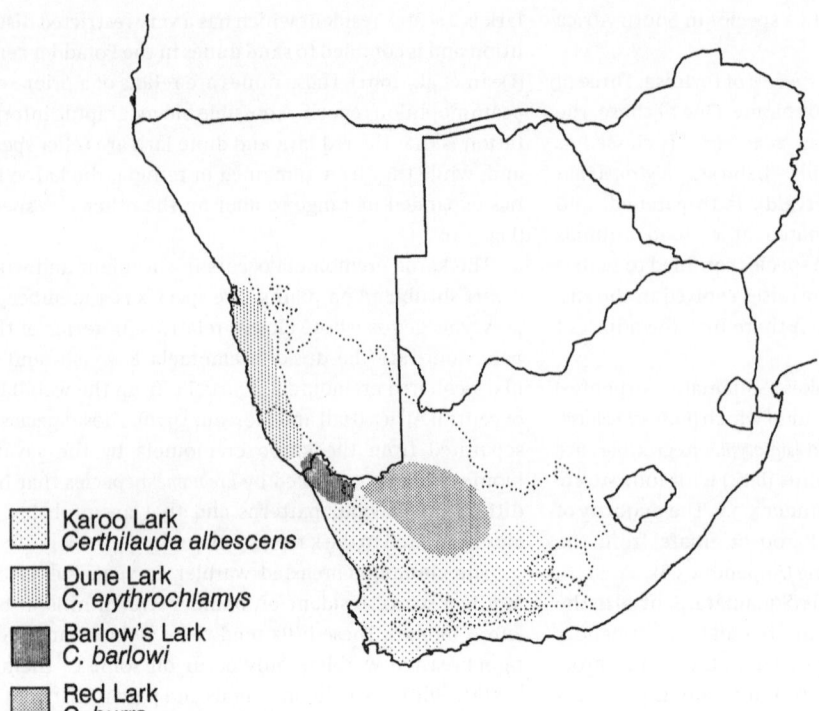

Figure 5.10 **The distribution of closely related species in southern Africa of the karoo lark *Certhilauda albescens* species complex**

Karoo Lark
Certhilauda albescens

Dune Lark
C. erythrochlamys

Barlow's Lark
C. barlowi

Red Lark
C. burra

species which undertake local movements when food resources are scarce. The tractrac chat and dusky sunbird occur equally in both the Nama-karoo and the succulent karoo. They are both resident and may have local movements.

There are 19 fynbos species, endemic to southern Africa, whose distribution includes the karoo. The majority of these (15 spp.) are found more widely in the succulent karoo than the Nama-karoo, suggesting a greater association between the fynbos and succulent karoo, than the fynbos and Nama-karoo. Cowling and Hilton-Taylor (this volume) point out that the relationship between the fynbos and succulent karoo floras is closer than that between the floras of the two karoo biomes.

Many of the southern African endemic species that occur both in the fynbos and the succulent karoo are members of polytypic genera. Two species, the black harrier *Circus maurus* and booted eagle *Hieraetus pennatus* (Accipitridae) occupy the ecotone between the fynbos, karoo and grassland biomes. The closest relatives of both appear to be in the Palaearctic. The thick-billed lark *Galerida magnirostris* is a succulent karoo counterpart of the savanna Sabota lark *Mirafra sabota*, but its closest relatives are found in North and West Africa (Fig. 5.11). Similarly, the closest relative of Layard's Tit-babbler *Parisoma layardi* (Sylviidae), common in the succulent karoo, is in East Africa even though the tit-babbler *Parisoma subcaeruleum* is a savanna biome counterpart. This

suggests that the closest relatives of such monotypic birds as the fairy flycatcher *Stenostira scita* and fiscal flycatcher *Sigelus silens* (Muscicapidae) may be in East Africa or the Palaearctic. It also questions the assumption that the closest relatives of polytypic genera such as the grey tit *Parus afer* (Paridae) and Karoo Robin *Erythropygia coryphaeus* are necessarily the adjacent savanna counterparts, ashy tit *P. cinerascens* and Kalahari robin *E. paena*.

There are 19 species of endemic grassland birds whose range extends into the succulent karoo. Ten species are found equally in both the Nama- and succulent karoo biomes, while nine species are found more widely in the succulent karoo. Three species, the grassbird *Sphenoeacus afer*, cloud cisticola *Cisticola textrix* (Sylviidae) and rock pipit *Anthus crenatus* (Motacillidae), are largely absent from the Nama-karoo, suggesting that the species have entered the succulent karoo both through the fynbos and directly from the grassland biome.

Mammalia: In the succulent karoo, there are 73 species of mammals (Appendix 5.11) of which 3 species are endemic. Among the endemics, *Cryptochloris wintoni* and *C. zyli* (Chrysochloridae) are insectivores, whereas *Bathyergus janetta* (Bathyergidae) is herbivorous, and all are fossorial and found along the coast. These mammals, plus the desert golden mole *Eremitalpa granti* (Chrysochloridae), form a faunal link between the fynbos, succulent karoo and desert biomes. The bush karoo rat *Otomys unisulcatus*

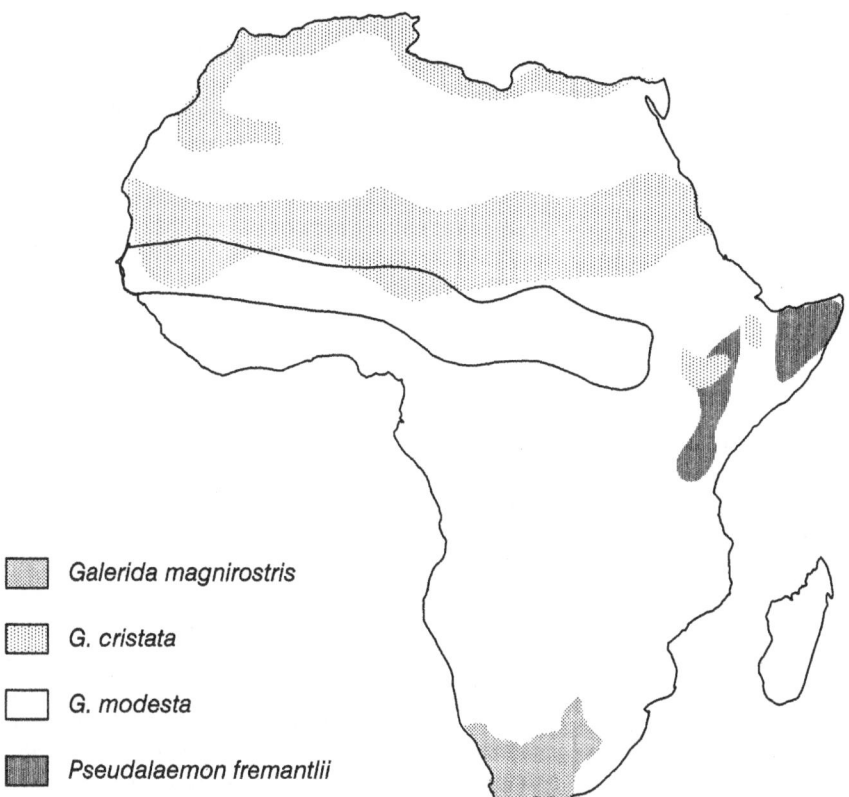

Figure 5.11 **The distribution of a karoo endemic species whose nearest relatives are in North and West Africa; the thick-billed lark Galerida magnirostris, and related crested lark Galerida cristata, sun lark G. modesta and short-tailed lark Pseudalaemon fremantlii**

Galerida magnirostris

G. cristata

G. modesta

Pseudalaemon fremantlii

(Otomyinae), endemic to the karoo *sensu lato*, occurs in the succulent karoo and extends into the adjacent half of the Nama-karoo and east to the Albany Centre. The spectacled dormouse *Graphiurus ocularis* (Gliridae) occurs throughout the succulent and Nama-karoo (Smithers, 1983) and is largely confined to the karoo, with its distribution extending from the Albany Centre to the Gariep Centre. Two other dormouse species occur marginally in the succulent karoo, the rock dormouse *G. platyops* in north-western Namaqualand and the woodland dormouse *G. murinus* along the southern edge of the succulent karoo.

5.5.   **Anthropogenic factors influencing biogeography**

In general, anthropogenic factors have severely masked or even modified ecological processes, the economy and the distribution of organisms in the karoo. These are discussed where they provide illustrations of biological processes in the karoo.

Important environmental changes have resulted from (a) imposing perennial pastoralism upon an environment which is either seaonal or episodic, (b) overstocking and

consequent accelerated erosion, (c) growing subsistence cereal crops on alluvial soils leading to the destruction of vleis and sponges, (d) the degradation of rangeland so that favourable episodic events replenish lower quantities of water over smaller areas than before. These problems had manifested by the 1870s (Shaw 1875), and in 1876 there was a commission of inquiry into the state of the karoo (Noble 1886) (see also Hoffman et al., this volume).

Tortoises: The Leopard Tortoise *Geochelone pardalis* is absent from the central parts of the Nama-karoo. This has been ascribed variously to anthropogenic, climatic and habitat factors. The tortoises may thus be readily extirpated from areas through a combination of predation by man (Branch, 1988a), or loss of food plants to domestic stock.

Birds: The birds show similar biogeographic patterns to those of the other faunal groups. However, these patterns are overwhelmed by the response of birds to anthropogenic factors. This simply illustrates the evolutionary principle that birds, like all organisms, attempt to maximize their fecundity and are only prevented from colonizing the world by environmental constraints.

Many species have extended their ranges into the karoo

because of anthropogenic factors (Macdonald, 1986a, 1986b, 1992; Macdonald et al., 1986). This phenomenon indicates the degree to which birds can capitalize upon opportunities, but imparts little information about the biogeography of the karoo. The presence of many species of birds, through their commensalism with man, depends upon man's exploitation of water resources which would not otherwise be available to those birds. The rate at which birds have increased in range consequent of man's colonization of the karoo in about the last 200 years is extremely rapid in evolutionary terms. The contraction in range or local extinction of avian commensals, when man abandons homesteads in the karoo, occurs very rapidly.

The Namaqua warbler is one karoo endemic that has been favoured by anthropogenic modification of the karoo. The increase in trees along drainage lines, and the oasis effect of homesteads have permitted the Namaqua warbler to increase in numbers.

However, at least three species of the grassland biome, bald ibis *Geronticus calvus*, Cape vulture *Gyps coprotheres* and wattled crane *Grus carunculatus*, have been extirpated from the karoo. Place names in the karoo like *aasvogelberg* (vulture mountain), *kalkoenkrans* (ibis cliff) and *kraanvoelkuil* (crane marsh) are mute reminders of the former presence of these species.

The bald ibis may have been shot out as it provided food for humans, but more likely may have declined through habitat modification and loss of food (Siegfried, 1966; Manry, 1985). The wattled crane went from the karoo at least 100 years ago, probably as a consequence of the loss of its wetland habitat (Brooke and Vernon, 1988). The extirpation of the wattled crane was presumably the result of pressures similar to those currently on the blue crane *Anthropoides paradisea* (Vernon et al., 1992).

The Cape vulture was probably resident in the fynbos and the grassland. From there, birds dispersed into the karoo biomes, more especially when conditions were favourable for the game herds. The replacement of the transient herds with permanent domestic stock stimulated the vultures to become resident in the karoo. With improved animal husbandry in the 1970s, the vultures were unable to obtain sufficient food and are no longer resident in the karoo (Boshoff and Vernon, 1980; Milton et al., this volume).

Mammals: The present suite of mammal species is a consequence of climatic events that have happened in the last 0.1 Myr. The extant larger karoo mammals, which are a remnant fauna almost replaced by man and his commensals, stands in contrast to prior epochs. Many mammals have long since been extirpated from the karoo and the adjacent fynbos biome (Skead, 1980, 1987; Stuart et al.,

1985; Rebelo, 1992; Macdonald, 1992). The general cause of their decline is anthropogenic and many were, and still are, inimical to the human colonists. By and large these mammals were widespread and their range contracted from the fynbos and karoo. The trend continues as though man and his commensals are displacing all other mammals.

However, the kudu *Tragelaphus strepsiceros* has recolonized drainage-line woodland in the Nama- and succulent karoo, and is said to be more common than before. The encroachment of trees along drainage lines has permitted the vervet monkey *Cercopithecus pygerythrus* to extend its distribution deep into the Nama- and succulent karoo (Macdonald, 1992). Some mammals regarded as vermin, such as the black-backed jackal *Canis mesomelas* and lynx *Felis caracal*, have survived in the karoo despite persecution.

Other species have decreased in distribution and abundance. The riverine rabbit *Bunolagus monticularis* has become rare owing to loss of its habitat of alluvial floodplains that have been transformed for agriculture (Macdonald, 1992). Species like the oribi *Ourebia ourebi* that occurred on the periphery of the karoo have been extirpated by the loss of their habitat. Anecdotal evidence maintains that oribi, along with many other ungulates, wintered in the wetlands at Vlekpoort in the Hofmeyr district of the eastern Nama-karoo. These wetlands were fed by water from the adjacent mountains in the grassland biome. The ungulates there either were extirpated by hunting or displaced by stock. The wetlands were later ploughed, inevitably eroded and now are a vast area of dongas. It is only the oral tradition and the presence of remnant freshwater mussel shells that provide evidence of a prior condition. These wetlands were ploughed and drained after 1850. Prior to the advent of the '75' plough (which had a superior mechanical ability) in about 1850, farmers had lacked the means to plough wetlands (W. S. Stretton, personal communication).

5.6.   **Conclusions**

The basic property of semi-arid ecosystems is that they exist as patchily dispersed alternate states of high and low productivity. Pulses of high rainfall, high productivity and accelerated nutrient cycling provide the 'fuel stores' for alternately occurring drought states of 'slow-burn' during which the systems merely 'tick over', and the environment becomes progressively drier until the next pulse of moisture and nutrients. Except for a few species that are able, through a variety of tactics, to reside permanently in the Nama-karoo, many species occur only in years of high resource levels, or only on the periphery of the Nama- and succulent karoo.

The fauna of the succulent karoo tends to have links with both the desert and the fynbos biomes, and only tenuously with the savanna. The fauna is rich in endemics, especially among the arachnids, hopliinid beetles, aculeate hymenoptera, reptiles and tortoises. The folding of the Cape mountains, the subsequent isolation of habitat patches and the extraordinary species-richness of plants (Cowling and Hilton-Taylor, this volume) have been major factors contributing to speciation among the animals of the succulent karoo. The high barriers along the boundary between the succulent karoo and the Nama-karoo, together with the relatively predictable rainfall in the succulent karoo have discouraged movement of animals between the two biomes, so that this species-richness has been largely contained. The fauna of the Nama-karoo tends to have strong links with that of the savanna and grassland biomes. It is not rich in endemics, and is characterized, among the birds and larger mammals, by species that are nomadic or partially migratory, utilizing resources that are patchy in space and time.

In dramatic terms, the Nama-karoo is a stage. The vegetation is the props, the climate is the effects and the animals are the actors. There is neither director nor script. Each actor enters and leaves at random, acting only in its own self-interest, exploiting and yet restricted by prevailing circumstances, and responding to cues that are poorly perceived and little understood by the audience. As the props change so does the scenery and the players. On close inspection, and over millennia, even the stage is being worn away. All is transient and we are reminded of Heraclitus 'we step but do not step twice in the same river'.

Appendices

Appendix 5.1. *Families and species of scorpions recorded from the succulent karoo, the Nama-karoo, the Namib Desert and the Kalahari Desert*

Family and species	Nama	Succ.	Namib	Kalahari
Bothriuridae				
Lisposoma elegans	+			
Buthidae				
Hottentota arenaceus	+	+		
H. conspersus	+			
Karasbergia methueni	+	+	+	
Parabuthus brevimanus	+	+		
P. calvus	+			
P. capensis	+			
P. distridor	+			
P. gracilis	+			
P. granulatus	+	+	+	+
P. kalaharicus	+			
P. kraepelini	+			
P. kuanyamarum	+	+		
P. laevifrons	+	+	+	+
P. namibensis	+			
P. nanus	+	+		
P. neglectus	+	+		

Family and species	Nama	Succ.	Namib	Kalahari
P. planicauda	+			
P. raudus	+	+	+	
P. schlechteri	+	+	+	
P. stridulus	+			
P. villosus	+	+	+	
Uroplectes carinatus	+	+	+	+
U. formosus	+			
U. gracilior	+	+		
U. lineatus	+			
U. longimanus	+			
U. marlothi	+			
U. otjimbinguensis	+			
U. schlechteri	+	+	+	
U. tumidimanus	+	+		
Ischnuridae				
Cheloctonus intermedius	+			
Hadogenes bicolor	+			
H. lawrencei	+			
H. minor	+			
H. phyllodes	+	+	+	
H. taeniurus	+			
H. tityrus	+	+		
H. zumpti	+	+		
Opisthacanthus capensis	+	+		
O. diremptus	+			
O. validus	+			
Scorpionidae				
Opistophthalmus adustus	+	+		
O. ammopus	+			
O. ater	+			
O. austerus	+	+		
O. capensis	+			
O. carinatus	+	+	+	+
O. chaperi	+			
O. coetzeei	+			
O. concinnus	+			
O. crassimanus	+	+		
O. fitzsimonsi	+			
O. flavescens	+			
O. fossor	+			
O. gigas	+	+	+	
O. granicauda	+			
O. granifrons	+			
O. haackei	+	+		
O. holmi	+			
O. intercedens	+			
O. intermedius	+			
O. karrooensis	+	+		
O. laticauda	+			
O. latimanus	+			
O. longicauda	+	+		
O. lornae	+			
O. macer	+			
O. nitidiceps	+			
O. opinatus	+	+		
O. pallidipes	+	+		
O. pattisoni	+			
O. penrithorum	+			
O. peringueyi	+			
O. pictus	+			
O. pygmaeus	+			
O. schlechteri	+			
O. schultzei	+			
O. wahlbergi	+	+	+	+

From Prendini (1995)

Appendix 5.2. *Some spider families and genera of the succulent karoo*

Family	Genus
Agelenidae	*Olorunia*
Ammoxenidae	*Ammoxenus, Rastellus*
Caponiidae	*Diploglena*
Eresidae	*Seothyra*
Heteropodidae	*Leucorchestris, Palystella*
Oecobiidae	*Uroecobius*
Oonopidae	*Salsula, Nephrochirus*
Palpimanidae	*Iheringia, Ikuma*
Prodidomidae	*Austrodomus*
Zodariidae	*Caesetius, Cyrioctea, Heradida, Mallinus, Palfuria, Psammoduon, Psammorygma*

Source: Lawrence (1964); Eryn Griffin (personal communication)

Appendix 5.3. *Solifugae of the karoo biomes*

Ceromidae		Solpugidae
Ceroma pictulum	*B. orangica*	*Solpuga alstoni*
C. sclateri	*B. robusta*	*S. brachyceras*
C. swierstrae	*B. rosea*	*S. cervina*
	B. schulzei	*S. chelicornis*
Ceromella pallida	*B. tricolor*	*S. ferox*
	B. uniquicornis	*S. spectralis*
Daesiidae		
Biton bernhardi	*Hemiblossia australis*	
B. cataractus	*H. bouvieri*	*Solpugassa furcifera*
B. crassidens	*H. etosha*	*S. kaokoensis*
B. gariesensis	*H. idioceras*	
B. hottentotus	*H. oneili*	*Solpugema*
B. kraekolpi	*Hemiblossia robusta*	*brachyceras*
B. leipoldti	*H. termitophila*	*S. coquinae*
B. lineata		*S. calycornis*
B. longisetosus	*Hemiblossiola*	*S. cycloceras*
B. namaqua	*kraepelini*	*S. derbiana*
B. ovambicus		*S. erythronotoides*
B. peasoni	**Gylippidae**	*S. genucornis*
B. schreineri	*Lipophaga schultzi*	*S. hostilis*
B. striata	*L. trispinosa*	*S. lateralis*
B. subulata		*S. scopulata*
B. tenuifalcis	**Hexisopodidae**	*S. spectralis*
B. triseriatus	*Chelypus barberi*	
Blossia alticursor	*C. hirsti*	*Solpugiba lineata*
B. brincki	*C. lennoxae*	
B. clunigera		*Solpugista hastata*
B. crepidulifera	*Hexisopus crassus*	*S. methueni*
B. echinata	*H. lanatus*	*Solpugopa fitzsimonsi*
B. falcifera	*H. reticulatus*	*S. villosa*
B. filicornis	*H. swarti*	
B. grandicornis		*Solpuguna alcicornis*
B. hessei		*S. collinita*
B. karrooica	**Melanoblossidae**	*S. organica*
B. laminicornis	*Lawrencega hewetti*	
B. lapidicola		*Zeria lethalis*
B. longipalpis	*Melanoblossia braunsi*	*Z. monteiroi*
B. namaquensis	*M. globiceps*	*Z. recta*
B. pallideflava	*M. namaquensis*	*Z. schlechteri*
B. pringlei	*M. tridentata*	*Z. schonlandi*
		Z. umbonata
		Z. venator

From Lawrence (1955) and Dean and Griffin (1993).

Appendix 5.4. *Termites that occur in the Nama- and succulent karoo, and their occurrence in savanna, fynbos or grassland biomes*

Species	Nama	Succulent	Savanna	Fynbos	grassland
Hodotermitidae					
Hodotermes mossambicus	+	+	+	+	
*Microhodotermes viator**	+	+			
Rhinotermitidae					
Psammoterminae					
Psammotermes allocerus	+	+	+ Arid		
Termitidae					
Amitermitinae					
Amitermes hastatus	+	+	+ Arid	+	
Microcerotermes sp.	+	+	+ Arid	+	
Macrotermitinae					
Allodontermes spp.	+	+			
Odontotermes sp.	+	+	+		
Nasutitermitinae					
*Baucaliotermes hainesi**	+	+			
*Fulleritermes mallyi**	+				
Trinervitermes trinervoides	+	+	+	+	
Trinervitermes spp.	+	+	+		
Termitinae					
Angulitermes spp.	+	+	+ Arid	+	
Lepidotermes simplex	+	+			
Promirotermes spp.	+	+			
Unicornitermes gaerdesi	+	+			

Arid = species that also occur on Kalahari sands in arid Savanna, * = species endemic to the karoo
Source: Coaton (1963); Coaton and Sheasby (1973–1978)

Appendix 5.5. *Amphibians that occur in the Nama- and succulent karoo biomes.*

Species	Nama	Succulent	Savanna	Fynbos	grassland
Pipidae					
Xenopus laevis	+	+	+	+	+
*X. gilli**	+	+			
Bufonidae					
*Bufo gariepensis**	+	+	+		
B. gutturalis	+	+			
B. garmani	+	+			
*B. rangeri**	+	+			
B. vertebralis	+	+			
Microhylidae					
Brevicipitinae					
*Breviceps namaquense**	+				
*B. macrops**	+				
Phrynomerinae					
Phrynomerus annectans	+	+			
Ranidae					
Raninae					
Pyxicephalus adspersus	+	+			
Tomopterna cryptotis	+	+			
Rana angolensis	+	+	+	+	
*R. fuscigula**	+	+	+	+	
*R. grayi**	+	+			
*R. montana**	+	+			
Phrynobatrachinae					
*Cacosternum boettgeri**	+	+			
*C. namaquense**	+				

* = endemic to South Africa

Appendix 5.6. **Fish that occur in the karoo biomes**

Anabantidae

Sandellia bainsii+ C
S. capensis+ F

Cichlidae

Tilapia sparrmanii O, C, L, T

Clariidae

Clarias gariepinus O, C, L, T

Claroteidae

Austroglanis sclateri+ O, Rare
A. barnardi+ F
A. gilli+ F

Clupeidae

Gilchristella aestuaria+ F, C

Cyprinidae

Mesobola brevianalis O, L, T
Barbus aeneus+ O
B. anoplus+ O
B. hospes* O, Rare
B. kimberleyensis+ O
B. pallidus+ O, C, L
B. paludinosus O, C, L, T
B. trimaculatus O, C, L, T
B. spp.+ F: 7 spp.
B. spp.+ C: 4 spp.
B. spp. C, L, T: 4 spp.
B. spp.+ L: 3 spp
B. spp. L, T: 9 spp.
B. spp. T: 19 spp.
Pseudobarbus spp.+ F: 6 spp.
Labeo capensis+ O
L. umbratus+ O, F
L. seeberi+ F
L. rubromaculatus+ C
L. quathlambae+ O, C

Galaxiidae

Galaxias zebratus+ F

From Skelton (1993).
* = confined to the karoo biome, + = South African endemic, O = occurs in the
Orange River system, C = occurs in rivers of the east coast, F = occurs in rivers
of the fynbos biome, L = occurs in the Limpopo River system, T = occurs in
tropical river systems, Rare = conservation status.

Appendix 5.7. **Tortoises of the karoo biomes**

Species	Nama	Succulent	Desert	Savanna	Fynbos	grassland
Testudinidae						
Homopus femoralis	+					
H. areolatus*	+	+				
H. boulengeri*	+					
H. signatus*	+					
H. bergeri*	+					
Geochelone pardalis	+	+	+	+	+	+
Chersina angulata*	+	+				
Psammobates oculifer	+	+ (arid)				
P. geometricus*	+	+				
P. tentorius*	+	+	+			

* = endemic to southern Africa

Appendix 5.8. **Snakes occurring in the karoo biomes and in other biomes**

Species	Nama	Succulent	Desert	Savanna	Fynbos	grassland
Typhlopidae						
Typhlops lalandei	+	+	+	+	+	
T. schinzi	+	+	+			
Leptotyphlopidae						
Leptotyphlops nigricans	+	+				
L. gracilior*	+					
L. scutifrons	+	+	+			
L. occidentalis	+					
Colubridae						
Lycodonomorphus rufulus	+	+	+	+	+	
Lamprophis fuliginosus	+	+	+	+	+	+
L. guttatus*	+	+				
L. aurora*	+	+	+	+	+	
L. fiskii*	+	+				
Lycophidion capense	+	+	+			
Pseudaspis cana	+	+	+	+	+	+
Prosymna sundevalli*	+	+	+	+	+	
P. bivittata*	+	+	+			
P. frontalis	+	+	+			
Dipsina multimaculata*	+	+	+	+ Arid		
Psammophylax rhombeatus	+	+	+	+		
Psammophis trigrammus	+	+				
P. notostictus	+	+	+	+ Arid	+	+
P. leightoni	+	+	+	+	+	+
P. crucifer*	+	+	+			
Atractaspis bibronii	+	+				
Aparallactus capensis	+	+	+			
Homoroselaps lacteus*	+	+	+	+		
Xenocalamus bicolor	+	+ Arid				
Philothamnus semivariegatus	+	+	+	+	+	
Dasypeltis scabra	+	+	+	+	+	+
Crotaphopeltis hotamboeia	+	+	+	+	+	
Telescopus semiannulatus	+	+	+	+	+	
T. beetzii*	+	+	+			
Dispholidus typus	+	+	+	+		
Elapidae						
Aspidelaps lubricus	+	+	+			
Naja nivea*	+	+	+	+	+	
N. nigricollis	+	+	+	+		
Hemachatus haemachatus*	+	+	+	+	+	
Viperidae						
Bitis arietans	+	+	+	+	+	+
B. caudalis	+	+	+	+		
B. cornuta*	+	+	+			
B. xeropaga*	+	+				
B. inornata*	+					
B. schneideri	+	+				
Amphisbaenidae						
Zygaspis quadrifrons	+	+	+			
Monopeltis capensis	+	+				

* = endemic to southern Africa

Appendix 5.9. *Lizards occurring in the karoo biomes and in other biomes*

Species	Nama	Succulent	Desert	Savanna	Fynbos	grassland
Scincidae						
Acontias gracilicauda*	+	+	+			
A. lineatus*	+	+				
A. litoralis*	+					
A. meleagris*	+	+				
Typhlosaurus gariepensis*	+					
T. lomii*	+					
T. meyeri*	+					
T. vermis*	+					
Scelotes sexlineatus*	+					
S. caffer*	+					
S. capensis*	+					
S. gronovii*	+					
S. kasneri*	+					
Mabuya acutilabris*	+					
M. capensis*	+	+	+	+	+	+
M. homalocephala*	+	+	+			
M. occidentalis*	+	+				
M. spilogaster*	+	+	+	+		
M. striata	+	+	+	+		
M. sulcata*	+	+	+	+		
M. variegata	+	+	+	+		
Lacertidae						
Heliobolus lugubris	+	+				
Meroles ctenodactylus*	+					
M. cueirostris*	+					
M. knoxii*	+					
Nucras intertexta*	+	+				
N. tesselata*	+	+	+			
Pedioplanis burchelli*	+	+	+			
P. laticeps*	+	+				
P. lineoocellata*	+	+	+	+	+	+
P. namaquensis*	+	+	+	+ Arid		
P. undata	+	+	+			
Cordylidae						
Cordylosaurus subtesselatus	+	+	+			
Gerrhosaurus typicus*	+	+				
Tetradactylus seps*	+					
T. tetradactylus*	+	+				
Cordylus cataphractus	+					
C. cordylus*	+	+	+			
C. lawrencei*	+					
C. macropholis*	+					
C. mclachlani*	+					
C. minor*	+					
C. namaquensis*	+					
C. peersi*	+					
C. polyzonus*	+	+	+			
Platysaurus capensis*	+					
Pseudocordylus capensis*	+					
P. microlepidotus*	+	+	+			
Varanidae						
Varanus exanthematicus	+	+	+			
V. niloticus	+	+				
Agamidae						
Agama aculeata	+	+	+	+		
A. anchietae	+	+				
A. atra*	+	+	+	+	+	+
A. hispida*	+	+				
Chamaeleonidae						
Brachypodion karroicum*	+					
B. pumilum*	+	+				
B. ventrale*	+	+				
Chamaeleo dilepsis	+	+				
C. namaquensis*	+	+	+			
Gekkonidae						
Afroedura africana*	+					
A. karroica*	+					
Chondrodactylus angulifer*	+	+	+	+		
Colophus wahlbergi*	+	+				
Lygodactylus bradfieldi	+	+				
Pachydactylus bibronii	+	+	+	+	+	+
P. capensis	+	+	+			
P. labialis*	+	+				
P. geitje*	+	+				
P. laevigatus	+	+	+			
P. maculatus*	+	+				
P. oculatus*	+					
P. mariquensis*	+	+				
P. namaquensis*	+	+	+			
P. punctatus	+	+	+			
P. rugosus*	+	+	+			
P. serval*	+	+				
P. weberi*	+	+				
Palmatogecko rangei	+					
Phelsuma ocellata*	+					
Phyllodactylus lineatus*	+	+				
P. peringueyi*	+	+				
P. porphyreus*	+	+				
Ptenopus garrulus*	+	+	+			

* = endemic to southern Africa

Appendix 5.10. **Bird species occurring in the karoo biomes. Separate lists are given for species occurring both in the karoo, sensu lato, and other biomes. Species that occur in particular habitats, or that occur only as a result of anthropogenic activities are listed separately**

Species endemic to the karoo
Eupodotis vigorsii
Certhilauda albescens
C. burra
Eremopterix australis
Spizocorys sclateri
Euryptila subcinnamomea
Eremomela gregalis
Phragmacia substriata

Species that occur in the karoo, desert and savanna
Falco rupicoloides
Meleirax canorus*
Poliohierax semitorquata N
Francolinus levaillantoides* N
Neotis ludwigii*
Eupodotis afraoides N
E. ruficrista N
Pterocles namaqua*
P. burchelli* N
Tricholaema leucomelas
Mirafra africanoides N
Mirafra sabota*
Spizocorys starki* N
Eremopterix verticalis*
Anthoscopus minutus*
Pycnonotus nigricans*
Erythropygia paena* N
Monticola brevipes* N
Myrmecocichla formicivora*
Malcorus pectoralis*
Parisoma subcaeruleum*
Prinia flavicans* N
Batis pririt*
Philetairus socius* N
Sporopipes squamifrons N
Serinus flaviventris*

Species that occur in the karoo and fynbos
Circus maurus*
Hieraeatus pennatus
Francolinus capensis* S
Eupodotis afra* S
Colius colius*
Galerida magnirostris*
Parus afer*
Pycnonotus capensis* S
Erythropygia coryphaeus*
Cisticola subruficapilla*
Parisoma layardi*
Prinia maculosa*
Sigelus silens*

Stenostira scita*
Tchagra tchagra* S
Promerops cafer* S
Nectarinia afra* S
Nectarinia violacea* S
Passer melanurus*
Pseudochloroptila totta* S

Species that occur in the karoo and grassland
Geronticus calvus* x
Tadorna cana*
Anas smithii*
Buteo rufofuscus*
Gyps coprotheres*
Francolinus africanus*
Anthropoides paradisea*
Eupodotis caerulescens* N
Cursorius rufus*
Centropus burchelli
Geocolaptes olivaceus*
Certhilauda curvirostris*
Chersomanes albofasciata
Mirafra apiata*
M. cheniana* N
M. chuana*
M. ruddi*
Spizocorys conirostris* N
Hirundo spilodera* N
Cercomela sinuata*
Monticola rupestris*
Oenanthe monticola*
Cisticola textrix S
Prinia hypoxantha*
Sphenoeacus afer S
Anthus crenatus* S
Macronyx capensis
Telophorus zeylonus*
Spreo bicolor*
Ploceus capensis
Serinus alario*

Species that occur in the karoo and forest
Turdus olivaceus
Zosterops pallidus
Nectarinia chalybea*

Species that occur in the karoo and that are widespread over southern Africa
Sagitarius serpentarius
Aquila verreauxi
Polemaeatus bellicosus

Falco biarmicus
F. tinnunculus
Ardeotis kori
Eupodotis senegalensis N
Burhinus capensis
Rhinoptilus africanus
Streptopelia capicola
Oena capensis
Bubo africanus
Apus barbatus
A. melba
Urocolius indicus
Merops apiaster
Calandrella cinerea
Hirundo cucullata*
H. fuligula
Parus cinerascens* N
Corvus senegalensis
Cercomela familiaris
Cossypha caffra
Oenanthe pileata
Saxicola torquata
Cisticola aridula N
Eremomela icteropygialis
Sylvietta rufescens
Motacilla capensis
Lanius collaris
Anthus similis
Onychognathus morio
Nectarinia famosa
Creatophora cinerea
Lamprotornis nitens
Passer diffusus
Plocepasser mahali N
Ploceus velatus
Amadina erythrocephala*
Serinus canicollis
Emberiza capensis*

Species that occur in rivers and dams in the karoo
Tachybaptus ruficollis
Phalacrocorax africanus
P. carbo
Ardea cinerea
Platalea alba
Ciconia nigra
Scopus umbretta
Alopochen aegypticus
Tadorna cana*
Anas undulata
Anas sparsa
Fulica cristata
Charadrius pecuarius

C. tricollaris
Vanellus armatus
Haematopus moquini*
Recurvirostris avocetta
Ceryle maxima
Hirundo albigularis
Riparia paludicola
Acrocephallus baeticatus
A. gracilirostris
Bradypterus baboecala S
Euplectes orix

Species that occur in the karoo mainly around homesteads and settlements
Columba guinea
Stretopelia semitorquata
S. senegalensis
Chrysococcyx caprius
Apus affinis
A. cafer
Colius striatus
Upupa epops
Sturnus vulgaris S
Passer domesticus

Species that occur in the karoo as a result of agriculture
Ardea melanocephala
Bubulcus ibis
Bostrychia hagedash S
Threskiornis aethiopicus
Elanus caeruleus
Coturnix coturnix
Numilda melagris
Vanellus coronatus
Burhinus capensis
Corvus albus S
C. albicollis
Cisticola juncidis
C. tinniens
Anthus cinnamomeus
Quelea quelea N
Estrilda asrild S
Vidua macroura

Species that occur in the karoo only as migrants
Ciconia ciconia
Buteo buteo
Circaeatus pectoralis
Falco naumanni

* = endemic or near-endemic to southern Africa, N = occurs mainly, or only in the Nama-karoo, S = occurs mainly, or only in the succulent karoo, x = no longer present in the karoo. List excludes feral introduced species.

Appendix 5.11. **Mammals occurring in the karoo biomes and in other biomes**

Species	Nama	Succulent	Desert	Savanna	Fynbos	grassland
Soricidae						
Myosorex varius	+	+	+			
Suncus varilla	+	+	+			
Crocidura cyanea	+	+	+	+	+	
Erinaceidae						
Erinaceus frontalis	+	+	+			
Chrysochloridae						
Cryptochloris wintoni*	+					
C. zyli*	+					
Chrysochloris asiatica*	+	+				
C. visagiei*	+					
Eremitalpa granti*	+	+				
Chlorotalpa sclateri*	+	+				
Macroscelidae						
Macroscelides proboscideus*	+	+	+	+		
Elephantulus rupestris*	+	+	+			
E. edwardii*	+	+				
Molossidae						
Tadarida pumila	+	+	+			
T. aegyptiaca	+	+	+	+	+	+
Vespertilionidae						
Miniopterus schreibersii	+	+	+	+	+	
Myotis seabrai*	+	+				
M. lesueuri*	+	+				
Myotis tricolor	+	+	+			
Eptesicus hottentotus	+	+	+	+		
E. melckorum	+	+	+			
E. capensis	+	+	+	+	+	
Nycteridae						
Nycteris thebaica	+	+	+	+		
Rhinolophidae						
Rhinolophus fumigatus	+	+				
R. clivosus	+	+	+	+	+	+
R. capensis	+	+	+			
R. denti	+	+				
Hipposideridae						
Hipposideros caffer	+	+				
Cercopethecidae						
Papio ursinus	+	+	+	+	+	+
Cercopithecus pygerythrus	+	+				
Manidae						
Manis temminckii	+	+	+			
Leporidae						
Lepus capensis	+	+	+	+Arid	+	
L. saxatilis	+	+	+	+	+	
Pronolagus rupestris	+	+	+	+		
Bunolagus monticularis*	+					
Bathyergidae						
Bathyergus janetta*	+					
Cryptomys hottentotus	+	+	+			
Georychus capensis*	+	+				
Hystricidae						
Hystrix africaeaustralis	+	+	+	+	+	+
Pedetidae						
Pedetes capensis	+	+	+			
Gliridae						
Graphiurus ocularis*	+	+	+			
G. platyops	+	+	+			
G. murinus	+	+	+			
Sciuridae						
Xerus inauris	+	+	+	+		
Petromuridae						
Petromus typicus	+	+	+	+		
Cricetidae						
Cricetinae						
Mystromys albicaudatus	+	+	+		+	
Gerbillinae						
Desmodillus auricularis	+	+	+	+		
Gerbillurus paeba	+	+	+		+	
G. vallinus	+	+	+			
Tatera leucogaster	+	+	+		+	
T. brantsii	+	+	+			
Dendromurinae						
Malacothrix typica	+	+	+		+	
Dendromus melanotis	+	+	+		+	+
Cricetomyinae						
Saccostomus campestris	+	+	+			
Otomyinae						
Parotomys brantsii*	+	+	+	+		
P. littledalei*	+	+	+			
Otomys saundersiae	+	+	+	+		
O. irroratus	+	+	+			
O. sloggetti*	+	+				
O. unisulcatus*	+	+				
Muridae						
Murinae						
Acomys subspinosus*	+	+				
Rhabdomys pumilio	+	+	+	+	+	+
Mus minutoides	+	+	+		+	
Praomys natalensis	+	+				
Thallomys paedulcus	+	+				
Aethomys namaquensis	+	+	+	+	+	+
A. granti*	+					
A. chrysophilus	+	+	+			
Hyaenidae						
Proteles cristatus	+	+	+	+		
Hyaena brunnea	+	+	+	+		
Felidae						
Panthera pardus	+	+	+	+	+	
Felis caracal	+	+	+	+	+	
F. lybica	+	+	+	+	+	+
F. nigripes	+	+				

Crocuta crocuta. Formerly widespread in the Nama- and succulent karoo, and still occurs as a vagrant in the northern Nama-karoo (Skead, 1980, 1987).

Acinonyx jubatus. Formerly occurred at a low density in the Nama-karoo (Skead, 1980, 1987), and may still occur as a vagrant in the northern Nama-karoo (Smithers, 1983).

P. leo. Formerly widespread in the Nama- and succulent karoo, but now extinct in the region (Skead 1980, 1987).

F. serval. Formerly occurred at a low density in the succulent karoo, but now extinct in the region (Skead, 1980, 1987).

Canidae

	1	2	3	4	5	6
Otocyon megalotis	+	+	+	+	+	

Lycaon pictus.   Formerly occurred at a low density in the Nama- and succulent karoo, but now extinct in the region (Skead, 1980, 1987).

	1	2	3	4	5	6
Vulpes chama	+	+	+	+	+	
Canis mesomelas	+	+	+	+	+	

Mustelidae

	1	2	3	4	5	6
Aonyx capensis	+	+	+	+	+	
Mellivora capensis	+	+	+	+	+	+
Poecilogale albinucha (rare)		+	+	+		
Ictonyx striatus	+	+	+	+	+	+

Viveridae

	1	2	3	4	5	6
Genetta genetta	+	+	+	+	+	+
G. tigrina	+	+				
Suricata suricatta	+	+	+	+	+	
Cynictis penicillata	+	+	+	+	+	
Galerella sanguinea	+	+	+	+		
G. pulverulenta	+	+	+			
Atilax paludinosus	+	+	+	+	+	

Orycteropidae

	1	2	3	4	5	6
Orycteropus afer	+	+	+	+	+	+

Elephantidae

Loxodonta africana.   Formerly widespread in the Nama- and succulent karoo (Skead, 1980, 1987), and a relic population still occurs in a protected area.

Procaviidae

	1	2	3	4	5	6
Procavia capensis	+	+	+	+	+	+

Rhinocerotidae

Ceratotherium simum.   Formerly occurred as vagrant to the Nama-karoo (Skead, 1980, 1987).

Diceros bicornis.   Formerly widespread in the Nama- and succulent karoo (Skead, 1980, 1987).

Equidae

Equus zebra.   Formerly occurred along mountain ranges and plains on the southern edge of the Nama-karoo, and in the southern and western succulent karoo (Skead, 1980, 1987). It now occurs at a low density only in protected areas.

E. quagga.   Formerly occurred in the central and eastern Nama-karoo, but now extinct as a species.

Suidae

Phacochoerus aethiopicus.   Formerly widespread in the Nama-karoo, and may still occur as a vagrant along the northern edge (Skead, 1980, 1987).

Hippopotamidae

Hippopotamus amphibius.   Formerly occurred in the southern and western succulent karoo, and in the Nama-karoo as a vagrant in high rainfall years, but now extinct in the region (Skead, 1980, 1987).

Bovidae
Alcelaphinae

Connochaetes gnou*.   Formerly occurred in the northeastern Nama-karoo (Skead, 1980, 1987), but now present only in protected areas.

C. taurinus.   Formerly occurred as a vagrant only along the northern and north-eastern edge of the Nama-karoo (Skead, 1980, 1987).

Alcelaphus buselaphus.   Formerly occurred as a vagrant in the Nama-karoo (Skead, 1980, 1987).

Cephalophinae

	1	2	3	4	5	6
Sylvicapra grimmea	+	+	+	+	+	+

Antelopinae

	1	2	3	4	5	6
Antidorcas marsupialis	+	+	+	+		

Reduncinae (right column species)

	1	2	3	4	5	6
Oreotragus oreotragus	+	+	+	+	+	
(Ourebia ourebi)	+	+				
Raphicerus campestris	+	+	+	+	+	+
R. melanotis*	+	+				

Peleinae

	1	2	3	4	5	6
Pelea capreolus	+	+	+	+		

Hippotraginae

Oryx gazella.   Formerly widespread in the Nama-karoo and western succulent karoo (Skead, 1980, 1987), but now extinct in the wild in the region.

Bovinae

Syncerus cafer.   Formerly widespread but at a low density in the Nama- and succulent karoo (Skead, 1980, 1987) and there is a relic population in a protected area in the eastern Cape at Addo Elephant National Park.

Tragelaphus strepsiceros

	1	2	3	4	5	6
	+	+	+	+		

Taurotragus oryx.   Formerly widespread in the Nama- and succulent karoo (Skead, 1980, 1987) and has been reintroduced to several protected areas.

Reduncinae

	1	2	3	4	5	6
Redunca fulvorufula	+	+	+	+		

* = endemic to southern Africa. List excludes feral introduced species. Species extirpated from the karoo have been listed.

Form and function

B. G. Lovegrove

Although a clear definition of an *adaptation* continues to elude evolutionary biologists (Williams 1966; Gould and Vrba 1982; Sober, 1984; Stearns, 1986, 1992; Brandon, 1990; Reeve and Sherman, 1993; Leroi et al., 1994), few would disagree that an adaptation can only be considered as such if it is associated with a specific *function* or task. Most would also agree that, for evolution to occur by natural selection, at least three necessary conditions need to apply; the *trait* under selection must be inheritable, the unit of selection must vary with respect to the trait, and the heritable trait must lead to differential reproductive success, i.e. enhanced *fitness* (Williams, 1966; Stearns, 1986, 1992; Harvey and Pagel, 1991). Individuals at all life-stages must therefore be able to survive the physical and biotic hazards of their environment and be capable of harnessing sufficient resources to grow and reproduce.

The purpose of this section of the book is to provide a synthesis of what we currently understand about the *form* and *function* of desert organisms. But what, exactly, do we mean by form and function? In terms of the adaptation concept, a form should be defined as any adaptive morphological, behavioural, or physiological phenotype or character serving a particular *function* or task which enhances the fitness of an individual. However, it should be emphasized that not all authors in this section have elected to define form and function in these terms. Given the confusion of meaning surrounding terms such as 'life form' and 'growth form' used to describe plant form, Midgley and Van der Heyden (this volume), define plant form as 'the four-dimensional structure of phytomass (time being the fourth dimension), made up of specific vegetative traits at the level of plant organ and above'. In other words, their definition of form describes the structure of a plant in space and time and disregards physiological and behavioural form. These authors define plant function as 'the flux of matter and energy through the plant body at the organ and sub-organ level'. Here function is clearly associated with plant physiology. It is important to appreciate that this alternative definition does not imply that a morphological adaptation is not dependent upon function or has no function, or that physiological and behavioural adaptations are not dependent upon morphology or are dissociated from morphology. On the contrary, as emphasized by Midgley and Van der Heyden, form and function are mutually dependent attributes with function perhaps more strongly constrained by form than form by function. Concerning animal form and function, I have gone with the conventional interpretation.

Our dichotomous approach to the concept of form and function is perhaps justified when one considers the paucity of empirical evidence confirming whether the fitness of desert organisms is indeed enhanced by the myriad of 'adaptations' identified in the literature. In this respect, we may even question whether many of these 'adaptations' do indeed represent true adaptations. They could, for example, represent exaptations, non-adaptations, or phylogenetic and/or developmental constraints (Williams, 1966; Gould and Vrba, 1982; Stearns, 1986; Brandon, 1990). Although some biologists maintain that the fitness accrued to an organism by adaptive traits should be measured directly (Reeve and Sherman, 1993; Leroi et al., 1994), others currently argue that the comparative, phylogenetic method can be used alternatively to determine correlated evolutionary traits within taxa from a known phylogeny (Felstenstein, 1985; Harvey and Pagel, 1991; Garland et al., 1992; Diazuriarte and Garland, 1996; Williams, 1996; Ward and Seely, 1996a). Either approach is lacking in studies on southern African desert organisms, although recently Ward and Seely (1996a) have taken the lead and have successfully adopted the latter approach to confirm and refute hitherto unchallenged claims concerning the adaptive nature of certain traits in Namib tenebrionid beetles. It is important, therefore, that in the following five chapters the reader place into context the interpretation of the term 'adaptation' which is, quite by perfunctory habit, used rather loosely to account for the adaptive traits of desert organisms. Our inability to be definitive about the adaptiveness of traits should serve as a clear warning to desert biologists that the time has arrived to tighten up our approach, be it comparative or otherwise, if we wish to obtain an equivocal understanding of how organisms are adapted to unpredictable environments and deserts.

The next five chapters describe the perceived adaptations of organisms to the deserts of southern Africa as they have been described in the literature to date. They concern adaptations which have permitted organisms to successfully colonize and inhabit some of the driest, hottest and oldest deserts on Earth. Considering the ecological complexity of the physical and biotic factors influencing communities in deserts (Noy-Meir, 1974; Polis, 1991a), it is refreshing that the adaptations of the organisms involved in these community interactions can be accounted for by relatively few environmental and biotic selection pressures. Foremost is rainfall, most importantly the temporal and spatial variability and unpredictability of rainfall, especially in the summer rainfall regions of southern Africa. Ambient temperature is also important in terms of seasonal and daily fluctuations within the absolute minimum and maximum limits capable of sustaining life. These two selective pressures are not mutually exclusive. They are inexplicably interdependent in processes such as photosynthesis, transpiration, osmoregulation, thermoregulation and production. In addition, the forms and functions of desert organisms have also been further moulded by interactions with other organisms in the same trophic level, or with those in higher or lower trophic levels, through competition and predation.

Low rainfall is a characteristic of all true, semi-tropical deserts and is caused by the persistent existence of high-pressure air masses generated by global air-pressure cells circulating between the equator and the semi-tropics (Tyson, 1987). Cool, dry air from the tropics descends from high altitudes over the semi-tropics and is heated by compression as it approaches the Earth's surface. Rainfall is rare because it can only be produced by the converse pattern, namely rising humid air cooling adiabatically. The latter air masses are kept out of desert regions by the prevailing high-pressure cells. Seasonally, however, the circulating pressure cells may be shifted towards the poles in summer, and towards the equator in winter. Deserts which occur more towards the poles (e.g. succulent karoo) may therefore be penetrated by frontal disturbances in winter producing widespread, gentle rain, whereas those lying more towards the equator, (e.g. arid savanna biome), are penetrated by humid, tropical air during summer bringing brief, torrential convectional rains. The unpredictability of move-ment of the high-pressure systems is further exacerbated in certain regions of the world by additional sources of unpredictable, global air-pressure perturbations, the most important of which are the El Niño Southern Oscillations (ENSO) (Shannon et al., 1986; Tyson, 1987; Philander, 1990). Although ENSO events generate three well-known, global climatic anomalies, namely above-average temperature and rainfall, and below-average rainfall, it is the latter which most severely influence arid regions. Of the major arid regions of the world, those affected by ENSO-related droughts fall within the Australian, Indo-Malayan and Afro-tropical zoogeographical zones (Philander, 1990). The latter zone encompasses sub-Saharan Africa, including Madagascar. Although, the ENSO effect is strongest on the eastern part of the continent where it influences monsoon rain systems, it has ripple or wave effects through the subcontinent when present.

In the simplest of terms, the survival of desert organisms depends most importantly upon the flux rates of matter and energy (water, temperature, nutrients and energy), in and out of desert organisms. These flux rates are influenced markedly by the unpredictability of desert ecosystems and are hence often difficult to manage (Hadley, 1975; Louw and Seely, 1982). Not surprisingly, though, many adaptations which do manage flux rates have evolved convergently in plants and animals (Hadley, 1972, 1981; Orians and Solbrig, 1977a; Louw and Seely, 1982).

The crucial importance of water and temperature fluxes in deserts can be identified at the first level of trophic energy flux, namely the photosynthetic fixing of CO_2 by plants. In fact, morphological and physiological photosynthetic adaptations, such as succu-lence and the relative utilization of C_3, C_4 and CAM metabolism, indirectly form the basis upon which the succulent karoo, Nama-karoo and Namib Desert biomes have been classified (Rutherford and Westfall, 1986) and illustrate important differences between plant and animal form and function. Unlike most animals, plants cannot easily 'escape' the physical conditions of deserts in the short-term and hence tend to display regional, climate-related life and/or growth forms. It is therefore easier to identify generalized categories of form and function in plants compared with animals. For example, succulence and CAM metabolism are synonymous with the winter rainfall deserts in southern Africa, whereas grasses and C_4 metabolism are synony-mous with the summer rainfall deserts (Midgley and Van der Heyden, this volume). It is difficult, if not impossible, to identify any such faunal distinctions associated with climatic zones and biomes in southern Africa.

The existence of obvious plant life forms within deserts does not necessarily imply that adaptations of desert plants rely upon tolerance rather than avoidance of abiotic stresses. On the contrary, annual plants epitomize the evolutionary solution to long-term escape from unfavourable desert conditions. This topic is dealt with by Van Rooyen (this volume). Here again, this distinctive therophytic plant form is synonymous with a particular climate zone, namely the winter rainfall deserts, in which rainfall is least variable spatially and temporally in the arid south-western zone. However, note that therophytic grasses are also a characteristic of the Namib Desert (Rutherford and Westfall, 1986). The annual plants also display convergent traits with those of the crustacean inhabitants of temporary water pools throughout the southern African deserts, namely very rapid growth rates during the short favourable season, and a long dormant 'seed' or 'egg' phase during the unfavourable season (Lovegrove, this volume).

Although plants naturally cannot rely upon behaviour to avoid the extremes of desert conditions as easily as animals can, certain behaviour in animals, such as daily retreat behaviour in the form of endogenous circadian activity rhythms, represents adaptations which are equally, if not more, important than the physiological and morphological adaptations governing the flux rates of water, heat and energy. Seely (1989) has emphasized this point and has argued that, whereas it is often assumed that physiological adaptations of desert organisms are somehow unique, many traits often differ quantitatively rather than qualitatively, and cannot be dissociated from a related behaviour essential for ensuring the particular function of the trait and enhanced fitness. For example, facilitated by physiological and morphological traits, tenebrionid beetles may indeed display some of the lowest rates of cuticular water loss ever measured in animals, yet their survival is dependent upon the ability to avoid the midday heat of the Namib dunes system either by burrowing into the sand or climbing up plants (Ward and Seely, 1996b).

Although desert organisms naturally need to survive for certain periods of time in desert conditions before they can reproduce, it is manifest that adaptations which optimize reproductive success will be subject to the strongest selection pressures. How this is achieved in plants forms the topic of the whole of chapter 8 (Esler, this volume), whereas for animals it forms the closing argument of chapter 9 (Lovegrove, this volume). Of course, adaptations need not necessarily be directly related to the reproductive process, but may involve aspects of efficient acquisition of the resources (e.g. energy, water) required for successful reproduction. For plants, competition for nutrients is important following rain when nutrients become soluble and are taken up by the root systems of desert plants (Midgley and Van der Heyden, this volume). Since rainfall is not a frequent event in deserts, competition for nutrients can be intense during short, favourable periods. Being mobile, animals are able to acquire resources by employing a wide diversity of foraging modes (Dean and Milton, this volume). Group co-operation and sociality is a considerable advantage in enhancing foraging success, as well as minimizing environmental stresses and threats from predators (Lovegrove, this volume).

6 Form and function in perennial plants

G. F. Midgley and F. van der Heyden

6.1. Introduction

The karoo flora contains an astounding variety of growth forms, and its plants range widely in size and shape, type and degree of succulence, leaf consistency and longevity, thorniness, woodiness and below-ground structure. Because environmental stress has strongly influenced selection of plant form in arid systems, it is likely that plant form attributes reveal much about their function. There is no doubt that appropriate growth form classifications are extremely useful in characterizing vegetation at a coarser level than the taxonomic, especially in speciose vegetation. At this stage though, plant growth form classification in arid ecosystems has a tenuous theoretical basis (sometimes as little as intuition alone), limited largely because of ignorance of the consequences of differences between growth form attributes for plant function. A clear understanding of the links between plant form and function is needed to assess the limits to the practical application of the growth form concept. Once established, this knowledge would be useful in identifying important growth form attributes, in understanding the determinants of growth form distribution and even their relative success, and thus in predicting their responses to a changing environment.

Discussion of plant form may be clouded by confusing use of terms. A good example is the apparently interchangeable use of the terms 'life form' and 'growth form' (Hoffman and Cowling, 1987). Life form is an abstract notion which attempts to capture the essence of the adaptive significance of plant structure. Growth form is concrete, and attempts to typify plants morphologically. Furthermore, it is seldom explicitly stated that the concept of plant growth form comprises more than the arrangement of living tissue in space, it also includes a time-dimension which is reflected in some grouping criteria in most accepted classification schemes (e.g. deciduousness, plant longevity). In this chapter, we avoid the abstraction of life form concepts, and refer rather to plant growth forms and important growth form characteristics. We define plant growth form as the four-dimensional structure of phytomass (time being the fourth dimension), made up of specific vegetative traits at the level of plant organ and above. We do this because the timing and duration of leaf cover is crucial for plant–water balance (Orshan, 1964), energy balance, nutrient balance and carbon balance. We define plant function as the flux of matter and energy through the plant body at the organ and suborgan level.

The apparent existence of plant growth form groups is possibly the result of optimal combinations of certain vegetative traits, and convergent evolution of form in edaphically and climatically similar sites in different continents supports this notion (Orians and Solbrig, 1977a). Plant growth form may be optimized not only with respect to environmental conditions, but also with respect to biotic pressures, such as competition and herbivory. It is difficult to discern the contribution of plant form to functional optimality without a powerful understanding of the relevance of certain traits (see Givnish, 1986). But even without a complete understanding, the growth form concept has proved useful as it provides a means of classifying plants into groups of like structure and assumed similar function, thereby characterizing vegetation in functional terms.

We have summarized some key ideas about growth form grouping in Table 6.1, not in order to develop a comprehensive classification system, but rather to provide a reference frame for the discussion which follows. Raunkiaer's life form system (Raunkiaer, 1934) provides a workable basis for categorizing major growth form groups, but we suspect that the power of this system lies more in the recognition of varying degrees of investment in woody tissue relative to photosynthetic tissue than in height of renewal buds, Raunkiaer's original idea. The minor groupings as listed in Table 6.1 fit uncomfortably within the classical Raunkiaer-type framework, and are often recognized as cohesive growth form groups with the same weight as the major groups (e.g. Cowling et al., 1994). The coarse groupings within a classical system necessarily ignore traits at the organ level, such as tissue succulence and leaf longevity, and these traits may be used to subcategorize groups (Cowling et al., 1994; Hoffman and Cowling, 1987). Developmental plasticity, modularity and adaptive plasticity in plants blurs the edges between groups, and complicates their application – dwarfism (minutism), for example, leads to hemicryptophyte-like characteristics in some species such as *Lithops* (stone plants) of the Mesembryanthema and some much reduced *Haworthia* species, and the bizarre woody hemicryptophyte-like gymnosperm *Welwitschia mirabilis*.

It is almost certain that plant root architecture is a key growth form attribute, especially in structurally diverse groups like shrubs, but the high degree of structural plasticity below-ground and lack of information limits the use of root structure in growth form categorization. However, recent advances in hydrogen isotope techniques are likely to lead to significant advances in identifying the water sources used by different life forms (Squeo et al., 1994), and this technique holds promise for improving understanding of this important plant characteristic.

Plant form and function are mutually dependent, but plant function may be more strongly constrained by form than vice versa. Examples are the key structural changes needed to allow the more advanced photosynthetic pathways in C_4 and CAM plants, namely Kranz anatomy and tissue succulence. It is this greater dependence of function on form which underpins 'functional type' classification schemes based on plant form parameters. In this chapter, we examine which of four main selective pressures, namely water availability, temperature, herbivory and nutrient availability have shaped the form and function of karoo plants. These four key selective pressures are not independent, but linked most strongly in two pairs, water availability with temperature, and herbivory with nutrient availability.

6.2.   Growth form abundance, distribution and diversity

One of the ways in which an initial assessment can be made of the functional significance of different growth

Table 6. 1. *The growth form grouping scheme on which the discussion in this chapter is based. This comprises commonly used major growth form groupings, and minor groupings less frequently used. We also provide a list of traits which either have been or could be used to distinguish subcategories within heterogenous groups. The table is arranged vertically in order of decreasing ratio of lignified : non-lignified tissue*

Major groups	Minor groups	Potentially useful traits for subcategorization
Trees (phanerophytes)		Stem succulence Leaf succulence Deciduousness Root architecture
Shrubs (chamaephytes)		Leaf succulence Stem succulence Deciduousness Root architecture
	Climbers/vines Woody parasites Herbs - (ferns, graminoids and other herbaceous non-grasses)	
Grasses (hemicryptophytes)		Presence of Kranz-anatomy
Geophytes (cryptophytes)		
	Root parasites	

forms is by studying their distribution and relative abundance in relation to climatic parameters. This should seed ideas for closer inspection and experimental testing, and identify groups which are poorly understood, show heterogeneous responses or require finer classification.

The relative abundance of different growth form characters in the succulent karoo biome has been studied by Orshan and co-workers (Orshan et al., 1984; Orshan 1989). Orshan et al. (1984) report that shrubs less than one metre tall (chamaephytes) dominate the perennial flora of sampled sites in the succulent karoo (79% of the species recorded). Roughly 60% of the perennial species were found to be evergreen, but less than 30% had leaves with a duration greater than 14 months. A high proportion (>60%) had succulent leaves. This analysis reveals a flora dominated by perennial shrubs which retain foliage year-round, but with an appreciable leaf turnover rate, regardless of the high frequency of succulence. A similar analysis of the flora of the Nama-karoo biome is not available, but Rutherford and Westfall (1986) showed that it differs from the succulent karoo biome in being more or less co-dominated by grass and shrub forms, a pattern confirmed by Cowling et al. (1994), who showed further a more variable and a more equitable growth form mix in Nama-karoo than succulent karoo sites.

A more detailed analysis of growth form types and their relation to climate has been carried out by M. T. Hoffman, G. F. Midgley and R. M. Cowling (unpublished data). At the sub-biome scale, they found that species-richness within particular growth forms may be related to cli-

Table 6.2. *Significance levels (P-values) for correlations between key climatic parameters and the relative contribution of growth forms to the total flora in sub-biome landscape types*

Growth form	Climatic parameter (relationship)	Significance level
Geophytes	PET (–)	$P < 0.02$
Perennial grasses	SAI (–)	$P < 0.001$
Perennial herbs	MAR (+)	$P < 0.001$
Shrubs	PET	NS
Trees	MAR	NS

PET is potential evapotranspiration, SAI is summer aridity index sensu Rutherford and Westfall (1986), and MAR is mean annual rainfall. The signs in brackets indicate positive or negative relationship.
Source: M. T. Hoffman, G. F. Midgley and R. M. Cowling, (unpublished data)

matic indices (see Table 6.2). The relationships seem even stronger when richness within a growth form category is expressed as a percentage of the total flora, an analysis which factors out climatic and other controls on total plant richness which are not well understood (Hoffman et al., 1994). Important climatic indices identified by Hoffman, Midgley and Cowling (unpublished data) included mean annual rainfall, potential evapotranspiration and summer aridity index, an index of water availability during the four hottest months of the year (*sensu* Rutherford and Westfall, 1986).

Relative richness within geophytes was negatively correlated with potential evapotranspiration, an index of energy loading and atmospheric demand for water, suggesting that high specific leaf area within this group may limit their success. Relative richness within perennial herbs was positively related to mean annual rainfall, a surprising finding that a rather coarse index of soil water availability sufficiently explains the success of this group. Relative richness of perennial grasses was negatively correlated with summer aridity index, revealing the strong dependence of this growth form on rainfall during the summer months. The latter finding is surprising in the light of Vogel et al.'s (1978) and Ellis et al.'s (1980) clear identification of different functional guilds within grasses and their changing proportional representation along climatic gradients. The patterns they identified occur at finer levels of scale, and these will be discussed later. The relative richness within two important groups, trees and shrubs, showed no relationship with climatic parameters. It is possible that relative tree richness is related more to the availability of ground water and thus site hydrology than to ambient climatic conditions. The same argument could be made for shrubs, but our suspicion is rather that the group 'shrubs' incorporates a diverse range of functional types, which obscures any clear relationship with climate. Both the succulent and Nama-karoo biomes are dominated by shrubs, so a clear need for a more detailed understanding of the group is identified by this analysis.

Is plant growth form of any importance in determining the structure and composition of karoo plant communities? Cowling et al. (1994) showed that growth form diversity in the karoo is positively related to an index of climatic temporal heterogeneity, which suggests that the distinct function of different growth forms may allow resource partitioning in time, thus reducing niche overlap (Cody, 1991). Greater growth form diversity in turn was found to promote a higher community species-richness (Cowling et al., 1994).

The relationships between plant growth form composition and climate discussed above strongly support the contention that plant growth form has a functional significance which influences relative plant success and causes compositional patterns to emerge at community and even sub-biome level. Furthermore, the finding that these facilitate temporal partitioning, at least, of resources (Cowling et al., 1994), suggests that karoo plant communities may be structured deterministically by trait-based community assembly rules (*sensu* Weiher and Keddy, 1995).

6.3.   Distribution and abundance of photosynthetic types

Because plant carbon gain is constrained by water and temperature limitations, and because photosynthetic carbon uptake is closely related to water loss, there has been much interest in the performance of different photosynthetic types in the karoo. Photosynthetic types encapsulate both water and carbon fluxes at leaf level which are clearly important determinants of plant function. In general, C_4 and CAM photosynthetic pathways have been shown to improve plant water-use efficiency, but not without costs – C_4 tends to have higher light and temperature requirements than C_3 photosynthesis, and CAM plants tend to have lower growth rates, and a capacity for photosynthate fixed by night which is limited by vacuolar storage of organic acids.

The approximate relative contribution of the different photosynthetic types to the karoo flora (Fig. 6.1), shows the floristic persistence of the ancestral C_3 form in the karoo. The derived C_4 and CAM photosynthetic types tend to be restricted to grass and succulent shrub growth forms respectively, but the converse is not true, that is, both grasses and succulent shrub species have many members with the ancestral C_3 pathway. Facultative CAM photosynthesis in succulent shrubs, the persistence of C_3 photosynthesis in many grass species, and the fact that halophytic shrubs of the Chenopodiaceae may possess C_4 photosynthesis (see Bond et al., 1994), complicates the interpretation of

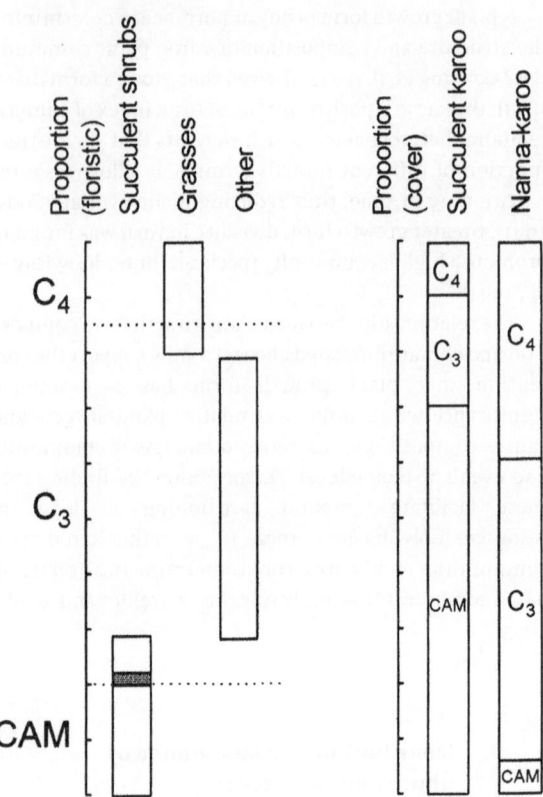

Figure 6.1 **The proportional floristic representation of different photosynthetic types in different growth form groups in the karoo region (stippled lines separate photosynthetic types), and the proportional representation in terms of plant cover abundance in the succulent and Nama-karoo biomes. Figures calculated by combining data from Werger and Ellis (1981), Mooney et al. (1977) and Vogel et al. (1978). The stippled region in the bar representing succulent shrubs indicates facultative CAM plants**

historical vegetation change studies based on carbon isotopic discrimination techniques.

In an analysis of 351 grass species, Vogel et al. (1978) showed that the South African grass flora is evenly distributed between C_3 and C_4 types, but in the karoo the contribution of these types in terms of plant cover diverges strongly in the succulent and Nama-karoo biomes (Fig. 6.1). Also, the cover dominance by CAM types in the succulent karoo clearly highlights the marked difference in plant function between these biomes. Werger and Ellis (1981) interpreted these results partly in terms of absolute water availability and rainfall seasonality. The demonstrated dominance of the CAM type in the west was ascribed to high water-use efficiency, rapid response to water input and the possible possession of reverse transpiration (water uptake from the atmosphere by leaves at night). The relative dominance of the C_4 type in summer rainfall areas was ascribed to its dependence on warm temperatures during the growing season.

These results follow qualitatively from a first principles

understanding of the relative advantages and disadvantages of the derived photosynthetic pathways, but this is not to say that the observed patterns would have been predicted quantitatively given a knowledge of the physiological differences. In the next sections, we focus on the relationship between growth form and resource availability, in an attempt to develop a more quantitative understanding.

6.4.   Water

The dependence of plants on water is many-faceted. Water is needed to support biochemical activity, for thermoregulation, to sustain growth (cell expansion through turgor pressure), and to facilitate nutrient uptake. Therefore plant–water relations (the aspects of plant function pertaining to water fluxes) are likely to be a key dependant of plant form in warm arid regions. The importance of water-related climatic indices in determining the relative success of many perennial growth forms in the karoo has been demonstrated above. At this point we ask two questions: (1) could differences between growth forms in terms of water relations provide a mechanistic understanding of growth form success and distribution? And (2), do water relations give us further insight with regard to growth form success, especially in the functionally diverse shrub growth form?

6.4.1.   Succulence

It is difficult to define tissue succulence clearly, but we adopt a pragmatic definition of Von Willert et al. (1992), namely that a succulent plant stores water for later use. A clear functional distinction must exist between succulent and non-succulent shrubs, as it is well known that this attribute distinguishes the dominant shrub forms of the two biomes of the karoo. It has been suggested that rainfall predictability may be one of the foremost driving variables in delineating succulent and Nama-karoo biomes (Hoffman and Cowling, 1987). The explosion of speciation especially within Mesembryanthema in the succulent karoo (Ihlenfeldt, 1994), but also within other succulent families, compels us to explore the functional significance of this attribute.

It has long been realized that an important benefit of increasing succulence is to increase volume : surface area ratio, as the succulent organ tends to approach a spherical or cylindrical shape. However, the size of the water-storing organ is also an important parameter, because increasing size also increases volume : surface area ratio, with important consequences for water storage capacity (Fig. 6.2).

We speculate about the consequences of increasing size for the drought survivorship, productivity, structural

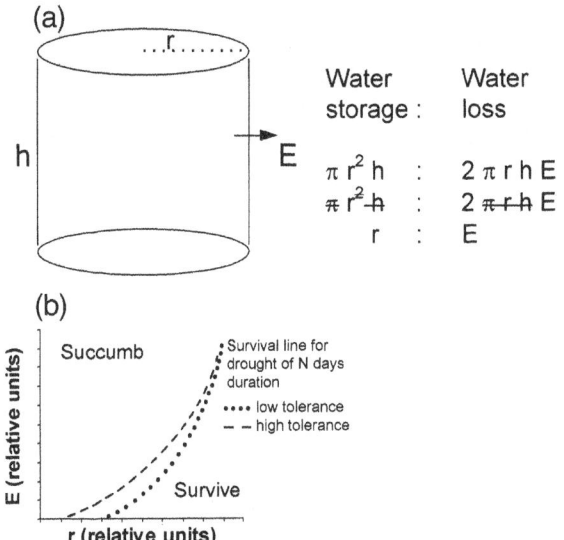

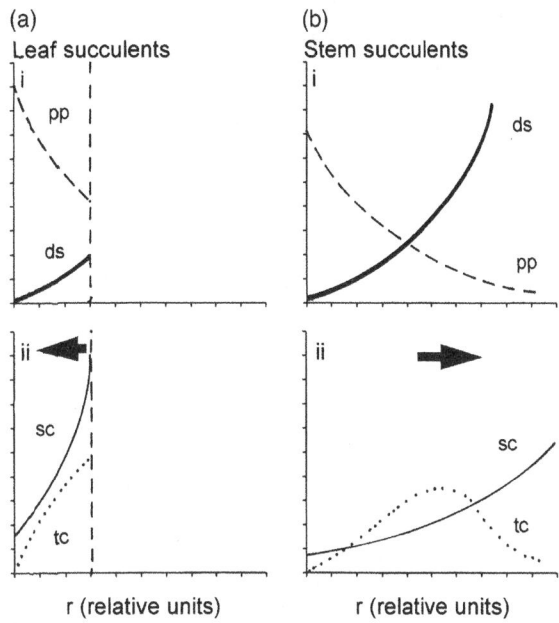

Figure 6.2 **(a): The drought survival potential of a cylindrical succulent organ of height h and radius r, with a per unit area transpiration rate E, once the plant roots are isolated from dry soil, depends on the ratio of water storage to water loss. This ratio can be simplified merely to the ratio of organ radius to transpiration rate per unit area. Thus (b), given a hypothetical drought of N days duration, organs with smaller radii need to maintain lower transpiration rates than more massive organs in order to survive the same drought. Conversely, given the same transpiration rate, organs with larger radii will tend to survive more extended droughts. This relationship may be complicated by inherent differences in tissue dessication tolerance (see high dessication tolerance and low dessication tolerance lines), and here we would speculate that less massive succulents would benefit more from tolerance of lower tissue water content than more massive ones**

Figure 6.3 **Four key aspects of succulent form and function (in relative units) in relation to organ radius, r. These are drought survivorship (ds), potential productivity (pp), structural costs (sc) and dependence on transpirational cooling (tc). Selection of leaf form in leaf succulents may act to reduce leaf size given favourable rainfall predictability (arrow in (a) ii), to gain from reduced cooling costs and productivity gains, but greater risk of mortality in longer drought periods would tend to favour greater radii in stem succulents (arrow in (b) ii)**

costs and cooling costs for stem and leaf succulents in Fig. 6.3. More massive succulents are able to survive longer droughts, given equal transpiration rates on a unit area basis. However, water storage above-ground has many other implications for plant function, including potential productivity, transpirational cooling costs and structural costs. It is possible that the predominance of leaf succulents with low volume : area ratios in the succulent karoo may be due to the low risk of extended drought in the biome, because this allows small-leaved succulents to take advantage of low structural costs, low dependence on transpirational cooling and potential productivity gains. With increasing aridity and decreasing rainfall predictability, low volume : area ratios become more risky, but a larger ratio carries cooling and support costs, an interesting conundrum which would make for fascinating and useful study. Leaf succulence becomes structurally non-viable above a certain radius limit for branched woody-stemmed plants (vertical dashed line in Fig. 6.3(a)), and the cylindrical photosynthetic stem-succulent design of *Euphorbias* or rosette leaf succulents such as *Aloe* species (some with non-photosynthetic succulent stems) may be more favourable. At high volume : area

ratios, heat load per unit mass of tissue (from radiation) is reduced, and cooling dependence on transpiration is lowered due to the thermal buffering capacity of the more massive water-filled structures and the vertical orientation of cylindrical stem succulents which reduces radiation interception at and near midday.

The functional significance of leaf dimension in leaf succulents has not been investigated in any detail. In a preliminary study on woody-stemmed leaf succulent species of the Mesembryanthema, R. Beukman (unpublished data) determined the relative growth rate and drought tolerance of the narrow-leaved *Ruschia caroli* and the thick-leaved *R. multiflora*. These species were found to occupy the relatively cool south- and relatively warm north-facing slopes, respectively, at the Worcester Veld Reserve. Under optimal water conditions, the relative growth rate of *R. caroli* was significantly greater than that of *R. multiflora*, but *R. multiflora* demonstrated greater diurnal acidity fluctuation, and daytime transpiration rate and stomatal conductance than *R. caroli* when water was withheld for up to 60 days. These findings support qualitatively the cost/benefit analysis in Fig. 6.3.

Leaf succulence appears to allow great physiological

flexibility. An intriguing and relatively well-studied phenomenon is C$_3$ CAM switching, or facultative CAM (the facultative CAM-plant *Mesembryanthemum crystallinum* is almost certainly the most studied non-agricultural species of southern Africa). Facultative CAM allows the plant capacity for greater daytime carbon fixation rates through the greater expression of the C$_3$ pathway when water is available, and greater water-use efficiency through CAM as soil dries. Three main types of CAM can be defined, based on patterns of CO$_2$ fixation and organic acid fluctuation, and these can be reversibly induced in some species (e.g. *Plectranthus marrubioides*, Herppich (1989) in Von Willert et al., 1992). CAM-cycling is closest to C$_3$ photosynthetic behaviour, typified by stomatal opening during the day and a C$_3$ carbon fixation pattern, but combined with fixation of respired carbon at night. 'True', or full CAM comprises mainly nighttime stomatal opening and appreciable organic acid accumulation. Finally, CAM-idling can be induced by extreme drought, and is characterized by no stomatal opening night or day, but detectable diurnal organic acid fluctuations which allow for endogenous carbon cycling and negligible water-loss rates. Mooney et al. (1977) found that 10 of 81 succulent species sampled in winter, summer and all-year rainfall areas in South Africa showed evidence of facultative CAM, and, by all similar accounts (Von Willert et al., 1992), it would seem that the leaf succulents of southern Africa include a wide range of C$_3$CAM types. Recent advances in application of chlorophyll fluorescence techniques to interpret photosynthetic activity in CAM plants (e.g. Herppich et al., 1998), combined with stable carbon isotope techniques, may help to develop a fuller understanding of the functional significance of CAM variations more productively than the use of laborious gas exchange techniques alone.

Some leaf succulents of the Mesembryanthema also have the ability (still relatively unstudied) to transport stored water between leaves along the growth axis. For example, the water content of young leaves and the apical bud of *Prenia sladeniana* dropped much more when older leaves were removed from the shoot than when older leaves were left in place before a drought treatment. The presence of older leaves also allowed improved photosynthetic characteristics in young leaves of droughted individuals of this species (Tuffers et al., 1995).

Older leaves on the growth axes of some Mesembryanthema show increasingly greater CAM expression, while younger leaves retain C$_3$ photosynthesis. In the annual *Mesembryanthemum crystallinum* (ice plant) the gradual increase of CAM activity as the plant grows might reserve water stored in older leaves which is able to extend the life of the plant into the drought season and fund the development of flowering parts and seed production even in the absence of rain, thus ensuring high fecundity in a precarious environment (Winter et al., 1978).

Even greater water savings in older leaves on a single growth axis can be effected by utilizing CAM-idling. In *Delosperma tradescantioides* (Herppich et al., 1996) older leaves showed large diurnal fluctuation of acidity, and younger leaves virtually none when well watered, but, when drought was imposed, acid accumulation was reduced in older leaves probably due to stomatal closure and CAM-idling (Fig. 6.4). Leaf area was able to increase unabated with no subsequent negative effect on whole shoot carbon uptake rate (Fig. 6.5). It is possible that similar mechanisms exist in facultative CAM plants of environments which experience greater fluctuations in water availability during the growing season than does *Delosperma tradescantioides*.

Thus CAM in older leaves might serve to improve their water retention and increase their usefulness as water stores to fund new growth, or even ensure the completion of the life cycle (e.g. *M. crystallinum*). The large varia-

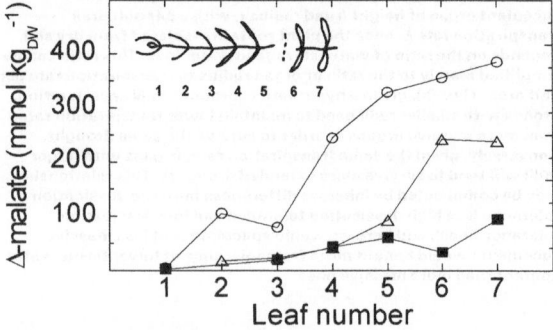

Figure 6.4 **Nocturnal malic acid accumulation of a sequence of leaf ages on the growth axis of the leaf-succulent *Delosperma tradescantioides* before (circles), during (squares) and after (triangles) a six-day drought. Leaves are numbered sequentially from the developing leaves at the shoot apex (1) through successive recently expanded leaves (2–5) to older leaves at the shoot base (6,7) (after Herppich et al., 1998)**

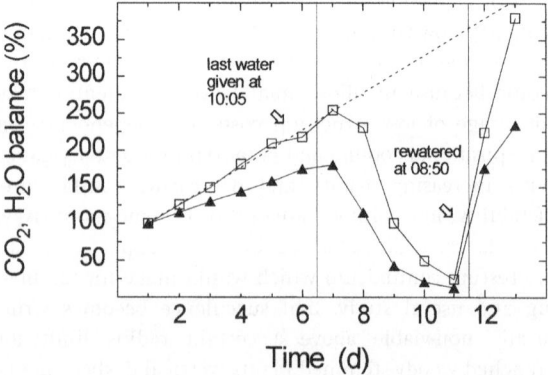

Figure 6.5 **Integrated diurnal shoot carbon uptake (squares) and water loss (triangles) (expressed as a percentage of that measured on day 1), and leaf area development (relative units, dotted line) for the leaf-succulent *Delosperma tradescantioides* during the development and release from a short-term drought (after Herppich et al., 1998)**

tions in shoot xylem pressure potentials of some Mesembryanthema (e.g. *Psilocaulon subnudosum*, Von Willert et al., 1992) suggest that this feature might be dependent on a mechanism of water compartmentation in the growth axis.

It is also interesting to speculate about the cost of succulent leaf construction – it is cheap in carbon terms to build succulent leaves, and succulent leaves of the Mesembryanthema have high leaf-specific areas (Von Willert et al., 1992). Therefore rapid leaf production and expansion when water is available increases photosynthetic leaf area at low cost, ensures the deprivation of that water from competitors, and gives the plant water-storage capacity for later use.

6.4.2.   Succulent vs. non-succulent shrubs

Throughout much of the karoo, succulent and non-succulent shrubs form an apparently stable matrix of relatively long-lived individuals. Modelling this type of community using a cellular automata approach suggests that the matrix may exist in a quasi-stable condition between large recruitment events which are driven by rainfall patterns (Wiegand et al., 1995). Plant lifespan, competitive interactions between species in the establishment phase, and between these and established shrubs, and gap availability seem to be the major determinants of community dynamics. We suggest here that the different function of these growth forms with respect to water limitation and response to water input would allow these insights to be further developed.

There is relatively little information on the drought survival of succulent and non-succulent shrubs of the karoo. A drought which occurred in the Richtersveld, an arid region in the northwest Cape which receives winter rainfall, killed around 80% of the succulent-dominated vegetation (Von Willert, 1985). The dessication-retarding role of CAM in leaf succulents is in question following reports by Von Willert (1985) that CAM succulents were more negatively affected by this drought than were C_3 succulents with similar form. This may have been due to greater ion accumulation in the CAM than non-CAM succulents during the desiccation process (Von Willert, 1985). This suggests that CAM leaf succulents may only be capable of surviving short-term drought conditions, and need water replenishment to avoid negative effects.

G. F. Midgley has carried out a field drought experiment in succulent karoo vegetation at the Worcester Veld Reserve, where two 36m^2 plots were covered by transparent fibreglass roofs and plastic-lined trenches dug to seal soil from lateral water movement. Results were calculated as LD_{50} (median lethal dose) values, that is, the number of days of drought required to kill 50% of the shoots of monitored individuals (see Table 6.3). On the whole, succulents

Table 6.3. *LD_{50} values (number of days of drought required to kill 50% of the shoots of monitored individuals) for a range of succulent and non-succulent species which had been deprived completely of water under fibreglass roofs constructed in the field at the Worcester Veld Reserve*

Species	growth form (n = no of individuals monitored)	LD_{50}-value (days of drought)
Succulent		
Ruschia caroli	(n > 10)	259
Euphorbia mauritanica	(n = 11)	343
Tylecodon paniculatus	(n = 5)	366
Euphorbia burmannii	(n = 4)	595
Non-succulent		
Protosparagus capensis	(n = 4)	413
Pteronia incana	(n = 4)	518
Lycium cinereum	(n = 3)	
Pteronia paniculata	(n = 4)	

were less drought-tolerant than non-succulents, and the leaf succulent *R. caroli* (Mesembryanthema) had an LD_{50} value lower by the significant margin of 100 days than all other species sampled. The narrow-stemmed *Euphorbia burmannii* was most drought-tolerant of all the succulent species, and was even more drought-tolerant than two of the non-succulent species. Two of the four non-succulents showed minimal shoot die-back during the course of the experiment. Water-potential studies suggested that the drought-deciduous *Lycium cinereum* had access to water at some depth in the soil, but that this was not the case for *Pteronia paniculata*, which retained leaf viability at shoot water potentials lower than −10 MPa for more than a year (the foamy nature of water expressed at the endpoints of the water-potential measurements suggested xylem cavitation and possibly xylem dehydration in this species). Thus it appears that this evergreen non-succulent has an astonishing level of drought tolerance which is akin to desiccation tolerance (see section 6.4.5).

A severe, natural drought of around 20 months (an event with an expected return time of 55 to 142 years) occurred during 1990 and 1991 in the Steytlerville district, a semi-arid region characterized by high year-to-year rainfall variability (Milton et al., 1995). This event reduced perennial vegetation cover from 45% to 21%, and both succulent and non-succulent forms suffered significant losses, though succulents had lower mortality (45%) than non-succulents (65%) (Milton et al., 1995). Post-drought recovery of established succulents was more rapid than that of non-succulents, and succulent seedlings established faster than did the seedlings of non-succulent species (Milton et al., 1995).

In a controlled greenhouse droughting experiment, Esler and Phillips (1994) showed that seedling drought survivorship of two non-succulent species was dependent on the timing of drought imposition, so that hardened seedlings which had experienced drought early had better survivorship than seedlings which were watered for longer before exposure to drought. The seedlings of the succulent shrub *Ruschia spinosa* were far more drought-tolerant than seedlings of the non-succulents, with only around 20% mortality after 400 days of drought, and they showed no hardening effects.

When considered together, these studies suggest that drought may cause major mortality of adult plants of both the succulent and non-succulent shrub growth form. On balance, it seems that adult succulents are more susceptible to drought-induced mortality than are adult non-succulents, but this may depend on the succulent subtype (stem or leaf succulent, CAM or C_3). Succulent seedlings, however, show significant drought tolerance, which assists the re-establishment from seedbanks following drought. Non-succulent shrubs display species-specific drought-induced mortality, their recovery from drought is slow, and their seedlings might be particularly drought susceptible.

Stem splitting is a phenomenon commonly displayed by non-succulent shrubs in many arid regions of the world (Ginzburg, 1973), but has received only limited attention in karoo species (Theron, 1964; Milton et al., 1997). It occurs when the growth axis of the developing shrub is separated by interxylary cork deposition. Over time, the apparently single axis develops into a bundle of 'daughter' axes, each joined independently to a separate part of the root system. This is thought to be a specific adaptation to drought (Danin, 1983) which allows plant axes with access to soil water to survive, while those without water die off. This anatomical adaptation effectively allows the shrub to sample the below-ground environment for favourable sub-habitats through trial and error, without risking the loss of the individual, and conferring the benefit of greatly enhanced persistence in the landscape.

Water inputs to the soil in semi-arid and arid systems occur predominantly as small events, wet soil dries quickly through evaporation and plants compete for water with each other, thus the response time of shrubs to water input may be an important functional attribute. Few studies have addressed the dynamics and timing of plant water uptake after rain in the karoo. The response of succulent and non-succulent shrubs to a simulated rainfall event has been investigated by G. F. Midgley, in an experiment conducted in succulent-dominated vegetation at the Worcester Veld Reserve. A rainfall event of 60 mm was simulated by irrigating 36m² plots with a microjet watering system prior to the onset of rains in April, after a prolonged dry summer. Both succulent and non-succulent shrubs responded within hours to the rainfall event (Fig. 6.6), showing increased water potential (succulents and non-succulents) and leaf water content (succulents).

Interestingly, both the leaf succulent *R. caroli* and the evergreen non-succulent *P. paniculata* increased leaf transpiration rates significantly shortly after water became available, showing that the water-stressed plants maintained viable leaves in a state of readiness for water input and demonstrating no hint of a 'water saving' strategy when soil was wet. The profligate stomatal behaviour of *P. paniculata* is reminiscent of the high stomatal conductances and transpiration rates, and low water-use efficiency of *Eriocephalus ericoides* and *P. incana* when well watered (Midgley and Moll, 1993), which suggests that these shrubs may use high water-loss rates as a means to reduce water availability to competitors.

Rapid responses to water input have also been recorded in the leaf succulents *Mesembryanthemum pellitum* and *Othonna opima* (both Mesembryanthema) by Eller et al. (1991). When invading natural vegetation in California, *Carpobrotus edulis* (Mesembryanthema) reduces water availability to indigenous shrubs (D'Antonio and Mahall, 1991).

In a detailed study on *Rhigozum trichotomum*, a facultatively deciduous non-succulent shrub, Moore (1989) found high transpiration rates when soil water was available. In general, as drought progressed, stomatal opening was limited to shorter periods of the day some 20 minutes after sunrise when air vapour pressure deficit was high. Air vapour pressure deficit was a key factor controlling stomatal closure, which led to pronounced midday stomatal depression in this species. Du Preez (1964) also found high rates of water loss in the non-succulent shrubs

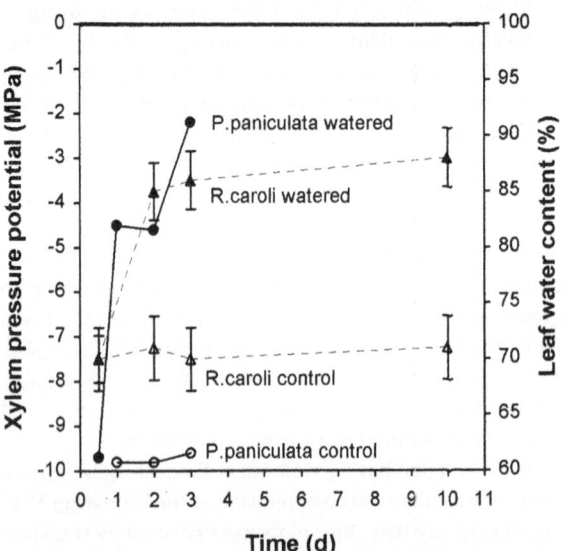

Figure 6.6 **Response of stem xylem pressure potential in the non-succulent *Pteronia paniculata* and leaf water content in the leaf-succulent *Ruschia caroli* to a simulated 60 mm rainfall event after an extended summer drought at the Worcester Veld Reserve, succulent karoo (bars represent standard deviations, *n* = 5 for each data point) (water added end of day 1)**

Phymaspermum parvifolium, Lycium arenicolum, Chrysocoma tenuifolia, Eriocephalus glaber and *E. spinescens*, when soil water was available.

We suspect that a combination of opportunistic water-use patterns with water-storage capacity (in succulents) or drought tolerance (in non-succulents) may have a powerful effect on karoo plant community structure and composition.

6.4.3.   Minutism

The quartz-strewn plains of the arid Knersvlakte and many other parts of the western succulent karoo are famous for their diversity of miniature forms, especially of the Mesembryanthema (*Lithops, Argyroderma, Conophytum*) and *Haworthia*. Although becoming much reduced in size seems to be self-defeating, especially in areas of moderate to high competitive intensity, the evolutionarily derived state of minutism seems to have a number of advantages where competitive interactions are minimized. Much of what follows in this section is anecdotal, due to limited availability of scientific findings.

It is well known that the western parts of the succulent karoo are fogbound at night and for a variable part of the day, depending on latitude. This source of water is trivial to large plants as it wets only the soil surface. However, the smaller the plant size, the more significant is this source of water! Thus a key advantage of minutism for west-coast species is the potential to derive a significant proportion of their water requirement from fog precipitation. The association of 'stone' or 'window' plants with quartz plains may be due partly to the condensation of dew on quartz pebbles.

The growth of minute Mesembryanthema comprises little more than the transfer of water from an existing leaf pair to a freshly expanding pair which simply replaces the old as it withers – here water transfer and recycling within the plant body, shown above to be an important aspect of Mesembryanthema function, reaches its ultimate state of efficiency.

Minutism is also likely to be favourable for obligate nurse-dependent plants. Conditions below the rather open canopies of drought-tolerant shrubs are much ameliorated in terms of soil and air temperature, soil nutrient status and even water status, as long as root systems of nurse and patient are spatially separate (see Beukman, 1991). Apparency to pollinators may be increased in this suite of species through the production of elongated flower-bearing stems, common, for example, in reduced forms of *Senecio, Haworthia, Adromischus, Crassula* and *Tylecodon*.

A further benefit of minutism is the evasion of tissue damage by wind-blown sand. Jürgens (in press) includes reduced Mesembryanthema in a larger group of psam-

mophorous growth forms. Geophytes are well represented in this group, and it is striking that two geophyte forms can be observed in the arid western karoo region, with either broader leaves very closely pressed to the soil surface, or with wiry, whiplike leaves. It is possible that these and other plant forms have been selected by high loads of wind-blown sand.

6.4.4.   Grasses

Vegetation of the succulent and Nama-karoo biomes can be distinguished in growth form terms on the basis of dominance by grass (Rutherford and Westfall, 1986). These biomes have distinct SAI ranges (summer aridity index see section 6.2), and this suggests a climatic restriction on grass dominance in the succulent karoo, which experiences intense summer droughts (SAI >5). A study of vegetation dynamics in an area with variable rainfall seasonality showed that the competitive balance between grass and shrub forms is controlled by the dependence of perennial grass on summer rainfall (Hoffman et al., 1990). Bond et al. (1994) have shown that the ecotonal zone between areas of grass/shrub codominance and shrub dominance has not shifted geographically through historical times, but they suggest that land use practices strongly influence grass cover in these transitional areas.

At a finer level of spatial scale than biomes, interesting distribution patterns of physiological subtypes of grasses exist within the karoo. Vogel et al. (1978) have demonstrated a change in dominance geographically in South Africa, from C_4 grasses in the eastern summer rainfall region to C_3 grasses in the western and southern winter and all-year rainfall region. This change does not explain the relative scarcity of grass in the south-western parts, which may be related more to soil texture (Rutherford and Westfall, 1986) and low soil nutrient status in the fynbos biome (Stock et al., 1997).

Differences at a fine level of leaf structure and physiological function seem to influence the distributional patterns of perennial grasses in semi-arid and arid Namibia (Ellis et al., 1980), where the C_4 type makes up more than 95% of the grass flora. C_4 grasses can be classified as either aspartate and malate formers, depending on which organic acid is formed in the initial fixation of atmospheric CO_2. Aspartate formers, in turn, can be classified into PEP-ck and NAD-me types depending on the enzyme involved in the decarboxylation of these acids in the bundle sheath. Each physiological type is associated with a particular leaf structure and cell arrangement around leaf vascular bundles. At mean annual rainfall (MAR) of around 500 mm (the high end of the distribution) aspartate and malate formers contribute equally to the C_4 grass flora, but aspartate formers predominate at lower MAR, and contribute more than 80% to the C_4 grass flora at MAR of

100 mm and below. Within the aspartate-forming group, PEP-ck types have a lower frequency than NAD-me types at MAR less than 350 mm. It would seem that different tissue desiccation tolerance characteristics or leaf water-use patterns which result from the biochemical or anatomical features of these subtypes might explain their differential dependence on soil water availability.

The distinct (but not disjunct) distribution of grasses of different photosynthetic types has been an important source of credibility to the application of ecophysiological research (e.g. Osmond et al., 1982). However, although it is possible to link ecophysiological performance to field distributions, no models exist which would predict distributions, relative success or species-richness of different grass photosynthetic types *a priori*. This is a tantalizing area of research which could yield great benefit from the attention of ecophysiologists, especially as these forms are amenable to study throughout their life cycle in controlled conditions. We suggest that comparative studies of the performance of several species of each photosynthetic subtype in relation to different combinations of temperature and water availability would be particularly rewarding.

The South African grass flora contains a priceless range of genetic and related functional diversity, comprising species suited to a wide range of rainfall and temperature conditions. This is a national resource of no small strategic value.

6.4.5.   Desiccation tolerance

Tissue desiccation tolerance would seem to be a favourable trait in the vegetative parts of arid-adapted plants, but desiccation tolerant plants (sometimes called 'resurrection plants') tend to be locally rare, and only occasionally form appreciable stands of vegetation (e.g. *Myrothamnus flabellifolia* on rocky slopes in the eastern Namib desert, G. F. Midgley, personal observations). The southern African flora has an appreciable representation of desiccation-tolerant species (Gaff, 1977), especially among monocots and ferns. Many of these species rehydrate rapidly after rain falls, but dewfall is not sufficient to cause rehydration (Gaff, 1977). Rehydration of several resurrection plant species occurs mainly through root water uptake, which implies refilling of xylem vessels by capillary action and possibly root pressure. This places a limit on the potential height of desiccation-tolerant plants – the tallest recorded by Gaff (1977) is 1.7 m high and none are trees.

The extreme stress tolerance of desiccation-tolerant species may place limits on their relative growth rate, thus restricting their ecological role to that of pioneer species (Gaff, 1977). However, the high degree of drought tolerance demonstrated in species such as *P. paniculata* suggests that a type of leaf tissue desiccation tolerance has been developed in higher plants which are capable of dominating vegetation, and this is an area of potentially fruitful research.

6.4.6.   Root architecture

The term 'root architecture' refers to the spatial configuration of the plant's root system (Lynch, 1995). Root architecture is of great importance to plant performance as soil resources are unevenly distributed horizontally and vertically, and subject to localized depletion. Competition for water, in particular, may be critical for the structure and composition of desert communities, and as a result plant packing may be dependent on root architecture (Yeaton and Cody, 1976). Root architecture constrains the plant's options for water uptake, and is thus a primary determinant of plant function in arid systems. It is difficult to define root system types because of the problem of plasticity of growth below-ground. Early comprehensive studies of root systems in desert plants in the western United States (Cannon, 1911) revealed a number of root system types, which could be classified into less than 10 main types (Cannon, 1949).

Relatively little is known about the root architecture of karoo growth forms. Early studies by Scott and Van Breda (1937a, 1937b, 1938, 1939) reveal extensive and deep root systems in the non-succulent shrubs *Elytropappus rhinocerotis* and *Galenia africana* (some excavated to more than 8 m depth). Two stem-succulent species (*Euphorbia mauritanica* and *E. burmannii*) had shallow root systems, but the leaf-succulent *Ruschia multiflora* had a concentration of shallow roots and a few roots penetrating to 4 m or more. It is important to note that most of these studies were carried out on deep soils. The gymnosperm *Welwitschia mirabilis* is known to possess a well-developed taproot which gives the plant access to water at some depth in the soil (e.g. Von Willert and Wagner-Douglas, 1994), and this is likely to limit its establishment to unusual periods of abundant rainfall to allow seedling taproot development.

However, much of the karoo is characterized by skeletal soils, often overlying fractured bedrock or shale beds (Watkeys, this volume). Seasonal water-potential changes in eight non-succulent shrubs growing in such a situation suggested that they could be broadly classed as shallow-rooted, intermediate-rooted or deep-rooted (Midgley and Bosenberg, 1990). While apparently deep-rooted forms did not access a water table, their relatively constant base water potentials through the year suggested that they were able to access soil water through the year, possibly from a perched water table (Midgley and Bosenberg, 1990).

Succulent species in North American deserts have been found to possess shallow root systems (Cannon, 1911; Franco and Nobel, 1990) with special water relations capabilities (Nobel and Sanderson, 1984). Rapid response to water input is ensured by the production of rain roots, which develop soon after the soil wets (Franco and Nobel, 1990). The rapid responses to water input in previously droughted plants described in section 6.4.2 suggests that this is a common feature in karoo plants.

In an exploratory study in Bushmanland (western karoo) (G. F. Midgley and A. V. Hall unpublished data), the root systems of three growth forms were found to be vertically separated (see Fig. 6.7). The leaf succulent *Ruschia robusta* had extremely shallow roots spreading widely horizontally, two grasses, *Stipagrostis brevifolia* and *S. ciliata*, had roots of intermediate depth and limited lateral extent, and the deciduous non-succulent *Lycium cinereum* had a taproot which penetrated until encountering a hardened soil layer which deflected the root system horizontally, and well-developed side roots extending horizontally between 10 and 30 cm depth. The base and mid-day water potentials of all three forms decreased with increasing rooting depth, showing the greater access of the shallower-rooted forms to soil water from recent rain.

The rooting pattern of the grass species is similar to that reported by Roux (1968a) for the grass *Tetrachne dregei*, found in the eastern karoo, namely a proliferation of relatively superficial roots with a herringbone pattern of side-root development, but no discernible taproot. Theron (1964) showed that the karoo shrub *Plinthus karooicus* allocated matter preferentially to root growth during the establishment period, and had a well-developed and relatively deeply penetrating root system.

This apparent vertical separation of roots among different growth forms implies niche partitioning similar to that encountered by Yeaton and Cody (1976), but may be deceptive. Moore (1989) found that the root system of *Rhigozum trichotomum* both explored surface soil (<200 mm depth) and had a number of taproots which penetrated to depth, similar to that of *Lycium cinereum* described above. *Rhigozum trichotomum* appeared to compete directly with established grasses for water, and greater densities of the shrub reduced grass production exponentially.

Du Preez (1964), working near Middelburg in the Nama-karoo, also found a well-developed taproot in the non-succulent *Lycium arenicolum*, but four other non-succulent shrubs had no such taproot, but had well-ramified relatively shallow root systems (generally less than 500 mm deep).

Vertical separation of roots of nurse and patient plants has also been described by Beukman (1991), and this was thought to reduce competition between these necessarily closely spaced individuals. Franco and Nobel (1990) have shown that a shallow root system may allow nursed individuals to retrieve up to 50% more soil water than exposed individuals, due to the reduction in soil surface evaporation in the shade of the nurse plant.

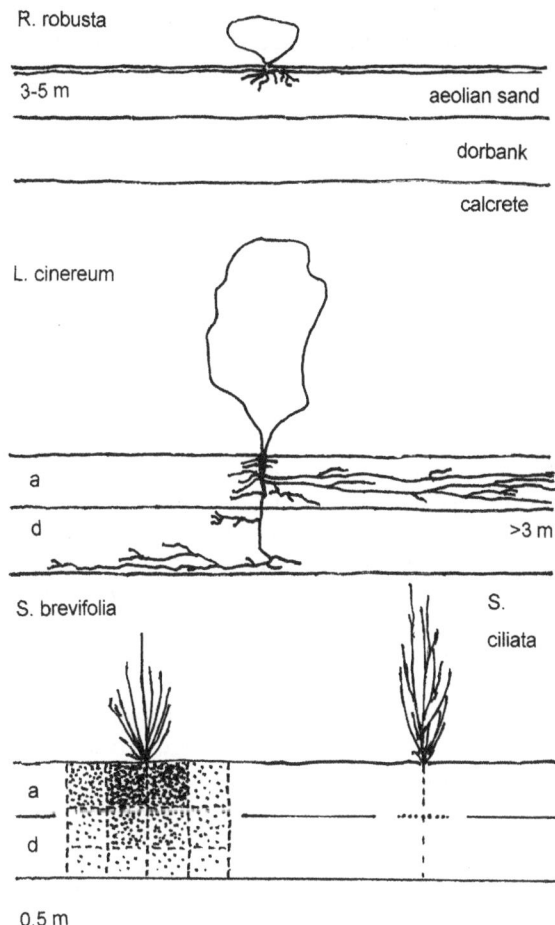

Figure 6.7 **Root systems of the leaf-succulent *Ruschia robusta*, deciduous non-succulent *Lycium cinereum* and the grasses *Stipagrostis brevifolia* and *S. ciliata* on the radiation waste dumpsite 'Vaalputs' in Bushmanland, ecotonal between Nama- and succulent karoo biomes (one indiviual of each species was excavated). The soil comprised three strata: a surface stratum of red aeolian sand overlying a compacted sand stratum (dorbank), also red, underlain by a lime-rich calcrete cemented layer. For the shrubs, the lateral extent of shallow roots is indicated. For the grasses, root system depth and width are indicated for *S. ciliata*, and densities in 12 excavated faces of soil beneath the plant indicated in *S. brevifolia* (heavy shading represents greater than 100 roots encountered, progressively lighter shading represents >25, >10 and <10 roots encountered per soil face)**

Few reliable conclusions can be reached about root architecture in karoo plant growth forms based on the information available, but it appears that root separation vertically in the soil profile may facilitate the coexistence of non-succulent and succulent shrubs, and encourage 'nursing'. Succulent shrubs and grasses may compete for water in the same horizontal plane. Succulents tend to have shallow rooting systems, allowing rapid response to water input, even from sources such as fog deposition. Non-succulent shrubs tend to have well-developed and deep root systems, but often retain a proliferation of roots near the soil surface. As mentioned in the introduction, advances in hydrogen isotope techniques provide the potential to advance our knowledge about the source of water accessed by different growth forms, as different water sources such as fog, rain, soil and ground water have different isotopic signatures (Squeo et al., 1994). This technique would be extremely useful in quantifying the intensity of competitive interactions in karoo plant communities.

6.5.   Temperature

The karoo environment is characterized by wide temperature fluctuations which occur diurnally and seasonally. As discussed above, the coincidence of rainfall with appropriate growing temperatures clearly limits the distribution of C_4 grasses, but do temperature extremes *per se* limit the success of growth forms?

At a biome scale, Rutherford and Westfall (1986) show that the desert, succulent karoo and Nama-karoo biomes have progressively cooler LMT (mean lowest minimum temperature for the coldest month), the desert above 2 °C, succulent karoo above –4 °C and Nama-karoo ranging from 0 to –9 °C. A temperature–moisture matrix comprising LMT and mean summer half-year rainfall clearly distinguishes the succulent from the Nama-karoo (Fig. 6.8). The succulent karoo and desert biomes are distant from the phanerophyte (tree) dominance area evident in the matrix (Fig. 6.8). Rutherford and Westfall (1986) speculated that the difference between succulent and Nama-karoo biomes in terms of dominance by leaf-succulent shrubs might be explained by cooler LMT in the latter. Werger (1983) suggested that minimum temperatures of –4 °C to –5 °C prevent dominance by succulents.

Stock et al. (1997) have shown that both low winter minimum and high summer maximum temperatures limit succulent richness. Only one study (Van Coller and Stock, 1994) on one nationally pandemic leaf succulent, *Cotyledon orbiculata*, has attempted to verify experimentally the cool temperature limits for succulent species. Lethal temperatures were found to range between –6.1 °C to –10.4 °C, which does not preclude succulent survival in the Nama-karoo, and does not show coincidence with the lower LMT in the succulent karoo biome. This study could not explain the observed survival of the species in highly frost-prone areas, and the authors cautioned against literal application of the results. Interestingly, a limited ability of the plant to 'harden' in pre-treatment of progressively colder temperatures was apparent, and hardening resulted in only a 2 °C lowering of the lethal temperature.

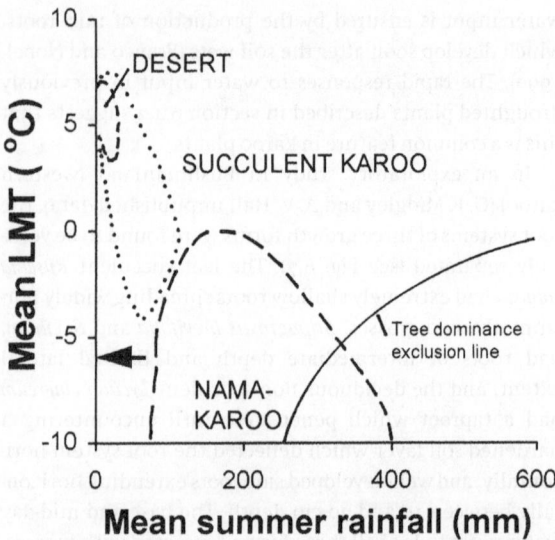

Figure 6.8 **Moisture–temperature matrix of summer rainfall and lowest minimum temperature (from Rutherford and Westfall, 1986), showing the distribution of Nama-karoo, succulent karoo and desert biomes. The experimentally determined minimum lethal temperature for the leaf-succulent *Cotyledon orbiculata* (Van Coller and Stock, 1994) is indicated by the arrow in the lower left quadrant of the matrix**

Hardening possibly occurs through the hydrolysis of fructan in the non-succulent shrub *Osteospermum sinuatum* (F. Van der Heyden, J. Ball and W. D. Stock, unpublished data), which leads to the reduction of osmotic potential and freezing-point (supercooling). It is possible that the effectiveness of this strategy is reduced in succulents with a relatively larger vacuole. The vacuoles of succulent plants build up organic acids at night which could reduce freezing-point, but acid accumulating enzymatic activity would be curtailed at low temperatures, thus leaving the vacuoles diluted and prone to freezing. Freezing tolerance may be dependent on the localization of ice-crystal formation in the apoplast, which prevents damage to symplastic organelles and membranes, but the relatively massive symplastic water storage in succulents may prevent this. Finally, freezing tolerance is dependent on changes in cell membrane properties (Krause et al., 1988), which may not be possible in succulent tissue.

Succulent species are also threatened by high daily temperatures because of their limited ability to cool by increasing transpiration rate. From the data of Jürgens (1991), Stock et al. (1997) concluded that succulent species-richness was curtailed above mean summer maximum temperatures of 32 °C. We have suggested that a combination of high rainfall variability and high temperature would not permit the survival of anything less than rather massive stem succulents. In support of this, we cite the report by Stock et al. (1997) of extreme leaf temperatures

(>50 °C) in large-leafed *Aloe* spp., which suggests that this kind of growth form requires protoplasmic heat tolerance. The leaves of leaf succulents may be exposed to the sun's rays through the entire day, but the naturally vertical orientation of stem succulents contributes to cooling by reducing exposure during the hottest midday period. Heliotropic movement of shoots of the C_3 leaf succulent species *Augea capensis* (G. F. Midgley, personal observations) seems to reduce exposure of its leaves to the sun and thus reduce cooling costs. Succulent leaf opacity is an important mechanism of temperature control in minute stone plants – clear-leaved species are associated with areas of lower radiation intensity than are opaque-leaved species (Turner and Picker, 1993).

Plant orientation has a further role to play in the succulent karoo biome, where plant growth may be limited by low temperature when water is available. The non-photosynthetic stem-succulent *Pachypodium namaquanum* appears to overcome this by orientating its growing point at the stem tip towards the equator, thus maximizing its temperature (Rundel et al., 1995)

There is not enough information available to come to a worthwhile conclusion about control of growth form distribution by temperature regime *per se.* Our suspicion is that temperature is an important parameter only in combination with rainfall regime, but it remains to be seen whether this holds for succulent plant distribution.

6.6.   Herbivory

A wide range of animal life forms feeds on the vegetation of the karoo, but relatively little is known of current and pre-colonial utilization pressures by indigenous fauna and by early human colonists. Birds such as the karoo korhaan (*Eupodotis vigorsii*) and the ostrich (*Struthio camelus*) feed on plant parts (leaves, flowers and seeds) of woody perennial, annual and leaf succulent plants (Boobyer and Hockey, 1994; Milton et al., 1994a; Dean and Milton, this volume). A multiplicity of insect species also feeds on karoo plants (Dean and Milton, this volume). The feeding ecology of insects has not been investigated in any great detail apart from studies on the outbreaks of the brown locust, *Locustana paradalina* (McKenzie and Longridge, 1988; Nailand and Hanrahan, 1993). During pre-colonial times it is likely that the vegetation was utilized extensively by migratory ungulates (Roux and Theron, 1987). These mammalian herbivores were replaced by enclosed domestic livestock (goats and sheep) at the time of European settlement in the karoo, made possible by the artificial provision of surface water. This marked the shift to excessive levels of plant utilization in the karoo. The

responses of karoo plants to these artificially high levels of herbivory have been addressed in many studies. In this section we use the findings from these rangeland ecology studies to examine whether herbivory has had a selective influence on plant form and function in the karoo.

Firstly, we compare the responses of the shrub growth form with those of the graminoid growth form. Since there have been almost no whole plant defoliation studies on indigenous karoo grasses, the studies on grasses native to other ecosystems are used for this comparison. This is followed by a comparison of the response of different functional types of karoo shrubs.

6.6.1.   Comparing shrubs with grasses

It has often been reported that total above-ground production of grasses is maximized at moderate grazing intensities (McNaughton, 1983a). The fast regrowth rates, or compensatory regrowth responses, following defoliation of grasses have not been observed for karoo shrubs (Van der Heyden, 1992; Van der Heyden et al., submitted), with clipped karoo shrubs not reaching control plant leaf biomass levels up to 26 weeks following severe clipping (Van der Heyden and Stock, 1995). It has been suggested that the regrowth of karoo shrubs is limited by the loss of apical meristems (removed during browsing) as well as the lack of secondary meristems (Van der Heyden and Stock, 1995). In contrast, the meristems of grasses remain protected when grazed, and new leaves can be produced fairly quickly following defoliation.

Severe defoliation of actively growing grasses generally results in the immediate reduction of non-structural carbohydrates in the remaining stubble and in the roots for periods between 4 and 6 days (Davidson and Milthorpe, 1966). However, in karoo shrubs, the period of decline in stored carbon levels ranges from 2 weeks under well-watered conditions (Van der Heyden, 1992), to periods exceeding 26 weeks (Van der Heyden and Stock, 1995). This indicates that karoo shrubs are dependent on stored carbohydrates for periods much longer than those quoted for grass species (Davidson and Milthorpe, 1966; Gonzales et al., 1989).

There is also a marked difference between the two growth forms with respect to the relative contributions of current photosynthates and reserve carbon to regrowth. The contribution (to regrowth) of stored carbon exceeded the contribution of photosynthetic carbon for only 2 to 5 days following defoliation of two *Agropyron* grass species (Richards and Caldwell, 1985). However, in a karoo shrub (*O. sinuatum*), most of the regrowth was derived from stored carbon (60%) up to 2 weeks following simulated browsing (Van der Heyden and Stock, 1996). Regrowth of the defoliated karoo shrub was therefore relatively more dependent (than grasses) on reserve carbon with respect to the duration of dependence, and quantities utilized to produce new leaves.

The carbon allocation responses of graminoids and shrubs outlined above illustrate important differences in the functional responses of these two growth forms to herbivory. The lower dependence by grasses on reserve carbon for regrowth, and their rapid production of new leaves, after defoliation, which are needed to maintain a positive carbon balance, may explain the poor competitive ability of grasses, and the persistence of shrubs in graminoid-shrub communities subjected to severe grazing pressures. Under conditions of severe and frequent defoliations, there would be few or no remaining leaves to produce photosynthates to facilitate the regrowth of grasses, and repeated defoliation would rapidly deplete their carbon reserves. Shrubs, which utilize mostly reserve carbon (and not photosynthates) for regrowth, would be at a competitive advantage. This may have led to the recent changes from grass-dominated to shrub-dominated communities in karoo rangelands subjected to severe grazing pressures (Acocks, 1955; Roux and Vorster, 1983; Bond et al., 1994). It is unlikely that browsing at relatively low frequencies and intensities (by large, indigenous mammalian herbivores) would have resulted in the large scale changes in community structure because perennial grasses and shrubs coexist in regions of the Nama-karoo where livestock is excluded, or on farms where low stocking rates are maintained. It is only under human-imposed livestock production systems, characterized by high defoliation intensities, that herbivory may have a selective influence on plant form and function, with the graminoid growth form being replaced by shrubs.

6.6.2.   Shrub functional types

Diet selection studies have indicated that livestock strongly differentiates among plant species, with some species being preferred above others and some species being avoided (Du Toit and Blom, 1995; Du Toit et al., 1995). These preferences cannot be explained by structural defences such as spines and thorns alone. There is evidence that carbon-based secondary compounds may explain acceptability and palatability of karoo shrubs. It has been shown that plants adapted to low resource environments, such as the karoo, generally do not show compensatory regrowth following browsing (Bryant et al., 1988; Coley et al., 1985). Photosynthetic processes continue under these conditions of slow regrowth, producing carbon which is not used for primary growth. It has been suggested that a range of carbon based compounds are formed from the excess photosynthates under these conditions (Baas, 1989). There is also evidence that tannins may be associated with palatability of karoo shrubs. *Osteospermum sinuatum*, a highly palatable karoo shrub, for

example, has naturally low tannin content, while a less palatable leaf succulent species, *Ruschia spinosa*, accumulates relatively high levels of tannins (Stock et al., 1993).

Van der Heyden and Stock (1995) and Stock et al. (1993) investigated the fate of excess photosynthate produced in karoo shrubs following browsing. In two Asteraceous shrubs, *Osteospermum sinuatum* and *Pteronia pallens*, excess carbon was stored as non-structural carbohydrates, and was not allocated to secondary carbon defence compounds. In *R. spinosa,* excess carbon was used for the production of secondary carbon defences, and was not allocated to storage of non-structural carbohydrates (Stock et al., 1993; Van der Heyden and Stock, 1995). These studies illustrate that karoo shrubs do not respond mechanistically in a similar fashion to herbivory, and that different functional responses occur among karoo shrubs.

It is also likely that the low specific leaf area of the many evergreen shrubs of the karoo, which is associated with the ability to tolerate low water potential, contributes to their unpalatability. Thus tolerance of tissue water deficit may simply confer the added advantage of low palatability. We are not aware of any test of this simple hypothesis, which might explain the progressive dominance in karoo landscapes of stress-tolerant shrubs with sub-optimal production and water-use characteristics. On the other hand, leaves of leaf-succulent species often have extremely high specific leaf area (Von Willert et al., 1992), and, as a highly apparent source of water in an arid landscape, would be expected to contain compounds which deter herbivory, such as tannins in the case of *R. spinosa*.

Studies to date suggest that herbivory has not provided a selection pressure for active chemical defence production in karoo shrubs. Stock et al. (1993), for example, recorded decreases in polyphenol and tannin concentrations in the stems and older leaves of *Pteronia pallens* in response to 80% clipping. Ras (1990) similarly recorded reductions in polyphenol and condensed tannin concentrations in leaves following simulated browsing of a leaf-succulent species, *Portulacaria afra*. In contrast, *Ruschia spinosa*, a slower growing leaf-succulent species responded to simulated browsing with elevated concentrations of these compounds in old stems and old leaves (Stock et al., 1993). The observed decreases in tannin concentrations following simulated browsing of some karoo shrubs (Ras, 1990; Stock et al., 1993), and the elevated concentrations in plant organs at low risk of being browsed in another species (Stock et al., 1993) do not provide convincing evidence for active chemical defence occurring in karoo shrubs. Stock et al. (1993) suggested that the observed changes in polyphenol and tannin concentrations were not active defence responses, but passive alterations in the chemistry of plants with a limited regrowth potential. Although this suggests that herbivory has not been an important selection pressure for tannin production in karoo shrubs, a far broader survey of secondary compound production, especially in the highly apparent succulent and deciduous shrub species with high specific leaf area is necessary to test this hypothesis.

One of the reasons advanced for the increasing dominance of unpalatable species (under livestock production systems) is the negative impacts of browsing on the recruitment of palatable shrubs (Milton, 1994a). It has also been postulated that variations in the inherent physiological abilities of different shrubs to tolerate browsing indirectly determine the changes in botanical composition (Roux and Vorster, 1983). The assumption that shrubs respond functionally in such dissimilar ways so that some shrub functional types are competitively more successful than others under livestock production systems was specifically tested by F. Van Der Heyden, J. Ball and W. D. Stock (unpublished data). They showed that the ecophysiological performance of the shrub *Eriocephalus ericoides* was unaltered by a wide range of stocking rates, but that drastic changes in the performance of *Pteronia tricephala* at high stocking rates included an improvement in plant–water relations and leaf nitrogen content, and an elevation in net photosynthetic rates. The carbohydrate status of heavily browsed *P. tricephala* plants remained unchanged, while that of *E. ericoides* plants decreased. This study illustrates that extreme intensities of herbivory may favour certain plant functional types which have compensatory mechanisms that operate only at intense levels of herbivory. This may confer a competitive advantage (high carbon reserves and elevated photosynthesis) on these shrubs when subjected to high stocking rates. Shrubs such as *E. ericoides,* without these compensatory mechanisms, may not be able to survive continuous heavy browsing. This illustrates that intense livestock production systems, which represent artificially high browsing pressures, presently influence the relative success of plant forms in the karoo, leading ultimately to the transformation of plant communities. The absence of appreciable changes in plant functioning at low and intermediate stocking rates suggest that herbivory by indigenous mammalian herbivores, at population densities expected for this vegetation type, may not have had such an influence historically.

6.7. Nutrient availability

Few studies have addressed plant nutrient relations in the karoo, and little is known of patterns of nutrient cycling in these semi-arid and arid systems. Soil fertility has been shown to be a significant driving variable of community structure in fynbos, and may be similarly important in arid zones of south-western Africa. Succulent and Nama-karoo vegetation appears to have high foliar nutrient levels, second only to levels in the savanna biome (Dean et al., 1994a). However, it is known that soil nutrient availability is variable regionally, and may be highly variable at small scales.

Functional and structural attributes such as deciduousness and spinescence appear to respond similarly to soil nutrient levels in the karoo (Milton, 1991), but it is difficult to isolate the nutrient effect from that of water in the 'natural experiments' available in the field. In a comparison of coenoclines in karroid and fynbos vegetation, leaf parameters such as succulence and consistency were shown to be related to soil moisture factors in karroid types, but more to soil fertility in fynbos (Cowling and Campbell, 1983).

Low soil nutrient levels may be found in the Worcester/Robertson Valley, where soils are highly acidic (Midgley and Musil, 1990). However, this condition is completely altered by mound-building termites, and mound soils are neutral to alkaline and have high nitrogen, phosphorus and macro-element concentrations. This may have led to the development of a distinct plant community type on mounds with representation of deciduous forms not found in surrounding off-mound soils (Midgley and Musil, 1990). Unfortunately, mound and non-mound soils are characterized by different disturbance regimes which have implications for species life histories (Esler and Cowling, 1995), and this confounds interpretations based on nutrient differences. It does seem, however, that even more subtle gradients of soil pH, nitrogen and phosphorus may be at least as important as interspecific competition in determining the relative success of three *Pteronia* species at Tierberg, Prince Albert (Esler and Cowling, 1993).

A detailed study of vegetation–substrate relationships was carried out in the all-year rainfall zone in Bushmanland by Lloyd (1985). The selected site has a range of soil types and plant communities dominated by different growth forms. Lloyd (1985) showed that grass-dominated vegetation was associated with acidic aeolian sands with low availability of phosphorus, leaf succulent (Mesembryanthema)-dominated communities occurred on saline soils of neutral to slightly alkaline pH, and non-Mesembryanthema leaf succulents (*Zygophyllum, Salsola*) dominated calcareous soils. It is clear from this study that small-scale patterns of plant distribution are controlled by soil texture, soil pH and, presumably, nutrient-availability characteristics. The role of simple soil features like these in limiting growth form success and distribution has yet to be assessed on a broader scale.

Mistletoes seem particularly limited by host nutrient status (Dean et al., 1994a), and are thus one group found in the karoo whose success is clearly nutrient-dependent. Mistletoe-richness tends to be higher in host genera with high nutrient status, and the number of mistletoes on any particular host in the field is significantly correlated with host nitrogen status in particular (Dean et al., 1994a). Mistletoes are not commonly found on succulent hosts, and it appears that succulents may limit mistletoe success through their low water availability in the xylem stream, as the mistletoes do not parasitize succulent water-storage tissue (Midgley et al., 1994).

6.8.   Conclusions

Plant form reveals much about plant function in karoo vegetation, but there is much that is still poorly understood. Water and temperature limitations have combined historically to shape the wide range of plant growth forms in this vegetation, so it is ironic that strong selection by herbivory is now acting to alter the relative dominance of plant species and growth forms.

It is difficult to come to any specific conclusions about the form and function of karoo plants, with so many identified unknowns, but at this point we suggest some key areas of research which would help to develop our understanding of the function of karoo ecosystems through an understanding of their component growth forms.

A major area is that of the timing and source of water acquisition by different growth forms. A better knowledge of water-use dynamics would improve our understanding of the intensity of competitive interactions between the diverse components of karoo vegetation. Use of isotope techniques would be highly productive in tracing seasonal water sources and water-use efficiencies of karoo growth forms. The highly drought-tolerant non-succulent evergreen growth form is particularly worthy of attention, due to its possible role in removing water rapidly from the soil profile.

The full implications of flexible physiology in the prolific group of leaf succulents needs to be realized, and their thermal relations more completely studied before we can appreciate the reasons for their exceptional abundance in the western karoo.

An understanding of the role of the obviously closely related physiological and morphological characteristics of grasses would be of much use in predicting their distributions and production potential.

We have also provided evidence that shrubs differ markedly from grasses in their functional responses to mammalian herbivory, findings which support reports of the transformation of grasslands to shrublands when karoo rangelands are subjected to intense browsing. Predictions of grazing intensity effects are limited by the weakly developed understanding of carbohydrate reserve dynamics in karoo grasses, and the secondary carbon metabolism of grass and shrub forms. It will be crucial to develop this area, as it is almost certain that artificially high intensities of herbivory, typical of current management strategies, will continue to alter community structure in karoo vegetation.

6.9.   Acknowledgements

G. F. M. acknowledges Professor D. J. Von Willert and Dr W. D. Herppich for their encouragement and inspiration during a sabbatical period spent at the Institute of Applied Botany, Münster, Dr W. D. Stock for sharing his insights regarding plant form and function, and the ongoing and generous support of the National Botanical Institute. We thank Norman Pammenter and Norbert Jürgens for helpful comments on a draft of the manuscript.

7   Functional aspects of short-lived plants

M. W. van Rooyen

7.1.   Introduction

Life in hot deserts is challenging for plants as they face severe physiological stress from drought and heat. Even during the rainy season, the availability of moisture is unpredictable in timing, amount and space (Mott and Chouard, 1979). There are two main survival strategies of plants growing under these conditions; firstly drought tolerance, which is usually exhibited by perennial plant species, and secondly, drought avoidance, a strategy common in short-lived species. In this chapter, short-lived plants include all those that complete their entire life cycle within one year and whose shoot and root systems die after seed production. To contrast these species with perennials (Midgley and Van Der Heyden, this volume), the term annual will be used in this chapter although most of the species complete their life cycles within a much shorter period.

Annual plants in hot deserts are frequently considered as paragons of adaptation (Mulroy and Rundel, 1977; Fox, 1992). The view that adaptation is important in the evolution of these plants is persuasive because these features are repeated among related taxa in deserts around the world (Gutterman, 1982). Moreover, the annual habit is thought to be a derived trait in most angiosperm taxa (Johnson, 1968; Axelrod, 1979).

Annuals constitute a large percentage of the flora of hot deserts, and this fraction tends to increase with environmental variability (Schaffer and Gadgil, 1975; Fox, 1989). The annuals, as a percentage of the flora of a number of sites in the karoo, are compared with the normal spectrum (Mueller-Dombois and Ellenberg, 1974) as well as with other desert areas in Table 7.1. The normal spectrum represents the world-wide distribution of species among the life form classes and was compiled by Raunkiaer to permit comparisons between different vegetation types. A prominent feature of all the spectra from arid areas is that annual plants make up a large percentage of the total flora. In contrast, the percentage annual plants in mesic montane fynbos at Swartboskloof (a Mediterranean climate shrubland) is exceptionally low (Van der Merwe, 1966). In Table 7.1, Namaqualand represents a winter-rainfall region, whereas rainfall at Whitehill is fairly uniform throughout the year and Hopetown lies in the summer-rainfall region (Werger, 1986). There is a progressive decrease of annual species in the flora along this rainfall gradient.

The unpredictable nature of the rainfall in arid areas is a strong selective force shaping the life-history patterns and affecting every stage in the life cycle of annual plant species. In the section that follows, the discussion is structured around the life cycle of an annual plant from seed to seed, outlining the morphological, anatomical, ecophysiological and demographic adaptations. Although seed dispersal and persistence in the seedbank are, strictly speaking, not part of the life cycle, these stages play crucial roles in the life-history strategies of annuals and are therefore included.

The discussion in this chapter will centre largely around annuals of Namaqualand in the succulent karoo biome. This area is renowned for the spectacular display of wild flowers of annual species which transforms the normally barren landscape into a wonderland of colour for a few weeks in spring, and is unique in being the only desert in the world to have such an extravagant and diverse spring flower display (Lovegrove, 1993).

Table 7.1. *Annuals as a percentage of the flora of some parts of the karoo compared with other areas with desert (BW) or steppe (BS) climates and fynbos with a Mediterranean (Cs) climate*

Locality	Climate	No of species	Annuals (%)
South Africa			
Namaqualand (succulent karoo)[a]	BW	582	28
Whitehill (southern karoo)[b]	BS	428	23
Hopetown (Nama-karoo)[b]	BS		18
Kalahari (arid savanna)[d]	BS	397	31
Swartboskloof (fynbos)[c]	Cs	448	4
Other deserts			
Death Valley, California[d]	BW	294	42
El Golea, Sahara[d]	BW	169	56
Ooldia, Australia[e]	BW	188	35
Annuals in normal life form spectrum[d]			13

[a] Van Rooyen et al. (1990)
[b] Werger (1986)
[c] Van der Merwe (1966)
[d] Cain (1950)
[e] Mueller-Dombois and Ellenberg (1974)

7.2.   Seedbanks

Populations of annual plant species have two components; the growing plants and the dormant seeds. In order to survive in environments where there is a high probability of a crop failure, annuals need a combination of a high seed yield in years when plants survive to maturity, seed longevity in the soil and a low annual germination fraction. As the environment becomes more unpredictable, greater seed carryover from year to year and increased dormancy are expected (Cohen, 1966, 1968; Venable and Lawlor, 1980).

Seedbank studies in the succulent karoo have not investigated the dynamics of the seedbank directly, and consequently the extent to which these predictions are realized in karoo annual plant species is largely untested. Research on annual seedbanks in the karoo has focussed on the spatial and temporal variation in seed densities, and the mechanisms by which seeds enter the soil and remain dormant.

7.2.1.   Mechanisms for seed longevity
Seed longevity is associated with annual species, and for many annuals the dormant fraction of the population is numerically far greater than the actively growing population. Seed persistence in the soil is determined not only by dormancy mechanisms, but purely physical attributes, such as seed size and shape, are also important. Small, compact and smooth seeds are readily buried and have been associated with longevity (Thompson et al., 1993). In general, the winter annuals of Namaqualand have small compact seeds, or if a pappus is present it usually becomes detached easily once the seeds land on the soil (Rösch,

1977; Van Rooyen, 1978). If seeds are to persist in the seedbank, predation and decomposition also have to be counteracted. In this regard, chemical defence mechanisms play an important role in maintaining seed persistence in the soil, probably by deferring or decreasing the rate of decomposition by microbes as well as defending against herbivory (Hendry et al., 1994).

7.2.2.   Spatial and temporal distribution of seed
Seedbank studies in the karoo reveal a high degree of spatial heterogeneity. On a geographical scale, the size of seedbanks varies over an order of magnitude. At Tierberg in the Nama-karoo, the size of the germinable seedbank ranged between *c.* 100 and 400 seeds m^{-2} in a shrubby vegetation with a low density of annual plants (Esler, 1993). Soil samples collected in the Koa Valley to determine the availability of food for red larks *Certhilauda burra* indicated that the size of the seedbank on the gravel plains ranged from *c.* 100 to 1500 seeds m^{-2} and were dominated by small seeds of succulent forbs and annual grasses, whereas in dune soils seedbank densities ranged between *c.* 300 and 4000 seeds m^{-2} and were dominated by large, smooth seeds of annuals (Dean et al., 1991). The size of the germinable seedbank was probably underestimated in the latter study as the soil samples were sieved, the seeds sorted and then counted. The minute seeds of many annual species are not recovered in this way. In strandveld in the succulent karoo, seed densities sampled in six communities (A. J. de Villiers, unpublished data) ranged between *c.* 100 and 10 000 seeds m^{-2} (Fig. 7.1).

Van Rooyen and Grobbelaar (1982) collected soil samples at five different localities in the Goegap Nature Reserve (formerly Hester Malan Nature Reserve) near Springbok in two successive years. Seed densities at Goegap ranged between 5000 and 41 000 m^{-2} (Table 7.2) and were of the same order of magnitude as those reported by Went (1979) for North American deserts. Seed density was higher in the sandy plains that have deeper sandy soils, are more subject to disturbance and have a low cover of perennial species, than in the less disturbed rocky soils of the hills and ridges that have a higher perennial plant cover (Table 7.2). The seedbank of all the localities at Goegap comprised mainly small, inconspicuous annual species, and seeds of perennial species were scarce. A similarity in species composition between the standing vegetation and the seedbank could only be demonstrated for the annual component and was absent in the case of the perennial component. As in the case of the Goegap Nature Reserve, the seedbank sampled in the strandveld was composed predominantly of annual species. In contrast, the germinable seedbank at Tierberg contained very few annual species, except along drainage lines (Esler, 1993).

On a microtopographical scale, it appears that the

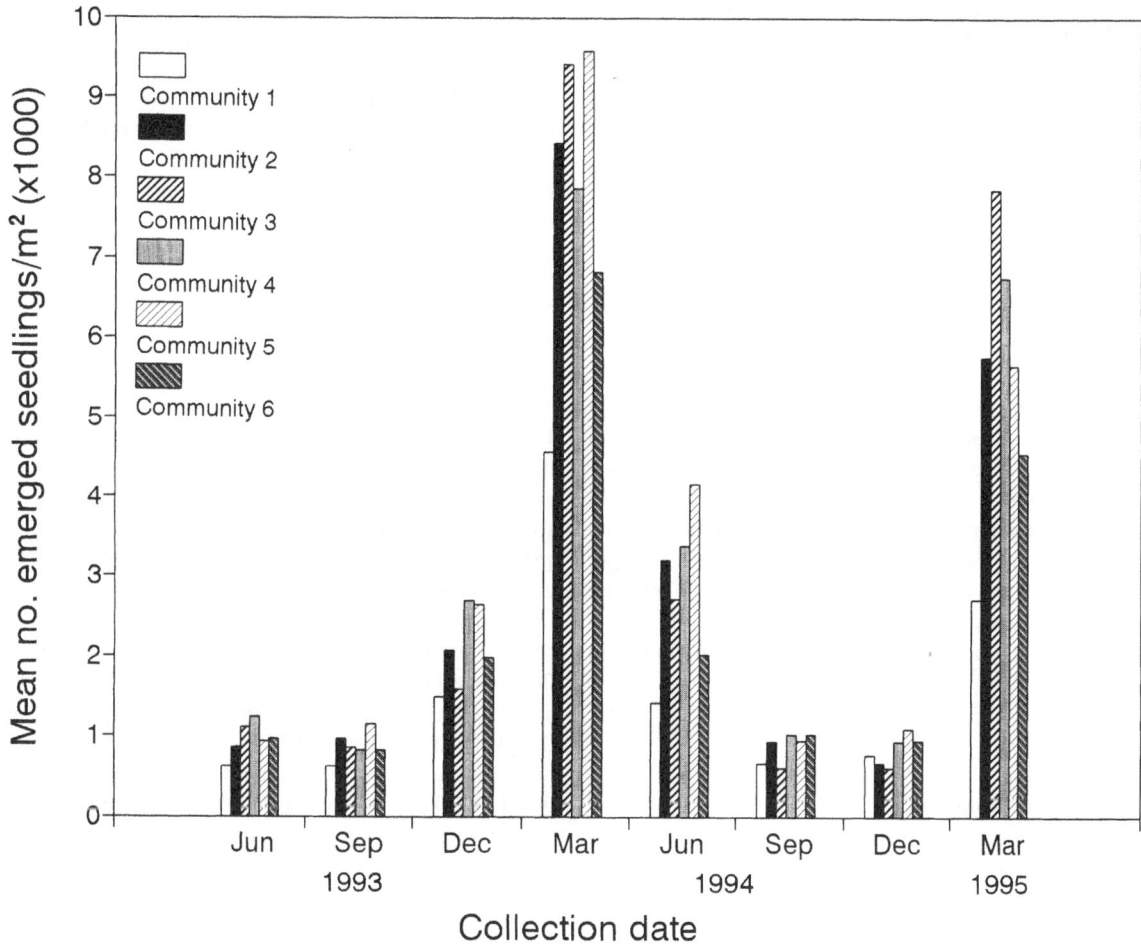

Figure 7.1 **Spatial and temporal changes in the size of the seedbank in the strandveld succulent karoo of the west coast. Community 1,** *Cladoraphis cyperoides – Odyssea paucinervis* **coastal strandveld. Community 2,** *Aloe framesi – Odyssea paucinervis* **dwarf shrub strandveld. Community 3,** *Babiana brachystachys – Odyssea paucinervis* **dwarf shrub strandveld. Community 4,** *Berkheya fruticosa – Osteospermum oppositifolium* **shrubland. Community 5,** *Ballota africana – Salvia lanceolata – Zygophyllum morgsana* **tall shrub strandveld. Community 6,** *Manulea benthamiana – Eriocephalus africanus* **tall shrub strandveld**

seeds are very patchily distributed in the soil. For example, a 197–fold variation in seed density was found for 100 cm³ soil samples from a single locality at Goegap (M. W. van Rooyen, unpublished data). Seeds in the soils are distributed mostly near the surface and there is a steep decline in seed density with soil depth (Van Rooyen and Grobbelaar, 1982; De Villiers et al., 1994a). The one locality where the layer, at 25–50 mm deep, was the richest in seeds was an old field where the cultivation of the loose sandy soil could have aided seed burial at greater depths (Van Rooyen and Grobbelaar, 1982). In addition to the spatial variability, there is great seasonal variability in the size of the seedbank, illustrated by results from the strandveld succulent karoo obtained by the emergence method (Fig. 7.1). The decrease in numbers of emerged seedlings from June to December does not, however, indicate a true reduction in

the size of the seedbank, but is due either to an endogenous rhythm or unfavourable temperatures for germination during these months. Duplicate soil samples which were kept at ambient temperatures yielded large numbers of seedlings when they were examined a few months later in March (A. J. de Villiers, unpublished data).

The size of the viable seedbank at Goegap did not differ significantly between the two consecutive sampling years. If it is taken into account that conditions prior to the second year's sampling were extremely unfavourable and that no annuals survived to a reproductive stage to contribute to the seed rain, it can be concluded that many annual species in this area have persistent seedbanks that are large in comparison with the annual seed additions and losses. Similar results were obtained by Westoby et al. (1982) in a desert region in Australia.

Table 7.2. *Size of the seedbank (m⁻²) in topsoil to a depth of 75 mm in the Goegap Nature Reserve*

Habitat	No. of viable seeds m⁻²
Sandy plains	
Old field	41 000
Plateau	23 000
Valley	13 750
Rocky soils	
Hillside	9 750
Ridge	5 000

Source: Van Rooyen and Grobbelaar (1982)

Besides functioning as reservoirs of genes and/or gene complexes and buffering populations from extinction (Kalisz, 1991), seedbanks are particularly important for annual plants as they can produce an age structure in the adult population. This change in generation time has the potential to alter population dynamics. A longer generation time slows the rate of population increase when a population is growing, but it also slows the rate of population decrease when the population is in decline.

7.3.   Germination

The success of annual species in desert environments depends largely on seeds germinating during periods when the environment is suitable for completion of the life cycle and remaining dormant when not. In one way or another, the mechanisms controlling germination of annuals are therefore adapted to harmonize the germination behaviour with the environment.

Some of these mechanisms that have evolved to reduce the risk of premature seedling mortality under unpredictable conditions are:

- After-ripening
- Polymorphism
- Selective response mechanisms to specific environments
- Endogenous rhythm.

7.3.1.   After-ripening

In many species, germination of freshly produced seeds is poor due to an after-ripening requirement. This dormancy usually wears off within six months, which is the duration of the hot and dry summer months in Namaqualand. Out of season rainfall will therefore not induce germination, but, as soon as the growing season approaches, the after-ripening requirement has been satisfied and germination

can take place. Fig. 7.2 illustrates the release of dormancy due to after-ripening in an annual species (Visser, 1993).

7.3.2.   Polymorphism

Fruit polymorphism, or heterodiaspory, the production of two or more morphologically distinct diaspore types by an individual plant has been described in many species found in unpredictable environments, such as disturbed habitats and arid environments (Zohary, 1937). The annual flora of Namaqualand is rich in polymorphic species (Rösch, 1977; Beneke, Von Teichman et al., 1992a, 1992b; Beneke et al., 1993c) and many of the species producing mass floral displays, for example *Dimorphotheca* spp., *Arctotis* spp., *Ursinia* spp. and *Leysera* spp., are polymorphic. The production of polymorphic diaspores is an adaptation improving the chances of survival of plant populations in arid environments as it enables the species to adopt two strategies when unsuitable conditions arise; an escape in space and an escape in time. Perhaps the best-known example, in Namaqualand winter annuals, occurs within the genus *Dimorphotheca* (the Namaqualand daisy), where the ray florets produce one type of achene and the disc florets another. These achenes differ not only in respect to dispersal but also in germination behaviour (Fig. 7.3). Even the plants grown from the different types of achenes have different traits (Beneke, Van Rooyen et al., 1992a, 1992b; Beneke et al., 1993a, 1993b). Disc achenes, which are dispersed widely and germinate readily, produce competitively superior plants, whereas ray achenes, which are not widely dispersed and do not germinate easily, produce plants which are competitively inferior. While disc plants are opportunistic and are responsible for the extension of the species' range, ray plants have adopted a more cautious strategy and are responsible for the maintenance of

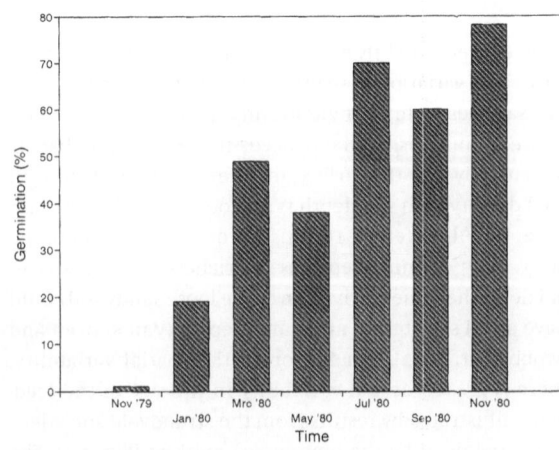

Figure 7.2 **Germination percentages of diaspores of *Heliophila pendula* at 22 °C/12 °C in the light over a period of 14 months after harvesting to indicate the release of dormancy (Visser, 1993)**

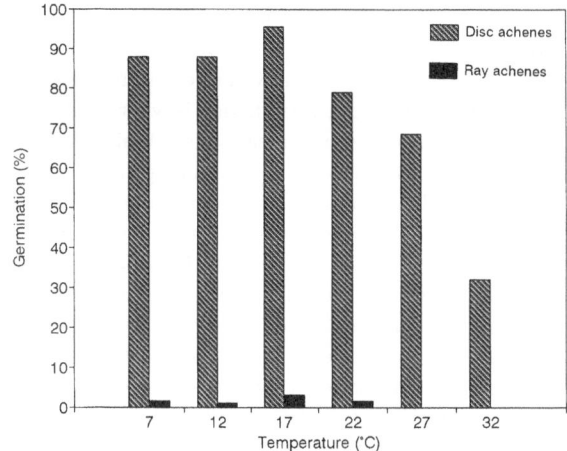

Figure 7.3 **Differences in germination behaviour of ray and disc achenes of *Dimorphotheca sinuata* at various temperatures in the light (Beneke et al., 1993c)**

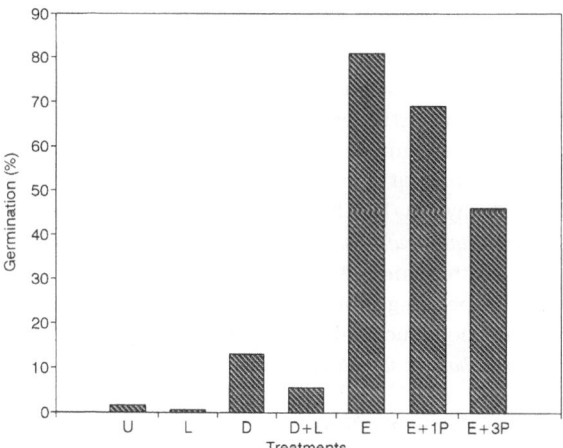

Figure 7.4 **The effect of dormancy-breaking treatments on germination of ray diaspores of *Dimorphotheca sinuata* at 17 °C after three weeks. U, untreated; L, leached; D, damaged; D + L, damaged and leached; E, excised embryos; E + 1P, embryos as well as their pericarps; and E + 3P, embryos plus three times as many pericarps than embryos (Beneke et al., 1993c)**

the species in the seedbank, thus ensuring the survival of the species under unfavourable conditions.

Several dormancy-breaking treatments were applied to ray diaspores of *D. sinuata* to increase their germination (Fig. 7.4). Diaspores that had been leached for three hours germinated poorly. Removal of one third of the pericarp, with or without subsequent leaching, improved germination slightly. Excising the embryos increased the germination to 81.5%, indicating that the pericarp offers mechanical and/or chemical impediment. When the pericarps were added to the embryos, the germination percentages

of the embryos decreased as the number of pericarps increased, possibly due to the presence of germination inhibitors in the pericarp. An anatomical study (Beneke, Von Teichman et al., 1992a, 1992b) indicated that the ray diaspores had a thicker layer of sclerenchyma in the pericarp than the disc diaspores and that tanniniferous substances were abundant in the ray diaspores.

Some of the other examples of polymorphism in the annuals of Namaqualand are less conspicuous and involve differences in the size of the pappus (*Leysera* spp.) or colour differences (*Ursinia* spp. and *Arctotis* spp.) (Rösch, 1977; Beneke, Von Teichman et al., 1992b).

Another form of staggered germination is found among synaptospermic plant species, such as *Tribulus* spp., *Grielum* spp. and *Didelta* spp. (Rösch, 1977; Van Rooyen et al., 1990). Synaptospermy is the phenomenon where two or more seeds are kept together in one compound diaspore until germination (Murbeck, 1920). Usually the seeds within a synaptospermic unit do not germinate simultaneously (Rösch, 1977) and thus the seeds produced by an individual are not all expended in a single attempt at establishment.

Germination polymorphism is also encountered among the annual Mesembryanthema. In the annual *Opophytum aquosum* the seed population consists of three types of seeds: a fast germinating, a slow germinating and a deep dormant portion (Von Willert et al., 1992). After a lag phase of more or less 10 days the fast germinating fraction germinates. After a second lag phase the slow germinating fraction germinates, quite well synchronized. However, even after this second phase of germination, a considerable portion of the seed population (the dormant fraction) remains ungerminated, forming a seedbank for future years. It has been proposed that the germination behaviour of a seed of the Mesembryanthema is determined by the relative position of the individual seed in the capsule (Gutterman, 1980/81).

7.3.3. Selective response mechanisms to specific environments

Selectivity creates response mechanisms to specific environmental signals with the result that the seed reserve is selectively utilized in specific environments and conserved in others. Temperature, light and water are the environmental factors most commonly used by the plant as indicators of environmental conditions. Light is often used as an indicator of depth of burial by seeds. The detection of soil depth is important in the case of small seeded species as these seedlings have limited resources and cannot emerge successfully if they germinate too deep in the soil. A study on the germination behaviour of 24 annual species from Namaqualand indicated that 18 of these species germinated better in the light and only 6 of the species had a higher maximum germination percentage in the dark (Visser, 1993). In a seedbank study it was also found that periodic mixing of the soil resulted in a large increase in the number of germinating seeds (Van Rooyen and Grobbelaar, 1982). The seeds germinating after mixing the soil were probably light-sensitive seeds that were brought to the soil surface where their light requirement was satisfied.

It is believed that the floristic composition of the annual vegetation in Namaqualand is determined by the interaction between temperature, on the one hand, and the timing and intensity of rainfall, on the other. Under laboratory conditions, the range of temperatures over which a species can germinate may appear too wide to ensure germination at a suitable time of the year, but, under field conditions, additional stress tends to narrow the range considerably. Temperature is probably the main factor distinguishing the germination response of annuals of the summer-rainfall karoo (Nama-karoo) and the winter-rainfall karoo (succulent karoo). The temperature response is particularly important where there are two distinct rainy seasons that differ in their average temperatures. In the transition zone between winter- and summer-rainfall regions in South Africa, Leistner (1991) distinguished three groups within the annual flora: winter annuals growing during the cooler months, summer annuals growing during the hotter months and a third group, the all-year annuals, which are able to germinate and grow throughout the year.

Biotic factors can also influence germination. It has been proposed (Inouye, 1980, 1991) that seed germination of desert plants could be inhibited by the presence of established seedlings and that this inhibitory effect increased with increased seedling density. The advantages of such a strategy would be to avoid competition and to conserve the seedbank. Van Rooyen and Grobbelaar (1982) however, demonstrated that established seedlings neither promoted nor inhibited the germination of the remaining seeds in the seedbank.

Water as a factor controlling germination is of survival value only if the mechanism is able to provide information on the environment other than the mere indication of the presence of water for hydration of the seed itself. In many deserts, studies have indicated a minimum amount of precipitation necessary to trigger germination and emergence (Juhren et al., 1956; Tevis, 1958a, 1958b; Beatley, 1974; Freas and Kemp, 1983; Gutterman, 1994). The amounts cited for other deserts range from 15 to 25 mm, to as little as 6 mm for some species. The minimum amount necessary for karoo annuals has not been determined. The effect of precipitation, however, depends not only on the total amount of precipitation, but also on the duration; a slow rain brings on more germination than a cloudburst yielding the same precipitation total.

One of the ways in which water detection is used to ensure precise timing of germination is when seeds contain water-soluble and water-leachable inhibitors. These seeds can meter the amount of water moving into the soil as well as its flow rate. Only after a rainfall of sufficient duration are the inhibitors washed out and germination can take place. Leaching was found to decrease dormancy significantly in two annual species of Namaqualand, namely *Grielum humifusum* and *Leysera tenella* (Visser, 1993).

High salt concentrations in the soil may act in a similar way to inhibit germination. De Villiers et al. (1994b, 1995a, 1995b) and Wentzel (1993) found that the germination and growth of annual species of the succulent karoo were inhibited by high salt concentrations. Heavy rains increase the osmotic potential in the upper soil layer and allow the seeds to germinate. In the saline soils of the Knersvlakte and strandveld this may be a very important mechanism of regulating germination.

7.3.4.   Endogenous rhythm

All the mechanisms mentioned above under selectivity were opportunistic, reacting directly to the environment becoming favourable and apparently not clocked in relation to solar time. In some species, however, an annual rhythm has been demonstrated, for example *Mesembryanthemum nodiflorum* from the Negev desert in Israel (Gutterman, 1980/1). An endogenous rhythm might also be operating in *M. nodiflorum* in the Knersvlakte as well as among such indigenous annual Asteraceae species as *Dimorphotheca sinuata* (Van Rensburg, 1978).

7.3.5.   Parental effects

The growing conditions of a parent plant may have a large effect on the degree of dormancy of its seeds (Gutterman, 1983, 1994; Roach and Wulff, 1987; Fenner, 1991). A study on *Dimorphotheca sinuata* and *Osteospermum hyoseroides* demonstrated that lower dormancy was associated with high temperatures during seed maturation (Visser, 1993). This decrease in dormancy was found regardless at what stage of maturation the seeds were subjected to the high temperatures (Fig. 7.5). Under field conditions this would mean that seeds maturing early in the season would have a higher degree of dormancy than those maturing late in the season. The effect of environmental conditions during seed maturation leads to differences in germination behaviour between seeds from different years, different populations within one year, between individuals within the same population and even within seeds on the same plant. Table 7.3 illustrates these differences for *Dimorphotheca sinuata* and also indicates that the germination of ray achenes is more sensitive to parental influences than disc achenes.

Table 7.3. *Comparison of the germination between different populations: different individuals in a population; and different collecting dates. Germination percentages are for* Dimorphotheca sinuata *disc and ray achenes at 17 °C in the light*

Origin	Germination	
	Disc achenes	Ray achenes
a. Different populations		
1	92	52
2	97	47
3	84	59
4	97	45
5	92	13
6	90	13
b. Different individuals (collected August)		
1	59	0
2	85	3
3	85	0
4	84	34
5	80	5
6	47	19
7	96	17
8	89	19
9	100	7
10	85	4
c. Different individuals (collected September)		
1	84	53
2	81	57
3	63	38
4	88	23
5	94	81
6	82	52
7	71	21
8	98	34
9	92	60
10	61	66

Adapted from Visser (1993)

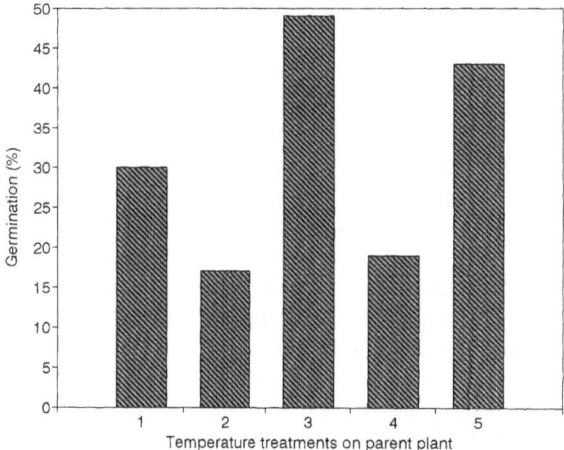

Figure 7.5 **Germination percentages of *Osteospermum hyoseroides* achenes that were subjected to different temperatures during seed maturation. 1 = motherplant kept at 25 °C/15 °C; 2 = motherplant at 25 °C/15 °C for first 16 days, then at 15 °C/8 °C; 3 = motherplant at 25 °C/15 °C for first 16 days, then at 30 °C/20 °C; 4 = motherplant at 15 °C/8 °C for first 16 days, then at 25 °C/15 °C; 5 = motherplant at 30 °C/20 °C for first 16 days, then at 25 °C/15 °C (Visser, 1993)**

7.4.   Vegetative growth

7.4.1.   Establishment

The most crucial stage in the life cycle of desert plants is establishment, during which stage mortality is high owing to the harshness of both biotic and abiotic factors. For the seeds to germinate and the seedlings to develop into mature plants in large numbers, a succession of specific environmental conditions is necessary. Owing to the erratic nature of some of these conditions, the seeds often fail to germinate. In some years many seedlings are produced, yet most perish before maturity (Rösch, 1977).

Seasonal fluctuations in density and composition of populations of desert annuals have been recorded in many arid regions (Went, 1948, 1949, 1955; Went and Westergaard, 1949; Juhren et al., 1956; Tevis, 1958a, 1958b; Beatley, 1967, 1974; Bykov, 1974; Bowers, 1987) and for the karoo (Van Rooyen, 1988; Dean and Milton, 1991a; Yeaton et al., 1993). Van Rooyen et al. (1979a, 1979b) observed the changes in composition in five permanent sampling plots of 1 m² from April to September 1974 in the Goegap Nature Reserve. Maximum seedling densities occurred in April and ranged between 115 and 1810 m⁻². Of the 1810 seedlings recorded at one site, 1558 belonged to a single species, *Cotula barbata*. These densities are not particularly high for the area, and densities in excess of 8000 m⁻² have been recorded in the same area for *Felicia australis*. The highest mortality rate was encountered at the seedling stage. Of all seedlings recorded in the five plots, 63% reached maturity, the range between plots being 48 to 74%. From studies in the Mojave Desert, Went (1955) concluded that desert annuals, once germinated, usually reach maturity, if only as depauperate and barely reproductive individuals. Figures provided by other researchers (Juhren et al., 1956; Tevis, 1958a; Beatley, 1967), however, indicate that it rarely happens that more than 50% of the seedlings survive.

7.4.2.   Growth rate

In order to exploit an environment which is only intermittently favourable for plant growth, most annuals have the capacity for high rates of dry-matter production. Relative growth rate values recorded for three winter annual species, grown under optimal conditions, were relatively high when compared with crop species (Van Rooyen et al., 1992a). Rapid establishment and growth are advantageous in an environment in which the period between rainfall events is unpredictable (Bell et al., 1979). Roots and leaves have the highest growth rates in the very young seedling (Fig. 7.6). Rapid root growth indicates the heavy early commitment to root growth, which enables roots to penetrate the soil and absorb moisture, whereas investment in leaf growth provides a large leaf area for photosynthesis,

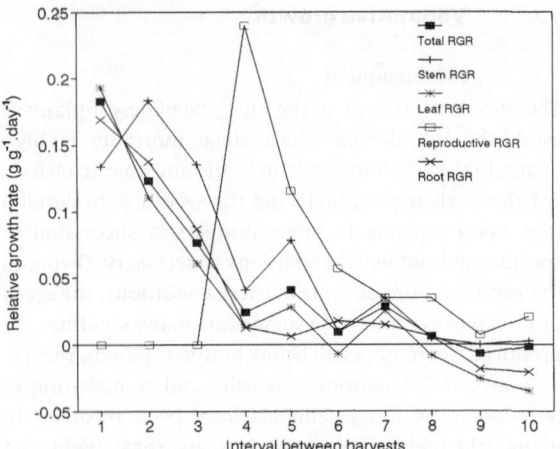

Figure 7.6 **Changes in the relative growth rate of the entire plant as well as the separate organs of** *Dimorphotheca sinuata.* **Plants were sown in early March (Van Rooyen et al., 1992a)**

which is essential for rapid development. Maximum stem growth rate occurs later in the life cycle, causing an increase in plant height, which allows for more effective pollination and dispersal. From the time that reproductive structures are initiated, their growth is more rapid than those of vegetative structures.

7.4.3.   Photosynthetic pathway
One of the attributes that has been recognized as important in linking a plant species to its environment is the type of photosynthetic pathway (Kemp, 1983). In many ways, C_4 plants may be ideally suited for an existence in warm, arid regions (DePuit, 1979; Ehleringer and Monson, 1993); their water-use efficiencies are generally greater than C_3 plants at high temperatures and high irradiances. As a result, C_4 photosynthesis is usually found among such summer annuals as *Setaria verticillata, Digitaria sanguinalis* and *Eleusine indica,* whereas winter annuals are predominantly C_3 plants. These include *Schismus barbatus* and *Bromus japonicus* (Vogel et al., 1986) and forbs of the Asteraceae, Brassicaceae and Scrophulariaceae. Although C_4-plants as a group have high photosynthetic rates, a C_3 winter annual has one of the highest photosynthetic rates ever measured (Mooney et al., 1976). Among the succulent annuals, C_3 photosynthesis is equally common among summer and winter species (e.g. *Trianthema triquetra*), whereas C_4 is found mainly among summer-growing species (e.g. *Zygophyllum simplex*) and CAM in winter-growing species (*Opophytum aquosum* and *Mesembryanthemum pellitum*) (Von Willert et al., 1992). The halophytic succulent annual *Mesembryanthemum cristallinum* belongs to an intermediate group and is able to undergo a transition from C_3 to CAM. At first it was believed that CAM was induced by salt

or drought stress. It has since been shown that even unstressed plants change to CAM as they age, but that the transition from C_3 to CAM occurs earlier in the life cycle of salt-stressed plants (Von Willert et al., 1992).

During their life cycle, winter annuals have to contend with large seasonal changes in temperature. During winter, temperatures may be extremely low and frost and even snow may occur, while temperatures towards the end of their lifespan may rise to summer levels. Studies have indicated that desert winter annuals have a larger capacity for temperature acclimation of photosynthesis in response to increased leaf temperature than do summer annuals (Seemann et al., 1986).

7.5.   Reproductive growth

7.5.1.   Timing of reproduction
The effects of photoperiod, temperature and moisture on development and flowering were investigated in three annual species of Namaqualand (Van Rooyen et al., 1991). Photoperiod did not have an effect on the time of flower initiation in any of the study species (Table 7.4). Similarly, no obligate photoperiodic requirements could be demonstrated for winter annuals from the Negev Desert (Evenari and Gutterman, 1966, 1985) or Australia (Mott and McComb, 1975a).

In all three species examined from the succulent karoo, flower initiation was promoted by lower temperatures (Table 7.4), although the time between flower initiation and anthesis was lengthened by low temperatures. Increasing the length of the vernalization period reduced the age at which primordia appeared, although no obligate vernalization requirement could be established. In general, vernalized plants were smaller, but allocated a larger percentage of their mass to reproduction. The longer the vernalization period the shorter the lifespan of the plants. An obligate vernalization requirement has been claimed for several winter annuals of the Sonoran Desert (Went, 1955, Beatley, 1967, 1974) although no such requirement could be demonstrated among the annual species from the Negev (Evenari and Gutterman, 1966) or Australian deserts (Mott and McComb, 1975a).

Although annuals are often classified as drought-avoiding, most of the annual species have a lifespan of several months, and over that period of time they are unable to avoid moisture stress completely. Plants growing under unfavourable moisture conditions are usually smaller than those growing under favourable conditions, and this led several authors, (Went, 1948, 1955; Solbrig et al., 1977; Rathcke and Lacey, 1985) to conclude that moisture stress promoted early flowering to increase the likelihood of

Table 7.4. *Number of weeks at which visible primordia appeared in different photoperiod and temperature treatments for three ephemeral species (mean of five replicates)*

Temperature & photoperiod	Species		
	Dimorphotheca sinuata	Ursinia calendulifolia	Heliophila pendulata
12° LD	8	9	9
12° SD	8	9	9
17° LD	9	12	9
17° SD	9	12	9
22° LD	11		9
22° SD	11		9
27° LD		13	13
27∞ SD		13	14

LD = Long days
SD = Short days
Source: Van Rooyen et al. (1991).

satisfactory seed set. Fox (1990b) however, points out that no convincing experimental evidence has yet been presented to show that drought does indeed stimulate flowering in any annual species. Both Fox (1990b) and Aronson et al. (1992) have argued against drought as a reliable signal for such a crucial phenological transition as onset of flowering in desert annuals. A drought-induced *delay* in time to flower initiation or anthesis has been reported for annuals from other deserts (Mott and McComb, 1975b; Fox, 1990a, 1990b) as well as for some winter annuals of the succulent karoo (Van Rooyen et al., 1991; Steyn et al., 1996b). In a detailed study on five annual species from Namaqualand, it was demonstrated that within a single species the reaction of the plant to water stress depended on sowing time (Steyn et al., 1996b). Water stress apparently interacted with other environmental cues (e.g. temperature and photoperiod) in determining the onset of anthesis. Plants from the early sowings (autumn), that have a potentially long growing season, delayed flowering under water stress apparently gambling on more water becoming available. In contrast, plants of the late sowings (mid-winter), which have a short growing season at their disposal, accelerated flowering under water stress apparently to increase the likelihood of successful reproduction before summer drought sets in.

The facultative nature of both photo- and thermo-induction of flowering reported for annuals of the succulent karoo has important advantages for their survival as it provides them with a flexibility which enables them to flower over a wide seasonal range in photoperiod and temperature, provided sufficient moisture is available. In an environment in which the start and duration of the growing season is unpredictable, this flexibility increases the likelihood of successful reproduction.

The reaction of winter annuals to temperature has

made it possible to develop a model to predict peak flowering time of Namaqualand annual plants (Steyn et al., 1996a, 1996c). The model is based on the thermal unit concept which stipulates that each species requires a specific amount of thermal units to reach a particular phenological stage. In short, this involves determining a cut-off temperature for each species experimentally. Cold units below this cut-off temperature are summed from germination to flower initiation, and from flower initiation to anthesis or peak flowering; heat units are summed, using the Growing Degree Days formula.

7.5.2. Phenotypic plasticity

Phenotypic plasticity in growth may be a critical life-history attribute allowing the persistence of annuals in a temporally variable environment. Since favourable conditions for seed germination in Namaqualand can occur over a long period, the growth period can last 6–8 months in a year when the rainy season starts early in autumn, whereas the growth period may last only 2–3 months when the rainy season starts in midwinter. This variability in duration of the growing season has profound effects on the allocation of energy to different organs. In years of below-average rainfall, plants can attain only a relatively small size, and would therefore produce few seeds if they did not have the ability to allocate proportionately more photosynthate to seeds when growth is limited.

As an example of the plasticity in growth form and biomass allocation, *Dimorphotheca sinuata* plants sown in autumn (early March) are compared to plants sown midwinter (June) (Fig. 7.7). Early commencement of the growing season allows the plants to grow for a few weeks at mild autumn temperatures and under relatively long day conditions. Flower initiation is delayed at high temperatures and only occurs after temperatures have dropped in winter (Van Rooyen et al., 1991). The plants, therefore, accumulate a large biomass early in the season, allocating energy to vegetative biomass and restricting allocation to reproduction (Van Rooyen et al., 1992a, 1992b). This strategy lengthens the potential lifespan of plants and increases the number of sites available for inflorescences to be produced. Such a commitment of energy increases the potential reproductive output if further precipitation occurs and there is enough time to mature the seeds. However, if no further rain falls, parts or even the whole plant could die without producing many seeds. It appears that precipitation events in Namaqualand are regular enough (Hoffman and Cowling, 1987) so that the investment in vegetative biomass toward lengthening the lifespan increases the long-term reproductive output.

When the growing season starts in midwinter, the relative growth rate of the young seedlings is slow due to low

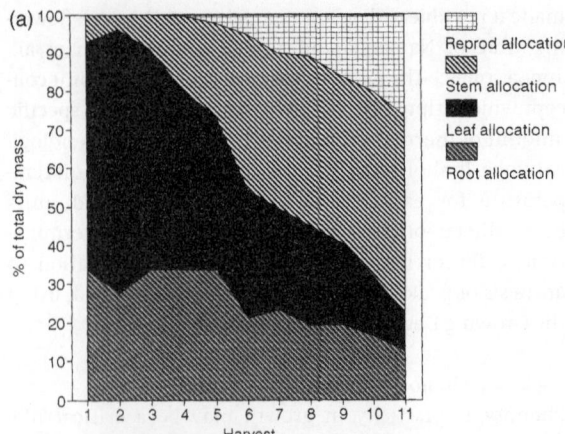

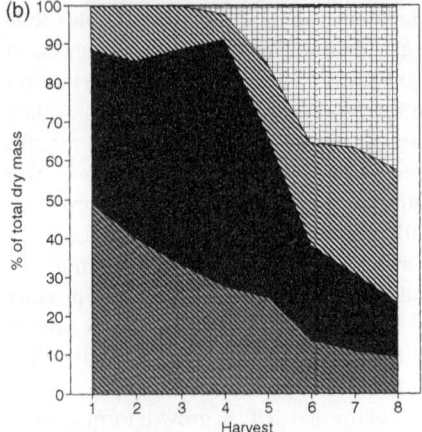

Figure 7.7 **Changes in biomass allocation of *Dimorphotheca sinuata* sown in (a) early March and (b) June (Van Rooyen et al., 1992b)**

Table 7.5. ***Dispersal spectrum (%) of the Goegap Nature Reserve***

Dispersal mechanism	Annuals (% of annual flora)	All life forms (% of total flora)
Telechoric		
Anemochory	83.0	66.3
Hydrochory	6.7	25.8
Autochory	7.9	19.6
Zoochory	15.2	14.8
Atelechoric	18.5	21.6
Antitelechoric		
Myxospermy	20.6	11.3
Trypanospermy	4.8	5.7
Synaptospermy	13.6	7.6
Heterodiaspory	8.5	3.6
Hygrochasy	1.8	9.3

temperatures and short day conditions. Low temperatures stimulate flower initiation, committing resources to reproduction at an early stage. This increases the probability of seedset. Although reproductive output of plants that develop from autumn-germinating seed exceeds that of winter-germinating plants, the reproductive allocation of the latter plants is greater (Fig. 7.7) (Van Rooyen et al., 1992a, 1992b).

7.6.   Seed dispersal

The telechoric or so-called long-range dispersal mechanisms of the different life-form classes for the Goegap Nature Reserve are given in Table 7.5 (Van Rooyen et al., 1990). Compared to the dispersal spectrum of all life forms, annuals have a higher percentage of species that are anemochorous (wind dispersed), while the percentage

of hydrochorous (water dispersed) and autochorous (self dispersed) species is less.

So-called antitelechoric mechanisms, or mechanisms restricting dispersal (Ellner and Shmida, 1981; Zohary, 1937; Van Rooyen et al., 1990) are well developed among the annuals at Goegap (Table 7.5). Compared with other life forms, the annuals are particularly rich in myxospermic and heterodiasporic species. Myxospermy refers to seeds or fruits that form a superficial layer of mucilage upon moistening (Murbeck, 1919). This mucilage increases the seed-soil contact and causes the seeds to adhere to the soil as soon as the mucilage dries out. Most of the myxospermic species belong to the Brassicaceae (*Heliophila* spp.) or Asteraceae (*Senecio* spp. and *Cotula* spp.). A heterodiasporic (or polymorphic, see section 7.3.2) plant produces different kinds of seeds, mericarps or fruits, the one type generally being telechorous whereas the other is atelechorous or antitelechorous. Synaptospermy, the formation of diaspores containing more than one fruit, is almost always associated with arid areas (Murbeck, 1920; Zohary, 1937) and at Goegap synaptospermic species were mostly annuals. The diaspores of trypanospermic species are equipped with a sharply pointed hygroscopic drilling apparatus or with a sharply pointed tip which enables them to penetrate the soil. Trypanospermy is typical among the Geraniaceae (e.g. *Pelargonium* spp. and *Erodium* spp.) and the Poaceae. Hygrochastic plants possess fruits that open when moistened to allow the seeds to escape and close again during dry conditions, so that dispersal is generally restricted in time. In hygrochastic plants, dispersal can only occur when water is available and conditions for germination should theoretically be suitable. In an area where moisture is a limiting factor, linking dispersal with water availability should be advantageous. Almost all the Mesembryanthema have hygrochastic capsules (Ihlenfeldt, 1971; Van Rooyen et al., 1980; Esler, 1993) but there are few annual Mesembryanthema. Atelechory, the absence of features specifically facilitating or restricting telechory, is also especially well represented among the annuals.

The generally accepted explanation for atelechory and antitelechory is that they are adaptive responses to the particular high mortality of dispersed seeds in deserts, and that they have evolved as mechanisms to reclaim the mother site (Zohary, 1937; Stopp, 1958; Van Der Pijl, 1982). Ellner and Shmida (1981) argue that atelechory in deserts is an adaptive response to the low benefit of long-range dispersal mechanisms in deserts rather than to any benefit arising from limited dispersal *per se*. They believe that antitelechory is disadvantageous and regard it as a side-effect of characters whose adaptive value is not directly related to dispersal. Antitelechory indicates the importance of adaptations in arid environments which spread the time of germination, provide suitable conditions for germination and maintain a reservoir of viable seeds.

Recent models shed new light on the advantages of reclaiming the mother site. Short-range dispersal leads to seed clumping, which facilitates coexistence (Green, 1989) due to spatial segregation leading to an increase in intraspecific competition relative to interspecific competition. Short-range dispersal allows some seeds to leave the overcrowded mother site, but to still remain in a neighbourhood where the probability of site suitability may be higher than random. When a seed establishes near its mother, the maternal site will be refilled the next year and available space will be filled concentrically and solidly around the population founders (Lavorel et al., 1994).

7.7.   Biotic interactions

As seen in the previous section, annuals provide many examples of unique adaptations to abiotic environmental stresses. Annual communities are, however, also strongly influenced by such biotic interactions as predation, disturbance and competition. These factors play significant roles in determining the species composition and abundance of annual communities. The relative importance of biotic interactions in the karoo varies considerably, both spatially and temporally (Van Rooyen, 1988; Dean and Milton, 1991a; Yeaton et al., 1993). For instance, it may be intuitively suggested that seed-eating birds that forage in flocks (Maclean, 1996) may have a large impact on seed consumption in one year and little or no effect in the next year.

7.7.1.   Disturbance
Annual species are excluded from habitats in the karoo where there is a dense cover of perennial species, and

occur mainly in sites where disturbance and physical stress inhibits the formation of a dense community of perennials (Dean and Milton, 1991a, 1995; Milton and Dean, 1991, 1992; Yeaton et al., 1993). Disturbances that reduce the perennial cover and provide opportunities for annuals include ploughing (Dean and Milton, 1995), grazing (Milton and Dean, 1992; Steinschen et al., 1996), burrowing animals (Dean and Milton, 1991a) and drought (Milton et al., 1995). Disturbances at all spatio-temporal scales are key elements of annual community dynamics. Small-scale disturbances created by burrowing rodents, porcupines, aardvarks or foxes or through antelope activity, provide patches that are favourable for colonization by annual plants. These patches often have increased nutrient status through deposition of faecal pellets and appear to be maintained by the activities of antelope and burrowing animals (Dean and Milton, 1991a).

Namaqualand has gained world-wide fame due to the spectacular floral display of the annuals during spring after a good rainy season. The annuals are especially prominent on old fields and otherwise disturbed rangeland (Van Rooyen, 1988; Smuts and Bond, 1995; Dean and Milton, 1995). As a result it is often contended that annuals occur only in abandoned fields and in overgrazed rangeland. This would imply that the spring aspect would disappear if rangeland condition could be improved. If this statement were true, then tourism and conservation would make conflicting demands on the management of the vegetation.

A comparison was made between (a) the number of species, (b) the number of individuals per species, and (c) the above-ground biomass in disturbed versus undisturbed vegetation at Goegap Nature Reserve (Table 7.6, M. W. van Rooyen, unpublished data). Although annuals were far more conspicuous in disturbed vegetation, no significant difference was found in the above-ground biomass between the two sites. Contrary to what was expected, the number of individuals m^{-2} was significantly higher in the undisturbed vegetation. Species-richness in the undisturbed vegetation exceeded that in the disturbed vegetation at both Goegap (M. W. Van Rooyen, unpublished data) and near Vanrhynsdorp (Steinschen et al., 1996).

Plant surveys on the Skilpad Wildflower Reserve in several old fields differing in the time since the last disturbance indicated that total plant species-richness steadily increased the longer the field had not been cultivated, although annual species-richness was not affected (Theron et al., 1993). A comparison between the invertebrate assemblages on old fields in the southern karoo (Dean and Milton, 1995) showed that invertebrate diversity was positively correlated to perennial plant diversity but negatively with annual plant diversity. The paucity of invertebrates on degraded rangelands, where the peren-

Table 7.6. **Comparison between disturbed and undisturbed vegetation in terms of dry mass per square metre, giving the density of individuals per square metre and number of species**

Species	Dry mass (g)		Number of individuals	
	Undisturbed	Disturbed	Undisturbed	Disturbed
Amellus strigosus	0.077	0.028	1.6	0.6
Atriplex lindleyi	0.002	0.297	0.2	6.8
Dimorphotheca sinuata	0.805	0.197	2.4	1.0
Foveolina albida	1.370	0.601	9.6	3.8
Galenia sarcophylla	1.602	0.005	1.2	0.2
Gazania lichtensteinii	0.059	0.912	1.2	2.8
G. tenuifolia	0.039		0.4	
Grielum humifusum	2.685	2.629	0.2	0.4
Heliophila variabilis	0.247	0.528	3.4	5.0
Ifloga paronychioides	0.041		2.0	
Karoochloa schismoides	0.329	0.067	7.2	1.0
Lasiospermum brachyglossum	0.023		0.2	
Lessertia diffusa	0.214	0.570	3.6	6.0
Leysera tenella	0.078	0.026	1.4	0.8
Lotonis brachyloba	0.003		0.2	
Mesembryanthemum guerichianum	0.009		0.2	
Oncosiphon grandiflora	4.424	0.332	22.0	0.8
Osteospermum pinnatum	33.984	46.189	89.6	28.2
Pentaschistis airoides	0.002		0.2	
Phyllopodium collinum	0.003		0.2	
Plantago cafra	0.05	0.105	0.4	1.6
Senecio arenarius		0.069		0.4
S. cardaminifolius		0.055		0.2
Zaluzianskya benthamiana	0.060		1.0	
Z. gilioides	0.022		0.6	
Z. villosa	0.073		1.4	
Total	46.196	52.610	150.4	59.6
Number of species	24	16		

nial cover had been replaced by annuals, was also noted by Milton and Dean (1992) and Smuts and Bond (1995).

Preliminary results to determine the effect of the type and frequency of disturbance on the density and species diversity of annual vegetation at Skilpad indicate that 'weedy' alien species, for example *Amsinckia menziesii, Fumaria muralis, Erodium* spp. and *Pelargonium* spp. increase in the season that a plot is tilled. Some species showed a marked response to either tilling or grazing. For example *Ursinia cakilefolia* was reduced in the season immediately following tilling, whereas *Arctotheca calendula* and *Amsinckia menziesii* were favoured by tilling. Grazing had a negative effect on the abundance of *Cotula barbata*, while *Dischisma spicatum* seemed to be favoured by grazing (Fig. 7.8). Athough mass flower displays are created by disturbances, the annual displays are at the same time threatened by grazing and, under certain conditions, the showy annual species may be outcompeted by invasive annual grasses (Steinschen et al., 1996).

7.7.2.   Competition

Plant populations respond to density through either plasticity or mortality or both (Harper, 1977). Results obtained from experimental field plots with two winter annual species (*Dimorphotheca sinuata* and *Senecio arenarius*) demonstrated an asymptotic relationship between density and above-ground plant yield (Van Rooyen et al., 1992c; Oosthuizen, 1994). Thus the total yield per unit area will increase with increasing density until a level is reached where yield remains fairly constant with further increases in density. The limit to this plateau yield will be set by the resource supplying power of the environment.

Competition, whether intraspecific or interspecific, reduces the reproductive output per plant (Cunliffe et al., 1990; Yeaton et al., 1993; Oosthuizen et al., 1996a, 1996b) although patterns of reproductive allocation do not respond uniformly to varying levels of intraspecific competition in different species. In the nine species examined, reproductive allocation was unchanged in five species; it was increased by increasing density in one species; and decreased in three species (Van Rooyen et al., 1992c). At the range of densities examined in these experimental plots (up to 1200 individuals m^{-2}), self-thinning did not have an important effect. It is not known whether the density-independent mortality observed in monocultures may become density-dependent in the presence of a second species.

With increasing density, the skewing of the frequency distribution of mass classes increased (Van Rooyen et al., 1992c; Oosthuizen, 1994). The mass of an individual in a population is probably determined by the timing of its emergence relative to the rest of the population. Once a size difference has been established between two individuals in a population it will be progressively exaggerated, especially under density stress. In species exhibiting seed polymorphism, differences in germination rate between the different forms would lead to the establishment of a size hierarchy in the population. If conditions in Namaqualand are favourable for consecutive germination events, those seedlings which establish first will capture a disproportionate amount of resources and deprive the seedlings which germinate later of their share (Van Rooyen et al., 1992c). The effect of the seedling that germinates first does not seem to be strongly species-specific (Oosthuizen, 1994).

A competitive hierarchy for 15 Namaqualand pioneer plant species was established by using the average phytometer mass (*Dimorphotheca sinuata*) when grown with a conspecific and in combination with fourteen other species (Rösch, 1996). The competitive effect hierarchy (Fig. 7.9) indicated which species were strong competitors (causing a low phytometer mass) and which species were weak competitors (causing a high phytometer mass). Opinions differ as to what particular plant traits confer a competitive advantage to a species. Regression of various plant traits (measured on plants grown singly) against

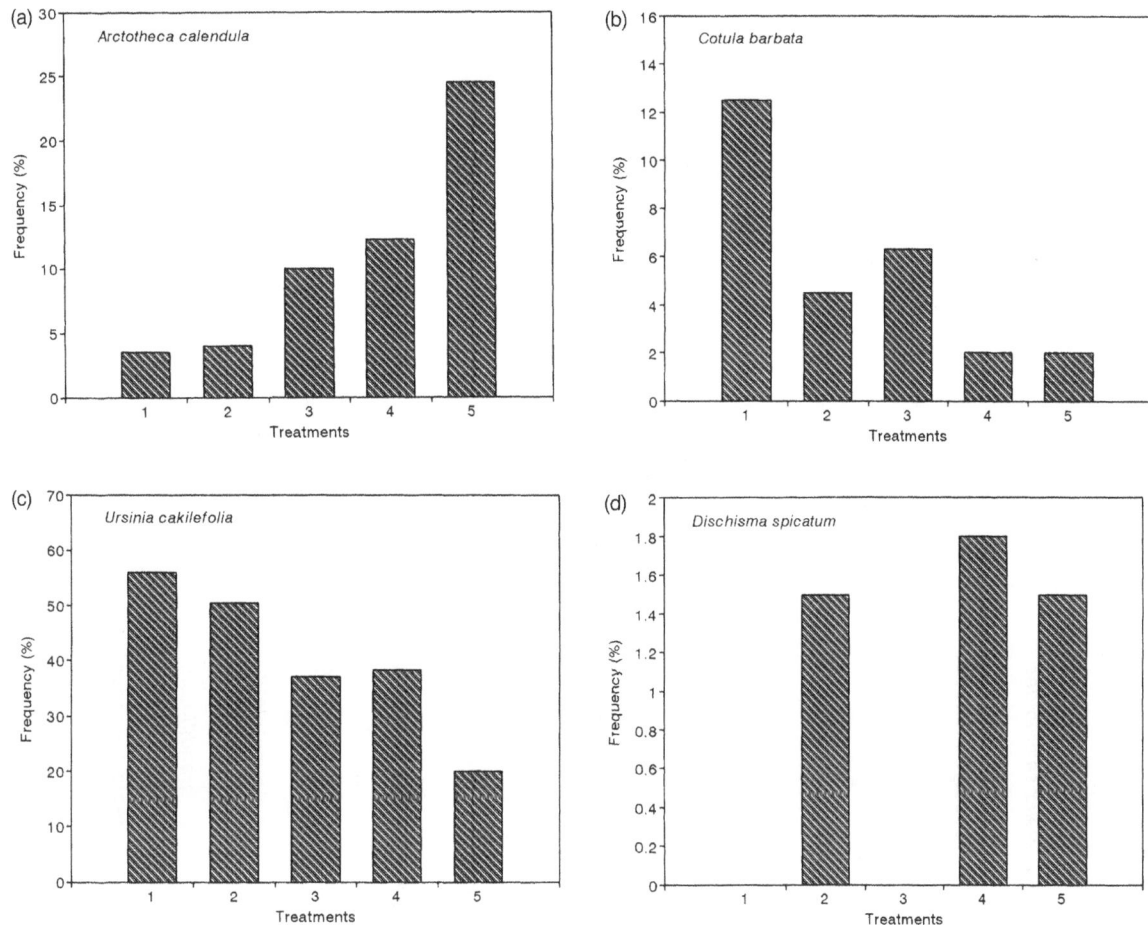

Figure 7.8 **Frequency of (a) *Arctotheca calendula*; (b) *Cotula barbata*; (c) *Ursinia cakilefolia*; and (d) *Dischisma spicatum* in various treatments. Treatment 1 = control; Treatment 2 = grazed; Treatment 3 = tilled; Treatment 4 = grazed and tilled; and Treatment 5 = grazed, tilled and sown**

phytometer biomass indicated which traits were significantly correlated. These traits, most of them being size related, were: maximum shoot mass ($r = -0.81$), total mass ($r = -0.80$), stem mass ($r = -0.74$), reproductive mass ($r = -0.68$), leaf area ($r = -0.66$), stem allocation ($r = -0.61$), specific leaf area ($r = -0.60$), vegetative height × diameter ($r = -0.56$), leaf area ratio ($r = -0.53$) and average number of days to flower initiation ($r = 0.52$). Maximum relative growth rate, which is usually considered a good predictor of competitive ability, was not significantly correlated with competitive effect ability. A forward stepwise multiple regression of the significant traits was used to determine an equation to predict competitive ability ($r^2 = 82.77$).

Only a single phytometer species was used in the above-mentioned study, so, to generalize about the competitive effect hierarchy that was obtained, it has to be assumed that neighbour hierarchies are consistent among phytometer species. This was indeed found in a subsequent study (Rösch, 1996) with four target species chosen over the whole spectrum of the hierarchy. It also has to be shown that the hierarchy obtained does not change over environments if generalizations are to be made on the validity of the hierarchy. A subsequent study of the competitive effect and response of 10 of these species at two nutrient levels (Rösch, 1996) demonstrated that the competitive effect hierarchy of these arid land species was concordant over nutrient levels. The status of the species was, therefore, not affected by nutrient level and the strong competitors at a high nutrient level were also strong competitors at a low nutrient level (Rösch, 1996).

On old kraals, it was found that dense stands of *Mesembryanthemum guerichianum* could prevent the recovery of the post-disturbance vegetation possibly through allelopathic effects on the growth of other annual species (Wentzel, 1993). *Mesembryanthemum guerichianum* (in common with other *Mesembryanthemum* spp.) is able to take up

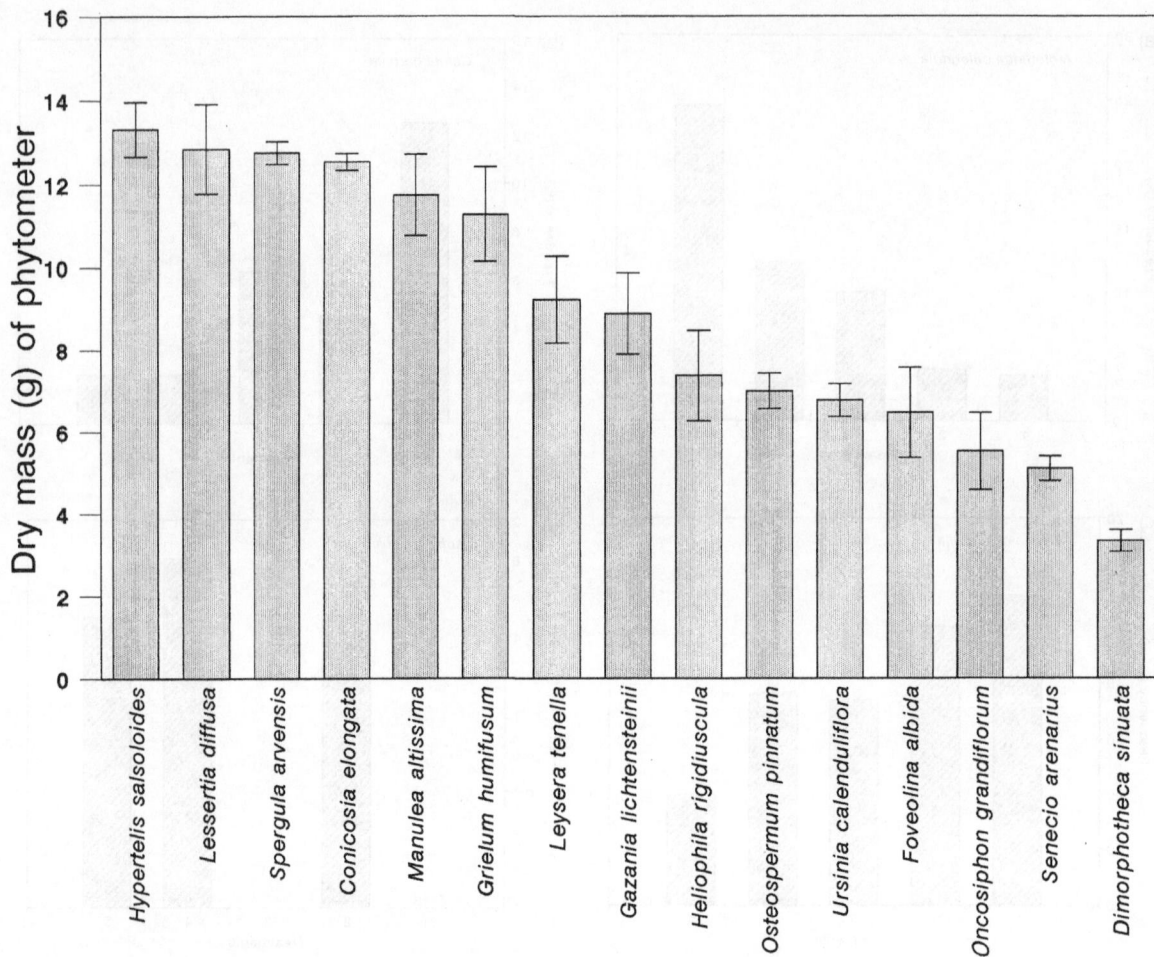

Figure 7.9 **Hierarchy of competitive effect of 15 Namaqualand pioneer plant species (Rösch, 1996)**

excess sodium and chloride throughout its lifespan and to accumulate them in its leaves (De Villiers et al., 1995a). When the plant dies, this salt is released into the soil again leading to salinization of the soil beneath the dead plants. The salt produces a detrimental osmotic environment preventing growth of non-tolerant species such as *Ursinia calenduliflora* and *Dimorphotheca sinuata* (Wentzel, 1993).

7.7.3.   Functional classification into guilds
Using 18 attributes of 30 pioneer species in an ordination (PCA), annual species could be clearly distinguished from perennial and facultative perennial species (Fig. 7.10(a), Rösch, 1996). Life-history strategies of perennial and facultative perennial species differ from those of the annual species in that their seedlings generally take longer to emerge. This delay in germination could be as a result of a more cautious germination strategy adopted by the perennials, whereas the annuals adopt a more opportunistic

germination strategy. In their first growing season, the perennials/facultative perennials are distinguished from the annuals by their low total dry mass values and in particular low stem mass values and consequently low stem allocation percentages. Most of the perennial and facultative perennial species flower later in the season, which could possibly be a mechanism to avoid pollinator competition.

An ordination, using only the annual species, produced three guilds (Fig. 7.10(b)). The species in the first group generally take between 3 and 6 days to emerge after sowing and are all large plants, with large stem masses, high stem allocations, but relatively small reproductive

Figure 7.10 (right) **(a) Distribution of 30 pioneer plant species from Namaqualand in a PCA ordination of 18 different plant attributes (Rösch, 1996). (b) Distribution of 23 annual plant species from Namaqualand in a PCA ordination of 18 different plant attributes (Rösch, 1996)**

(a)

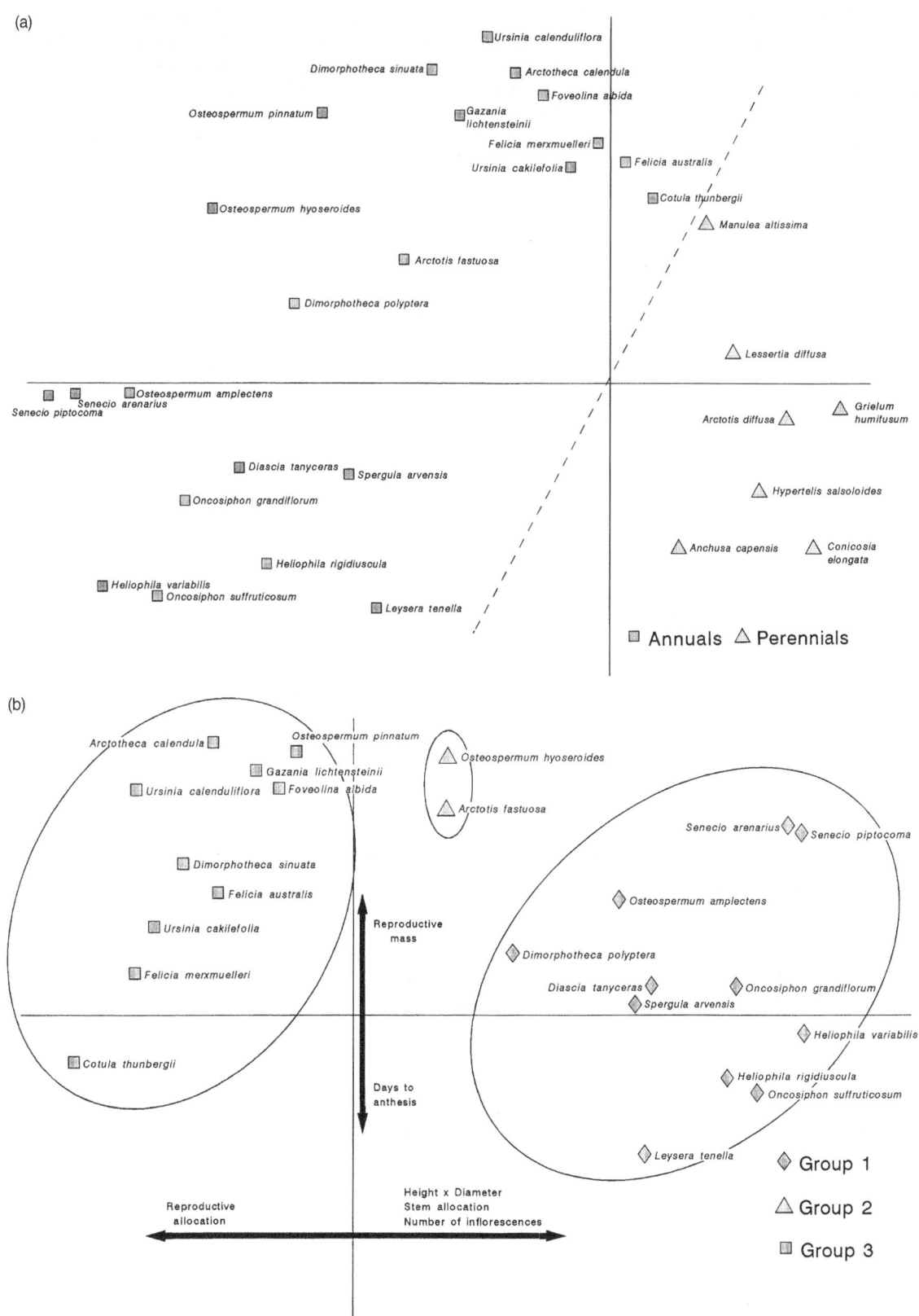

masses and reproductive allocations. Species in the second group are of intermediate size and, although they take relatively long to emerge (on average 6 to 7 days), they are among the first to reach anthesis. Species in the third group are small, with small stem allocations, but reproductive allocation of these species is high. Although the vertical stratification in the annual community is not as pronounced as in other ecosystems, for example forests, these three groups form different vertical layers. In the field, many annual species coexist not only in different vertical layers but also in the same layer. It is possible that vertical stratification brings about stability of the community as a whole through asymmetric competition between species in different layers (Hara, 1993). On the other hand, the species in the same vertical layer undergo symmetric competition (Oosthuizen et al., 1996a) and hence species diversity is realized in each layer (Hara, 1993).

7.8.   Conclusions

The temporal variability of rainfall in the karoo environment has been a strong selective force shaping the demographic characteristics of individual annual species in terms of their life-history patterns. The life-history pattern of an annual affects its population dynamics and determines its competitive success and, thus, its continued coexistence in its environment.

A large gap in our knowledge concerns the interactions between annuals and other groups of organisms. Perennials have significant direct and indirect effects on the distribution and density of annuals, for example by way of competition for water, space and nutrients. Conversely, dense stands of annuals prevent the establishment of perennial species. Processes governing the interactions and feedbacks between annuals, perennials and animal groups such as invertebrates and rodents have not been investigated in depth.

At this stage, our knowledge of the functioning of the annual component of the vegetation is not sufficient to allow the effective management and conservation of this natural asset. There is a need for relatively long-term manipulative studies to uncover the way in which biotic interactions shape the biology of annual plant communities.

7.9.   Acknowledgements

The support of the Department of Environmental Affairs and Tourism, the Foundation for Research Development and the University of Pretoria is gratefully acknowledged. I thank Guillaume Theron, Annelise le Roux and Noel van Rooyen for sharing their insight of the Namaqualand system.

8   Plant reproductive ecology

K. J. Esler

8.1.   Introduction

The turnover of plant populations is not well understood for the karoo, but is critical in the understanding of the relative importance of the regeneration phase. For annuals and perhaps many of the short-lived leaf succulent shrubs in the succulent karoo, germination and recruitment are the critical life-cycle stages. The factors that determine and influence successful reproduction therefore play a major role in the overall dynamics of these species and are an important consideration in the development of a predictive understanding of plant community structure and function. The longer-lived, deeper-rooted evergreen and deciduous shrubs associated mainly with the Nama-karoo are persisters in the system and, although they do produce seeds and occasional seedlings, the selective pressures to avoid death of the modular organism/bud bank are likely to be as or more critical. Seed biology is an important research priority for species that are not strong persisters, while understanding growth and regeneration phenology is critical for those species that do persist (Table 8.1).

The stages in the process of reproduction of karoo species are dealt with here in a chronological sequence, from flowering phenology and pollination through to seed banks and seedling recruitment. All of these stages are subject to different selective pressures which ultimately influence regeneration. Obstacles are posed by the abiotic (environmental stresses) and biotic (competition, predation, disease) environments. The consequent trade-offs in allocation (e.g. more resources allocated to growing = less resources allocated for seed production) due to these obstacles result in sets of characteristics exhibited by plants that maximize the chance of successful reproduction. It is these characteristics that differentiate the reproductive strategies and life-histories of plants.

Throughout the chapter, the underlying theme is the comparison between guilds of plants with different life-histories. The succulent and Nama-karoo biomes are distinguished by distinct climates and species guilds, so there is a natural comparison between these biomes (see Desmet and Cowling, this volume; Cowling and Hilton-Taylor, this volume). The definition of seeds in this chapter includes all seeds and certain fruits.

Table 8.1. *Research questions differ depending on the life-history strategies of the species under study. Some species are virtually 'immortal' (in ecological time) because they resprout readily. For these species, traditional reproductive biology questions are irrelevant. An understanding of poor persisters, in contrast, requires an understanding of plant reproductive biology*

Characteristics	Good persisters	Poor persisters
	Herbaceous resprouters	**Weak sprouters**
	Poor at making seedlings	**Good at making seedlings**
Example	*Portulacaria afra*	most Mesembryanthema
Directed research questions	1. Are viable BUD banks present?	1. Are SEED banks present?
	2. How does sprouting ability change with plant age?	2. How do seedbanks accumulate over time?
	3. What causes buds to sprout?	3. What influences seed germination?
	4. How does disturbance influence bud survival?	4. How does disturbance influence seedling survival?
	5. What determines the survival and vigour of sprout growth?	

Adapted from Bond (1996)

8.2.   General phenology

The variability in phenological patterns in the karoo reflects the variability of climatic factors (rainfall/temperature combinations) and the diversity of species. Since growth phenology influences the timing of flowering and the production of seeds, a brief discussion is included.

8.2.1.   Phenology

Although some perennials begin growth at the end of summer in the succulent karoo, most species begin to grow after the first significant rains in early winter (Van Rooyen et al., 1979a). What appears to be unique about the succulent karoo is that growth continues through winter because of moderate minimum temperatures (Esler et al., 1994; Esler et al., this volume). Some perennial plant species flower during winter, but the main flowering period for both annuals and perennials is in spring (Van Rooyen et al., 1979a). Apart from the species released and dispersed by water, most seeds are dispersed during the dry summer months (Rösch, 1977). These either accumulate in a seed bank or germinate with the onset of the autumn rains.

Growth and reproduction in the summer-rainfall region is less seasonal and is associated with unpredictable rainfall and the growth form composition of the vegetation. For example, it may occur at any time of the year depending on rain events in *Sarcocaulon vanderietae* (Hobson et al., 1970). Generally, however, most shrub species flower and/or fruit in autumn and spring (Roux, 1966), because extreme temperatures in winter and summer limit growth (Vorster and Roux, 1983). Rainfall in the form of thunderstorms is more spatially variable, so adjacent areas in otherwise homogeneous environments might differ in the amount of rain received, and therefore have distinct phenological activities and possibly distinct growth form compositions (Hoffman and Cowling, 1987). This is particularly relevant for C_4 grasses, whose abundance in space and in time is closely correlated with incidence of summer rainfall (Hoffman et al., 1990).

8.3.   Flowering phenology

The patterns of growth and the timing and amount of flowering have important repercussions for the recruitment of plant species in the karoo (Wiegand et al., 1995) through their effects on reproductive processes such as pollination and timing of seed dispersal. Timing of flowering may be influenced by a variety of selective factors including the availability of pollinators, competition for pollinators, moisture availability and/or conditions for seeding recruitment. In addition, flowering time may be conservative within a phylogenetic lineage. Since there is substantial variation between species in their flowering response to climatic variables, the reproductive success of co-occurring species may differ in any particular blooming season. This variation may be important for the coexistence of species, particularly in the succulent karoo, where turnover is rapid (i.e. poor persisters) and there are many ecologically equivalent species (Cowling et al., 1994).

8.3.1.   Timing of flowering

There are substantial differences in the flowering phenology of succulent and Nama-karoo communities. In the winter rainfall region (succulent karoo), flowering occurs mostly in showy, multi-species displays in spring (August and September), followed by seed ripening in late spring and early summer (Le Roux et al., 1989; Struck, 1992, 1994a). Annuals are usually the first to flower, their performance (i.e. abundance of flowers) varying depending on moisture availability (Van Rooyen et al., 1979b, 1991), and, towards the end of spring, many geophytes (although not all, Amaryllidaceae for example) and a number of the succulent species have their main flowering season (Rösch, 1977; Van Rooyen et al., 1979a). Insufficient moisture is probably the major factor limiting summer flowering in the winter-rainfall regions. In the less predictable, summer-rainfall regions of the Nama-karoo, flowering appears to more flexible and, although it is more common in summer, a proportion of the flora can flower during any month of the year. *Lycium cinereum* and *Pentzia incana* flower opportunistically in response to rainfall (Hoffman, 1989b), but those species with storage organs or deep roots do not (Werger, 1986). Management decisions, however, have largely been based on the idea that different growth forms flower and set seed at predictably different times of the year (average growth-cycle model; Roux, 1968b). Mass flowering, synchronized among species, in these areas is not a notable feature, except in years of unusually high rainfall.

Why is mass flowering a succulent karoo phenomenon? The answer may partially lie in the growth form mix of the region. Mass flowering displays in Namaqualand are usually associated with annuals, short to medium-lived succulent perennials and geophytes, which are uncommon in the summer rainfall areas (Van Rooyen, this volume). These growth forms rely more on seeds for population perpetuation than do shrubs, hence abundant flowering. In addition they may also have design constraints which could reduce the flowering time (for example, annuals are vegetatively present only during the growing season, whereas geophytes usually produce a single inflorescence with a few buds). Finally they may have to compete for pollinators. In contrast, the dominant species of the Nama-karoo are persisters that allocate more resources to survival than reproduction. Abundantly branched shoots in long-lived shrubs allow for flower-bud development over an extended period.

In studies in the winter-rainfall region of the succulent karoo, where seasonal precipitation is relatively predictable, Hoffman and Cowling (1987) have shown that flowering is markedly synchronized compared to other regions with similar climates (Orshan et al., 1989). Le Roux et al. (1989) show that 90% of the shrubby perennials and virtually all of the annuals flower during late winter to spring in Namaqualand. Duration of flowering varies roughly between 20 and 80 days in this region (Struck, 1992). Despite this synchrony, a number of species flower outside the main season. Some petaloid monocotyledons and succulents that have storage organs flower outside the main season by uncoupling flowering and growth. These include *Haemanthus* spp. and other Amaryllidaceae which generally flower in autumn (March–April) in the winter-rainfall region, and in summer (October–February) in the summer-rainfall region (Snijman, 1984; Du Plessis and Duncan, 1989). The production of flowers at times when the surrounding vegetation is dormant presumably enhances pollination, but also enables the fleshy seeds of this group, which have no dormancy period (Markotter, 1936) to germinate almost immediately, making optimal use of the rainy season for seedlings to establish themselves. *Dactylopsis digitata* (Aizoaceae, Mesembryanthema) and *Tylecodon pygmaeus* (Crassulaceae) flower in summer in Namaqualand where most other species flower in winter and spring (Struck, 1995) (Fig. 8.1). Staggered flowering times in communities are often explained in terms of competition for pollinators, although many of these patterns can also be explained by the diversity of phylogenies in a community (for example, characteristic autumn flowering in Amaryllidaceae). Long-term phenological data are required to distinguish between proximate (ecological) and ultimate (evolutionary) explanations for flowering phenology patterns in communities, and these do not exist for the karoo.

8.3.2.   Triggers for flowering
What then, are the triggers for flowering? The answer is dependent on the species, and the information available is sketchy. Unfortunately, community-level studies seeking

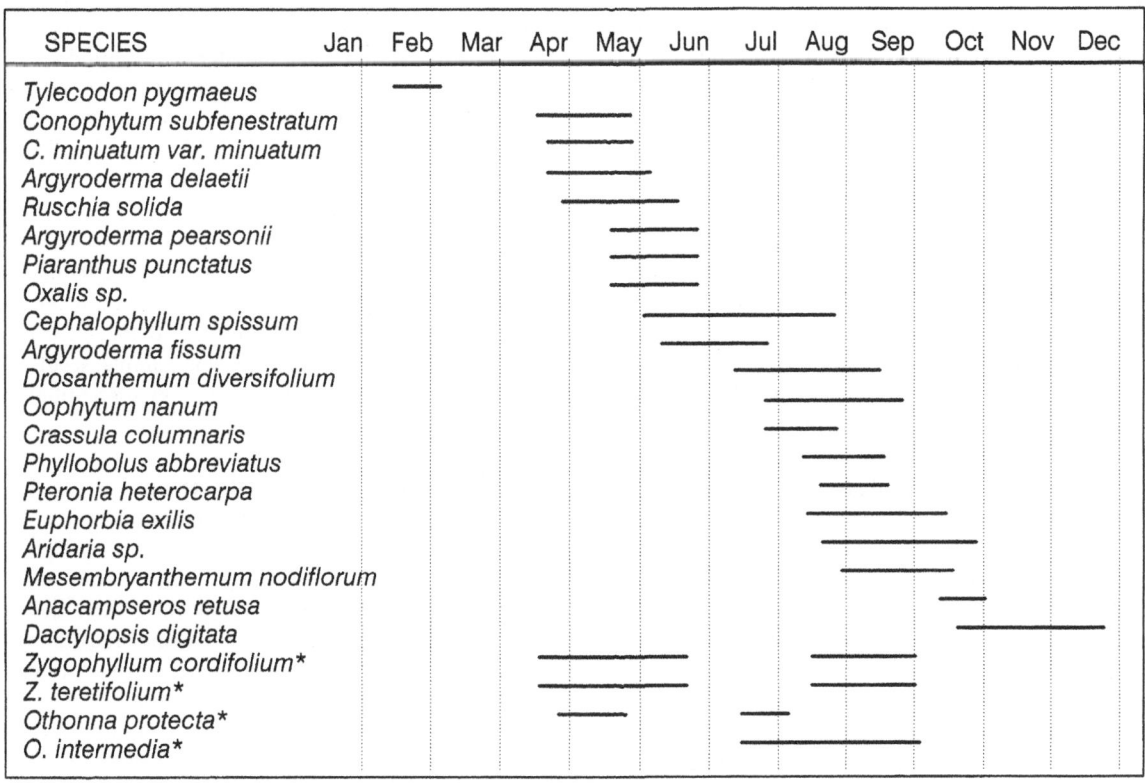

Figure 8.1 **Generalized flowering phenologies of selected species from the central Knersvlakte. In response to the prevailing climatic conditions, the onset of the blooming season is subject to considerable variation, nevertheless, the flowering sequence remains consistent between the years. *Variable in response to moisture availability (adapted from Struck 1995)**

correlations with climatic factors are lacking, since it is difficult to generalize when dealing with plants differing in flower developmental time and in the source of resources used in flower development. For example, such succulents as *Tylecodon* spp. appear to 'pump' water from leaves to developing inflorescences and are usually leafless during anthesis. These species may therefore be largely independent of rainfall prior to anthesis. Other succulent species, such as *Mesembryanthemum crystallinum* (Von Willert et al., 1992) flower only when there is no longer sufficient moisture to grow. Flowers of *Pteronia* spp. take six months to develop from bud to anthesis stage while most mesembs only take a few weeks, and thus the correlation of anthesis with rainfall may vary considerably depending on the bud development time of the species concerned.

Struck (1994a) infers that the timing of flowering of ephemerals in the winter-rainfall karoo is indirectly influenced by the first significant drop in autumn temperatures (beginning of growing period) and/or temperature increases in spring (beginning of dry period). The actual flowering time appears to be delayed by water stress (Struck, 1994a; Van Rooyen et al., 1991; Van Rooyen, this volume). Experiments by Steyn et al. (1996a, 1996b, 1996c) also show that temperature can be used to explain phenological differences within and between annual species occurring in the winter-rainfall region as well as to predict the timing of certain phenological stages. Interestingly, temperature declines in winter due to the passage of cold fronts are correlated with decreases in barometric pressure and there is anecdotal evidence that mass flowering of some geophytes appears to coincide with strong frontal systems (barometric trigger perhaps) (N. M. du Plessis, personal communication). While changes in temperature and day length may cue flowering in some species, rainfall patterns can influence flowering times. Snijman (1984) reports that when rainfall is patchy and of short duration in the winter-rainfall region, only isolated *Haemanthus* spp. populations flower, but, when rainfall at the beginning of autumn is widespread, most *Haemanthus* spp. flower simultaneously throughout their ranges. The pattern and amount of seasonal precipitation may therefore influence species composition and the numbers of individuals flowering in any one year or at any particular site (Struck, 1992, 1994a).

Flowering responses to rainfall may be related to the morphological and physiological characteristics of the species. For example, Struck (1994a) shows that, in the winter-rainfall region of Namaqualand, non-succulent shrubs produce the greatest number of flowers in moister years (most Asteraceae, *Hermannia* spp., *Lebeckia sericea*), while succulents flower more prolifically in dry years (Mesembryanthema and *Euphorbia* spp.). The abundance of flowering annuals appears to be related to the amount of

rainfall two months prior to the flowering 'season' (Struck, 1994a). The mean number of flowers produced on new growth by *Osteospermum sinuatum* (a widespread, drought-deciduous shrub that flowers up to four times annually) was closely correlated to rainfall three months prior to anthesis, but was unrelated to rain that occurred during the four-week period when flowers were developing, or to day length (Milton, 1992b). The cues and mechanisms explaining these observations are not well understood, although the implications of this data for explaining vegetation dynamics using the lottery mechanism for coexistence could be significant.

It has been shown that individuals given additional water (Milton, 1995a), or those growing on such nutrient-rich patches as *heuweltjies* (Midgley and Musil, 1990) and harvester ant nest-mounds (Dean and Yeaton, 1993b) or released from competitive effects of neighbouring plants (Milton, 1992b), tend to produce significantly more flowers (Table 8.2). A direct correlation has been shown between number of capitula and basal diameter in *Pteronia* spp. (Milton, 1995a), individuals of which are strongly influenced in size by both intra- and interspecific competition (Esler and Cowling, 1993).

This suggests that release from competition for resources indirectly influences the numbers of flowers produced, and emphasizes the importance of good rain years for flower (and therefore seed) production. Unfortunately, the management practice of thinning-out unpalatable species to promote the flowering and establishment of palatable species (Joubert and Van Breda, 1976) can have negative community implications. Although there are no studies which document these effects, one would predict that reduction of cover would increase runoff and soil erosion. The resulting reduction of species/growth form diversity could potentially reduce the resilience of plant communities to the effects of drought, hail and insect damage. For example, seedlings of the grass *Tetrachne dregei* are highly susceptible to destruction by sheet erosion (Roux, 1960).

8.3.3.   Landscape patterns of flowering

Local and regional differences in soil moisture and nutrient status in the karoo result in functional differences across landscapes. Seasonality in growth and flowering of communities in the southern karoo was greater among plants on runoff sites (plains) than in washes and drainage lines (Milton, 1990). The proportion of actively growing species in plains communities averaged 26% in summer and 58% in winter, but in drainage-line communities 41% of species were active in both seasons. *Heuweltjies*, common in the succulent karoo, and other nutrient-rich patches, also influence phenological patterns across landscapes (see section 8.3.2 and Milton, 1990).

Table 8.2. *The number of capitulae and seed capsules ('fruits') counted in 150 mm diameter 'plots' on plants growing on harvester ant (Messor capensis) mounds and on plants in intermound spaces at Tierberg*

Species	On mounds		Intermound	
	n	Fruits	*n*	Fruits
Brownanthus ciliatus	27	54.9 ± 19.6	104	31.6 ± 23.2
Drosanthemum montaguense	34	13.8 ± 6.3	112	7.1 ± 5.1
Ruschia spinosa	27	33.5 ± 19.8	96	8.4 ± 9.8
Hereroa latipetala	8	2.1 ± 1.3	20	1.0 ± 1.5
Ruschia approximata	5	16.2 ± 5.9	20	9.4 ± 6.3
Rhinephyllum macradenium	10	13.0 ± 11.3	32	9.0 ± 8.6
Pteronia pallens	104	36.6 ± 13.8	341	16.1 ± 10.9
P. cf empetrifolia	21	35.9 ± 11.6	73	30.7 ± 10.7
Osteospermum sinuatum	36	13.6 ± 5.0	116	9.5 ± 6.8

Data are means ± standard deviations
From Dean and Yeaton (1993b).

8.4.   Pollination biology

A large amount of anecdotal information about pollination biology exists in the karoo, but ecological and evolutionary interpretations are less common. Table 8.3 provides a summary of some of the known pollination syndromes of some karoo families.

8.4.1.   Flowering phenology and competition for pollinators

Despite observed variability in the onset and abundance of flowering, a constant flowering sequence of species between years has been observed (Struck, 1992, 1994a). This suggests that rainfall is not the overriding environmental factor controlling timing of flowering but that more predictable triggers mediated through temperature changes may also play a role (Van Rooyen et al., 1991; Struck, 1994a). It has been shown in many systems that flowering phenologies of coexisting plant species are related to pollinator availability and to the avoidance of competition for pollinators (in order to maximize seed set). Temperature may therefore be a cue for regulating both flowering and pollinator activity. A cautionary note is that, while patterns have been attributed to the avoid-

Table 8.3. *Pollinators recorded for a range of karoo plant families*

Family	Pollinator(s)	Reference
Acanthaceae	Bees	Struck (1992, 1994b)
Aizoaceae	Short-tongued visitors; bees	Struck (1992, 1994b)
Aizoaceae (Mesembryanthema)	Short-tongued visitors; bees, Bee-flies, monkey beetles, Masarine wasps, moths	Struck (1992, 1994b), Gess and Gess (1989, 1980a,b), Gess (1992), Whitehead (1984)
Amaranthaceae	Bees	Struck (1992, 1994b)
Amaryllidaceae	Bees, butterflies	Snijman (1984)
Apiaceae	Aculeate wasps, bees	Gess and Gess (1991a,b)
Asphodelaceae	Bees	Struck (1992, 1994b)
Asteraceae	Short-tongued visitors; bees, Bee-flies, monkey beetles, Masarine wasps	Struck (1992, 1994b), Gess (1992), Whitehead (1984), Midgley (1993), Picker and Midgley (1996)
Brassicaceae	Bees	Struck (1992, 1994b)
Campanulaceae	Bees	Struck (1992, 1994b)
Capparaceae	Birds	Skead (1967), Rebelo (1987)
Crassulaceae	Bees, birds	Struck (1994b)
Euphorbiaceae	Short-tongued visitors	Struck (1992, 1994b)
Fabaceae	Bees, birds	Skead (1967), Struck (1992, 1994b)
Geraniaceae	Bees, long-tongued flies	Struck (1992, 1994b), Manning and Goldblatt (1996)
Hyacinthaceae	Bees	Struck (1992, 1994b)
Iridaceae	Bees, flies, long-tongued flies, Butterflies, sphingid moths	Struck (1992, 1994b), Manning and Goldblatt (1996), Goldblatt (1991), Goldblatt et al. (1995)
Liliaceae	Birds, bees	Hoffman (1988a)
Loranthaceae	Birds	Skead (1967), Rebelo (1987)
Molluginaceae	Bees	Struck (1992, 1994b)
Montiniaceae	Bees	Struck (1992, 1994b)
Oxalidaceae	Bees	Struck (1992, 1994b)
Rosaceae	Bees	Struck (1992, 1994b), Whitehead (1984)
Santalaceae	Bees	Struck (1992, 1994b)
Scrophulariaceae	Bees, butterflies, masarine wasps	Struck (1992, 1994b), Whitehead et al. (1987), Gess (1993)
Solanaceae	Bees, birds	Skead (1967), Rebelo (1987), Struck (1992, 1994b)
Sterculiaceae	Bees	Struck (1992, 1994b)
Zygophyllaceae	Bees	Struck (1992, 1994b), Whitehead (1984)

ance of competition for pollinators, recent studies suggest that they can equally well be explained by the diversity of phylogenies in a community. Staggered flowering times may simply reflect a wide range of phylogenies represented in a community (Johnson, 1992).

8.4.2.   Insect pollination

Insect pollinators are a key to the persistence of many self-sterile species in the karoo (Jacobsen, 1960) (e.g. species in the Asclepiadaceae, Asphodelaceae, Mesembryanthema and Oxalidaceae). Struck (1994b) reports that the majority of species at a site in Namaqualand (Goegap Nature Reserve) show entomophilous pollination syndromes. Insect species visit more than one host plant and most plant species attract a variety of floral visitors.

Many of the insect flower visitors such as social and solitary bees, wasps, butterflies and bee-flies (Gess and Gess, 1989; Gess, 1992; Struck, 1992; Whitehead, 1984) are endemic to southern Africa and the karoo (Vernon, this volume), possibly because their distributions are restricted by their host plants, many of which also show a high degree of endemism (Cowling and Hilton-Taylor, this volume). Variations in the abundance of floral resources between blooming periods favour opportunistic use of a broad range of host plants rather than specialist use of a few (Neff et al., 1977), which may explain why so many endemic karoo pollinators are generalists. The dominance of generalist pollinators is reflected by the preponderance of plant species with large, open flowers which are accessible to a range of insects. Most flowers of the Mesembryanthema exhibit large, showy yellow, orange, purple and pink petals, and have nectaries which act as generalist insect attractants (Hartmann, 1991; Ihlenfeldt, 1994), as do the families Apiaceae, Celastraceae and the subfamily Mimosoideae in the Nama-karoo (Palmer and Hoffman, 1997).

Masarine wasps are most diverse in the Mediterranean and temperate arid areas of southern Africa (c. 145 species, all endemic; Gess, 1992), their distribution being strikingly similar to the nodes of species-richness associated with the Mesembryanthema (Gess and Gess, 1980a; Gess, 1992). Masarine wasps tend to be oligolectic (collect pollen from flowers of plants of a single family or genus), although the species associated with the Mesembryanthema (in the genus *Ceramus*) tend to be more widely distributed than those species associated with Asteraceae and Papilionaceae (Gess, 1992). Monkey beetles (Coleoptera: Scarabaeidae: Hopliini) are endemic to southern Africa (Scholtz and Holm, 1985) and pollination by this group is important in the arid areas of the south-western Cape and Namaqualand where adaptive radiation of the subfamily has occurred. This is of particular interest since

beetle pollination is globally uncommon (Picker and Midgley, 1996). Monkey beetles emerge during the spring flowering peak. Colour discrimination by the beetles influences their host plant selection, enabling the separation of beetles into guilds (Picker and Midgley, 1996) and avoiding to some extent pollinator overlap. Bee-flies (Bombyliidae) show a centre of diversity in Namaqualand (Hesse, 1938). The proportion of bee species endemic to the dry areas of southern Africa (95% in regions receiving less than 500 mm/yr) is higher than in the rest of the subcontinent (75%) (Struck, 1994a).

There are a number of described mechanisms to avoid pollinator overlap (spatially, temporally and mechanistic) in karoo species. Despite the synchronized flowering of Mesembryanthema, and frequent hybridization in cultivation, hybrids in nature are rare (Hammer and Liede, 1990; Liede et al., 1991). A combination of spatial (habitat) and temporal (flowering time) segregation and/or different flower structures probably reduce pollinator overlap (Liede et al., 1991). Although the anthophilous insect fauna is largely dominated by generalists (Struck, 1994b), there are some specialist pollinators. In the winter-rainfall karoo, for example, 28 species of Iridaceae and Geraneaceae have intensely coloured purple to crimson flowers with exceptionally long corolla tubes which attract long-tongued flies (Goldblatt et al., 1995; Manning and Goldblatt, 1996). Pollen contamination is avoided by differential placement of pollen on the insects' body, and the advantage to the plant species of having a dedicated pollinator include increased pollination success and decreased pollen contamination and loss (Manning and Goldblatt, 1996). The narrow flowers of *Lycium cinereum*, *Hermannia cuneifolia* and *Conophytum* spp. in the Knersvlakte restrict entry to generalist insects but attract bees and butterflies with long mouth parts using nectar rewards (Struck, 1995). Long proboscid *Celonites* spp. (Masarinae) are pollinators for the specialist flowers of *Aptosimum* and *Peliostomum* (Scrophulariaceae) which occur throughout the karoo (Gess and Gess, 1989; Gess, 1992, 1993). Oil-collecting bees specialize on annual *Diascia* spp. (Scrophulariaceae) and there is a correlation between flower and bee density (Whitehead and Steiner, 1985). Carrion flies are attracted by the foetid scent of sub-canopy stapeliad flowers (Bayer, 1978) as well as *Ferraria* spp. and *Moraea lurida* in the Iridaceae (Goldblatt, 1991). Flies are also attracted by the honey-like floral fragrance of *Piaranthus punctatus* (Meve, 1994; Meve and Liede, 1994). The fruity scents of such white and yellow flowers as *Hesperantha* spp. and *Moraea* spp. that open in the early evening (Goldblatt, 1984, 1991) and *Lithops* spp. that open at night (Hammer, 1991) attract moth pollinators.

Table 8.4. **Seed losses in the period from ovule formation to the development of ripe seeds in the family Mesembryanthemaceae. Data are means ± standard errors**

Species	Ovules/capsule[a]	Seeds/capsule[b]	Percent aborted	Seeds/plant[c]
Hereroa latipetala	610.8 ± 47.5	387.4 ± 69.3	36.6	659
Mesembryanthemum crystallinum	518.5 ± 17.6	120.1 ± 20.5	76.8	6004
Ruschia approximata	131.6 ± 6.2	94.0 ± 6.4	28.6	2819
Rhinephyllum cf. graniforme	128.6 ± 4.1	65.6 ± 11.2	50.0	931
Drosanthemum montaguense	114.8 ± 5.4	81.1 ± 6.2	29.4	3809
Malephora lutea	84.2±9.1	86.5 ± 7.3	0	493
Drosanthemum hispidum	60.2 ± 10.2	30.6 ± 5.0	49.2	1497
Brownanthus ciliatus	28.6 ± 4.0	26.9 ± 2.0	6.0	5021

[a] Unpublished data, K. J. Esler. N = 5 flowers
[b] Data from Esler and Cowling (1995). N = 20 capsules with no visible signs of predation.
[c] Unpublished data, K. J. Esler

8.4.3.   Other forms of pollination

Mesembryanthema may not rely entirely on insects for pollination. Bittrich (1987) states that many Mesembryanthema species produce copious amounts of dry, powdery pollen, suggesting that wind pollination may play a role in these species. Pollination by birds is less common than insect pollination in the karoo. Ornithophily, which is widespread in biomes adjacent to the karoo (Rebelo, 1987) was not observed at Goegap Nature Reserve in Namaqualand (Struck, 1994b). Most of the ornithophilous species have vivid flower colours (red, yellow, green), robust floral parts, and perches to support heavier avian visitors, mainly malachite (*Nectarinia famosa*), lesser double-collared (*N. chalybea*) and dusky (*N. fusca*) sunbirds (Skead, 1967). In the Richtersveld (succulent karoo), arborescent succulents (*Aloe dichotoma, A. pillansii, Pachypodium namaquanum*) are among the few bird-pollinated species in the area (Midgley et al., 1997). In the eastern Cape, the arborescent succulent *Aloe ferox* flowers in winter and is pollinated by sunbirds (*Nectarinia* spp.) and honeybees (*Apis mellifera*), whereas the majority of karroid plants in the area flower after rain during the warm to hot months and are pollinated by solitary wasps and bees (Hoffman, 1988a). Other examples of bird-pollinated genera are *Cadaba, Lycium, Loranthus, Lachenalia* and *Cotyledon* (Skead, 1967; Rebelo, 1987).

8.4.4.   Pollinator abundance and seed set

Are plants pollinator limited or resource limited? This is an important question which needs to be addressed in the karoo. Although what is needed are manipulative experiments involving bagging and artificial pollination, an indirect measure is to observe seed set and seed production through counts. Counts of the number of ovules per capsule versus the number of seeds per capsule in the Mesembryanthema show that seed set is highly variable, ranging from no, or low, ovule loss in some species (the perennials, *Malephora lutea, Brownanthus ciliatus*), through to abortion levels of 77% (the only annual in the survey,

Mesembryanthemum crystallinum) (Table 8.4). Since all species were from the same site (Tierberg), and since counts on predation-damaged capsules were avoided, mortality of the ovules was unlikely to be due to predation or to resource deficiency (all species presumed to have equal availability of resources).

Alternative causes of ovule mortality could be pollination failure or developmental failure. Excess ovules may be functional – for example as testing sites for optimal genetic combinations. Whatever the reason, there is a need to distinguish between proximate (ecological) and ultimate (evolutionary) explanations for low seed set. Abortion levels may not be significant at the level of the whole plant, since, in the examples given, abortion levels did not correlate with seed production per plant (Table 8.4).

8.4.5.   Associations among modes of pollination, seed dispersal and establishment

Little is known about the interface between pollination biology and seed and seedling biology. In most cases, we have no idea how the processes involved in the production of seed relate back to population viability. Shaanker et al. (1990) argue that morphology of flowers is influenced not only by adaptation to pollination, but also by the modes of seed and fruit dispersal and seedling establishment. This, they suggest, is because of the morphological continuity of flowers and fruits, and the possibility that simultaneous adaptation for the dispersal of gametes (pollen) and seeds would result in more economic usage of reproductive resources. They found that more often than expected, wind-pollinated species tend to have wind-dispersed seeds and insect-pollinated species to have either passive or animal-dispersed seeds. Although similar studies have not been conducted in the karoo, it is my opinion that these patterns are unlikely to hold here. For example, the majority of species in plain, wash and drainage habitats at Tierberg (southern succulent karoo) are wind and water dispersed (Milton, 1990), but belong to genera known to be insect pollinated (Gess, 1992; Struck, 1994b).

8.5.   **Pre-dispersal hazards**

Changes in plant species composition in the karoo have been attributed partly to lowered seed set and seed production (since seed availability is said to determine future vegetation composition) of palatable species (Acocks, 1953). A consideration of the pre-dispersal hazards facing seeds is therefore essential, although this argument may only hold true for short-lived species that do not resprout. Seed set is generally far lower than flower production for a variety of reasons, including pollination failure, resource deficiency, damage by predators or pathogens or merely because the flowers were acting as pollinator attractants or pollen 'donors'. Although 'persisting' versus 'seeding' represent different life-histories, both groups do produce seeds and seedlings, and the extent to which they are affected depends on the scale (both in time and space) of hazards.

8.5.1.   Pollen loss

Adaptations that appear to reduce pollen loss have not been addressed in the majority of karoo species and there is a need for appropriate experimentation. Pollen theft may be a strong selective (and ecological) force. Unfortunately, very little is known about the impacts of pollen predators on seed production. It has been hypothesized that the dark areas on ray florets of the self-incompatible annual *Gorteria diffusa* mimic beetles that feed on the pollen of other annual Asteraceae, giving the impression that their flowers are already occupied (Hutchinson, 1946; Midgley, 1991). More recently, however, it has been suggested that these black spots are not to repel beetles, but rather to attract fly pollinators through visual sexual mimicry (Midgley, 1993; Johnson and Midgley, 1996a, 1996b). Dark-centered flowers have evolved independently in a large number of succulent karoo genera. Picker and Midgley (1996) note that the relatively hairless guild of hopliinid beetles (*Scelophysa, Heterochelus, Gymnoloma* and *Pachycnema* spp.) in the south-western Cape that tend to embed in flowers of Mesembryanthema and Asteraceae, probably have a limited role in pollination, but, as they feed extensively on ray florets, ovules and pollen of species in these families, they may have considerable impact on the reproductive biology of some Namaqualand plants.

8.5.2.   Florivory

Although there are a number of studies on pre-dispersal predation of karoo species, these studies do not unambiguously demonstrate the population or community consequences of predation. Pre-dispersal predation might have acted as a major selective force in the evolution of allocation patterns of karoo species. Mass flowering, for example, might not be entirely for pollinator attraction,

but might also confer an advantage, though 'masting' effects, to those species that would normally suffer the impacts of pre- and post-dispersal predation (Janzen, 1976). Pre-dispersal predators include invertebrates (tephritid fly larvae (Milton, 1995b), lepidopteran larvae, pentatomid bugs and chrysomelid beetles) and vertebrates (mammalian herbivores, birds and reptiles), and their impacts vary depending on how generalist/specialist their feeding habits are. Ostriches (*Struthio camelus*) feed on the green shoots (39%), leaves (12%), flowers (17%), fruits and seeds (4%) and annual grasses and forbs (28%) (Milton et al., 1994a). Flowers of certain species of Asteraceae, Brassicaceae and Mesembryanthema and the flowers and seeds of many annual species are common in the diet of the karoo korhaan *Eupodotis vigorsii* (Boobyer and Hockey, 1994). The flowering and fruiting of karoo shrubs is frequently depressed after defoliation (Venter, 1962), especially in such species as *Galenia fruticosa* and *Pteronia* spp. that develop flowers once a year over many months (Milton and Dean, 1990b). Grazing and browsing by sheep has been shown to reduce the reproductive outputs of preferred forage species, including *Osteospermum sinuatum* (Milton, 1992b), *Rhigozum obovatum* (Milton and Dean, 1988) and *Themeda triandra* (O'Connor and Pickett, 1992), thus reducing the abundance of their seeds and seedlings relative to the weedy species (Roux, 1968a; Milton, 1994a; Gatimu, 1995). Milton (1995a) showed (for two *Pteronia* species occurring on the southern edge of the Nama-karoo/succulent karoo interface) that the proportion of the seed crop lost to pre-dispersal predators (in this case sheep and tephritid fly larvae) was not constant, but that it increased during years and in places where fruit production was low. This means that heavy grazing during wet years could reduce the reproductive potential of favoured (palatable) species, and that pre-dispersal seed/flower predators could potentially alter the competitive interactions between plant species. It has been suggested that protection of palatable shrubs during flower development in wet years may reduce the impact of introduced domestic stock on preferred forage species (Milton, 1995a). Defoliation does not necessarily reduce flowering; moderate grazing has been shown to increase seed production in such species as *Osteospermum sinuatum* that flower two to three times per year on new growth (Milton and Dean, 1990b).

8.5.3.   Resource availability

One of the key limiting factors in arid environments is the availability of water, and this in turn limits nutrient uptake. Below-average rainfall after flowering may influence seed set through resource availability, although no data are available for karoo species. The seed production of some species is, and others is not, affected by variation in rainfall (Esler et al., 1992). It is not known whether the reduction in seed production occurs at the pre- or post-flowering stage.

8.6.   Seed production

Questions relating to whether the turnover of karoo vegetation is seed-limited, are intriguing and certainly have direct management implications. Theoretically, changes in vegetation composition may be a function of seed availability (Liljelund et al., 1988) in that the species producing the most seed would have the greatest probability of occupying regeneration 'gaps'. In a three-year study of the growth, flowering and recruitment of shrubs in grazed and protected rangelands in the southern karoo, Milton (1994b) showed that changes in the demographic structure of shrub populations in arid rangelands can be a function of seed availability. However, producing more seeds does not necessarily mean eventual domination of a community, since many other factors also play a role in recruitment, as will be shown later in this chapter.

Indirectly, seed production of individual species may play an important role in livestock management. Seeds of *Augea capensis*, a winter-rainfall succulent herbaceous shrub (Van der Merwe, 1991) are highly nutritious (Marloth, 1925) and are produced in vast quantities (up to 500 kg ha[-1]). They provide a valuable source of fodder during prolonged drought in some areas (C. A. Van der Merwe, personal communication).

8.6.1.   Seed numbers (per plant)
Not surprisingly, seed production per plant in the karoo is highly variable (Table 8.5).

Since seed production represents a large investment in energy and water, species-specific tradeoffs between energy conservation and reproduction occur and this is reflected in the thresholds of rainfall that are required before seed production (as well as germination and seedling survival) can occur (Esler, 1993; Wiegand et al., 1995; Milton, 1995a). Variation between species depends largely on the resource allocation patterns of the plant, and this is tied to life-history. Resources for producing seeds are limited, and because seed size is variable and depends on the compromise between pressures that select for large (better chance of recruitment) or small (greater numbers translate into fitness) seeds, relationships between seed size and seed number should be apparent. Because seed size and abundance is correlated with taxonomic affiliation, these relationships may only be apparent when species in the same family or genus are compared. This is the case for perennial karoo

Table 8.5. *Seed production per plant and seed mass (mg) of some Nama- and succulent karoo species. Region refers to Nama-karoo (NK) or succulent karoo (SK)*

Species		Seeds/plant	Seed mass
Perennial grasses			
Aristida congesta	NK	6 620[a]	0.53[a]
Cymbopogon excavatus	NK	2 160[a]	0.8[a]
Eragrostis plana	NK	6 000[a]	0.25[a]
Hyparrhenia hirta	NK/SK	2 451[a]	1.2[a]
Sporobilis capensis	NK	23 520[a]	0.16[a]
Tetrachne dregei	NK	31 858[b]	0.34–0.58[b]
Themeda triandra	NK/SK	378[a]	3.9[a]
Perennial non-succulent shrubs			
Augea capensis	SK	3 410[c]	8.3[c]
Galenia fruticosa	NK/SK	3 680[c]	0.62[c]
Geigeria passerinoides	NK/SK	1 855[d]	0.49[d]
Osteospermum sinuatum	NK/SK	50[c]	9.1[c]
Plinthus karooicus	NK/SK	2 274[e]	0.19[e]
Pteronia empetrifolia	NK/SK	665[c]	3.7[c]
Pteronia pallens	NK/SK	200[c]	3.3[c]
Senecio retrorsus	SK	2 417[f]	1.13[f]
Succulent shrubs			
Aridaria noctiflora	SK	826[g]	0.64[h]
Brownanthus ciliatus	SK	5 021[g]	0.20[h]
Drosanthemum hispidum	SK	1 497[g]	0.03[h]
Drosanthemum montaguense	SK	3 809[g]	0.10[h]
Hereroa latipetala	SK	659[g]	0.11[h]
Malephora lutea	SK	493[g]	0.20[h]
Psilocaulon utile	SK	3 106[g]	0.23[h]
Rhinephyllum macradenium	SK	513[g]	0.90[h]
Ruschia approximata	SK	2 819[g]	0.66[h]
Ruschia spinosa	SK	476[g]	0.16[h]
Sphalmanthus brevifolius	SK	1 101[g]	0.62[h]
Annuals			
Mesemb. crystallinum	NK/SK	6 004[g]	0.24[h]

[a] Papendorf, in Theron (1964)
[b] Roux, in Theron (1964)
[c] Milton and Dean (1990b) (estimations from lightly grazed habitat)
[d] Van der Schijff, in Theron (1964)
[e] Theron (1964)
[f] Brynard, in Theron (1964)
[g] Esler and Cowling (1995)
[h] K. J. Esler, unpublished data

grasses, and succulent shrubs in the family Aizoaceae (Mesembryanthema) (Table 8.5). Variation in seed production within species (in space or in time) depends on a wide variety of factors. All of the pre-dispersal factors influencing plant reproduction discussed in this chapter are likely to contribute to the variability of seed production per plant in space and time (see sections 8.1 to 8.8).

8.6.2.   Seed numbers (landscape level)
Given the above, it is not surprising that the number of seeds produced per square metre in the arid regions of southern Africa is highly variable, and depends not only on favorable climatic conditions (timing and amount of rainfall) and levels of pre-dispersal loss but also on the mix of species present on a site (Esler, 1993; Milton, 1995a). In areas where perennials are dominant and annual species are rare, overall seed production is comparatively low. For

example, seed production at Tierberg, under a light grazing regime in a shrubland where annuals comprise less than 1% of the biomass at any time, is about 3000 seeds/m^{-2} or 34 kg ha^{-1} yr^{-1} (Milton and Dean, 1990b). Allocation patterns associated with perennial versus annual life-history patterns would account for much of this variation, although there are many exceptions to the rule. A single tuft of the pioneering perennial grass, *Stipagrostis ciliata*, for example, can produce up to 30 000 seeds during a favorable year (Skinner, 1965). Geophytes, which are unusually common in the succulent karoo, tend to produce quantities of seed which appears to be necessary for perpetuation of such geophyte species as Iridaceae (Goldblatt, 1991).

This variation has important implications for vegetation dynamics and has been modelled in detail for the southern edge of the Nama-karoo/succulent karoo interface by Wiegand et al. (1995). The model indicated that the dynamics of this shrub community are characterized by events (episodic and discontinuous) that change vegetation composition. This is driven by the timing and amount of rainfall which influences seed production and seedling recruitment and establishment.

8.7.   Dispersal

Dispersal and dormancy are common mechanisms through which plants deal with spatially and temporally patchy environments (Levin et al., 1984). Escaping in space (via dispersal) or in time (via dormancy) is critical for many plant species in the unpredictable karoo environment. The diversity of dispersal mechanisms represented in the karoo has long received attention (Marloth 1894), although general patterns and ecological processes of dispersion are less well understood. Recently it has been shown that variation in the dispersal characteristics of co-occurring species can influence the outcome of competitive interactions and hence the overall dynamics of vegetation (Yeaton and Esler, 1990; Wiegand et al., 1997).

8.7.1.   Dispersal mechanisms and habitat
Dispersal mechanisms vary depending on seed size, shape and energy allocation (Harper, 1977). Since the costs associated with producing seeds vary with seed size, seed dispersal characteristics are closely linked to life-history strategies, which would explain the variability of dispersal strategies in relation to habitat. Seed dispersal mechanisms in the karoo vary with differing selective pressures between habitats (Milton, 1990).

Animal dispersal of seeds by ingestion of fruits tends to be uncommon in dry habitats (Howe and Smallwood, 1982) although in the karoo, species occurring on nutrient

rich (*heuweltjies*) or mesic patches (washes and drainage lines), do more commonly exhibit interrelationships with mammals, birds and insects (Milton, 1990; Milton et al., 1990; Milton, 1992b). These specialized habitats comprise a very small part of the land surface, and thus component species require effective means of interpatch dispersal (Milton, 1990). For example drainage-line species, less limited by water stress, produce succulent fruits which are dispersed by frugivorous birds (Knight, 1988; Dean et al., 1993) and omnivorous mammals (Kuntzsch and Nel, 1992) or fruits with adhesive barbs or spines dispersed on the pelts of browsing mammals (Milton et al., 1990). The seeds of *Galium tomentosum* (Rubiaceae), a climbing plant that is restricted to drainage lines in the karoo, are presented on long, woolly peduncles that are developed only by the females of this dioecious species. The peduncles, with attached seeds, are dispersed into trees in drainage lines by tree-nesting bird species that use them to construct the framework of nests (Dean et al., 1990; Dean and Milton, 1993). *Cucumis humifructus* is the only species in the Cucurbitaceae which bears its fruits underground (Geocarpy). This annual species, which is found in the Northern Province, appears to have an unusual mutualistic relationship with the Aardvark (*Orycteropus afer*). In addition to its usual diet of ants and termites, the Aardvark also eats the fruits of *C. humifructus*, dispersing the seeds to disturbed areas near its burrows (Meeuse, 1958; Mitchell, 1965; Hollmann et al., 1995). Colonizer species also require a means of long-distance dispersal. Many of the food items of the karoo korhaan are endozoochorous seeds of species associated with intensive grazing and other forms of disturbance (Boobyer and Hockey, 1994). This suggests that the karoo korhaan is an important disperser of such colonizer and nurse species as *Lycium* spp. (Beukman, 1991), and that it may play an important role in the reclamation of degraded areas and the spread of certain 'increaser' species into vegetation types bordering the karoo biome (Boobyer, 1989). Endozoochory by tortoises, rodents and aardvarks has also been reported for the families Aizoaceae (Mesembryanthema) and Molluginaceae (Jump, 1988; Dean and Yeaton, 1992; Milton, 1992b). The genus *Carpobrotus* (Mesembryanthema) has fleshy indehiscent fruits that are embedded in a mucilaginous substance attractive to birds, mammals and humans (D'Antonio, 1990; Zedler and Scheid, 1988). This genus comprises species with extraordinary colonizing attributes and it is a successful invasive weed in many countries (D'Antonio et al., 1993).

8.7.2.   Short-distance dispersal and seed retention
In the more common xeric vegetation matrix, selective pressures are towards short-distance dispersal and/or seed

retention. It has been suggested that this strategy has a selective advantage over long-distance dispersal because seedlings have a higher probability of survival close to where the parent plant has survived to reproduction (mother site theory; Ellner and Shmida, 1981). Seeds in the xeric matrix of the succulent and Nama-karoo are mostly wind- or water-dispersed (Hoffman and Cowling, 1987; Milton, 1990; Van Rooyen et al., 1990). Species which colonize open, exposed sites usually have small water-dispersed seeds (<1 mm in diameter) that are trapped by soil particles (Yeaton and Esler, 1990; Esler, 1993). Although some larger seeded species are also effective colonizers of disturbed areas because their seeds exude gelatinous or sticky substances which 'glue' them to wet soil (e.g. *Euryops multifidus*, *Lepidium* spp. and *Augea capensis*), most large-seeded, wind dispersed species with pappuses or wing-like appendages are ultimately trapped and establish under 'nurse' plants. These seeds are well represented in the families Asteraceae, Portulacaceae and Zygophyllaceae and tend to be from groups which have broad distributions. Interestingly, *Eriocephalus* spp. (Asteraceae) which are widely distributed throughout the karoo, have light, fluffy, wind-dispersed seeds that are used by birds as nest lining material (Dean et al., 1990). The shade-loving species of Mesembryanthema, that disperse by shedding either the entire (*Fenestraria*) or parts of (*Apatesia, Conicosia, Herrea, Hymenogyne*) the large and light-weight seed capsules (Ihlenfeldt, 1971; Hartmann, 1982; Hartmann and Gölling, 1993) are also included in the wind-dispersed category although this type of dispersal is the exception rather than the rule in the family.

8.7.3.   Dispersal in Mesembryanthema

The predominantly water-dispersed dissemination system of the succulent karoo family, Mesembryanthema (98% of the species of this family possess hygrochastic capsules; Ihlenfeldt, 1983) is particularly well studied (Van Rooyen et al., 1980; Hartmann, 1988; Hartmann, 1991; Ihlenfeldt, 1971, 1983). Seeds usually ripen towards the end of the growing period, but are retained (canopy stored) in capsules that remain closed until they are soaked by rain (hygrochastic) (Esler and Cowling, 1995). Seeds (0.1 to 0.5 mm in length) are therefore released when there is a higher probability that moisture conditions are favourable for germination and seedling establishment (Garside and Lockyer, 1930; Lockyer, 1932; Ihlenfeldt, 1971; Van Rooyen et al., 1980; Hartmann, 1991). Different capsule types have evolved in this family to 'promote the extension of dispersal in time' (Hartmann, 1991). Varying degrees of seed retention are found between species (Volk, 1960; Ihlenfeldt, 1983), which to some extent is related to habit and habitat. Esler and Cowling (1995) show that long-lived woody species growing in areas receiving low

levels of disturbance germinate rapidly, have low overall seed dormancy and low levels of seed retention, which is a viable strategy where competition for recruitment sites is strong. Species occurring on adjacent disturbed areas (*heuweltjies*, drainage lines) have low overall germination and high seed retention, indicating an opportunistic life-history strategy (Esler and Cowling, 1995). Interestingly, Hartmann (1989) reports opposite trends to these when comparing the genera *Argyroderma* (long-lived) and *Oophytum* (short-lived) in the Mesembryanthema.

Dispersal mechanisms in the Mesembryanthema have received some functional and evolutionary attention due the interesting trends within and between genera (Straka, 1955; Ihlenfeldt, 1983). In the Dorotheanthinae, for example, the seven types of capsules represented in this group follow a topocline which runs from south to north. As aridity increases, seed retention decreases (as capsules tend to be simpler) and seeds are distributed before the growth period, thus forming a soil seedbank (Fig. 8.2) (Struck, 1989).

Durable seed capsules spread seed release over several rain events thereby reducing the risk of recruitment failure and enabling plants to 'sample' the size and intensity of a rain event (more seeds tend to be released after heavier rain events; personal observations). It is possible that seed retention may reduce the risk of seed release into an 'adult-saturated' population, although the time period of seed retention is generally orders of magnitude shorter than the average lifespan of a perennial plant (Esler and Cowling, 1995). Where rain events are infrequent, the seed retention adaptation is defunct in this group. Sampling different types of rain events is no longer necessary, since all rain events are likely to be exploited. One would predict that species at the arid end of the topocline would have some form of seed dormancy which would still spread the risk of failed recruitment. What is needed here is a study of the germination/dormancy characteristics of the species along the topocline.

8.7.4.   Dispersal distances

Few estimates of dispersal distances exist for the karoo. Seeds can be dispersed from the capsules of Mesembryanthema in the karoo up to 1.65 m away from the parent plant depending on the capsule design (Lockyer, 1932; Garside and Lockyer, 1930; Volk, 1960), but secondary dispersal by sheet flow is thought to increase dispersal distances significantly. Large tumble seeds of fynbos Proteaceae are moved by wind over greater distances (10–40 m) on unvegetated smooth rather than vegetated or stony ground (Bond, 1988). It is likely that karoo tumble seeds (e.g. *Pteronia pallens, Osteospermum sinuatum, Tetragonia spicata*) have similar dispersal distances, but are

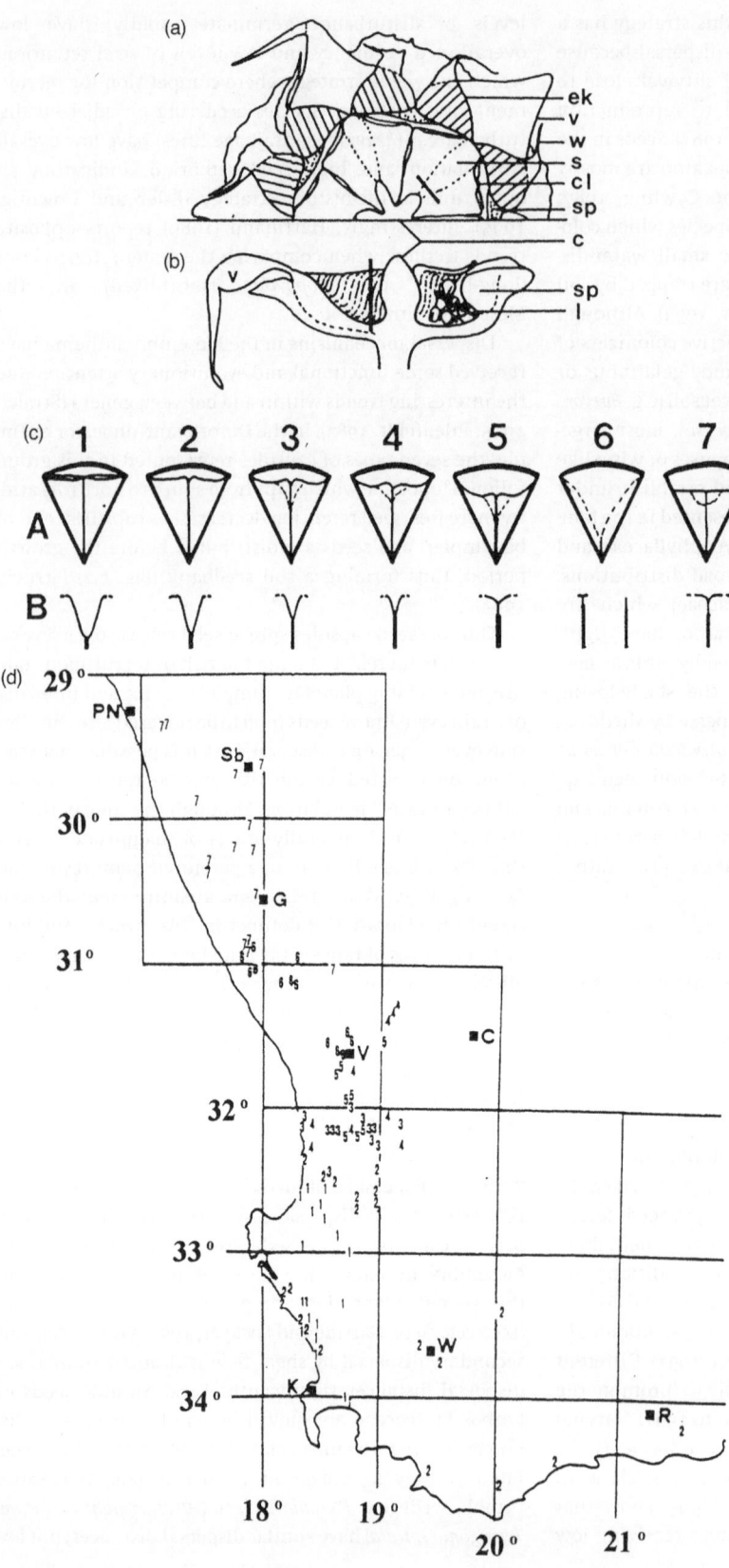

Figure 8.2 **Capsule types of Dorotheanthus spp. (Mesembryanthemaceae). (a) Opened capsule in top view showing expanding keels (ek), valves (v), wings (w), septas (s), cell-lids (cl), sealing protuberances (sp), and column (c). (b) Opened capsule in longitudinal section showing the sealing protuberance (sp) and a locule (l) filled with seeds. (c) Morphology of the covering membrane complex with (A) diagrammatic representation of a locule in top view of each fruit type showing the reduction of the cell lids and (B) diagrammatic representation of the septas and cell lids in tangential sections (see solid base in (a) and (b)) showing the receding splitting of the septas. (d) Distribution of the capsule types. The numbers indicate the capsule types according to (c). The succession of capsule types from south to north can be interpreted as an adaptation to increasingly drier localities (reprinted from Struck (1989), with permission of the Botanical Society of South Africa)**

eventually trapped by other shrubs. Leopard tortoises (*Geochelone pardalis*) can potentially disperse seeds of Chenopodiaceae, Crassulaceae, Cyperaceae, Fabaceae, Mesembryanthema, Poaceae and Scrophulariaceae uphill or between inselbergs in the southern karoo (Milton, 1992a), since the tortoises have home ranges of 1–2 km^2 and move from rocky hillside habitats to valley bottoms in search of water.

8.7.5.   Dispersal and seed polymorphism

Many succulent karoo species have polymorphic seeds (Rösch, 1977; Von Willert et al., 1992; Beneke et al., 1993a,b,c) which enable plants to spread dispersal through space and time. Seeds from the same plant or inflorescence can vary in size, seed coat anatomy, morphology or mechanism and timing of release, influencing not only dispersal, but also germination behaviour and growth (Van Rooyen, this volume).

8.7.6.   Other forms of dispersal

Dispersal in some species is not confined to seeds. In light, sandy soils, the corms of geophytes can be moved up to 60 m by molerats (Bathyergidae) (Lovegrove and Jarvis, 1986). It has been suggested that corms that readily separate from one another when disturbed may be a coadapted specialization in response to molerat predation (Lovegrove and Jarvis, 1986). In the southern Namib desert, weathered branches of *Geigeria alata* (Asteraceae) break off and act as tumble weeds, dispersing seeds which remain for a time on the branches in serotinous containers. In addition, some seeds are not dispersed and remain at the parent site in basal seed containers (Günster, 1994). There are similar dispersal mechanisms in *Aloe* (Liliaceae) and *Hessia* (Amaryllidaceae) with candelabra-like inflorescences that detach and roll in the wind, as well as in dead branches of *Psilocaulon* spp. and *Mesembryanthemum* spp. that break off at ground level (S. J. Milton, personal communication). The latter are often fugitive, requiring mobility (big wheels travel further than little wheels over rough ground).

8.8.   Seedbanks

Interactions among seed dispersal patterns, seed dormancy and seed germination influence the turnover and spatial distribution of seeds in the seedbank. These characteristics are tightly bound to the life-histories of the individual species. Since short-lived species most commonly spread the risk of germination through space and time, seedbanks are a characteristic of environments where these species are common.

8.8.1.   Soil seedbanks

Only three detailed studies on the spatial and temporal variation in seedbanks exist for the karoo; one in the southern karoo, where annuals comprise less than 1% of the biomass and cover at any time (Esler, 1993), and the others in Namaqualand (Van Rooyen and Grobbelaar, 1982; De Villiers, 1993) where annuals and short-lived perennials are an important component of plains vegetation, covering 30–40% of the soil surface in the growing season (Steinschen et al., 1996). These studies appear to conform with the general pattern described above. Seed densities in Namaqualand of the succulent karoo range from 5 000 (scattered perennials and a few annuals) to 41 000 seeds per m^2 (in sandy bottom lands dominated by annuals) (Van Rooyen and Grobbelaar, 1982), while in the southern karoo, estimates are substantially lower (17 to 426 seeds per m^2) (Esler, 1993). In the latter area, there is high seasonal variability in the size and composition of the seedbank as a whole. These differences reflect the life-history strategies of the component species. Seedbanks of large-seeded, non-succulent shrubs in the southern karoo (*Pteronia* spp., *Osteospermum sinuatum*) are transient, being recorded temporarily in the seedbank in summer after dispersal, but absent following rains in winter (Esler, 1993). Artificially sown seeds placed in the field in summer germinate only in the first year (Esler, 1993; Milton, 1994b), indicating that these seeds do not remain viable in the field for long periods.

8.8.2.   Canopy seedbanks

While seed dormancy is one way to deal with environmental uncertainty, delay in seed release (seed retention) is an alternative strategy. Seed retention is common in the Mesembryanthema (see section 8.7.2). While long-lived Mesembryanthema tend to have short-lived canopy stored seedbanks (low seed retention), species occurring in disturbed sites (drainage lines, *heuweltjies*) appear to maintain larger canopy seedbanks (Esler and Cowling, 1995). Serotinous plants are also an important component of Namib desert edge vegetation, although seed longevity in these desert plants might not be as long as previously assumed (Günster, 1994).

8.8.3.   Spatial distribution of seeds in seedbanks

Seed distributions in the karoo are highly clumped. There is a 3.5 fold variation in the numbers of seeds between open sites and those under shrubs in the southern karoo (Esler, 1993). Surface microtopography plays an important role in the distribution of seeds in the soil seedbank and these differences are largely a result of variation in dispersal among individual species (Esler, 1993). Larger wind-blown seeds are commonly associated with adult plant cover, whereas water-dispersed seeds are found in a variety of habitats including open areas between shrubs.

Soil depressions on hard-capped karoo soils are created by such digging mammals as aardvark, porcupine, *Hystrix africaeaustralis*, bat-eared fox, *Otocyon megalotis* and cape fox, *Vulpes chama*, and provide sites where seeds collect (Dean and Milton, 1991b). Wind-dispersed seeds tend to accumulate in litter because they are lightweight and have similar attributes to litter (Table 8.6).

Accumulations of seeds in depressions and in litter can result in increased seed germination because of a more favourable moisture environment. However, these are not necessarily the best microsites for seedling establishment (Dean and Milton, 1991b). Seeds accumulated and buried by such food hoarders as the harvester ant (*Messor capensis*) may escape destruction. *Messor capensis* is a granivore and stores viable seeds in its nest mounds. Some of these seeds are returned to the seedbank by mammalian predators of ants which periodically excavate the mounds while foraging (Dean and Yeaton, 1992).

Many seeds have specific morphological adaptations (awns, barbs and so on) that influence seed movement into suitable microsites for germination. These can influence not only the burial of individual seeds but also species distributions. *Stipagrostis ciliata*, for example, only occurs on deep, well-drained sandy apron veld in the arid western karoo. The physical qualities of the soil, in combination with dispersal characteristics of the awned seeds, determine the distribution pattern of this species (Skinner, 1964). Seeds of *Stipagrostis* spp. are negatively affected by extreme surface summer temperatures, so burial appears to confer a germination advantage (Skinner, 1964). Selective pressures for seed burial are different for different species, and it has been suggested that the hygroscopic awns of species in the tribes Paniceae and Andropogoneae (Poaceae) do not confer any germination advantage via burial but have a dispersal function (e.g. *Themeda triandra*; Adams and Tainton, 1990). Burial may also have a distinct advantage as a post-dispersal predation avoidance mechanism, although no examples could be found for karoo species. Efforts to reseed degraded karoo vegetation have been vastly improved by coating seeds with mud, indicating that 'burial' does have an ecological advantage (Van Breda, 1939). Barnard (1987), in a study of nine winter-rainfall karoo species showed that sowing depth has a definite influence on seed germination. Depending on seed size and structure, the species studied germinated more successfully when covered by at least 2–20 mm of soil (Barnard, 1987). Not all species gain advantage from burial. Some karoo species do not germinate when buried (e.g. *Geigeria passerinoides* and *Plinthus karooicus*, Theron, 1964). Germination trials indicate that these species have low levels of germination in the dark (Theron, 1964).

Although data are limited on seed distributions within soil profiles, patterns in karoo vegetation do appear to correlate with those found in arid systems elsewhere (Kemp, 1989). De Villiers et al. (1994b) found that seed numbers decrease significantly with increasing soil depth, and that most seeds are found in the top 50 mm layer of soil. To some extent seed burial depths are related to seed size and to soil texture. Skinner (1964) notes that depth of burial of *Stipagrostis ciliata* (11–14 mm) and *Stipagrostis obtusa* (8.5–11 mm) is related to the length of the unbranched feathered awn (of the lemma) plus the length of the seed.

8.8.4.   Correspondence between the seedbank and vegetation

Soil seed densities do not appear to correlate strongly with densities of adult plants in the southern karoo (Esler, 1993). This discrepancy may be partially explained by the highly patchy distribution of seeds in the landscape, owing to non-uniform dispersal, or merely to short dispersal distances combined with the patchy distribution of plants (Yeaton and Esler, 1990). The discrepancies may also be explained in a successional context, since later successional species tend to produce seeds with limited longevity and dispersal abilities than do early successional species, although this relationship is not absolute. In a study at Tierberg, Esler (1993) found that those species which were proportionally over represented in the seedbank (i.e. more opportunistic species associated with heuweltjies, drainage lines and washes) were species identified as early successional by Yeaton and Esler (1990).

8.9.   Post-dispersal hazards

Even when in the seedbank, seeds are not safe from predators or pathogens which could theoretically affect recruitment and distributions of species as well as the composition of plant communities. This is particularly relevant for shrub species that do not maintain persistent seed banks.

Plant seeds are a primary food resource for granivorous animals that can impact the species composition of vegetation by altering the reproductive potential of plants. Kerley (1991) measured the removal rates of non-native seeds by small mammals, birds and rodents at two

Table 8.6. *Observed and expected (based on the percentage cover of the different microsites) seedling frequencies in relation to type of dispersal at Tierberg*

	Litter	Open	Cover	Dispersal
Expected frequency	5.3	64.5	30.1	
Observed frequency				
Galenia fruticosa	34.8	48.3	16.9	Passive
Mesembryanthemaceae spp.	30.0	56.1	13.9	Water
Pteronia spp.	84.6	0	13.2	Wind

Source: K. J. Esler (unpublished data).

Table 8.7. *Germination characteristics of a selection of species from the family Mesembryanthemaceae*

Species	Longevity	Germination Rate	Max(%)	Treatment	Reference
Aridaria noctiflora	Medium	delayed	23	10 °C night/20 °C day	K. J. Esler
Brownanthus ciliatus	Short	tardy	5	10 °C night/20 °C day	K. J. Esler
Carpobrotus edulis	Short	delayed[a]	96	Through gut of mammal	D'Antonio et al. 1993, Zedler and Scheid (1988)
Cleretum papulosum	Short	delayed	1	scarified and 10 °C night/20 °C day	Pierce et al. (1995)
Conophytum litorale	Long	rapid	95	21 °C night/24–27 °C day	Distefano (1990)
C. declinatum	Long	rapid	100	21 °C night/24–27 °C day	Distefano (1990)
C. obcordellum	Long	rapid	100	21 °C night/24–27 °C day	Distefano (1990)
Dinteranthus spp	Long	delayed[b]	100	60–100 hrs exposure to 65 °C	Visser et al. (1976)
Drosanthemum hispidum	Short	delayed	2	10 °C night/20 °C day	K. J. Esler
D. montaguense	Medium	rapid	60	10 °C night/20 °day	Esler et al. (1992)
D. speciosum	Medium	tardy	48	10 °C night/20 °C day, smoke	Pierce et al. (1995)
Herreroa latipetala	Long	rapid	100	10 °C night/20 °C day	K. J. Esler
Lampranthus aureus	Medium	delayed	19	10 °C night/20 °C day, smoke	Pierce et al. (1995)
L. haworthii	Medium	delayed	10	10 °C night/20 °C day, smoke	Pierce et al. (1995)
Malephora latipetala	Short	delayed	6	15 °C night/30 °C day	K. J. Esler
Mesemb. crystallinum	Short	delayed	0		K. J. Esler
Pleiospilos leipoldtii	Long	rapid	90	21 °C night/24–27 °C day	Distefano (1990)
Psilocaulon utile	Short	delayed	7	15 °C night/30 °C day	K. J. Esler
Rhinephyllum graniforme	Long	rapid	78	15 °C night/30 °C day	K. J. Esler
R. macradenium	Long	rapid	77	10 °C night/20 °C day	Esler et al. (1992)
Ruschia approximata	Long	rapid	98	15 °C night/30 °C day	K. J. Esler
R. caroli	Medium	rapid	70	10 °C night/20 °C day, smoke	Pierce et al. (1995)
R. multiflora	Medium	delayed	23	10 °C night/20 °C day, smoke	Pierce et al. (1995)
Sceletium tortuosum	Medium	delayed	43	10 °C night/20 °C day, leaching	K. J. Esler
Sphalmanthus brevifolius	Medium	delayed	17	15 °C night/30 °C day	K. J. Esler

Max (%) refers to the maximum percentage germination recorded. Short = short-lived (<1–3 yrs); Medium = Intermediate longevity (4–15 yrs); Long = Long-lived (>15 yrs). Note that short- and medium-lived species tend to 'bet-hedge' by having delayed germination (this implies either very specific germination requirements (for example *Carpobrotus edulis*) or seed dormancy.
[a] unless animal dispersed
[b] unless exposed to heat

localities in the karoo (Steytlerville and Tierberg). He observed that ants were the most important seed removers, with rodents removing lesser amounts of seed and birds the least (although this technique was biased against birds). Ants tend to harvest more seed during the warmer seasons than in winter when their activities are depressed by the cold (Kerley, 1991). The harvester ant *Messor capensis* takes about 10% (5.2 kg ha^{-1}) of the annual seed production at Tierberg (Milton and Dean, 1993), concentrating on the most abundant of the larger seeds available. Up to 60% of the annual seed output of preferred species can be removed. Despite being an important granivore, this species might also play a vital role in the process of regeneration. If nutrient-rich *Messor capensis* mounds are disturbed (see section 8.8.3), seeds can escape predation, and recruit into a highly favourable environment (Dean and Yeaton, 1992) where resources are abundant and competition from adult plants is reduced.

8.10.   Germination

It is very difficult to summarize and compare karoo germination data because of the wide variety of techniques used

and the inadequate information provided on sample sizes and germination conditions. However, a few generalizations can be made. Bet-hedging tends not to be a strategy of long-lived species (although there are exceptions), and seeds generally exhibit high percentages and rates of germination. In contrast, seeds of short-lived and opportunistic species tend to exhibit delayed germination and seed dormancy.

8.10.1.   Germination and life-histories
The Mesembryanthema are a good group for an overall survey of germination characteristics (Table 8.7) because a wide variety of life forms (annuals, short-lived perennials, long-lived perennials) occurs in the family. The family is best represented in the succulent karoo.

In a comparative survey of the life-history characteristics of Mesembryanthema occurring in disturbed (*heuweltjies*) and undisturbed (flats) habitats in the southern karoo, Esler and Cowling (1995) found distinct differences in germination characteristics. Species occurring on *heuweltjies* have low overall germination and high seed retention (i.e. they maintain canopy seedbanks) compared with those species occurring off *heuweltjies*. Disturbance levels on *heuweltjies* are high, and consequently adult population turnover is rapid (i.e. plants tend to be short-lived).

Opportunistic life-history characteristics confer an advantage to these species, because occasions for establishment are spatially and temporally variable and the probability of death after establishment is high (Esler and Cowling, 1995). In contrast, for the long-lived species occurring on the flats, the chances of encountering favourable recruitment sites are very low (Yeaton and Esler, 1990; Esler, 1993), but, once seedlings develop into adult plants, they have a higher chance of survival. Flats species are therefore geared to exploiting suitable rain events. Seeds tend to be released during the first rain events of the season and they exhibit rapid germination and low seed dormancy (Esler and Cowling, 1995). Although seedling mortality rates are very high after germination events (Esler, 1993), adult longevity allows plants to risk the high probability of seedling mortality. These patterns within the Mesembryanthema (see also Von Willert et al., 1992) accord with anecdotal data for germination of species found in other families. For example, seeds of the long-lived stem-succulent, *Pachypodium namaquanum* (Apocynaceae) in the Richtersveld, exhibit 95% germination within two days of sowing (Retief, 1978), although in the field this species produces very few seeds because of severe parasitism (Retief, 1988). Of the 85 winter-rainfall region perennial shrub species described by Van Breda and Barnard (1991), 77% germinate within 10 to 15 days of sowing and only 13% require some sort of dormancy breaking treatment (e.g. scarification, seed maturation). Vegetative propagation is recommended for 3% of the species, and it is not known how the remaining 7% of the species regenerate (Van Breda and Barnard, 1991).

8.10.2.   Germination and intrinsic factors

In the field, non-dormant seeds of long-lived plant species in the karoo do not persist in seedbanks due to seed decay (pathogens) and losses to granivory or germination (Esler, 1993; Roux, 1960). Under laboratory conditions, however, seed can remain viable for several years after collection. Hartmann and Stüber (1993) report 8 years for Mesembryanthema, and Roux (1960) reports at least 11 years for the grass *Tetrachne dregei. Lithops* spp. also tend to be long-lived, and, on average, the percentage germination of new seed is about 85% (Du Plooy, 1993). These seeds deteriorate by approximately 4.35% per annum over the first 15 years until all seeds are non-viable (Du Plooy, 1994). The decrease in viability of laboratory-stored seeds has been noted for other species too (Table 8.8), and this has important implications for gene bank collections. Van Jaarsveld (1994) noted that the viability of *Gasteria* spp. seed decreases rapidly after one year.

Dormant seeds are those that will not germinate under normal environmental conditions and must first undergo a period of after-ripening or embryo maturation

Table 8.8.   *Germination of laboratory stored seeds of* Pteronia pallens, Ruschia spinosa *and* Brownanthus ciliatus *from the Tierberg Karoo Research Center, South Africa*

Species	Months after seed collection				
	0	6	12	18	24
Pteronia pallens	20.8 ± 2.5	16.8 ± 2.1	16.0 ± 2.1	12.5 ± 7.6	8.0 ± 1.4
Ruschia spinosa	21.5 ± 2.4	24.5 ± 0.6	20.8 ± 3.0	21.3 ± 4.1	25.0 ± 0.0
Brownanthus ciliatus	6.3 ± 3.4	1.0 ± 1.4	4.0 ± 1.6	2.5 ± 1.7	2.0 ± 1.4

Seeds were tested from 0 to 24 months after storage at room temperatures in the laboratory. Data are means ± standard deviations, $n = 4$ petri dishes, each containing 25 seeds.
Data from Esler (1993).

or be exposed to some sort of environmental stimulus (e.g. scarification, temperature fluctuations, leaching). Essentially, dormancy allows seeds to disperse in time. This risk-spreading strategy is an advantage when seeds are the only mechanism for long-term persistence and where environments are variable and unpredictable.

Seed of the Nama-karoo shrub, *Plinthus karooicus* require a two year after-ripening period before germination can occur (Theron, 1964). The only technique (out of a variety applied) which broke dormancy of seeds younger than two years was the application of concentrated H_2SO_4, suggesting that dormancy in this species is controlled by the seed coat (Theron, 1964). The widely distributed grasses, *Cenchrus ciliaris, Stipagrostis ciliata* and *S. obtusa* that occur throughout the Nama-karoo have substances present in the structures surrounding the caryopses which inhibit germination (Skinner, 1964; Venter and Rethman, 1992). Only when these have been leached out by sufficient rainfall, will germination occur, as is also the case for *Augea capensis* (Zygophyllaceae) (C. A. Van Der Merwe, personal communication). An ecological advantage of leaching is that germination only occurs when moisture conditions (after successive rains) are favourable for successful recruitment. Zoochorous species have higher percentage germination after passing through the digestive systems of animals, as is the case in *Geigeria africana* (Steyn, 1934; Van der Schijff, 1948), *Carpobrotus* spp. (D'Antonio et al., 1993) (Fig. 8.3) and *Tetragonia spicata* (Van Breda and Barnard, 1991). The former two species are opportunistic, readily colonizing disturbed vegetation.

Polymorphism for germination requirements is a common bet-hedging strategy in arid zones (Harper, 1977) and has been reported for a number of ephemerals in Namaqualand (Van Rooyen, this volume). Ihlenfeldt (1983) reports that, although some seeds from Mesembryanthema capsules germinate very quickly (after 24 hr), it takes others from the same capsule a longer time to germinate (30 days). He suggests that the seed population consists both of fast-germinating individuals which are able to make the maximum use of the short growing

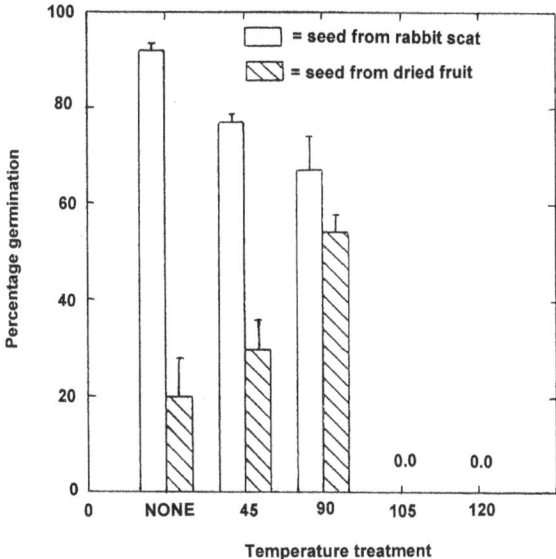

Figure 8.3 **Germination of *Carpobrotus* sp. seeds from scat or fruit at varying temperatures. Seeds were subjected to preset temperatures for 5 minutes and then spread on wet sand for 3 months to observe germination. Bars represent means plus one standard error. *N* = 5 (reprinted from D'Antonio et al. (1993), with permission from Springer-Verlag)**

period and of slowly germinating individuals that avoid the risk of insufficient rain (e.g. *Malephora* spp.). These patterns have also been described by Hartmann (1991).

8.10.3.   Germination and extrinsic factors

Some degree of temperature control prevents seed germination during extreme summer months, and it has been noted in many studies that germination percentages in the karoo tend to be higher under regimes that simulate spring/autumn conditions (10 °C night/20 °C day) when there is a higher probability of rain (Esler and Cowling, 1995; Esler et al., 1992; Hartmann and Dehn, 1987; Henrici, 1935a; Theron, 1964) (Table 8.7). The temperature optima for perennial grasses (*Stipagrostis* spp.) from the winter-rainfall region, peak at 20 °C (Skinner, 1964). However, germination of *Tetrachne dregei*, a grass of the summer-rainfall region, peaked at 30 °C (Roux, 1960). Maximum germination of the perennial shrubs *Geigeria africana*, *Senecio retrorsus* and *Plinthus karooicus* are at 25 °C (Van der Schijff, 1948), 20 °C (Van Brynard in Theron, 1964) and 21 °C respectively (Theron, 1964). After leaching to remove a water-soluble inhibitor(s) associated with the structures around the embryo, seeds of *Augea capensis* germinate more readily when temperatures are low (i.e. 20 °C; C. A. van der Merwe, personal communication). The overall germination of a number of Nama-karoo shrubs is higher at 20 °C than at 30 °C (Henrici, 1935, 1939), suggest-

ing that above average winter rains might favour shrub recruitment. However, this has not been documented at a community level. Despite a degree of innate dormancy, higher germination percentages at cooler temperatures have also been noted for Namaqualand pioneer species (De Villiers et al., 1994b).

Although fire is not a major driving variable in the karoo except in the eastern karoo mountains (Roux and Vorster, 1983), many seeds of karoo Mesembryanthema are stimulated to germinate by plant-derived smoke extract (Pierce et al. 1995). These results cast some doubt on the ecological significance of smoke as a fire-related cue in the adjacent fire prone fynbos biome. It is inferred that the chemical component in smoke is merely acting on a general germination inhibitor (Pierce et al., 1995).

Trials that varied osmotic potential of water-using salts (NaCl) or sugars (mannitol), have shown that the germination of *Tetrachne dregei* (Roux, 1968a) and *Plinthus karooicus* (Theron, 1964) is limited at low water potentials. This could partly explain why these species are absent on 'brak' (saline) soils.

8.11.   Seedling establishment

8.11.1.   Timing of recruitment

The most limiting factor in arid environments is moisture availability, and it is not surprising therefore that seedling establishment is directly related to rainfall timing and amount. Shrubs in the southern karoo mostly germinate in the autumn months (Van Breda and Barnard, 1991; Esler, 1993; Dean and Yeaton, 1992; Milton, 1995a) when ambient and soil temperatures have declined after hot summers, rainfall is more predictable, and there is a greater probability of follow-up rains. Rains in autumn coincide with the period of greatest seed abundance after dispersal in late summer/autumn (Hoffman, 1989b; Milton, 1992b), since seedbanks are transient (Esler, 1993). Rain events occurring out of season do not result in germination possibly because high temperatures prevent seedling germination and survival (Hartmann, 1983; Dean and Yeaton, 1992; Esler, 1993; Milton, 1995a) (Fig. 8.4).

A seedling emergence experiment conducted in summer on a Nama-karoo grass, *Tetrachne dregei*, indicated that more than 60 mm of rain within a three-week period would be needed to provide sufficient moisture for the germination and establishment of this species. It was concluded that seedlings were more likely to establish readily during autumn when temperature conditions are more favourable for germination, and when the soil dries out more slowly (Roux, 1968a).

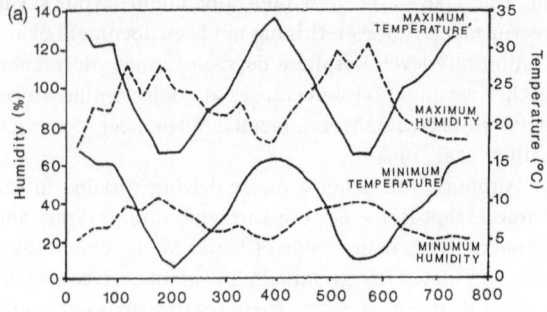

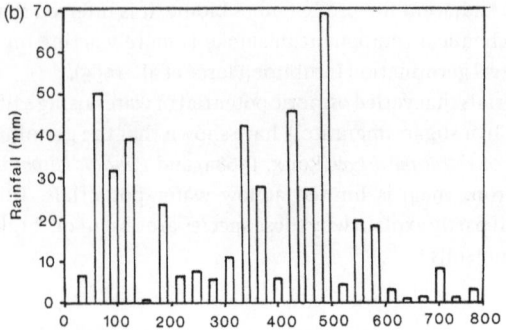

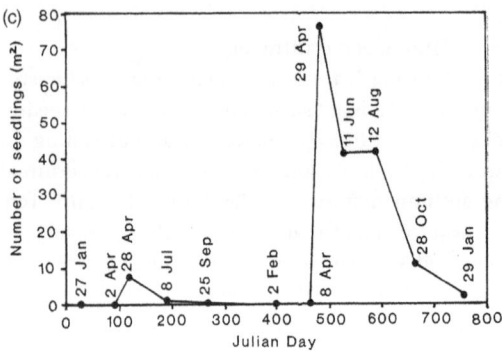

Figure 8.4 **(a) Temperature and humidity, (b) rainfall and (c) number of seedlings per m² recorded over a two-year period at the Tierberg Karoo Research Centre in the southern succulent karoo (from Esler 1993)**

8.11.2.   Size of recruitment events

The size of any germination event is directly related to the availability of seed, which can be influenced by a number of factors already discussed in this chapter, as well as the size, duration and timing of the rainfall event. Follow-up rain in the six months after emergence, however, is crucial for seedling establishment (Esler, 1993; Milton, 1995a). Few seedlings reach reproductive maturity in undisturbed arid shrublands where long-lived species dominate (Milton, 1994a), and seedling survival ranges from 1–5% of emergent seedlings in average years to 20–30% in years

with ample post-germination rain in winter and spring (Milton, 1995a). After good follow-up rains (>130 mm between May and December) in the winter and spring of 1991 in the southern karoo, seedling survivorship was 25% (Milton 1995a). A simulation model that incorporates southern karoo rainfall patterns, indicates that rare recruitment opportunities are further reduced by root competition for water (Wiegand et al., 1995). This highlights the slow turnover of species in southern karoo rangelands. In contrast, seedlings of short-lived species and ephemerals have higher probabilities of survival. In Namaqualand, recorded survival rates of annuals are highly variable within and between species at different sites and range from 47 to 74% of emerging seedlings (Van Rooyen et al., 1979b). This is generally higher than for karoo perennials (Milton, 1995a).

8.11.3.   Microsites and recruitment

The potential benefits of seed dispersal for successful plant recruitment depend on the microsites to which the seeds are dispersed. Field observations indicate that seedlings are not uniformly distributed with respect to microhabitat, and that these patterns are largely the result of the dispersal biology of individual species (Table 8.9) (Esler, 1993). Variance to mean ratios indicate that seedling distributions for some species are more heterogeneous than the distribution of adult plants at the same scale, and that temporal heterogeneity of seedlings is high, although not necessarily greater than spatial heterogeneity at the microhabitat scale (Table 8.10) (Esler, 1993).

The nurse-plant phenomenon has been described for a variety of semi-arid environments (McAuliffe, 1988), including the succulent karoo (Beukman, 1991). Beukman (1991) and Dean and Yeaton (1992) suggest that the occurrence of seedlings under nurses is primarily a method of avoiding abiotic stress at the seedling stage. Temperatures, for example, on open soil surfaces in the karoo can reach up to 61 °C in summer (Dean, 1992a), although other possibilities, such as escape from herbivory (Yeaton and Esler, 1990) and simple 'trapping' (Esler, 1993) also exist. Differences in the relative tolerances of seedlings to water stress have been demonstrated (Esler and Phillips, 1994), giving additional support to the stress-avoidance hypothesis. Seedlings that have large cotyledons tend to be more vulnerable, and establishment below canopies possibly provides them with an added advantage against herbivory and/or water stress. For example, the seedlings of *Osteospermum sinuatum* and *Pteronia pallens*, commonly found under shrubs (Esler, 1993) are more vulnerable to water stress than those of *Ruschia spinosa* (Esler and Phillips, 1994) that are dispersed into open areas (Esler, 1993). Whatever the reason for seedling establishment of some species under others,

Table 8.9. *The numbers of seedlings per m^2 (counted in 10, 100 × 1 m transects) in seven microhabitats at the Tierberg Karoo Research Centre in the southern karoo*

Species	Microhabitat						
	Open	Mound-side	Litter	Mesemb shrub	Mesemb mat	Dead mesemb	Shrub
Mesembryanthema	40.560	1.513	8.210	2.837	4.465	0.038	2.081
Galenia fruticosa	4.654	2.005	2.573	0.870	0.416	0.038	0.492
Osteospermum sinuatum	0.003	0	0.012	0	0.009	0.001	0
Tetragonia spicata	0.002	0.002	0.050	0.008	0.060	0.001	0.006
Pteronia spp.	0.004	0.025	1.729	0.001	0.568	0.045	0.115
Lotononis sp.	0.681	0.303	0.341	0	0.114	0.076	0.114
Augea capensis	0	0	0.114	0	0	0	0
Other spp.	0.567	0.038	0.529	0.076	0.265	0	0.227
Total	46.47	3.89	13.56	3.79	5.90	0.2	3.03

Seedling abundances were monitored approximately 20 days after a rainfall event in April 1990. Explanation of microhabitats are as follows: Open = in open areas on fine sands or clays between adult plants; Mound-side = at the side of soil mounds accumulated around adult plants; Litter = in litter accumulated on the soil surface; Mesemb shrub = under the canopy of adult Mesembryanthema spp. with erect, shrub-like growth forms (woody branches and long internodes); Mesemb mat = under the canopy of adult Mesembryanthema spp. with mat-forming growth forms; Dead mesemb = under dead Mesembryanthema spp.; Shrub = under the canopies of other woody shrubs (not Mesembryanthema)
Data from Esler (1993).

Table 8.10. *Spatial and temporal heterogeneity of adult plants and seedlings (as measured by variance to mean ratios) at the Tierberg Karoo Research Centre in the southern karoo. Seedlings were counted immediately after a recruitment event on 29 April 1990*

Species	Spatial heterogeneity		Temporal heterogeneity
	Adults	Seedlings	Seedlings
	Transect[a]	Micro-habitat[b]	Transect[c]
All species	14.76	0.89	969.25	–
Mesembryanthema	10.74	26.03	665.95	634.06
Galenia fruticosa	20.12	4.58	49.28	1047.30
Tetragonia spicata	2.72	10.92	39.34	31.38
Osteospermum sinuatum	4.30	2.42	5.89	8.75
Pteronia spp.	4.36	212.37	1188.24	824.90

[a] Variance to mean ratios of adult plants and seedlings in ten 100 × 1 m strip transects at one recording period (29 April 1990)
[b] Variance to mean ratios of seedlings among seven distinct microhabitats at one recording period (29 April 1990)
[c] Variance to mean ratios of total numbers of seedlings counted over five sampling periods from 29 April 1990 to 29 January 1991. All of the seedlings were from the same original cohort which germinated immediately prior to the first recording period. These ratios reflect the relative survivorship of each seedling species
Data from Esler (1993).

these patterns certainly influence the composition and arrangement of vegetation in the karoo (Yeaton and Esler, 1990; Milton and Dean, 1995a).

8.11.4.   Disturbance and recruitment
Large-scale disturbances often precede major recruitment events in the karoo. Droughts (Danckwerts and Stuart-Hill, 1988; Milton et al., 1995), hail storms (Powrie, 1993) and intensive trampling (Roux, 1980) can significantly reduce adult plant cover, leaving 'gaps' in the vegetation where competition for moisture, and possibly allelopathic effects are reduced. Prolonged drought is often followed by significant rainfall events (Hoffman and Cowling, 1987), which provide favourable recruitment conditions

(Milton, 1995a). Although vegetation clearing can increase the amplitude of soil moisture oscillations at 50 mm below the soil surface, soil moisture depletion at 150 mm below the surface is reduced (Milton, 1995a).

Small-scale disturbances (<1 m^2) generate spatial heterogeneity in the landscape (Dean and Milton, 1991a,b). These disturbances, mainly due to the action of digging and burrowing mammals, create patches where nutrients, water and seeds accumulate. Seedling survival is greater in disturbed or nutrient-enriched microsites (Dean and Milton, 1991a,b; Dean and Yeaton, 1992, 1993b). In seeding trials with *Osteospermum sinuatum* (Joubert and Van Breda, 1976), and *Ehrharta calycina* (Muirhead, 1962) no establishment occurred on undisturbed soil, while good establishment occurred when soil was disturbed and/or potential competitors were removed. Similarly, seedling recruitment in natural vegetation was greater on 25 m^2 plots from which established plants had been cleared than in adjacent undisturbed vegetation (Fig. 8.5) (Milton, 1995a). Van Rooyen and Grobbelaar (1982) report a large increase in the number of Namaqualand annuals in soil that had been periodically mixed.

Despite the predominant longevity of plant species (especially in the Nama-karoo), small-scale disturbances create 'gaps' which allow the maintenance of shorter-lived species in the environment (and therefore increase species diversity). In addition, they permit the continual, low-density recruitment of perennials outside major recruitment events. The effects of small-scale disturbances on the temporal and spatial dynamics of karoo shrub communities have been modelled by Wiegand et al. (1997), who showed that differences in scales and intensities of disturbances may partly account for the mixture of regeneration strategies and longevities found in apparently superficially homogeneous vegetation such as that in the southern succulent karoo.

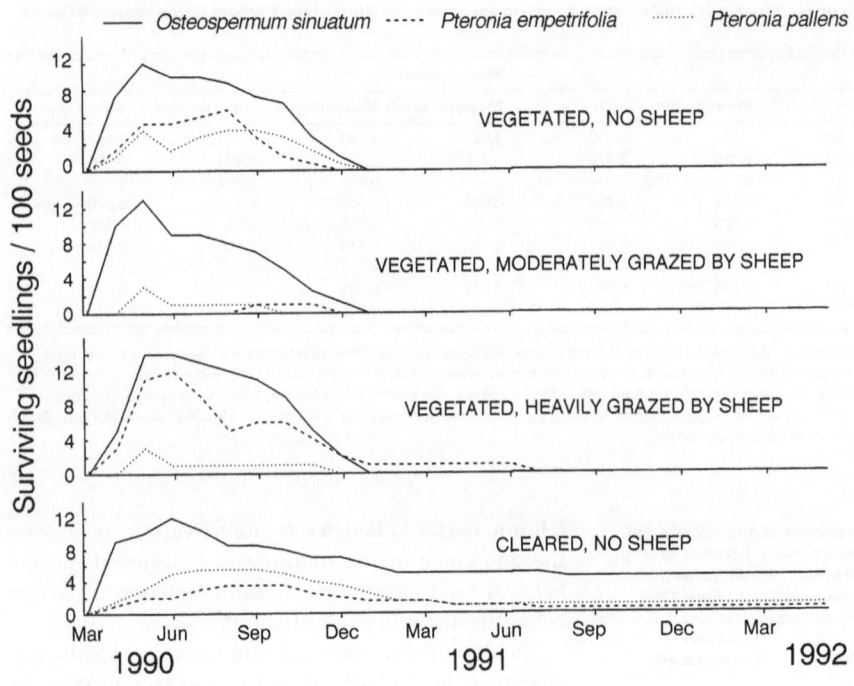

Figure 8.5 **Emergence and survival of three species of shrub seeds sown in vegetated plots at the Tierberg Karoo Research Centre (TKRC) (no sheep), Tierberg Ranch (moderately grazed by sheep) and Argentina Ranch (heavily grazed by sheep), and in cleared plots at TKRC (no sheep). (Reprinted from Milton (1994b), with permission of the Grasslands Society of Southern Africa and Blackwell Science Ltd.)**

Arborescent succulents (*Pachypodium namaquanum, Aloe dichotoma* and *A. pillansii*) are conspicuous elements of the arid flora of the succulent karoo. Recently, adult mortality due to disturbance by baboons (*Papio ursinus*) and porcupines has been reported for some of these species (particularly arborescent aloes in the Richtersveld) (Midgley et al. 1997), and it has been suggested that these agents may account for the absence of seedlings of these species in some areas. This is cause for concern because the arborescent succulents are keystone species, providing nest sites for birds and vantage sites for raptors, as well as being an aesthetically pleasing element of the stark arid landscape.

8.11.5.   Mycorrhizal status and recruitment

Most terrestrial plants probably form mychorrhizal associations between their roots and certain fungi (Brundrett, 1991). Interestingly, the Crassulaceae, Aizoaceae (Mesembryanthema) and Zygophyllaceae, all speciose families in the succulent karoo, appear to be non-mycorrhizal (Berliner et al., 1989; Allsopp and Stock, 1993, 1994), but other families, such as the Euphorbiaceae, are highly mycorrhizal. It is suggested that the absence of mycorrhizal association may provide a key to the understanding of the dynamics that shape this biome. The turnover of plants in the succulent karoo is believed to be rapid relative to the Nama-karoo (Cowling et al., 1994), and, since seedling establishment of species in the above families do not rely on the presence of certain fungi, this might allow rapid exploitation of newly available habi-

tats. It is not surprising that the Mesembryanthema are often dominant in post-disturbance colonizing floras (Acocks, 1953).

8.12.   Seedling mortality

Attempts to reseed degraded karoo vegetation are generally not successful (Van Breda, 1939; Roux and Vorster, 1983), despite good results when forage plants are grown in cultivation (Barnard, 1987; Donaldson, 1989; Van Breda and Barnard, 1991). Apart from predation (prior to germination) and germination failure (Van Breda, 1939), seedling mortality contributes significantly to these losses. Reasons for mortality include frost damage (Donaldson, 1989), drought, herbivory (Gatimu, 1995) and competition from existing vegetation (Milton, 1994b). Virtually nothing is known about the influence of soil-living parasitic nematodes, but it has been suggested that these animals could play a significant role in the prevention of establishment of grass seedlings that are shaded by larger shrubs or mature grass tufts (Roux, in Theron, 1964). Such root feeders as nematodes have been reported to cause drought stress in plants (Dropkin, 1989). Although reducing the density of competing perennials appears to improve the survival of forage plant seedlings (Milton, 1994b), artificial reduction in plant cover should be approached with caution until more is known about the potential adverse effects (Milton, 1994b; Milton and Dean, 1995a).

8.13.   Seed vs. vegetative reproduction

Vegetative reproduction, particularly common in succulent families with leaves and stems that store water and carbohydrates, is prominent in the succulent karoo, although not confined to the biome. Hartmann (1983), in a study of the population dynamics of *Cephalophyllum* spp. (Mesembryanthema) in the succulent karoo noted distinct tradeoffs between runner formation (vegetative reproduction) and seedling recruitment. Runner production appeared to be at the cost of sexual production. Van Jaarsveld (1994) suggested that vegetative reproduction in *Gasteria* spp. is a means to survive herbivory, and also noted that there was an inverse correlation between sexual and vegetative reproduction in this genus. *Gasteria* species that grow prolifically from the base tend to have fewer inflorescences with fewer flowers, while the inverse is true for species like *G. excelsa* which is not prolific from the base but which can produce large inflorescences bearing up to 2 700 flowers (Van Jaarsveld, 1994). Vegetative reproduction is also common in the Crassulaceae that shed their leaves during dry periods to be dispersed by wind or water into suitable microsites under shrubs and in litter (Von Willert et al., 1992; Milton et al., 1997). Seeds of Crassulaceae species germinate rapidly (Tölken, 1977) indicating that there is less need to bet-hedge via seed dormancy when the back-up of vegetative reproduction exists. Some karoo Asteraceae (*Helichrysum dregeanum*, *Pentzia incana*) root where branches touch the soil surface, a habit that leads to clonal spread as well as anchoring the soil around the mother plant (Van Breda and Barnard, 1991).

8.14.   Concluding remarks and future research

The literature on seed and seedling biology for the karoo is mostly descriptive, comprising many natural history type and anecdotal observations. This makes it difficult to formulate broad-based generalizations about how processes at the seed and seedling stage influence ecological processes at the population and community level.

It is clear that marked differences exist between plants with different life-history strategies. The proportion of seeds lost between the seed production and seedling establishment phases is proportionally higher for long-lived species compared to short-lived species. The losses

incurred may or may not have a corresponding impact on recruitment, but the significance of these losses to population dynamics is difficult to determine. This is especially so in semi-arid environments where successful recruitment events are difficult to detect.

The question is, what is unique, or worth noting about the karoo with respect to plant reproductive ecology? What follows is a brief description of some of the highlights.

The phenomenon of mass-flowering, particularly in Namaqualand, is notable. These flower displays are of great economic value because of their ecotourism attraction, but still, little is known about what environmental factors trigger a good flowering year.

It is fascinating that insect pollinators are a key to the persistence of many species in the karoo. The paradox is that insect abundance is generally low and it is possible that mass floral displays are the result of stiff competition for pollinators. What is especially intriguing is that many of the variety of insect pollinators are generalists. The succulent karoo in particular has witnessed the adaptive radiation of pollinator groups which are uncommon in the rest of the world. What ecological features have structured such radiation? There is considerable lack of knowledge in the field of pollination biology in the karoo and there are many unanswered questions about the diversity of phenological patterns and floral characteristics that exist here. For example, to what extent is reproduction in different plant species guilds limited by pollinator availability? How strongly does pollination influence seed production? Would a population decline with the extinction of a specialist pollinator? Why are so many of the endemic insect pollinators generalists?

Seeds are an important component of the life-history of most karoo species and, relative to other vegetation types (savannas, temperate vegetation types), there is very little vegetative reproduction. With so much dependence on seeds, it is surprising that seedbanks are relatively small, especially in a semi-arid environment. Is this because our environments are relatively predictable compared to other semi-arid regions?

Certainly, the western succulent karoo has a predictable rainfall pattern associated with low-pressure systems in winter. This appears to have strong implications for the turnover of species (Cowling and Hilton-Taylor, this volume; Esler et al., this volume) and therefore for seed and seedling biology. One of the groups important in this rapid turnover (in time and space) is the non-mychorrhizal Mesembryanthema. Can the lack of mycorrhiza alone explain why the group is able to exploit newly available habitats so rapidly? More research on mychorrhizal associations at the seedling stage is required.

It is clear that much still needs to be learned about

karoo seed and seedling ecology, especially if we are to understand the processes and dynamics driving karoo vegetation. Only long-term monitoring and such manipulative experiments as seed additions will show whether seed losses have a significant effect on population dynamics and if, indeed, reproduction by seed is the limiting factor for population persistence in the karoo.

8.15.   Acknowledgements

I am most grateful to Richard Cowling, Phil Rundel and Dick Yeaton for never tiring of enthusiastic discussion and critical debate. Thanks are also due to Sue Milton, Richard Dean and Steve Johnson for providing valuable input into a first draft of this chapter.

9   Animal form and function

B. G. Lovegrove

9.1.   Introduction

It is very important to appreciate at the outset the huge regional disproportionality of ecophysiological research which has been conducted on desert animals in southern Africa over the past four decades. The research output concerning Namib and Kalahari animals from scientists who have been associated, respectively, with the Desert Ecological Research Unit at Gobabeb, Namibia, and, to a lesser extent, the former Nossob research facilities in the Kalahari Gemsbok National Park, is considerable compared with the overall output for karoo species. This is unfortunate because, although it is possible to borrow conceptually from the Namib and Kalahari studies to a certain extent, many physical, biotic and environmental characteristics are unique to the succulent and Nama-karoo biomes. Wherever possible I have obviously reviewed karoo studies in preference to analogous Namib or Kalahari studies, but, for the sake of holism and completeness, I have been compelled to employ non-karoo studies where data are lacking. Had I restricted this chapter to karoo studies only, the chapter would, in all likelihood, have had a more limited educational and philosophical appeal due to the fragmentary nature of available information. It is therefore important that, where extension of logic is required between karoo and non-karoo affinities, this be undertaken with educated caution.

Animal form and function is best appreciated by reviewing the three types of adaptations; morphological, physiological and behavioural, which have evolved in desert animals in response to the selective pressures of desert environments. Invariably, these adaptations aid desert animals in maintaining water, thermal and energy balances in the face of the common physical and ecological stresses characteristic of arid ecosystems: limited water and food resources and the extremes of ambient temperature. These stresses, in addition to those of predation, have also played an important role in the evolution of a wide variety of social and reproductive systems in desert organisms, and in the frequency distribution of body sizes.

Climatic unpredictability is also an important selective force accounting for many of the movement and reproductive patterns of desert organisms. Although in many adaptations the influence of unpredictability is manifestly obvious, it is arguable that the full extent and importance of its role remains largely underestimated. This is hardly surprising considering that scientists have only recently, in the past decade or so, begun to identify and understand climatic phenomena, such as the El Niño Southern Oscillations (ENSO), implicated in climatic unpredictability in the southern hemisphere.

9.2.   Water balance

That water is the most limiting resource to all desert animals is well emphasized by one definition of deserts as 'water-controlled ecosystems with infrequent, discrete, and largely unpredictable water inputs' (Noy-Meir, 1974). Apart from its absolute physiological importance as the universal solvent, the availability of water also markedly influences primary production and hence food availability. Like all animals, desert animals need to maintain a regulated water balance in order to achieve homeostasis of body fluid concentrations. Animals in mesic habitats can easily maintain water balance by replacing losses (e.g. respiratory losses) with the high water content of their food or simply by drinking free-standing water. This balance is, however, not easily achieved in desert animals given the scarcity and unpredictable availability of free water and the prevailing high water-vapour saturation deficits and consequent high rates of evaporation from moist surfaces (Louw, 1993; Hadley, 1994). Moreover, evaporative cooling is the most effective avenue of heat loss open to animals. Yet, many desert animals, especially small ones, cannot afford to 'waste' water in order to defend their body temperatures (T_b) against hyperthermia in hot conditions. Instead, they resort to alternative means of thermoregulation, such as retreating temporarily from the harsh desert environment by being nocturnal or retiring to refugia. Nevertheless, many adaptations which serve to minimize losses and maximize water gains are often closely associated with thermoregulatory considerations.

9.2.1.   Water loss

Water is lost from animals via four important routes: evaporation from the skin or body surface (cutaneous losses), evaporation from the lungs (respiratory losses), in the urine, and in the faeces (Schmidt-Nielsen, 1983; Wharton, 1985; Hadley, 1994). In the short term, the easiest way to avoid high rates of water loss is (a) to escape the desiccating conditions of the heat of day by retreating to cooler refugia, such as burrows and nests, where relative humidity (RH%) is often elevated above that of the ambient, resulting in small water-vapour saturation deficits (Louw and Seely, 1982; Louw, 1993), and (b) to restrict activity to the night for the same reasons (see 9.5 Retreat).

Respiratory water loss is one of the biggest avenues of water loss from desert animals. In arid-adapted insects, such as the tenebrionid beetles of the Namib Desert, these losses are minimized (Nicolson et al., 1984b) through the evolution of sunken spiracles (Ahearn, 1970; Hadley, 1994), the fusion of the elytral wing coverings to form the sub-elytral space (Ahearn, 1970; Dawson, 1989), and discontinuous ventilation cycling (Bartholomew et al., 1985; Louw et al., 1986; Lighton, 1991, 1996). These adaptations either minimize the time period during which evaporation of body fluids can occur, or increase the distance and therefore the width of the water-vapour pressure boundary layer between the desiccating outside air and moist respiratory surfaces (Hadley, 1994). Compared with insects, desert vertebrates incur much higher rates of respiratory water losses, but these loses are nevertheless minimized through efficient nasal cooling of saturated air exhaled from the lungs. Nasal cooling reduces the water-vapour density of the expired air and thus the amount of water lost on each exhalation (Withers, 1992).

The ostrich (*Struthio camelus*) is one of the few animals besides the camel (*Camelus dromedarius*) that is capable of exhaling unsaturated air (Withers et al., 1981). Although the mechanism has yet to be explained, the bird is capable of exhaling air with 85 RH% via the nares. This extraction of water vapour recovers about 35% of the water from the lung air resulting in a saving of about 30% of the ostrich's daily water requirements. This is one of several adaptations which account for the success of ostriches throughout the arid and semi-regions of southern Africa.

The evaporation of water from the surface of animals is another major avenue of water loss and can be very substantial in certain animals. Desert arthropods have the lowest cuticular rates of water loss, due mostly to their waterproof cuticles (Edney, 1971; Hadley, 1974, 1981, 1994; Nicolson et al., 1984b). Waterproofing is greatly enhanced by the secretion onto the cuticle surface of long-chain saturated hydrocarbons (Hadley 1994) and wax blooms, such as those observed in tenebrionid beetles (Hanrahan et al., 1984; McClain, 1984; McClain et al., 1984, 1985). Rates of cuticular water losses from Namib tenebrionid beetles, scorpions and fishmoths rank among the lowest ever recorded in terrestrial animals (Edney, 1971; Nicolson et al., 1984b; Louw, 1993; Hadley, 1994). Equally impressive are the high resistances to water loss across the skin of southern African reptiles (Louw and Seely, 1982; Louw, 1993).

In reptiles, birds and insects, excretory water losses are reduced by the excretion of waste nitrogenous products in the form of uric acid. Being barely soluble in water, uric acid is excreted as a white paste by precipitation in the cloaca in birds and reptiles and in the rectum in insects. Precipitation of uric acid is induced by the re-absorption of salts (active transport) and water (passive osmosis). Uricotelic animals therefore do not need to waste water in the production of urine, which accounts for much of the success of uricotelic animals in deserts. In birds and reptiles, the excess salts that are re-absorbed are excreted by the salt glands. The disadvantage of uricotely in vertebrates is that the potential production of concentrated urine increases the likelihood of precipitation of uric acid crystals within the kidney. Consequently, birds and reptiles cannot concentrate their urine more than about two-to-three times that of the blood. Nevertheless, the ostrich is capable of producing urine with an osmolality of 800 mOsm kg^{-1} (Louw et al., 1969; Withers, 1983b), one of the highest values ever measured in birds. Interestingly, salt-gland function in the ostrich appears to be negligible (Gray and Brown, 1995). In mammals, which excrete urea, urinary water losses can be substantial. However, their efficient kidneys minimize losses considerably. By maximizing the counter-current multiplier effect, the relatively long Loops of Henle in the medulla of the kidney are capable of producing urine as much as 20 times more concentrated than that of the blood (Table 9.1). Gerbils (*Gerbillurus* and *Desmodillus* spp.) produce the most concentrated urine of the southern African mammals. Certain of these gerbils (*Desmodillus auricularis*, *Gerbillurus paeba*,

Table 9.1. *The maximum urine concentrating ability of desert mammals in southern Africa*

Species	Urine concentration. (mOsm kg⁻¹)	Urine/plasma osmolal ratio*
Rodents		
Gerbillurus vallinus[a]	7144	20.4
G. paeba[a]	6997	20.0
G. tytonis[a]	6129	17.5
Desmodillus auricularis[b]	6102	17.4
Gerbillurus setzeri[a]	5398	16.4
Thallomys nigricauda[c]	5364	15.3
Tatera leucogaster[d]	5000	14.3
Aethomys namaquensis[d]	3725	10.6
Rhabdomys pumilio[b]	3294	9.4
Elephant shrews		
Elephatulus myurus[e]	5210	14.9
Ungulates		
Madoqua kirkii[f]	4300	12.3
Antidorcas marsupialis[g]	3000	8.6
Oryx gazella[g]	2700	7.7

* Assuming a plasma osmolality of 350 mOsm kg⁻¹ for all mammals tabled.
[a] Downs and Perrin (1991)
[b] Christian (1978)
[c] J. Frean, Downs and Lovegrove (1998).
[d] Buffenstein et al. (1985)
[e] Du Toit (1993)
[f] Hofmeyr and Louw (1987)
[g] Maloiy (1973)

Tatera leucogaster) are also exceptional in their ability to produce allantoin in the urine (Buffenstein et al., 1985). Being less soluble than urea, allantoin production reduces urinary water losses by about 20%, albeit at a higher cost of production than that of urea.

As mentioned, birds, reptiles and insects reduce the amount of water which is lost in the faeces by producing a fairly dry faecal pellet following water reabsorption in the cloaca or rectum. Several mammals, such as rodents (e.g. ground squirrel *Xerus inauris*) and ungulates (e.g. Damara Dik-dik *Madoqua kirkii*), also minimize faecal water losses by reabsorbing water in the large intestine to produce dry faecal pellets (Maloiy, 1973; Haim et al., 1987). This adaptation probably applies to karoo ungulates as well.

Ruminants recycle 'waste' urea (from the kidney) via the saliva or the general circulation to the rumen where micro-organisms convert it into microbial protein (Le Bars, 1967; Maynard et al., 1979). Although the microbial protein is a valuable source of protein to the ruminant, urea recycling, and hence a lower water requirement for its excretion, is an important physiological mechanism for reducing urinary water loss in ruminants. However, this topic has not been investigated in desert ungulates in any detail in southern Africa.

9.2.2. Water gain

There are four sources of water available to desert animals: free-standing water, water in food, metabolic water, and atmospheric water (Louw and Seely, 1982; Schmidt-Nielsen, 1983; Wharton, 1985; Louw, 1993; Hadley, 1994). Free-standing water is rare in deserts, and those animals which rely upon it travel large distances, often daily, to water holes and other water sources (Knight, 1989). Being strong fliers, sandgrouse (*Pterocles* spp.) are well-adapted in this respect. They employ highly specialized chest and abdominal feathers to absorb large quantities of water (Winterbottom and Rowan, 1962; Cade and Maclean, 1967; Maclean, 1983; Cade et al., 1986). After soaking the feathers, adults fly back to the chicks which employ a 'feather-stripping' behaviour to extract the water from the feathers.

For most desert animals, food is the only reliable source of water. In fact, it is arguable that a true desert animal is one which is able to survive indefinitely without access to free-standing water. Deviations from regular or expected diets to more succulent diets to maintain water balance, especially when water-stressed, are not uncommon in desert animals. For example, the hairy-footed gerbil (*G. paeba*) in the Namib feeds on the leaves of *Trianthema hereroensis*, one of the few succulent plants in the Namib (Louw, 1972). Many rodents also readily include insects (c. 70% water content) in their diet. In the southern Kalahari Desert, about 50% of the diet of lactating brown hyaenas (*Hyaena brunnea*), which are predominantly carrion feeders, comprises the succulent fruit of Tsama melons (*Citrillus lanatus*) and gemsbok cucumbers (*Acanthosicyos naudinianus*) (Mills, 1990). These melons are most abundant during the cold and hot dry seasons in the Kalahari and are relied upon heavily by many animals, including birds, rodents, carnivores and ungulates, as an alternative water source to the free-standing water in pans during the hot wet season (Knight, 1995). The tubers of *A. naudinianus* are also a valuable source of water to mammals capable of excavating them and are often dug up in enormous numbers by gemsbok (*Oryx gazella*) during very dry years (Williamson, 1987; Lovegrove and Knight-Eloff, 1988; Knight, 1991). Troops of Chacma baboons (*Papio ursinus*) inhabiting the dry Kuiseb River bed in the Namib Desert are unique among terrestrial primates for their relative independence of drinking water. The baboons spend periods in excess of 11 days away from the few sources of drinking water available, but manage to balance their water budget by feeding on the fruits of *Salvadora persica* (69% water content) and, to a lesser extent, on figs (80% water content) of a few widely spaced *Ficus sycomorus* trees (Brain, 1988).

Many desert plants increase their water content at night by absorbing atmospheric water and fog. For example, perennial *Stipagrostis uniplumis* grasses are very hygroscopic and can increase their water content from 9% during the day to 26% at night (Louw, 1972). Consequently, ungulates tend to feed at night to maximize their food water uptake (Knight, 1991).

Fog is an invaluable water source for many Namib and succulent karoo animals, and the unusually high diversity of invertebrates in the Namib in particular can largely be attributed to the high incidence of fog along the west coast of southern Africa's deserts (Louw, 1972; Louw and Seely, 1982). Elaborate fog-capturing behaviour has evolved in several animals whose distributions are largely restricted to the fog-belt of the Namib Desert. Tenebrionid beetles, and some karoo tortoises (Boycott and Bourquin, 1988) employ 'fog basking' (*Onymacris unguicularis* and *O. bicolor*) and 'trench digging' (*Lepidochora discoidalis*) to indirectly access this atmospheric water vapour (Hamilton, 1976; Seely and Hamilton, 1976; Hattingh, 1984). When sprayed with a fine mist of water, Péringuey's Adder (*Bitis peringueyi*) immediately flattens its body to increase its dorsal surface area and moves its head up and down the body licking up the water droplets (Louw, 1972). It periodically raises the head to facilitate swallowing by gravity. Several insect, reptile and mammal species have been observed licking water droplets condensed from fog off vegetation and stones (Louw, 1972; Louw and Seely, 1982).

The ability to absorb atmospheric water directly from the air via a highly specialized cryptonephric system in the rectum is well-documented in the larvae of Namib tenebrionid beetles (Coutchie and Crowe, 1979). This morphological and physiological adaptation which relies upon powerful osmotic gradients to extract water from unsaturated air (85 RH% and above), compensates for the inability of the worm-like larvae to practice the fog-capturing behaviour of the adults.

The ability of certain animals to store excess water in their bodies is an adaptation which permits them to survive for extended periods without access to free-standing water. It is, however, surprisingly rare, and seems to be restricted to dune-dwelling animals within the Namib fogbelt. The lizard *Aporosaura anchietae* licks condensed fog off vegetation and the sand surface and can store as much as 10% of its body mass in water in its caecum following heavy fogs (Louw, 1972).

Studies on north African desert ruminants (camels, Bedouin goats, Zebu cattle) suggest that the rumen plays an extremely important role in storing water (Silanikove, 1994). Compared with monogastric animals and nondesert ruminants, these animals can withstand severe dehydration (18–40% of initial body weight), a capacity related to adaptations permitting the gradual use of rumen water (50–70% of total water loss) during dehydration. Unfortunately, very little is known about this ability in southern African desert ungulates. The risk which dehydrated ruminants face when imbibing a large amount of water in one session is that a huge rumen-blood osmotic

gradient of 200–300 mOsm kg^{-1} is generated which can cause a sudden drop in the blood osmolality and hence water intoxication. Silanikove (1994) has shown that desert ruminants overcome this problem by a) dramatically increasing hypotonic saliva secretion, and b) marked retention in the kidney of water, Na$^+$ and carbonic acid. Increased salivation recycles a considerable portion of the water absorbed by osmosis by the blood, whereas enhanced solute retention replaces ions, especially Na$^+$, drained from the blood during salivation. These adaptations permit the rumen to act as a water-storage vessel which is gradually depleted over the next dehydration cycle in response to the osmoregulatory demands of the animal. It would be interesting to establish the extent to which southern African ungulates, such as gemsbok, springbok, steenbok and eland display these adaptations. For example, to what extent do these animals rely upon hydration–dehydration cycles, or are they able to maintain continuous water balance by selective feeding (i.e. food water) alone?

Some gerbils, such as *D. auricularis* and *G. paeba*, the pygmy rock mouse, *Petromyscus collinus*, and the Namaqua rock mouse, *Aethomys namaquensis*, are able to balance their water budgets on a diet of air-dried seeds when deprived of water (Christian, 1978; Withers et al., 1980; Skinner and Smithers, 1990). These gerbils rely very heavily upon metabolic water formed during respiration. It has been calculated that metabolic water can comprise as much as 70% of the daily water requirements of *D. auricularis* (Christian, 1978).

9.3.   Thermal balance

Desert organisms face two main temperature problems: (a) extremes of temperature, both hot and cold, and (b) the rapid fluctuations of ambient temperature which can often occur within a short period of time (hours). For both ecto- and endotherms, adaptations permitting tolerance of extreme hot and cold temperatures are an obvious advantage.

9.3.1.   Thermal tolerance

Physiological tolerance of extreme temperature, e.g. freeze and heat tolerance, has not been studied in any detail in southern African desert organisms. Yet, certainly in the case of heat tolerance, some insects routinely tolerate extremely high body temperatures (T_b) which are normally fatal to most insects. For example, the critical thermal maximum (the T_b at which 50% of insects become inactivated by heat) of several ants in the Namib Desert exceeds 50 °C (Curtis, 1985c; Marsh, 1985b; Marsh, 1987a).

For most thermophilic insects, heat tolerance maximizes the chances of prey capture. Ant-lion (*Cueta trivirgata*) larvae tolerate very high body temperatures (LD_{50} = 53.4 °C) and remain vigilant in their pits despite sand surface temperatures reaching 65 °C (Marsh 1987a). The pits are slightly cooler than the sand surface temperatures. Their most important prey items (65%) are *Ocymyrmex robustior* ants (Marsh, 1987a). These thermophilic ants are active during the hottest hours of the day preying on heat-stressed insects. The ants are active at sand surface temperatures ranging from 27–68 °C with most individuals active at 50–55 °C (Marsh, 1987a). These behavioural and physiological heat tolerance adaptations therefore also represents feeding adaptations in both species since they could and do easily retreat beneath the sand surface when temperatures become extreme.

9.3.2.   Heat gain

Apart from the obvious behavioural response of retreating to thermal refugia during the day (see 9.5 Retreat), most adaptations to high ambient temperatures involve behavioural, physiological and morphological avenues of minimizing heat gain by the body. Physically, heat is gained by radiation and conduction.

The composition, architecture and colour of the body surfaces of many desert organisms are adapted to maximize the reflection of incidence radiation from the sun. Examples include the waxy blooms of tenebrionid beetles (McClain et al., 1985; Turner and Lombard, 1990), and the ability of reptiles, such as the Namaqua chamaeleon (*Chamaeleo namaquensis*), to lighten skin colour under high radiation loads (Burrage, 1973).

For small desert mammals and birds, the interaction between metabolic rate, insulation and body temperature determines the thermal extremes which they can tolerate (Lovegrove et al., 1991b). The range of thermal tolerance is also strongly influenced by the availability of water for evaporative cooling. Perhaps the best 'model' of thermal tolerance against which karoo rodents can be gauged is the black-tailed tree rat (*Thallomys nigricauda*), an arid-adapted arboreal rodent most frequently associated with camelthorn trees (*Acacia erioloba*). Being arboreal, this rodent is subjected to the full range of daily and seasonal temperature fluctuations and extremes. It does not retreat to thermally buffered burrows like most desert rodents. Black-tailed tree rats are capable of maintaining normothermia between temperatures of −8 °C to 43 °C in summer and −18 °C to 43 °C in winter (Lovegrove et al., 1991a). Even at temperatures of 43 °C they are able to maintain their body temperatures below 40 °C. This remarkable thermal tolerance is made possible by (a) a low basal metabolic rate (BMR), (b) a low thermal conductance (i.e. good insulation), and (c) access to sufficient food water. A low thermal con-

ductance not only prevents hypothermia at very low ambient temperatures (such as occurs in exposed trees in winter) by minimizing heat loss, it also retards heat gain at high ambient temperatures. However, a low thermal conductance can also potentially retard heat loss at high ambient temperatures which can lead to hypothermia even at moderate air temperatures (32–38 °C). This eventually is, however, offset by the low BMR permitting the animal to 'store' heat without adverse hyperthermia. In effect, a low BMR shifts the lower critical limit of thermoneutrality to a higher ambient temperature (Lovegrove et al., 1991a). At the highest ambient temperatures *T. nigricauda* must rely heavily upon evaporative cooling to avoid hyperthermia. It is presumed that the consequent high water demand is met by food water obtained from young *Acacia* leaves on which these animals feed.

Generally speaking, most nocturnal karoo rodents display low BMRs (see 9.4.1) and low thermal conductances for the same reasons. The low conductances permit the rodents to maintain normothermia on the ground surface at night during winter even during the coldest months and also help to maintain body temperatures in the mammalian range despite low BMRs. The biggest differences between these rodents and *Thallomys nigricauda* concerns their tolerance of high ambient temperatures. Without access to reliable sources of food water or free water, they cannot depend upon evaporative cooling and seem physiologically incapable of avoiding lethal hyperthermia at ambient temperatures above 38 °C (Lovegrove et al., 1991a). At this temperature, the body temperature of *Aethomys namaquensis* exceeded 41 °C (Lovegrove et al., 1991a). More water-conservative species, such as *G. paeba,* become hyperthermic at ambient temperatures as low as 35 °C (Buffenstein, 1984).

Certain diurnal, burrowing mammals, such as suricates (*Suricata suricatta*) and ground squirrels tend to display high thermal conductances in association with low BMRs (Marsh et al., 1978; Müller and Lojewski, 1986; Haim et al., 1987). These sparsely haired, small mammals are thus easily able to offload stored body heat rapidly by conduction to cooler sand surfaces within burrows by 'belly basking'. A high conductance also promotes conductive heat flux between huddling animals at night, and is probably important in rapidly dissipating locomotory heat by radiation and convection during escape dashes from predators. The disadvantage of a high conductance is that heat gain by the body is more rapid. However, in the case of *X. inauris*, this problem is offset by their behavioural use of the broad, bushy tail as an 'umbrella' to shield incident solar radiation from the body (Louw and Seely 1982).

For large desert mammals, such as ungulates, which cannot escape the heat of the day by entering burrows, adaptations which minimize heat gain are important. In this respect, the springbok is arguably the best adapted to desert conditions and especially to those of the karoo where shade trees are rare. In terms of heat flux, its pelage is unique (Hofmeyr and Louw, 1987). The white belly fur and long 'pronk' hairs on the rump reflect 72% of the incident radiation from the sun. The belly fur thus minimizes heat gained ventrally via reflectance off the desert surface, whereas the pronk hairs minimize dorsal heat gain. Behavioural orientation of the long axis of the body towards the sun's rays is important in this respect, as the fawn and brown fur on the flanks of the springbok reflect only about 40% of the incident radiation. Continuous observations of herds of springbok show that >60% of animals in a herd orientated the long axis of the body towards the sun at any one time. In contrast, only 32% of animals adopted this behaviour on clouded days (Hofmeyr and Louw, 1987). Indeed, most desert ungulates exposed to high radiation loads during the day orientate the long axis of their bodies parallel to that of the sun's rays to minimize total area exposed to solar radiation and hence irradiation (Wilson, 1989). In general, the thermal conductance of the Springbok and Steenbok (*Raphicerus campestris*) is much higher than that of all other African ungulates investigated to date (Hofmeyr and Louw, 1987). This can be attributed to the fur on the flanks which, for the body size of the animal, is considerably shorter than it is for other ungulates. The explanation offered for the high conductance is that it facilitates the rapid convective loss of large amounts of heat generated by locomotion during frantic escape dashes from predators.

The thermal significance of animal colouration, specifically the significance of black body colouration and heat gain, has been a subject of much debate (Hamilton and Heppner, 1967; Walsberg et al., 1978; Turner and Lombard, 1990; Louw, 1993; Walsberg and Wolf, 1995). Whereas some authors de-emphasize the importance of black body colour in desert beetles (Turner and Lombard, 1990), others recognize its importance (Hamilton and Heppner, 1967; Hamilton, 1973; Wharton, 1980). In the Black Homeotherm Pigmentation Hypothesis, Hamilton and Heppner (1967) suggested that endotherms with black pelages or plumages conserve a component of obligatory thermoregulatory energy in the morning and evenings by absorbing the equivalent energy component from the sun's radiation through the enhanced heat loading of black bodies by short-wavelength electromagnetic radiation. This hypothesis has received some support from studies in the Negev Desert which have shown that black Bedouin goats in winter used 25% less energy per day than white goats (Louw, 1993). The disadvantage of higher heat loading in summer may be offset by the reliability of afternoon winds in deserts and the observation that at wind speeds higher than about 3 m s^{-1}, black animals absorb less heat than white ones (Walsberg et al., 1978). In this respect, birds (males mostly) which may benefit from a black plumage in the karoo, Kalahari and Namib are ostriches, black korhaans (*Eupodotis afra*), black-eared finchlarks (*Eremopterix australis*) and grey-backed finchlarks (*E. verticalis*). On the other hand, black is a visually conspicuous colour in desert landscapes, and may alternatively serve an aposematic function (Cloudsley-Thompson, 1979) since the vast majority of desert birds are light-coloured for crypsis (Maclean, 1996).

It has been argued that all ruminants, camels, canids, and felids, possess the ability to store heat in the body by selective brain cooling; keeping the brain temperature below that of the body (Taylor, 1969; Laburn et al., 1988; Kuhnen and Jessen, 1991; Kuhnen and Jessen, 1994). This is made possible by a carotid rete countercurrent blood-flow system below the brain which cools the hot arterial blood from the body before it enters the brain (Laburn et al., 1988). The blood is cooled by venous blood returning from the moist nasal regions. Oryx (*Oryx beisa*), for example, permit their body temperatures to rise to 45 °C, whereas that of the brain hardly exceeds 40 °C (Taylor, 1969). The gemsbok is likely to display a similar capacity to store heat (Wilson, 1989). The advantage of storing heat in the body is that water is not wasted on evaporative cooling to maintain a constant body temperature. Heat storage can therefore be employed by the larger desert mammals that warm up more slowly during the course of the day than small animals because of their relatively smaller surface-area-to-volume ratios. Stored heat is then offloaded passively at night by convection, for example by retiring to high ridges or dunes to maximize convective cooling by breezes. It has also been pointed out that selective brain cooling is crucial for all large mammals which possess a fur coat, and which need to sprint to escape predators or to catch prey (Hofmeyr and Louw, 1987; Louw, 1993).

However, recent telemetric studies of the blood and brain temperatures of free-ranging black wildebeest (*Connochaetes gnou*) confirm that not all African ungulates rely upon heat storage as an alternative to evaporative cooling (Jessen et al., 1994). Black wildebeest display a circadian body temperature pattern consistent with an endogenous T_b rhythm with an amplitude not exceeding 1 °C, and possess an efficient capacity to dissipate heat by panting and respiratory evaporative cooling. The long snout of both species of wildebeest facilitate nasal evaporative cooling (Jessen et al., 1994).

The ostrich, although flightless, has well-feathered wings which, when held forward and to the side effectively shield the body from incident radiation (Louw et al., 1969; Louw, 1993). The shade provided is also crucial for young chicks. The ostrich can effectively manipulate the sparse plumage on its dorsal surfaces. At high ambient temperatures the feathers are erected permitting lateral air movement across the skin surface and hence convective cooling.

9.3.3.   Heat loss

Heat is lost from the bodies of animals by convection (forced and free), conduction, radiation and evaporation. Forced convection by winds and breezes, is the most important avenue of heat loss available to desert insects (Turner and Lombard, 1990). However, small, ground-dwelling desert insects, such as ants, are particularly susceptible to high ambient temperatures because they spend much of the time within the thermal boundary layer between the ground and the air and thus cannot easily rely upon forced convection to offload heat. Thermophilic ants, such as *Ocymyrmex barbiger* (4 mg) from the Namib Desert, overcome this problem by practising 'thermal respite' behaviour (Marsh, 1985b). They forage for heat-stressed insects during the hottest hours of the day when sand surface temperatures are as high as 67 °C. As their body temperatures approach critical levels, they climb onto objects such as pebbles which lifts them out of the boundary layer and they can then offload heat by convection (Marsh, 1985b). As sand surface temperatures increase, the frequency of respiting increases to about eight respites per minute. Because the ants have such a large surface area to volume ratio, they can offload heat in 5–80 sec.

Many insects lift their bodies as far away from the hot sand surface by employing stilting behaviours; the full extension of some or all legs (Dreisig, 1990). In southern Africa this behaviour is perhaps best illustrated by the tenebrionid beetle *Stenocara phalangium* which inhabits the hot, gravel interdune 'streets' of the Namib. This beetle cannot retire below the ground by burrowing or retreat beneath shade plants as do most tenebrionid beetles in the Namib dune sea. It is almost entirely dependent upon convective cooling in full sunshine throughout the day. *S. phalangium* has the longest legs relative to its body size of all the tenebrionid beetles (Henwood, 1975). By bending or extending its legs and thus moving the body closer or further away from the ground surface, it can regulate heat gain from the ground and heat loss out of the boundary layer, depending on its body temperature (Henwood 1975). During the very hottest hours of the day it climbs onto pebbles and practices a motionless stilting behaviour pointing the lighter-coloured posterior abdomen towards the sun thus minimizing absorbance of incident radiation (Henwood, 1975). One beetle of the Namib dune sea, *Onymacris plana*, maximizes convective heat loss not by seeking out breezes, but by creating its own – it is one of the fastest runners of all land insects (Nicolson et al., 1984a).

Other ectotherms, such as lizards, also rely heavily upon forced convection for cooling. For example, a specialized behaviour known as 'breezing' is employed by the Namib day gecko, *Rhoptropus afer* (Odendaal 1979). Whenever breezes occur during the hottest hours of the day, the geckos assume the breezing posture on the highest point on a rock and lift their bodies as high as possible off the hot rock surface. They also orientate the line of the body with respect to the sun's rays thus minimizing irradiation (Odendaal 1979).

For endotherms (birds and mammals) which maintain constant body temperatures, heat can be dissipated from the body by convection provided that the T_b is greater than T_a, i.e. $(T_b - T_a) > 0$. In this respect birds are preadapted to high ambient temperatures simply because their body temperatures are about 2–6 °C higher than those of mammals (Prinzinger et al., 1991; Maclean 1996). Higher body temperatures maintain a greater temperature gradient between the body and the ambient air $(T_b - T_a)$ and hence a greater 'driving force' for non-evaporative conductive and convective heat loss. Theoretically, birds are therefore capable of tolerating higher ambient temperatures than mammals. However, desert birds do not display higher T_bs than their mesic counterparts (Prinzinger et al., 1991), so selection for high body temperatures in birds is not considered to have been influenced by thermoregulatory considerations (Maclean, 1996). For mammals, the ability to dissipate heat is a problem if ambient temperatures approach body temperatures, which they regularly do on a daily basis in the southern African deserts, especially in summer. If small mammals have no recourse to evaporative cooling to avoid hyperthermia they are obliged to be nocturnal and/or live in burrows or thermal refugia (see 9.5). Large mammals rely upon heat storage and regional heterothermy discussed earlier.

Although birds benefit from a higher $T_b - T_a$ gradient, feathers have a lower conductance than mammalian fur and so heat dissipation is slow (Schmidt-Nielsen, 1983; Withers, 1992). Desert birds overcome this problem by employing a variety of behavioural and morphological adaptations which effectively create high conductance 'thermal windows' to facilitate heat dissipation by convection and radiation (Maclean, 1996). Typically, heat-stressed birds droop their wings and hold them away from the body to expose the more sparsely feathered parts of the body normally covered by the wings (Thomas and Maclean, 1981; Louw and Seely, 1982; Maclean, 1996). When holding its wings laterally, the Ostrich exposes the lateral thermal windows of the thorax (Louw et al., 1969; Louw, 1993). At low ambient temperatures the dorsal feathers are interlocked, air is trapped between the skin surface and the feathers, and conductance and hence heat loss is reduced.

Mammals can less easily create thermal windows with postural adjustments, but small diurnal mammals such as suricates and ground squirrels promote heat loss by conduction by practising belly-basking in cool, shaded sand (Marsh et al., 1978; Van Heerden and Dauth, 1987). Indeed, all rodents employ thermoregulatory behaviour at high ambient temperatures by lying flat on the ground in a stretched out posture (Gordon, 1993). This behaviour increases the surface areas of the rodent relative to its volume thus facilitating conductive heat loss from the ventral surfaces as well as convective heat loss in general (Gordon, 1993).

At ambient temperatures close to or exceeding body temperatures, evaporative cooling represents the sole remaining avenue of heat loss. Animals which cannot afford to 'waste' water on cooling must either retreat to thermal refugia (see 9.5.4), or they will become hypothermic and die. As discussed earlier (see 9.3.2), the efficiency of evaporative cooling is therefore entirely dependent upon the water resources available to the animal.

As a rule, insects do not employ evaporative cooling (Hadley, 1994). However, the Sonoran desert cicada, *Dicerprocta apache*, is one of the few insects which is known to maintain its T_b below that of the ambient by sweating via specialized sweat pores on the cuticle (Toolson, 1987; Toolson and Hadley, 1987; Hadley, 1994). These insects can thus remain active during the hottest hours of the day, provided that they simultaneously suck sap from desert plants to replace the lost water. No similar examples have been reported from the southern African deserts. Although desert reptiles lose heat mostly by shuttling to shade, they also rely upon evaporative cooling generated by panting, mostly at ambient temperatures above 40 °C. Birds promote evaporative cooling using gular flutter; the rapid oscillation of the floor of the mouth which rapidly pumps air in and out of the mouth cavity (Louw and Seely, 1982; Maclean, 1996). For example, the ostrich increases the frequency of oscillations from 4 min^{-1} at T_a = 36 °C in windy conditions to 40 min^{-1} at T_a = 35 °C in windless conditions (Louw, 1993). The frequency of fluttering can be as high as 1000 min^{-1} in small birds in extreme heat (Louw, 1993). For small ungulates, the important avenue of evaporative heat loss (EHL) is also respiratory, via panting. The springbok increases its respiration rate from 40 min^{-1} at 22 °C to 275 min^{-1} at 60 °C (Hofmeyr and Louw, 1987).

9.4.   Energy balance

Desert animals often walk an energetic 'tightrope', having to balance the energy requirements of a daily existence,

thermoregulation, activity and locomotion, and reproductive costs, with the meagre energy resources available in their low-productivity arid environments. Selection for adaptations which minimize energy costs is strong in desert animals, and is mostly the topic of this section.

9.4.1.   Basal metabolic rates
All desert mammals of the world have low metabolic rates which minimize maintenance costs (Bradley et al., 1974; McNab, 1979; Maloiy and Kamau, 1982; Lovegrove, 1986, 1996; Withers, 1992). In the Palaearctic and Nearctic zoogeographical zones the basal metabolic rates (BMR) of desert mammals is markedly lower than those of mesic counterparts (Lovegrove, 1996, unpublished data). This large difference does not exist between the desert and mesic mammals of the Indo-Malayan, Afrotropical and Australian zones (unpublishd data). The hypothesis for these regional differences is that the latter zones are subjected to prolonged, negative rainfall anomalies (droughts) caused by phenomena such as the El Niño Southern Oscillations (ENSO), which have a more widespread influence over the southern continents (mesic and desert regions) than they do in the northern hemisphere (Lovegrove, 1996). Prolonged droughts represent ecological bottlenecks during which primary production and hence food availability is very low. It is argued that, for small mammals in particular, which cannot migrate out of affected areas (see 9.5) and which are burdened with high mass-specific metabolic rates (McNab, 1970; McNab, 1974, 1983; Schmidt-Nielsen, 1983; Hayssen and Lacy, 1985), strong selection for low metabolic rates may have occurred during ENSO events as a means of reducing overall daily energy expenditure. Low metabolic rates also significantly reduce respiratory water losses.

9.4.2.   Circadian metabolic rhythms
As a single measure of an animal's metabolic rate, the BMR is, however, not very informative. BMR is typically measured in resting animals during the rest phase. It has been argued that a low BMR is more indicative of a larger-than-average fluctuation of metabolic rate between the animal's rest and active phase (Lovegrove and Heldmaier, 1994). In other words, normothermic desert mammals permit their metabolic rates to drop to lower levels during the rest phase, and hence benefit from greater overall energy savings on a daily basis (Lovegrove and Heldmaier 1994). The amplitude of these circadian fluctuations are easily measured by monitoring circadian body temperature rhythms. Desert rodents such as *Thallomys nigricauda* and *Aethomys namaquensis* display circadian amplitudes of T_b which are nearly double those expected for their body sizes (Lovegrove and Heldmaier, 1994).

9.4.3. Torpor

Daily torpor is an extreme exaggeration of the above response (Lyman et al., 1982; Berger, 1993; Kilduff et al., 1993). The metabolic rate is substantially reduced during the animal's rest phase resulting in normothermic body temperatures being abandoned. The resultant adaptive hypothermia leaves the animal in a physiological state in which it is incapable of normal locomotory function but in which it remains perceptive to some, but not all, external stimuli. It is capable of arousal by endogenous heat production generated by non-shivering thermogenesis (NST) immediately prior to the onset of the activity phase.

Although torpor is regularly employed by many Australian, Neotropical, Palaearctic and Nearctic birds and mammals in response to cold and food deprivation (Lyman and O'Brien, 1974; Prinzinger et al., 1986; Bozinovic and Rosenmann, 1988; Prinzinger and Siedle, 1988; Caviiedes-Vidal et al., 1990; Hiebert, 1990; Snyder and Nestler, 1990; Bozinovic and Marquet, 1991; Brigham, 1992; Bucher and Chappell, 1992; Ruf and Heldmaier, 1992; Withers, 1992; Geiser and Broome, 1993), the incidence of torpor use by Afrotropical mammals has received little attention and is restricted to a handful of studies. The pouched mouse (*S. campestris*), which occurs throughout the semi-arid regions of southern Africa, employs daily torpor in response to low temperatures (Ellison and Skinner, 1992; Ellison, 1993) and food shortage (Lovegrove and Raman, 1998). A recent study on the round-eared elephant shrew (*Macroscelides proboscideus*) has shown that this omnivorous (Kerley, 1989, 1995) karoo mammal employs torpor in response to low ambient temperatures, and the frequency of torpor use increases markedly with food deprivation (M. Lawes, L. Roxburgh and B. G. Lovegrove, unpublished data). Daily torpor is also used extensively by several genera of golden moles (*Amblysomus*, *Eremitalpa* and *Chrysochloris*) (Kuyper, 1979; Fielden et al., 1990; Bennett and Spinks, 1995), the woodland dormouse, *Graphiurus murinus* (Ellison and Skinner, 1991), the spectacled dormouse, *Graphiurus ocularis* (personal observations), and has been reported in the field in *Petromyscus collinus* and *Aethomys namaquensis* (Withers et al. 1980). Daily torpor is probably also employed by several bat species, although this has not been adequately documented. Although adequate data are lacking, sunbirds are also suspected of employing daily torpor.

It is likely that torpor is much more prevalent among southern African small mammals and birds than currently appreciated. For example, it is realistic to expect that some of the smallest species of gerbils of the genera *Gerbillurus*, *Tatera* and *Desmodillus* might employ torpor in response to food deprivation and perhaps even water deprivation. Also, given the high incidence of torpor use by non-African nectarivorous and insectivorous birds (Prinzinger et al., 1986; Prinzinger and Siedle, 1988; Hiebert, 1990; Brigham, 1992; Bucher and Chappell, 1992; McNab and Bonaccorso, 1995), it is reasonable to expect similar physiological adaptations to have evolved in the Afrotropical ecological counterparts (e.g. sunbirds, nightjars).

There are, perhaps, important factors, such as diet, which may prevent certain species from entering torpor in captivity. For example, certain mammals can access stored fat energy reserves during torpor only if the melting point of the fat is decreased below the normal melting point (*c.* 25 °C) by the presence of significant amounts of polyunsaturated fatty acids, such as linoleic and α-linolenic acids in the cell membranes (Frank, 1992; Geiser et al., 1992; Geiser and Kenagy 1993; Frank 1994). Also, the incorporation of polyunsaturated fatty acids into the cell membranes increases membrane fluidity thus maintaining cellular physiology at low body temperatures. Since mammals cannot synthesize polyunsaturated fats themselves, hibernating mammals must ingest plant foods rich in polyunsaturated fatty acids prior to hibernation events. Therefore, if captive animals are not fed a diet with the appropriate polyunsaturated fat content, they are probably unlikely or unable to enter torpor successfully. This is especially relevant if the animal relies upon the energy of stored fat during torpor.

In this latter respect, it is interesting that *Macroscelides proboscideus* includes a significant proportion of plant material in its diet (Kerley, 1989, 1995). Until recently, these animals were thought to be exclusively insectivorous. Interestingly, the morphology of their digestive systems shows no obvious adaptation which could promote cellulose digestion. However, since these animals employ torpor in response to unpredictable food availability (M. Lawes, L. Roxburgh and B. G. Lovegrove, unpublished data), it is possible that their vegetarian habits may be partly related to the need for polyunsaturated fatty acids.

At least one karoo rodent, the spectacled dormouse, that inhabits mountainous regions (often above the snow line), is a true hibernator (personal observations). This animal hibernates for periods of up to 13 days without arousal. Hibernation is rare in African mammals, and is thought that it has evolved in the spectacled dormouse in response to low temperatures and the high probability of snow cover, and hence poor food availability, during winter.

9.5. Retreat

Problems balancing water, thermal and energy budgets necessitate that animals retreat behaviourally and physiologically from prevalent desert conditions (Louw and Seely, 1982; Lovegrove, 1993). These retreats can be categorized as either short-term (hours or days), or long-term (weeks, months, whole seasons, years).

9.5.1.   Short-term retreat

Short-term retreat is particularly important to small animals and ectotherms (e.g. insects, spiders, reptiles). They have higher surface-area-to-volume ratios, and therefore experience greater rates of water, heat and energy fluxes per unit of body mass than larger animals (Louw and Seely, 1982; Schmidt-Nielsen, 1983; Louw, 1993). Ectotherms do not possess the efficient thermoregulatory mechanisms of the endotherms (such as evaporative cooling by sweating) to avoid extreme body temperature fluctuations. Small endotherms also cannot rely upon evaporative cooling to maintain constant body temperatures; rodents weighing 10–100 g would loose 15–30% of their body water per hour under hot desert conditions (Schmidt-Nielsen, 1983).

Various forms of daily physiological retreat such as large amplitudes of circadian metabolic rhythms (9.4.2) and torpor (9.4.3) have already been discussed. The easiest behavioural escape from stressful desert conditions other than the construction of a nest structure is to retreat under the ground or under stones. Most small southern African desert animals are therefore either good burrowers themselves, or routinely occupy burrows excavated by other animals. Scorpions of the Buthidae and Scorpionidae are abundant throughout the arid regions of southern Africa (Lamoral, 1979). Generally, buthid scorpions which have slender claws (thick tails), tend to live under rocks and do not excavate their own burrows, whereas the Scorpionidae have large, powerful pedipalps (slender tails) and are generally excellent burrowers (Hadley, 1974). All 24 species of *Opisthopthalmus* (Scorpionidae) in Namibia construct their own burrows, whereas species of the genera *Buthotus*, *Parabuthus*, *Uroplectes* and *Karasbergia*, all of the family Buthidae, live in narrow scrapes under rocks (Lamoral, 1979). Burrow entrances of scorpions are easy to distinguish from those of other arthropods because they are all oval in cross-section and have a fan-shaped mound of sand or soil radiating away from the entrance (Lamoral, 1979). The average configuration of the burrows of *Opisthopthalmus* spp. is a vertical spiral burrow (one to three turns) starting 10–20 cm from the burrow entrance and ending in a chamber large enough to enable the scorpion to turn around (Lamoral, 1979).

Most ant species are subterranean nesters in the southern African deserts, although several also nest under stones. For example, at Tierberg in the southern karoo, ants selectively choose quartzite, sandstone, Dwyka tillite and dolerite stones under which to nest (Dean and Turner, 1991). In summer, the stones reduce the daily fluctuation

of temperatures compared with the soil profile (Dean and Turner, 1991). Other insects exploit the thermal benefits of the ant nests and the protection afforded by the ants. For example, there are 29 species of butterflies belonging to the family Lycaenidae which are restricted to the Nama- and succulent karoo biomes (Clark and Dickson, 1971). Of these, 17 species pupate or spend part of their larval stages within formicine ant nests (e.g. *Anoplolepis custodiens*, *Camponotus* spp. and *Crematogaster* spp.), a proportion which is significantly higher than in other more mesic biomes (Clark and Dickson, 1971). Bees, especially solitary species, and wasps, are also excellent burrowers in arid regions (Gess and Gess, 1993). Wasps of the Masarinae and Eumeniae tend to excavate burrows in hard, non-friable soils, whereas species of Nyssoninae prefer sandy soils (Gess and Gess, 1993).

For rodents which have access to limited food water, for example those that include a significant proportion of seeds in their diet (there are no strictly granivorous rodents in the southern African deserts), avoidance of daytime desiccating conditions is essential. Seeds generally have a much lower water content than green plant material or fruits (Schmidt-Nielsen, 1983). These animals are therefore obliged to be nocturnal burrowers and, without exception, include all three genera of gerbils (*Desmodillus*, *Gerbillurus* and *Tatera*) in southern Africa. They are morphologically adapted to a nocturnal life style, typically possessing large eyes and well-developed ear bullae (Skinner and Smithers, 1990). Owing to its small size (c. 25 g) and the problems of water loss associated with being small (Schmidt-Nielsen, 1983), the burrowing behaviour of *Gerbillurus paeba* is highly specialized to minimize evaporative water losses within the burrow. These gerbils always plug the burrow entrance after entry, which serves to minimize water losses to the outside air by maintaining through moist, expired air in the burrow a high RH% and hence a low water-vapour saturation deficit.

Of the larger mammals, the most prodigious burrowers are the aardvark (*Orycteropus afer*), and the springhare (*Pedetes capensis*) (Skinner and Smithers, 1990). At least 17 other mammals species are known to inhabit aardvark burrows (Skinner and Smithers, 1990). Several mammals, such as the aardwolf (*Proteles cristatus*), and the bat-eared fox (*Otocyon megalotis*), are good burrowers themselves, but will readily alter aardvark and springhare burrows to their needs. The ant-eating chat, *Myrmecocichla formicivora*, nests in the roof of burrows originated by aardvarks (Maclean, 1993), and may be dependent upon the presence of aardvarks for successful breeding.

9.5.2.   Burrowing adaptations

Morphological adaptations for burrowing in desert arthropods and reptiles has been well reviewed by

Cloudsley-Thompson (1991) and will thus receive brief treatment here. The limbs of psammophilous insects, lizards and geckos are adapted for digging as well as for weight support on loose sand. *Aporosaura anchietae*, for example, has exceptionally long toes lined with broad scale fringes permitting it to move very fast across the loose sand of dunes (Branch, 1988a). The web-footed gecko (*Palmatogecko rangei*) is unique in having the toes joined by webbing on the fore- and hindlimbs. Although these 'sand shoe' feet must certainly aid the animal in moving over loose sand, the webbed feet have evolved primarily to aid in digging burrows and excavating large volumes of loose sand (Russel and Bauer, 1990). The dune cricket of the Namib Desert (*Comicus* spp.) displays several ultrapsammophilous adaptations (Scholtz and Holm, 1985). The tarsal joints on the legs bear long, fleshy projections which markedly increase the surface area of the limb in contact with the sand surface. These crickets are also apterous, unpigmented, and have well-developed hind limbs for jumping and predator escape.

In certain ultrapsammophilous insects (e.g *Onymacris unguicularis*), reptiles (e.g. *Aporosaura anchietae*, *Angolosaura skoogi*) and mammals (*Eremitalpa granti*), sand-diving behaviour has evolved which permits these animals to 'swim' through the loose, dry sand of dunes to escape predators and desiccating conditions (Louw and Seely, 1982). Morphological adaptations which enhance the efficiency of this behaviour by increasing streamlining are smooth body surfaces, shovel- or wedge-shaped snouts (lizards and moles), and reduced appendages (e.g. lack of ear pinnae in moles). For *A. skoogi* it has been suggested that activity periods in excess of about four hours per day can potentially result in negative water and energy balances (Clarke and Nicolson, 1994). In these herbivorous lizards, evaporative water loss increases tenfold when the lizards are active on the sand surface. However, they have a reliable plant food source (*Acanthosicyos horridus*) which has a high water content and so they do not need to spend long hours foraging to balance energy and water budgets (Clarke and Nicolson, 1994).

The strong spatial niche separation of lepismatid fishmoths in the Namib dunes is emphasized by obvious differences in morphology (Watson, 1989). Compared with the familiar long, thin bodies of surface foraging fishmoths such as *Ctenolepisma pauliani* and *Ctenolepisma terebrans*, subsurface burrowers such as *Namibmormisma muricaudata* and *Hyperlepisma australis* have broad heads, a tapered humpbacked body and short appendages (Watson, 1989).

9.5.3.   Circadian rhythms of retreat

The timing of the active and rest phases in most animals is controlled by endogenous circadian 'clocks' which synchronize the animal to cycles of predator risks, physical environmental variables and food availability, thus optimizing fitness (Aschoff, 1964; Enright 1970; Holm and Edney 1973; Daan and Aschoff, 1982; Aschoff, 1984; Marsh, 1988). These endogenous circadian rhythms are arguably particularly well developed in desert organisms where daytime physical conditions are harsh (Cloudsley-Thompson, 1991; Applin et al., 1993). Whereas most desert species display nocturnal activity rhythms, there are a surprisingly large number of diurnal species. Diurnal species are generally well adapted to heat . Whether diurnal of nocturnal, the endogenous pattern and expression of many circadian activity rhythms is influenced by season, surface temperature, wind and competition for food (Holm and Edney, 1973). One common seasonal adjustment is that bimodal (usually crepuscular) diurnal patterns in summer become unimodal in winter (centred around midday) in tenebrionid beetles, ants and fishmoths (Wharton, 1980; Curtis, 1985b; Marsh, 1988). Activity peaks thus avoid extreme heat during midday in summer, but shift to the midday hours in winter as temperatures are less extreme (conversely, an avoidance of cool mornings and evenings).

Species-specific emergence patterns are not, however, always predictable. The harvester termite (*Microhodotermes viator*) minimizes predation risks by emerging to forage for plant material on the surface at unpredictable times during the day, irrespective of temperature, humidity or rainfall patterns (Dean, 1993). Unpredictable foraging precludes predators from synchronizing their circadian feeding periods with those of the termites. Also, the intense interspecific competition for food accounting for the spatial niche separation patterns of lepismatid fishmoths in the Namib dunes has also selected for strong temporal niche separation patterns (Watson, 1989). The six species which inhabit the dunes fall into one of three general patterns of activity; diurnal to crepuscular, early nocturnal and late nocturnal (Watson, 1989).

9.5.4.   Nest shelter retreats

Elaborate nest-building behaviour, not seen in many mesic species, has evolved in several karoo vertebrates. Although these nests are utilized for breeding and predator escape functions as well, they maintain a favourable physical internal environment (buffered temperatures and RH%) which reduces water losses. Certain rodents construct grass nests above burrow entrances (*Aethomys namaquensis*), or twig nest 'lodges' either on the ground surface (*Otomys unisulcatus*) or in trees (*Thallomys nigricauda*) (Skinner and Smithers, 1990; Lovegrove 1993). In the nests of *O. unisulcatus*, temperatures are buffered against diel ambient temperature fluctuations and fall within the thermoneutral environment of the animal. Water-vapour pressures in the burrow are higher than those of the ambient, in both winter and summer (Du Plessis and Kerley, 1991; Du Plessis et al., 1992). Relative humidities of 64–74% in summer and 56–83% in winter were recorded inside the nest structure (Du Plessis et al., 1992).

Nest-building behaviour is taken to the extreme by the sociable weaver (*Philetairus socius*) that builds huge communal grass nests in trees or any suitable, available structure (Maclean, 1973, 1993). These nests also buffer daily temperature extremes maintaining internal temperatures above and below mean ambient temperatures in winter and in summer, respectively (Bartholomew et al., 1976; White et al., 1981; Avery, 1990b; Lovegrove, 1993). Pygmy falcons (*Polihierax semitorquatus*) are entirely dependent upon these nests for roosting and breeding (Maclean, 1993).

9.5.5.   Long-term retreat (diapause, migration and nomadism)

In invertebrates, long-term retreat from deserts takes the form of diapause, whereas, in vertebrates, it involves seasonal and predictable movement out of the area (migration) or nomadic movement associated with localized rainfall patterns (nomadism).

During diapause, the development of invertebrate eggs, embryos and larvae is arrested during unfavourable periods, sometimes for many consecutive years. Diapause is induced and terminated by environmental factors such as photoperiod and ambient temperature, although rainfall, relative humidity, soil moisture content, water salinity, water oxygen concentration and light intensity invariably terminate diapause in very unpredictable desert environments (Belk and Cole, 1975; Matthee, 1978; Louw and Seely, 1982; Friedländer and Scholtz, 1993; Higuchi, 1994; McNamara, 1994). Although many desert arthropods employ diapause, it is particularly important to the inhabitants of temporary or ephemeral pools and pans which occur throughout the deserts of southern Africa (Belk and Cole, 1975; Louw and Seely, 1982). The invertebrates which inhabit these depressions – anostracans, conchostrachans, cladocerans, copepods, ostracods and notostracans (Rayner and Bowland, 1985a, 1985b; Day, 1990) – have extremely fast growth rates and are tolerant of large fluctuations of water temperature, salinity and oxygen content (Belk and Cole, 1975; Louw and Seely, 1982). The egg of temporary pond crustacea is actually a shelled embryo capable of cryptobiosis or ametabolism for considerable periods of time during dormancy (Belk and Cole, 1975; Lovegrove, 1993). The shell serves a protective rather than a waterproofing function, for the embryos can withstand extreme dehydration (Belk and Cole, 1975).

Migrations can only be undertaken by animals which have a low cost of locomotion, limiting the option to large

mammals and birds in the deserts of southern Africa. For small mammals, the cost of locomotion is prohibitive (Taylor et al., 1970; Louw, 1993; Langman et al., 1995). Prior to the rinderpest epidemic in 1896 which markedly thinned ungulate herds in southern Africa, herds of springbok, reportedly numbering millions of animals, undertook periodic 'emigrations' from the summer rainfall regions of the Nama-karoo and southern Kalahari Desert (Cronwright-Schreiner, 1925; Eloff, 1959a, 1959b, 1961, 1962; Lovegrove, 1993; Skinner, 1993). From historical accounts (Cronwright-Schreiner, 1925), it seems that these movements (*treks*) were mostly unpredictable and aseasonal. Recent interpretations (Lovegrove, 1993; Skinner, 1993) suggest that they probably coincided with poor summer rangeland condition in the Nama-karoo caused by (a) large population numbers and overgrazing, and/or (b) periodic droughts as a component of the rainfall cycles in southern Africa (Tyson, 1987), and/or those caused by ENSO events. There were three general patterns of emigration: (a) southward, out of the southern Kalahari Desert and into the Nama-karoo, (b) westward, out of the Nama-karoo and into the winter rainfall regions of the succulent karoo biome, and (c) eastwards, out of the southeastern karoo towards the Eastern Cape (Skinner, 1993).

The option of nomadism is also determined by the cost of locomotion (kJ g^{-1} km^{-1}) and is thus common in desert birds – flight is a considerably cheaper option of movement than either walking, running or hopping (Withers, 1992). Typical nomads of the karoo, Namib and Kalahari are the various desert-dwelling larks (red-capped *Calandrella cinerea*, pink-billed *Spizocorys conirostris*, Stark's *Eremalauda starki*, grey-backed and black-eared finchlarks) and the Namaqua dove (*Oena capensis*), which move from one region of localized rainfall to another in search of food (Willoughby, 1971; Maclean, 1970a, 1971, 1984, 1993, 1996; Dean and Hockey, 1989). Among the larger animals, the well-known nomads are ostriches, springbok, gemsbok and eland (Berry and Siegfried, 1991; Skinner and Smithers, 1990).

9.6.   Sociality and co-operative breeding

Although sociality, eusociality and co-operative breeding have evolved in a number of diploid, haploid and haplodiploid animal groups for a number of debatable reasons (Andersson, 1984; Tyson, 1984; Rubenstein and Wrangham, 1986), it is fair to argue that the incidence of social organization is high in desert animals. Moreover, the degree of sociality is particularly well developed in some animal groups. For example, the African subterranean rodents (Bathyergidae) are the most social of all the mammals and aridity gradients appear to be the driving force behind selection for the social systems observed in these rodents (Lovegrove, 1991; Jarvis and Bennett, 1993). It is possible to identify a number of desert constraints which can impede an individual's fitness but which can be overcome through cooperation and group living to optimize the inclusive fitness of individuals within groups. These include food acquisition, nest building, predator avoidance, territory defence, thermoregulation and even the maintenance of water balance. Examples of each are provided.

Ants and termites are probably no more social in deserts than they are in more mesic habits, but their abundance throughout the desert regions in southern Africa (Coaton, 1958, 1963; Coaton and Sheasby, 1974, 1975; Willis et al., 1992) confirms the success of sociality in these insects. The benefits probably include all those listed above. Common termites in the karoo, such as the harvester termites *Microhodotermes viator* and *Hodotermes mossambicus*, are preyed upon by a wide variety of animals including arthropods, lizards, birds and mammals (Cooper and Skinner, 1979; Haarhoff, 1982; Dean, 1988, 1989; Bauer et al., 1989; Kok and Hewitt, 1990; Willis et al., 1992; Milton et al., 1993). Although harvester termites are generally defenceless when foraging on the surface (Dean, 1989), cooperation in nest building minimizes losses to predators. The termites emerge from foraging tunnels at unpredictable times of the day (see 9.5.3) in great numbers, but only remain on the surface for about an hour. The advantage of sociality in this respect is that large amounts of plant material can be harvested in a short period of time, thus minimizing predation risks. The dominant ant in the dune sea of the Namib is *Camponotus detritus* (Curtis, 1985a, 1985b, 1985d; Curtis and Seely 1987). The large body size of this ant coupled with its extreme intra- and interspecific aggression has effectively eliminated all other ants species from the dune sea. *Camponotus detritus* vigorously defend their food plants against invaders and predators of scale insects (Curtis, 1985a).

The high degree of sociality within the Bathyergidae has evolved in response to the high individual risks of foraging by burrowing for widely dispersed and/or patchy subterranean food resources such as tubers, bulbs and corms (Jarvis, 1978, 1981; Lovegrove and Wissel, 1988; Lovegrove, 1991; Jarvis and Bennett, 1993). Increasing sociality and group size within the Bathyergidae tends to follow a pattern of increasing aridity with decreasing density of geophyte storage organs, but increasing sizes of the organs (Lovegrove and Wissel, 1988). In the most social of the Bathyergidae, the naked molerat, *Heterocephalus glaber* of East Africa, and the Damara molerat, *Cryptomys damarensis* of the Kalahari Desert, there is a single breeding queen molerat, and several castes of workers, intermediate workers, and non-workers (Jarvis, 1981; Bennett and Jarvis, 1988; Bennett, 1990; Jarvis and Bennett, 1990; Jacobs et al., 1991; Sherman et al., 1992). Reproductive suppression of females appears to be maintained through aggression from the queen and non-worker males (Abbott et al., 1989; Faulkes et al., 1990).

Among the large carnivores, territory defence and efficient food acquisition are important variables selecting for social organization. The social system of the spotted hyaena *Crocuta crocuta* is related to competition for clumped, high-quality food resources that are defendable (Mills, 1983, 1990). This observation is, to a lesser extent also true for the brown hyaena (Mills, 1990). In both species the social organization consists of female-bonded groups. Whereas related and unrelated brown hyaenas within clans forage alone (Mills, 1978), they co-operate in raising cubs (Mills, 1990). Related spotted hyaenas forage cooperatively, but do not cooperate in raising cubs (Mills, 1985, 1990). For both hyaena species in the Kalahari Desert, territory size was correlated with the average distance moved by individuals between significant food items (Mills, 1990). Group size was not related to territory size, but was related to food patch richness in brown hyaenas (number of large carcasses), and mean herd size of prey in spotted hyaenas (Mills, 1990). Unfortunately, little is known about the social organization of other large, desert-dwelling carnivores in southern Africa.

In contrast to hyaenas, the group size of karoo and Rüppell's korhaans (*Eupodotis vigorsii* and *E. rueppellii*) increases with territory size and decreasing rainfall (Viljoen, 1983; Hockey and Boobyer 1994). In the karoo korhaan, as territories become larger, sub-adult males remain with the breeding pair and assist with territory defence. The sub-adults thus gain the experience of territory defence and a territorial existence before reaching maturity (Hockey and Boobyer, 1994). In the Kaokoland in the northern Namib, the largest group size recorded in Rüppell's korhaan was eight birds (Viljoen, 1983).

Vigilance against predators is implicated in the herd sizes of ungulates (Siegfried, 1980), the flock sizes of ostriches (Bertram, 1980) and perhaps other birds, and the social organization of suricates (Lovegrove, 1993). A minimum group size permits animals to minimize the risk of attack by predators while maximizing feeding time. For example, it has been shown that a herd size of 20 springbok is the optimum for fastest feeding in the Etosha National Park (Siegfried, 1980). Sub-adult suricates share vigilance duty thus permitting reproductive females and juveniles to optimize foraging time and minimize the risks of aerial attacks (personal observations).

Although the adaptive significance of flocking in birds remains debatable, there is general agreement that individuals minimize the risks of aerial predation by joining flocks (Thompson et al., 1974; Siegfried and Underhill, 1975). Desert birds such as sandgrouse and Cape turtle doves (*Streptopelia capicola*) forage individually or in pairs during the day, yet congregate in large flocks prior to drinking at water holes in the Kalahari Desert (Siegfried and Underhill, 1975; Cade et al., 1986; Lovegrove, 1993). Flocking behaviour minimizes the risks of attacks from aerial predators such as lanner falcons (*Falco biarmicus*) that are reluctant to risk injury attacking large flocks of panic-stricken birds (Siegfried and Underhill, 1975).

Social behaviour may also minimize physiological stresses in desert animals. Certain examples of birds and mammals which co-operatively build and maintain large nest structures which serve as important thermal refugia have been discussed elsewhere (see 9.5.4). Presumably the nests also afford the inhabitants a measure of protection against predators and, in the case of the sociable weaver, they certainly enhance breeding success (see 9.8). It has recently been suggested that gregarious behaviour has evolved in a species of tenebrionid beetle (*Parastizopus armaticeps*) as an adaptation to reduce evaporative water loss (Rasa, 1994). The group sizes of these beetles within burrows increases during droughts.

9.7.   Predator–prey adaptations

The sparse vegetation of deserts makes it difficult for animals to conceal themselves and take refuge from predators. Likewise, predators cannot rely upon vegetation cover for concealment. The low population numbers of prey items also necessitate that some predators range considerable distances in search of food. Consequently, various behavioural, morphological and physiological adaptations have evolved in desert species to aid them in avoiding predators and finding and securing prey.

9.7.1.   Behaviour
Several descriptions of how sociality minimizes predation risks through vigilance and group defence have been detailed elsewhere (see 9.6). Moreover, although endogenous circadian activity rhythms (see 9.5.3) emphasized the importance of emergence times in terms of physiological stresses, the synchrony of predator–prey activity rhythms are also crucial in minimizing the risks of predation (Aschoff, 1964; Enright, 1970; Daan and Aschoff, 1982). Here I discuss some of the more unusual and generalized behavioural observations of predator avoidance and hunting behaviour.

The behavioural diversity of predator avoidance and hunting behaviour in desert spiders is impressive. One dune-dwelling spider in the Namib Desert, *Carporachne*

aureoflava, employs a highly unusual form of escape behaviour – it cartwheels down sand dunes to escape predators (Henschel, 1990). The spiders' main predators are pompilid wasps which actively excavate the spiders out of their burrows in the dune slopes. Spiders which initially escape the wasps roll themselves into wheels and roll down the dune slope at speeds of $0.5–1.5$ m s^{-1} (Henschel, 1990).

Corolla spiders, *Ariadna* spp., employ 'tool use' on the gravel plains of the Namib Desert to enhance prey capture (Henschel, 1995). The spiders select 7 or 8 stones of a preferred size, shape and composition (quartz crystals), and arrange them in a circular, daisy-flower pattern around the lip of their vertical burrow entrances. The stones are connected via silk threads to the burrow entrance, and are used by the spiders to detect when prey touch or brush past any of the stones. The spiders thus extend their foraging range with the use of the stones. In the succulent karoo, bauble spiders (*Achaearanea* spp.) construct an ingenious spiral retreat which protects the spiders from heat, rain and predators, yet also serves as a platform from which to hunt ants (Henschel and Jocqué, 1994). The bauble-like retreat structure is comprised of silk, sand grains and pebbles and is suspended under overhanging rocks where it is protected from direct solar radiation. The spiders reside within spiral tubes within the bauble. Ants are trapped with sticky silk threads constructed at night from the ventral surface of the bauble to the substrate below. Once trapped, the ants are hauled up towards the bauble by the spiders and consumed.

In general, most desert reptiles have a fast burst of speed and escape predators by fleeing, whereas others, such as the girdled lizards (Cordylinae), are slower and have opted for heavy armour in the form of spiky scales which not only protects them from certain predators, but also enables them to anchor themselves tightly within crevices of rocks and trees. For example, the giant girdled lizard (*Cordylus giganteus*), the largest of the girdled lizards in southern Africa, retreats backwards out of its burrow towards danger lashing its spiny tail from side-to-side (Branch, 1988a). Cordylid lizards are very wary, seldom straying far from safe shelters. They retreat at the first sign of danger, and jam themselves in crevices. The karoo girdled lizard (*Cordylus polyzonus*) protectively curls its tail over its head after retreat, whereas the armadillo girdled lizard (*Cordylus cataphractus*), an endemic of the succulent karoo biome, is well known for its habit of rolling into a ball and biting its tail to avoid being eaten by predators.

The propagation of sound at frequencies >10 kHz in desert air is poor, and hence many desert mammals depend upon low frequency sounds (<2 kHz) for effective detection and communication of the presence of predators (Eisenberg and Kleiman, 1977). Apart from the morphological implications (see 9.7.4), this constraint has led to the evolution of seismic communication by footdrumming which serves to warn conspecifics against predators (Kenagy, 1976; Eisenberg and Kleiman, 1977; Randall and Stevens, 1987). In southern Africa, footdrumming or stamping as a warning communication has been observed in the Cape hare (*Lepus capensis*) and Brants' whistling rat (*Parotomys brantsii*) (Skinner and Smithers, 1990), elephant shrews (Rathbun, 1979) and molerats (Narins et al., 1992). With the exception of the Cape hare, these species either live permanently in burrows, or retire to them when threatened.

Ungulates display one of two responses when attacked by predators; they either flee or they stand and defend themselves. Fast ungulates such as springbok, blue wildebeest and red hartebeest which flee from predators, tend to be ignored by most large carnivores such as spotted hyaenas, unless they are breeding. In this respect, synchronized oestrous and breeding is an important behavioural adaptation minimizing lamb mortalities, especially in migratory and nomadic species which either cannot, or do not, employ group defence (Estes, 1976). In the Kalahari Desert it is common in large herds of springbok and blue wildebeest (Estes, 1976; Sinclair, 1977; Berry, 1980; Knight, 1991; Heske et al., 1994). Synchronized calving minimizes calf mortalities by providing a glut of food for predators only during the first few days that it takes the neonates to keep up with the herd (Estes, 1976).

Although the significance of the characteristic stotting or 'pronking' behaviour in alarmed springbok is unclear, the most likely explanation is that it represents antipredator behaviour. In the typically scrubby vegetation of the favoured karroid habitats of springbok, the behaviour may serve to confuse predators, gain a better visual view of concealed, stalking predators, and advertise danger to conspecifics. In the latter respect, the concomitant raising of the 'pronk' of stiff white hairs on the rump (Skinner and Smithers, 1990) probably enhances the warning signal.

Mills (1990) has made careful observations of the behaviour of larger ungulates to attacks by spotted hyaena clans in the Kalahari Desert. When in herds, gemsbok and eland employ group defence fairly successfully against predators. Calf mortalities are understandably the highest. Both species swing their horns viciously sidewards in frontal attacks. When isolated, individual gemsbok often protect their rear by backing into a thorn bush. Eland, on the other hand, often form a defensive circle with their heads together and kick out vigorously with their hind legs.

9.7.2.   Crypsis and colouration
The adaptive significance of body colour in desert animals remains contentious (Hamilton and Heppner, 1967; Willoughby, 1969; Hamilton, 1973; Cloudsley-Thompson, 1979). Nevertheless, few biologists today doubt that

many desert animals depend heavily upon crypsis for concealment from predators and prey (Willoughby, 1969; Cloudsley-Thompson, 1979). Willoughby (1969) adopted the term 'desert colouration' to describe a specific form of crypsis; the pale ventral and dorsal body colours which harmonize with the colour, texture and even shapes of desert sand, gravel and rocks. Desert colouration can be seen in insects, spiders, lizards (e.g. Lacertidae), snakes (e.g. Viviperinae), birds and mammals, in both prey and predators (Cloudsley-Thompson, 1979).

Several species of toad grasshoppers of the genus *Batrachotetrix* in the Namib and karoo are unusually short and squat and display cryptic patterns which mimic perfectly the colours, shapes and sizes of pebbles (see photos in Louw and Seely, 1982; Lovegrove, 1993). Ground-dwelling birds are particularly cryptic, especially the larks of the family Alaudidae (Willoughby, 1969; Maclean, 1993). For example, it has been shown that eight species of larks breeding in the Kalahari Desert fall naturally into two colour groups: 'red' and 'grey' (Maclean, 1983). The 'red' larks (fawn-coloured *Mirafra africanoides*, clapper *M. apiata*, spike-heeled *Chersomanes albofasciata* and pink-billed larks, and black-eared finchlark), inhabit mainly the red sand dunes, whereas the 'grey' larks (Sabota *M. sabota* and Stark's larks, and grey-backed finchlark) inhabit the light-coloured, calcrete along the banks of the Nossob and Auob Rivers. Birds with desert colouration in the Namib Desert (Rüppell's korhaan, two-banded courser, karoo lark, spike-heeled lark, Gray's lark *Ammomanes grayi*, trac-trac chat *Cercomela tractrac* and karoo chat *C. schlegelii*) tend to be sedentary, endemic species or subspecies, whereas those displaying more 'generalized cryptic colouration' (e.g. greater kestrel *Falco rupicoloides*, Ludwig's bustard *Neotis ludwigii*, crowned plover *Vanellus coronatus*, Burchell's courser *Cursorius rufus*, Namaqua sandgrouse *Pterocles namaqua*, Stark's lark, grey-backed finchlark) tend to be widely distributed outside the Namib (Willoughby, 1969).

9.7.3.   Mimicry
There are several interesting examples of Batesian and Müllerian mimicry in desert arthropods and reptiles in southern Africa. Jumping spiders of the genus *Cosmophasis* mimic the ants *Camponotus detritus* and *C. fulvopilosus* (Curtis, 1988). They apparently feed on the ants only to acquire the ants' odour to avoid being attacked while moving among the ants in the colony, possibly to feed on the ants brood (Curtis, 1988). Curtis also suggests that, by mimicking the aggressive formicine ants, the spiders gain protection from predators who have learnt to avoid the ants.

Adult arid savanna lizards (*Heliobolus lugubris*) are cryptic like all other lacertid lizards in the Kalahari Desert and eastern Namib. However, their hatchlings mimic the black and white aposematic colours and patterns, and even the walking gait, of 'oogpister' (eye-squirter) beetles of the genus *Anthia* (Huey and Pianka, 1977). When threatened, the beetles squirt a strong jet of formic acid from the pygidial glands which can cause blindness in birds and mammals. Young arid savanna lizards forage actively during the daytime and walk jerkily on stiff legs with their backs arched in much the same way as oogpisters dart around foraging for prey.

Some beetles in the Namib Desert mimic the speed and agility of others (Holm and Kirsten, 1979). *Scarabaeus rubripennis* has an aposematically coloured orange elytra which apparently advertises its fast flight speed and agility and clearly distinguishes it from the normally black-coloured, flightless scarabs in the Namib. The aposematic orange elytra code to predators is reinforced by another larger, but slower beetle, *Drepanopodus proximus*. This mimicry is thus part Müllerian and part Batesian. One apterous scarab beetle, *Pachysoma denticolle*, is a full Batesian mimic of *S. rubripennis* where these two species are sympatric in the Namib, but is non-mimetic in allopatry.

9.7.4.   Morphology and body shape
One of the distinctive morphological features of nocturnal desert rodents is the large size of their inflated middle-ear cavities reflected externally in the expansion of the mastoid bullae (Eisenberg and Kleiman 1977). Given the rapid loss of energy of high-frequency sounds in deserts (see 9.7.1), these auditory adaptations increase sensitivity to low-amplitude sounds by focusing energy at the meatus (Eisenberg and Kleiman 1977). Indeed, within a single species of gerbil (*Meriones*), increased bulla expansions were correlated with increased aridity and reduced vegetation cover (Lay 1972). In southern Africa, all desert gerbils (Gerbillinae), but particularly *D. auricularis*, *Gerbillurus tytonis* and *G. setzeri*, posses well-developed ear bullae (Griffin 1990; Skinner and Smithers 1990).

Several desert mammals also have very large ears which aid them in locating food or detecting predators, rather than serving the alternative function of thermal windows for heat dissipation, the basis of Allen's Rule (Louw and Seely 1982; Withers 1992). For example, as its name implies, the large-eared mouse (*Malacothrix typica*), a small, nocturnal mouse that inhabits calcrete pans throughout the karoo, has exceptionally large ears which aid in the detection of aerial predators in its exposed habitat (Knight and Skinner, 1981). Bat-eared foxes rely upon their acute hearing senses facilitated by their enormous ears to locate subterranean prey items, mostly beetle larvae, termites and scorpions (Skinner and Smithers, 1990).

The risks of predation during foraging are reflected in the morphology (Huey and Pianka, 1981) and locomotory performance (Huey et al., 1984) of lacertid lizards of the Kalahari Desert. 'Sit-and-wait' lizards, such as *Pedioplanis lineoocellata*, are stocky and have short tails, whereas widely foraging lizards, such as *Nucras tesselatta*, are sleek, streamlined and generally have longer tails (Huey and Pianka, 1981). Widely foraging lizards also tend to have a greater locomotory and sprint endurance, but lower burst speed, than sit-and-wait lizards (Huey et al. 1984). Lizards associated with rocky habitats, such as the Cordylinae, tend to be spiny (see 9.7.1).

9.8.   Reproduction

In terms of fitness and energy requirements, reproduction is risky in deserts where rainfall, and hence the resources for reproduction, are temporally and spatially unpredictable (Noy-Meir, 1973, 1974; Wiens, 1977, 1991; Louw and Seely, 1982). Phenomena such as the ENSO reduce the regularity of summer rainfall in sub-Saharan Africa (Tyson, 1987; Seleshi and Demaree, 1995; Elfatih and Eltahir, 1996) adding markedly to the unpredictability of the already meagre rainfall of arid and semi-arid regions (Noy-Meir, 1973). Moreover, rainfall in the form of convective storms, the prevalent summer rainfall in southern Africa's arid regions, is extremely patchy spatially in desert ecosystems (Noy-Meir, 1973). Areas separated by only a few kilometres are capable of receiving sufficient rain (*c.* 20 mm) to stimulate breeding in insects, mammals and birds, whereas surrounding areas may receive no rain at all during the same period (Sharon, 1981; Louw and Seely, 1982; Berry and Siegfried, 1991).

9.8.1.   Cues and Zeitgeber for reproduction
The cues which initiate reproduction in desert animals are therefore more complex and elaborate than those seen in seasonal breeding animals from predictable ecosystems (Immelmann, 1971; Louw and Seely, 1982; Maclean, 1996). Proximate or primary environmental factors, also termed *Zeitgeber,* stimulate recrudescence of the male and females gonads and prepare the animal physiologically for reproduction (Immelmann, 1972; Louw and Seely, 1982). In temperate species, the most common and reliable *Zeitgeber* is photoperiod or daylength (Immelmann, 1971; Louw and Seely, 1982). In desert regions, however, summer rainfall, although predictably seasonal, may not be sufficient to sustain the reproductive efforts of most desert species. Desert species cannot risk reproduction until assurance that sufficient rain has indeed fallen has been registered by the neuroendocrine axis. Therefore, most desert species in addition rely upon secondary *Zeitgeber* to release them

from a 'sub-threshold sexual activity' initiated by the primary cue to fully initiate reproduction (Immelmann, 1971, 1972; Louw and Seely, 1982). This signal may be rainfall or some aspect of vegetation quality such as protein content. As stated by Maclean (1996), 'primary *Zeitgeber* provide stimuli which initiate gonadal development, while secondary *Zeitgeber* operate at a later stage of gonadal development and maintain the congruence between the internal rhythm and the environment'. Together, the primary and secondary *Zeitgeber* interface precise environmental information with the neuroendocrine system of the animal to synchronize reproduction and permit them to rapidly exploit unpredictable flushes of resource availability when arid regions respond to rainfall events. Interesting exceptions to this generalized rule occur wherever local environmental conditions, such as the regular occurrence of fog in the Namib Desert and parts of the succulent karoo, decrease the unpredictability of resource availability.

Reproduction in desert insects is not as risky as it is for birds and mammals. They have the capacity to endure long periods in diapause and require only an emergence stimulus, most often associated with rainfall.

Desert reptiles display varied patterns of reproductive traits and life histories, dependent mostly upon distribution, foraging mode and thermoregulatory constraints. In some regions breeding is highly seasonal. For example, the burrowing, legless skink *Typhlosaurus lineatus* in the Kalahari Desert shows a clear testicular cycle reaching a peak in August/September when mating occurs (Huey et al., 1974). These viviparous termite specialists produce a single small brood (average of 1.6 young) in summer each year (Huey et al. 1974). The sympatric species *Typhlosaurus gariepensis* has an even more conservative reproductive output, giving birth to a single baby after a five-month gestation period each year (Huey et al. 1974).

In contrast to the burrowing skinks which are all known to be viviparous, all lacertid lizards in the desert regions of southern Africa are oviparous. In these Old World lizards, relative clutch mass (clutch volume divided by body mass) is dependent upon foraging mode. 'Sit-and-wait' foragers have greater relative clutch masses compared with lizards which are fast-moving and forage widely, mostly for termites (Huey and Pianka 1981). It has been suggested that these characters (foraging mode and clutch mass) have co-evolved in response to the relative risks of predation and predator avoidance associated with the two foraging modes (Huey and Pianka, 1981).

Thermoregulatory constraints on winter breeding activity, as well as summer rainfall, restrict the egg-laying and hatching period of most karoo and Kalahari lacertid lizards to early summer and late summer, respectively (Branch 1988a). Six species of geckos in the Kalahari Desert

also showed breeding activity restricted to the summer months (Pianka and Huey, 1978). However, wide-ranging species, such as *Meroles suborbitalis*, breed throughout the year in the Namib Desert, but are summer breeders in the Kalahari Desert (Branch 1988a). Indeed, several other lacertids confined to the Namib Desert, typically *Aporosaura anchietae* and *Meroles cuneirostris*, are also aseasonal breeders (Goldberg and Robinson 1979; Robinson 1990). Aseasonal breeding in these lizards is presumably made possible by the predictable availability of moisture in the form of coastal fog in the Namib, and the consequent year-round availability of insects and wind-blown seeds (Murray 1984; Murray and Schramm 1987; Robinson 1987).

Whereas many rodent communities in Africa breed opportunistically after rain or in response to the consequent flush of green vegetation (Delany 1986), the breeding of several desert-adapted rodents in the Namib Desert is independent of rain. For example, *Petromys typicus*, *Petromyscus collinus* and *Aethomys namaquensis* inhabiting rocky outcrops in the Namib Desert displayed a highly seasonal breeding period in January and February which was not correlated with rainfall (Withers 1983a). These rodents may rely instead upon photoperiod as the primary *Zeitgeber*, although it has also been suggested that breeding may be initiated by the increased frequency of coastal fog after September (Withers, 1983a). *Petromys typicus* and *Petromyscus collinus* are physiologically well adapted to food and water stresses (Withers et al. 1980). The more dense vegetation of rocky outcrops and inselbergs may capture more moisture from fog than the surrounding gravel plains which may benefit the reproduction of these rodents. On the gravel plains of the Namib Desert, rodents such as *Rhabdomys pumilio* and *Gerbillurus paeba* also showed seasonal increases in population numbers, whereas *Desmodillus auricularis* tended to breed over a protracted period (March to September) (Christian, 1980).

Certain generalizations concerning the association between reproductive patterns, phylogeny and physiological traits have been identified for desert rodents in southern Africa with relative assurance (Christian, 1978, 1979, 1980; Withers, 1979, 1983a; Withers et al., 1980; Delany, 1986; White and Bernard, 1996). Firstly, desert-dwelling murid rodents (*Aethomys namaquensis*, *Rhabdomys pumilio*) are more recent invaders of the arid regions of southern Africa than are the cricetid rodents (*Petromys*, *Petromyscus*, *Desmodillus*, *Gerbillurus* spp.). The murids display less advanced physiological adaptations to food and water stresses than do the cricetid rodents, and hence tend to breed seasonally during favourable conditions and opportunistically after rain. They have high reproductive outputs and mortality rates and can increase population numbers dramatically following good rainfalls. Cricetid rodents, on the other hand, display well-developed physiological adaptations to aridity (concentrated urine, allantoin production, low BMRs and WTRs, use of torpor), maintain fairly stable populations, have lower reproductive and mortality rates, and can breed over protracted periods throughout the year.

Desert lagomorphs and ungulates (steenbok, springbok, gemsbok and eland), on the other hand, breed throughout the year (Skinner and Smithers, 1990). Although springbok lambing tends to peak during spring (September and October), their year-round breeding capability contrasts sharply with the highly restricted breeding season of their mesic counterparts, impala (*Aepyceros melampus*), which lamb predictably within two months only in summer every year (Skinner et al., 1977, 1984; Skinner and Van Jaarsveld, 1987). The ability to breed throughout the year, depending upon food availability, permits rapid increases in the population sizes of arid-adapted ungulates. It has been pointed out that, theoretically, given no mortality, springbok ewes can breed 6.5 times in three years, during which time the population can treble (Skinner 1993). Springbok apparently rely upon odours carried by the prevailing wind to detect rainfall many hundreds of kilometres way (Skinner et al., 1984). The secondary *Zeitgeber* for oestrus behaviour in springbok is thought to be the flush of green foliage following rain (Skinner et al., 1984). Although this *Zeitgeber* does not necessarily ensure good grazing when the lambs are born, it facilitates synchronous lambing and the swamping of predators during the lambing period (Estes, 1966; Kruuk, 1972; Estes 1976; Skinner et al., 1984; Skinner and Van Jaarsveld, 1987; Mills, 1990). The breeding success of springbok in the arid regions of southern Africa can thus be attributed to their nomadic behaviour and ability to move vast distances to isolated regions where rain has fallen and the growth response of succulent ephemeral grasses, such as *Enneapogon desvauxii*, is very rapid (Skinner et al., 1984).

Not surprisingly, large carnivores, such as brown hyaenas, spotted hyaenas, leopards and cheetahs, also breed throughout the year in the arid regions of southern Africa (Mills, 1990; Skinner and Smithers, 1990) and can thus benefit from the non-restrictive breeding patterns of desert ungulates.

For many desert birds, rainfall itself may act as the secondary *Zeitgeber* and breeding takes place opportunistically after significant rainfall events (Winterbottom and Rowan, 1962). Indeed, in species such as the sociable weaver, the time lag between a rainfall event of 20 mm or more and the laying of the first egg is six days (Maclean, 1973). This short period is facilitated by the existence of a permanent nest structure maintained throughout the year, illustrating one of the advantages of sociality in

these birds (Maclean, 1973; Lovegrove, 1993). The rufous-eared warbler (*Malcorus pectoralis*) also responds rapidly to rain, laying the first egg within seven days (Maclean, 1996). The onset of breeding by all larks in the Kalahari Desert is also stimulated by rainfall (Maclean, 1970a, 1970b, 1970c, 1996). Although high thermoregulatory costs are presumed to restrict white-backed mousebirds (*Colius colius*) from breeding during the coldest winter months in the succulent karoo, breeding during summer is correlated with rainfall (Dean et al., 1993).

The breeding of dune larks (*Certhilauda erythrochlamys*) and the double-banded courser (*Rhinoptilus africanus*), which inhabit extremely arid regions, is independent of rainfall and occurs throughout the year (Maclean, 1967; Boyer, 1988). It is suspected that the same factors which permit year-round reproduction in Namib insects and lizards, namely reliable sources of wind-borne resources and moisture from coastal fog, account for the non-restrictive breeding pattern of dune larks. Aseasonal breeding by the double-banded courser, on the other hand, appears to be related to 'multiple-brooding' (Maclean, 1967). The bird lays a single egg, and may lay a second egg while the chick of the previous brood is still being fed. Maclean (1967) has suggested that the combination of a single-egg clutch and sequential laying imposes little energetic stress on the birds thus permitting them to raise several chicks successfully throughout the year. Several larger, ground-dwelling birds, such as Rüppell's korhaan and, in the western limits of its distribution on the edge of the Namib Desert, the karoo korhaan, remain seasonal breeders yet also lay a single egg (Maclean, 1993). Their breeding season coincides with the best chances of rain in summer. The larger body size and hence higher total energy requirements of these birds presumably restrict year-round breeding in and around the Namib Desert.

Certain intraspecific life history attributes differ between species of birds breeding in arid and mesic regions of southern Africa (Siegfried and Brooke, 1989). For example, several insectivorous passerines have relatively larger clutches and shorter incubation and fledging periods in arid regions than they do in the eastern, mesic regions (Siegfried and Brooke, 1989).

9.8.2.   The brown locust

Perhaps the one species which personifies reproductive adaptation to the unpredictability of the karoo is the brown locust (*Locustana pardalina*). From an agricultural perspective, it is best known for its periodic outbreaks of huge, migratory swarms, the damage it causes to crops, and the controversies surrounding its control (Jago, 1987; Hanrahan, 1988; Hockey, 1988). From a biological perspective, the characteristics of its reproduction, emergence and polymorphism remain intriguing (Faure, 1932; Lea,

1964; Botha et al., 1974; Bouaichi et al., 1995). Like other members of the suborder Caelifera, which includes the desert locust (*Schistocerca gregaria*) and red locust (*Cyrtacanthacris septemfasciata*), the species can occur in three phases or forms: a cryptic, solitary phase or *solitaria*, perhaps the 'normal' phase of the locust similar to that of ordinary grasshoppers; an aposematic swarm phase or *gregaria,* during which the locusts aggregate to form huge migratory swarms; and an intermediate *transiens* phase (Uvarov, 1921; Faure, 1932; Scholtz and Holm, 1985). The phases differ markedly in colour, morphology, behaviour and physiology. Density-dependent 'crowding' of *solitaria* hoppers stimulates the transition to the *gregaria* phase and consequent swarming (Faure, 1932; Bouaichi et al., 1995).

The eggs laid by *solitaria* are small and capable of diapause, whereas *gregaria* eggs are large, extremely drought-resistant, and do not go into diapause (Matthee, 1950; Botha, 1967; Venter and Potgieter, 1967; Petty, 1973a, 1973b, 1974). The hatching cue for *solitaria* eggs is probably a combination of photoperiod and humidity, since the eggs do not hatch once diapause has commenced in late summer. In contrast, the hatching cue for *gregaria* eggs is rainfall. The contrasting life-history attributes of the *solitaria* and *gregaria* phases are thus well adapted to hedge bets and maximize fitness in desert regions where seasonal rainfall patterns are unpredictably interrupted by droughts which may last for many consecutive years as a consequence of climatic phenomena such as the ENSO. The migratory *gregaria* phase, in particular, is capable of spreading the gene pool over considerable areas of southern Africa during the favourable periods which follow the termination of droughts.

10 Animal foraging and food

W. R. J. Dean and S. J. Milton

10.1. Introduction

Resources in the karoo, desert, and the arid savanna are generally patchy in time and space, lacking the seasonality typical of more mesic winter and summer rainfall biomes. The southern and western parts of the karoo are poor in species and abundance of grasses (Acocks, 1953; Gibbs Russell et al., 1990), and lack the large grazing herbivores of the grasslands and mesic savannas (Skead, 1980, 1987) or granivorous rodents characteristic of northern-hemisphere arid systems (Abramsky, 1983; Kerley, 1991). Plants bearing juicy fruits are spatially restricted (mainly to drainage lines), so this resource is only available to animals of these habitats, or animals able to move economically between drainage lines (Acocks, 1976). Seeds are abundant but patchy, so granivores must either move between patches, as the granivorous birds do, or store seeds during a time of plenty, as the ants do. Small mammals are not abundant (Kerley, 1992a) and, in general, carnivores tend to take a wider range of prey sizes and species in the karoo than they do elsewhere (Palmer and Fairall, 1988; Boshoff et al., 1990). Even such species as black eagle *Aquila verreauxii*, thought to be relatively stenophagous, take a wide range of prey species (Boshoff et al., 1991). Below-ground, however, there is an abundance of food, and, for those animals that are fossorial, or are able to dig into hard soils, there are rich rewards in the form of underground plant parts as well as invertebrate and vertebrate prey.

Food resources for animals in the karoo can be classified along an availability (or predictability) gradient as dependable, seasonal or unpredictably scattered in time and space. We show, in this chapter, that animals in the karoo are adapted in a number of ways to exploit these resources. Our focus is on foraging behaviour, food, and food handling that differ from those in other southern African biomes. Carnivores in general are not dealt with here, because they apparently have no special adaptations for life in the karoo. We have not discussed the aquatic food resources in the karoo for the simple reason that no detailed studies have been done on the ephemeral water bodies of the central and western Nama-karoo and the north-western succulent karoo.

10.2. Dependable resources

Foods that are available throughout the year comprise plant detritus, foliage of (mostly) toxic and distasteful plants, roots and bulbs, and the animals that eat such food. Some general principles that apply to animals feeding on dependable resources in the karoo are:

- The animals should be relatively specialized in their diet, particularly well-adapted to finding and obtaining particular foods, and able to survive (thrive?) on toxic, distasteful or low quality foods. This may be combined with short-term food storage because of environmental constraints on foraging activity, or by the unpredictable or irregular activity patterns of prey animals.

- The animals should be resident, territorial and at low densities.

Few food resources are available above-ground all year round in the karoo. Leaves, seeds and fruits are markedly seasonal or are produced in response to unpredictable and patchy rainfall. Similarly, most invertebrates and reptiles, as potential prey, are not active throughout the year and do not provide a dependable source of food. Animals that are able to make use of chemically defended foliage or such plant parts as stems and roots or that feed on non-seasonal insects, however, may be able to survive throughout the year without switching their diet or storing food. Food storage may not only involve gathering the food and placing it in a safe site. Many species that are resident are territorial, and these species are, in effect, storing food for their own exclusive future use. Territoriality is particularly marked among such birds as karoo korhaan *Eupodotis vigorsii* (Hockey and Boobyer, 1994), long-billed lark *Certhilauda curvirostris* and karoo chat *Cercomela schlegelii* (W. R. J. Dean, personal observations) that defend territories year-round, and sunbirds (*Nectarinia* spp.) that defend ephemeral patches of flowers with high levels of nectar.

10.2.1.   Below-ground foragers

Temperature and humidity underground vary less than they do on the soil surface (Du Plessis, 1989; Lovegrove and Knight-Eloff 1988; Lovegrove, this volume). For this reason, a number of plants store water and energy resources underground (Midgley and Van der Heyden, this volume), and some invertebrates feed or nest below the soil surface.

Herbivores

The molerats (Bathyergidae), are an important group of below-ground herbivores, feeding on roots, corms and bulbs (Skinner and Smithers, 1990; Lovegrove and Painting, 1987; see *10.3.3.*). The common molerat *Cryptomys hottentotus* is widespread, but patchy (Lovegrove and Siegfried, 1989) in the Nama- and succulent karoo, the Cape molerat *Georychus capensis* occurs in sands in the western karoo and marginally along the southern edge of the succulent karoo, and the Namaqua dune molerat *Bathyergus janetta* is restricted to dune sands in the north-west succulent karoo. Molerats may feed on only one species of food plant depending on the soil type of the habitat patch. For example, in arid savanna, individual groups of common molerats living in dune sands eat roots of *Acanthosicyos naudinianus*, and groups living in alluvial flood plains eat bulbs of *Dipcadi gracilimum* (Lovegrove and Painting, 1987). The energetic costs of foraging for geophytes increases in dry sands because of the difficulty of excavating sand (Lovegrove and Painting, 1987).

Carnivores

Lycaenid butterflies (Lycaenidae) are widespread in southern Africa and of the 125 species 51 occur in the karoo (Clark and Dickson, 1971). The larvae of some genera in this family associate with formicine ants and their nests, may spend part of their larval stage underground feeding on detritus or young ant larvae (latent carnivores), and generally pupate deep inside ant nests. Only the ants *Anoplolepis custodiens*, *Camponotus* spp. and *Crematogaster* spp. are listed as hosts by Clark and Dickson (1971). Henning (1983) gives *Acantholepis capensis* (widespread in the karoo) as a host to *Aloeides dentatus* in the Transvaal. It is likely that *A. capensis* hosts other lycaenid species in the karoo. Lycaenid butterflies are generally at a low density in the karoo (W. R. J. Dean and S. J. Milton, personal observations).

Some solifuges (*Chelypus* and *Hemiblossia* spp.) occur in termite mounds and forage underground, feeding almost entirely on termite workers and larvae (Lawrence, 1963). Driver ants *Dorylus helvolus* feed on carrion, live termites, and other prey below-ground (personal observations), including such small mammals as the pouched mouse *Saccostomus campestris* (Ellison, 1988).

Two species of burrowing snake (Typhlopidae) and four species of thread snake (Leptotyphlopidae) occur in the Nama- and succulent karoo. Delalande's blind snake *Typhlops lalandei* and the beaked blind snake *T. schinzi* feed on termites and other insects below-ground. Delalande's blind snake is widespread in the Nama-karoo, and the beaked blind snake occurs in the north-western succulent karoo (Branch, 1988a). The thread snakes *Leptotyphlops nigricans*, *L. gracilior*, *L. scutifrons* and *L. occidentalis* similarly live underground, and are highly adapted to feeding on termites, since they can produce pheromones that repel termite soldiers (Branch, 1988a). Both Typhlopidae and Leptotyphlopidae may be restricted to sandy or loamy soils in the karoo.

Similarly, fossorial, insectivorous golden moles (Chrysochloridae) may be restricted by soil type in the karoo, since they occur only on the southern and western edges of the Nama- and succulent karoo, with an isolated population of *Chrysochloris visagiei* in the central western Nama-karoo (Vernon, this volume). Desert and karoo golden moles feed on lizards as a main part of their diet, whereas savanna and forest moles apparently eat only insects (Skinner and Smithers, 1990).

10.2.2.   Above-ground foragers

Detritivores

As primary production decreases with aridity, detritus becomes more important as the basis of the food chain and provides a reliable source of food for those organisms that are able to use it. Concomitantly, more predators apparently rely on detritivores as food. Detritus therefore allows top predators to occupy habitats that have little or no primary production. In the most arid parts of the Namib Desert, imported detritus (wind-blown grass stems, leaves and awns) is the only source of energy (Louw and Seely, 1982), and is eaten by termites, thysanurids and tenebrionid beetles, which in turn are eaten by solifuges, spiders, lizards, chameleons, crows and black-backed jackals (Louw and Seely, 1982). In the karoo, detritivores, including termites and tenebrionid beetles, are major components of the diet of many invertebrate predators (Lawrence, 1963; Dean, 1988, 1989), reptiles (Branch, 1988a), birds and mammals (Kok and Hewitt, 1990; Kok and Louw, 1994).

Termites *Microhodotermes viator* collect plant detritus and store it underground (Coaton and Sheasby, 1974a,b,c). These termites are selective detrivores, and nitrogen and phosphorus content of plant detritus are more important in determining food selection than energy content (Scarola, 1995). Plant material is collected by the termites and stockpiled at the entrances to foraging ports before being moved underground. This temporary storage may be because the termites must forage and gather food in as short a space of time as possible to avoid a build-up of predators (Wilson and Clark, 1977; Dean, 1989). *Microhodotermes viator* colonies are regularly spaced throughout their distribution in the Nama- and succulent karoo (Lovegrove and Siegfried, 1989; Milton et al., 1992a), suggesting that there is competition between colonies.

Tenebrionid beetles are a conspicuous element of the fauna in parts of the Nama- and succulent karoo (Dean and Milton, personal observations), but have been studied only in the Namib Desert. Species for which there are food and foraging data show that *Physadesmia globosa*, *Stenocara gracilipes* and *Onymacris rugatipennis* are omnivorous detritivores, but not indiscriminate detritivores, since they eat flowers and dry cuticular fragments of insects, apparently selecting the nutrient-rich components of the available detritus (Hanrahan and Seely, 1990). *Parastizopus armaticeps* apparently feeds mainly on detritus from *Lebeckia* sp. (Fabaceae) in the southern Kalahari (Rasa, 1995). *Physadesmia globosa* has also been recorded eating termites, and *Rhammatodes longicornis* feeds on seeds, owl pellets, and dead mosquito larvae, as well as plant detritus (Wharton and Seely, 1982).

Not all detritivores in the south-western arid zone are invertebrates. The lizard *Angolosaurus skoogi* feeds on vegetable matter in the Namib Desert, including components of dry wind-blown detritus; it obtains water from eating the endemic perennial cucurbit *Acanthosicyos horrida* (Pietruszka et al., 1986).

Herbivores

Plants that are available above-ground throughout the year in the karoo have adaptations for conserving water and reducing herbivory (Midgley and Van Der Heyden, this volume). These traits include waxy or hairy cuticles, tannins, alkaloids and phenolic compounds. Consequently, resident herbivores need to be tolerant of these defences or highly selective for undefended plant parts.

The snail *Trigonephrus haughtoni* in the southern Namib Desert includes a toxic plant (*Othonna sparsiflora*) in its diet (Curtis, 1991). Insect herbivores in the karoo include, *inter alia*, the karoo caterpillar *Loxostege frustalis* that feeds mainly on *Pentzia* spp. (Annecke and Moran, 1977). Larvae of this species spin webs on food plants and remain inside the webs while feeding. Webs spun by the caterpillars render the plants unpalatable to sheep (Annecke and Moran, 1977) suggesting that the webs are an effective defence by the caterpillars against accidental ingestion by (indigenous) mammalian herbivores.

The food of the leopard tortoise *Geochelone pardalis* (Milton, 1992a) and serrated tent tortoise *Psammobates oculifer* (Rall and Fairall, 1993) includes several species of geophytes and Crassulaceae which contain toxic chemicals such as acids, alkaloids and glycosides. These plants are avoided by antelope, sheep and goats. Tortoises, as well as certain desert lizards and rodents, are apparently able to denature or excrete the toxins (Milton, 1992a). Leopard tortoises are dependent on water, and are territorial with home ranges of 1–2 km² (Branch, 1988a). Serrated tent tortoises were not observed to drink water in arid savanna, suggesting that this species is able to survive without access to free water (Rall and Fairall, 1993). However, the related tent tortoise *Psammobates tentorius* is able to use condensed fog and rain water by elevating its rear and channelling the water towards its mouth (Boycott and Bourquin, 1988). Home range and territorial behaviour are not known in serrated tent tortoises.

Typical, resident, herbivorous mammals in the karoo tend to be solitary and territorial and, in general, support Hobbs and Swift's (1988) hypothesis that unproductive systems offer less benefits to gregarious grazers than do productive systems. This suggests that large herds of grazers could occur in the karoo only during periods of high primary production.

The distribution of large grazing mammals in the karoo is unlikely to have been homogeneous over the region, since the composition of the vegetation is not grassy in the southern and western Nama-karoo and the succulent karoo. Vegetation in the southern and western karoo *sensu lato* is not very structured, so that herbivores must forage either low down (in shrubs) or high up (in drainage-line vegetation), and this too is an effective filter for certain herbivore species. It is possible that the scarcity of grass (and the aridity) in the southern and western karoo also influences the species-richness and abundance of rodents in these areas.

Whistling rats *Parotomys brantsii* and karoo bush rats *Otomys unisulcatus* are central-place-foragers living in small family groups and feeding on the most common plants in the vicinities of their burrow systems or lodges

(Vermeulen and Nel, 1988; Du Plessis, 1989; Brown and Willan, 1991). These are usually the toxic stems and foliage of *Pteronia pallens*, and the saline and generally unpalatable leaves and stems of *Malephora lutea* and *Augea capensis*, as well as the foliage of palatable, but spiny, *Lycium* spp. There is a great deal of dietary overlap between whistling rats and karoo bush rats (Du Plessis et al., 1991). However, whistling rats burrow into bare, sandy patches, and karoo bush rats build lodges of sticks in tangled thorn bushes in drainage lines, so competition is avoided by spatial separation (Du Plessis, 1989). Neither species shows any evidence of seasonal diet change, and both species feed opportunistically on annual plants (Brown and Willan, 1991; Du Plessis et al., 1991). Whistling rats and karoo bush rats collect plant stems and succulent leaves, usually enough for their daily needs, and stockpile food outside burrow and lodge entrances respectively (Vermeulen and Nel, 1988; Du Plessis, 1989; Brown and Willan, 1991). Such stores, however, are seldom used the following day, and the rats collect fresh food daily, regardless of how much food remains in the temporary store. Whistling rats and karoo bush rats are preyed on, *inter alia*, by pale chanting goshawks *Melierax canorus* (Malan, 1995). Making temporary stores reduces the exposure of the rats to predators by substantially limiting time away from shelter. This may only lower the probability of predation outside the burrow system, since burrows are sometimes shared by Cape cobras *Naja nivea* and predation within the burrow system cannot be excluded. Whistling rat colonies and karoo bush rat lodges are regularly spaced in suitable habitat (Vermeulen and Nel, 1988; Brown and Willan, 1991; Milton et al., 1992a), suggesting that intraspecific competitive interactions are maintaining the density of colonies and lodges of both species.

Information on the food of most resident herbivorous mammals in the karoo suggests that diets are highly plant-part selective. Published food data for the Cape hare *Lepus capensis* are available for only one site in the southern Nama-karoo. The Cape hare is a browser at this locality, feeding on foliage and flowers of forbs and shrubs, and on one occasion eating seed capsules of Mesembryanthema (Kerley, 1990). Succulent and woody species eaten included plants generally considered distasteful or toxic to domestic livestock. It is not known how frequently seed capsules of succulent plants are eaten by Cape hares, but one direct observation of a Cape hare eating seed capsules of *Malephora lutea* and numerous records of capsules that had been eaten (presumably by Cape hare) suggest that it may not be infrequent (W. R. J. Dean, unpublished observations). Seedlings of *Drosanthemum* sp. and *Trichodiadema* sp. (Mesembryanthema) germinated in dung pellets of Smith's red rock rabbit feeding on a rocky ridge in the southern karoo (S. J. Milton, unpublished

data), showing that seed capsules of mesembs are also eaten by this species.

At two sites in the succulent karoo, klipspringers *Oreotragus oreotragus* fed on flowers and fruits when these were available, selectively removing only these parts of the plant (Norton, 1984). The klipspringers generally ate plant parts that were low in fibre and high in nutrients and moisture. Kudu *Tragelaphus strepsiceros* select patches of habitat within the landscape matrix (Fabricius, 1994), presumably not only for shade and shelter, but also for the quality of the forage within patches.

Other resident herbivores in the karoo include porcupine *Hystrix africae australis*, three lagomorphs (scrub hare *L. saxatilis*, Smith's red rock rabbit *Pronolagus rupestris* and riverine rabbit *Bunolagus monticularis*) and four antelope (common duiker *Sylvicapra grimmea*, steenbok *Raphicerus campestris*, Grysbok *R. melanotis* and grey rhebok *Pelea capreolus*).

Rock hyraxes *Procavia capensis* do not conform to Hobbs and Swift's (1988) hypothesis, but are resident in harem groups averaging 16 individuals within 'colonies' (see Milton et al., this volume). Rock hyraxes feed selectively on new growth of forbs and shrubs in their rocky habitat, although they take a wide range of species (Lensing, 1983), including such alkaloid-defended plants as *Pteronia pallens* that are toxic to other herbivores (S. J. Milton, personal observations).

Termitivores

Termite (*Hodotermes*, *Microhodotermes* and *Trinervitermes* spp.) workers are eaten by a number of invertebrate and vertebrate predators in the karoo, including solifuges (Lawrence, 1963), carabid beetles (Dean and Milton, personal observations), lizards and snakes (Branch, 1988a), and birds and mammals (Skinner and Smithers, 1990; Kok and Hewitt, 1990; Maclean, 1993). For some species, feeding on workers and alates of termites may be opportunistic, as in some gekkonid lizards (Bauer et al., 1989), but, for other species (see below), termites are an important part of their daily food intake. Southern African vertebrate termitivores are larger than their Australian counterparts (Milewski et al., 1994), because termites appear to be more abundant, and thus more reliable as a food source in southern Africa than in Australia. Animals large enough to dig into termite nests are generally able to feed on termites daily and throughout the year in the karoo.

Ponerine ants (*Pachycondyla (Ophthalmopone) hottentota*) prey exclusively on small harvester termites *M. viator* in the southern karoo (Dean, 1989). The termites forage dynamically and unpredictably (Dean, 1993) and the ants catch workers returning to foraging ports. The ants are not able to dig deeply for prey, but are able to exhume termites that are active just below the soil surface at termite-nest soil dumps. Prey capture by the ants thus depends on two sets of activity by the termites, neither of which is constant. Since the termites are not active all the time, the ants catch and store termites during brief periods of termite activity. No data are available on the length of time that termites are stored by the ants, but it is likely that it is not for extended periods, since we have observed that ant larvae share the storage space with the immobilized termites. *Ammoxenus* spp. spiders (Ammoxenidae) similarly prey on *M. viator* during brief termite activity periods and bury them in temporary storage sites (Wilson and Clark, 1977; Dean, 1988). *Ammoxenus* spiders catch and eat termites of a specified size range, testing potential prey by rapid manipulations and either rejecting or accepting the prey. These spiders also draw their prey beneath the soil surface and feed in this position (Wilson and Clark, 1977). Pit-trap data collected at Tierberg suggest that the activity patterns of *Pachycondyla (Ophthalmopone) hottentota* and *Ammoxenus coccineus* match the activity patterns of *M. viator* workers fairly closely (Fig. 10.1).

Many lizards prey on termites in addition to other prey (Branch, 1988a). No lizards appear to feed exclusively on termites, but the bulk of the food may, on occasion, be made up by termites (Bauer et al., 1989). Other termitivores, to whom termites vary in importance, include the bat-eared fox *Otocyon megalotis* (Lourens and Nel, 1990) and aardvark *Orycteropus afer* (Skinner and Smithers, 1990). Both bat-eared foxes and aardvarks are widespread in the karoo. Aardvarks are generally resident, solitary and appear to be at a low density in the karoo. They may be territorial, although Van Aarde et al. (1992) were unable (through constraints in the study) to show territoriality in this species. Aardvark home ranges in the north-eastern

Nama-karoo are up to 3.5 km² (Van Aarde et al., 1992). Aardvarks are well adapted to dig deeply into termite and ant nests in hard soils, and may switch diet from mainly termites to mainly ants at certain times of the year (Willis et al., 1992). Location of termite and ant colonies by aardvarks may be by a combination of sound and smell (Melton, 1975) or local knowledge, as it has been shown that aardvarks preferentially dig into the west side of termite nest-mounds in the Nama-karoo (Dean and Siegfried, 1991).

Bat-eared foxes appear to be permanent residents elsewhere in the arid zone of southern Africa (Mackie and Nel, 1989; Lourens and Nel, 1990; Nel, 1990). They forage in groups (family parties), and are not strongly territorial, with home ranges in the north-eastern Nama-karoo of about 1 km² and group sizes of 2.68 ± 1.27 (n = 214) (Mackie and Nel, 1989). In coastal western succulent karoo, home ranges are <0.5 km² with group sizes of 2.3 (n = 10) (Lourens and Nel, 1990). Territory boundaries may be fluid and dependent on rich food patches (Nel, 1984). In the southern karoo, a group of foxes generally occupies a burrow for only a few months before moving to another shelter or foraging area. Group sizes in bat-eared foxes are influenced by rainfall and amount of food available (Nel et al., 1984) so are likely to differ between areas and between seasons. Termites form a small to large part of their varied diet in the karoo, which includes ants and other insects, lizards and, particularly in the winter, fruit (Nel and Mackie, 1990; Kuntzsch and Nel, 1992).

The aardwolf *Proteles cristatus* is an obligate termitivore, occurring marginally in the Nama-karoo. The population dynamics of aardwolf is determined by the availability of their prey, the nasute harvester termite *Trinervitermes*

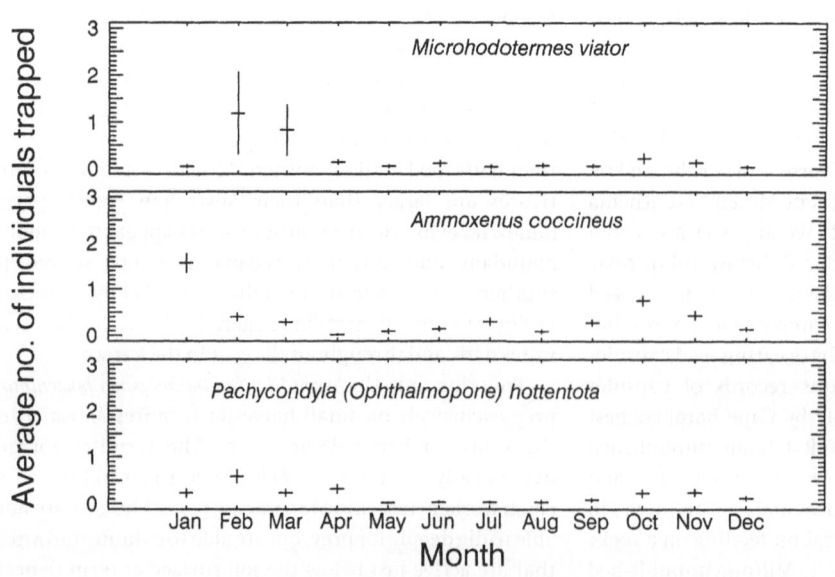

Figure 10.1 **Average numbers (horizontal bar indicates mean, vertical lines indicate the standard error) of *Microhodotermes viator* workers (prey), *Ammoxenus coccineus* spiders and *Pachycondyla (Ophthalmopone) hottentota* foragers (predators) trapped in pit traps (n = 40) set for 24 hours once monthly (for 30 months) at Tierberg, southern Nama-karoo (W. R. J. Dean, unpublished data)**

trinervoides. Specialization on this small prey with low nutritional value means that aardwolfs need to occupy territories that contain about 3000 termitaria and to spend 6–8 hours nightly licking up about 250 000 of these insects from the soil surface (Richardson 1987a). With such constraints, the female cannot intensify milk production, but must sustain lactation for 15 weeks before cubs reach weaned weight (Van Jaarsveld et al., 1995). Births of 2–4 cubs per litter in October coincide with the seasonal peak in termite activity (Richardson 1987b). Scarcity of termites during winter months causes cub mortalities and weight losses in adults.

Formicivores

Ants provide a dependable, though distasteful, food resource for many other animals, including other ants. Predation on ants in arid areas may be highly specific, as it is in *Cosmophasis* spiders (Salticidae) that mimic *Camponotus detritus* and *C. fulvopilosus* (Formicidae) respectively (Curtis, 1988), but it may be opportunistic by other species. The ant *Ocymyrex robustior* is adapted to high soil surface temperatures (up to 70 °C) and in the Namib Desert preys on other insects, including ants, that are stressed by heat, dehydration or that are unable to locate thermal refuges at critical times (Marsh, 1985c). Whole ants and parts of ants, including *Monomorium* spp., *Tetramorium* spp., *Messor capensis, Anoplolepis steingroeveri* and *Acantholepis capensis*, were the most frequent food items brought to a nest of *Ocymyrmex cilliei* (44 of 98 prey items) in the southern Nama-karoo. Other prey consisted of termites *Microhodotermes viator* and *Amitermes* sp., bees, wasps, weevils, roaches, 1st instar grasshoppers, coccids and hemipterans (W. R. J. Dean, unpublished data). *Ocymyrmex cilliei* is active at soil surface temperatures of 29.6–61° (mean 45.7°) (Dean, 1992a) and probably occupies a similar thermal and biological niche in the southern karoo to that occupied by *O. robustior* in the Namib Desert and by *Cataglyphis* spp. in the Saharan Desert (Marsh, 1985b).

Ocymyrmex robustior is itself preyed on by pit-building ant-lion larvae. Marsh (1987a) found that 65.4% of the biomass consumed by the larvae of *Cueta trivirgata* (Myrmeleontidae) consisted of *O. robustior* in the Namib Desert.

A number of birds in the karoo feed on ants, but it is not known whether birds in the karoo include a higher percentage of ants in their diet than do similar-sized congeners in the savanna. Birds recorded eating ants in the karoo include karoo korhaan (Boobyer and Hockey, 1994), several lark spp. (Alaudidae), ant-eating chat *Myrmecocichla formicivora*, karoo chat *Cercomela schlegelii*, karoo robin *Erythropygia coryphaeus*, Namaqua warbler *Phragmacia substriata* and rufous-eared warbler *Malcorus*

pectoralis (Dean and Milton, unpublished data). The ground woodpecker *Geocolaptes olivaceus* is a major formicivore throughout its range in South Africa, including the karoo (Oatley et al., 1989; T. B. Oatley, personal communication).

Among other small insectivorous mammals that forage above-ground, elephant shrews (*Macroscelides* and *Elephantulus*) occur widely in the karoo, including the most arid areas. Shrews (*Myosorex, Suncus* and *Crocidura*), however, are not widespread in the karoo, and tend to occur along the edge of the region in more mesic habitats. A major difference between the two groups is that all species of elephant shrews include termites and ants in their diet, whereas shrews are not known to eat ants, although admittedly their diet is generally poorly known (Skinner and Smithers, 1990). We suggest that the ability to eat ants may contribute to the success of elephant shrews in the karoo.

The Pangolin *Manis temminckii* occurs marginally in the northern Nama-karoo, and feeds almost exclusively on ants, digging into ant nests to do so (Skinner and Smithers, 1990). More widely distributed bat-eared foxes take an increased proportion of ants during the summer (November–March) in the karoo, particularly in shrublands grazed by domestic livestock (Kuntzsch and Nel, 1992).

Other insectivores

The spider- and caterpillar-hunting wasps are an important group of invertebrate predators (Weaving, 1988). Sympatric *Ammophila* spp. differ in families and species of caterpillars taken as prey, indicating some partitioning of the resources, chiefly by the wasps hunting in different habitats, including one species taking prey from beneath leaf litter (Weaving, 1988). *Ammophila* wasps initially search on the ground for faecal pellets to locate prey in overhead vegetation, and Weaving (1988) suggests that this may have led to the evolution of hunting on the ground by some species. Weaving further speculates that ground-hunting by wasps may be more prevalent in arid areas, because caterpillars or other prey species may shelter in leaf litter or other sheltered sites on the soil surface.

10.2.3   Parasitoids and parasites

Root-feeding scale insects (Hemiptera: Coccoidea) are common on roots of certain succulent Aizoaceae in the southern Nama-karoo (S. J. Milton, personal observations) but no studies have quantified occurrence, abundance and range of food plants in these insects in the karoo. The energy flow through populations of *Lasius flavus* ants feeding on root-feeding aphids and honeydew in Europe is high ($310 \, kJ \, m^{-2} \, yr^{-1}$) and compares favourably with the energy flow through above-ground ant populations (Woodell and King, 1991). It is likely that root-feeding aphids are exploited by nectarivorous ants in the karoo, but no studies have yet shown an association between root-feeding scale insects and ants in the karoo.

The nymphs of cicadas (Homoptera: Cicadidae: *Quintillia, Henicotettix, Munza, Platypleura, Masupha* spp.) in the Nama-karoo feed on root xylem sap throughout their lengthy (2–3 year) pre-adult phase and are not constrained by seasonal food shortages. The supply of root xylem sap, however, may be affected by drought and herbivory (Midgley and Van Der Heyden, this volume), and may in turn influence the growth rates of nymphs. Although resident, densities of cicadas do not provide support for the hypothesis that such animals should be at a low density, and the density of *Quintillia* spp. adults that emerged at one site in the Nama/succulent karoo interface was calculated at 7400 ± 1036 cicadas ha^{-1} with a biomass of 2.516 kg ha^{-1}, greater than the (wild) mammalian herbivore biomass at the same site (Dean and Milton, 1991c).

Some data on parasitoids are available from studies on aculeate hymenoptera in the karoo (Gess, 1980a; Gess and Gess, 1975, 1976, 1980b, 1982). A number of parasitoid wasp families occur in the region, attacking the larvae and pupae of Lepidoptera, Hemiptera, Coleoptera, Diptera, Hymenoptera, coccids and spiders (Scholtz and Holm, 1985). The levels of parasitism on hymenoptera varies between hosts and parasites. Only one of 60 nests of *Dichragenia pulchricoma* (Hymenoptera: Pompilidae: Macromerini) examined was parasitized by *Ceropales punctulatus* (Pompilidae: Ceropalinae) on the karoo edge north of Grahamstown (Gess and Gess, 1974), but 6 of 9 nests of *Euodynerus euryspilus* (Hymenoptera: Eumenidae) were parasitized by *Chrysis hoplites* (Hymenoptera: Chrysididae) at the same locality (Gess and Gess, 1991a).

Bee-fly larvae (Diptera: Bombyliidae) are parasitic on the larvae or pupae of other insects, including other Diptera, Coleoptera, Lepidoptera and Hymenoptera, although the adults are generally nectar or pollen feeders (Scholtz and Holm, 1985). Host-mimetic conopid flies (Diptera: Conopidae) are parasitic on aculeate hymenoptera in dry areas, and other species of conopids have been recorded parasitizing dung-breeding Muscidae elsewhere (Scholtz and Holm, 1985).

There are several groups of blood-sucking flies present in the karoo, of which mosquitoes (Diptera: Culicidae) are the most well known. Tabanids (Diptera: Tabanidae) are common along water-courses throughout the karoo, feeding on blood of mammals, but also reptiles, amphibians and birds (Scholtz and Holm, 1985). Blood-sucking simulid flies, including *Simulium chutteri* along the Orange River, and the salt-pan midge *Leptoconops karteszi*, widespread in the karoo, are increasing in abundance, due, in part, to the increase in storage dams and flood-control systems along rivers (Bath, 1978). Bombyliid and tabanid flies, as well as asilid, syrphid and sarcophagid flies are preyed on by such wasps as *Bembix* spp. (Gess, 1986). Mosquitoes have been extensively studied in southern Africa (Scholtz and Holm, 1985), but very little has been published on blood-sucking flies in the karoo.

The larvae of a blood-sucking fly, *Passeromyia heterochaeta*, feed exclusively on nestling birds (Ledger, 1969). Although not known to be widespread in the karoo, *P. heterochaeta* does occur in the eastern karoo (Taylor, 1949), and larvae have been found in the southern karoo in a nest of the spotted prinia *Prinia maculosa* (W. R. J. Dean, personal observations).

Ticks (Acaridae) are widespread in the karoo and hosted by mammals, birds, reptiles and tortoises (Theiler, 1959, 1965; McLatern, 1987). *Amblyomma* spp., essentially parasites on mammals, has been found on several species of non-passerine birds, including the ostrich. Similarly, *Hyalomma* spp. have been widely recorded on birds, including passerines, and also parasitize such small mammals as hares (Theiler, 1959). Other examples are *Rhipicephalus sanguineus*, which is associated with human habitation and parasitizes small mammals and larger non-passerine birds, including the ostrich and kori bustard *Otis kori*, and *Aponomma* spp., parasitic on reptiles (Theiler, 1959). *Amblyomma hebraeum* and *A. marmoreum* are hosted by leopard tortoises, and may play a central role in the transmission of heartwater fever in the eastern Nama-karoo (McLatern, 1987).

10.3.   Seasonal resources

Some food resources in the karoo, such as deciduous leaves, flowers, fruits and seeds, and the animals that eat such food, are available predictably but only at certain times of the year.

Some general principles that apply to animals feeding on seasonal resources are:

- If the animal is resident, it should change its diet seasonally or be able to switch diet.
- If a specialist feeder, it may store food and live for part of the year on food stores.
- The animal may only be active at those times of the year when the resources are available.
- The animal may have a short lifespan, so that the life cycle is completed during the time that the resources are available, or it needs a diapause phase in the life cycle.

10.3.1.   Opportunism and diet switching

There is no evidence that any animal in the karoo has a true seasonal diet switch. Most apparent dietary switches

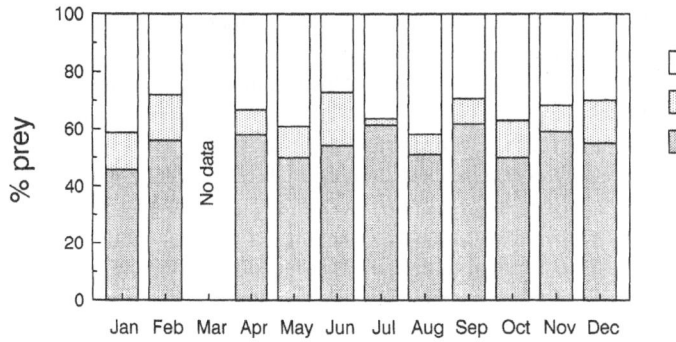

Figure 10.2 **Prey taken by the pale chanting goshawk *Melierax canorus* near Calitzdorp, southern succulent karoo (G. Malan, unpublished data)**

can be ascribed to opportunism, but a number of omnivorous species do take a higher proportion of (usually) plant foods, including leaves, flowers, fruits and seeds, during the winter.

Lizards are generally considered to be insectivorous (Branch, 1988a), but the southern rock agama (*Agama atra*) feeds on flowers of *Gazania krebsiana* in early spring, before insects reach peak abundance. Both *Angolosaurus skoogi* and *Aporosaura anchietae* eat dry seeds in addition to invertebrates (Pietruszka et al., 1986; Murray and Schramm, 1987; Robinson, 1987) in the Namib Desert. There was a shift from predominantly invertebrate food to predominantly plant food in midwinter to early summer in a 'dry' year, but not in a 'wet' year in *A. anchietae* (Robinson, 1987).

Several species of birds that occur in the karoo, other than the herbivorous ostrich (Milton et al., 1994a), include in their mixed diet leaves and flowers of shrubs and trees (Winterbottom, 1973; Milewski, 1978; Boobyer and Hockey, 1994), young leaves and basal nodes of grasses (Willoughby, 1971), unripe seed capsules of Mesembryanthema, seedlings of forbs and grasses, and the bulbs of geophytes when available (Broekhuysen, 1963; Maclean, 1993; Boobyer and Hockey, 1994). Resident white-backed mousebirds (*Colius colius*) that are entirely vegetarian (Rowan, 1967) vary the proportions of flowers, fruits and leaves in their diets in response to resource availability (W. R. J. Dean and S. J. Milton, personal observations). The omnivorous karoo korhaan opportunistically feeds on flowers of Asteraceae, Brassicaceae and Aizoaceae (Mesembryanthema) and leaves of the halophyte *Zygophyllum retrofractum* (Zygophyllaceae), as well as a wide variety of invertebrate and vertebrate prey (Boobyer and Hockey, 1994).

The pale chanting goshawk *Melierax canorus* is an opportunistic predator, taking both invertebrates and vertebrates (Maclean, 1993). The patchy mosaic of habitats near Calitzdorp in the succulent karoo has led to a low diversity of prey and large spatial differences in prey abundance (Malan, 1995), but little seasonal variation in diet (Fig. 10.2). However, the proportion of tent tortoise

Psammobates tentorius, angulate tortoise *Chersine angulata* and leopard tortoise prey in pale chanting goshawk diets peaked in autumn when hatchlings emerged (Malan and Branch, 1992), indicating some opportunism by the goshawk.

Opportunistic foraging by raptorial birds on road casualties is common in the Nama-karoo. Many species of mammals, birds and reptiles are killed on the roads in the karoo (Siegfried, 1965a), providing food for a number of scavenging birds including eagles and kites (Steyn, 1982), jackal buzzard *Buteo rufofuscus* (Steyn, 1982; Schmitt et al., 1987), pale chanting goshawk (Malan, 1992), rock kestrel *Falco tinnunculus* (Siegfried, 1965b), black crow *Corvus capensis*, pied crow *C. albus*, and white-necked raven *C. albicollis* (Siegfried, 1963a; Winterbottom, 1975; Macdonald and Macdonald, 1985), and smaller species such as the fiscal shrike *Lanius collaris*. In some areas, road kills may contribute to increases in populations of scavengers along major roads (Siegfried, 1963a).

Elephant shrews (*Macroscelides proboscideus*), are considered to be obligate insectivores (Skinner and Smithers, 1990), but Kerley (1989) found that foliage formed as much as 60% of the food of this species at both winter and summer rainfall sites in the karoo. There was a seasonal trend in food, with a higher percentage of plant food, including seeds, eaten during the winter. Similarly, the hairy-footed gerbil *Gerbillurus paeba* at the same sites consumed relatively more foliage during the winter, with seeds dominant in the food during summer and autumn, and insects dominant during spring (Kerley, 1989).

Seeds are eaten by some of the small mammals in the karoo, but do not form as large a component of the diet in *G. paeba* and *Desmodillus auricularis* as previously thought (Kerley, 1989). Seed formed only 36% of the diet overall in *G. paeba*, and was the largest component of the diet in summer. Of the other species investigated, *Mus minutoides* ate seeds, foliage and insects, and *Rhabdomys pumilio* ate mostly seeds, with some foliage and insects. For these two species, this finding does not differ substantially from results obtained in other ecosystems (Kerley, 1989).

Predation in sparse vegetation may spatially and temporally limit foraging by seed-eating rodents in the Namib Desert. Both diurnal *R. pumilio* and nocturnal *Gerbillurus tytonis* limit their foraging activity to the immediate vicinity of islands of spinescent grass (*Stipagrostis sabulicola*) or shrubs (*Acanthosicyos horrida*). Foraging effort decreased on moonlit nights (Hughes et al., 1994). Reduction of populations led to an increase in foraging effort, presumably in response to increased food rewards.

Plant material (mainly juicy and dried fruits) formed a significantly higher proportion of bat-eared fox food in winter in both protected areas and areas grazed by livestock, but the proportion of termites eaten remained constant through the year (Kuntzsch and Nel, 1992). These shifts are not only seasonal, but also related to aridity, and the proportion of beetles and fruit in the diet of the bat-eared fox apparently increases from mesic to arid sites (Macdonald and Nel, 1986; Kuntzsch and Nel, 1992; Nel and Mackie, 1990). The consumption of dried, fallen fruits of *Diospyros lycioides* in the southern karoo allows bat-eared foxes to extend their fruit diet beyond the fruiting season (W. R. J. Dean and S. J. Milton, personal observations).

The chacma baboon *Papio ursinus* is an opportunist omnivore, and throughout most of its range in southern Africa is dependent on water (Skinner and Smithers, 1990). Consumption of fruits such as figs (*Ficus* spp.), together with changes in behaviour patterns to conserve moisture, allowed one baboon troop in the Kuiseb River canyon to go for up to 11 days without drinking (Brain, 1988). In periods of food and water shortage, Namib Desert baboons consumed alternative plant foods, including plants toxic to humans and other animals, and plants with a high moisture content not usually eaten (Hamilton, 1986). During this time, one baboon troop made three excursions to particular fig trees at least 5 km away from their base in the Kuiseb River canyon, implying a high degree of local knowledge. Similarly, the behaviour of a troop of *c.* 30 baboons observed at Tierberg in the southern Nama-karoo, suggested that they knew when and where they could obtain resources within their territory. After a rain shower of 12 mm, the troop moved from a rocky ridge to a patch of moist ground where they excavated and ate the bulbs and corms of geophytes. These underground plant parts are usually not available to the baboons because of the hard soils. However, the baboons were able to dig into damp soil for a short period following rain. More than 200 bulbs and corms of a select range of species were removed and eaten by the baboons from a 0.5 ha patch in about 2 hrs (W. R. J. Dean and S. J. Milton, unpublished observations). At Tierberg, chacma baboon troops forage over areas of *c.* 30 km², taking about 30 days to cover the area. This movement is, in effect, trap-lining, or a short-distance migration, since the baboons roost at the same sites at regular intervals and spend 2–3 days foraging in the vicinity of the roosts before moving on to the next camp-site (W. R. J. Dean and S. J. Milton, unpublished observations).

10.3.2.   Storage and caching of food

Resident animals feeding on durable food items (for example, seeds, nectar, pollen and corms) may store food, whereas species feeding on perishable food (e.g. foliage, animal prey and carrion) are less likely to do so. Perishable foods are likely to be temporarily cached, rather than stored, when superabundant. Short- and long-term storage of food is frequent in the karoo, but in most cases this does not differ from similar tactics used by the same or related species in other biomes in southern Africa. Aculeate wasps and bees store invertebrate prey and pollen and nectar in nests to provision young (Gess, 1981; Weaving, 1988); in the honey-bee *Apis mellifera*, this storage extends to food for both adults and young, and stores, depending on the rainfall and security, may not be tapped for some years (W. R. J. Dean and S. J. Milton, personal observations). Ants, *Messor capensis*, *Tetramorium* spp., *Monomorium* spp. and *Pheidole* spp., among others, harvest and store seeds of a variety of plant species (Marsh, 1987b, 1990; Dean and Yeaton, 1992; Vorster et al., 1992; Milton and Dean, 1993). The ants generally store more seeds than they eat, so that seed stores may reflect seed abundances of particular species in past years. Seeds are often seen to be carried out of stores by *M. capensis* after rain and later collected, and it is possible that this species manages its stores to prevent mildew or germination.

Messor capensis, a major seed-harvesting ant, apparently collects food only during the warmer months of the year (Milton and Dean, 1993). An estimated 4.1 million seeds or (5.2 kg) are collected ha⁻¹ yr⁻¹ by *M. capensis* in the southern karoo (Milton and Dean, 1993). The ants gather and store seeds, especially when seeds are superabundant following episodic rainfall events. *M. capensis* often collects wind-dispersed seeds from accumulations in mammal diggings. Such accumulations may allow the ants to gather a larger number of seeds from a particular seeding event than would otherwise be possible, and appear to play an important role in the foraging economy of the ants. Harvester ants can form 96% of ant biomass (Marsh, 1990) in the Namib Desert, but no comparable estimates for the karoo are available. *M. capensis* nests are regularly spaced at Tierberg (8 nests ha⁻¹, Dean and Yeaton, 1992), but occur at a relatively high density where shrubland has been degraded because of greater availability of seed of toxic shrubs, a biennial succulent and ephemeral forbs (Milton and Dean, 1993). In semi-arid grassland, however, the situation is reversed, and *M. capensis* nest-mounds occur at a lower density in degraded than in climax vegetation (Vorster et al., 1992) probably because of an overall reduction in the size of available seeds. *M. capensis* gather the most abundant, medium-sized seeds (Milton and Dean, 1993) regardless of plant species.

The transport and storage of seeds with elaiosomes, common in the fynbos biome (Bond and Slingsby, 1983) is rare in the karoo (Breytenbach, 1988), although a few species of plants with elaiosomes on their seeds occur in the genera *Polygala* and *Euphorbia*.

The fiscal shrike *Lanius collaris* caches invertebrate and vertebrate prey items. Prey items too large to be consumed in one meal, or temporarily superabundant, are impaled on thorns or barbed-wire fences and shrikes return to feed on them later (Maclean, 1993), but this behaviour is not specific to shrikes in the karoo biomes.

Food storage in the karoo is frequent in mammals, and ranges from scatter-hoarding in small rodents (Skinner and Smithers, 1990), to longer-term storage in common molerats (Lovegrove, 1993). Nel (1984) notes, without details, that black-backed jackals *Canis mesomelas* and Cape foxes *Vulpes chama* cache food, but whether this is done only in semi-arid regions is not stated.

In parts of the karoo and Kalahari Desert, molerats collect bulbs and corms and store them for some months in large underground storage chambers. Not all molerats in the karoo store food, and the storage of bulbs and corms may be related as much to local sizes and abundances of food species as it is to the moisture status of the soil in which the animals live (Lovegrove, 1993). It is also related to the size and density of food items; storage is thought to be advantageous where food items are small and at high densities (Lovegrove and Painting, 1987). Stores of bulbs

may be managed by the molerats to prevent sprouting (Lovegrove, 1993). Scatter-hoarding by the short-tailed gerbil *Desmodillus auricularis* and pouched mouse has been recorded in laboratory studies (Pettifer and Nel, 1977) and presumably occurs in the wild. Both species are central-place-foragers, and the short-tailed gerbil is known to store food in blind tunnels in its burrow systems, both in the laboratory (Christian et al., 1977) and in the wild (Nel, 1967).

10.3.3.   Seasonally active and short-lived animals

Animals with short lifespans and those capable of going into prolonged torpor, are able to time their activity to coincide with that of their food plants or prey species. In this way, they avoid the problems associated with fluctuations in food availability caused by cold, heat and drought.

Snails *Trigonephrus haughtoni* appear to be active only in winter in the southern Namib Desert (Curtis, 1991), and feed only at night and the early part of the day. Between feeding bouts, snails bury themselves in sand, with the apex of their shells 5–15 mm below the surface. Presumably, snails remain buried throughout the summer in these sites.

The activity of all invertebrates (Fig. 10.3) and ants in particular (Fig. 10.4) is markedly seasonal in the southern Nama-karoo (W. R. J. Dean, unpublished data). Many aculeate wasps and bees in the karoo show marked seasons of activity, and overwinter in a pre-pupal or pupal stage, emerging as adults in spring or summer (Gess, 1981). Some species of carpenter bees and social wasps

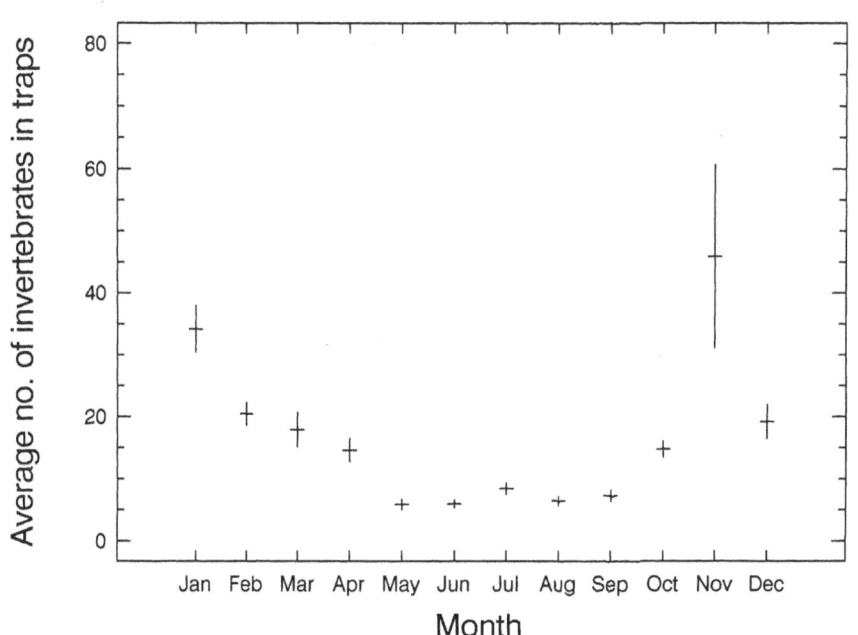

Figure 10.3 **Average numbers (see Figure 10.1) of all invertebrates trapped in pit traps at Tierberg, southern Nama-karoo (W. R. J. Dean, unpublished data). Pit-trap details as Figure 10.1**

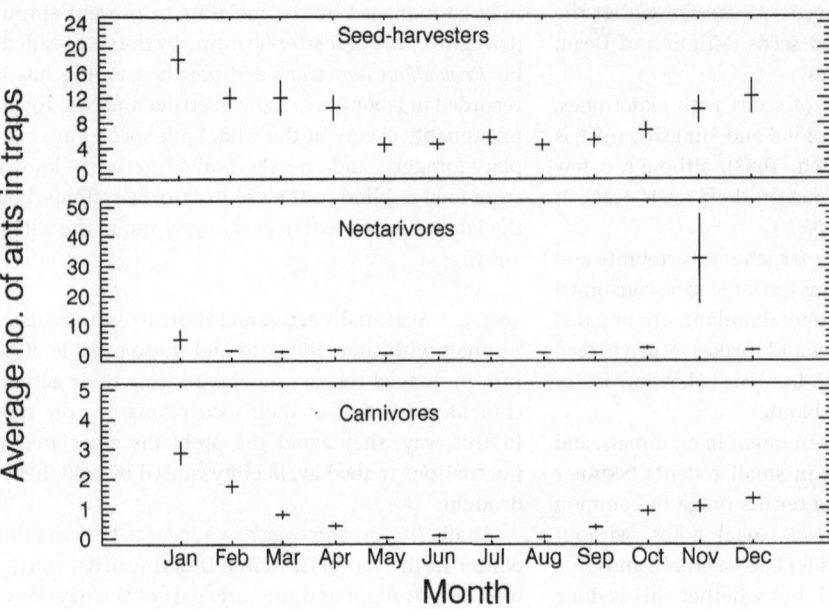

Figure 10.4 **Average numbers (see Figure 10.1) of ants (Formicidae) trapped in pit traps at Tierberg, southern Nama-karoo (W. R. J. Dean, unpublished data). Pit-trap details as Figure 10.1**

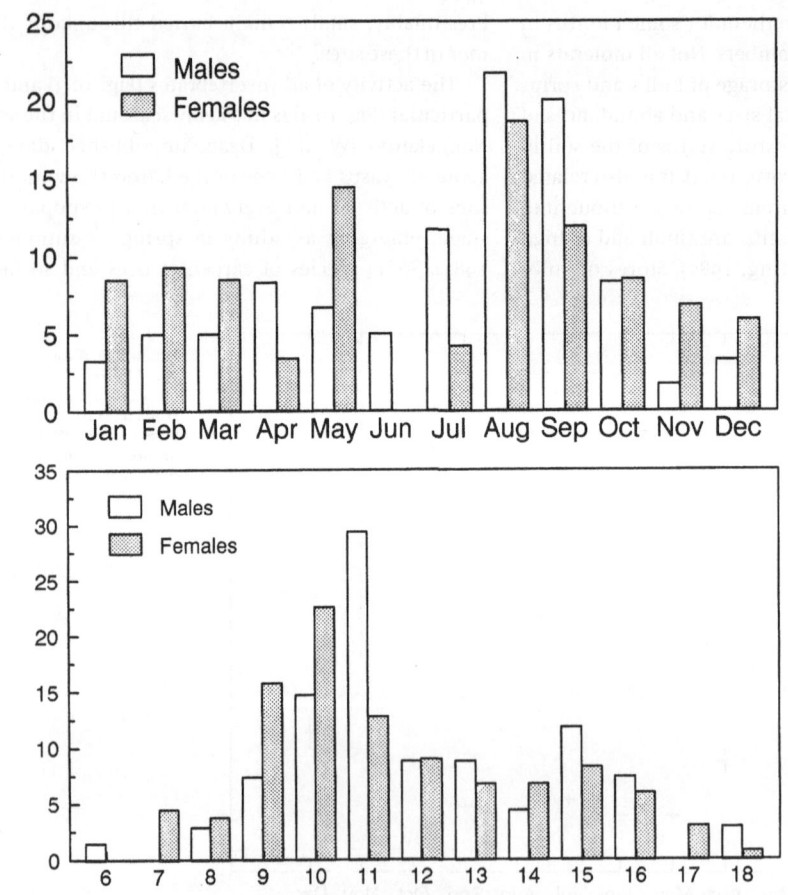

Figure 10.5 **Annual and diurnal activity patterns of adult tent tortoises *Psammobates tentorius* in the southern Nama-karoo (W. R. J. Dean and S. J. Milton, unpublished data)**

emerge as adults before the end of summer and then over-winter as adults (Gess, 1981).

Male tent tortoises show a peak of activity in late winter and early spring, but females are more active through the summer and autumn (Fig. 10.5). Females are also more active in the early part of the day than males (Fig. 10.5) so there may be some partitioning of resources between the sexes.

10.4. Unpredictable resources scattered in time and space

(i) Some foods are dependably available after rain, but the rain itself is unpredictable. Foods include foliage, flowers, nectar, pollen and seeds of plants that leaf out and seed after rain, and the animals that eat such foods. These can be exploited by either a short adult life cycle, or nomadic tracking of patches of a particular resource by a specialist animal.

An example of a short adult life cycle cued by rain is that of the pollen wasp. Development of imagoes in their subterranean cells is stimulated by rainwater penetration which synchronizes their emergence with the flowering of their forage plants (Gess, 1996).

Birds, with their extreme mobility, are able to use resources that are patchy in space and time. Such movements are made not only by birds that are truly nomadic, but also by so-called resident species that may gather to exploit superabundant food patches such as locusts *Locustana pardalina* or caterpillars *Loxostege frustalis*. Nomadic birds include aquatic species that breed opportunistically on ephemeral wetlands in the karoo.

Granivores

Some resources, for example seeds in summer- and winter-rainfall regions, have a relatively high degree of predictability. The Namaqua sandgrouse *Pterocles namaqua* moves seasonally over long distances within the Nama- and succulent karoo from west to east. These movements, which do not involve the whole population (partial-migration), appear to track seed crops in autumn in the summer rainfall area of the eastern Nama-karoo, and seed crops in the western Nama-karoo and northwestern succulent karoo in the spring (Malan et al., 1994). Ludwig's bustard *Neotis ludwigii* is a winter migrant to the succulent karoo, and migrates seasonally from the Namib Desert to the Namibian escarpment (Allan, 1994). There is evidence that this is a partial migration as there is no decrease in relative abundance of the birds in the Nama-karoo during summer (Allan, 1994). No data are available on seasonal shifts in foraging or food in this species, and movements appear to track rainfall.

Most nomadic bird species in semi-arid regions throughout the world are granivorous (Maclean, 1974, 1996; Davies, 1984). In the karoo, regional nomads (species that may move over hundreds of kilometres to find rainfall patches) are mainly granivorous, whereas local nomads (species that remain in one general area) tend to be mainly insectivores or nectarivores (Dean, 1995).

There is some difference in the kinds of seeds that are eaten by resident, nomadic and locally nomadic avian granivores in the karoo. Summarized data on the seed components of the food of 24 resident, nomadic and locally nomadic species from the karoo is presented in Table 10.1. Resident species feed on the seeds of forbs, shrubs and legumes. The small nomadic larks and buntings feed mainly on the seeds of perennial and annual grasses (*Brachyaria glomerata*, *Stipagrostis* spp. and *Schmidtia kalihariensis*), sometimes on the seeds of ephemeral Asteraceae and occasionally legumes, whereas the nomadic or partially migratory Namaqua sandgrouse feeds largely on legumes (Table 10.1). Nomadic species eat a significantly larger proportion of grass seeds (Kruskal–Wallis test, $p = 0.016$) than do resident species.

Small birds (<25 g) eat significantly more grass seeds than larger birds in the karoo (Mann–Whitney U-test: $z = -2.925$, $p < 0.003$). The samples in Table 10.1 are too small to test differences in the proportion of legume seed eaten by resident and nomadic birds. Legumes appear to be important only for some of the resident larks and the nomadic Namaqua sandgrouse. Dixon and Louw (1978) found that Namaqua sandgrouse in the Namib Desert feed almost exclusively on legume and forb seeds. Lists of carefully identified seeds from stomachs and crops of many of the granivorous species in southern Africa are not available, so the conclusions drawn from the above analysis are tentative. The results do support the findings of Morton and Davies (1983), who found that small birds (<100 g) in the Australian arid zone eat proportionately more grass seeds than other seeds. It would be of interest to test the generality of this in all arid ecosystems.

The advantages of eating grass seeds, particularly for nomadic birds, are that the seeds are often abundant after rain, are easily harvested and handled, and lack the tannins and alkaloids often present in the seeds of forbs and shrubs (Watt and Breyer-Brandwijk, 1962). Seeds of grasses are produced relatively quickly after rain (*c.* 4 weeks from rain to seeding in *Stipagrostis ciliata*, S. J. Milton, personal observations). There are few data available for the karoo on numbers of grass or shrub seeds on the soil surface. A study in progress suggests that forb and shrub seeds may be abundant (up to 6 g m^2) on sandy soils in the northern Nama-karoo (Lloyd et al., 1994).

Table 10.1. **The proportion of different kinds of seeds in combined stomach and crop contents of some granivorous or partly granivorous birds in the karoo. Only the seed component of the diet of individual birds is given here**

Species	Mass	n	Grass	Others	Legumes	Total seeds
Resident						
Fawn-coloured ark	23	2	0.54	0.46	0	270
Sabota lark	25	9	0.69	0.27	0.04	99
Long-billed lark	50	3	0	1	0	15
Karoo lark	31	33	0.13	0.77	0.10	3501
Red lark	37	12	0.27	0.73	0	832
Spike-heeled lark	27	8	0	0.97	0.03	252
Red-capped lark	26	2	0.02	0.98	0	64
Thick-billed lark	44	4	0.04	0.93	0.03	295
Cape sparrow	27	1	0	1	0	5
Masked weaver	26	1	0	1	0	29
Cape canary	15	2	0	1	0	40
Cape bunting	21	1	0	1	0	53
Rock bunting	15	1	1	0	0	32
Nomadic						
Namaqua sandgrouse	180	4	0	0.63	0.37	16 067
Namaqua dove	40	3	0	1	0	6 843
Sclater's lark	20	6	0.99	0	0.01	157
Stark's lark	19	16	0.90	0.10	0	377
Grey-backed finchlark	17	30	0.99	0.01	0	869
Black-eared finchlark	15	8	0.67	0.33	0	360
Scaly-feathered finch	11	1	1	0	0	63
Black-headed canary	12	3	0.85	0.15	0	200
Yellow canary	17	6	0.05	0.95	0	263
White-throated canary	27	7	0	1	0	242
Larklike bunting	15	6	0.91	0.09	0	138

Herbivores

It has been suggested that mobility is the secret of success in ostriches in arid parts of Africa. Ostriches, which select high-quality forage (Milton et al., 1994a), follow water-courses and move nomadically between rainfall patches (Sauer and Sauer, 1966; Berry and Siegfried, 1991; Williams et al., 1993).

The springbok 'treks' (Skinner, 1993), in which many hundreds of animals moved on a roughly circular route through the karoo, possibly were away from adverse veld conditions because of drought. However, in some instances the treks were in the direction of localized rainfall patches and new plant growth (Skinner, 1993). In the succulent karoo, springbok may have moved seasonally to remain in the most productive rangelands on the summer–winter rainfall interface (Skinner, 1993). The movements also imply that the springbok populations built up to levels which could not be supported by the available forage. However, why there was a mass movement in one direction by this species, instead of dispersing in small groups to recent rainfall patches, has not been explained satisfactorily. The movements may be the ghost of migration past, evolved in a wetter, more productive, more sharply seasonal environment when the springbok moved regularly between the winter-rainfall and summer-rainfall areas of the karoo. Fencing constrains the movement of these animals, and there have been no mass movements of springbok in the karoo since the early 1900s (Skinner, 1993).

Formerly, other large herbivores, including elephant *Loxodonta africana*, black rhinoceros *Diceros bicornis*, mountain zebra *Equus zebra*, quagga *E. quagga*, black wildebeest *Connochaetes gnou*, red hartebeest *Alcelaphus buselaphus*, gemsbok *Orys gazella* and eland *Taurotragus oryx* periodically occurred in the karoo (Skead, 1980, 1987; Vernon, this volume), but are now extinct in the region or survive only as relict populations in protected areas. These large herbivores are likely to have only occurred in the karoo during high-rainfall periods, when there was above-average primary production.

(ii) Other foods are patchily available in space and time. Great distances or several years may separate places and periods of abundance in these resources. These superabundant resources (e.g. locusts and karoo caterpillars), are utilized by many species, even herbivores, but only highly mobile animals can use such patchy food resources efficiently.

Locusts *Locustana pardalina* are exploited by a diversity of animals, including baboons (Hamilton, 1986), other smaller and larger mammals, birds, reptiles, invertebrate predators and parasites (Erasmus, 1988). The large numbers of locusts are able to eat and recycle an immense amount of vegetation, mainly grasses, and it has been suggested that locusts play an important role in nutrient cycling in the karoo. Locusts eat an estimated 171.4 kg ha^{-1} of vegetation in the core areas of abundance in the Nama-karoo (Boshoff, 1987). A conservative estimate suggests that locusts could have deposited 1.4 g m^{-2} of frass containing 0.65% nitrogen over an area of 237 300 km^{-2} in the Nama-karoo during one outbreak (Boshoff, 1987). The amount consumed by other locust and grasshopper species in the karoo is not known, but is likely to be significant. It has been estimated that grasshoppers consume more vegetation (94 kg ha^{-1} yr^{-1}) than all other herbivores combined in more mesic savanna at Nylsvley (Gandar, 1988).

It is possible that in the later stages of a locust outbreak the predators may depress the locust population, and then may be crucial in reducing the numbers of solitary-phase locusts, helping to delay the onset of the next outbreak. The removal of the predators of locusts by man, either directly (through problem animal control) or indirectly (through poison spraying for locusts) may have had the effect of causing the increased frequency of locust outbreaks observed since 1910 (De Villiers, 1988).

10.5.   **Food webs**

Some examples of food webs in the karoo are only briefly discussed here. Food chains in the karoo may be short, as in the chain in which the ant-lion *Cueta trivirgata* is the top predator, preying on the detritivore and predator *Ocymyrmex robustior* (Marsh, 1987a), or they may be more complex mutual predation webs. For example, *O. cilliei* interacts with *Anoplolepis steingroeveri* in a mutual predation web, in which the top predator oscillates between *O. cilliei* at high soil surface temperatures, and *A. steingroeveri* at lower soil surface temperatures. In this interaction, *A. steingroeveri* is active at soil surface temperatures (T_s) of 10.2–58.9° (mean 210.5°), and *O. cilliei* is active at T_s 29.6–61° (mean 45.7°). At temperatures <40°, *A. steingroeveri* catches and kills individuals of *O. cilliei*, but at temperatures >40°, *O. cilliei* preys on *A. steingroeveri* (W. R. J. Dean, unpublished data). There may be a similar interaction between sympatric *O. weitzeckeri* and *A. custodiens*, since there are similar differences in preferred temperatures in these species, but if so, it was not noted by De Bie and Hewitt (1990).

Loops in food webs in the karoo may be frequent. Tephritid flies, particularly *Desmella anceps* (Diptera: Tephritidae) are seed parasites of *Pteronia pallens* and other karroid shrubs (Milton, 1995b). Adult *D. anceps* lay eggs in seed capsules and larvae feed on the developing seeds, and the larvae are themselves parasitized by small chalcidonid wasps (Eurytomidae: *Eurystoma* spp. and Torymidae: *Antistropholex* spp.) (Milton, 1995b). Some of the scorpions are cannibalistic ('self loop') in the karoo, to the extent that a large female might subsist almost entirely on her young for a time (L. Prendini, personal communication). Cannibalism may be common in carnivorous ants. Alates of *A. custodiens* and *A. steingroeveri* that emerge during rain or moist periods and that delay their departure from the nest, are caught and eaten by workers of the same species (W. R. J. Dean, personal observations).

Food webs generally have been poorly studied in the karoo, and this may be a rich field for further studies. Available evidence suggests a similar complexity in food webs as in other arid ecosystems (Polis, 1991b, 1991c). Specialization in foraging, as in the ant *Pachycondyla (Ophthalmopone) hottentota*, that preys only on *Microhodotermes viator* in the southern Nama-karoo (Dean, 1988) or sphecid wasps that prey only on halictid wasps or single ant genera (Gess, 1980b) represent one end of a foraging spectrum, and such species as baboons or the large bustards, that feed on a wide range of animal and vegetable foods, represent the other. The karoo is nutrient-rich but may be limited in standing biomass and production because of aridity, resulting in fewer large herbivores and concomitant carnivores. The spatially and temporally patchy distribution of primary, and, concomitantly, secondary production leads to opportunistic diet switches, omnivory and ephemeral use of resources by animals and consequently to complexity of food webs.

Dynamics

P. A. Novellie

The study of rangeland dynamics has been orientated largely towards the important objective of preventing range degradation or desertification. This study has relatively recently entered an exciting phase. Confidence in the adequacy of conventional rangeland paradigms, which has prevailed for a long time, is progressively giving way to a succession of challenges from alternative theories. There is a growing conviction on the part of a number of range scientists, in southern Africa and elsewhere, that the conventional approach is not succeeding in its ultimate aim, and that it has not had a discernible positive effect on rangeland management (e.g. Hoffman, 1988b; Behnke and Scoones, 1993). It is therefore appropriate to start this overview with a brief introduction to the conventional 'mainstream approach', and its perceived shortcomings, before going on to the alternative paradigms.

Perhaps because of its complexity and relative lack of empirical data, the field of rangeland dynamics tends to be more strongly influenced than other disciplines by paradigms or models. One such model, appropriately termed the range succession model because of its roots in Clementsian succession theory (Westoby et al., 1989), has in the past largely dominated perceptions in range dynamics. This constitutes the conventional or 'mainstream approach' (Behnke and Scoones, 1993) of range scientists virtually throughout in the world. This model holds that communities are characterized by an equilibrium or climax state. Overgrazing causes the system to regress along a predictable pathway to earlier successional stages whereas a reduction in grazing pressure allows a return to the climax.

This model forms the basis of the widely applied botanically based system of range condition assessment. Range condition is judged from the relative proportions of plant species that are characteristic of an early 'overgrazed' successional stage in relation to plant species that are characteristic of more advanced stages. In this approach, condition is simply the technical term for the position of the vegetation along the presumed successional continuum. A system rangeland condition assessment along these lines was developed for the eastern karoo by Vorster (1982).

It is particularly in arid and semi-arid environments that the range succession model has proved to be inadequate. The model assumes a consistent and predictable relationship between the numbers of animals on the range, the biomass and species composition of the vegetation, and the output of the system in terms of animal products. In view of the high variability of the rainfall in arid regions, this is unlikely to hold. Successions of dry years drastically reduce forage availability, forcing herbivores to either migrate elsewhere or suffer catastrophic mortality. The rapid plant recovery and growth occurring in runs of wet years proceeds in advance of herbivore population growth, resulting in periods of superabundance of forage. Under such conditions, the state of the vegetation is unlikely to be closely linked to numbers of grazers. The difficulty of distinguishing rainfall-induced from herbivore-induced vegetation changes is a fundamental problem in rangeland monitoring in arid regions.

Early indications that the notion of a climax was inappropriate for the karoo was given by Roux (1966) who showed that variations in the proportion of summer to winter rainfall influenced the abundance of grasses relative to karroid shrubs. Later studies (Hoffman et al., 1990; Novellie and Bezuidenhout, 1994) confirmed that summer rainfall promoted the rapid establishment of grasses but rejected Roux's generalization that winter rain promoted the dwarf shrubs. The implication of Roux's work was that the karoo was characterized by long-term fluctuations in the abundance of grass rather than an enduring climax state.

To some extent, these rainfall-induced fluctuations may be interpreted in terms of the range succession model. In the successional interpretation, rainfall is held to produce trends similar to those of grazing, with drought retarding the succession and higher rainfall promoting an advance to the climax. However, it has been widely observed in arid systems in general that neither increased precipitation nor withdrawal of grazing pressure invariably lead to the anticipated advance along presumed successional sequences. Instead, relatively sudden and apparently irreversible changes in vegetation state may occur. The prevalence of such anomalies led Westoby et al. (1989) to propose the state-and-transition model, which is more appropriate than the succession model for systems in which change is precipitated largely by rare events, such as a succession of abnormally wet seasons or extended droughts. This model describes a particular vegetation in terms of different alternative states that may exist, and different possible transitions between the states. Complexities of different events (e.g. grazing pressure, rainfall patterns, episodes of seed production or of seed predation) may combine to produce a transition in state. Unless this complexity is understood, the role of stocking rate in the transition may be misunderstood.

The state-and-transition model implies a totally different approach to research and management. Testing of the state and transition model requires experimentation, and the variability of the environment requires that such experiments be opportunistic, designed to take advantage of rare combinations of circumstances that may play a role in a transition between states (Westoby et al., 1989). Successful management under such circumstances depends on an understanding of a wide range of factors that may influence the demography of the key plant species. The need for such an understanding in the karoo was emphasized by Hoffman (1988b), who urged for research into physiological and demographic responses in order to gain predictive insights. The numerous recent studies that have focused on these aspects are dealt with by Milton et al. and Palmer et al. in this volume.

Another characteristic of conventional rangeland thinking has been a tendency to focus on average conditions across landscapes and to neglect the patterns of spatial dynamics within landscapes. Soil erosion has been a major preoccupation, to the point of overlooking the fact that soil is not only eroded but also transported and deposited. The spatial dynamics of soil transport and accumulation, and the resulting relocation of primary productivity and redistribution of nutrients by grazing animals, are natural and inevitable features on a landscape scale. The fact that soil moves does not necessarily imply degradation. Models that are insensitive to these spatial variations are likely to be unrealistic (Stafford Smith and Pickup, 1993). Incorporation of spatial variation is particularly necessary for arid areas where precipitation varies as much over space as over time. These spatial variations have critical implications for stock management (Behnke and Scoones, 1993).

Whereas the study of temporal dynamics has been a major preoccupation in the karoo, spatial variation on a landscape scale has only recently received critical attention. One of the most important concerns is patch selective grazing. Fuls (1992) and Kellner and Bosch (1992) view patch selective grazing as a key factor promoting range degradation in the semi-arid regions of southern Africa. These studies note that recommended stocking rates are determined on the basis of overall range conditions and do not take patch selection into account. Their concern is that patch over-grazing degrades the most favoured parts of the habitat, whereas neglected patches represent a waste of resources. Kellner and Bosch (1992) note that setting stocking rates at levels that would avoid localized degradation is probably not economically feasible.

Is this a real concern, or does it represent a failure to understand natural spatial dynamics at the landscape scale? Patch selective grazing is an extremely widespread phenomenon in grassland ecosystems. The availability of nutrients important for plant growth typically varies in space, because of edaphic gradients or because of zoogenic effects such as termitaria (McNaughton, 1988). Many large herbivore species tend to congregate at nutrient-rich sites, and accumulations of their excreta accelerate the nutrient-cycling rate by bypassing the more lengthy process of litter decomposition (Ruess and Seagle, 1994). This increases the rate of production of certain plant species that are adapted to conditions of high nutrient availability (Roux, 1969; Day and Detling, 1990). The resulting positive feedback system favours the large herbivores and creates a mosaic of vegetation communities, thus playing a vital role in maintaining biodiversity (McNaughton, 1983b; Lovegrove, 1993).

Far from being nuclei of degradation, these nutrient-rich patches may be critical determinants of the capacity of the range to support herbivores. The neglected patches, rather than being wasted resources, may represent important forage reserves in times of food scarcity. In the karoo, there have been few attempts to study both temporal and patch dynamics in an integrated way, and hence it is not clear whether patch selective grazing is a universal cause for concern or whether it is an entirely natural consequence of spatial variation in nutrient availability. An understanding of nutrient transport on a landscape scale and of the origin and dynamics of the nutrient-rich/heavily grazed patches is much needed to resolve the question, and this forms the subject of chapter 12.

Westoby et al. (1989) emphasized the need for opportunistic experiments to understand the nature of state-and-transition type systems. Experimentation has an important role to play, but management decisions can seldom await the outcome of opportunistic experiments. As demonstrated by Nailand and Hanrahan (1993) for outbreaks of the brown locust, simulation modelling can be an effective aid to decision-making in an uncertain environment. Data on the probability of occurrence of key events can be used in conjunction with a knowledge of demography and the physiology of species to produce simulation models capable of identifying the types of transitions that may occur. There is a particular need for models that integrate spatial and temporal components (Stafford Smith and Pickup, 1993). Model results can be used to design appropriate experiments or to formulate management procedures to guard against undesirable transitions. This is the subject of the final chapters of the section on dynamics (chapters 13 and 14).

To fulfil its function, it is necessary that our understanding of dynamics be sufficient to identify the likelihood of range degradation and to contribute to its prevention. This is difficult in environments subject to wide spatial and temporal variation. Range degradation is defined as an effectively permanent decline in the rate at which land yields livestock products under a given system of management (Abel and Blaikie, 1989). The emphasis is therefore on permanent change, or at least change which is impractical or uneconomical to reverse. Temporary localized declines in productive capacity, resulting from droughts or episodes of intensive grazing, do not constitute degradation. Degradation can also only be defined in relation to the objectives of specific systems of management (Bell, 1982a). The definition of what constitutes degradation is all important.

The major criticism of the indices of rangeland condition derived from the range succession model is that they are not always clearly relevant to whatever is defined as degradation. Unless the indices of rangeland condition do in fact provide early warnings of permanent loss of capacity to achieve defined management objectives, they are irrelevant. There is reason to doubt claims that indices of rangeland condition do, in fact, achieve this in African rangelands, and their relevance to permanent loss of economic capacity is dubious (Behnke and Scoones, 1993). The new paradigms of rangeland dynamics will, in the end, be judged by the role they play in apprehending and avoiding range degradation.

11 Population level dynamics

S. J. Milton, R. A. G. Davies and G. I. H. Kerley

11.1. Introduction

Short-term variability in rainfall patterns in hot and cold deserts is thought to be reflected by fluctuations in population sizes of plants and animals (Caughley and Gunn, 1993). The periodicity of these fluctuations is likely to be tied to mobility and life-histories, and therefore to differ greatly among species.

This chapter reviews the patterns and causes of plant and animal population fluctuations in time and space in the arid zone of southern Africa. Wherever possible, we compare the effects of environmental stochasticity and cyclicity on populations with those of competition, disease and predation. We also consider the possible role of refuges and immigration on the resilience and persistence of small populations in a variable environment. The population dynamics of plants, invertebrates, reptiles, birds and mammals is dealt with separately because the resolution and focus of the information available differs among these groups. In conclusion, we indicate research directions that could improve current predictions based on the few autecological studies and limited long-term data series available for arid zone biota.

11.2. Structure and dynamics of plant populations

11.2.1. Ephemerals and annuals

The mass flowering of annual plants in the winter-rainfall western karoo is a major tourist attraction, and the prediction of the flowering time and abundance of annual plants is consequently of economic importance (see Van Rooyen, this volume). Namaqualand ephemerals grew larger and produced more flowers when growing at low temperatures (17 °C) than at higher temperatures (22–27 °C). Moisture-stressed plants were smaller than well-watered plants, and allocated less biomass to reproduction (Van Rooyen et al., 1991). Flowering in ephemerals may also be depressed by competition and grazing. Densities of the wind-dispersed annual *Gorteria diffusa* (Asteraceae) in Namaqualand were higher near perennial shrubs which trapped their seeds, but flower production was eight-fold greater among isolated individuals of this species (Cunliffe et al., 1990). Similarly, in the southern karoo, seed production in *Tetragonia echinata* (Aizoaceae) was significantly greater in plots from which shrubs had been removed, than in undisturbed vegetation (Milton, 1994b). Selective flower grazing (florivory) by sheep changes the reproductive output of some ephemeral species. For example, the number of capitulae per plant of *Senecio arenarius* was reduced by tenfold in a sheep camp relative to an ungrazed road verge, whereas no significant intersite difference in flowering occurred in *Gorteria diffusa* which is poisonous to sheep (Görne, 1995).

Seedbank densities of ephemerals in Namaqualand differed between habitats, and ranged from 5 000 m^{-2} on ridge tops to 41 000 seeds m^{-2} in sandy bottomlands (Van Rooyen and Grobbelaar, 1982). Germination of seeds of annuals and ephemerals is often staggered (Von Willert et al, 1992; Beneke et al., 1993c) as a result of dormancy mechanisms (see Van Rooyen, this volume) that prevent the loss of an entire seed population when germination is followed by drought. Germination of annual plants from the arid winter-rainfall region of Namaqualand was stimulated by cool temperatures (17 °C), and emergence was faster in planted seeds subjected to a fine water spray for three hours, than in unleached seeds or soil. Established seedlings did not inhibit further seed germination (Van Rooyen and Grobbelaar, 1982).

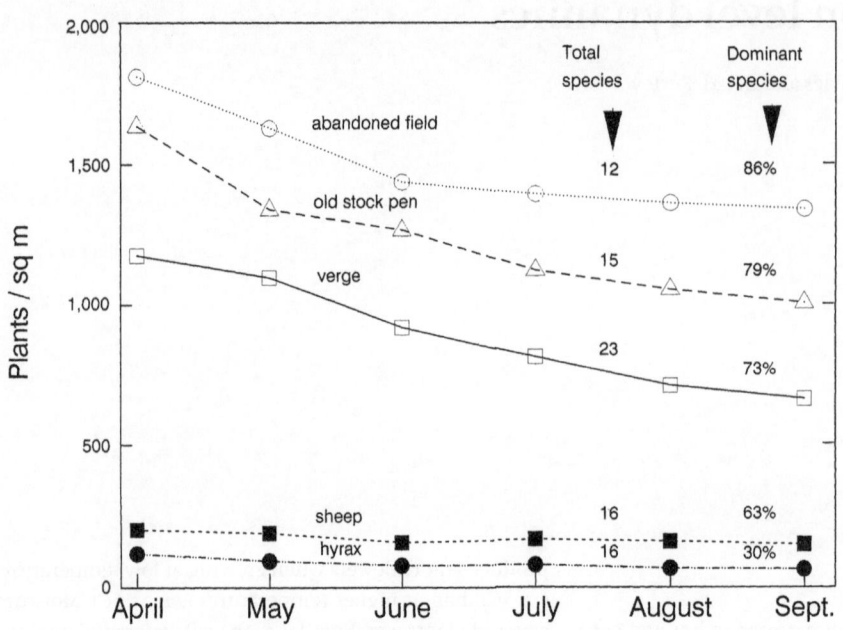

Figure 11.1 **Changes in densities of annual forbs in permanent plots at five sites in Namaqualand. Species-richness is the total number of species recorded over 6 months, % dominance refers to the proportion of the population comprising a single species (Rösch, 1977)**

Under field conditions in Namaqualand, 51–74% of ephemeral seedlings survived to maturity after emerging at densities of 115–1810 m^{-2} (Rösch, 1977; Van Rooyen et al., 1979b). Survival varied among species and sites (Fig. 11.1). Mortality was attributed to moisture stress, but herbivores selectively reduced densities of some ephemeral species. For example, rock hyraxes *Procavia capensis* removed 95% of *Dimorphotheca sinuata* (Asteraceae) from a fixed plot but did not feed on the related *Osteospermum amplectens*, which was present at similar initial densities (Rosch, 1977). Ostriches in the southern karoo reduced densities of *Galenia papulosa* and *Gazania lichtensteinii* but had little influence on *Leysera tenella* (Gatimu, 1996).

Under the hyper-arid conditions of the Namib, patches of annual grass (*Stipagrostis* spp.) appear following local rain events. Grass cover was correlated with measured rainfall for individual stations. However, grass grew more frequently in drainage lines and around inselbergs where rainwater collected (Günster, 1995). In the Richtersveld, the timing of rainfall (whether in autumn or midwinter) influenced the relative abundance of species of annual succulents surviving to reproductive maturity in successive years (Von Willert et al., 1992).

In higher-rainfall regions of the karoo, perennial plants appear to control the abundance of ephemerals through competition for water. This results in spatial and temporal heterogeneity in the density and performance of ephemerals within a landscape. Improved survival, growth and seed production of ephemerals occurs when and where large or small disturbances, such as drought, grazing, vegetation clearing and soil excavation, kill or reduce perennial grasses or shrubs (Dean and Milton, 1991a; Yeaton et al., 1993; Milton et al., 1995). However, ploughing and grazing generally increase densities of a few ephemeral plant species (Fig. 11.1), at the cost of species-richness (Rösch, 1977; Dean and Milton, 1995). The almost monospecific stands of Namaqualand daisies *Dimorphotheca sinuata* that blanket old fields in Namaqualand are well-known examples of this phenomenon (Smuts and Bond, 1995). Similarly, the inconspicuous *Stipa capensis*, a winter growing annual grass normally restricted to *heuweltjies* (disturbed, nutrient-rich soil patches overlying nests of the small harvester termite *Microhodotermes viator*) in well-vegetated parts of southern Namaqualand (Acocks, 1953), becomes dominant on abandoned fields and overgrazed rangeland (Steinschen, 1995).

11.2.2. Grasses

Populations of perennial desert grasses appear to be structured by rain-driven recruitment events and subsequent mortality, whereas competition and herbivory may override rainfall effects in structuring perennial grass populations in higher-rainfall regions of the karoo.

Rainfall events >15 mm cause widespread germination of *Stipagrostis sabulicola* on mobile dunes in the Namib Desert. In the absence of follow-up rainfall, however, which rarely occurs, all seedlings die (Seely, 1990). Site-specific differences in seedling survival, rather than seed availability, appear to determine the patchy distribution of perennial grass on these desert dunes. Following the unusually high rainfall (>100 mm) of 1976 and 1978 *S. sabulicola* established at similar densities (50 to 128 ha^{-1}) on all

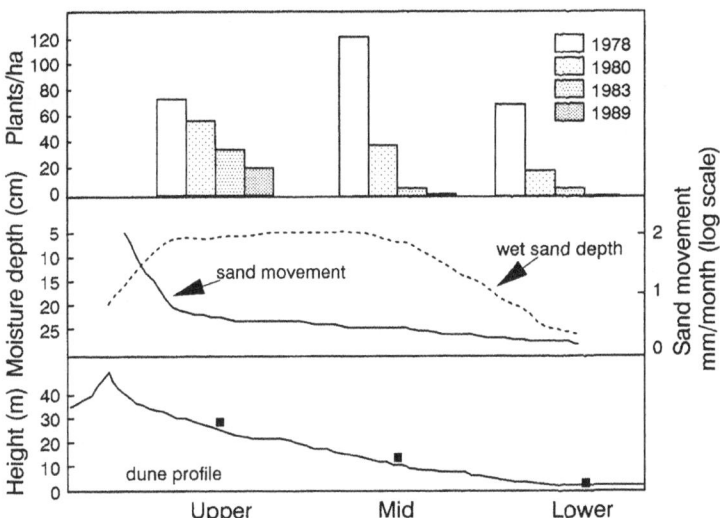

Figure 11.2 **Changes in density of *Stipagrostis sabulicola* over 11 years at 3 sites on a Namib dune slope (data from Seely, 1990) in relation to soil moisture and sand movement profiles (data from Yeaton, 1988)**

parts of the dune except the very mobile crest and slipface (Seely, 1990). After 11 years, 29% of the population on upper slopes survived compared with 0.6% of plants on mid to lower dune slopes (Fig. 11.2). Survival of perennial grasses on the upper slopes was apparently facilitated by the proximity of moisture to the soil surface in this part of the dune (Yeaton, 1988).

Large rain events (>15 mm) that stimulate germination of seeds of the desert grass *Stipagrostis sabulicola* may occur once or twice per year, but recruitment has been recorded only twice in this century. Despite seed losses to seedling mortality and to continual post-dispersal seed predation by beetles, lizards, birds and rodents, seedbank depletion does not occur because established plants, sustained by fog moisture, set seed annually in summer (Seely, 1990).

Under the more mesic conditions of the Stormberg Plateau in the eastern karoo, *Tetrachne dregei* stands comprise dense populations of large tufts and there is little seedling establishment. Roux (1960) suggested that reduced seed production caused by summer grazing, together with the need for moist conditions and a suitable, competition-free seedbed, limited population renewal in this species. On tilled soil, emergence and recruitment of *Tetrachne dregei* occurred only in wet years or on irrigated fields. Best germination (90%) took place at 35 °C, decreasing at both higher and lower temperatures and mid-summer sowing was unsuccessful in the field (Roux, 1960). Seedling establishment failed on unworked and unstable soil surfaces. Two years after emergence, the seedlings were tolerant of frost and drought, and 10% were reproductively mature. From the association of

established seedlings with shrubs (*Lycium* sp.) and the tufted grass *Eragrostis lehmanniana*, it may be inferred that *T. dregei* seedlings benefited from protection provided by established plants.

Seed set in *Tetrachne dregei* is better in years when rainfall coincides with development of the fruits during mid summer. Pure stands produce an estimated 7500 seeds m^{-2}, which are dispersed up to 12 m by wind and up to 32 m by runoff water, but are not known to be mammal dispersed. Seed loss is mainly to grazing mammals, birds and termites before dispersal and to the granivorous ant *Messor capensis* after dispersal. Although germinability of dry-stored seeds remained high after 11 years (98%), and seeds survived dry temperatures ranging from –50 °C to 65 °C without reduction in viability (Roux, 1960), *T. dregei* has highly germinable fresh seeds, so that it is unlikely that large reserves of dormant seeds are retained in the soil.

Life-histories of karoo grasses range from ephemerals with permanent seedbanks, small-seeded tufted or creeping perennials with seed dormancy, to large-seeded perennials with transient seedbanks. Large-seeded perennials generally occur in more mesic parts of the karoo (Van Breda and Barnard, 1991), or on rockier and sandy soils than small-seeded annuals and perennials (Novellie and Bezuidenhout, 1994). Populations of short-lived and small-seeded grasses, such as *Aristida congesta, Eragrostis lehmanniana* and *Tragus koeleroides*, respond rapidly to drought-breaking rains (Henrici, 1940; Fourie et al., 1987; Novellie and Bezuidenhout, 1994; Milton et al., 1995), but later decrease as slower-growing, longer-lived, large-seeded

186   S. J. Milton, R. A. G. Davies and G. I. H. Kerley

grasses and shrubs increase in cover and competitive ability (Fig. 11.3).

Large-seeded perennial grasses that set few seeds and have transient seedbanks, depend on regular production of fresh seed for population renewal (O'Connor, 1991). Inflorescence destruction by grazing consequently reduces the resilience of these grass populations to stochastic events such as drought (O'Connor, 1995). In many livestock ranching areas of the Kalahari, Bushmanland and the karoo there is a tendency of large-seeded perennial grasses (*Stipagrostis ciliata*, *Themeda triandra*, *Tetrachne dregei*) to be replaced by shorter-lived, small-seeded grasses (*Eragrostis lehmanniana*, *Aristida* spp.) that maintain dormant seedbanks (Roux, 1960; Danckwerts and Stuart-Hill, 1988; Dean et al., 1991). Repeated summer grazing almost eliminates all non-stoloniferous, perennial grasses (O'Connor and Roux, 1995).

11.2.3. Succulents

The arid zone succulents of southern Africa can broadly be separated into two guilds, namely shade-loving and sun-loving. The first group is confined, either during establishment or throughout their lifespan, to patches of deep shade beneath shrubs or among rocks. This group includes all leaf- and stem-succulent plant families of southern Africa except the Mesembryanthema and Zygophyllaceae. These two leaf-succulent families are able to colonize open, unshaded sites, and are therefore important in determining pattern in some types of karoo vegetation (Yeaton and Esler, 1990).

Seeds of the succulent Asclepiadaceae, Asteraceae, Asphodelaceae, Crassulaceae and Euphorbiaceae are released as soon as the capsules dry and split, and seeds are dispersed autochorously or by wind. None of these shade-loving succulents is known to maintain a persistent soil-stored seedbank, but most can multiply vegetatively by suckers and runners and rooted leaves (Jacobsen, 1960; Von Willert et al., 1992; Van Jaarsveld, 1994). Through clonal expansion, a single plant is able to monopolize the limited space available beneath a host plant or rock shelter.

Seeds of sun-loving Mesembryanthema are generally retained in hygrochastic capsules on the parent plant, and released onto wet soil when capsules open during rain (Volk, 1960; Ihlenfeldt, 1971). The numerous (30–900) small seeds (Esler, 1993) are ejected over short distances (20 to 90 cm) by the impact of raindrops (Volk, 1960). Runoff water is likely to disperse the seeds further over smooth soil surfaces, until they are trapped by loose soil particles. Uphill movement is made possible through dispersal in dung of rodents (Jump, 1988), tortoises (Milton, 1992a) and other herbivores (Smith, 1966; S. J. Milton unpublished data). Capsule durability varies within and between genera (Struck, 1989), and more durable capsules, which close again when dry, presumably spread seed release over time, reducing the risk of total recruitment failure (Ihlenfeldt, 1971). A few species of Mesembryanthema also expand vegetatively by adventitious rooting from decumbent (*Malephora lutea*, *Cephalophyllum* sp.) or reflexed (*Drosanthemum* sp., *Leipoldtia* sp.) branches. Zygophyllaceae are larger-seeded than Mesembryanthema, but some, such as *Augea capensis*, have seed coverings that become mucilaginous and sticky when wet, enabling seeds to adhere to smooth soil

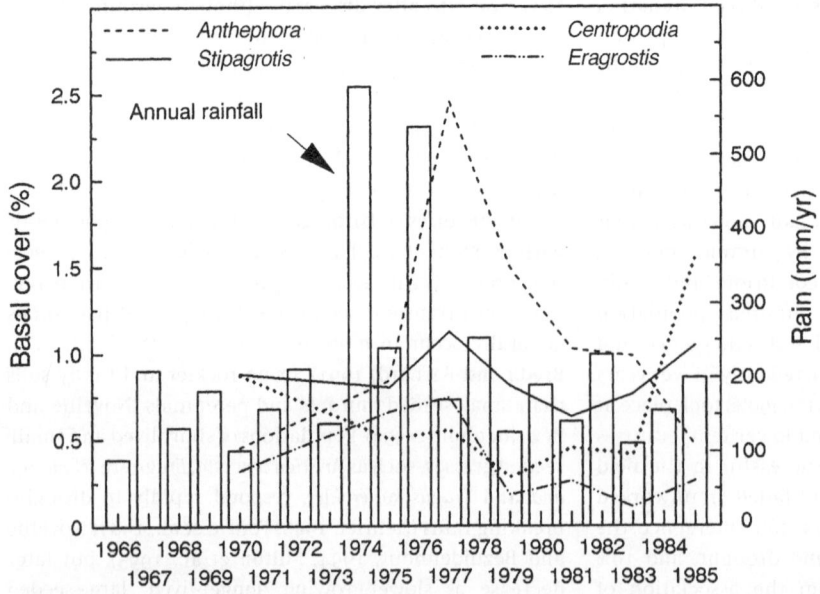

Figure 11.3 **Small-seeded, weakly perennial grasses *Centropodia glauca* and *Eragrostis lehmanniana* decrease in basal cover (abundance) as large-seeded, slower-growing grasses *Anthephora argentea* and *Stipagrostis* spp. increase** (data from Fourie et al., 1987)

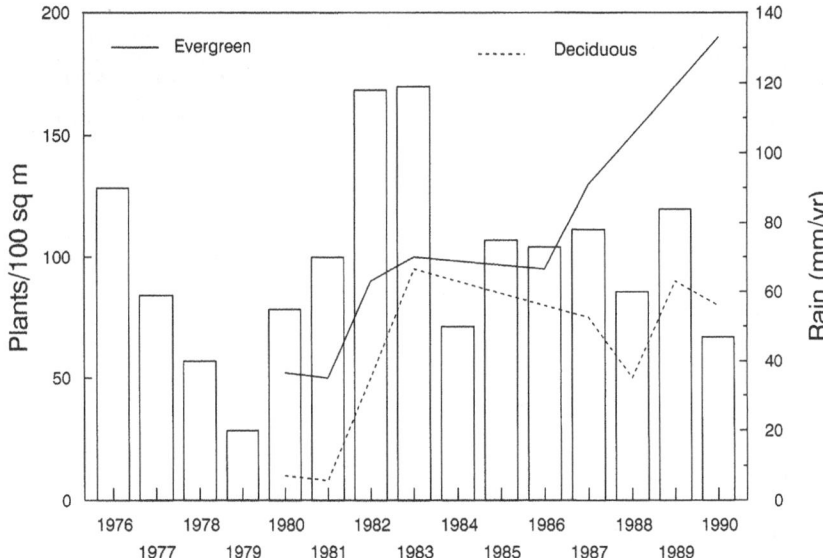

Figure 11.4 **Post drought recovery of populations of Mesembryanthema in the Richtersveld. Populations of deciduous species recovered rapidly but decreased in dry years, whereas evergreen species recovered slowly but were more drought-tolerant (data from Jürgens et al., 1997)**

surfaces and to colonize open ground (Van der Merwe, 1987).

Seed dormancy among Mesembryanthema varies from enforced to a combination of enforced and innate (Ihlenfeldt, 1971; Von Willert et al., 1992), and intraspecific differences in germinability of fresh seeds may occur between sites (Hartmann, 1983). Short-lived Mesembryanthema that colonize frequently disturbed habitats, such as *heuweltjies*, maintain larger seedbanks than species of relatively undisturbed sites that store only the current year's seed (Esler et al., 1992; Esler and Cowling, 1995). Seedling emergence of leaf-succulent shrubs is in autumn and winter (Esler, 1993; Milton, 1995a). In the Richtersveld, there was no emergence of succulents in response to spring (October) irrigation, but abundant emergence followed March rain (Von Willert et al., 1992).

Once established, seedlings of the evergreen leaf succulents are very tolerant of aridity. Approximately 80% of *Ruschia spinosa* (Mesembryanthema) seedlings, deprived of water seven days after emergence, survived for 400 days without water under greenhouse conditions (Esler and Phillips, 1994). The potential longevity of succulents varies from a few months for the mesophyllous ephemerals to many decades for arborescent aloes and some embedded leaf succulents (Cole, 1988). When mortality is continuous rather than episodic, the proportion of dead individuals in a population can provide an index of turnover and longevity. Esler et al. (1992) reported that, in the southern karoo, 8% of the population of a woody leaf succulent *Drosanthemum montaguense* was dead as compared with 22% and 24% in the non-woody leaf succulents *Rhinephyllum macradenium* and *Brownanthus ciliatus*.

Mortality of leaf succulents, at patch or local population scale, may be episodic in response to drought (Von Willert et al., 1992; Jürgens et al., 1997) or hail (Milton and Collins, 1989; Powrie, 1993). Population fluctuations appear more frequent in fast-growing mesomorphic than xeric species of leaf succulents. For example, drought in the Richtersveld in 1979 killed about 80% of a predominantly leaf-succulent plant community (Von Willert et al., 1985). Exceptionally high rainfall from 1981 to 1983 initiated a recovery in the vegetation, which was then monitored for 10 years (Jürgens et al., 1997). Results showed that populations of Mesembryanthema with thin-walled epidermal cells and drought-deciduous leaves recovered rapidly after drought but decreased again in moderately dry years. Populations of evergreen succulents in the same observation plot recovered slowly, remaining stable in dry years, but increasing in wet years over a period of 10 years (Fig. 11.4). No comparable data are available for shade-succulents, but, on the basis of their less exposed habitats, ability to reproduce vegetatively, and the absence of plant- or soil-stored seedbanks, it may be predicted that their population sizes vary less over time than those of leaf succulents that colonize open ground.

11.2.4. Non-succulent dwarf shrubs

Seedlings of *Pteronia* spp., *Osteospermum sinuatum* and other dwarf shrubs at Tierberg in the southern karoo emerged in autumn (March–May) in each of five consecutive years (Esler, 1993; Milton, 1995a). Interannual differences in emergence densities were related to the timing and amount of rainfall, with high densities being correlated with autumn rainfall. *Plinthus karooicus* from the eastern karoo germinated mainly in spring and autumn and ger-

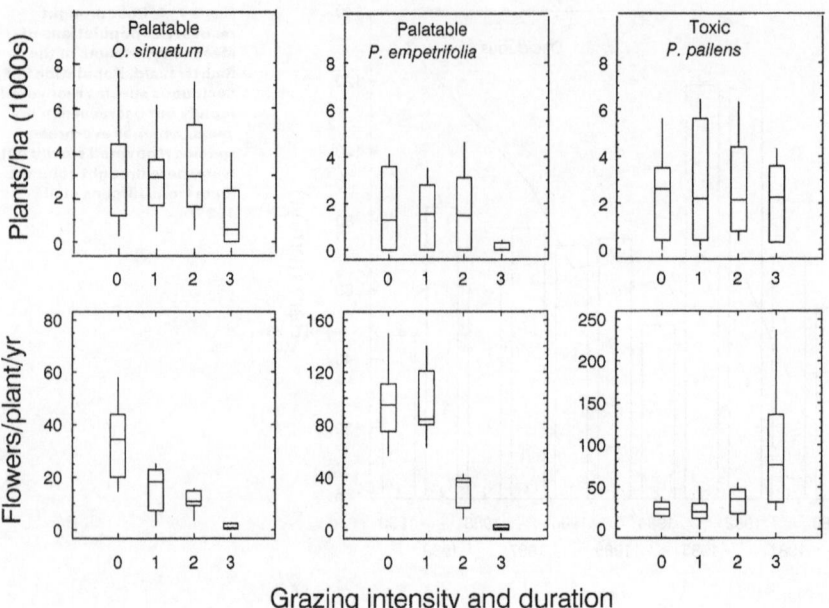

Figure 11.5 **Density and annual flower production of palatable (*Osteospermum sinuatum*, *Pteronia empetrifolia*) and toxic (*P. pallens*) dwarf shrubs in herbivore exclosures (0), under light continuous grazing by small indigenous herbivores (1), under moderate rotational grazing by sheep (2), and under continuous heavy grazing by sheep (3) (S. J. Milton, unpublished data)**

Grazing intensity and duration

mination was suppressed by temperatures below 18 °C or above 21 °C (Theron, 1964), a pattern reported for other karoo shrubs (Henrici, 1935a; Hoffman and Cowling, 1987). Although some dwarf shrub seeds (*P. karooicus*, *Galenia* spp.) require a period of after-ripening, few species maintain persistent soil seedbanks (Theron, 1964; Esler, 1993). Population renewal therefore depends on the availability of fresh seed when conditions are suitable for seedling germination and recruitment.

Some karoo shrubs are toxic to domestic livestock, and therefore have a competitive and reproductive advantage in rangelands. Consumption of flowers by sheep can reduce seed production of palatable dwarf shrubs by 80–90% (Fig. 11.5) relative to protected plants (Milton and Dean, 1990b; Milton, 1992b, 1995b). Field observation of population structures of palatable plants in rangeland (Milton, 1994a), and models of palatable shrub population responses to seed reduction when growing with freely seeding neighbouring species (Wiegand et al., 1995; Jeltsch et al., this volume), indicate that karoo shrub populations may become seed limited under repeated heavy browsing.

Pre-dispersal seed losses to florivorous insects, particularly endophageous tephritid fly larvae, vary between years but may destroy 60–80% of seeds of some shrubs species in dry years when seed crops are small (Milton, 1995b). Granivorous ants are major post-dispersal predators of karoo shrub seeds (Kerley, 1991), taking up to 60% of seed production of some species annually (Milton and Dean, 1993). However it is unlikely that ants influence vegetation composition because they collect seeds in proportion to their availablity (Milton and Dean, 1993).

Spatial variation in the densities of emerging seedlings of four dwarf shrub species in the southern karoo was correlated with seed production at the paddock scale (100 ha) and with densities of parent plants at the patch (25 m²) scale (Milton, 1995a). Smaller-scale spatial heterogeneity in seedling densities appears to be generated by patchy dispersal. Seedlings of tumble-seeded plants were associated with seed traps, including mat-forming succulents (Yeaton and Esler, 1990), plant debris (Esler, 1993; Milton, 1995a) and soil pits dug by foraging mammals (Dean and Milton, 1991b), as well as with disturbed nest-mounds of seed-harvesting ants (*Messor capensis*) that store viable seeds (Dean and Yeaton, 1992). In contrast, small-seeded, non-succulent Aizoaceae (*Plinthus karooicus* and *Galenia fruticosa*) establish in relatively exposed sites (Theron, 1964; Esler, 1993; Milton, 1995a). Shading of *P. karooicus* increased the incidence of fatal root damage by nematodes (Theron, 1964).

Initial growth of deep-rooted dwarf shrubs is slow. During the first two months after emergence, *P. karooicus* invests 85% of its biomass in roots (Theron, 1964). Survival of seedlings recorded at Tierberg over six years was generally poor (1–5%), but 20% of new seedlings survived for 12 months in a year when rains occurred frequently in the winter and spring following autumn emergence (Milton, 1995a). Seedlings of non-succulent shrubs appear to be less drought-tolerant than leaf succulents. All seedlings of non-succulent Asteraceae (*Pteronia pallens*, *Osteospermum sinuatum*) died within 160 days when deprived of water under nursery conditions (Esler and Phillips, 1994). It is probable that dependence on moist years and sites for

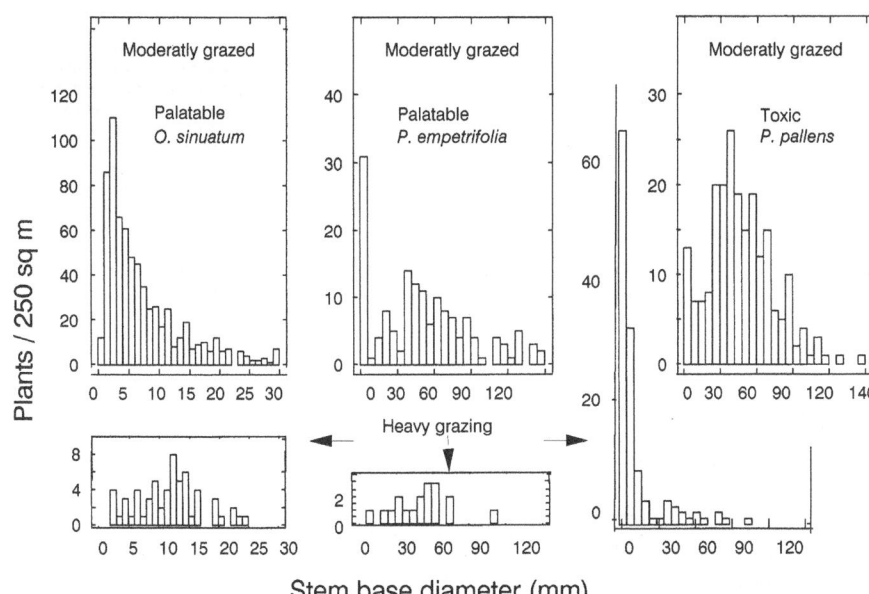

Figure 11.6 **Palatable shrub populations (*Osteospermum sinuatum, Pteronia empetrifolia*) which show continuous or episodic recruitment on a moderately stocked sheep ranch, occur at lower densities and are skewed towards old plants on a ranch with a 50 year history of heavy and continuous grazing by sheep. Toxic shrub *P. pallens* populations show recent recruitment on the overgrazed site (Milton, 1994a)**

establishment results in clumped distributions of shrubs in the more arid landscapes, and in pulsed recruitment events.

The availability of microsites sheltered by living or dead plants can increase seedling emergence of shrubs with tumble seeds (Esler, 1993; Milton, 1994a), but seedling survival may be limited by neighbouring plants. No increase in plant density was observed five years after litter addition to undisturbed vegetation (S. J. Milton, unpublished data). Seedling survival was improved by the death or removal of established neighbours, which prolonged post-rain moisture saturation 150 mm below the soil surface from 4 to 12 days (Milton, 1995a). Similarly, seedling recruitment was more frequent on overstocked rangeland with a relatively sparse cover of shrubs than in an adjacent, well-vegetated rangeland (Milton, 1995a). However, seedlings of shrubs avoided by domestic livestock predominated at the overgrazed site. Differences in the population structures of palatable and unpalatable shrubs on adjacent ranches differing in management history (Fig. 11.6) suggest that population growth integrates a number of variables including seed density, microsite quality and moisture availability in response to rainfall, runoff and transpiration.

There is little information on longevities of the non-succulent dwarf shrubs of the karoo. Ring-dating is complicated by the stem splitting behaviour characteristic of shrubs of arid areas (Theron, 1964). In a sample of *Pteronia pallens* stem bases (*n* = 16) collected at Tierberg, Prince Albert, 5–45 growth rings were counted in stems 5–90 mm in basal diameter. Ring widths were positively correlated

with cool-season rainfall, from which it was inferred that *P. pallens* forms annual growth rings in active sections of the stem during the growing season (Milton, Gourlay and Dean, 1997). Growth rings averaged 0.9 mm in width, so that basal diameter increments of *Pteronia pallens* on plains at Tierberg are not expected to exceed 2 mm yr^{-1}. Assuming that the relationship between stem age and basal diameter obtained for this small sample held for the population, *P. pallens* at this site probably lives for up to 130 years.

Low annual rates of turnover were recorded in marked populations of *Osteospermum sinuatum* (3.0%), *Pteronia empetrifolia* (3.4%) and *P. pallens* (2.2%) at Tierberg, and most of the recorded mortalities (88%) occurred among pre-reproductive plants (Milton, 1994a). Occasional mortality of individual established plants is caused by the foraging activities of insectivorous mammals excavating ants and termites (Dean and Yeaton, 1992). Karoo shrublands rarely burn and the effects of fires on plant populations have not been documented (Forrester, 1988). However, drought in the karoo, like fire in other biomes (Bond and Van Wilgen, 1995) can cause large-scale mortality of shrubs and bring about temporary dominance by herbaceous plants. At Baroe near Steytlerville, 65% of karoo bushes died when only 70 mm of rain (26% of long-term mean, probability 1:70 years) fell during a 12 month period (Milton et al., 1995). Populations of the dominant karoo bushes *Eriocephalus ericoides* and *Pentzia incana* were reduced by 95% and 78% respectively (Fig. 11.7). During the 1933 drought at Fauresmith (54% of mean annual rainfall, probability 1 : 20 years), Henrici (1940) estimated that 25% of

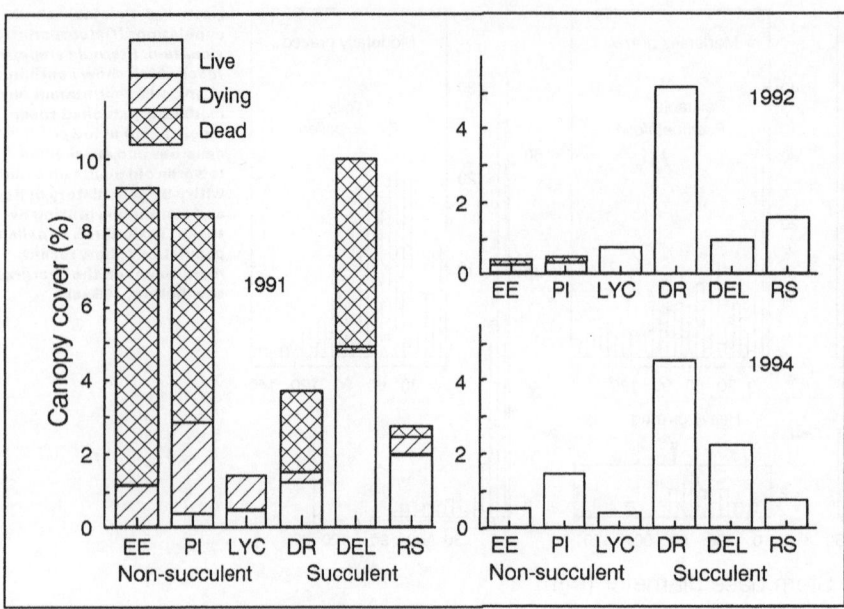

Figure 11.7 **Changes in cover of non-succulent and succulent dwarf shrubs following the 1990–1 drought near Steytlerville in the eastern karoo.** EE = *Eriocephalus ericoides*, PI = *Pentzia incana*, LYC = *Lycium* sp., DR = *Drosanthemum montaguense*, DEL = *Delosperma* sp., RS = *Ruschia spinosa* (reproduced with permission from Milton et al., 1995)

the plants died and mortality was greatest among non-succulent karoo bushes *P. incana* and *Rosenia humilis*.

11.2.5.   Shrubs and trees

Periodicity of population fluctuations among woody plants varies with potential longevity and recruitment responses to drought and above-average rainfall. In the Kalahari Gemsbok National Park, above average rainfall during the period 1974–6 was followed by eight years of average to below-average rainfall. This weather sequence had little effect on populations of trees and large, deep-rooted shrubs (*Acacia erioloba*, *Rhigozum trichotomum*, *Lycium oxycarpum*), although it caused populations of non-succulent dwarf shrubs (*Plinthus sericeus*, *Chrysocoma obtusata*, *Sericorema remotiflora*) to expand and then contract (Van Rooyen et al., 1984).

Large shrubs and trees of the karoo generally belong to families associated with subtropical thickets. In the arid extremes of their ranges, they are restricted to drainage lines and rocky outcrops where there are shaded, relatively moist establishment sites. Most produce fleshy fruits, and their large seeds (3–15 mm diameter) are dispersed by frugivorous birds (Dean et al., 1993), baboons, humans and canids, particularly bat-eared fox *Otocyon megalotis* (Kuntzsch and Nel, 1992). Bignoniaceae and Sapotaceae have wind-dispersed seeds, and leguminous trees of the karoo and Kalahari (*Acacia Parkinsonia*) are dispersed by wind, water and large herbivores (Story, 1952; Leistner, 1961; Hoffman et al., 1989). Fresh seeds of non-leguminous karoo trees are unlikely to maintain a durable, soil-stored seedbank because they are ger-

minable when fresh (Van Breda and Barnard, 1991; Joffe, 1993).

It is probable that regeneration of the tree and shrub species of subtropical thicket species at the arid extremes of their ranges is episodic, and occurs during exceptionally wet or warm–wet periods (Midgley and Cowling, 1993). In 1990, seedlings of *Euclea undulata* were absent from 68 plots in 10 western and eastern Cape sites where the trees were common (Midgley and Cowling, 1993). But, in 1996, following the unusual spring rainfall sequence of 1995–6, high densities of seedlings of *E. undulata* and other thicket species were recorded in drainage lines in the southern karoo (S. J. Milton, personal observation). Population structures of some trees and shrubs indicate pulsed recruitment, possibly associated with high-rainfall years, local flooding or disturbance events. Even-sized cohorts of *Acacia erioloba* were recorded on a cattle pen (*kraal*) abandoned for 20 years at Riemvasmaak in the Richtersveld (Hoffman et al., 1994) and on disused agricultural land along the Molopo River (Milton and Dean, 1995b). During the 1980s and early 1990s, *Acacia erioloba* populations sampled in the Namib (Theron et al., 1985), Kalahari (Van Rooyen et al., 1994) and northern Cape (Hoffman et al., 1994) included few saplings, possibly reflecting the low to average rainfall experienced throughout southern Africa from 1977 to 1994. Height distribution of a *Schotia afra* population sampled at Riemvasmaak in the northern Cape during 1994 was bimodal, the most recent recruitment phase apparently having been during the wet years from 1972 to 1976 (Hoffman et al., 1994). *Pappea capensis* showed a similar trend at this site.

Longevity of *Acacia erioloba* has been inferred from trunk diameter and annual diameter increments to be about 400 years (Van Rooyen et al., 1994). *Acacia karroo*, a common pioneer of karoo river banks, reaches its maximum height in 20 years and develops only 25–40 annual growth rings (Barnes et al., 1996). The multi-stemmed form of some karoo trees and their ability to coppice from roots following above-ground damage, complicate age estimations. Two *Rhus lancea* in karoopoort were sketched by Burchell in 1811 and photographed in 1928 (Hutchinson, 1946). Such anecdotal evidence suggests that hardwood trees of karoo areas are slow growing and may live for centuries. The impacts on riverine woodland of harvesting such hardwoods as *Rhus lancea* for fence construction on ranches in the early 1900s are unknown. Longevity and rare, episodic recruitment makes the status of tree populations in arid southern Africa difficult to assess. Little is known of the causes of mortality in arid zone trees and shrubs. Mortality of *Acacia erioloba* in the Kalahari during droughts was greater among saplings than trees (Van Rooyen et al., 1984, 1994), however, mortality during a wild fire which killed 75% of *A. erioloba* trees in the Nossob River Valley in 1976 was skewed towards large, moribund trees (Van der Walt and Le Riche, 1984).

11.2.6.   Principal drivers of karoo plant populations

There is growing evidence that both mortality and recruitment may be episodic in many arid zone plant species. Droughts, or fires that follow unusually abundant rainfall, are major causes of plant mortality. Rainfall sequences that favour seedling survival probably result in recruitment events when they coincide with seed availability and reduced competition from established plants (Wiegand et al., 1995). Depending on species-specific drought tolerance, longevity and competitive ability, pulses of mortality and population renewal may occur at intervals of years, decades or centuries.

The conspicuous influence of rainfall and drought on vegetation and plant populations in arid climates including the karoo (Fairall and Le Roux, 1991; Novellie and Bezuidenhout, 1994; Hoffman et al., 1995; Milton et al., 1995; O'Connor and Roux, 1995) led to the development of the disequilibrium concept of vegetation dynamics (Walker, 1993). The logical extension of this notion is that herbivores, when unconfined and unsupported by forage supplements, have little influence on the dynamics of populations of arid zone plants (Scoones, 1992; Helldén, 1991). In southern Africa, grazing land is a limited resource, and both domestic livestock and indigenous herbivores are now confined by fences and provided with water and supplements during prolonged droughts (Hoffman et al., this volume). Under these conditions, herbivores override, or act synergistically with, weather patterns to influence the

dynamics of plant populations in karoo rangelands (Roux, 1960; Danckwerts and Stuart-Hill, 1988; Milton and Hoffman, 1994; O'Connor and Roux, 1995). Seed reduction together with alteration of establishment sites are likely mechanisms for herbivore-driven changes in the abundance and structure of arid zone plant populations (Milton and Dean, 1990b; Milton, 1994a; Milton et al., 1994b; Stokes, 1994; Jeltsch et al. this volume).

11.3.   **Invertebrates**

11.3.1.   Brown locust *Locustana pardalina*

The brown locust *Locustana pardalina* is one of the few karoo insects for which life-history data and some index of variation in abundance over the past 200 years is available (Fig. 11.8). Population research has been motivated by a perceived need to predict the time and place of outbreaks of this economically important species. The brown locust is a grass-eater, and the relatively grassy, late summer-rainfall regions of the karoo are centres of brown locust outbreaks (Faure, 1937; Vernon, this volume). It is from these areas that locust swarms move northwards into the maize-growing areas.

Lounsbury (1915), who reviewed five locust plague periods in southern Africa between 1797 and 1909, concluded that periods of high locust density (plagues) lasted on average for 13 years and were separated by periods of low locust density that lasted about 11 years. Between 1910 and 1967, there were four periods of high locust density each lasting 7–11 years (Lea, 1968). From 1964 to 1976 locust density was generally high, although over this period there were high and low density spells that lasted for one to two years. Low locust densities during drought years from 1977 to 1984 were followed by an irruption extending from 1985 into the 1990s (De Villiers, 1988; Nailand and Hanrahan, 1993). Reduction in the duration and increase in frequency of high-density phases of locust populations since the 1950s have been attributed to the use of insecticides to reduce the impact of swarming locusts on crop production areas to the north of the karoo (Lea, 1968; De Villiers, 1988). However, the apparent change in cycle length may also reflect changes in the methods used to quantify locust abundance.

Rainfall has long been considered a major factor in the quasi-cyclic behaviour of locust populations. Although locust outbreaks do not correlate well with rainfall cyclicity in South Africa (Fig. 11.8), and rainfall totals and seasonal distribution vary more between the locust outbreak regions than between years (Lea, 1968), more detailed analyses of locust life-history responses to rainfall lead to fairly accurate predictions of outbreak years and areas.

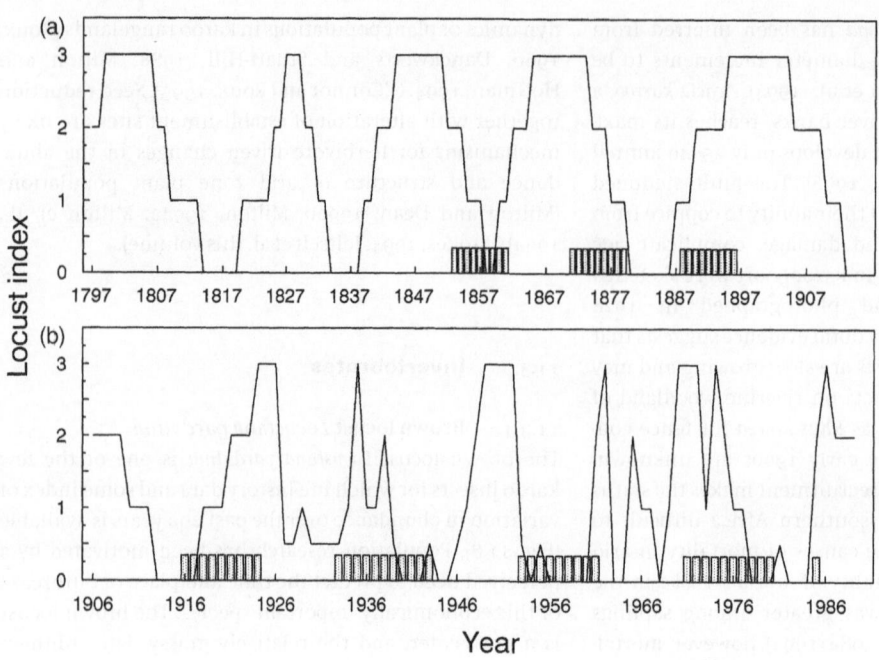

Figure 11.8 **Incidence of swarming in the brown locust in South Africa. Incidence classes are 0 = none, 1 = low, 2 = moderate, 3 = high. Shading indicates periods of above-average rainfall over southern Africa (from Tyson, 1987). (a) 1797–1909 swarm data based on anecdotal evidence (Lounsbury, 1915), (b) 1910–91 swarm data based on expenditure on control (Lea, 1968) and weight of insecticides used to control swarms (De Villiers, 1988)**

Locusts take two months to complete their life cycle (from egg to egg). Locust eggs survive better in dry than wet winters, and warm, wet conditions trigger development and hopper emergence. This led Smit (1941) to predict that hopper emergence would be greatest in years when five dry winter months were succeeded by good rains at two-monthly intervals in summer. Locust populations would increase rapidly under this scenario which allows three generations in one summer. Two such seasons would cause low-density *solitaria* populations to change to high-density *gregaria* swarms. Verification of a simulation model (Nailand and Hanrahan, 1993), which uses the 12-day running averages of soil moisture, derived from rainfall data, to predict the timing of mass emergence of hopper bands at the local scale, further supports the notion that the sequence of dry and wet periods can be used to predict locust irruptions. The spatial distribution of locust swarms in the karoo can also potentially be anticipated from remote sensing of winter soil moisture and summer vegetation productivity (greenness), an index of locust food quality (Nailand, 1993).

Lea (1968) proposed disruptive selection as an alternative explanation for the observed population cycles. The brown locust has two morphologically and behaviourally distinct phases, *solitaria* and *gregaria*, the former being sedentary and the latter forming mobile swarms. He suggested that, in growing sedentary populations, predation favours the more fecund, density-sensitive individuals.

These continually disperse out of crowded populations. However, with further population growth, dispersing individuals are forced to aggregate, becoming easy prey for storks, raptors, carnivorous mammals and specialized parasitic flies (*Stomorhina lunata*) and protozoa (*Malamoeba locustae*). Higher mortality levels in the dispersing genotype would eventually favour the less fecund sedentary form that remains in the permanent karoo breeding grounds. Locust populations thus decline and stabilize until population growth is again triggered by low mortality and suitable rainfall sequences. Other factors, such as lagged, density-dependent predation and parasitism, or high stress levels associated with crowding, might also promote cyclicity.

More recently, Owen-Smith (1988) suggested that locust outbreaks and collapses could be explained by changes in the nutritional quality of their food plants, as follows. Drought restricts plant growth, conserving soil nutrients (see Whitford, this volume). When drought is broken by spring rains, quiescent locust eggs hatch and grass resprouts. Dry spells during summer lead to high concentrations of proteins but low carbohydrate and tannin levels in moisture-stressed grass leaves. This promotes rapid development of locust nymphs, and locust populations increase and swarm. In years when rains are uniform and plants are unstressed throughout the karoo, locust nymphs, stressed by low protein availability, grow slowly. Nutritional stress, combined with heavy losses to parasites and predators, cause locust populations to collapse. As yet, there is inadequate information to reject this hypothesis or any of the others proposed to explain locust outbreaks.

Clearly, brown locusts are arid region specialists, benefiting from periodic droughts and triggered by summer rain to exploit the protein-rich leaves of resprouting plants. However, the mystery of locust population irruptions has not yet been unravelled completely.

11.3.2.   Other insects

The karoo caterpillar *Loxostege frustalis* (Lepidoptera: Pyralidae), which feeds on about 20 species of small shrubs of the Asteraceae, and in particular on the genus *Pentzia*, has shown sporadic population outbreaks in the karoo over the past century (Annecke and Moran, 1977). It is of economic importance because *Pentzia incana* and related species are important forage plants for sheep in large areas of the karoo, particularly during winter. The caterpillars spin webs and defoliate the plants in late summer, making them unacceptable to sheep until spring regrowth. Karoo caterpillar populations are usually maintained at fairly low levels by 12 indigenous species of parasitoids (Diptera and Hymenoptera) which destroy about 41% of the caterpillars and 47% of the pupae (Annecke and Moran, 1977).

Karoo caterpillar epidemics are known to have occurred in the late 1800s, from 1937 to 1941 and in 1985 (Annecke and Moran, 1977), all of which were periods of above-average rainfall over the karoo. During the 1985 outbreak at Middelburg, caterpillars occurred at densities of around 800 000 ha^{-1} and moths at 100 000 ha^{-1} in species-poor vegetation dominated by large *Pentzia* plants (Donaldson, 1986). Caterpillar populations are thought to irrupt after droughts, when a rapid increase in food supply and quality is combined with a temporary release from predation as a result of the lag in the population responses of parasites (Pretorius, 1989). It has been found that the karoo caterpillar has a density-dependent response to its host plant population (Möhr, 1982). Consequently, its populations reach higher densities and cause greater economic losses in *Pentzia*-dominated vegetation than in areas where there is a mixture of grasses and shrubs. Large, ungrazed bushes support more caterpillars than heavily browsed bushes (Donaldson, 1986). Irregular and patchy population irruptions have been recorded in other folivorous insects including Chrysomelid beetles (Milton, 1993).

Populations of cicadas in the genus *Quintillia* (Dean and Milton, 1991c) and of small sapsuckers (Coccidae, Cicadellidae, Psyllidae, Tingidae) also undergo sporadic outbreaks on karoo shrubs and trees (Webb, 1977; S. J. Milton, personal observation). Psillid (*Acizzia russellae*: Homoptera) densities on *Acacia karroo* show regular spring

peaks and winter troughs related to nitrogen levels in the host foliage (Webb, 1977). Population densities of this sucking insect were 10 times greater on regenerative foliage of pruned trees than on controls, and declined less rapidly with the onset of winter (Webb and Moran, 1978). Host plant quality appeared to have a greater influence on the local abundance of the psyllid than did its parasites.

Population densities of some karoo insects have been found to vary with land use and vegetation structure. High densities of termitaria of grass-eating snouted termites (*Trinervitermes* spp) are associated with overgrazing and winter burning of rangeland in the north-eastern karoo (Coaton, 1948). When grass is scarce, the termites build supplementary mounds linked by subterranean passages to a central termitarium. These temporary shelter and larder facilities, which enable them to increase their foraging area, are abandoned when grass becomes plentiful. Large termite populations are therefore not necessarily correlated with high densities of termitaria (Coaton, 1948). The harvester termite *Hodotermes mossambicus* is also associated with overgrazed rangeland. Coaton and Sheasby (1975b) suggested that grazing practices that denude the soil during the summer swarming season favour population increases because alates select bare rather than vegetated ground for settlement. Populations of harvester termites (*H. mossambicus, Microhodotermes viator*) are further thought to be limited by intercolony competition, drought (Nel and Hewitt, 1969), scarcity of suitable forage (Hewitt and Nel, 1969; Nel et al., 1970) and by natural enemies. Management practices that reduce such specialist termite predators as aardvark *Orycteropus afer* (Van Ark, 1967) and *Pachycondyla (Opthalmopone)* ants (Dean, 1989) and *Ammoxenus* spp. spiders (Dean, 1993), might allow termite populations to increase.

In their research on the life-histories of aculeate wasps, F. W. Gess and S. K. Gess have shown that anthropogenic modification of physical habitats and plant communities can increase or exterminate local populations of endemic insects. For example, the building of small impoundments greatly increases populations of the potter wasp (*Ceramius socius*), whereas vegetation destruction by ploughing reduces populations of specialized wasps that foraged on perennial flowers (Gess and Gess, 1993). Similarly, population densities of adult cicadas (*Quintillia conspersa*) in the southern karoo were positively correlated with shrub cover. Cicada densities were lower on overgrazed ranches, dominated by annual rather than perennial plants, than on adjacent, good-condition rangeland (Milton and Dean, 1992).

11.3.3.   General patterns in invertebrate populations

Despite the economic importance of the brown locust and karoo caterpillar in the karoo, and research on the factors controlling the abundance of these species that spans 200 years, the current understanding of their dynamics remains hypothetical. Correlative evidence supports the notion that irruptions are related to rainfall variability which regulates population growth rate through the quantity and nutritional quality of food. Population crashes seem to be brought about by resource depletion and exponential increases in parasites and disease organisms. Some karoo species appear more sensitive to resource availability and habitat quality than predation, but there is too little empirical data to progress beyond speculations about their population dynamics.

11.4.   Reptiles and birds

11.4.1.   Tortoises

Nine species of tortoises (Testudinidae), of which six are endemic, occur in the karoo (Branch et al., 1995). Although up to five species may be found within an area of 300 square kilometers (as is the case in the Karoo National Park), they use different vegetation or topographic units within the landscape. In the Calitzdorp area, angulate tortoise *Chersina angulata* densities were highest in dense, closed succulent thicket whereas tent tortoise *Psammobates tentorius* was restricted to the interface between dwarf succulent shrubland and shrubland with patches of thicket (Malan and Branch, 1992). In the Great Karoo, near Prince Albert, leopard tortoise *Geochelone pardalis* occurs on rocky ridges and along wooded water-courses, *C. angulata* in dense shrubland and *P. tentorius* in sparse, succulent dwarf shrubland (Milton, 1992a). The availability of habitat patches that offer suitable hibernation, nest sites and food probably imposes limits on population sizes, while territoriality may regulate density.

Factors that could negatively influence karoo tortoise population sizes are habitat destruction by ploughing and overgrazing, and direct persecution by landowners who fear that tortoises are vectors for tick-borne diseases and that they contaminate the water at stock watering points (Boycott and Bourquin, 1988; Milton, 1992a). Man-made barriers, including railway lines, roads and electric fences hamper movement and cause mortality through injury and electrocution (Burger and Branch, 1994). Chelonians are potentially long-lived and well defended as adults. However, there is heavy predation on eggs and hatchlings by raptors and mammalian carnivores (Boycott and Bourquin, 1988; Boshoff et al., 1991; Malan and Branch, 1992). It is probable that the impact of predation on tortoise populations may be greater in rangelands in which heavy grazing or trampling by ostrich and other livestock

has reduced the vegetation cover, making hatchlings more conspicuous to their natural predators.

11.4.2.   Spatial and temporal patterns in bird abundance

Man-made structures, habitats, introduced plants and predator control or extermination over the past 200 years have apparently changed the population sizes and distribution of a number of karoo birds. For example, black crow *Corvus capensis* and pied crow *C. albus* populations have increased, probably in response to roads that supply abundant carrion in the form of road kills, and telephone lines that provide perches and nest sites (Siegfried, 1963a; Macdonald and Macdonald, 1985). Alien trees have allowed pied barbet *Tricholaema leucomelas* and masked weaver *Ploceus velatus* to extend their distributions into formerly treeless shrublands (Macdonald, 1986b; Macdonald, 1990). Similarly, pylons may have expanded the breeding ranges of martial eagle *Polemaetus bellicosus* in the karoo (Boshoff, 1986). The abundance of the laughing dove *Streptopelia senegalensis*, blue crane *Anthropoides paradiseus* and black-shouldered kite *Elanus caeruleus* have increased in disturbed and cultivated areas, where they preferentially feed on resources (weed seeds and grain, rodents) that were once absent or scarce in the arid zone (Dean, 1975; Dean, 1995; Allan, 1995).

Seasonal and interannual fluctuations in bird populations have been monitored in the Kalahari and southern karoo. Four patterns in the temporal abundance of bird populations can be distinguished (Dean, 1995), namely stable populations, regular fluctuations, irregular fluctuations and irruptions (Fig. 11.9). Resident, territorial species, including insectivorous passerines such as karoo chat *Cercomela schlegelii*, anteating chat *Myrmecocichla formicivora*, chat flycatcher *Melaenornis infuscatus* and rufouseared warbler *Malcorus pectoralis*, (Maclean, 1970d; Earlé and Herholdt, 1988; Dean, 1995), the omnivorous karoo korhaan *Eupodotis vigorsii* (Boobyer, 1989), and such raptors as black eagle (Davies, 1994), martial eagle (Boshoff, 1986), and pale chanting goshawk *Melierax canorus* (Malan, 1995), have fairly stable populations that vary less over time than between habitat types. Seasonal and interannual fluctuations in these populations appear to reflect recruitment of juveniles into the populations.

Birds that move seasonally between the summer- and winter-rainfall zones of the karoo, include the blue crane *Anthropoides paradiseus* (Allan 1995), Namaqua sandgrouse *Pterocles namaqua* (Malan et al., 1994; Dean, 1995) and Ludwig's bustard *Neotis ludwigii* (Allan, 1994). Irregular movement among patches of favourable habitat is characteristic of the nomadic and predominantly granivorous passerine birds that feed their young on insects. This group includes red-capped lark *Calandrella cinerea*, grey-backed finchlark *Eremopterix verticalis*, black-eared

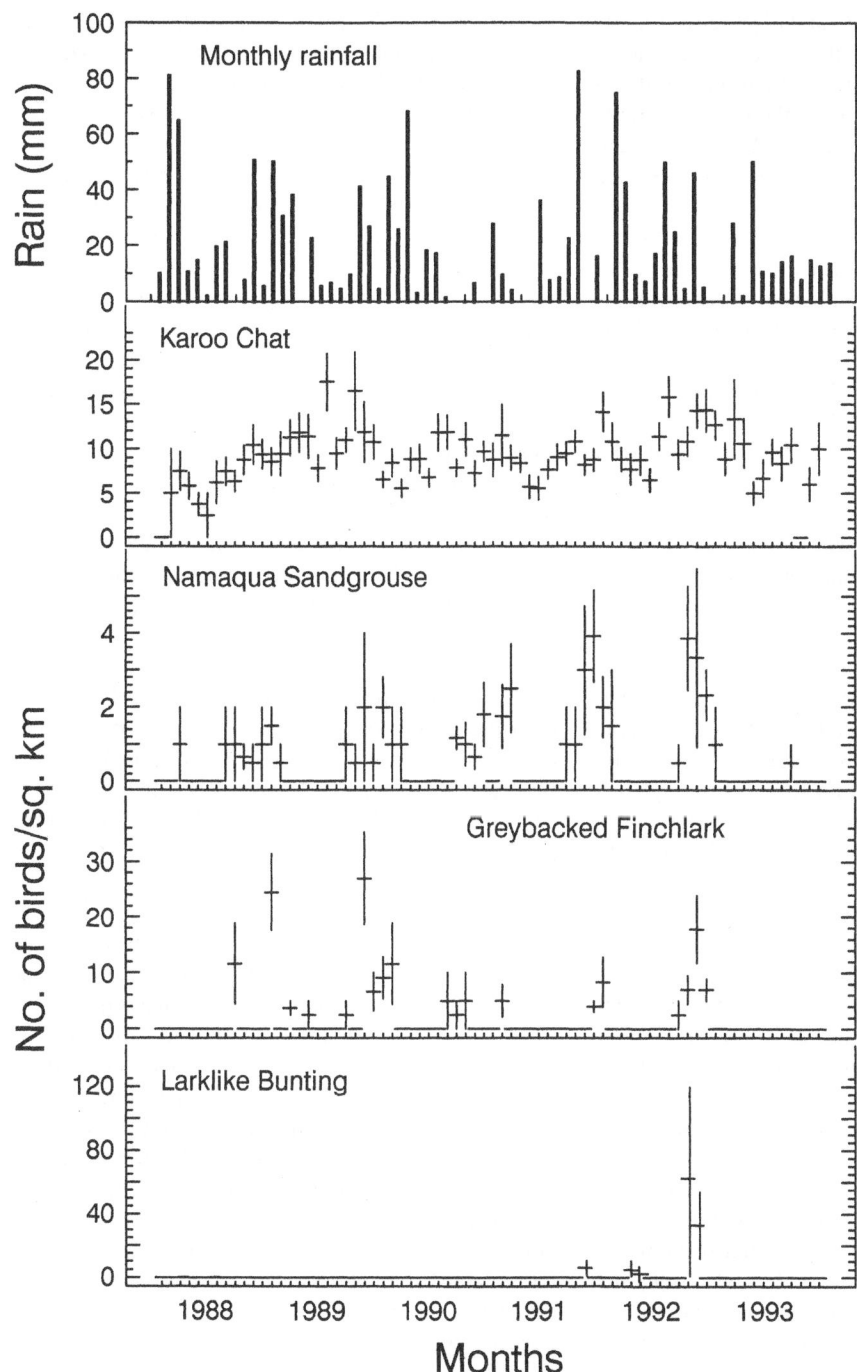

Figure 11.9 **Bird counts along walked transects at Tierberg in the southern karoo illustrate four patterns in the temporal abundance of bird populations in relation to monthly rainfall. Resident karoo chat had fairly stable populations, partially migratory and nomadic Namaqua sandgrouse showed regular fluctuations, nomadic grey-backed finchlark underwent irregular fluctuations and nomadic larklike bunting occurred sporadically in large flocks (Dean, 1995)**

finchlark *E. australis* and black-headed canary *Serinus alario* (Maclean, 1970d; Dean, 1995). Some nomadic species (including larklike bunting *Emberiza impetuani* and wattled starling *Creatophora cinerea*) show irruptive behaviour, aggregating rapidly in very large flocks to feed and breed (Liversidge, 1961; Skead, 1960).

Plant and insect population data reviewed earlier in this chapter have identified post-drought rainfall events as important determinants of primary productivity, forage quality and arthropod abundance in the Karoo–Kalahari region. Changes in the temporal abundance of resident bird populations and in the spatial distribution of nomadic birds are thus expected to reflect rainfall patterns. That the patchy and unpredictable rainfall of the karoo, and the seasonal variation in the timing of rainfall across the arid zone of southern Africa exercises a powerful influence on reproduction, immigration and emigration of birds is evident in both the Kalahari (Maclean, 1970d; Liversidge, 1980) and the karoo (Dean, 1995). Commenting on temporal patterns in bird species abundance in the Kalahari, Maclean (1970d) said that it was possible to see what was happening, but very difficult or impossible to understand why it was happening. The scarcity of autecological information available for most arid zone bird species still makes patterns of abundance difficult to interpret. We summarize here the few detailed studies that have sought to unravel the processes influencing population dynamics and spatial distribution of karoo birds.

11.4.3.   Resident, raptorial birds

Populations of large raptorial and scavenging carnivorous birds have generally decreased in the karoo and Kalahari with the expansion of commercial ranching of domestic livestock (Dean, 1975; Boshoff and Vernon, 1980). Between 1968 and 1980, total raptor densities were 4–30 times greater in the Kalahari Gemsbok National Park (20–30 raptors 100 km^{-2}) than on livestock ranches (1–3 raptors 100 km^{-2}) to the south of the park, and differences were greatest among large raptors and vultures (Dean, 1975; Liversidge, 1984). Human activities rather than weather patterns, predation or disease appear to explain the timing and spatial pattern of vulture and raptor population declines in the karoo.

The Cape vulture was widespread and numerous throughout the karoo during the nineteenth century, but began to decrease rapidly in the late 1800s. From about 1900 until 1950, Cape vultures were absent from the karoo, but after 1950 their populations began to expand south-westward again until the mid 1970s when declines were again reported. The reasons for these fluctuations in Cape vulture populations are not clear. Boshoff and Vernon (1980) consider that the extermination of game herds in the 1800s accounted for the first documented decreases in vulture numbers. It has also been suggested that vultures migrated northward in response to the carrion provided by the rindepest epidemic. The most recent decline in the karoo population of Cape vulture has been attributed to a decrease in food supply through farming methods that greatly reduced livestock losses on karoo ranches between 1950 and 1970 (Boshoff and Vernon, 1980). From 1972 to 1983, a colony of Cape vultures at Potberg in the southern Western Cape, decreased from 40 to 20 breeding pairs. Calcium deficiency, collision and electrocution on power lines, persecution and the use of strychnine and organochlorine based insecticides on farms (Boshoff and Vernon, 1980; Boshoff and Robertson, 1985; Boshoff, 1986; Boshoff and De Kock, 1988; Allan, 1990) may be accelerating their decline.

The distribution and abundance of some raptors is limited by specialized prey and nest sites. In the Great Karoo and Namib Desert, territories of martial eagles are about 250 km^{2} per breeding pair (Boshoff, 1986). The large territories, and cliff or large-tree nest sites required by this and other large raptors, apparently leave little scope for increases in the breeding population. Despite their low population densities in the karoo, a questionnaire survey indicated that 130 black eagles and 25 martial eagles were shot, trapped or poisoned annually on sheep farms in the Laingsburg District over an area of about 3300 km^{2} (Siegfried, 1963b).

The pale chanting goshawk *Melierax canorus*, a perch-hunter that feeds on a wide variety of prey (Maclean, 1993), is probably an exception among karoo raptors in that it appears to have benefited from overgrazing and can compensate behaviourally for the limits on population growth normally imposed by territoriality. In the Little Karoo, where this species specializes on otomyid rodents, it occurs at unusually high densities of one bird per 2–5 linear kilometres on ostrich farms in the Calitzdorp district, compared with the one bird per 13–200 km elsewhere in the karoo and Kalahari (Malan, 1995). High densities of pale chanting goshawks were correlated with abundant prey in habitat that offered good prey visibility and suitable perches.

Birds in such habitat produced more offspring, but juvenile birds had less opportunity to establish territories. The social behaviour of pale chanting goshawks in high-density populations differed from that in less dense populations. In dense populations, juvenile dispersal was delayed, allowing non-breeders to share in the returns of co-operative hunting, and related male birds bred in polyandrous trios rather than monogamously, improving their survival and fitness (Malan and Crowe, 1996). This behaviour compensated for the loss in reproductive fitness caused by the limited availability of territories. Consequently, breeding populations in high-quality habitat remained stable (<8% cv) over five years despite large fluctuations in the population densities of their otomyid prey, and territory holders tended to be replaced by their own offspring.

11.4.4.   Resident, omnivorous birds

Habitat quality, including soil colour, vegetation structure and food availability appear to control the distribution of the red lark (*Certhilauda burra*) endemic to Bushmanland on the red sand dunes south of the Orange River (Dean et al., 1991). Red larks occur at densities of 1–18 birds km^{-2}, the higher densities being on dunes with a good cover of *Stipagrostis ciliata* and where large-seeded ephemerals were abundant. Red larks were absent or scarce on bare dunes and where the vegetation was shrubby (*Rhigozum trichotomum, Lycium* spp) rather than grassy. Heavy grazing of dune grasslands by cattle is thought to make the habitat unsuitable for red larks by reducing the availability of nest sites (beneath spreading grass tussocks) and food (large seeds, insects).

The sociable weaver *Philetairus socius* is colonial, roosting and breeding in large nests (2–4 m diameter, 1–2 m thick) incorporating nest cavities of up to 500 birds. The sheltered conditions in the nest reduce energy and water expenditure (White et al., 1975; Bartholomew et al., 1976; Williams and du Plessis, 1996). The southern distribution limits of the sociable weaver are probably determined by the scarcity of trees, such as *Acacia erioloba*, robust enough to support the nest mass (Maclean, 1973). Man-made structures, such as telephone poles, that provide suitable support for nests, have extended the distribution of the sociable weaver southwards into the treeless grasslands of Bushmanland. In addition to being nest-site limited, there is evidence of density-dependent population regulation in this species (Maclean, 1973). Neighbouring colonies are at least 0.8 km apart even when suitable trees are abundant, indicating intercolonial territoriality. Breeding activity may occur up to five times per year, and is triggered by large rain events (>20 mm), which are reliable predictors of insect abundance. Chicks are fed on invertebrates rather than seed, and the number of birds in a colony varies with breeding success.

The karoo korhaan, a small (1.2–1.7 kg) endemic bustard that feeds on plants (90%) and invertebrates, is social (2–5 birds per group) and territorial (0.6–3.3 km² per group) (Boobyer and Hockey, 1994; Hockey and Boobyer, 1994). The size of social groups and territories are adjusted in response to spatial and temporal differences in food availability. Group sizes and territories increase in disturbed or very arid areas where food is patchy in space and time, and territories shift in the landscape as the availability of resources changes (Boobyer, 1989).

Hunting-bag records can provide long-term indices for karoo game-bird populations. One such example is the helmeted guineafowl (*Numida meleagris*), an indigenous gallinaceous bird with a live mass of 1.3–1.5 kg. Despite large clutch sizes of 6–19 eggs, breeding success is low in dry years. This may be because drought either reduces vegetation cover increasing predation of the precocial chicks, or reduces the availability of food to chicks. In the eastern karoo, helmeted guineafowl fed on insects (Orthoptera, Coleoptera, Lepidoptera), large seeds and berries, flowers, young grass leaves, and corms (Skead, 1962), the availability of which is tied to warm season rainfall (Crowe, 1978). At Rooipoort near Kimberley, in the ecotone between karoo and Kalahari savanna, the numbers of guineafowl bagged daily by hunters (an index of population size) were closely correlated with number of raindays per year (Fig. 11.10). Population responses to rainfall were density-dependent, being greater at times when population densities were low (Crowe, 1978). Breeding populations of guineafowl were clustered around water points and, when populations were large, resident males spent more time aggressively excluding competitors from prime territories. Such competitive interactions, together with high parasite loads (Crowe, 1978) may explain why high-density guineafowl populations did not grow as rapidly in wet years as low-density populations.

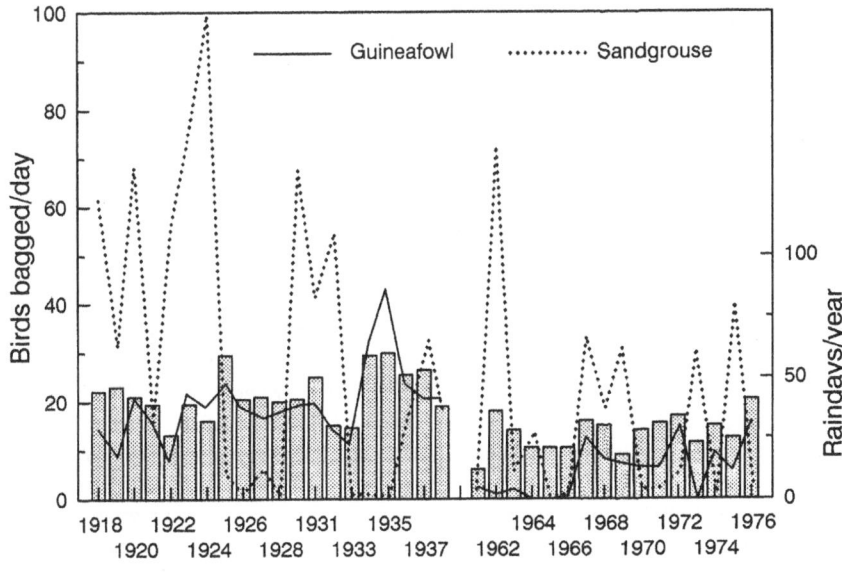

Figure 11.10 **Numbers of helmeted guineafowl and Namaqua sandgrouse bagged per hunting day at Rooipoort and the total number of raindays (bars) for the years 1918–38 and 1961–76 (data from Crowe, 1978; Little et al., 1996)**

11.4.5.   Local migrants and nomads

Among the local migrants and nomadic birds of the karoo, there appears to be no evidence for the density-dependent regulation of populations through the territoriality that characterizes resident bird species. Populations of local migrants and nomads show high temporal and spatial variability that has thus far proved difficult to predict.

Granivorous Namaqua sandgrouse, which are partial migrants in southern Africa, apparently track potential seed availability. They move eastwards into the summer rainfall region of the Kalahari in autumn, then back westwards in spring to breed in the winter-rainfall Bushmanland area of the karoo (Malan et al., 1994). The birds are dependent on high-quality drinking water for their own needs and those of their chicks. Sandgrouse populations have a fairly low potential growth rate because less than half of the population is reproductively active at one time; they lay 2–3 eggs per clutch and have a reproductive success of only 16–45% (Malan et al., 1993). Hunting-bag data (1907–92) for Namaqua sandgrouse at Rooipoort in the east of their range (Fig. 11.10), indicate interannual fluctuations in abundance as well as an apparent population decline from 1950 to 1992 (Little et al., 1996). There was no relationship between sandgrouse abundance and rainfall frequency, but sandgrouse bags were high in years when above-average rain fell in autumn following a dry summer. The explanation for this pattern is that delayed germination of annual legumes enables more sandgrouse to remain in the eastern extreme of the range during the winter hunting season. Since sandgrouse are migratory, models for predicting huntable sandgrouse numbers should probably also include indices of resource availability in the south-western part of their range. The long-term decline in sandgrouse numbers has been attributed to the increase in stock watering points (from 5 per 100 km² in 1930 to 27 per 100 km² in 1960), which possibly led to greater dispersal of sandgrouse populations. An alternative explanation is that improved ranching practices have increased vegetation cover, reducing the abundance of annual legumes on which sandgrouse depend for seed (Little et al., 1996).

Unlike the smaller nomadic birds of the arid zone, the large (3–7 kg) Ludwig's bustard is omnivorous, feeding on large insects (mostly locusts) and small vertebrates as well as flowers and seeds. Part of the population apparently moves from the summer-rainfall karoo and Namibian escarpment westwards into the winter-rainfall Namib and succulent karoo during winter (Allan, 1994). On a more local scale, the density of Ludwig's bustards is positively correlated with total rainfall in the preceding three months (Allan, 1994). Unusually large rainfall events in arid regions (such as the Tanqua Karoo or the Gariep) attract more bustards than good rainfall in a mesic region.

The nutritional quality hypothesis proposed by Owen-Smith (1988) to explain locust outbreaks, may be a good general predictor of arthropod or small vertebrate population increases large enough to support the bustards. Ludwig's bustards concentrate where prey is abundant, as in the case of a flock of about 230 of these birds seen feeding on ovipositing locusts near Rietbron (W. R. J. Dean, personal observation). In the Little Karoo and Noorsveld, Ludwig's bustard densities are consistently low, suggesting either that prey is scarce in these habitats or that tall succulent thicket vegetation reduces their foraging success.

Ostrich *Struthio camelus* are well adapted to unpredictable and patchy rainfall. Where they are not restrained by fences, ostrich are able to emigrate from drought-stricken areas and rapidly repopulate resource-rich patches in the mosaic drought and drought-recovery landscapes of the karoo and Namib (Berry and Siegfried, 1991). Their success in deserts can be attributed to a combination of longevity (maximum 40 years), mobility (daily foraging range 19 km, Williams et al., 1993) and rapid population growth under post-drought conditions. Temporary freedom at such times from nest predation (70% of nests predated) allows ostriches to realize the high reproductive rate that results from polygamy, large clutches (20–30 eggs per nest) and crèching of chicks (Bertram, 1992). Ostriches require high-quality forage, feeding mainly on ephemeral plants and flowers (Milton et al., 1994a). Fencing and artificial water points, which prevent or delay emigration from desiccated vegetation patches, has been responsible for starvation of up to 10% of the adult ostrich population during prolonged droughts in the Namib–Naukluft and Kalahari–Gemsbok National Parks (Kok, 1980; Knight, 1995a).

11.5.   **Mammals**

11.5.1.   Lagomorphs, rodents and shrews

Small mammal populations, in both the winter-rainfall succulent karoo and the summer-rainfall Nama-karoo, are generally greater in autumn and winter, than in spring and summer (Jooste and Palmer, 1982; Kerley, 1992b). However, peak densities were observed during one summer season in the Nama-karoo (Kerley and Erasmus, 1992). The reproductive patterns of small mammals in the karoo differ among species. Bernard et al. (1996) found that the round-eared elephant shrew *Macroscelides proboscideus* (Macroscelididae), which feeds on insects (46%) and foliage (48%) is an aseasonal breeder in the Nama-karoo. Spermatogenically active males and pregnant females were recorded throughout the year. In contrast, White et al. (1996) working at the same site over the same period, found that, although males of the hairy-footed gerbil *Gerbillurus paeba* were spermatogenically active all year round, pregnancies were observed only in summer months (October–May). Similarly, a laboratory population bred only under an extended daylight regime, despite abundant food being available (Ascaray, 1986). However, in arid savanna, *G. paeba* is an aseasonal breeder (Smithers, 1971). The restricted breeding period observed in southern karoo populations may be related to a winter scarcity of insects, a preferred food of this omnivorous rodent (Skinner and Smithers, 1990; Kerley, 1992c), or to other, as yet unknown, extrinsic factors.

Changes in day length are generally recognized as proximal cues that stimulate breeding in small mammals of temperate regions, and may also influence karoo species. For example, a laboratory population of the short-tailed gerbil *Desmodillus auricularis* bred successfully under an extended daylight regime (16L:8D), after failing to breed under natural-light conditions despite availability of adequate food (Keogh, 1973). However, under the unpredictable and low-rainfall regime of the karoo, short-lived taxa are more likely to benefit from opportunistic reproduction. Reproductive activity in some female rodents is stimulated by 6-methoxybenzoxazolinone (6-MBOA) in fresh green leaves (Linn, 1991), but this has not been investigated in karoo rodents.

Although there has been very limited research on the population dynamics of small mammals in the karoo, there are no published accounts of rodent populations ever reaching 'plague' levels. Anecdotal accounts do, however, indicate that populations of a few species of rodents (particularly pygmy mouse *Mus minutoides*, striped field mouse *Rhabdomys pumillio* and hairy-footed gerbil) become locally abundant for a few months when good rains follow droughts or when grasses are abundant (Leistner, 1967; Kerley and Erasmus, 1992; Milton et al., 1995). The hairy-footed gerbil is the most common small mammal in the karoo, and its temporal abundance varies markedly. At Tierberg, its population density varied between 2 and 30 individuals ha^{-1} within a year, while, at a site 60 km further east, the small population (2 ha^{-1}) of this species disappeared and no small mammal species were recorded in the area during three trapping seasons (Kerley and Erasmus, 1992).

Some rodents, such as the short-tailed gerbil, rarely occur at densities greater than one individual per hectare, and others appear ephemerally in certain habitats. For example, the big-eared desert mouse *Malacothrix typica* was recorded in the open plains of the southern karoo during only three months of a four year sampling programme (Kerley 1991, 1992c; Kerley and Erasmus, 1992). The animals were trapped from February to May, apparently in response to a 60 mm rainfall event during late spring. Pygmy mice were similarly recorded during only half of the trapping sessions at this site, in periods following good rains. Such species probably maintain populations in specific habitat islands from which individuals emigrate as populations fluctuate or habitat conditions vary over time. Similar responses of rodent populations to rainfall and productivity have been inferred from studies in the Chihuahuan Desert (Whitford, 1976).

Folivorous otomyid rodents (whistling rat *Parotomys brantsii* and karoo bush rat *Otomys unisulcatus*) at times reach densities of 3 ha^{-1} and 13 ha^{-1} respectively in some vegetation types in the Little Karoo (Malan, 1995). In peak rodent years, the combined biomass of these rodents in open shrubland with scattered trees was 210 kg km^{-2}, compared with 48–73 kg km^{-2} in sparse dwarf shrubland and thickets. Populations periodically declined to 20–50% of peak levels (Malan, 1995), and, as reported from the Kalahari, these population cycles appeared to be uncoupled from rainfall (Nel and Rautenbach, 1975). Population explosions in some small mammal species in the karoo and Kalahari may be the result of successful breeding and predator swamping during favourable seasons, whereas crashes in populations may be caused by the rapid spread of diseases, such as bubonic plague *Pasteurella pestis* by house-rat fleas *Xenopsylla brasiliensis*. Davis (1953) monitored a population crash in plague-infected highveld gerbils (*Tatera brantsi*) in the Free State from 1940 to 1942 and found that the estimated population decreased from 1250 to 125 and only two of 34 gerbil colonies persisted after nine months.

Habitat destruction has driven riverine rabbit *Bunolagus monticularis*, one of seven endemic karoo mammals (Vernon, this volume), to near extinction over the past century. This is the only African lagomorph confined to tall, dense *Salsola* sp. thickets on deep alluvium. Virtually all alluvium in the karoo has been ploughed, because arable land with access to water is at a premium in this region (Hoffman et al., this volume). Effects of habitat loss are exacerbated by subsistence hunting by farm labourers. Small litters (1 kit per litter) and territoriality limit rates of population growth even in protected habitat (Duthie, 1989). Efforts to re-establish the remnant riverine rabbit population through captive breeding (Dippenaar and Ferguson, 1994) have not yet achieved success because of the high mortality rate in captivity and of deaths of rabbits released into unsuitable habitat (Matthews, 1995).

11.5.2. Rock hyrax *Procavia capensis*

The rock hyrax *Procavia capensis* is well suited to a consideration of parameters affecting karoo mammal populations because its reproductive biology has been studied (Millar, 1971; Steyn, 1980; Fourie, 1983) and its population dynamics have been mathematically modelled (Swart et al., 1986; Davies, 1994). Research was prompted by the economic significance of rock hyrax outbreaks in the karoo biome where hyraxes sometimes compete with domestic livestock for the available forage. The total rock hyrax population size of the karoo may exceed 4.4 million animals. These medium-sized herbivores (c. 3.5 kg) are especially well adapted to the arid, rocky habitats of the karoo. They live in groups comprising a matriline monopolized by a single territorial male, with young of both sexes (Fourie and Perrin, 1987).

Additions to rock hyrax populations

Hyrax birth pulses occur in early summer (November to December) in the karoo. In the Mountain Zebra National Park, 92% of births occurred within four weeks of the first birth for that season (Fourie, 1983) and, in the Noorsveld, 81% of hyrax were born in November (Millar, 1971). Modal parturition date in the Karoo National Park was mid-November (Davies, 1994). Heavy predation pressure on the juveniles may explain why rock hyrax reproduction is seasonal in a relatively unpredictable climate. Their lengthy gestation period (230 days), which might be an evolutionary throw-back (Sale, 1970), precludes hyrax from rapid reproductive responses to rainfall events in the karoo.

Recruitment of young is affected by population composition. Only about 45% of the naive juveniles survive their first year. Sexual maturity is usually attained by month 17 (second rut) for females and by month 17 or a year later for males. In exceptionally good conditions, up to 33% of females attain sexual maturity at their first rut (5 months old) and this can contribute significantly to population growth (Millar, 1971). Most males disperse from their natal colony before they are sexually mature. Adult males are only tolerated by the resident territorial male if they remain in a sexually quiescent state (Hoeck et al., 1982; Van der Merwe and Skinner, 1982), but one or two mature males may be present on the periphery of colonies. Dispersal of females is less obligatory and occurs later.

Survivorship of hyrax in their second year is relatively low (60%) partly as a result of risks incurred during dispersion. Survival of both sexes from two years on, as adults, is good (80–85%), until eight years old when dental attrition first becomes manifest (Fourie, 1983). From this age on, survivorship drops markedly, especially for males. Females generally live longer than males and can attain ages of 13 years (Fourie, 1983).

Recruitment is more closely related to the composition of mature females in the population than to that of males, because a single male is able to monopolize up to 17 females. In hyrax population models it is therefore assumed that males are in sufficient supply and that recruitment is a function of the age distribution and body condition of the mature females. Fecundity rates vary with female age, and usually 80–90% of mature females fall pregnant. Average litter sizes range from two in young females to four (exceptionally 5–6) in older females (Fairall et al., 1986). The average birth pulse is equivalent to about 35% of the hyrax population in the karoo, but reproduction is influenced by environmental conditions. In the Karoo National Park at Beaufort West, birth pulses ranged from 22% to 42% of the total population size. A number of studies have demonstrated a correlation between the size of the birth pulse and rainfall patterns (Millar, 1971; Fourie, 1983; Davies, 1994). Particularly large birth pulses are associated with seasons of high rainfall, and particularly small birth pulses with drought. In the karoo, where resources fluctuate according to rainfall, this results in a highly dynamic population structure. As these abnormal cohorts survive through the age classes, their influence on population structure may be seen in samples taken from the population for up to a decade.

Underrepresentation of hyrax cohorts born during drought years was evident in a shot sample from the karoo National Park (Davies, 1994). Age-determination of a much larger sample of hyrax cranial material recovered from beneath black eagle eyries at this site revealed the same trend and indicated a positive correlation between representation of cohorts and rainfall in the summer preceding their conception (Davies, 1994). It is assumed that good rains enhance the breeding condition of rock hyrax during the rut by providing a plentiful supply of forage. Follow-up rains appear to be just as crucial. Extensive annual counts in the Karoo National Park indicated that hyrax recruitment was closely tied to rainfall during the gestation period (Fig. 11.11), and that rainfall in the month prior to the birth pulse could explain as much as 84% of the variation in recruitment (Davies, 1994). Most hyrax colonies do not have access to free water, so it is likely that the onset and process of lactation would be dependent on moisture in food plants.

Analyses of hyrax age structures, in black eagle prey collected from a variety of regions in the former Cape Province in different decades by Boshoff et al., (1991), also revealed patterns which related to rainfall regimes over time and space (in Davies, 1994). For instance, predominance of less profitable juvenile hyrax in prey captured during the 1970s suggests that black eagles were preying on small hyrax populations which were growing in response to good rains. In the 1980s, black eagles took more surplus elements (e.g. males and old individuals) from hyrax populations that were thought to be larger and declining due to drought.

Wet conditions in the karoo appear to lead to an increase in the mass of individual hyrax as well as an increase in their numbers (Davies, 1994). The dimensions

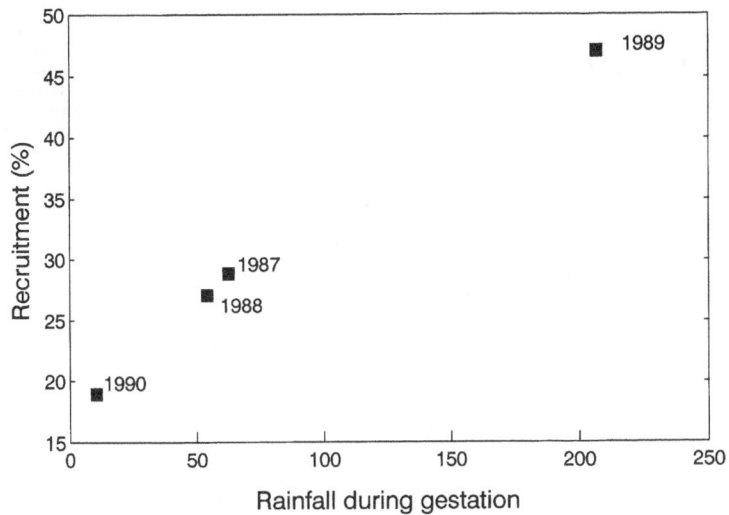

Figure 11.11 **The relationship between hyrax recruitment (no. juveniles at a particular birth pulse compared with total population size at the start of the preceding year) and the amount of rain falling during the gestation period (data from Davies, 1994)**

of hyrax maxillae recovered beneath black eagle eyries in the eastern karoo were significantly larger for animals born during the wet 1970s than for animals born during the arid 1960s (Fig. 11.12), and scaling suggests that they would probably have been 12.5% larger by mass. Over one era, there were statistically significant differences between maxillae dimensions recovered from different regions of the karoo. Hyrax from arid Namaqualand were probably 12% smaller by mass than those from the central karoo, whereas hyrax from the more mesic eastern Cape karoo were probably 20% larger by mass than those from the central karoo. It would appear that body mass of rock hyrax from arid southern Africa is more directly linked to patterns of precipitation than to patterns of latitude, temperature or vegetation type.

There is circumstantial evidence suggesting that sex ratio of juvenile rock hyrax is linked to rainfall patterns in the karoo (Davies, 1994). Significant correlations were demonstrated between the proportion of male hyrax in cohorts and arid conditions at the time of their birth in both the shot sample and collections of hyrax prey remains from black eagle eyries in the Karoo National Park. Foetal sex ratio in pregnant females from the shot sample did not differ significantly from parity, but was biased towards females in the wet year of 1989. It is not known whether sex ratios were influenced before or just

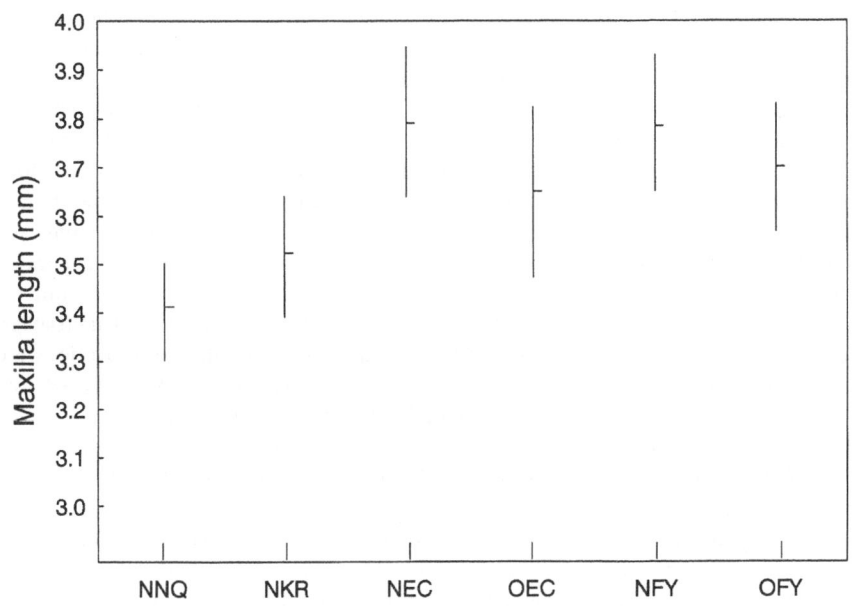

Figure 11.12 **Variation in the mean size (length) of adult male hyrax maxillae from black eagle prey remains collected from different regions of the Cape Province, and at different times. NNQ: Namaqualand ($n = 41$), NKR: karoo 1980s ($n = 50$), NEC: eastern Cape 1980s ($n = 50$), OEC: eastern Cape: 1970s ($n = 50$), NFY: fynbos 1980s ($n = 37$), OFY: fynbos 1970s ($n = 10$). Bars show SD (measurements taken from hyrax skulls collected by Boshoff et al., 1991, and Davies, 1994)**

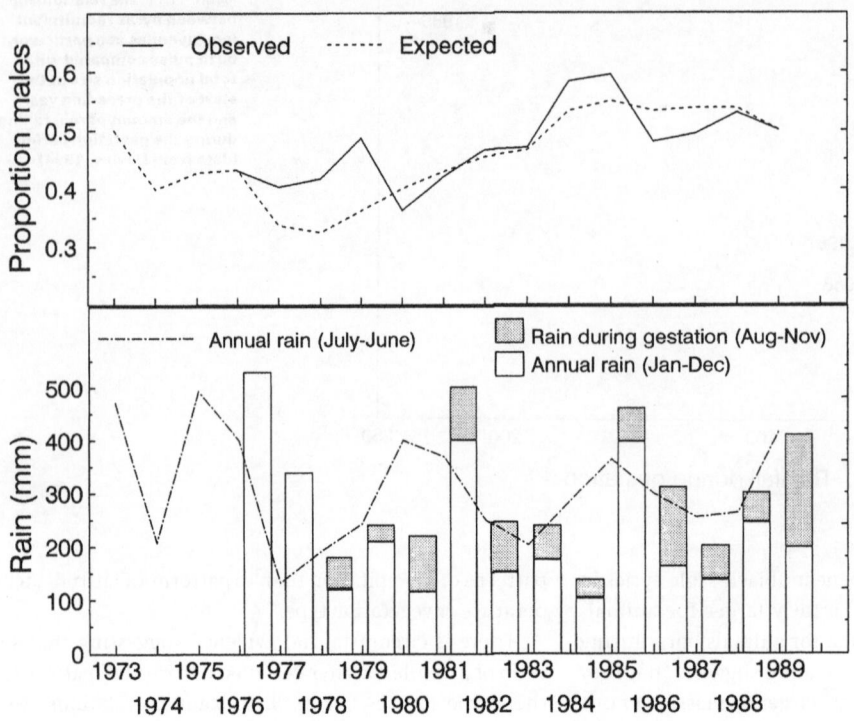

Figure 11.13 **Observed and expected contribution of males to hyrax cohorts in relation to rainfall patterns in the years during which the cohorts were born. Observed values were determined from fresh skulls or maxillae in the prey remains collections. Expected values were calculated for eagles predating a stable hyrax population with a constant and equal production of male and female juveniles each birth pulse (data from Davies, 1994)**

after birth. Analyses of hyrax maxillae from black eagle eyries suggested that a build-up in the male component of cohorts occurred during the two dry spells of 1977–9 and 1983–4 (Fig. 11.13). Both periods culminated in a sudden shift to a female biased cohort, and these shifts roughly coincided with peaks in summer rainfall.

In mammal societies where males disperse and females tend to remain with the group, it has been suggested that females refrain from producing female offspring if local resources are diminishing (Clark, 1979). To gain an understanding of the phenomenon, Armitage (1987) argues that we should rather compare the reproductive potential of male or female offspring. Female hyrax that produce more male offspring during karoo droughts would be likely to benefit from more second-generation offspring provided that reproductive opportunities for territorial males increase during population declines.

Losses to rock hyrax populations

Some hyrax are lost to the population by abortion and resorption of foetuses before they are born (Millar, 1971; Fourie, 1983; Davies, 1994). During drought, pre-natal mortality can claim up to 20% of foetuses in older female hyrax (Fourie, 1983), contributing to the observed correlation between recruitment and rainfall during gestation. Aborting and resorbing foetuses, when conditions might make lactation impossible, represents a major energy sav-

ing for female hyraxes, and allows hyraxes to adjust their litter sizes to prevailing conditions.

Juvenile hyrax suffer about 55% mortality in their first year. Drought years may claim up to 67% of juveniles (Davies, 1994). Fourie (1983) considered juvenile mortality to be the most important demographic factor affecting hyrax population change in the karoo, and ascribed it to heavy parasite burdens, low fat reserves, low thermolability, greater susceptibility to disease, and a reluctance to forage away from natal colonies. Juvenile mortality should not be considered as separate from predation. Young rock hyrax are naive to predators and, when the alarm is sounded, may take refuge in bushes that afford inadequate protection against black eagles rather than in rock crevices. In addition to large eagles and large cats, juvenile rock hyraxes are vulnerable to many generalist predators in the karoo, and predation may account for 40% of juvenile deaths in the Karoo National Park (Davies, 1994). Extremely rapid hyrax population recoveries were witnessed on karoo farms where the generalist predators had been largely exterminated (Davies, 1994).

Refuge has an important bearing on hyrax losses through predation. Behavioural observations of hyrax groups in the Karoo National Park revealed that individuals spend 96% of their day within 5 m of shelter (crevices or thorn bushes) and were never observed >15 m from shelter. There appears to be a safe feeding zone for rock hyraxes in the vicinity of shelter (Davies, 1994), reflected in distinct vegetation zones up to 15 m from isolated rocky outcrops. The zones are thought to result from a behavioural trade-off between foraging return and the risk of predation (Davies, 1994). The area of safe feeding habitat is limited by the near permanent distribution of the refuges and is a measure of carrying capacity for rock hyraxes in the karoo landscape. Protection of hyrax within their safe feeding habitat, and their vulnerability to predation beyond this, imposes some stability on their populations. However, whilst the safe feeding area is fixed, the number of hyrax that it can support varies with vegetation growth and consumption, and hyraxes must move farther and become more vulnerable to predation during drought (Davies, 1994).

The most important predators of adult rock hyrax in the rocky habitats of the karoo are black eagles and caracals (Palmer and Fairall, 1988; Davies, 1994). The leopard *Panthera pardus*, another potential predator, has now been eradicated from most of the karoo. Rock hyrax populations suffer losses to black eagles and caracals in very different ways. Black eagles are specialist predators of rock hyrax, and take their prey by using cover, surprise and co-operative hunting even when small populations of hyrax are well protected within rocky habitat. In the Karoo National Park in the 1980s, black eagles accounted for two of three deaths of adult hyraxes (Davies, 1994). Unlike black eagles, caracals can respond to population increases of rock hyraxes by a dramatic increase in density and by a rapid change in diet (Davies, 1994). Swart et al., (1986) modelled the population processes by which caracal numbers might track hyrax numbers in the Mountain Zebra National Park. Population modelling for the Karoo National Park indicated that, while black eagles take a steady toll on rock hyrax, caracals can remove hyrax surpluses as and when they occur, and a range of predation is necessary to maintain balance between hyrax and their changing food supplies in the vicinity of shelter (Davies, 1994).

At high densities, agonistic encounters between hyraxes may lead to deaths when wounds, inflicted by fighting amongst males, become infested with ixodid ticks and susceptible to secondary bacterial infection (Fourie, 1983). Parasite burdens increase during high-rainfall periods (Fourie, 1983). Falls and other accidents are an additional source of hyrax mortality (Davies, 1994). Rock hyrax should be very susceptible to a disease epidemic because they are 'close-contact' social animals. Hyraxes are affected by outbreaks of bubonic plague (Kingdon, 1971), sarcoptic mange (Hoeck, 1982) and pulmonary tuberculosis (Wagner et al., 1958; Wagner and Bokkenheuser, 1961). This raises speculation that the major hyrax population declines observed in the early 1980s may have been caused by disease, although no sick animals were observed in the

Karoo National Park during this period (N. Fairall, personal communication). It does seem likely that disease might be involved in hyrax population declines, but it is probably encouraged by other circumstances such as high densities and drought. Predators may remove sick individuals, thereby reducing the rate of disease spread. Although disease might be involved in hyrax population declines, decreases may be accelerated by high densities and drought.

Synchronous population declines of other herbivore populations in the karoo during the early 1980s have been ascribed to the low quantity and quality of available forage (Davies, 1985). Drought conditions may increase hyrax mortality through starvation, disease, competition and fighting, foetal loss and greater vulnerability to predation. Improved survival of hyrax colonies near water sources in the Karoo National Park (N. Fairall and R. A. G. Davies, personal observations) during the 1980s decline suggests that moisture in food plants may also have been limiting. Thus, population declines in hyrax can be attributed to the collusion of many factors.

Key factors affecting rock hyrax populations

The influence of refuges and the marked effect of rainfall on the karoo system were incorporated in the population model of Davies (1994). This was achieved through a lower trophic model which simulated the interaction of rainfall, vegetation and herbivory within the safe feeding habitat. Phytomass was considered as 'capital' which could accumulate or depreciate over time within realistic limits, and fresh growth was considered as the 'interest' on that capital determined by rainfall. This simple model predicted phases of increase and decline for the hyrax population which were synchronous with those observed in the field (Davies, 1994). Such a model may find wider application in the simulation of rainfall – karoo rangeland – herbivore dynamics. The following conclusions are drawn largely from manipulations of this model.

Additions to the hyrax population only affect the first age class, whereas mortalities can affect all age classes. Modelling indicated that recruitment had far more influence on population structure than on population size. Key factor analyses showed that losses through the upper trophic level (predation) could account for most of hyrax population change and that losses through the lower trophic level (variation in recruitment and non-violent mortality) were only partly deterministic during the increase phases. Losses through the lower trophic level were insufficient to maintain close tracking of the carrying capacity of the safe feeding habitat by the hyrax population. By contrast, black eagles and especially caracals (which were realistically handled in the model) had the capacity to remove all surplus hyraxes in excess of the carrying capacity of the safe feeding habitat, even after the largest birth pulses.

Hyrax populations in the karoo are effectively regulated by their predators around the number of protected prey, which is determined by food supplies in the vicinity of shelter. A high degree of stability is resonant in this system and can be ascribed to effective refuge and effective predation. Long-term studies have shown that the order of magnitude variation in rock hyrax populations does not exceed a factor of four (Hoeck, 1989; Davies, 1994; N. Fairall, personal communication). Populations of other similar-sized mammals regularly vary by factors of up to 30 times. This stability is surprising in light of previous accounts of hyrax population explosions in the karoo.

Karoo hyrax populations are disturbed by eradication of predators and by rainfall patterns. Reports of rock hyrax irruptions in the karoo coincided with predicted decline phases in the hyrax population following major peaks. Good rains, as experienced in the late 1970s, enable hyrax populations to build up within their rocky habitat. If these conditions are followed by severe drought, as occurred in the early 1980s, large hyrax populations deplete the food supplies around shelter and are forced away from their rocks to find food. Although the populations are already declining at this stage, their habitat shift makes them much more conspicious, leading to the perception of population explosions. Long-term cycles in rainfall patterns, as reported for the summer-rainfall region of southern Africa (Tyson, 1987), can be expected to encourage these disturbances of hyrax populations. Most hyrax irruptions reported in the literature can be attributed to food shortages during drought, and a few of them coincided with springbok treks reported in the karoo at the same time. Intensive predator eradication probably contributed to a population irruption of rock hyrax witnessed during the 1940s, which otherwise would not have been predicted by the population model. The cost of hyrax surpluses to the farmer in the absence of effective predation can be considerable (Davies, 1994).

11.5.3   Large herbivores
Under natural conditions, populations of large herbivores in hot and cold deserts undergo long-term aperiodic fluctuations, which Caughley and Gunn (1993) show are a mathematical consequence of short-term weather effects on forage biomass. Although most large herbivore populations in the karoo are now managed rather than regulated by environmental factors, populations can be expected to fluctuate in large nature reserves.

Large karoo herbivores respond to drought by attempting to emigrate to better watered areas. The springbok Antidorcas marsupialis 'treks' or mass migrations of hundreds of thousands of animals, described by travellers and early European settlers in the karoo in the eighteenth and nineteenth centures are the best-known examples of this drought-evasion response (Skead, 1980; Skinner, 1993). Breeding is opportunistic, apparently being cued by rain and favoured by the availability of high-quality forage. In the winter-rainfall karoo at Robertson, springbok mate in autumn. Energy indices (kidney fat, bone-marrow fat) of pregnant ewes remain high throughout winter until parturition in spring, but fall off during the dry summer (Vorster, 1994).

Springbok populations grow rapidly when forage is abundant because lambing percentages are high (>85%), but, when confined, are subject to sudden crashes if food quality is reduced by drought. For example, a population of 7 springbok released into a 109 ha camp in karroid vegetation near Victoria West increased to 130 animals over 10 years, but 2 years later, towards the end of the severe El Niño-related drought of the early 1980s, the population crashed, leaving only 8 (female) animals (Skinner et al., 1987). Mortalities of springbok during drought were lower on this farm, where they were stocked with merino sheep that removed lignified grass and shrubs (Davies, 1985). An unmanaged springbok population on the 4543 ha Hester Malan Nature Reserve in Namaqualand (Fig. 11.14) fluctuated between 20 and 110 animals (4–24 animals 10 km^{-2}) in response to varying rainfall over a period of 15 years, showing a clear response to the early 1980s drought (Fairall and Le Roux, 1991).

Most confined antelope populations in the karoo are managed by culling or live capture to control population growth, to avoid drought-induced population crashes (Norton, 1989) or to optimize harvests (Skinner et al., 1987). However, Fairall and Le Roux (1991) suggest that it may be desirable (in terms of natural selection and vegetation dynamics) to permit large population fluctuations and drought-induced mortality of antelope on karoo nature reserves. Drought apparently had less impact on the gemsbok Oryx gazella population in the Hester Malan Reserve (Fig. 11.14), and it is likely that a policy of benign neglect could lead to considerable habitat change before reducing populations of this and other hardy, roughage-feeding antelope.

Access to water may alter seasonal movement and habitat use by captive antelope on ranches and nature reserves, in some cases exacerbating the impact of drought on vegetation and animal condition. For example, in October 1985, the sixth consecutive year of the 1980s drought, 3400 carcasses were counted during an aerial survey in the Kalahari National Park. These represented 35% of the eland Taurotragus oryx, 19% of the wildebeest Connochaetes taurinus, and 12% of the red hartebeest Alcelaphus buselaphus populations of the park. All are migrant species, and the high mortality is thought to be related to provision of

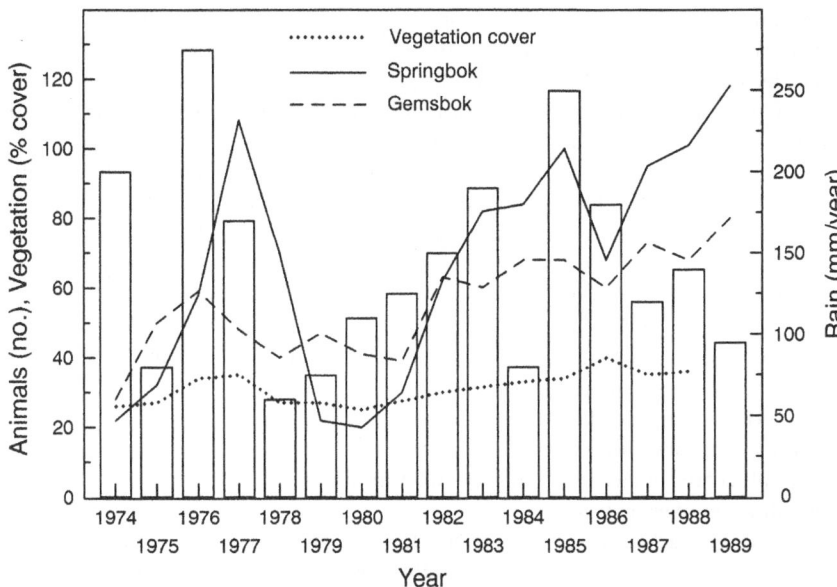

Figure 11.14 **Numbers of springbok on Hester Malan Nature Reserve in relation to annual rainfall (bars) and vegetation cover (redrawn from Fairall and Le Roux, 1991)**

artificial water sources which concentrated the animals in their summer range during the dry winter (Knight, 1995).

In the period of European settlement in southern Africa, mass slaughter of game for meat and sport, competition for grazing from domestic livestock, and fencing of rangeland have greatly altered the sizes of karoo herbivore populations (Skead, 1980; Vernon, this volume). A number of large herbivores, including eland, black rhinoceros *Diceros bicornis* and mountain zebra *Equus zebra* survive only as small captive populations in the karoo. One species, the endemic quagga *Equus quagga*, was apparently hunted to extinction between 1830 and 1870 (Skinner and Smithers, 1990), but the demise of this grass-eater may have been accelerated by overgrazing or climate-induced changes in vegetation (Palmer et al., this volume).

11.5.4. Primates and carnivores

Baboons are omnivorous and opportunistic, but their distribution, breeding success and survival is limited by water availability (Skinner and Smithers, 1990). In the most arid parts of the karoo and in the Namib Desert, baboon *Papio ursinus* troops are confined to the vicinity of permanent water (Hamilton, 1985; Brain, 1988).

Baboons have learned to exploit agricultural produce in the karoo, and may frequently be seen feeding in orchards and wheat fields. For this reason, and because they also kill poultry and small-stock, foul water and damage installations (Brand, 1995), baboons have no legal protection in South Africa. Baboon troop sizes at 10 localities in the southern and Little Karoo ranged from 22–90 individuals. Cages baited with food removed between 46 and 95% of a troop before the remaining individuals became

trap-shy (Brand, 1995). Despite such control measures, baboons remain common throughout the karoo, and are particularly abundant in rocky terrain that provides shelter from climatic extremes and predators (Lloyd and Millar, 1983). Their success may possibly be attributed to the superabundance of once limiting food and water resources now available on karoo farms, as well as to the eradication of leopards from most karoo areas.

As is the case with baboons, carnivorous mammals have long been viewed as pests by karoo livestock farmers (Van der Walt, 1949; Marincowitz, 1992). Recent surveys show that farmers attribute losses of 1–2% of their livestock herds to indigenous predators (Brand, 1992). Despite their perceived importance to agriculture, their probable control of rodents, hares and hyraxes (Swart et al., 1986; Palmer and Fairall, 1988), and the role of some (*Cynictis pencillata*, *Otocyon megalotis*) in the spread of rabies (Keightley et al., 1987) or in the dispersal of melons and fleshy-fruited trees (Kok, 1996), there have been no rigorous long-term studies of factors influencing the dynamics of karoo carnivore populations.

Prey availability influences the timing and success of breeding in arid zone carnivores. Breeding in black-backed jackal, the most carnivorous canid in the karoo, occurs in midwinter (July-August). The birth of pups thus coincides with peak availability of ungulate carcasses towards the end of winter (Bernard and Stuart, 1992). In contrast, the spring birth peak of the Cape fox coincides with the reproductive season of its rodent prey species, and the early summer (October-November) births of aardwolf *Proteles cristatus* and bat-eared fox are timed to exploit maximum abundance of invertebrate prey (Richardson 1987b; Bernard and Stuart, 1992).

Dietary specialization in the aardwolf necessitates large territories and prolonged nocturnal foraging activity. This constrains annual fecundity and milk production, leading to high male and female investment in parental care (Van Jaarsveld et al., 1995; Dean and Milton, this volume). Aardwolf population dynamics are thus particularly sensitive to management actions that reduce the abundance of their prey or kill adult animals.

Hunting and electric fencing have influenced the density, distribution and behaviour of the black-backed jackal *Canis mesomelas* in the karoo (Stuart, 1981; Heard and Stephenson, 1987). Black-backed jackals maintain long-term pair bonds, and successful weaning of pups (1–5 per litter) depends on provisioning by both parents and the previous year's offspring (Skinner and Smithers, 1990). The lower incidence of site or material records of jackals from karoo rangeland than from nature reserves may be partly attributed to behavioural adaptation. Black-backed jackals are diurnal, but become nocturnal in areas were they are intensively hunted (Stuart, 1981). Jackals also learn to avoid traps, coyote-getters and poison bait (Brand et al., 1995), and, despite the annual sale, in karoo areas, of enough strychnine to kill 68 000 black-backed jackals (Allan, 1990), the species still persists at low densities on karoo rangeland.

Anecdotal evidence indicates that jackal populations are currently increasing in some areas. In the Steytlerville District, jackal have been trapped frequently in the 1990s after an absence of nearly 30 years (R. B. Kirkman, personal observations). Increases are attributed to the deterioration of the jackal-proof fences that were erected in the early 1900s (see Hoffman et al., this volume) but have been poorly maintained in the past decade since government subsidies fell away. In the Prince Albert magisterial district, for which Cape Nature Conservation had no sight records for black-backed jackal (Stuart, 1981), one farmer reported killing 12 black-backed jackal and 52 caracal *Felis caracal* on an 8000 ha farm over six years from 1985 to 1990 (C. P. Marincowitz, personal communication).

Leopard *Panthera pardis* populations have been exterminated in most parts of the karoo. The total population of leopard in South Africa is estimated to be only 2000–3000 animals (Norton, 1990), and in the karoo they are now restricted to mountains on the southern and north-western margins of the region (Norton, 1986). Despite protective legislation, hunting with gin traps, poison and dogs destroys about 28 leopards annually in the karoo (Esterhuizen and Norton, 1985). Population growth is further limited by slow maturation (3–4 years), low rates of increase (2 young/female/year) and territoriality (Skinner and Smithers, 1990). Caracal populations, however, are considered to be increasing in some parts of the karoo (Novellie, 1988). Caracal produce larger litters (2–4 young)

than leopard, but are also solitary. Their increase is thought to be a response to reduction in densities of competing sympatric carnivores or to habitat changes (aridification) brought about by farming practices (Grobler, 1987).

A host of smaller, non-target mammals, birds and reptiles are killed by 'problem animal control' activities (Allan, 1990; Kok, 1996) that include poisoning, the use of coyote-getters, trap-guns, baited gin traps, fall-traps and hunters accompanied by packs of dogs. The impact of hunting mortality on non-target populations, and on those of their prey, has not been quantified.

11.6.   Conclusions

This review has shown repeatedly that rainfall quantity and seasonality have a marked influence on plant and animal populations in the karoo. Rainfall cyclicity over the karoo is probably driven by temperature oscillations (El Niño effect) in the southern oceans (Tyson, 1987; Mason, 1990). Whatever the cause, rainfall patterns influence the grassiness of karoo vegetation, and consequently the abundance of locusts, movements of nomadic birds and the ratios of grazing to browsing herbivores (Novellie, 1988). Periodic droughts reduce forage quality in the short term and cause large-scale mortality in grass and shrub populations, leading to migrations or crashes in populations of springbok and other herbivores. The soil nutrient reserves available to plants during drought-breaking rains are rapidly exploited by ephemeral plants and resprouting perennials (Whitford, this volume). It has been suggested that this pulse of productivity frequently results in insect outbreaks and rapid increases in populations of small mammals as well as immigration of birds and larger mammals (Owen-Smith, 1988).

Rainfall-driven fluctuations in resource availability cannot explain all the mortality and recruitment events documented in karoo biota. The uneven structure of many perennial plant populations is indicative of episodic recruitment that might be related to periods of either unseasonal or above-average rainfall. Reduced competition following mortality caused by hail, frost, fire or local disturbances by humans and other animals may also facilitate recruitment. Replacement of indigenous herbivores by domestic livestock, as well as the introduction of alien plants, modification of drainage and the lowering of the water table, are likely to affect the frequency and size of recruitment events for long-lived plant species.

Apart from rainfall-driven fluctuations in food quality, major determinants of population trends in animals include infectious diseases and human alteration of habi-

tats and communities. Many large herbivore and carnivore populations in the karoo are directly controlled by
humans through fencing, hunting and breeding programmes, and such manipulation of the upper levels of
the food chain are likely to affect population processes
lower in the hierarchy. Addition of crops, drinking water,
alien trees and man-made structures have increased the
ranges of baboons and some birds, while loss of riverine
habitat through ploughing, erosion and construction of
impoundments has led to the virtual extinction of the
riverine rabbit.

Despite the need to predict and manage trends in karoo
plant and animal populations on ranches and in protected
areas, we conclude that the current understanding of population dynamics, of even the most common karoo plants,
insects and mammals, is inadequate for this purpose. In
addition to long-term population monitoring, a predictive
understanding of karoo populations will require experimental manipulation of predation, disease, competition,
mutualism and habitat quality.

12 Community patterns and dynamics

A. R. Palmer, P. A. Novellie and J. W. Lloyd

12.1. Introduction

Patterns in the distribution of the biotic components are a function of the interaction between the biota and environmental variables. The vegetation and its associated biota co-evolved under common environmental conditions. These units are recognizable at different scales and will be discussed at the levels of biomes, landscapes and patches. Using geographic information systems and their associated spatial modelling capabilities, it has become possible to describe the environmental conditions associated with larger units in the landscape and to discriminate between units on the basis of the environmental correlates. The landscape may be viewed as units of differing size, and it is vital that scale be defined before discussing pattern. Early researchers were aware of the role that the environment plays in determining the spatial arrangement of the biotic components, and their results can be integrated into the interpretation of pattern in the landscape.

Following a strongly descriptive bias in research design, early interpretations (Marloth, 1908; Bews, 1916; Cannon, 1924) emphasize the diversity of growth forms, but do not relate growth form to abiotic conditions, or interpret patterns in the biome. Rutherford and Westfall (1986) divide the arid and semi-arid region of southern Africa into three biomes. The Desert biome is discriminated from the succulent and Nama-karoo biomes on the basis of the former's high summer aridity and low mean annual rainfall (<70 mm). In defining the boundaries between the Nama-karoo and the succulent karoo, Palmer and Hoffman (1997) attribute the differences in growth form characteristics of the biomes to a lower co-efficient of variation in annual rainfall, a stronger winter seasonality and a smaller temperature variation in the succulent karoo.

At the landscape scale, a structural approach is evident, with authors defining 'desert shrub' (Pole-Evans, 1936), 'karroid scrub' (Dyer, 1937), 'arid bush' (Adamson, 1938), 'dwarf and succulent shrub' (Edwards, 1970) and 'dwarf shrub steppe formation' (Martin and Noel, 1960). In the most intensive study of the landscapes in the biome, Acocks (1953) used elevation and rainfall, in combination with associated floristic and structural information, to discriminate between 'veld types' (vegetation units differing in structure and agricultural potential), 21 of which occur in the karoo biome *sensu lato*. Acocks attributed the floristic composition and structure of nine of the veld types to the impact of herbivory by sheep and cattle in the post-colonial period. This notion was never fully tested at landscape scale, and was based primarily upon the impact of herbivory on paddock-scale experiments. The validity of interpolating paddock-scale results to the landscape has been questioned (Stafford-Smith, 1996). The interpretation of pattern as a function of recent herbivory by domestic stock was further perpetuated after the Drought Commission Report (Anon, 1923) by, amongst others, Tidmarsh (1936, 1948), Werger (1980), Vorster (1985) and Roux and Theron (1987). In an effort to establish realistic expectations of community distribution along the topo-moisture gradients in the biome, Palmer (1991a) and Palmer and Cowling (1994) modelled the distributions using GIS. Following further analysis of data collected by Werger (1980), Palmer and Cowling (1994) suggest that the presence or absence of taxa along the gradient is a function of the climatic environment. Differences in species proportions in topo-moisture classes were not investigated. The contemporary vegetation types (*sensu stricto*, see Werger, 1980) correlate well with those described and mapped by Acocks (1953). This result indicates that changes pre-date 1953, or that little change in species presence/absence has occurred at the landscape level. Even after applying a modelling approach, the absence of historical vegetation data or sites of known disturbance history make it difficult to evaluate the nature and extent of vegetation change within the karoo biome.

12.2. Vegetation patterns: landscape scale

12.2.1. Topo-moisture gradients, rainfall seasonality and uncertainty

Studies of vegetation pattern at the landscape scale in the biome have focussed on defining units using environmental variables such as elevation, rainfall and geology (e.g. Acocks, 1953; Vorster, 1982; Hoffman, 1996; Palmer and Hoffman, 1997). These studies attempt to relate pattern to macro-climatic gradients in the biome. The most objective basis for this approach has been provided by Dent et al. (1989), who recognized 87 homogeneous climate zones in the South African portion of the biome. These climate zones, based primarily on a cross classification of rainfall and elevation response surfaces, provide the basis for defining units which have the same potential or expected vegetation (Palmer, 1991a). Elevation, a surrogate for a range of environmental variables (Austin et al., 1983), is an important determinant of pattern within the biome. Palmer and Hoffman (1997) and Hoffman (1996) divide the karoo biome into three subregions using elevation and associated macro-climatic variables.

The largest region is Griqualand-West and Bushmanland, with elevation ranging from 550 to 1300 m asl; rainfall varies from 60–200 mm; and the co-efficient of variation is high (75%). July minimum temperatures are moderate (>2 °C) with high January maxima (>35 °C). The region contains patches of transitional succulent dwarf shrublands (Lloyd, 1989a), where the annual rainfall CV is 36–38%.

The arid shrublands or Orange River Nama-karoo (Hoffman, 1996) are synonymous with Orange River Broken Veld (Acocks, 1953) and parts of the Namaqualand Broken Veld where it occurs along the Orange River. They occur at moderate elevation (800–1100 m) on soils derived from ancient basement granites and gneisses of the Namaqualand complex. Dominant taxa are *Rhigozum trichotomum*, *Zygophyllum suffruticosum* and *Acacia mellifera*. Where the terrain becomes hilly, and elevation exceeds 1100 m, *Aloe dichotoma* and *Euphorbia avasmontana* occur. Annual rainfall varies from 150 to 200 mm. The CV of annual rainfall is high (50–60%) and January maximum temperatures exceed 34 °C.

At the eastern and south-eastern edge of the region, an arid grassy dwarf shrubland occurs. This is synonymous with the Stipagrostion (Werger, 1980) and parts of Arid Karoo (Acocks, 1953). Elevation range is small (1200–1300 m), annual rainfall is low (150–200 mm), rainfall variability is relatively low (43–45%) and mean January maximum temperature is 31 °C. Communities include the grassy dwarf shrubland dominated by caespitose grasses *Stipagrostis ciliata* and *S. obtusa*, on soils derived from Jurassic-age dolerites, and the succulent dwarf shrubland dominated by *Ruschia spinosa*, occurring mainly on soils derived from mudstone and sandstone. Other shrubs include *Pentzia incana*, *Lycium prunus-spinosa*, *Felicia* spp., *Pteronia* spp. and *Eriocephalus* spp. These grassy dwarf shrubland communities are transitional to the dwarf shrublands of the upper karoo.

The extensive spiny caespitose *Stipagrostis* spp. grasslands of Bushmanland are unique. They have strong affinity with the desert grasslands of the southern Namib (Yeaton, 1988). Where dolerite or sandstone are overlain by Kalahari sand, a community dominated by *Stipagrostis obtusa*, *S. ciliata* and *Enneapogon desvauxii* occurs. The arid grasslands are described more fully by Werger (1980).

The second region, the Great Karoo and central lower karoo, extends in elevation from 550 to 1500 m and temperature varies accordingly. Winters are cold with regular frost, and summers are moderately hot. Rainfall is low (60–200 mm), but is more certain (CV <50%) than sites north of the Great Escarpment. The rain-bearing south-westerly winds are influenced by the Swartberg mountain range to the south of the Great Karoo and then the Nuweveldberge to the north. In the rain-shadow of the Swartberg, the rainfall is extremely low (c. 175 mm yr^{-1}) and uncertain. At the foot of the Nuweveldberge, precipitation increases as the rain-bearing clouds release their moisture. Larger woody shrubs are supported and dwarf shrubland is replaced by grassland. The region comprises two major topo-moisture classes: the western plain, or Great Karoo and the eastern plain or Central Lower Karoo (Acocks, 1953). Both are dominated by dwarf shrublands with grass increasing in abundance towards the east.

In the third region, the upper karoo and eastern Cape Midlands, elevation is from 900–1300 m, annual rainfall from 200–400 mm, and co-efficient of variation in annual rainfall from 40–50%. July minimum temperatures are <2 °C and January maximum temperatures range from 30–35 °C. The structure and floristic composition of the vegetation of this sub-biome has been surveyed at regional (Acocks, 1953; Werger, 1973, 1980; Vorster, 1985; Palmer, 1991b) and local scale (Jooste, 1980; Van Der Walt, 1980; Werger, 1973; Palmer 1988, 1989). Three formations have been recognized on the basis of broad structural and floristic attributes.

The arid montane grasslands occur on the steeply sloping landforms. Rainfall varies from 350 to 400 mm, with a unimodal peak in late summer. Geology varies from sandstone to a mixture of blue and grey mudstone, with dolerite rocks. This formation is synonymous with karroid *Merxmuellera* mountain veld (Acocks, 1953) and *Merxmuellera disticha* is the dominant species of these high-elevation grasslands. Other differential species include *Cymbopogon plurinodis*, *Felicia filifolia*, *Pentzia globosa*, *Elytropappus rhinocerotis* and *Nenax microphylla*. Species found towards the mesic end of the gradient are *Themeda triandra*, *Tetrachne dregei*, *Passerina montana* and *Melica decumbens*.

Dwarf shrublands occur on the gently sloping, arid pediments between the mountain ranges. Structureless to weakly structured soils are a feature of these pedologically young landscapes. Soils have generally developed *in situ* from colluvium, with lime present in the entire landscape, and the Mispah and Glenrosa forms predominate (Ellis and Lambrechts, 1986). Low, irregular precipitation patterns, and the increasing aridity from east to west contribute towards the mosaic of communities.

In Namaqualand, Van Blerk (1987) (using unpublished data from J. P. H. Acocks), found that there were topographic and geological gradients reflected in the plant communities. The topographic gradients changed from mountainous areas and rocky koppies with sandy slopes to sand plains and sand dunes.

12.2.2.   Soil types, depth, stoniness, texture and chemistry

Of the communities described from Bushmanland (Lloyd, 1989a), several are associated with variations in geology and soil. Physiognomy varies from dwarf semi-open shrublands on saline and calcareous soils, to low open grasslands on deeper, strongly acid aeolian sands. Two communities (*Aptosimum procumbens* var. *procumbens*–*Salsola tuberculata* dwarf/low open shrubland and *Aptosimum procumbens* var. *procumbens*–*Ruschia muricata* low semi-open/moderately closed shrubland) are associated with calcrete and calcareous soils. One community (*Aptosimum procumbens* var. *procumbens*–*Brownanthus ciliatus* subsp. *ciliatus*–*Ruschia levynsiae* dwarf semi-open shrubland) is associated with shallow saline soils, and another (*Eberlanzia armata* low open/semi-open shrubland) with shallow soils overlying granite-gneiss basement or dorbank. The *Stipagrostis ciliata* var. *capensis* low open grassland is associated with deeper, strongly acid aeolian sand. Exchangeable sodium, exchangeable potassium, soil pH, exchangeable calcium, exchangeable magnesium and soil depth were, in order of importance, the best discriminating variables between these communities.

Although community-level descriptions of the Great Karoo and the Central Lower Karoo are rare (Acocks, 1953), three formations are described by Rubin and Palmer (1996):

- Situated along the edge of the escarpment are the grassy shrublands on rocky surfaces. Grasses include *Aristida diffusa*, *Digitaria eriantha*, *Heteropogon contortus* and *Themeda triandra*. Bush clumps are dominated by *Rhus burchellii*, with *Grewia robusta*, *Maytenus polyacantha*, *Carissa haematocarpa* and *Acacia karroo*. The substratum consists of dolerite boulders overlying sandstone. Eroded surfaces are rare, and soil between the rocks is fertile with high Al, Mg and organic carbon levels. The environment is mesic due to the precipitation deposited by the south westerly fronts. Elevation varies from 900 to 1250 m asl. There is a steep aridity gradient from the edge of the escarpment to the pediments below.

- The dwarf shrublands of the Great Karoo and Central Lower Karoo comprise the *Rhigozum obovatum*-dominated communities and a *Ruschia spinosa* complex. The drainage lines are occupied by distinctive woody communities. Regular flooding events ensure that drainage lines experience a high disturbance hydrological regime. Taxa from many non-riparian communities are represented, with woody species being most successful and obvious. These are dominated by the spinescent taxa *Acacia karroo*, *Lycium oxycarpum* and *Maytenus polyacantha*, but may include *Grewia robusta*, *Diospyros lycioides* and *Rhus lancea*. The stoloniferous grass *Cynodon incompletus* and perennial bunch grasses such as *Cenchrus ciliaris*, *Fingerhuthia africanus*, *Stipagrostis namaquensis* and *Hyparrhenia hirta* occur in scattered, well established clumps. Cover varies both spatially and temporally, as moisture levels fluctuate, and the deep, sandy alluvium provides an ideal germination environment for many short-lived taxa such as *Atriplex lindleyi*, *Delosperma* sp. and *Atriplex* spp. This community dissects the landscape and appears uncoupled from the controlling influence of the topo-moisture gradients of the region.

- Le Roux (1984) distinguished three vegetation communities in the Namaqualand Broken Veld that are associated with rocky hills, three communities in the flatter, sandier strandveld of the West coast that have less or no rock cover, and one community in the Arid Karoo that is flatter, sandier and has a greater proportion of summer rainfall. The *Zygophyllum morgsana* – *Eriocephalus ericoides* communities of the Namaqualand Broken Veld are further sub-divided into the *Eriocephalus ericoides*–*Elytropappus rhinocerotis* community, the *Eriocephalus ericoides*–*Montinia caryophyllacea* community and the *Eriocephalus ericoides*–*Pentzia incana* community. The *Zygophyllum morgsana*–*Othonna cylindrica* communities of the

Strandveld of the West coast are further sub-divided into the *Othonna cylindrica–Hermannia multiflora* community, the *Othonna cylindrica–Tetragonia fruticosa–Chrysanthemoides monilifera* community and the *Othonna cylindrica–Tetragonia fruticosa–Stipagrostis ciliata* community. The *Eriocephalus africanus–Stipagrostis brevifolia* community is that of the Arid Karoo.

The vegetation types of the Namaqualand coastal plain reflect differences in substrate (Boucher and le Roux, 1993). These are strandveld, succulent karoo and sand plain fynbos. The strandveld vegetation is associated with calcareous sands, with the shorter forms of strandveld vegetation associated with exposed calcretes and coastal granitic or shale rocks. Taller vegetation is associated with deeper sands. Short strandveld occurs over shallow soils with little soil moisture and contains a considerable succulent element. The succulent karoo vegetation occurs on local patches of quartz-rich granite, exposed silcretes and compacted saline soils, and the sand plain fynbos is associated with leached acidic sand that has a lower nutrient status than the other sands in the region.

12.3. Vegetation patterns: local scale

12.3.1. Nutrient enrichment patterns

In African ecosystems there is a general interrelationship between precipitation, vegetation biomass and nutritional value which is of fundamental importance in determining the use of the landscape by herbivores. Rainfall tends to be positively correlated with vegetation biomass, but negatively correlated with soil fertility and forage nutritional value. This is true of the karoo and surrounding savanna and grassland biomes (Du Toit et al., 1940) as it is elsewhere in Africa (Bell, 1982b; Ruess and Seagle, 1994). In mesic regions, it is particularly during the dry season that nutritional value of forage tends to be low, contrasting with the high dry season value of forage in arid regions. In South Africa, the former type has long been known as sourveld and the latter as sweetveld (Scott, 1959). Climate is, however, not the exclusive determinant of sweetness or sourness. Parent material that gives rise to nutrient-rich soil tends to be sweetveld in regions where one would have expected sourveld on the basis of climate alone. Ellery et al. (1995) conclude that sourness is related to the degree to which the environment promotes carbon assimilation relative to nutrient mineralization and assimilation, an explanation that is consistent with the resource availability hypothesis (Bryant et al., 1983). Cool, wet conditions favour carbon assimilation, while hot, dry conditions favour nitrogen assimilation.

These relationships are clearly evident on a landscape scale in the eastern karoo. Overall conditions favour sweetveld in most habitats, but on cooler mountain tops and south-facing slopes tend to support grasses such as *Merxmuellera disticha* or *Elionurus muticus* that are characteristic of sourveld (Ellery et al., 1995). On a general landscape and regional scale, the 'sweet/sour' dichotomy often results in herbivores being presented with a 'choice' between sparse vegetation of high quality, on the one hand, and more abundant vegetation of poorer quality, on the other. The influence of this dichotomy on animal dynamics is described elsewhere (12.5.1).

Within each of the regions, resources are non-randomly distributed in the landscape. The riparian zone bisects the landscape, providing refugia for many species (Milton, 1990). Geochemical features provide further patchiness; for example the sandstone plateaus in the central lower karoo which protect underlying, nutrient-rich (phosphate) mudstone layers. Geochemical patches may be important in regulating photosynthetic patterns and possibly herbivory (Palmer and Van der Heyden, 1997) (Fig. 12.1).

The distribution patterns of plants and animals are influenced by geology, soil colour, soil particle size and soil nutrient status (Dean, 1995) with infiltrability and

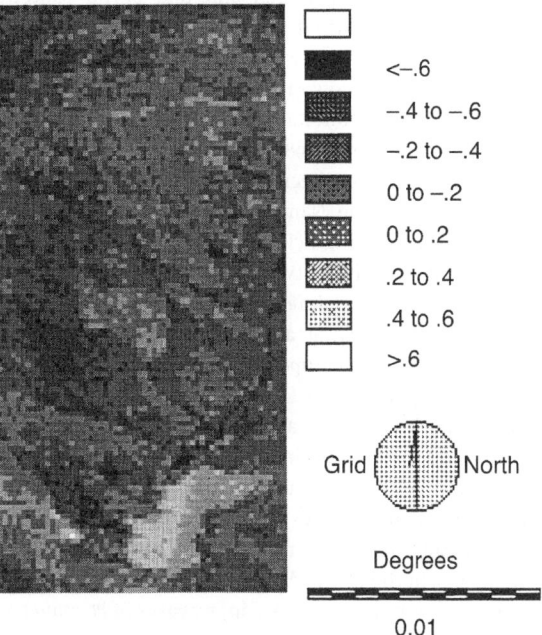

Figure 12.1 **A normalized difference vegetation index (NDVI) derived from a SPOT HRV image (25 December 1989) of the Karoo National Park, Beaufort West, showing an area of high active greenness on a raised geochemical mound. The site has low basal cover in comparison to adjacent rangeland and is dominated by actively growing dwarf shrubs (*Rhigozum obovatum* and *Eriocepahlus ericoides*)**

water retention capacity being influenced by soil texture, soil depth, slope (Wilcox et al., 1988) and plant cover (Snyman and Fouché, 1991). A combination of sandy soil and moderate rainfall results in a high subsoil : topsoil moisture ratio that favours woody species (Walker and Noy-Meir, 1982). Soils of washes and cracked soils have the highest absorption rates, followed by the soils of nest-mounds of the harvester ant *Messor capensis* (Dean, 1992b). Soils rich in organic matter, such as those of ant nest-mounds or *heuweltjies* (low mounds of nitrogen-rich soil, 5–15 m in diameter and about 1 m in height – see below, and section 12.5.2., *Termites and ants*), markedly increase absorption ability of the soil, although the water repellency in some of the *heuweltjie* soils may be related to the amount of termite frass present in the soil. Differences in absorption rates and infiltration on *heuweltjies* may be related to salinity or the presence of algal crusts which are thought to impair infiltration. Emergence holes of adult cicadas (Homoptera: Cicadidae) were also found to increase the absorption rate and infiltration of water into karoo soils after rains, particularly in the hard, bare inter-shrub soils (Dean, 1992b).

Soil nutrient status affects vegetation patterns. The soils are generally well supplied with most of the important plant nutrients, with the exception of the soils along the west coast which are nutrient-poor. Soils in the pans of Bushmanland have very high zinc and boron contents (Ellis, 1988). The nutrient status depends largely on the geological material from which it is derived, for example low nutrient status, low nitrogen, low phosphate in soils derived from sandstones and quartzites (Wild, 1978; Du Plessis and Shainberg, 1985; see Watkeys, this volume). Broad associations between dominant plant species and 10 soil forms in the Nama-karoo were determined by Vorster (1986), although climatic influence on these patterns cannot be excluded (Vorster, 1985). For example, the woody dwarf shrubs *Galenia fruticosa*, *G. procumbens*, *Kochia pubescens*, *Salsola aphylla* and *S. glabrescens* and grasses *Panicum maximum* and *Stipagrostis namaquensis* are associated with deep soils of the Dundee form.

High species turnover along environmental gradients is recorded for succulent karoo communities (Cowling et al., 1994; Jürgens, 1986). Turnover is often abrupt, resulting in narrow boundaries between communities (Lloyd, 1989a, 1989b; Smitheman and Perry, 1990; Yeaton and Esler, 1990). In the Prince Albert area, on the Nama/succulent karoo boundary, there are many cases of *Pteronia* spp. replacing each other over apparent edaphic discontinuities (Esler and Cowling, 1993). Edaphic variables account for much of the variation in the species data. Species on the colluvial plain are separated from those on tillite slopes along a soil texture gradient. A separation of species in a second axis indicates that soil phosphorus, pH, mag-

nesium and calcium levels are also important in determining species distribution patterns (Esler and Cowling, 1993). *Pteronia* distribution patterns in the succulent karoo result from the combined effect of soil physico-chemical factors and competition. If edaphic specialization alone was responsible for the distribution patterns of *Pteronia* spp., one would expect there to be reduced competition at the boundaries of species overlap (since each species would be physiologically adapted to exploit resources differentially, thus minimizing niche overlap). This was not the case, and nearest-neighbour analysis indicated that there were strong inter- and intraspecific competitive interactions at *Pteronia* community boundaries (Esler and Cowling, 1993).

Evidence of localized zoogenic effects in the Nama-karoo biome is provided by the so-called *heuweltjies* or mima-like mounds found largely in the succulent karoo biome (Lovegrove, 1991, 1993; Midgley and Musil, 1990). These are regularly spaced round or oval patches of vegetation which are conspicuously different from the surrounding area. Regions in which *heuweltjies* are abundant have a characteristic pock-marked appearance (Lovegrove, 1993).

There is little doubt that the *heuweltjies* owe their origin to termitaria (Moore and Picker, 1991), those south of the Orange River being constructed mainly by *Microhodotermes viator* (Coaton, 1958). Termite activity results in a raised mound of finer textured, alkaline soil with increased water content and higher water-holding capacity. These soil changes favour a different plant community, increasing between patch diversity and providing habitat for other animal species. The characteristically uniform spacing of the mounds is probably due to competition between different termite colonies for resources.

The raised *heuweltjies* are less prone to flooding in the wet season. This favours burrowing animals, for example common molerat (*Cryptomys hottentotus*) which use the mounds to store bulbs or corms (Lovegrove, 1993). When the mounds dry out in spring after the winter period of waterlogging, the mole rats burrow outwards, thus contributing to mound development by pushing soil towards their nests.

Numerous other species may contribute to the nutrient enrichment of *heuweltjies*: they are frequently used as dung middens by aardvark (*Orycteropus afer*) and Steenbok (*Raphicerus campestris*), as colony sites by the herbivorous rodent *Parotomys brantsii* and as preferred grazing sites by sheep (Armstrong and Siegfried, 1990; Milton and Dean, 1990a).

Nest-mounds of other termite species, as well as burrow systems of some mammals, also exert an influence on the patch selective grazing which is characteristic of both domestic and indigenous ungulates. De Jager and Joubert (1968) report that sheep, introduced into a camp, tend to focus their grazing around the margins of termitaria. These localities act as nuclei from which patch selective grazing develops, until finally the camp becomes a mosaic of grazed and ungrazed patches. The tendency to graze the margins of termitaria is also characteristic of blesbok (*Damaliscus dorcas phillipsi*), with results similar to those described for the sheep (Du Plessis, 1972). Burrowing mammals such as gerbils (*Desmodillus* and *Gerbillurus* spp.) and ground squirrels (*Xerus inauris*) also contribute to plant species diversity in arid regions through alteration of the substrate (Louw and Seely, 1982), and such sites occasionally become the focus of patch selective grazing by wild ungulates (P. Novellie, personal observations).

Apart from compounding the effects of termites or burrowing mammals, indigenous ungulates can independently exert a considerable influence on nutrient enrichment of the soil and on vegetation transformation on a landscape scale (Du Plessis, 1972; McNaughton, 1983b, 1988; Ruess and Seagle, 1994). In the Serengeti grasslands, localized concentrations of ungulates can be explained in terms of availability of soil nutrients resulting from edaphic variations (McNaughton, 1988). The ungulate concentrations and resulting accumulations of excreta strongly influence soil microbial activity and accelerate the nutrient cycle by short-circuiting the more lengthy process of litter production and decomposition (Ruess and Seagle, 1994).

Soil enrichment is associated with a change in plant species composition, which often takes the form of replacement of tall tufted grasses by prostrate creeping species (McNaughton, 1979; Stuart-Hill and Mentis, 1982; Coughenour et al., 1985). Those plant species that take over under intense localized grazing are adapted to conditions of high nutrient availability (called 'greedy feeders' by Roux, 1969). Nutrient enrichment through excreta enables these plants to effectively increase the availability of nutrients to the herbivores and this can contribute to improved animal performance (Roux, 1969; Day and Detling, 1990). Patch selective grazing constitutes a positive feedback system resulting in the creation of grazing lawns that are consistently favoured by the herbivores. This process is viewed as an important agent of plant community diversity in Serengeti grasslands (McNaughton, 1983b).

A take-over by greedy feeders (*Cynodon* spp. and *Eragrostis* spp.) can be achieved entirely by fertilization of so-called climax veld, in the absence of defoliation by large herbivores (Roux, 1969), which shows that the process is primarily driven by nutrient availability, rather than selective defoliation.

A number of indigenous ungulate species focus their grazing on specific patches: blesbok, black wildebeest (*Connochaetes gnou*) and springbok (*Antidorcas marsupialis*) (Von Richter, 1971; Du Plessis, 1972; Mentis and Duke, 1976; Mentis, 1980) and habitat they occupy becomes a mosaic of grazed and ungrazed patches. The males of these species occupy small territories or leks during the rutting season (Von Richter, 1971; Lynch, 1974), and localized concentrations of soil nutrients in such territories are likely to play an important role in attracting females to graze there (McNaughton, 1988). The territorial males tend to defaecate at specific localities, so creating conspicuous dung patches. These patches are normally surrounded by lawns of creeping grass (*Cynodon* spp.) which is particularly associated with fertilized soils (Roux, 1969). Milton et al. (1992b) found that mineral nutrient levels in a soil sample taken from beneath a springbok dung patch were three to six times higher than that of surrounding soil, and that the carbon and nitrogen content of the soil was respectively 6 and 14 times that of adjacent soil.

The dung patches have been interpreted as olfactory and visual territorial markers (Lynch, 1974). It is not improbable, however, that the adaptive value of this behaviour is to assist with the cultivation of nutrient-rich grazing lawns. The fact that individual males may remain attached to the same territory in several successive rutting seasons (Estes, 1969; David, 1975) implies that their self-maintained grazing lawns may reap rewards in the form of enhanced mating opportunities.

Blesbok and black wildebeest occurred in the eastern and northern parts of the karoo in historical times (Skead, 1987) as they do in protected areas today. Their impact on the habitat, the creation of grazing lawns of *Cynodon* spp. and the typical mosaic of grazed and neglected patches is evident in habitat occupied by these species in the eastern Cape midlands (Novellie, 1990a, 1991). It is likely that, before the advent of European farmers, the wild ungulates, together with termites and burrowing mammals, were as important agents of patch diversity as they are in the Serengeti grasslands (McNaughton, 1983b).

There is evidence to suggest that early man may have occupied sites which have become degraded (Sampson, 1986; Smith, this volume). Evidence of the impact of contemporary human disturbance is profound, with roads, windpumps, stock watering troughs and dwellings, all providing foci for change. Enrichment, salinization and physical compaction (Roux and Opperman, 1986) of substrate occurs around all watering points and stock-handling facilities (including human dwellings). Species dominance shifts favour plant species with well-developed anti-herbivory mechanisms. Infra-red gas analysis reveals high CO_2 assimilation rates at these sites. In the interdune troughs of Griqualand West, *Rhigozum trichotomum* attains single dominance status around homesteads and surface-water points, and these piospheres are clearly visible on satellite images (Fig. 12.2). Fence contrasts between

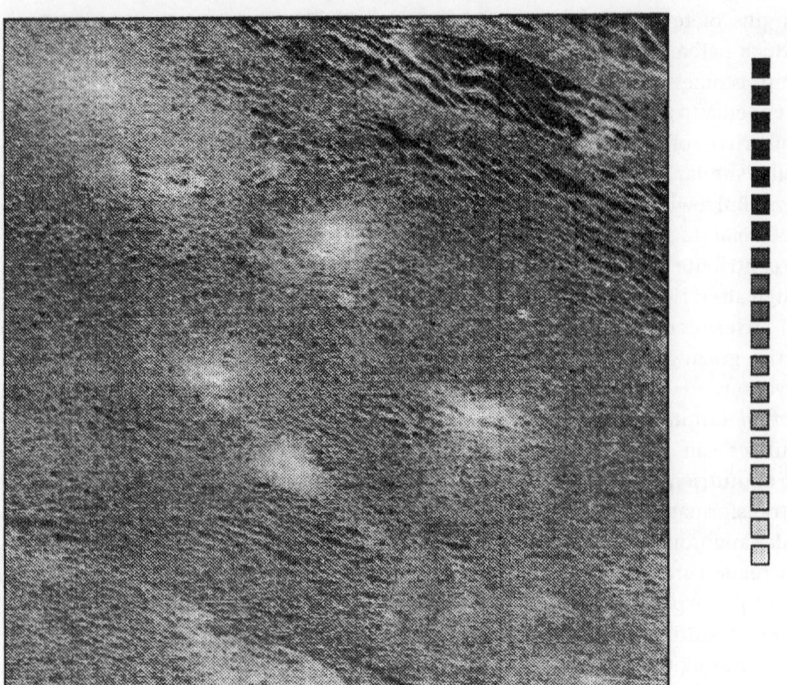

Figure 12.2 **An NDVI of disturbed (over-grazed) patches around water points in the southern Kalahari. Patches in the dune troughs are dominated by *Rhigozum trichotomum*, with the dune vegetation having reduced grass cover and containing clumps of woody shrubs (*Acacia haematoxylon, Acacia mellifera, Boscia albitrunca*). Lighter tones indicate higher NDVI (actively growing plants)**

SCALE: 1 : 140000

26:15S

REGION: 20:05E   20:15E

26:25S

different farms and camps further reveals the impact of contemporary human disturbance.

12.3.2   Nutrient transport patterns

With the exception of descriptive work on the distribution of soil types within the biome (Ellis and Lambrechts, 1986; Ellis, 1987), little is known about nutrient transport and movement. In a brief analysis of landscape-scale soil patterns, Palmer (1991c), suggests that nutrients may be transported from the dolerites of the escarpment to the pediments. In semi-arid rangelands in Australia, Tongway and Hindley (1995) report that landscape degradation is accompanied by a 'leaking' of nutrients from enriched patches into the surrounding landscape. By implication, a landscape comprising many well-defined patches should be more productive. Few studies have addressed the apparent patchiness of nutrients in the karoo (Walters, 1951; Milton et al., 1992b) and an evaluation of this, following Tongway and Hindley (1995), would prove extremely valuable. Patches are created (Martens et al., 1990) and maintained (Roux, 1992) by domestic herbivores at the paddock scale, with sheep spending a higher proportion of their time in the modified patch. The creation of patches is not generally adequately tested, and herbivory has been cited as the cause of rangeland degradation (Acocks, 1953; Hurt

and Bosch, 1991). These studies fail to recognize the importance of nutrient enrichment and salinization in the process. These events, all consequences of the greater time spent by the herbivore near the water point or in a particularly grazing treatment, are seldom, if ever, cited as the reason for changes in vegetation state.

12.3.3.   Photosynthetic patterns

Remote sensing and the preparation of vegetation indices (e.g. normalized difference vegetation index or NDVI) provides the opportunity to evaluate landscape patchiness, as sites of high moisture and nutrient status support more actively growing plants. These actively growing plants reflect more light in the NIR (near-infra-red) region and provide an index of the patches or fertile islands in the system. During a study of photosynthetic patterns in the Karoo National Park, Palmer and Van der Heyden (1997) explored the assimilation rates of three taxa at two sites. Two taxa (*Eriocephalus ericoides, Rhigozum obovatum*) showed higher CO_2 assimilation rates when growing on a raised geochemical mound, than plants of the same species off the mound. They attribute these differences to elevated P levels in the rocks below the mound. The stronger NDVI (Fig. 12.1) signal could be attributed largely to the higher woody biomass and photosynthetic activity of *Rhigozum obovatum.* Following an analysis of Landsat TM images from the southern Kalahari, higher NDVIs are associated with the presence of clumps of actively growing shrubs or trees. Nodes of rangeland degradation (for example, stands of *Rhigozum trichotomum* and *Acacia mellifera* in the southern Kalahari) also exhibit high NDVIs (A. R. Palmer, unpublished data) and these results have been used to identify sites of desertification (Ringrose, 1987)

High surface albedo has been used to identify erosion sites using monochromatic aerial photography. These early remote-sensing methods were used extensively in farm planning and erosion sensitivity analysis. However, highly reflective sites in the visible spectrum often display extremely high NIR reflectance. On shallow substrata, these high values are a function of the actively growing dwarf shrubs that provide nutritious plant material to the selective grazer. On aeolian sand, this is a function of woody shrubs or dwarf shrubs which are able to utilize the elevated moisture stored in the sand. The saline pans, which have high albedos in the dry season, also display high NIR reflections, particularly after wet events.

The low NDVIs of upper- and middle-slopes in the Karoo National Park are an indication of the stress associated with these landscape units. Low NDVIs are also recorded in old, unburnt or senescing montane grasslands and montane shrublands (*Elytropappus rhinocerotis*-dominated communities). Satellite imagery of wet season states in the biome continues to provide enormous potential for understanding ecosystem function.

12.3.4.   Diversity patterns in the landscape

Although floristic diversity has been described for the succulent karoo (Cowling and Hilton-Taylor, 1994) and Nama-karoo (Cowling et al., 1994; Cowling and Hilton-Taylor, this volume), landscape diversity patterns have only recently been explored using floristic samples (quadrats) and satellite derived data (Fig. 12.3) (Tanser, 1997). These patterns suggest a weak relationship between floristic diversity and the NIR (band 3) pixel diversity derived from a SPOT HRV image of the Karoo National Park.

12.4   **Plant community dynamics**

12.4.1   Rainfall variability

Bond et al. (1994) examined C13/C12 ratios in the soil profile at sites across the biome. The results suggest a significant change from grass to shrubs in the False Upper karoo in recent times (±50 yr BP). Geographic analysis of this area reveals a higher co-efficient of variation in annual rainfall (Fig. 12.4) relative to other regions experiencing the same mean annual rainfall, suggesting that the False Upper Karoo experiences a less certain rainfall regime and is susceptible to shifts in dwarf shrub/grass dominance. In other dwarf shrublands, shifts between perennial grasses and dwarf shrubs in response to rainfall variation, are documented (O'Connor and Roux, 1995). The impact of high annual rainfall CV on vegetation response has been reported elsewhere in Africa (Ellis and Galvin, 1994; Le Houérou et al., 1993).

12.4.2   Fire

There are few studies on effects of fire on karoo vegetation. In the Mountain Zebra National Park, Novellie (1990b) shows that *Eriocephalus*, *Chrysocoma* and *Felicia* are killed by fire and gaps are filled by grasses (*Aristida*, *Melica*, *Eragrostis*) and forbs in much the same way as in post-drought succession. Fire may be used to maintain grassiness in the montane shrublands of the eastern Cape midlands (Roux and Smart, 1977, 1979); and to reduce the dominance of *Elytropappas rhinocerotis* and *Euryops* spp., but should be avoided in the dwarf shrublands as post-fire recovery is extremely slow.

12.4.3.   Grazing

Mechanisms for grazing-induced change
Investigation of the impact of herbivory on the vegetation by domestic animals has occurred at the paddock scale (Roux, 1992; O'Connor and Roux, 1995). Changes in plant species composition of paddocks is primarily driven by rainfall variation, with the influence of grazing treat-

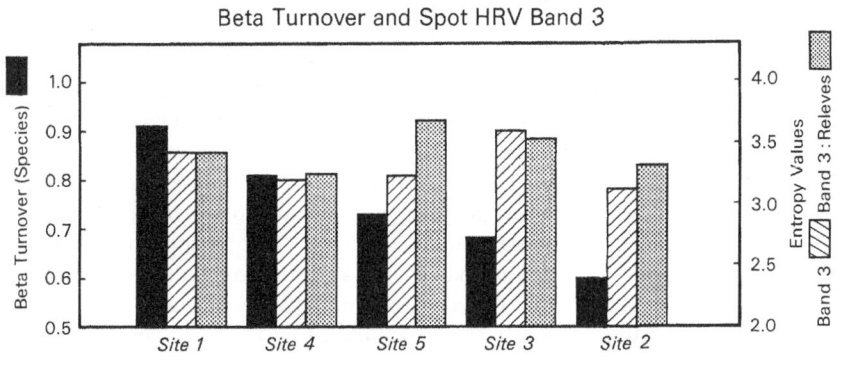

Figure 12.3 **The results of an analysis of diversity patterns at the landscape scale in the Karoo National Park. Diversity indices were applied to (i) total floristic data from 10 replicated plots; and (ii) pixel diversity in band 3 of a SPOT HRV image from 25 December 1989 (Tanser, 1997)**

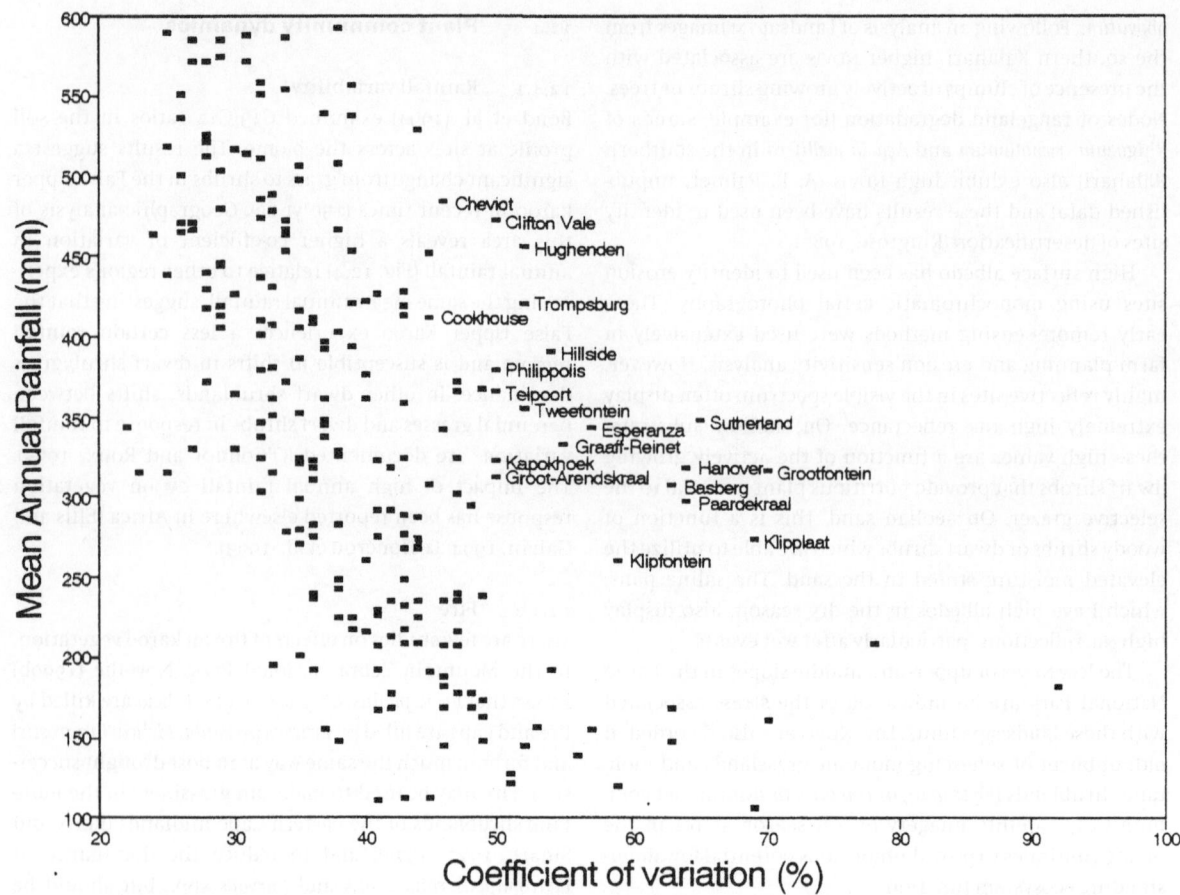

Figure 12.4 **Relationship between the co-efficient of variation in annual rainfall and median annual rainfall for stations with 40 years and more rainfall records. Many of the stations which are outliers to the general function are situated in the False Upper Karoo (Acocks, 1953) and these sites are subject to great interannual variation in grass/dwarf shrub cover**

ments on longer-lived plants becoming more important over a longer time period (O'Connor and Roux, 1995). Extrapolation of these results to the landscape, implying landscape-scale vegetation change, must be done with caution. Tidmarsh (1948) and Acocks (1953) suggest a trend towards the dominance of dwarf shrubs. It was further suggested that dwarf shrubs were expanding into the eastern grasslands at some 2.0–3.5 km per year (Jarman and Bosch, 1973), but these spatial patterns recognized from satellite-derived imagery (Landsat MSS) are inconclusive. Satellite-derived indices are a function of infra-red reflectance within a 800 m^2 area and are influenced by rainfall seasonality, phenology and even the presence of a single actively growing tree or shrub. The reflectances vary between seasons, and interpretation of change requires careful ground referencing which was not applied by Jarman and Bosch, (1973). Efforts to improve remote-sensing techniques to monitor change continue (Palmer

and Van Rooyen, 1995), but these must still be applied at the biome scale.

Land management policies at farm scale can affect the states towards which different landscape units shift (Milton and Hoffman, 1994). In the mesic regions of the biome, changes in management policy can reverse the degradation process. Some superficial evidence for this is found in the national road (N1) purchase area between Colesberg and Bloemfontein. Unlike narrower road reserves, where elevated run-off and nutrient enrichment influence species composition and cover, on the N1 there is a wide (>100 m) purchase area situated on formerly commercial rangeland. This area is now protected from herbivory and is mowed regularly to prevent accidental fires. The absence of herbivory, in conjuction with annual harvesting, has, over 15 yr, transformed the vegetation from the 'degraded' condition of adjacent grazed rangeland to a grassy sward dominated by *Themeda triandra* and other pyrophilic species. In the more arid parts of the karoo, post-disturbance recovery is less rapid, with old fields on alluvium in the succulent karoo taking >50 yr to recover despite the availability of species of all successional stages in the vicinity (Dean and Milton, 1995).

Further evidence of landscape-scale transformation is the success of *Rhigozum trichotomum* and *Acacia mellifera* in the arid savanna and in the Orange River valley. Both taxa appear to be limited by their fundamental niches and are not competitively successful outside of their ranges. Within their fundamental niche space, both taxa dominate a transformed landscape. The mechanism of transformation in the *Rhigozum trichotomum* scenario is thought to be a competitive interaction for available soil moisture between grasses and the shrub (Moore, 1989). When grasses are removed by herbivory, the shrubs are able to access the soil moisture and finally suppress grass production. The result is a change from a multispecies, productive landscape into a low-production single species or co-dominant community.

From the above narratives, it is clear that plant species performance in the karoo cannot be explained by a single model, and expectations of the performance of species at different locations in the biome need to be established. The capacity to develop niche models (Austin et al., 1983) has increased with the development of GIS techniques and these have been applied to *Portulacaria afra* (Gibson, 1995) with favourable results. Application of these techniques to establish realistic expectations for other taxa at the landscape scale is necessary.

The establishment and growth of vegetation, at a point in the landscape, occurs in response to the availability of seed, and three basic resources: water, nutrients and solar energy (Stafford-Smith and Pickup, 1993). The availability of these resources is influenced by a complexity of environmental factors: soil, topography, aspect, rainfall, fire, competition with other plants and other organisms. Grazing can act in combination with any of these factors in a wide variety of ways, to either augment or diminish their effects on the vegetation. Consequently, the role played by grazing as a determinant of vegetation change cannot be considered in isolation from other factors, and it is extremely difficult to make any wide generalizations. Nevertheless, there are a number of hypotheses for the mechanism of vegetation change induced by large herbivores that have been considered in the karoo and semi-arid grasslands of southern Africa.

Hypothesis 1. Vegetation change is caused by the selective removal of palatable plants or selective reduction of the reproductive potential of palatable plants, thereby allowing for increase in cover of less palatable species.

There is considerable evidence that prolonged herbivory can reduce the fitness of palatable plant species to the point where unpalatable species have a competitive advantage, and changes in species composition of the vegetation result (O'Connor, 1991). In the karoo, one of the most dramatic demonstrations of this can be seen at the Grootfontein Research Centre, Middelburg where a paddock subjected to summer grazing by sheep during a 30–year trial is now dominated by karroid shrubs, while paddocks subjected to winter grazing are dominated by perennial grasses (Roux, 1981). This presumably resulted either from the seasonal change in grazing preference of the sheep, which favour grasses in summer and dicotyledons in winter (Meyer, 1992; Du Toit and Blom, 1995), or from a differential response by grasses to summer (growing season) versus winter (dormant season) grazing (Tainton, 1981).

Relatively few studies have analysed the way in which grazing can influence the population processes of palatable plant species. In the southern karoo, Milton and Dean (1993), Milton (1994a) and Stokes (1994) report that grazing by sheep could promote unpalatable species by reducing seed abundance in palatable species. Although grazing did not affect the survival of seedlings or established plants during a three-year study, abundant recruitment of unpalatable species occurred on degraded rangeland, probably because of the effect of grazing in depressing the abundance of seed of palatable species (Milton, 1994a). Grazing by sheep thus determines the relative abundance of a new generation of seedlings by selectively reducing the sizes and reproductive success of preferred forage plants.

None of the larger, more long-lived, shrub species studied by Milton (1994a) in the southern karoo had soil stored seedbanks, increasing the probability of extinction of palatable shrubs if they are subjected to heavy grazing for prolonged periods. In contrast, short-lived ephemerals have seedbanks and can irrupt quickly after suitable rainfall in some seasons (Milton and Hoffman, 1994).

The population processes of plant species characteristic of the grassier eastern karoo are not well known (Milton and Hoffman, 1994), but perennial grass cover can decrease substantially under the combined influence of drought and grazing (O'Connor and Roux, 1995). Because perennial grasses show little intrinsic dormancy, and because grazing depresses seed production, these are also prone to extinction as a result of prolonged drought and severe grazing (O'Connor 1991). As in the southern karoo, irruptions of short-lived plant species may occur in the eastern karoo as a result of favourable rainfall, probably from soil stored seeds (O'Connor and Roux, 1995; Novellie and Bezuidenhout, 1994). Grazing may therefore result in the extinction of the perennial component and a transition to short-lived species with seeds that have the capacity for lengthy dormancy.

> **Hypothesis 2.**   Local enrichment of soil nutrient and base status by accumulations of urine and dung where large herbivores congregate changes plant communities by favouring nitrophilous or halophytic species and fast-growing more productive plants.

In mesic grasslands of South Africa, it has long been known that the changes in species composition brought about by grazing animals can, in some instances, be duplicated by applying fertilizer to the veld, even if grazing is totally absent (Roux, 1969). This takes the form of a replacement of perennial grass species characteristic of undisturbed veld by the nitrophilous grasses *Cynodon* spp. and *Eragrostis* spp. Such changes result from the fact that accumulations of dung and urine, as well as nitrogenous fertilizer, accelerate the nutrient-cycling rate by bypassing the more lengthy process of litter decomposition, resulting in conditions that favour nitrophilous plant species (Roux, 1969; Ruess and Seagle, 1994). This transformation can increase the rate of production of nutrients available to herbivores (Day and Detling, 1990) and can actually enhance animal performance (Roux, 1969).

It appears likely that nutrient enrichment plays an important role in the karoo, although experimental verification of this is lacking. *Cynodon* and *Eragrostis* spp are associated with antelope dung patches in the eastern karoo (Novellie, 1990a, 1991) and the fact that these patches carry high herbivore densities for extended periods (Novellie, 1991) suggests that nutrient enrichment may result in an enduring enhancement of grazing capacity. Nutrient enrichment also appears to play an important role as an agent of vegetation change in the nutrient-poor desert soils of the western parts of the karoo biome (Dean and Milton, 1991a). In arid grassy dunes, antelope shelter sites below trees are characterized by accumulations of antelope dung. This nutrient enrichment, together with trampling, evidently promotes high densities of annual plants at the expense of perennials (Dean and Milton, 1991a).

> **Hypothesis 3.**   Vegetation composition is influenced by changes in hydrology including accelerated runoff on fine soils that favours more drought-tolerant, xerophytic species.

As postulated by Roux and Theron (1987), thinning of the vegetation by grazing leads to increased runoff, higher soil temperatures, increased soil moisture evaporation and accelerated erosion. The general result is a decrease in the effectivity of rainfall, decreased soil moisture status, harshening of the microclimate, and an increase in short-lived xerophytic species. There is no doubt that, as plant cover declines, runoff and soil erosion increase (Snyman and Van Rensburg, 1986) and that regular trampling by ungulates can reduce soil absorption (Dean, 1992b). However, there is practically no information on the rate and magnitude of changes in the soil and the microclimate in the karoo biome (Roux and Theron, 1987).

> **Hypothesis 4.**   Interactions between grazing and other disturbances such as drought and fire lead to changes in vegetation

Drought and heavy grazing tend to have similar effects on the vegetation, but the former tends to dominate the latter in terms of magnitude of influence (Stokes, 1994; Novellie and Bezuidenhout, 1994; O'Connor and Roux, 1995). Directional trends resulting from heavy grazing can therefore only be properly determined if a study is continued for an extended period, probably about 40 years in the karoo (O'Connor and Roux, 1995). Post-drought rain in the karoo favours the rapid establishment of annuals or weak perennials (Novellie and Bezuidenhout, 1994; O'Connor and Roux, 1995; Milton, 1995a; Milton et al., 1995). These short-lived species tend to survive droughts in the form of seeds (O'Connor, 1991). In the eastern karoo, perennial grasses may be severely depressed during droughts, but show a strong tendency to recover during extended periods of favourable summer rainfall (Novellie and Bezuidenhout, 1994; O'Connor and Roux, 1995). With successive years of good rainfall, the increasing cover of perennial grasses may suppress the annuals that irrupted during the initial post-drought period (Novellie and Bezuidenhout, 1994). However, this suppression may not occur if persistent heavy grazing restricts the growth of perennials. The grazing pressure prevailing during and after drought periods may, in fact, be a critical factor determining vegetation change. Perennial grasses or palatable long-lived shrubs do not have dormant seeds, and may become locally extinct if disadvantaged by persistent heavy grazing (O'Connor, 1991; Milton and Hoffman, 1994). This can lead to a sudden shift in favour of unpalatable species.

Fire occurs regularly in the mountain veld of the central and eastern karoo, but its effects have been little studied. In other veld types, occasional fires may occur after successive years of good rainfall in combination with light grazing have resulted in an increased fuel load. Regrowth after fire tends to attract herbivores, and in karroid *Merxmuellera* Mountain Veld (Acocks, 1988) there is a marked post-fire increase in defoliation of both grasses and dicotyledons (Novellie, 1990b). Fire can eliminate certain karroid shrub species (*Felicia filifolia* and *Chrysocoma ciliata*), but others (*Pentzia globosa* and *Nenax microphylla*) may resprout vigorously after being burnt (Novellie, 1990b). It appears that fire can result in sudden transitions in karroid vegetation, generally in the form of a loss of part of the dwarf shrub component. However, fire is likely to play a role only in situations where grazing has been low and rainfall high for extended periods, currently an unlikely combination of circumstances except in mountain veld.

Implications of grazing-induced change

The dynamics of karoo vegetation in response to herbivory has been given scant attention by the scientific community. With suggestions of reduced production due to the invasion of unpalatable dwarf shrubs, and a reduction in perennial grasses, some attention was focussed on the processes affecting species composition and structure. The 'average growth cycle' model (Roux, 1981) attributes the contemporary flora to the impact of domestic herbivory, with the timing and intensity of herbivory having the greatest effect on vegetation, runoff and erosion. This model follows the results of an intensive long-term grazing trial at the camp scale. The results have not been scaled up and their application to landscape-scale processes remains uncertain. A further numerical analysis of these data (O'Connor and Roux, 1995) detects possible system run-down in certain grazing treatments, with a detectable increase in unpalatable dwarf shrubs. Perennial grasses are related to rainfall pulses in the 30 years of climatic data.

Following Westoby et al. (1989), Milton and Hoffman (1994) provide a catalogue of states and transitions for the karoo biome. The routes for transition may be either rapid through overgrazing to a low-cover pavement dominated by annual grasses (*Aristida* spp.) and succulents, or by slow system run-down to a system dominated by unpalatable dwarf shrubs (*Pteronia* spp., *Chrysocoma ciliata*). An alternative model of the dynamics of semi-arid systems is presented by Novellie (1990a) who demonstrates that patchiness in the distribution of nutrients is the fundamental cause of so-called 'degraded' conditions in the system. Nutrient-rich patches are created and maintained by both wild (this chapter) and domestic herbivores (Martens et al., 1990). These patches account for most of the nutritious production and receive most attention by grazing herbivores. Anthropogenically induced patches occur around stock management facilities (water points, fences and gates) and are most obvious on sheep ranches. If this process were to be responsible for complete landscape transformation, then degraded patches would have to coalesce and form mega-patches (Dean et al., 1995). Coalescing of patches does not appear to occur, with the impact of domestic herbivory declining with distance from enriched sites near water points and stock-handling facilities. The vegetation furthest from the centre of the degraded patches deviates substantially from the patch, and contains many so-called 'decreaser' or desirable species.

The concept of potential or expected vegetation (Palmer, 1991a) does not provide a full explanation for the fence-line contrasts which are evident throughout this semi-arid landscape. Structure and composition of the rangeland can be radically influenced by the timing and intensity of grazing events, with intense summer herbivory on the dolerite dykes transforming the landscape into one dominated by phreatic woody shrubs. The rehabilitation of this landscape to a condition approaching the potential is seldom achieved by the removal of herbivory. Suggested reasons for this are the removal of certain key species in the degradation process (Yeaton and Esler, 1990). The absence of these taxa prevents colonization of associated species, and the vegetation remains in the transformed state.

Reversal of grazing-induced changes

The reversal of the grazing-induced change is a controversial topic in the biome. Earlier research suggested that resting of rangeland from herbivory was sufficient to arrest change and initiate rehabilitation. After a lack of success with this approach in recently established nature reserves (Karoo Nature Reserve) and national parks (Karoo National Park), it was suggested that the absence of certain keystone species from the degraded condition delayed the re-establishment of desirable taxa (Yeaton and Esler, 1990). In addition, long-lived shrubs had colonized the system and there was a lag phase (similar to that described for savanna (Walker, 1993) before the shrubs died and made space for other species (Milton et al., 1995).

In the succulent dwarf shrublands of the central lower karoo, the enduring changes to soil chemistry, particularly elevated salinity, appear to prevent recovery. Other changes in soil chemistry, including elevated nutrient status and pH (Novellie, 1990a; Perkins and Thomas, 1993; Beukes et al., 1994; Wentzel et al., 1994), further exacerbate the problem.

Schlesinger et al. (1990) suggest that desertification can best be understood by exploring changes in the spatial and temporal distribution of soil resources. It is argued that desertification occurs when the relatively uniform distribution of resources is replaced by an increase in resource heterogeneity. This change may occur as a result of the invasion of grasslands by woody dwarf shrubs, and subsequent concentration of soil resources under the dwarf shrubs. Evidence for increased heterogeneity in degraded environments adjacent to the Nama-karoo has been presented using the moving standard deviation index (Tanser, 1997) of Landsat TM band 3 (red to near infra-red). Sites in degraded environments (e.g. communally managed rangeland with high dwarf shrub cover) have much higher standard deviation indices than adjacent well-grassed rangeland. There are suggestions that numerous sites in the Nama-karoo have undergone transformation of the soil surface, resulting in a more arid environment with accelerated runoff and soil capping, all of which preclude the establishment of the original plant species (Roux, 1968a; Roux and Theron, 1987; Beukes et al., 1994).

An alternative reason for the slow rate of recovery is presented by Wiegand et al. (1995) and Jeltsch et al. (this volume), who suggest that there are rare opportunities for recruitment of original species as a result of stochastic rainfall regimes of karoo areas.

12.4.4. Time scales in plant community dynamics
The timescale for change in the karoo can be from minutes to hundreds of years. In the case of sudden events, such as fire or flooding, dwarf shrubs and larger woody taxa can be rapidly removed (Novellie, 1990a, 1990b) from a site in the landscape. Less rapid changes in the composition of the short-lived component occur due to seasonal climatic events (O'Connor and Roux, 1995), and sudden reductions of perennial shrub cover have been associated with drought (Milton et al., 1995). Gradual decrease in perennial grass cover and palatable shrubs has occurred over many years under certain grazing regimes (O'Connor and Roux, 1995). The rate of post-disturbance recovery of the vegetation of ploughed and cleared sites under arid conditions is extremely slow (Dean and Milton, 1995). The idea would be to distinguish between short-term fluctuations, gradual successional type changes and more or less permanent changes in vegetation state (for example, salinization, nutrient enrichment, bush encroachment).

12.5.   Animal community dynamics

12.5.1.   Indigenous ungulates
The use of space by the large herds of indigenous ungulates present in the karoo in historical times is unknown, but a tentative reconstruction can be derived from the following sources: migratory patterns in ecosystems that have remained lightly impacted by man (For example, the Serengeti and the Kalahari Desert), behaviour of present-day ungulate populations in reserves in the karoo, and historical records of movement patterns. The latter are sparse and unreliable except for the dramatic springbok treks which tended to draw comment from observers (Skinner, 1993). Additional evidence comes from the distribution

pattern of plant species which carry structural defences against large herbivores and plants which are adapted for using these animals as seed dispersers (Milton et al., 1990). The contribution of such plants to the flora indicates the extent to which it was utilized by large herbivores during the course of its evolution.

Migratory grazing ungulates, such as the Serengeti wildebeest, in grassland and savanna ecosystems spend the wet season in short grass habitats in semi-arid regions and in the dry season move to taller grass habitat in more mesic regions (McNaughton, 1985; Sinclair and Fryxell, 1985; Fryxell and Sinclair, 1988). These movements do not simply track seasonal changes in the absolute abundance of food because the animals move away from the area of highest resource abundance in the wet season and return in the dry season. Fryxell and Sinclair (1988) conclude that variations in food quality are more important than absolute abundance of food in determining movement patterns. It is also probable that sward structure, which influences the rate of food intake (Owen-Smith, 1985), also plays a role. Thus the short grass of arid areas could provide sufficient quality and quantity of forage in the wet season but quantity may be lacking in the dry season. This would influence the animals to return to the mesic regions in the dry season where forage is more abundant.

A similar pattern of movement was evident in blesbok and black wildebeest in the Mountain Zebra National Park in the eastern karoo (Novellie, 1991). These species tended to favour short-grass areas dominated by creeping grasses in the wet season, but in the dry season showed increased use of areas dominated by taller tufted grasses. The overall annual rainfall also influenced this tendency: in wet years the animals favoured the short-grass communities, but in drought years spent more time in the taller-grass patches. The grazing lawns, with abundant dung patches and associated *Cynodon* spp. were heavily used almost without respite from one year to the next. However, during drought years and during dry seasons, the intensity of use of at least some of the grazing lawns decreased, and the generally neglected patches of taller grass appeared to act as an important forage reserve (Novellie 1990a, 1991 and unpublished data). Thus the trade-off between quantity and quality of forage seems to be the critical factor.

In historical times, grazing antelope probably penetrated the eastern and central karoo from the more mesic grasslands of the southern Free State and Eastern Province during wet seasons and moved back to the more mesic grasslands to the north and east in dry seasons. During extended periods of summer rainfall, when grass was more abundant in the eastern karoo (Roux, 1966) large herds of grazing antelope may have remained in the central and eastern karoo year round. Possibly, they tended to show seasonal movements on the landscape scale between patches of short and long grass as they do currently in the Mountain Zebra National Park.

Historical reports indicate a more or less regular seasonal movement of springbok from Bushmanland, where they spent the summer, westwards and southwards to the succulent karoo and the Koue Bokkeveld in winter (Skinner, 1993). Rain in the latter regions falls in winter with less between-season variation than in the interior of the country, and this may have promoted a regular pattern of movement. In other areas, springbok treks appeared to be irregular, the only discernible pattern being that the general direction of movement was from an arid to a more mesic region (Skinner, 1993). It is probable that the treks were initiated when local forage supplies ran out and the animals moved *en masse* to different areas, possibly attracted by distant flashes of lightning (see Vernon, this volume; Lovegrove, this volume).

In the Kalahari, major migrations of ungulates are not regular annual events, but tend to occur when the quality of the habitat drops below a certain minimum (Knight, 1991; Spinage, 1992). However, the pattern often followed by the wildebeest is to move to more mesic regions in the north and east of the Kalahari in the dry season (Spinage, 1992).

The use of pans in the Kalahari Gemsbok and Vaalbos National Parks by springbok and blue wildebeest (*Connochaetes taurinus*) is, in some respects, similar to that described above for grassland ecosystems (Knight, 1991, and unpublished data). The pans, which generally support short grasses, are particularly favoured in the wet seasons. However, the use of the surrounding Kalahari sands, with their taller, stemmier grasses, tends to show a relative increase in dry seasons. The pans are more nutrient-rich than the sands (Milton et al., 1992b), but in the late dry season the quantity of vegetation available on them is extremely limited and this may account for the increasing use of the surrounding dunes (Knight, 1991).

In historical times, regular seasonal movements of large numbers of indigenous grazing ungulates would have been a feature of the eastern and north-eastern fringes of the karoo where it intergrades with the grassland biome (Skead, 1987). Regular movements of springbok appear to have occurred on the interface between the Nama- and succulent karoo. As indicated by Milton et al. (1990), penetration by large ungulates into the southern and western karoo would have been sporadic, possibly tracking rainfall events that were extremely variable in both time and space. In the grassier northern and eastern karoo, thorny trees and shrubs and barb-fruited grasses are abundant, whereas, in the arid south-western karoo, such plants are largely confined to drainage lines. This indicates that vertebrate herbivores of the south-western karoo were mostly small mammals and avian herbivores, with large mammals generally being restricted to drainage lines (Milton et al., 1990).

12.5.2.   Invertebrates

Insects, amongst which the most important order is the Hymenoptera that feed on nectar, pollen and other parts of flowers (Kevan and Baker, 1983), assist in maintaining patchiness. The specialist feeders, with a low diversity of floral choice, promote the process of pollination, and succulent taxa, such as *Aloe ferox*, gain some advantage by growing in clumps (Stokes and Yeaton, 1995). In a study of aculeate wasps and bees as flower visitors and pollinators throughout the karoo, Gess (1981) and Gess and Gess (1994) found that the majority of these insects have an adult phase synchronized with the flowering of the flora. Visitors to the flowers of 35 families of plants have been sampled. Preliminary results show an overall high diversity of choice for non-masarine wasps, an overall high diversity of choice for bees and a relatively low diversity of choice for masarine wasps (those wasps which, bee-like, provision their young with pollen and nectar). At the family level, their results show high diversity of choice for the bee families Anthophoridae, Megachilidae and Halictidae, and relatively low diversities of choice for Colletidae, Melittidae and Fideliidae (Gess, 1992). This association between anthophilous insects and the flora suggests that their activities may further promote patchiness. Such plant families as Apiaceae, Celastraceae and Mimosaceae are mostly generalists, attracting generalist feeders. Others, such as Asteraceae, though as a rule generalists, attract not only generalist, but, in addition, specialist feeders. Though more dependable visitors where they occur, specialists may have a narrower geographical distribution than their forage plants. Papilionaceae are generally visited principally by Megachilidae, Anthophoridae and the honey-bee *Apis mellifera* (Apidae) (Gess and Gess, 1994). Examples of specialist flowers are the gullet flowers of *Aptosimum* and *Peliostomum* (Scrophulariaceae), which require their pollinators to be able to reach the nectar produced at the base of a very narrow tube, and to be large enough when inside the flower to receive a pollen load dorsally. Throughout their range in the western, northern, southern and eastern karoo, flowers are almost solely attended by *Celonites* (Masarinae) spp., themselves specialists, which are equipped with long tongues with which they are able to reach the nectar after entering and which fulfil the requirements of fit and behaviour to make them dependable pollinators (Gess and Gess, 1989, 1991b; Gess, 1992). Other recent work includes Hoffman's (1988a) study of the pollination ecology of an important succulent, *Aloe ferox*. He showed how both birds and honey-bees are involved in its pollination. It is of note that this plant, unlike the majority of karroid plants which flower after rain during the warm to hot months, flowers in the winter, when the solitary wasps and bees are not active.

Termites and ants

Climate, vegetation and soil interact to influence the distribution and abundance of termites (Lee and Wood, 1971). Soil texture affects water-holding capacity and binding properties, which indirectly affect food availability for termites. The most important soil feature for mound-building termites is the proportion of sand, silt and clay and the distribution of these throughout the profile. A certain amount of clay, or other colloidal material, is needed to cement particles together, which explains the general absence of termite mounds on pure sands (Lee and Wood, 1971). Similarly with subterranean termites, where the proportions of sand, silt and clay are important in determining the stability of subterranean galleries (Lee and Wood, 1971).

Low mounds of nitrogen-rich soils (heuweltjies) are fairly evenly distributed over the Tierberg Karoo Research study site in the southern karoo at a density of c. 2.1 ha^{-1} (Milton et al., 1992a). These mounds are probably formed by the activities of termites (Moore and Picker, 1991) and their soils are further altered by burrowing mammals and antelope (Milton and Dean, 1990a). Heuweltjies may influence the distribution of harvester ant Messor capensis nest-mounds (Dean and Yeaton, 1993b). Harvester ant-mounds appear to be most abundant in the plains community and on the edges of heuweltjies. They occur less frequently in minor washes and are rare in the major drainage lines (Dean and Yeaton, 1992).

Although high in most plant nutrients, soils in the karoo are generally poor in nitrogen and water infiltrability due to the low organic-matter content (Ellis and Lambrechts, 1986; see Watkeys, this volume). Ants and termites concentrate organic matter in patches, or contribute to the movement of organic matter underground, thereby influencing the local distribution of moisture and plant nutrients in the soil.

Soils of the harvester ant Messor capensis nest-mounds in the southern karoo have a higher infiltrability and contain significantly more organic matter, phosphorus, potassium and nitrogen than soils of intermound spaces (Dean and Yeaton, 1993a). The nest-mounds are conical or elongated heaps of organic matter and soil which vary in size from 0.3 m to 2.9 m in length, 0.26 m to 1.6 m in width and 0.07 m to 0.42 m in height (Dean and Yeaton, 1992). The organic matter in these nest-mounds is derived from seed husks, pappuses, wings from wind-dispersed seeds, pods from legumes, insect remains, faecal matter and ant-frass, the latter also increasing nitrogen levels (Dean and Yeaton, 1993a). Ant-nest soils in a semi-arid area such as the karoo provide a nutrient-enriched environment for seedlings and mature plants. Harvester ant nests in the karoo may be important in raising the level of plant-available nutrients, particularly phosphorus, provided that there is rain to increase access to nutrients (Dean and Yeaton, 1993a). Woody shrubs growing in nest-mound soils have higher levels of nitrogen than plants growing in intermound soils (Dean and Yeaton, 1993a).

The high nutrient status nest-mound soils are more favourable for seed germination than the intermound soils in cases where leaching is not required to remove salts and inhibitors from certain seeds (Dean and Yeaton, 1992). In the biome, ants are faced with extremes of heat and cold and the shallow, gravelly or hard clay soils impose constraints on digging nest holes (Dean and Turner, 1991). Existing crevices or cavities under stones may be used (Skaife, 1979) or nest holes may be excavated under stones by ants, the stones providing relatively safe nest sites from predation by ant-eating mammals (Dean and Turner, 1991). It was found that ant nests in the southern karoo are not randomly dispersed with regard to available stones. The analysis of sites on stone type alone suggests that white quartzite or black dolerite are selected for, but a preference index shows that nests tend to be under large stones and that grey Dwyka tillite was the most favoured stone type (Dean and Turner, 1991).

The presence or absence of stones may, therefore, determine the species-richness and abundance of the ant assemblage in an area, and this will, in turn, cause local cascade effects to run through the local plant and animal community. The abundance of ants will influence the amount of seed harvested in that area, and the levels of predation of other invertebrates. The presence of harvester ants Messor capensis, that often found (begin) colonies under stones (Dean and Turner, 1991), subsequently building their nest-mound over the stone, will influence infiltration rates (Dean, 1992b) and local soil nutrient levels.

12.6.   Conclusions

Plant community patterns in the Nama-karoo do not appear to differ from those of other arid ecosystems, with a strong association between growth form, and nutrient and base status. Perennial dwarf shrubs and annual grasses predominate in nutrient-rich, high-base status patches, and perennial grasses predominate in low-nutrient status soils. Woody shrubs are successful in the phreatic conditions of rocky ridges and drainage lines. Disturbance of the herbaceous layer in the phreatic sites favours woody shrubs.

Concern has been expressed that patch selective

grazing represents a major threat to the grazing resource in the semi-arid regions of South Africa (Fuls, 1992; Kellner and Bosch, 1992). The reason for concern is that the neglected patches may constitute a waste of resources while heavily grazed patches are susceptible to progressive degradation that is exacerbated by droughts. On the other hand, the above review shows that patch selective grazing is a natural component of African grassland ecosystems utilized by indigenous herbivores, and is an important agent of plant community diversity. Little is known of the long-term dynamics of heavily grazed patches, but their usefulness to indigenous herbivores can persist for extended periods without evidence of vegetation degradation (at least six years: Novellie, 1991). The major question is: do these patches represent useful and productive self-maintaining systems capable of sustaining continuous use, or are they nuclei of degradation that will lead to loss of the grazing resource? Are the neglected areas a waste of resources, or important reserves of forage

in periods of drought? Answering these questions requires an understanding of the processes of formation on nutrient-rich patches, and the factors affecting the useful lifespan of such patches.

There is little doubt that woody, deep-rooted species (dwarf shrubs, shrubs and small trees) are more successful once the perennial grasses have been removed by herbivory. On the shallow (<0.3 m) soils which dominate the karoo landscapes, the removal of perennial grasses promotes dwarf shrub dominance. For the grazier this is a desired state, as, in contrast to grasses, dwarf shrubs provide nutritious fodder outside of the growing season. On the other hand, the removal of grass from the deeper, sandy soils of Bushmanland and the Kalahari, and along certain rocky ridges, promotes the growth of woody shrubs such as *Rhigozum trichotomum* and *Lycium* spp. that have little value for domestic livestock. The success of these woody shrubs in some areas may be regarded as a serious management problem.

13 Modelling populations and community dynamics in karoo ecosystems

K. Kellner and J. Booysen

13.1. Introduction

Hall and Day (1977) define a model as any abstraction or simplification of a system and extension of scientific analysis by other means. A model should have some important functional attributes of the real system, although not all attributes are necessary, and is generally simpler than the real system. Modelling is used to aid the conceptualization and measurements of complex systems and also to predict the consequences of an action on the real system. According to Goodman (1990), models are ways to predict the behaviour of complex, poorly understood entities from the behaviour of parts that are already well understood. Models must, however, be checked frequently against the real world to ensure that the presentation is accurate. If the model and the real world disagree, then one or the other, or both, are imperfectly known and, by finding points of difference, the understanding of the real world or the model system will be increased. A principal use of models is, therefore, to test the validity of field measurements and assumptions made from the data. The ultimate function of modelling is to know more about the structure and behaviour of natural systems, both now and in the future. Models are one tool of many that can aid us in this process, especially in interdisciplinary research. Models can incorporate known spatial heterogeneity of biological, physical and environmental components, such as the vegetation, soils and climatic parameters. Major drawbacks are the numerical complexity and large number of input parameters which often require long-term research (Keating et al., 1995). Different aspects influencing the dynamic process can be explained in several sub-models. Spatial models are at different scales and can be combined to address a given problem and identify the key factors that influence the behaviour of the ecological system (Richardson, Hahn and Wilke, 1991). Models help in the identification of research questions and in future decision-making processes.

In summary, the advantage of models is that they:

- generate hypotheses;
- simulate treatments as an alternative to costly field experiments;
- are faster and less destructive than field experiments;
- synthesize available information and theory;
- identify new directions for research;
- provide guidelines for management.

In this chapter, we present brief descriptions and results of a variety of models that have been used in the karoo *sensu lato*.

13.2. Models to predict the outbreak of problem animals

Predictive models have been used in the karoo to guide management decisions relating to the protection of rangeland and crops. In the arid parts of the central Nama-karoo, outbreaks of the brown locust, *Locustana pardalina*, is a perceived threat, not only to the extensive rangelands, but also to the croplands of neighbouring provinces and countries. Predictions of brown locust swarming can be used as an early warning system in order to practise effective integrated control measures. By using an age- and stage-structured Leslie matrix model, Nailand and Hanrahan (1993) developed a rule-enhanced simulation model in which rainfall patterns, daily temperature and wind conditions are key factors in the transitions between the life-cycle stages of the locust population, thereby predicting population increases and movement patterns. Changes in locust population dynamics conducive to swarming are event driven. Exogenous factors, such as good rains in early summer that end an extended dry period, are the most significant trigger mechanisms. Data generated by the Leslie matrix indicate that endogenous factors, such as a high growth rate, cannot account for irruptions of brown locusts. This is supported by studies that correlate rainfall and locust irruptions in the natural outbreak regions of the Nama-karoo. For large-scale population increases of brown locusts to occur, there should be an extended dry period of several months in winter with no rain of more than *c.* 6 mm on any day. This apparently ensures that both diapause and quiescence in the majority of overwintering eggs is broken. Good rains in early summer (October, November or December) will precipitate large-scale, wide-spread hatching and facilitate population development. The model suggests that a wet winter is detrimental to locust population increase, but light winter rainfall enables eggs to survive the normally dry winter period in the summer rainfall area of the Nama-karoo. The rainfall may bring about the end of diapause and quiescence and prevent desiccation, but the low temperatures throughout the winter will certainly inhibit development.

13.3. Models to study animal population dynamics

A modelling approach helps to focus on the dynamics rather than simply the structure of populations (Norton, 1994). Often it is more important to understand the process of change than just population size. Starfield and Bleloch (1986) state that 'it does not suffice to hypothesize vaguely about what causes a population to decline; we have to build a model that can actually make the population change'. Starfield and Bleloch (1986) further suggest that the greatest strength of modelling is not so much its ability to simulate reality, but rather that it can be used to 'think usefully about the problem'. Norton (1994) used a simple deterministic, spreadsheet model incorporating events, such as rainfall, in order to provide guidelines for the management of the mountain reedbuck (*Redunca fulvorufula*) population in three nature reserves in the karoo. Since decisions on game population management are usually made annually, the time steps for the model are also annual. The model is based on field data on fecundity, recruitment and mortality of the mountain reedbuck obtained from these reserves. Results show that there is a clear birth peak in early summer, and that juvenile mor-

tality is much higher than adult mortality, although the adult mortality increases after the age of seven years. Most of the population dynamics are driven by variation in rainfall. Body condition of the animals varied with available grass biomass. The animals are not affected equally by drought. A significantly lower condition index suggests that males are subject to greater stress than females. Sensitivity analysis of the different components of the model suggest that peri-natal and first year survival are the population parameters that have the greatest effect on population growth. This population study suggests that mountain reedbuck probably have to be culled on many small reserves, depending on the quality of the grazing and the management objectives of the reserve. A cycle of a long population build-up followed by a major cull is likely to be closest to the natural cycle of the mountain reedbuck, especially if the timing of the cull is coupled with periods of low rainfall.

According to Van der Hulst (1980) we should study total ecosystem dynamics and not only, as in most cases, vegetation dynamics. It is, however, true that most faunal models and especially those predicting population changes, are based on a framework of vegetation dynamic models. Novellie and Strydom (1987) used vegetation parameters to predict the performance and grazing habits of large herbivores. A stochastic variance analysis type of model was used to study the implications of climate induced changes on the performance and grazing habits of large herbivores. Several vegetation parameters, such as canopy spread cover, diversity and grazing height, were monitored. The study indicates that the Cape mountain zebra (*Equus zebra zebra*), and the grazing of other larger herbivores has no apparent effect on the cover and density of the vegetation in the Mountain Zebra National Park. In fact, the grass-tuft density and canopy-spread cover of the shrubs showed a five-fold increase within one year following good summer rains. It was not clear how much the cover and height of grass need to decrease before the zebra population would be affected. An integrated monitoring programme of the interrelationships between the species composition and structure of the grass sward and dynamics of the zebra population would be necessary to improve confidence in the predictions of the model (Novellie and Strydom, 1987).

Habitat suitability models (HSM), which can be arithmetic models, mathematical models or rule-based decision support systems, can also be used to predict the quantity or quality of habitat for animals. An HSM model in which multivariate statistics and expert system approaches were combined, was used by Fabricius and Mentis (1992) to predict suitable habitat for kudu (*Tragelaphus strepsiceros*) in the arid savanna of South Africa. Plant species were classified into functional and structural groups, rather than species, and food and shelter attributes, mainly influenced by seasonality, were taken into account. Plant attributes, such as phenophase, spinescence, leaf size and palatability were used. The expert system calculated a habitat-suitability score, based on discriminant functions. Although most variables are assumed to be linearly related, which is unrealistic, the linear additive effect of the discriminant functions was counteracted by the expert system without sacrificing their quantitative properties in this model.

By combining the results of both the multivariate statistics with the expert systems, the kudu model retained both flexibility and objectivity. A disadvantage of expert systems is that they can impart a false sense of security to the user if they are not based on quantitative data, since, wherever expert systems are used in habitat suitability models, the modeller has to rely on personal opinions or value judgements in formulating the rules of the model.

Models have been used to establish faunal niches and to understand the population dynamics of birds and other, smaller herbivores in the karoo. Dean (1995) describes the temporal and spatial variability of bird densities in response to changes in plant species-richness and cover. The latter is influenced by the grazing patterns which causes, together with the rainfall variability, a mosaic of vegetation patches. These data were used by Fahse (1994) in a Lefkovitch matrix type of approach in the construction of a metapopulation model. With the aid of different submodels based on the vegetation dynamics and spatio-temporal characteristics of the rainfall patterns, all which form part of a broader range condition model, suitable protected areas for the breeding and feeding habits of nomadic lark species over the long term are predicted or simulated (Fahse, 1994)

Little et al. (1996) used a stepwise multiple regression analysis approach to investigate long-term population trends and environmental correlations in the development of a predictive model for the population fluctuations of the Namaqua sandgrouse Pterocles namaqua. This species is endemic to the semi-arid regions of south-western Africa. Because the birds congregate in large flocks at traditional watering points, particularly during the nonbreeding season, the species has attracted considerable attention from bird hunters. This creates opportunities for the gamebird hunting industry which serves both national and international tourism clients. The model was therefore constructed to identify key ecological factors which influence annual variations in gamebird populations, in order to maintain sustainable populations over the long term and to forecast the offtake for commercial hunting. However, "the model was more useful for correctly predicting" low sandgrouse abundance (i.e. seasons that will offer poor hunting bags) than for predicting seasons of high sandgrouse abundances (Little et al., 1996). Negative correlations between annual sandgrouse abundance indices and December rainfall, and positive correlations with March rainfall, are probably mediated through the effects of rainfall on preferred food plants. The breeding area of the sandgrouse populations is on the transition zone between the winter rainfall in the southwest and summer rainfall in the northeast. The relationship between sandgrouse and rainfall is due to the combined effects of low December rainfall and heavy March rainfall, prior to the hunting season in winter. An apparent decline in numbers of sandgrouse might reflect the influence of a qualitative change in grazing pressure and/or increase in available watering sites. The increase in small carnivores (black-backed jackal Canis mesomelas, Cape fox Vulpes chama and caracal (lynx) Felis caracal) probably due to the removal of larger carnivores, may also have contributed to the decrease in the sandgrouse populations, as predation on nests can be as high as 80%.

The hyrax model (Davies, 1994) predicts population responses of this small mammalian herbivore to rainfall, available forage and predation by black eagles Aquila verreauxii. Using a spreadsheet model, the population composition and age-specific rates of mortality and fecundity of the rock hyrax, Procavia capensis, was studied in the karoo. The age-specific rates of mortality and fecundity used in the model in a given year were derived from the ratio of P : K, where P equals hyrax population size within the black eagle territory after the birth pulse; and K equals the number of hyrax that can be accommodated by food resources within the 'safe feeding area' as determined by the rocky habitat of the study area. Peaks in hyrax numbers correspond to periods of high winter phytomass, while declines correspond with droughts. During drought periods, hyraxes have less shelter, making them more vulnerable to predation when foraging. Good rains lead to increased vegetation cover, ensuring protection from predators. The model predicts that major phases of increase and decrease in rock hyrax populations are correlated with rainfall, but moderated by predation. From very simple rainfall data, the model makes robust predictions about hyrax population changes which concur with field observations. Predictions of hyrax populations fell within the correct range only when predators were included in the model. Dramatic differences in hyrax populations were obtained when the model was used without predators (see Milton et al., this volume).

Henwood (1975) used a dynamic deterministic approach with much shorter iterations (minutes), to construct a thermoregulation model for two diurnal Namib tenebrionid beetles. Onymacris plana is a large, shiny tenebrionid with a round, pancake-shaped body. Stenocara phalangium is smaller, with a distinct pear-shaped body. Observations showed that the beetles attain relatively high and constant sublethal body temperatures for several hours each day. At steady-state, an organisms temperature is dependent on the rate of heat gain from, and loss to, the environment, plus any metabolic contribution. A thermal energy model was developed using beetle heat exchange parameters and combined with microclimate data to predict the range of body temperatures available to the insects in their different environments. Observations of thermoregulatory behaviour, when coupled with the effect of behaviour on modelled body temperatures, make it possible to evaluate the efficacy of various thermal strategies and microhabitat selection by which the beetles maintain preferred temperatures. Conclusions gained from the model are, for example, that the activity of the beetles is not only influenced by the temperature and wind speed that prevails in their habitat on the dune slopes, but also differs between sexes.

13.4. Production and drought prediction models

The simplest type of model that can be used to predict the effects of the grass and shrub basal cover fluctuations on production in response to climate or management is the regression analysis model. Using a simple model, Van den Berg (1983) shows that there is a positive correlation between rainfall and forage production. These regressions can also be used to predict the available biomass of forage for herbivores over the short term and can estimate the grazing capacity of rangeland in the long term. One of the major constraints of such a model is that the calculated grazing capacity can seldom be extrapolated to other areas and the results are only as reliable as the seasonal rainfall data given over the short- or long-term.

Larger, more complex, dynamic models for predicting production are the PUTU rangeland models by Fouche (1984, 1992). The model is a hybrid of empirical and deterministic algorithms, commonly referred to as an intermediate model. It is based on the physical and biological aspects of the environment and simulates water balance and plant growth on a daily basis. These models are used to predict the production potential of the herbaceous layer for an area. They quantify droughts, using scientific norms for drought characterization. They can be used as a tool for predicting the numbers of stock that can be kept on the rangeland without a reduction in the fodder production. The PUTU II model takes the basal cover and carbohydrate reserves of the herbaceous layer into account. The build-up and translocation of the carbohydrates in the plant is a function of the growth stage, maximum growing rate and environmental stress at any particular stage, of which the

most significant is water stress. The latter not only depends on the water balance in the soil and the soil–water status, but also climatic variables, such as the rainfall, minimum and maximum temperature, evaporation rate, daily radiation and other soil physical factors. The PUTU model indicates that drought periods, calculated from the dry-matter production per unit area, increase with higher stocking rates. It is evident that the inputs for such a complex model are quite substantial and, consequently, limits it to cases to which it can be applied. Although this model was developed for the climatic climax grasslands of southern Africa, it has often been used in other semi-arid areas because there are no other production models available.

Other drought-related models include the development of a relative drought resistance index (RDI) for the karoo region, developed by Roux (1993), whereby the adaptability of the phenological stages of the plant species according to the standard deviation (STD) of the monthly long-term mean of the rainfall figures are taken into account. The drought resistance of the natural vegetation of an area is related to rainfall variability. In drier regions, plants tend to be better adapted to drought stress and an erratic rainfall pattern, whereas, in higher rainfall areas with more evenly distributed seasonal rainfall, the vegetation is less drought resistant. RDI values are proposed for parts in the karoo region and can aid the national drought committee of the government in the drought relief scheme for farmers.

The Palmer drought index (PI) is frequently referred to in drought studies and is one of the best-known examples of an objective, rational measure of drought severity (Palmer, 1965). In this dynamic deterministic type of model, soil type and monthly rainfall and temperature indices are used to calculate an index of drought prediction. This model is, however, a hydrological index allowing comparison between different areas with different climatological characteristics. The relationship between these indices and explicit ecological activities are not well defined and needs further investigation. Although the Palmer drought-analysis procedure has been adapted for South African conditions, certain application problems are still encountered. As noted, one of these limitations is that it is a climatological index and does not necessarily reflect the drought impact on the natural vegetation. This is shown by poor correlations established between the PI and grassland biomass. Venter (1992) tried to address this problem by a numerical index of drought severity that was calculated in a simple mathematical model using actual monthly precipitation as well as monthly averages of daily maximum temperature, compared with rainfall probabilities generated by the ZA model (Zucchini 1992; Zucchini et al., 1992). In this model, a single parameter (h)

represents the rate of decline of plant available water. It shows how long-term averages of maximum daily temperature can be used to estimate possible variation in the available water in the plant, which can in turn be used to calculate a shrub production index using dry-matter yield of dwarf shrubs within a growing season. The drought impact on the shrubland with respect to the calculated drought index can be quantified. Models of this nature can be used to decide whether aid for drought should be paid, and could play a major role in management decisions at farm level (Roux et al., 1995). They can also be used to calculate drought risk and the economic consequences of selling or feeding livestock during drought. Using models in decision-making, even 'unsophisticated' empirical models, can be considered a step forward in drought prediction and vegetation production when compared with previous methods where decisions were based on rather arbitrary comparisons of average rainfall figures.

Production simulation models are also used for the prediction of animal performance in a livestock production systems (Richardson et al., 1991) and for assessing the economic viability of a proposed scheme. The Òptimization Decision Making model for small-stock farming (Herselman, 1989) integrates the product prices, input costs and management practices (type and number of animals, nutritional requirements of each, and so on), with the ever-changing environmental variables to establish an optimal utilization production model. This can aid the farmer in the so-called 'tactical' decision-making process for the implementation of strategies with short-term effects. Various management alternatives can also be evaluated and compared. Quantitative biological information is integrated with economic considerations which will provide a sound basis for decision-making by the farmer in order to manage the natural resources in a sustainable manner (see Jeltsch et al., this volume).

13.5.   Models for vegetation dynamics and sustainable rangeland management

Nailand (1992) suggests that, when studying community dynamics, rather concentrate on the 'how' and 'why' than the 'what' stages of the factors contributing to the changes in the community. More emphasis is placed on the vegetation community dynamics of the arid karoo, where the spatio- and temporal variability of plant species, influenced by the environmental conditions during the successional process, are being studied. The majority of rangeland management decisions in the karoo are based on conceptual models, explaining the directional, cyclic or stochastic life-histories and population dynamics

influenced by the intrinsic and extrinsic factors of the environment. Existing conceptual models of vegetation dynamics adequately explain the impact of domestic livestock on karoo vegetation. One such conceptual model (Milton, 1992c) suggests that selective herbivory affects interactions among plant species and between plants and animals, influencing the relative abundance of shrubs differing in their acceptability as forage. The karoo vegetation however has a backdrop of climatic variability, influencing the population dynamics, biomass productivity and species change (Dean, Hoffman et al., 1995).

Models of ecosystem dynamics in the past were unable to provide the functional and predictive insights necessary for the development of management programmes. They were mostly based on conceptual and deterministic views of the ecosystem and ignored the stochasticity that occurs in such an arid region, focussing instead on the 'averages' as a basis for their rationale (Hoffman, 1988b). Milton (1992c) and Milton and Hoffman (1994) recognized this constraint and suggested that the state-and-transition model approach by Westoby et al. (1989) would be appropriate for karoo vegetation. Directional types of models, which mainly include conceptual, cyclic and deterministic models, concentrate more on the Clementsian continuous, climax (mature) succession, at an equilibrium state type of approach, whereas stochastic type of models are more flexible, emphasizing more the non-equilibrium and state-and-transition approach to describe the changes in ecosystem dynamic processes (Milton, 1992c; Nailand, 1992; McCook, 1994; Milton and Hoffman, 1994). Succession paradigms mainly emphasize equilibrium states and are unable to provide a predictive basis for community changes in the climatically unpredictable environments such as the karoo. A distinct shift towards the non-equilibrium theory, as expressed by the state-and-transition model, is shown by rangeland managers. These models also include the unpredictable climatic or disturbance factors, which can be incorporated in transition matrix and Markov chain type of prediction models, which are often used to explain the spatio- and temporal dynamic changes in population dynamics (Nailand, 1992; O'Connor, 1993; Kellner, 1995). State-and-transition models are also more suitable to assess and describe rangeland dynamic changes and consist of a set of discrete 'states' of vegetation and a catalogue of 'transitions' which attempt to explain the changes that take place from one 'state' to another. With this stepwise type of model, where each step represents a certain state in which successional changes may occur, the dynamics of vegetation change are better understood, leading to more appropriate management strategies. An example is the proposed state-and-transition model for the eastern karoo (Fig. 13.1).

In addition to differences in opinion on the mecha-

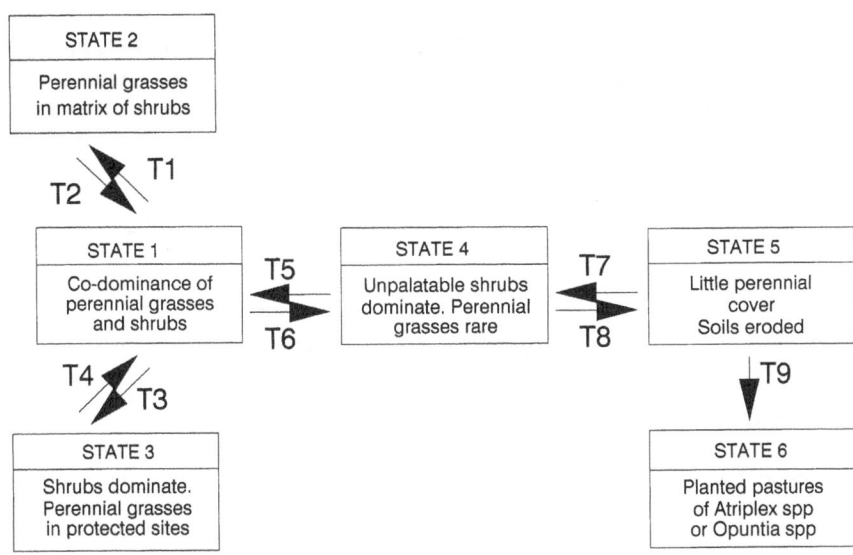

Figure 13.1 **Example of a proposed state-and-transition model for the grassy eastern karoo. States (1–6) and transitions (1–9) (redrawn from Milton and Hoffman (1994), with permission from the Grassland Society of Southern Africa)**

nisms for changes in plant species composition in karoo rangelands (Hoffman et al., 1990; Milton, 1992c, 1995a; Milton and Hoffman, 1994; Milton et al., 1994b), Acocks's (1953) proposal that karroid shrubland is advancing north-east into grasslands has been challenged by alternative models. The 'desertification' process has been much debated (Bosch, 1988, 1989; Hoffman, 1988b; Hoffman and Cowling, 1990b; Hoffman et al., 1990; Dean, Hoffman et al., 1995). Using a simple regression analysis approach, Hoffman et al. (1990) examined fluctuations between the grass and shrub basal cover in the eastern karoo. Results of this study suggest that an increase in grass basal cover may negatively influence shrub basal cover and that the summer rainfall, which increases the grass cover, is the driving variable for community dynamics in this system. Competition for resources between grasses and shrubs may lead to a decrease in shrub cover. With the failure of average summer rainfall, grass cover decreases, enhancing shrub recruitment into open spaces, a process favoured by winter rain. A matrix-modelling approach with stage-structured Lefkovitch models can simulate vegetation change when data on the response of demographic parameters to herbivory and climatic events are available (O'Connor, 1993). The results of such matrices can also be incorporated into Markov chain models that have predictive ability and are helpful, not only in interpreting the replacement of plant species types, but also in interpreting views on the mechanism of ecological community succession, contributing to the change in rangeland condition (Usher, 1981; Nailand, 1992; McCook, 1994; Kellner, 1995). The heuristic Markov chain approaches are also suitable in predicting states in the state-and-transition model (Van der Hulst, 1980; Usher, 1981; McCook, 1994),

despite their disadvantages, such as the need to define states, furnish a large amount of data and consider the historical background, and the fact that the predictive states are not explanatory but are based on empirical estimates of the replacement probabilities.

Management models designed for understanding and managing changes in rangeland conditions for livestock can be incorporated into expert systems to facilitate their use by the decision-maker. The comparisons of management models can include conceptual ideas, historical data and quantitative data. Decision Support Systems (DSS) are compiled so that the research information and knowledge can be transferred to land managers in a usable form (Bosch and Booysen, 1992). These systems include various modelling approaches, but are based mainly on multivariate analytical procedures. Using multivariate statistical analysis, range condition models can be constructed which are used to interpret the changes and incorporate thresholds between alternative states in the ecological system (Bosch and Gauch, 1991; Bosch and Kellner, 1991; Friedel, 1991; Van Rooyen et al. 1994). These qualitative or quantitative models generally assume that deterministic succession occurs in nature. A residual value calculated in the Degradation Model Construction (DMOC) analysis indicates changes in states, and degradation values, which are successional changes along a gradient. DMOC model construction uses various ordination methods, namely standardized principal component analysis (PCA), centered PCA and reciprocal averaging (Bosch and Gauch, 1991). Gradients can further be subdivided into thresholds on a subjective scale, supporting state-and-transition principles. Grazing capacities are deterministically calculated and adjusted by means of numerical calculations within an expert system. Management strategies selected are based on the previous information by means of a rule-based forward chaining expert system (Bosch et al., 1992). An integrative approach, based on new or existing scientifically sound understanding of the nature, dynamics and production potential of the rangeland, is incorporated in the ISPD (Integrated System for Plant Dynamics), decision support system. With multivariate analysis, deterministic degradation models (DMOC) are constructed for relatively homogeneous grazing areas and, by means of an expert system approach, grazing capacity and management models are used to define appropriate management strategies for a certain area (Bosch et al., 1992).

Decision support systems based on simulation models, together with long-term field studies, could also help to determine management strategies and stocking rates appropriate for given states of rangelands.

13.6.   Conclusions

The construction of models and their application improves insight into ecosystem dynamics, furthering our understanding of the processes that cause changes in the system. The main value of a model is probably the experience gained during its construction. A model in the process of development is perhaps more valuable than the final product. This emphasizes the importance of the validation of models constructed for specific objectives. Models need to be tested with independent data sets before predictive understanding of the ecosystem dynamics can be gained. The danger of models is that they may have been developed for a particular aim or vegetation community, but may be used for other applications without proper validation. The constraints of some models, such as the state-and-transition model, is that they need too much input in the form of data collected from long-term trials. Detailed knowledge of physiological and demographic responses to vegetation changes and climate is needed to provide a greater predictive insight into ecosystem dynamics. On the other hand, long-term trials are needed for the verification of conceptual and other types of models in vegetation community studies in such arid environments as the karoo. It must always be borne in mind that, no matter how good a model may be, it cannot replace field observations.

14 Spatially explicit computer simulation models – tools for understanding vegetation dynamics and supporting rangeland management

F. Jeltsch, T. Wiegand and C. Wissel

14.1. Introduction

A class of spatially explicit models, the grid-based simulation models, has proved valuable for dealing with problems on large temporal scales, and where complex interactions depend on coincidences in time and space (Menaut et al., 1990; Wiegand et al., 1995; Thiéry et al., 1995; Jeltsch et al., 1996, 1997a; Wiegand and Milton, 1996). Most of the grid-based models applied to vegetation dynamics in semi-arid systems represent an advanced form of cellular automata models (Wolfram, 1986). These models typically subdivide a modelled area by a grid of spatial subunits, so-called 'cells'. Each cell is characterized by its location and by one or more discrete ecological states which may change in the course of time due to the influence of (1) neighbouring cells, (2) the previous state of the cell itself and (3) of such external factors as climate, disturbance or management actions. The size of these spatial subunits is determined by the initial question to be addressed by the model, and is usually based on typical biological scales of the modelled system, for example the size of individual plants, characteristic distances for seed dispersal or typical ranges of plant interactions (Jeltsch and Wissel, 1994; Wiegand et al., 1995; Jeltsch et al., 1996, 1997a).

Grid-based models such as those of Wiegand et al. (1995) and Wiegand and Milton (1996) focus on the processes and mechanisms that drive community dynamics at the level of individual plants. Although there are usually few long-term field data available on the dynamics of arid plant communities, such rainfall-dependent life-history attributes as growth, seed production, germination, recruitment and mortality factors, and seed dispersal and interactions between individual species, are relatively easy to observe on shorter timescales. The basic idea of this bottom-up approach is to incorporate such short-term knowledge in the form of rules into a computer simulation model. In order to investigate community dynamics, the model simulates the fate and the interactions of individual plants within the community, the sum of which is community dynamics (Wiegand et al., 1995; Wiegand and Milton, 1996). In this way, a model extrapolates from the behaviour of individual plants to long-term community dynamics. However, if plant life forms of distinctively different sizes are to be modelled, it is often necessary to base the model on plant assemblages instead of individuals (Jeltsch et al., 1996; Stephan et al., 1996). An important advantage of this type of grid-based simulation model is the inclusion of the necessary biological information for the modelled processes in the form of rules rather than mathematical equations. In more complex problems this allows the direct inclusion of expert knowledge that is not necessarily restricted to hard data (for this rule-based approach see also Jeltsch and Wissel (1994), Wiegand et al. (1995) and Jeltsch et al. (1996, 1997a).

Here, we present three different spatially explicit simulation models using a grid-based approach. This has proved to be a useful way to investigate selected aspects of karoo vegetation dynamics and rangeland management. Two of the examples focus on the vegetation dynamics of certain areas of the karoo, using a bottom-up approach based on individual plants or small plant assemblages. By contrast, the third example uses a top-down approach based on forage production to model the ecological and economic system of a ranch.

14.2.   **Simulation models**

14.2.1.   Vegetation dynamics of a shrub ecosystem in the karoo

Wiegand et al. (1995) and Wiegand and Milton (1996) developed a grid-based simulation model based on field data for a typical semi-arid ecosystem at the Tierberg Karoo Research Centre (TKRC) near Prince Albert in the southern karoo, South Africa. The aim of the model was to capture the processes and mechanisms that determine the temporal and spatial dynamics of common plant species on a large temporal scale and to calculate probabilities and timespans for transitions between different vegetation states.

Detailed information about interannual variation in rainfall, and on rainfall-dependent plant attributes, were included in the model. Five shrub species dominate the plains vegetation at TKRC. These are *Brownanthus ciliatus*, a mat-forming stem-and-leaf succulent, and *Ruschia spinosa*, an evergreen leaf succulent (both Mesembryanthema: Aizoaceae), and three non-succulent species, semi-deciduous *Galenia fruticosa* (Aizoaceae), deciduous *Osteospermum sinuatum* and evergreen *Pteronia pallens* (both Asteraceae). These shrubs differ in their life-history attributes and in acceptability to domestic sheep. The dominant species do not reproduce vegetatively, *R. spinosa*, *P. pallens* and *O. sinuatum* have no seedbank, and *G. fruticosa* and *B. ciliatus* appear to have a short-lived seedbank (Esler, 1993).

The five dominant species can be divided into two functional groups. Seedlings of 'colonizer species' (*B. ciliatus*, *G. fruticosa* and *R. spinosa*) need large gaps in open vegetation to establish, while seedlings of 'successor species' (*P. pallens* and *O. sinuatum*) establish in shaded sites under the canopy of colonizer plants. The authors considered less common species in the model only as occupiers of space, and termed them 'fixed plants'. The life-histories of the plants were not considered, and the plants remained at fixed densities throughout simulated time, with their only function being to prevent colonization of cells by pioneers.

The space was divided into a grid of cells which represent mature plant sites. The local dynamics (succession) within a cell were given by the sequences ('empty' → 'colonizer plant' → 'successor plant' → 'empty') or ('empty' → 'colonizer plant' → 'empty'). For a given cell, the pathway followed and the duration (in time steps) of each state was determined by the variables which characterized the state of a cell, and the rule-set which determined how these variables change in the course of time depending upon the states of neighbouring cells, and on the external factors rainfall and management (grazing, clearing of unpalatable plants).

The rule-set contained rules on seed production, germination, seedling survival, seed dispersal, safe sites, competitive interactions, establishment, growth and mortality. Detailed descriptions of the rule-set are given in Wiegand et al. (1995) and Wiegand and Milton (1996). Although the output for the spatial and temporal simulations was in annual time steps, processes such as seed production, germination and survival depended on rainfall seasonality. For this purpose, a submodel internally calculated, on a monthly basis, the total number of seeds produced and dispersed, as well as those that germinated and survived, and summed these values for one year. The cell dynamics for a single iteration (one year) then proceeded by determining effects of neighbouring plants and competition on seedling survival, and deleted all dispersed, non-surviving seeds other than those in the seedbank. The annual iteration was concluded once time, weather and management effects on plant size, reproductive maturity and survival had been considered.

Wiegand et al. (1995) used a parameter set for ungrazed vegetation and started their simulations with a species composition typical for rangeland in good condition (Milton and Dean, 1990b). Modelling results showed that, in this case, all five species could coexist for a simulation period of some centuries. However, relative densities of component species did not reach a state of equilibrium (Fig. 14.1). Instead, an episodic, event-driven behaviour occurred, with quasi-stable periods interrupted by sudden, discontinuous changes in species composition. Sudden increases in density of colonizer species occurred when rains suitable for germination and recruitment followed long periods with rainfall not favourable for recruitment. Failure of plant populations to replace natural mortality during these prolonged periods led to a decrease in the density of established plants, and consequently to an increase in the size and abundance of gaps that served as safe establishment sites for colonizers. Large recruitment events occurred only when timing and amount of rainfall over the year facilitated seed production, seed germination and post-germination survival of seedlings (Fig. 14.2(a)), and secondly, if safe sites were available to the dispersing seeds (Fig. 14.2(b)).

The coincidence of rainfall conditions suitable for reproduction and availability of safe recruitment sites was such a rare event that large recruitment events were likely to occur only 2–5 times per century in these arid shrublands.

The stochastic and unreliable rainfall resulted in unpredictable driving events. For this reason, the future development of the plant community could be described only in terms of probabilities. To deal with this problem, Wiegand and Milton (1996) conducted subseries of 100 simulation runs for different scenarios using a different sequence of rainfall data (with the same monthly mean

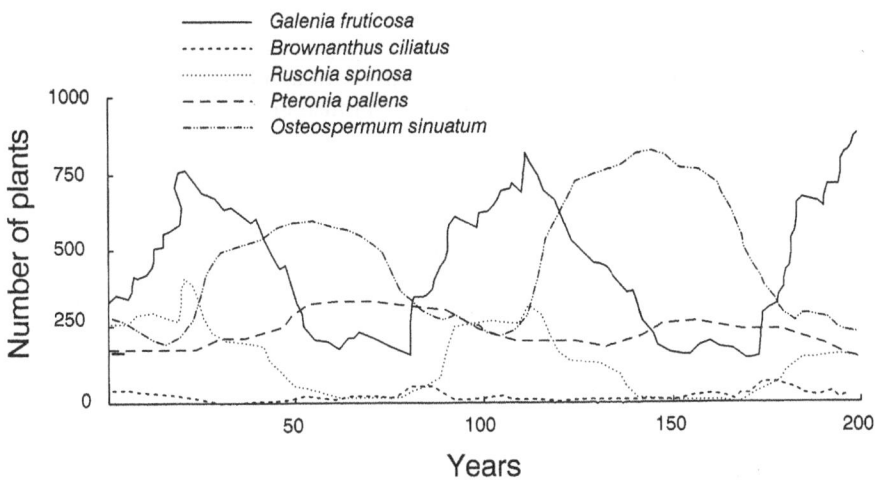

Figure 14.1 **Time series showing the mean abundance of the five plant species**

- —— *Galenia fruticosa*
- - - - - - - *Brownanthus ciliatus*
- .... *Ruschia spinosa*
- - - - - · *Pteronia pallens*
- -··-··-·· *Osteospermum sinuatum*

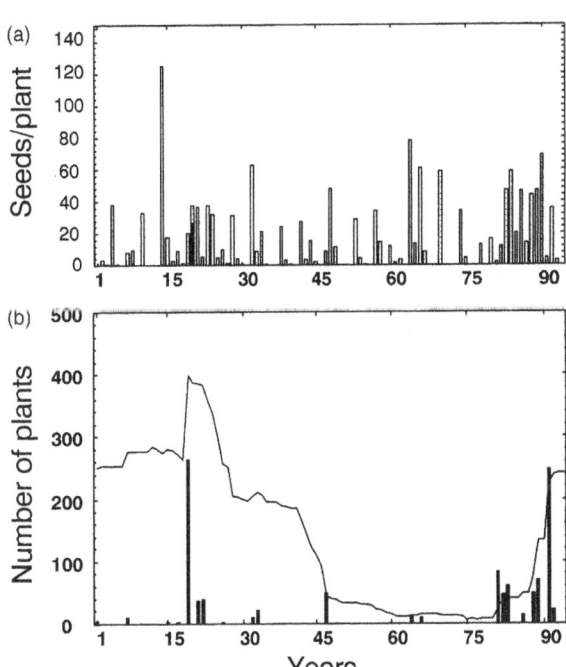

Figure 14.2 **(a) Seed production, germination and number of seedlings which would survive in safe sites of the species *Ruschia spinosa* calculated with the submodel SEED over the 93 years of the Prince Albert rainfall scenario. Bar shading shows seeds that fail to germinate (white), seeds that germinate but die (hatched) and surviving seedlings (black). (b) Time-series (92 years) for *R. spinosa* showing the number of seeds (bars) per adult plant that survive as seedlings in safe sites**

long-lived plants. The vegetation state was characterized through a grazing potential index that sums up the densities of all species in the simulation grid weighted with a sheep utilization index (Milton and Dean, 1993).

Simulation results (Fig. 14.3, Table 14.1) indicated that little improvement in rangeland condition was likely during a period of 60 years. After 60 years there was a 54% probability that the overgrazed rangeland would remain in an overgrazed condition or even deteriorate, and only a 7% probability that there would be a substantial improvement in its condition. Even such active management, as (simulated) clearing of unpalatable shrubs, resulted in only a 66% probability that degraded shrubland would be in good condition after 60 years resting.

To investigate timescales for vegetation change due to overgrazing, Wiegand and Milton (1996) conducted a subseries of simulation experiments in which 80 years of heavy, continuous grazing of rangeland in good condition was simulated. Continuous grazing was modelled by reducing the seed production of palatable species by a certain factor.

The results of this simulation (Fig. 14.4) showed that the rangeland remained (on average) in good condition for 20 years after the initiation of heavy grazing. Thereafter, the average range condition declined almost linearly, and after 50 years the probable range of the grazing potential varied from degraded to good. After 70 years of this treatment, the average range condition had declined to that of an overgrazed rangeland.

14.2.2.   Shrub encroachment in the north-western karoo/southern Kalahari

In contrast to *14.2.1.*, this model investigated the vegetation dynamics of more than one life form (a grass–shrub mix). The increase in shrub dominated areas, i.e., shrub encroachment, is thought to partly explain the decreasing

and variance as the original rainfall data) for each run. They simulated resting of rangelands in (1) good condition, (2) in overgrazed condition and (3) in degraded condition. To test the hypothesis that recovery of grazing potential may possibly be accelerated by partial clearing of long-lived, unpalatable shrubs, they simulated resting of (4) overgrazed rangeland with removal of unpalatable, stocking rate of karoo and other semi-arid grasslands (Dean and Macdonald, 1994). Because shrub encroachment is a slow process, and because domestic livestock are stocked at low densities in the karoo, field experiments for determining of stocking rates that avoid shrub encroachment under various rainfall scenarios are almost impossible to replicate. Jeltsch et al. (1997a, 1997b) therefore used a spatially explicit grid-based simulation model to investigate the shrub–grass dynamics of the north-western karoo/southern Kalahari under various realistic rainfall scenarios and stocking rates of domestic livestock. They analysed the formation of vegetation zones around artificial waterholes (Jeltsch et al., 1997a) and addressed the following questions: (1) does simulated cattle grazing lead to shrub encroachment, (2) over what timescale does the process take place and (3) are the dynamics of vegetation change continuous or abrupt?

In these simulation models an area of approximately 50 ha was subdivided into a grid of 5 × 5 m cells. Only the dominant life forms in each cell were modelled, and cells were characterized as being dominated by shrubs, perennial grasses and herbs, annuals and combinations of these. On the basis of these grid cells and a one-year time step, four different submodels were created, (1) simulating top and subsoil moisture, (2) vegetation dynamics and biomass production, (3) grazing by cattle and (4) grass fires. In the soil moisture submodel, the effective moisture content in a top and subsoil layer were calculated for each grid cell as a function of the frequency and quantity of precipitation and water losses, for example evaporation and runoff. In the vegetation dynamics and biomass production submodel the moisture levels calculated in the previ-

ous submodel were further reduced by the life forms in each grid cell separately. The available soil moisture determined changes in life form occurrence and levels of biomass production. In the grazing submodel the spatially explicit reduction of herbaceous biomass by cattle was simulated. The grazing pressure on the herbaceous component in a grid cell eventually influenced the potential biomass production and the local occurrence of the herbaceous vegetation. Finally, in the grass-fire submodel, the amount of remaining grass fuel determined the probability at which grass fires occurred. Grass fires are modelled to eventually change the potential biomass production and occurrence of life forms.

Simulation results indicated that the shrub–grass dynamics depended on the quantity and sequence of rainfall. Simulated cattle grazing led to bush encroachment under all rainfall scenarios, once stocking rates exceeded a threshold determined by long-term mean annual rainfall (Fig. 14.5). However, the stocking rate threshold for shrub encroachment was clearer under mesic than xeric climatic scenarios. This is because either competition from the herbaceous layer or rain could limit the establishment of woody plants. In relatively mesic scenarios, where shrub encroachment was limited mainly by grass competition, grazing of grasses beyond a certain threshold led deterministically to an increase in shrub cover. However, under xeric climates, where rainfall was lower and more stochastic, the rate of shrub encroachment, in response to a given intensity of grazing, became less predictable. The results also indicate the formation of distinct vegetation zones (piospheres) around artificial waterholes only at the higher rainfall sites. Under all stocking

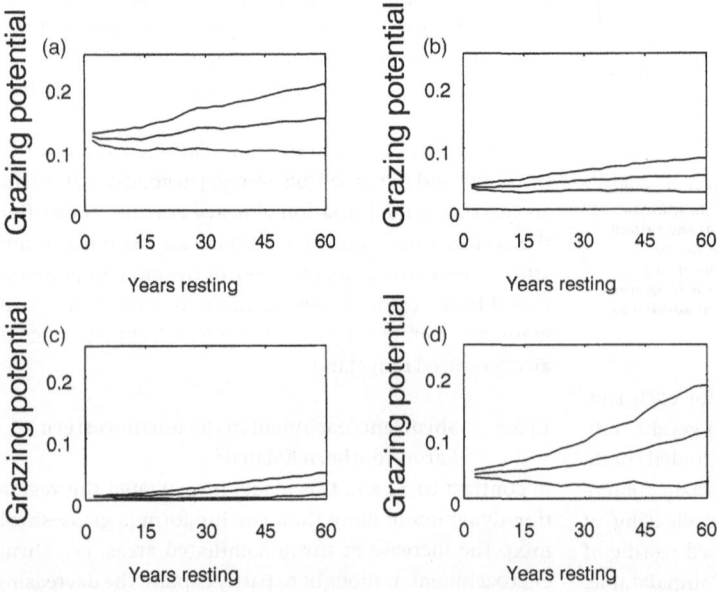

Figure 14.3 **The temporal development of the mean grazing potential index and its mean variation for the first four simulation experiments. (a) Simulation experiment 1, (b) simulation experiment 2, (c) simulation experiment 3, (d) simulation experiment 4**

Table 14.1. *The probability that the state of rangeland will change to a given condition after 10, 30 and 60 years*

Grazing potential	After 10 years			After 30 years			After 60 years		
	good	overg.	rem.	good	overg.	rem.	good	overg.	rem.
Worse	0.00	0.00	0.00	0.05	0.03	0.01	0.13	0.10	0.00
Stable	1.00	0.91	0.78	0.71	0.62	0.23	0.44	0.44	0.09
Better	0.00	0.09	0.22	0.24	0.35	0.55	0.43	0.39	0.25
Good	0.00	0.00	0.00	0.00	0.00	0.21	0.00	0.07	0.66

Left columns: resting of rangeland in good condition (good), middle columns: resting of overgrazed rangeland (overg.), right columns: resting of overgrazed rangeland and removal of unpalatable shrubs (rem.). Worse: grazing potential <0.8 initial grazing potential, stable: grazing potential (>0.8, <1.5) of initial grazing potential, better: grazing potential (>1.5, <3) of initial grazing potential and good: grazing potential >4 of initial grazing potential

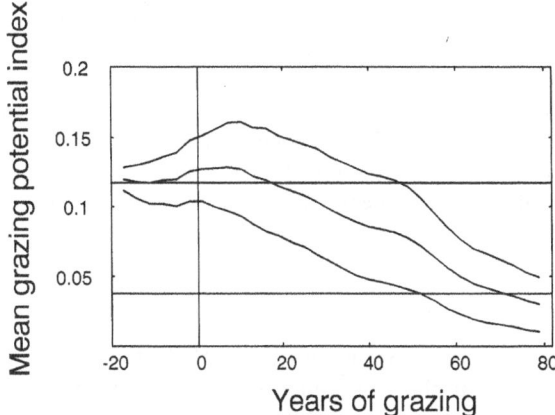

Figure 14.4 **The temporal development of the mean grazing potential index and its mean variation for a rangeland in good condition under heavy, continuous grazing. The rangeland was rested for the first 20 years (years –20 to 0), and heavy continuous grazing started at year 0. The upper horizontal line gives the value of the grazing potential index of a rangeland in good condition, the lower line corresponds with an overgrazed rangeland**

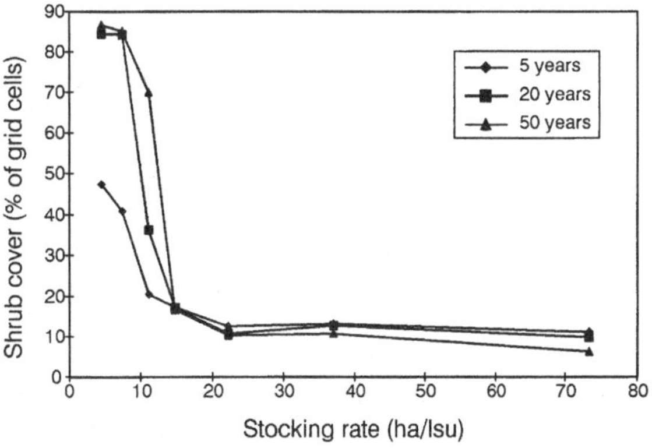

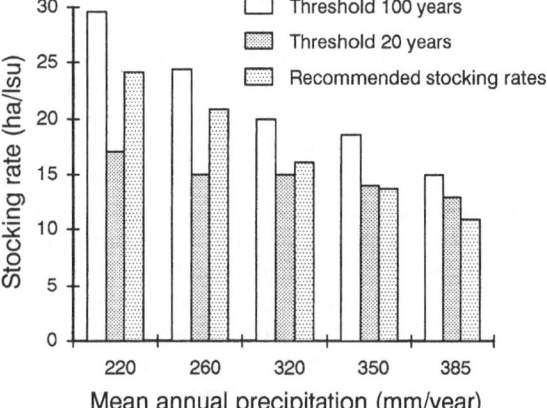

Figure 14.5 **Shrub cover (given in % of grid cells) after 5, 20 and 50 years, versus stocking rate (given in ha per livestock unit) for a sample location with a mean annual rainfall of 385 mm. The shrub cover shows a typical threshold behaviour with increasing grazing pressure (i.e., decreasing values at the abscissa)**

rates tested in the model, distinct zones of bare soil and woody shrubs developed and expanded outwards at a rate correlated with the grazing pressure. At the lower rainfall sites, the zone development was limited and strongly influenced by rainfall.

The most significant finding of the simulation experiments was that, although the stocking rates currently recommended by pasture scientists were unlikely to lead to shrub encroachment within 20 years, they had a high probability of bringing about shrub encroachment within a century (Fig. 14.6). The recovery potential of shrub encroached zones after withdrawal of cattle was negligible in a timespan of 100 years.

14.2.3.   Analysing a sophisticated grazing strategy at a ranch level

In contrast to the previous examples, Stephan et al. (1998) used a top-down approach to investigate the sustainable management strategy of a sheep ranch with the help of a simulation model. The ranch 'Gamis' is located about 250 km southwest from Windhoek at the edge of the

Figure 14.6 **Recommended stocking rates for various locations with differing mean annual precipitation compared with model results of thresholds of shrub encroachment. Stocking rates below the given threshold values show on average a doubling of shrub cover within timespans of less than 100 years or less than 20 years, respectively. In a wide range of mean annual precipitation, the recommended stocking rate is between the two threshold values. Thus, shrub encroachment can, on average, be avoided in timespans below 20 years but is likely to occur in longer periods**

Namib (mean annual precipitation 180 mm). Extensive pastoral agriculture is thought to be the suitable type of land use in this ecologically sensitive region where much of the land is now degraded due to overgrazing in the past. At Gamis, Karakul sheep are kept for production of pelts. The rangeland on the ranch (30 000 ha) is divided into 98 paddocks, 60 of which are used for grazing for up to 3 production flocks. A sophisticated grazing rotation system has prevented land degradation and at the same time ensured economic survival of the ranch. The simulation model was constructed in order to understand the basis of this successful strategy and to apply this knowledge to other situations. Instead of modelling the vegetation dynamics in detail, the model simulated rainfall and the response of the vegetation on a paddock basis according to long-term observations of the rancher. This expert knowledge as well as the grazing strategy were included in the model in the form of rules.

Stephan et al. (1998) distinguished 4 different life forms (trees, shrubs, annuals and perennial grasses) and 5 habitats with different soil types (river, slope, brackish, stones and slate). One paddock could contain all 5 soil types. Typical distributions of the life forms on the different habitats under different rainfall conditions were known from long-term observations. Trees, shrubs and perennial grasses were described in terms of their herbaceous biomass and vitality, but annuals were characterized only by their herbaceous biomass (see Jeltsch et. al., 1996). The vitality described the potential of the vegetation to produce herbaceous biomass and depended on the rainfall and on the intensity of grazing during the past years. The vitality represented the 'memory' of the vegetation and considered, for example, that production of herbaceous biomass would be lower after a long period of stress (drought, heavy grazing) than after a less stressful period. Grazing was modelled as a reduction of the vitality. The model kept track of the herbaceous biomass and the vitality of the different life forms in each paddock and simulated their temporal development in accordance to rainfall and grazing intensity. The model considered a spatial explicit distribution of rainfall and distinguished between low, average and high rainfall for each year and paddock. Rules for the yearly growth of the vitality and the production of herbaceous biomass under different rainfall conditions and different grazing intensities were based on long-term observations by the rancher.

The grazing system was rotational and comprised two weeks grazing followed by (at least) two months resting of the paddocks. Additionally, one third of all paddocks remained rested during the growing season (September–May). In the case of insufficient rainfall, certain modifications of these basic rules were employed. If rainfall was not sufficient for one year, 'spare' paddocks

were used, resting time was decreased and production of lamb was ceased or moved to rented rangeland. After two years of low rainfall, numbers of adult sheep were reduced and all paddocks were grazed without resting to distribute the grazing pressure equally over all paddocks. In the case of severe droughts (more than two years of low rainfall), all sheep were moved to rented rangeland.

Variations of the grazing rules facilitated the comparison of different strategies. Model results indicate that this strategy was able to cope with aridity and typical periods of drought. On increasing the grazing pressure in the simulations by 15%, it took 40 years before the degradation of the vegetation became apparent (Fig. 14.7). However, degradation was severe after that. Comparing alternative grazing strategies in the simulation experiments showed that from an economical point of view the present management is more successful than continuous grazing or strategies which were not adapted to the occurrence of long-time droughts (Fig. 14.8).

14.3.   **How do the models contribute to the understanding and management of karoo vegetation?**

Applied ecology disciplines such as range management are based on models of ecosystem functioning (Westoby et al., 1989). These models are a philosophical system of concepts, generalizations or assumptions rather than qualitative models that guide management strategies or collections of data. With the quantitative simulation model, Wiegand et al. (1995) were able to test the two main concepts existing for rangeland ecology, the 'range succession model' and the 'state-and-transition' model. The plant community at Tierberg Karoo Research Centre showed the essential properties described by the state-and-transition concept. These were (1) event-driven change, (2) long timescale of changes, (3) demographic inertia and lag-effects, (4) unpredictability of vegetation change and (5) huge variances in fodder production. The authors identified the rare recruitment events to be the key processes driving the dynamics of the plant community. Just the occurrence of one big establishment event after a sequence of dry years can drive the dynamics of the plant community in a completely different direction and can determine the composition of the plant community for many years. Such events are opportunities for management to influence vegetation change in a desirable direction (for example, to maintain a viable population of palatable species within the plant community) that may occur once in 10 or 20 years. To recognize these rare opportunities, managers need to monitor the vegetation and esti-

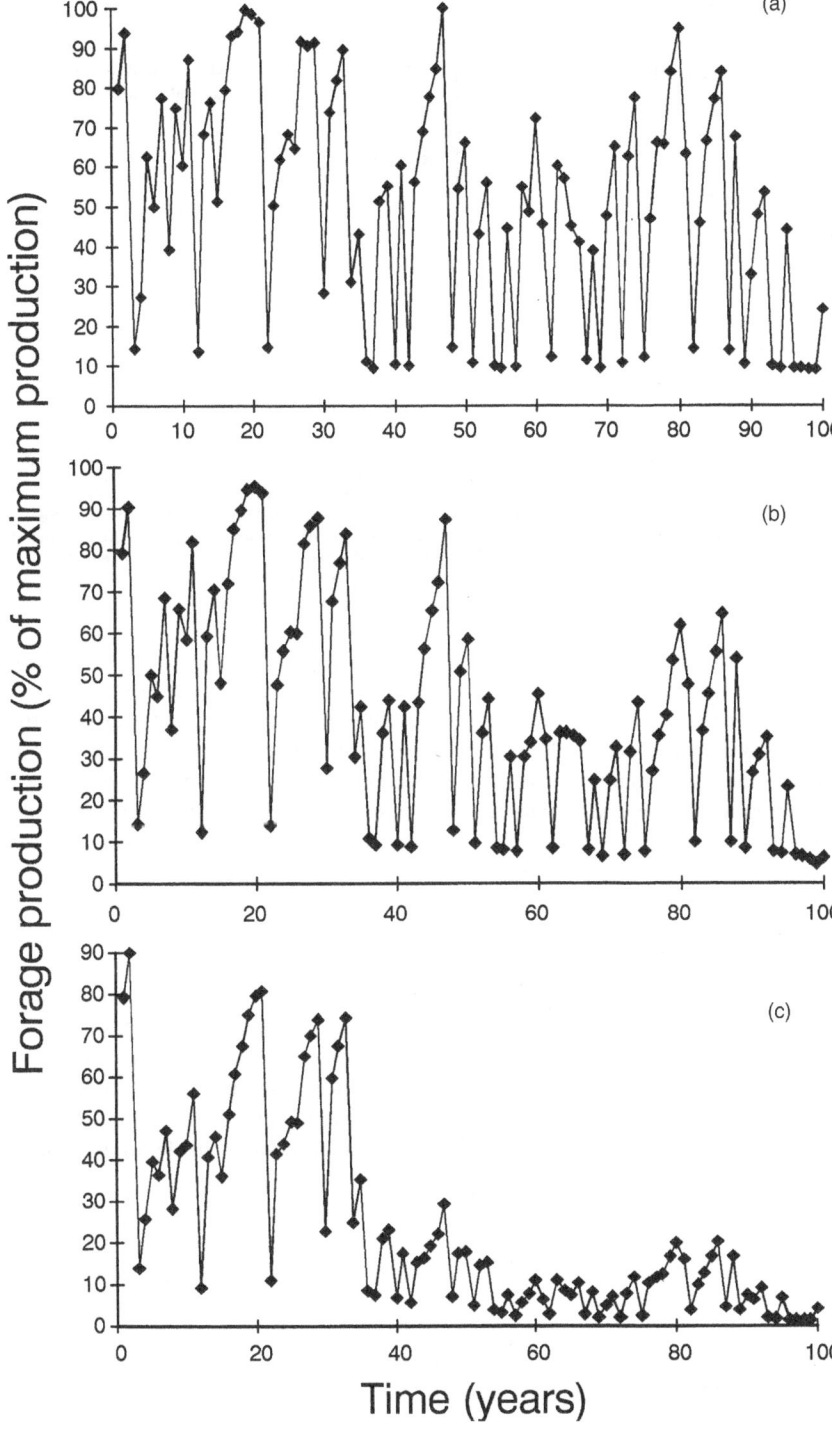

Figure 14.7 **Comparison of the effects of different grazing intensities. Sample time series of the forage production of perennial grasses produced by the simulation model. (a) no grazing pressure. (b) sustainable grazing strategy as realized at Gamis ranch. (c) a 15% higher grazing intensity than in (b). The y-axis shows the percentage of the maximum production reached without grazing in year 47**

mate the densities of safe sites for the different species or functional groups of the rangeland to assess whether a large recruitment event may be possible. Wiegand and Milton (1996) showed that changes as a result of bad management are only visible after decades, but, once overgraz-

ing becomes obvious, rehabilitation is almost impossible within economic timescales. For that reason, management should be planned for the long term. This requires details of the history (accurate records) of the rangeland and knowledge about the age-structure of the plant community. The unpredictability of events and an uncertainty and lack of precise knowledge make management necessarily risk-based (Walker, 1993). Both, 'bad periods' and 'good periods' are innate to the system. 'Bad periods' and 'good periods' are evenly distributed over centuries, but not over decades which is the economic timescale of the rancher. Even resting rangeland does not ensure an improvement in carrying capacity. Thus, managing semi-arid rangeland always involves a considerable risk, and flexibility to respond rapidly to an event is needed ('event-orientated management').

The simulations of Jeltsch et al. (1997a, 1997b) demonstrated that, even at or below recommended stocking rates, continual grazing can eventually cause a marked increase in shrub cover in a grass–shrub mix. Since the dynamics of the system seem to involve discontinuous changes, the concept of a potential carrying capacity should be replaced by a dynamic 'state response strategy' (Abel, 1993). In this context, a 'state' should be defined as a point in the 3-dimensional space of present and previous years' vegetation condition and distribution, rainfall, and disturbances (including fire and grazing). Stephan et al. (1998) show that such a strategy can be ecologically and economically successful. If, as results of the arid savanna model suggest, shrub encroachment shows a threshold behaviour, production might be economically maintained under heavy stocking rate for a number of years. But, even though effects of grazing are variable in the short term (years), over the longer term (decades) the economic cost of controlling the increase of woody cover will drastically increase. In the Gamis ranch model, as well as in the Tierberg shrubland model, the effects of overgrazing became visible only after a delay of decades. Using the traditional but outdated range succession concept, where change is perceived as gradual and reversible, it would be difficult to find a good measure of how much utilization is ecologically acceptable (Westoby et al., 1989). But, as the savanna model results emphasize, gradual, reversible changes only occur when stocking rates are below a threshold. This limit could be defined as a 'state' within

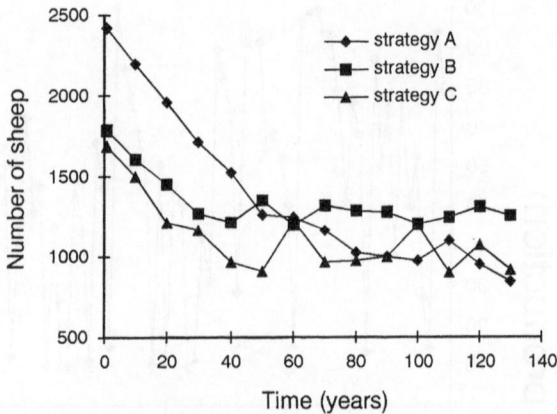

Figure 14.8 **Analysis of the effect of three different grazing strategies on the long-term carrying capacity of the Gamis ranch (given in number of sheep that can be stocked). Strategy A: continuous grazing. Strategy B: in years of sufficient rainfall one third of the paddocks are rested, in years of low rainfall all paddocks are grazed. Strategy C: one third of the paddocks are rested every year. The values are averages for 20 simulation runs**

the 'state-and-transition' concept (Westoby et al., 1989). In this line of argument, transition corresponds to the stocking rate threshold at which rapid changes towards a shrub-dominated state occur. Unfortunately, the high stochasticity of the process of shrub encroachment, particularly in more arid regions, makes it difficult to identify and distinguish 'states' and 'transitions' for a given concrete example. Analysing such successful management strategies as used at the Gamis ranch could be a promising approach to sustainable landuse in the karoo. Decision support systems based on simulation models (Teague and Smit, 1992; Abel, 1993), together with long-term field studies, could help to determine management strategies and stocking rates appropriate for given states of rangelands in various localities, thereby improving the utilization of the biomass produced in high rainfall years and avoiding overutilization under conditions likely to precipitate change.

Human impacts

W. R. Siegfried

The following three chapters treat some of the basic and major prehistoric and modern land use economies and their impacts in the karoo. In keeping with the theme of this book, the impacts are expressed essentially in ecological contexts, rather than in terms of economics or sociology. The chapters do not constitute a complete and comprehensive account of human impacts on karoo ecosystems.

The thesis running through much of the discourse in the following chapters stems from a premise that the environment influences people, and people influence the environment. In this context, the two primary environmental factors in the karoo are climate and geology or, more particularly, soils. Summary propositions on these two factors follow.

The environment

The soils of the karoo are not deficient in basic elements and minerals (Watkeys, this volume); contrasting, for example, with the condition of the neighbouring fynbos biome in which nitrogen and phosphorus are in scarce supply (Stock and Allsopp, 1992). As a generalization, it is fair to say that primary production is not limited by infertile soils in the karoo.

The ambient temperature regime for much of the karoo is temperate (Desmet and Cowling, this volume). Prolonged extremes of cold (below 5 °C) and heat (above 35 °C) are generally abnormal. The relatively low incidence of persistent frost appears to be important in the karoo's extraordinary development of succulent vegetation (Cowling and Hilton-Taylor, this volume). In short, ambient temperatures are relatively unimportant in limiting primary production, and human occupation, in the karoo.

The availability of water is of overriding importance in influencing production, both primary and secondary, and human dispersion in the karoo. The average total annual rainfall ranges from less than 100 mm in the west to 500 mm in the east. The karoo is an arid to semi-arid area, with parts receiving less than an average of 100 mm per annum being classed as desert. Moreover, the karoo is subject to a relatively high incidence of persistent wind. This enhances desiccation, with significant consequences for the type of vegetation and its production (see, for example, Grace, 1977, and the review by Ennos, 1997).

The bald statistics of low rainfall alone are far from adequate in exemplifying the major environmental factor influencing production and shaping human subsistence in the karoo. After all, dryland wheat and other cereal crops are cultivated successfully at rainfall levels below those of the karoo and as low as 300 mm per annum in several regions of the world. This is only possible in those regions because their co-efficient of interannual variation in rainfall is significantly lower than that of the karoo, and because of the karoo's relatively low threshold of non-equilibrial ecosystem dynamics. All that this means is that the availability of water in much of the karoo is highly variable and unreliable (cannot be predicted with much confidence) in both space and time. It is this fundamental external forcing factor, more than any other, that sets the character of the karoo, expressed in its primary and secondary production, the subsistence of its people and their impacts on the ecosystems they exploit.

In passing, it should be noted that, in the karoo, unlike in its neighbouring biomes (fynbos, savanna and grassland) and in semi-arid Australia, fire (caused either naturally or artificially) is not an important determinant of ecosystem form and functioning.

The people

Before the arrival of European settlers, the people of the karoo, for hundreds of thousands of years, subsisted by exploiting a largely unpredictable, dynamic mosaic of primary and secondary resources (Smith, this volume). These prehistoric people did not cultivate crops. Instead, they moved and adjusted their population densities in response to, and in accordance with, the spatial and temporal dynamics of wild resource availabilty. This is not to say that all and everything was ephemeral and nomadic. Some degree of permanency, or at least semi-permanency and migration (movement between fixed points) was involved as well. This would have been promoted by the fact that much, perhaps the greatest part, of primary production in the karoo occurs underground (itself an adaptation to the fundamental physical factors driving the ecosystems). In this context, it is likely that the digging stick was of greater importance to the subsistence of these people than the bow and arrow. In other words, roots, bulbs and other subterranean storage organs of plants formed the staple food of the hunter-gatherer communities. The karoo, or at least parts of it, is known to have extraordinarily diverse and abundant geophyte assemblages (Cowling and Hilton-Taylor, this volume).

People (*Homo sapiens*) herding domestic livestock probably have been present in the karoo for some thousands of years, as opposed to the hundreds of thousands of years of occupation by hunter gatherers. The prehistoric herders practised a nomadic life style. This pastoral life style was copied by the first Europeans in the region, but extensive pastoralism of a nomadic, as well as as of a migratory, nature had all but ended in the karoo at the onset of the twentieth century.

Impacts

It is popularly believed that the prehistoric people of the karoo lived in a state of 'perfect harmony' with their environment. More importantly, however, is a recognition that, as far as is known, the impacts of these people on their environments did not cause any irreversible damage to either the form or functioning of major ecosystems. The same cannot be said for modern inhabitants during the last 200 years in the karoo. Permanent changes to ecosystem composition (e.g. biotic extinctions) are widespread, varied and many, and many more appear to be inevitable in the future if current trends are maintained. What is less clear, indeed it is not at all clear, is whether these changes have resulted generally in a reduction in the production capacity of the karoo under present systems of exploitation. In reviewing the available evidence (of which some is treated in the chapters of this book), it is remarkable how little research has been aimed at resolving what amounts to a fundamental and crucial set of problems: what is the production capacity; what can be exploited sustainably and what is not sustainable, and over what periods; and what is acceptable ecologically, economically and sociologically?

The changes (at least within several human generations), both locally and more widespread, that have occurred in the functioning of karoo ecosystems stem basically from a superimposition of alien land use practices over inherently incompatible indigenous ecosystems. More particularly, the colonists' imperative to establish regularly repeatable, constant patterns of settled subsistence, characteristic of northern European environments, caused the initial and subsequent impacts of change. These impacts would not have been as widespread and severe as they have become, had the initial attempts at inducing and reproducing stability of extraction on an unstable base been left to fail, as fail they must if left to their own devices. Conceivably, such early failures would have left the ecosystems with enough capacity to recover eventually to their former full states. Instead, an array of long-term (decades of) financial subsidies from the state encouraged settlers to continue to practise land use that all the time enhanced the level of mismatch between indigenous ecosystem capacity, on the one hand, and superimposed alien extraction, on the other. In short, without importing foreign capital, it is impossible to have an enduring stable system of extraction based on an inherently unstable system of production whose performance is largely unpredictable in space and time.

15 Hunters and herders in the karoo landscape

A. B. Smith

15.1. Introduction

When the first trekboers moved into the karoo in the eighteenth century they were entering an environment that had already been impacted by human use for over three million years. While the impact would have been minimal for most of this time, increasing skills in modifying and manipulating the environment, such as driving game with fire, meant that humans played a greater and greater role in the distribution of plants and animals. Sampson (1986) has noted that around prehistoric hunter campsites in the karoo surface erosion is evident, and alterations have occurred to plant communities. The latter was evident in the consistent growth of *Lycium* sp. on sites, as well as seasonal growth of *Arctotheca calendula*, *Salvia verbenaca*, *Oxalis depressa* and *Ifloga paronychioides*, all of which are considered to be weeds. According to Sampson (1986: 41) this damage amounts to 5.6% of the rangeland (excluding dolerite ridges and hills). The greatest damage would have been around important resource loci, such as waterholes.

15.2. Human adaptation to arid lands

While the karoo appears to be a dry and difficult environment today, and would have possibly been even more so without underground water being available, in the past this was not a stumbling-block to human habitation. People knew where water was to be found, even if they had to dig in dry river beds or for underground water-bearing plants to get it. Water is the least available necessity, so presumably controlled to a considerable extent where people would have been found before underground aquifers were tapped by wind-driven pumps at the end of the nineteenth century.

Since most of the interior of the Cape is dry, any rainfall is a bonus, and creates micro-environments for animal and plant life when it occurs. This 'event-driven' system (Walker et al., 1986) could only be used on an opportunistic basis by highly mobile human groups taking advantage of highly localized conditions. Resources would thus tend to be sporadic and patchy. Patchy resources are some of the constraints of an arid environment which require an 'r-strategy' response for exploitation (see Dyson-Hudson, 1984, for application of this biological concept to humans). Limiting group size for mobility is part of this strategy, so has an effect on social organization. It is also known that hunters of arid Southern Africa have a low reproductive rate (Kolata, 1974).

Exploitation areas would have to be large enough to permit families to find enough food throughout the year. These areas would need to allow for localized rainfall events, as well as contain several permanent waterholes. They would also contain the raw materials for making implements, such as fine-grained cryptocrystalline rocks for knives and arrowheads.

Plant foods constituted the bulk of the food eaten by the aboriginal people of the karoo. While a complete compendium of useful plants has never been compiled for the different areas of the karoo, Archer's (1994) important work on the Richtersveld can be used as a guide to indigenous botanical knowledge. She has three categories of useful plants: edible (70 species), medicinal (40 species) and other useful plants (43 species + 14 used for firewood). Table 15.1 gives the breakdown by parts utilized (different parts of some plants can be used, and some plants can have a wide variety of uses). From this regional database we can extrapolate much of the knowledge by indigenous hunters and herders over the entire karoo area, albeit with different species. Richard Lee (1969: 59) has noted that the Kalahari Ju/'hoansi of northern Namibia and Botswana differentiated over 200 plant and 220 animal species, of

Table 15.1. *Useful plants of the Richtersveld*

Plant parts	Edible	Medicinal	Other utilization
Gum	2		2
Stems	12	2	4
Flowers	6		2
Roots	10	15	
Leaves	11	22	6
Fruits	15		
Tubers	7		
Corms	5	1	
Seeds	2		1
Pods	3		
Bark		4	3
Latex		2	1
Caudex	1		
Wood			6
Branches			12
Reeds			5
Juice			1
Firewood			14

After Archer (1994)

which 85 plants and 54 animals were considered edible. In the case of the Dobe Ju/'hoansi, they had a staple, the Mongongo nut *Ricinodendron rautanenii* which accounted for up to two thirds of the annual vegetable diet by weight. For the karoo it is unlikely that a similar staple was available, as there does not appear to have been a single food resource that was concentrated in such quantity.

Territories would have varied in size, depending on resources, but estimates of the areas used by the /Xam Bushmen (see endnote) of Bushmanland at the end of the nineteenth century would suggest they ranged from approximately 400 to 1000 km^2. This overlaps with Lee's (1979: 334) estimate of the exploitation areas of the Ju/'hoansi as being between 300 and 600 km^2, although the smaller areas may be due to the staple described above. The Ju/'hoansi prioritize their foodstuffs. Those more desirable foods would be eaten first, and gradually, as the season wore on, they would revert to less desirable foods, and during periods of drought famine foods would be the last resort.

The Ju/'hoansi people could be mobile during the rains because of temporary pans filling up. They would fall back on the permanent waterholes as the dry season progressed, and their main social activities, such as reciprocal *hxaro* exchange, would take place at these aggregation points (Lee, 1979: 446).

By contrast, the G/wi Bushmen of the central Kalahari (Silberbauer, 1981) did not have permanent waterholes. They would tend to aggregate during the wet season when there was plenty of water for larger groups, and disperse during the dry season when they had to rely on underground water-bearing roots or *tsama* melons for their moisture.

The deep permeable sands that restricted artesian water in the central Kalahari would have been less of a

problem in the karoo, where artesian springs would have permitted a pattern probably similar to that of the Ju/'hoansi of aggregation around waterholes in the dry season. An archaeological observation has been the discovery of ostrich eggshell caches where water was stored by burying the eggs in the sand. At Drögrond, 50 km south of Kakamas, the farmer uncovered six of these eggs, each with a small 1.5 cm hole at one end, when he was bulldozing a new dam. These eggshells presumably would have had fibre 'stoppers' when they were used as containers (see Wannenburgh et al., 1979, plate 35 for an example). Such caches would have allowed the owners use of a wider area away from water in an emergency.

Such emergencies could have arisen when a group of men were tracking a wounded animal which wandered for several days before it died. The hunting techniques used in the karoo included those of the Ju/'hoansi: relatively weak bows, relying on poisons to immobilize the prey. We can say this on the basis of similarities in the arrows used in both areas. The arrow was made in three sections: (a) the point (this might be a long, slender polished bone, or a wooden piece with a metal or stone tip. The tip was not poisoned, only the lower part of the point); (b) a bone or wooden link shaft (separating the point from the main shaft); (c) the reed shaft (with flight feathers and a nock for the bow string at the end). The principle was to allow the poison to work its way through the animal. Even if the prey ran through thick bush and the shaft was dislodged, the tip would separate at the link shaft and stay in place.

Such equipment required the hunter to get quite close to his prey. They did this by use of camouflage, such as wearing the skin of an ostrich (Fig. 15.1) or vegetation (Fig. 15.2). In addition to hunting with bows and arrows there are records of pit-falls being used (Fig. 15.3). These were often used in conjunction with brush fences that channelled the animals towards the trap.

Information about the living conditions of hunters in the area under study is rare, since the way of life was sadly gone by the end of the nineteenth century. We are fortunate that, in 1872, the geologist Dunn travelled across Bushmanland. He passed a hunter's camp and described it thus (1873: 35):

> The *werfs* were of the usual description. A circle of bushes with a fire-place in the centre. Stuck into the bushes are the digging sticks of the women, the deadly arrows tipped with puffadder poison, and the bow made of karee wood, famed for its strength and toughness. The string is made of twisted springbok sinews, is very strong, but unserviceable in damp weather. Curious culinary utensils: a dish made from the upper shell of a tortoise; a brush made from the long hairs of a hyena – it is used for

Figure 15.1 **Watercolour sketch by Charles Davidson Bell, c. 1835, showing ostrich camouflage used by Bushmen in hunting wildebeest. UCT Manuscripts Division, Bell Sketchbook no. 11:C27 (with permission of the John and Charles Bell Heritage Trust, University of Cape Town)**

'whipping' ostrich eggs in the shell, after breaking a hole in one end by being twisted rapidly round, for eating soup or milk, and also if the weather be hot as a handkerchief for wiping perspiration from the face; spoons made from ribs of gemsbok, used for extracting cucumbers; dishes made from ostrich leg-bones. Half a dozen springbok *zakkies* await an enterprising purchaser, while a bran [sic] new leather gown, doubtless of the most fashionable cut in these parts, lies on top of an adjoining bush. Many of the old ladies wear a kind of cowrie shell fastened by threads to the centre of their foreheads. These ladies are given to painting [their faces].

On the latter point, Dunn (1872: 380) also came across a pit where the Bushmen had been following a vein to extract soft red ochre to use as paint.

15.3.   Early humans in the karoo

The earliest traces of human occupation in the karoo goes back to at least 3 million years. This is the estimated date,

based on comparable associated fauna from East Africa, for the Taung skull (*Australopithecus africanus*) recognized and named by Raymond Dart in 1924. There is no reason to doubt that human use of the karoo landscape was ever interrupted from then on. Acheulean hand axes and cleavers have been found at a number of localities (Fig. 15.4), often on the surface (see Sampson, 1974: Fig. 40), but *in situ* in several sites, for example Montagu Cave (Keller, 1973), Pniel (Beaumont, 1990a), Rooidam (Fock, 1968), Doornlagte (Mason, 1966), Kathu Pan (Beaumont et al., 1984), Equus Cave (Klein et al., 1991), etc. attesting to widespread occupation by *Homo erectus*.

While the economies of these early Stone Age inhabitants of the karoo are hard to gauge, we do know that meat was obtained from a wide range of large mammals. The faunal record from Pniel (Klein, 1988), for example, has given us a list which includes elephant, rhino, hippo, *Megalotragus* (giant alcelaphine) and *Pelorovis* (giant buffalo), as well as a number of smaller animals like pigs and bushbuck. There has been a great deal of discussion in the literature about how the meat was obtained (Shipman and Phillips-Conroy, 1977; Shipman, 1983; Blumenschine and Cavallo, 1992). The general consensus is that it was probably scavenged, but *H. erectus* was aggressive enough to be

246   A. B. Smith

Figure 15.2 **Companion sketch to Figure 15.1 showing a camouflaged Bushman hunter stalking springbok. UCT Manuscripts Division, Bell Sketchbook no. 11:C26 (with permission of the John and Charles Bell Heritage Trust, University of Cape Town)**

capable of chasing most other carnivores off their kills, so had access to good 'cuts' of meat.

The transition from *H. erectus* to archaic *H. sapiens* has been notoriously difficult to date, as it occurred during a period beyond the range of radio-carbon dating (i.e. 50 000 years). Several other dating techniques are being devised, and a consensus of around 250 000 years ago would seem to be the best fit. Middle Stone Age occurrences are even more ubiquitous across the degraded landscape of the karoo. Almost any farm will produce at least some surface material, and, close to raw material sources, these surface indications can be considerable. The differences between early Stone Age and middle Stone Age lies in the biological development of archaic *H. sapiens* and changes in technology, seen in the disappearance of hand axe forms, to be replaced with a technology based primarily on flakes and blades. This has ramifications about how meat was obtained. These flake tools were now hafted onto a wooden shaft to create the first compound tools. Now humans were capable of actually hunting and immobilizing animals bigger than themselves, since these weapons could be thrown, and the dangers of hunting large game considerably diminished.

The good preservation of bones on archaeological sites means that meat-eating, and thus the male side of food procurement, tends to dominate the vision of the diet of early people. Plant foods, collected by women, most probably were more important. A study of the potential for exploitation of underground plant foods by Vincent (1985) has shown that, once a simple digging stick was devised adequate plant food would have been easily available. While her study is in the savanna zone of East Africa where there are many tubers, there is no reason to doubt that similar early exploitation could have taken place in the karoo for underground parts of geophytes.

Although most of the caves containing middle Stone Age levels, and thus long sequences, are along the south and west coasts (Thackeray, 1992), there are a few in the area under study, e.g. Wonderwerk (Beaumont, 1990b), Diepkloof (Parkington and Poggenpoel, 1987), Elands Bay Cave (Parkington, 1992), Montagu Cave (Keller, 1973), Zoovoorbij Cave (Smith, 1995), as well as a number of *in situ* open sites: Kathu Pan (Beaumont, 1990c), Orangia (Sampson, 1968), Zeekoegat (Sampson, 1968), Florisbad (Kuman and Clarke, 1986), Bundu (P. Beaumont, personal communication), many of which have associated faunal remains.

Figure 15.3 **Watercolour sketch by Charles Davidson Bell showing rhinoceros trapped in a Bushman pitfall. UCT Manuscripts Division, Bell Sketchbook no. 11:C29 (with permission of the John and Charles Bell Heritage Trust, University of Cape Town)**

The attributes of this early form of *H. sapiens* most probably included speech, and we have clues from pieces of ground ochre found in south-coast sites that red pigment was used, probably for decoration. It is also within this period that intentional burial of the dead took place, although this has yet to be demonstrated in Southern Africa. Some of the earliest remains of *H. sapiens* come from South Africa, a fact which has contributed to the debate on the origins of modern people, and supported the hypothesis that Africa was the source (see Wilson et al., 1992).

The appearance of modern people, *H. sapiens sapiens,* in southern Africa heralded a new phase of development. This probably occurred around 40,000 years ago, and it is where we first see the small stone tools characteristic of the later Stone Age. Along with this went behaviour and cultural activity that was essentially modern. This would have included social organization based on family life, sexual division of labour and equal distribution of product. Based on studies of modern Bushmen of the Kalahari (Lee, 1979; Lee and Devore, 1976; Marshall, 1976) no leaders are accepted except to perform specific tasks, such as hunting. This egalitarianism cross-cuts gender, because women produce more than men in the daily foraging activities. Such a model of hunting society is supported by studies of simple mobile band communities elsewhere who obtain their food from the environment and use it immediately (Woodburn, 1988). This contrasts markedly with more sedentary hunters whose reliable resources allow them to stay in one place for most of the year and which result in food storage. Such economic wealth can create surpluses, especially as population density increases on the most productive land. This can result in population pressures building up (see Keeley, 1988) and surpluses being used to create social and political hierarchies through control and redistribution of products.

Roughly 2000 years ago, the first agricultural people appeared in southern Africa, bringing with them domestic animals, plants and pottery. The impact on aboriginal hunters was a reduction in their traditional prey species as domestic animals used up parts of the grazing. In addition, the new immigrants brought with them a structured society with leaders and a concept of private property, either in the form of herds, or control over land for fields. While the initial impact may have been slow, gradually hunting people would have been pushed more and more

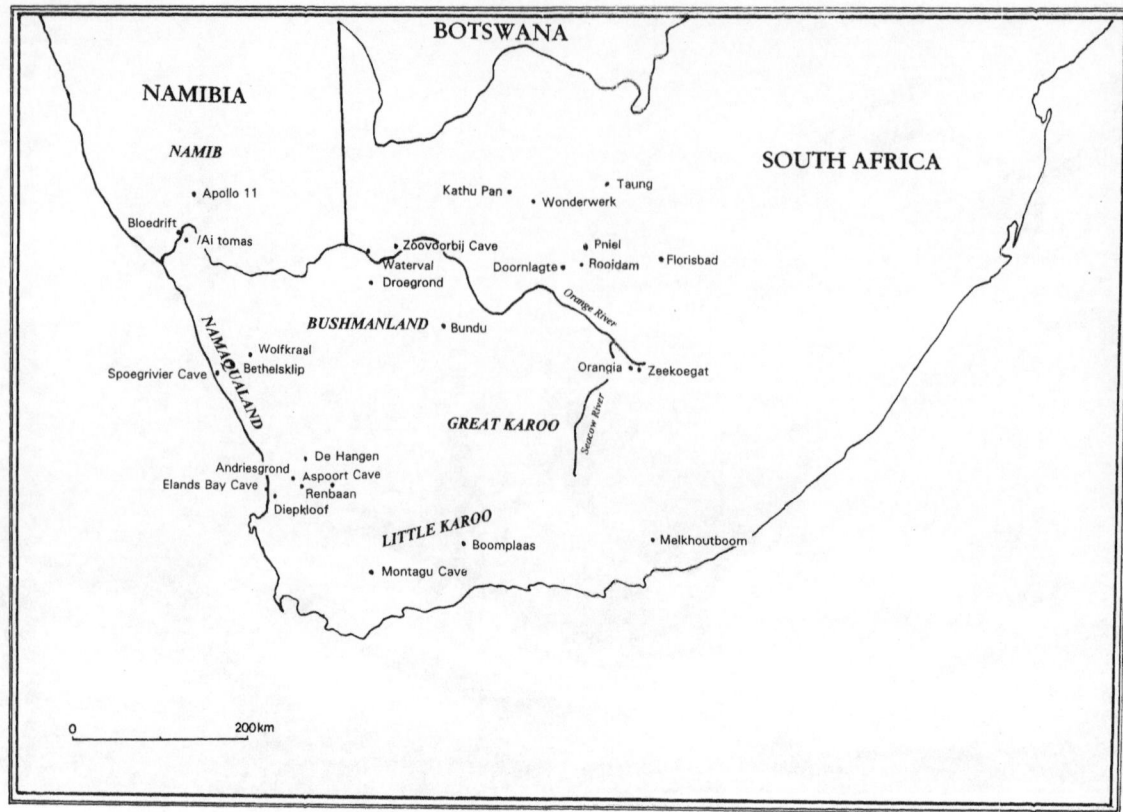

Figure 15.4 **Map of South Africa showing sites mentioned in the text**

on to marginal grazing lands. With the rise in importance of cattle herding and wealth stored in these animals, hunters would have found themselves even more on the periphery of the dominant society, either working for them as hunters, honey gatherers, rain-makers etc., or as stock thieves raiding the herds. Since much of the Great Karoo and Bushmanland was marginal for traditional stock-keeping, hunters were only slightly interfered with in these areas by herders until Europeans began taking land for farms in the eighteenth century. So low was the European population density until well into the nineteenth century that the Bushmen held their own, and were able to raid with impunity.

15.4.   **The later Stone Age**

The later Stone Age archaeology of the karoo biome should be divided up into a number of regions: the Cape west coast, Namaqualand, northern Namaqualand and the Namib, Bushmanland and the Middle Orange River, the Cape Fold Belt (west) and the Tanqua Karoo, the Cape Fold Belt (south) and the Little Karoo, and the Great Karoo. Each of these regions has its own special resources that were exploited by aboriginal people, so should be looked at separately.

15.4.1.   **The Cape west coast**
A great deal is known of the later Stone Age of this region, in particular the Verlorenvlei which has been studied intensively over the past 25 years (Parkington, 1976, 1981, in press; Parkington et al., 1988). The work has looked at human adaptation to later Pleistocene and Holocene environmental changes by reconstruction of the resource base over this period. The coastline has significantly altered, causing variable estuarine conditions at the mouth of the Verlorenvlei as the sea rose to its present level. This is evident, not only in the geomorphology (Miller et al., 1993), but in the species of fish caught by the early inhabitants of Elands Bay Cave (Poggenpoel, 1996). All marine resources were tapped by people living along the coast. In particular, seal exploitation goes back to middle Stone Age times, and would have been a regular source of fat for people who regarded this substance as having greater than purely nutritional value. Lewis-Williams (1981: 48–50) notes that

among the Kalahari Bushmen fat and sex are linked, even as far as saying 'to eat fat' is to have sexual intercourse. The fat of the eland is especially highly regarded, as the eland is the 'rain animal' and its fat has potency in ritual (Lewis-Williams, 1981: 51). The coastal site of Dunefield Midden, in addition to seal bones and other marine foods, has produced the bones of several eland (Nilssen, 1989). One reason for fat's importance to southern African hunters is the dearth of this substance in most of the game animals. The eland is similar to domestic cattle and is an exception, which gave it even greater value and potency.

Elands Bay Cave has a long sequence covering most of the period since the Last Glacial Maximum, c. 18 000 years ago. Of note is a hiatus in occupation between 8000 and 4000 years ago, which is repeated at other sites nearby. Why this should have existed can only be surmised: possibly due to a significant dry period that meant there was little fresh water anywhere in the vicinity, especially if the Verlorenvlei opened up to the sea. At Steinbokfontein, north of Elands Bay, occupation did exist during this period, perhaps because the 'fontein' remained active.

In addition to a 'wide' range of marine resources found at Elands Bay Cave – seals, marine birds, molluscs, fish etc. – are terrestrial species, the most abundant being the angulate tortoise, *Chersina angulata*. An interesting feature noted by Parkington (1992) is that tortoise bone density peaks coincide with the first appearance of limpets as the shoreline reached its present level at the end of the Pleistocene before 10 000 BP. In addition, large numbers of ostrich eggshell fragments (with relatively few beads, suggesting the eggs were predominantly a food source), as well as increased numbers of neonate *Raphicerus* spp. bones, appear at this time. This may be an indicator of stress and adjustment to the new conditions people were faced with as sea levels rose flooding much of the coastal plain. An alternative, but complementary, interpretation has been offered by Parkington (in press) who suggests that the cave was being used for domestic purposes at the beginning of the Holocene, where previously it had been a hunting camp. This view implies that whole families began using the cave at this time and women were doing the collecting of the small 'packages', such as limpets and tortoises.

The coastal plain offered a range of wild plant foods as well as terrestrial mammals. Among the plants exploited were underground corms of the Iridaceae used seasonally as an important carbohydrate source. Table 15.2 lists the plants identified from excavations at Elands Bay Cave and Diepkloof. The berries of *Nylandtia spinosa* (skilpadbessie) were another seasonal resource, as were the fruits of *Carpobrotus edulis*. In addition to food plants were those used for medicinal and other purposes, such as the dried seaweed *Zostera capensis* used as bedding in caves, and *Ecklonia maxima* used in its dried form to store oil.

Table 15.2. *Plant species identified from Western Cape sites*

Species	Elands Bay Cave	Diepkloof
Submerged aquatic plants		
Ecklonia maxima	x	
Suhria vittata	x	
Zostera capensis	x	
Reeds/grasses		
Ehrharta cf. *calycina*	x	
Willdenowia sp.	x	x
Geophytes		
Veltheimia glauca	x	
Boophane sp.	x	x
Moraea cf. *fugax*	x	
Babiana sp.	x	x
Gladiolus sp.	x	x
Antholyza plicata	x	
Watsonia sp.	x	x
Oxalis sp.	x	
Succulents		
Aizoon sp.	x	
Carpobrotus cf. *edulis*	x	x
Malephora sp.	x	
Ruschia cf. *maxima*	x	x
Creeper		
Cysticapnos vesicarius	x	
Shrubs		
Myrica cordifolium	x	
Leucadendron sp.	x	
Nylandtia spinosa	x	x
Ricinus communis	x	
Helichrysum sp.	x	
Chrysanthemoides sp.	x	
Trees		
Ilearia argentea	x	
Rhus sp.	x	
Cassine parviflora	x	
Euclea sp.	x	x
Diospyros sp.	x	
Olea sp.	x	x

After Liengme (1987)

Analysis of the different plant residues from Diepkloof (Parkington and Poggenpoel, 1987: Table 15.3) showed that 10% by weight came from Iridaceae waste (e.g. corm tunics of geophytes), 16% from *Nylandtia* seeds, and 10% from *Leucadendron* seeds. The remaining 64% were mostly woody parts and charcoal from trees, and grasses.

15.4.2. Namaqualand

This area comprises the coastal forelands south of the Orange River. The mountains of the Kamiesberg allow good rains to develop in winter, and the granite substrate produces good quality grazing. Coastal resources were utilized intensively by foraging people, as demonstrated by the excavations at Spoegrivier Cave (Webley 1992b). The bottom of the sequence is dated to 2100 BP (Sealy and Yates 1994), which is an accelerator date on sheep bones, the earliest in the Cape. The fauna tends to be dominated by seal bones, with small antelope next in quantity. The range of animals identified covers most of the known mammalian fauna from the area, both terrestrial and marine. Shellfish were an important part of the diet, as were plant foods. Berries from *Rhus* sp. and *Euclea tomentosa* were two of four types identified. The remains of a poisonous geophyte with medicinal properties, *Boophane disticha* were also recovered. Although this plant is not currently used by people in the Namaqualand Reserves (Webley, 1992b: 7), according to Smith (1966: 228) the inner scales of the bulb were used to make arrow poison by indigenous people. These data suggest the Spoegrivier Cave was occupied by hunters with access to some domestic stock.

In the interior sandveld towards the mountains few sites have been found, but what we know of the later occupation by the Namaqua, a Khoikhoi group described in some detail by Gordon in 1799–80 (Smith and Pheiffer, 1992), this was good pasture country. The site of Bethelsklip in the Leliefontein Reserve gave a lower date of *c.* 800 BP, from which has come the earliest-known record of goats on the west coast. Included in the fauna from this site are the bones of smaller antelope, cf. *Raphicerus* or *Sylvicapra*, but dassies (*Procavia capensis*) and tortoises predominate. Corm tunics and bases of *Cyanella hyacinthoides* and *Oxalis* sp. were found at the site (Webley, 1984, 1986).

Further east is another site, Wolfkraal, which has an upper part of its sequence similar to Bethelsklip, but goes back in time as far as the middle Stone Age. As it is a smaller site, the later occupation is regarded as a stopover between the winter-rainfall area of Namaqualand and the summer-rainfall area of Bushmanland. Just such sojourns were noted by Gordon in his journal entry for 27 September 1779 (Raper and Boucher, 1988:298) when he described a Bushman named 'Uijsita' who had travelled to Namaqualand from one *graafwatertje* to another. He was probably a Grassveld Bushman or Taaneina (Grass Bushman, as per Deacon (1986a) from the Kubiskouberg near Loeriesfontein (see Raper and Boucher, 1988: 214).

15.4.3.   Northern Namaqualand and the Namib

This area across the Orange River into southern Namibia is much drier than Namaqualand proper. Within the great bend of the Orange, near its mouth, lies the Richtersveld, the largest of the Namaqualand reserves. In spite of its aridity, its soils are relatively shallow, so plants react quickly to even small amounts of rain. Many of the species are good foraging plants for animals, and, with the river as a lifeline, it has been used by hunters and herders during the later Holocene,

About 50 km from the Atlantic coast is the site of /Ai tomas. Excavation was carried out at the base of one group of boulders, producing a dense scatter of pottery and stone artefacts (Webley, 1992a). A date of 1980 BP showed the site first to have been occupied around the time domestic animals appeared in Namaqualand. Most of the occupation, however, probably came from the period between 400 and

300 years ago. Ovicaprid bones were found in most levels, as were the bones of the smaller antelope, especially *Oreotragus*. Goat and cattle bones were found in the upper levels. Plant remains included the bases of Iridaceae, as well as the seeds of *Diospyros* sp. and *Rhus undulata*, all of which are edible during spring to early summer.

Along the lower Orange River, at Bloedrift are a number of geometric rock engravings. There is no way of dating these, but they may well have been tied into the religious activities of hunters in the area.

In southern Namibia, work has been done on another site producing rock art. Apollo 11 has a long sequence going back to middle Stone Age times, but it has become famous for the earliest known *art mobilier* in Africa, in the form of several painted stones dated to between 27 000 and 25 000 BP (Wendt, 1976). Evidence for nara melon use going back to around 8000 BP has also been found here. (Deacon, 1984), and further north in the Namib at Mirabib *c.* 6500 BP (Sandelowsky, 1977). Nara seeds (*Acanthosicyos horrida*) are found as much as 100 km from the present known distribution of the plants, indicating that historical records of people specifically going to the coast to collect nara melons from the interior has great antiquity. The season of harvesting nara is between December and June, and the seeds can be used as a dry-season food if conditions get rough. Ethnographic information suggests that individuals or groups had inherited rights to melon patches (Budack, 1983). Nara knives made of bone have come from several archaeological sites in the Kuiseb (!Khuiseb) Delta (Kinahan, 1991).

Of interest is the incidence of whale-bone structures used as huts on the Namibian coast (Smith and Kinahan, 1984). The layout of these archaeological sites conforms to that depicted by Gordon at the mouth of the Orange River in 1779 (Raper and Boucher, 1988: plate 52) with whale-vertebrae seats around a hearth, and discarded shells close by. A sixteenth century Spanish silver coin was found on one of the more degraded of these sites (Kinahan and Kinahan, 1984), which might be an indicator of age. According to Gordon, the size of the group he met living in these kind of structures was 11 people. Gordon wrote in his journal on 20th August 1779:

> There was a large hut . . . with two high doors . . . open
> to the E, of wood from cast up trees, and
> Noordcaper or whale bones covered with grass and
> vegetation, and very hot. In it were nine or ten
> sleeping-places where the skins of rock rabbits and
> jackal lay . . . In the huts hung sacks of hide and
> canna, or Cape eland horns with buchu and fat, and
> a baked earthenware pot, as well as many ostrich
> eggshells, some empty, others filled with a supply
> of water . . . In front of the door they had planted

dry cast-up trees, on the branches of which hung pieces of raw whale-meat which had been cut off [and] which they roast or cook as food. We also found two beautiful tanned sealskins.

(Raper and Boucher, 1988:269)

15.4.4. Bushmanland and the Middle Orange River

While a number of surveys have been carried out in this area, few sites have been excavated. One of the surveys (Beaumont *et al.*, 1995) collected organic material from 26 sites and produced 59 radio-carbon dates ranging over the past 4600 years. On the basis of analyses of surface materials, it became obvious that within the last 2000 years two different, but coeval archaeological signatures could be recognized. The first, called Swartkop, continues from the pre-2000 year period, and has over 60% of the finished stone tools in the form of backed blades used as arrow tips, as well as coarse undecorated pottery with grass temper. The sites are located around pans or stream-bed margins, near springs and soaks (*graafwater*), hollows on dunes and on the flanks or crests of low kopjes.

A second and distinct archaeological assemblage is that of the Doornfontein industry, found in the vicinity of the Orange River, or along more-or-less permanent water sources close to the river. This industry is characterized by a consistently large number of thin-walled potsherds, with thickened bases and lugs, bosses, spouts and decorated necks and rims. There are few, if any, formally retouched stone tools. All the stone flakes are irregular, with a high incidence of quartz. The ostrich eggshell beads tend to be larger than on Swartkop sites.

By analogy with sites in the southwestern Cape, the suggestion is that Doornfontein sites contain the residues of herder activities, while Swartkop assemblages extend back in time before domestic stock arrived in southern Africa, so come from hunters living in Bushmanland, and later within the last 2000 years at the same time when herders occupied the better-watered area along the Orange River.

Of the few excavated sites in Bushmanland, Drögrond is perhaps indicative of the Swartkops industry. The site is on a low rise immediately next to a shallow aquifer from which sweet water can be obtained with relative ease. The site produced many small flaked tools (particularly crescents made in crystal quartz) and grass-tempered pottery. The larger fauna consisted of springbok (*Antidorcas marsupialis*), steenbok (*Raphicerus campestris*) and *Equus* sp. (probably zebra) (Smith, 1995). While the faunal assemblage was not large, the animals are indicative of those available to hunters who targeted migratory game. Large quantities of ostrich eggshell fragments were also collected that showed the eggs were an important food resource. Three

glass beads were excavated, and along with the radio-carbon date of 400 BP indicated that external trade had penetrated Bushmanland at least by the sixteenth century.

On the farm Waterval (Augrabies Falls) a surface scatter of pottery and other artifacts was mapped (Smith 1995). Included was the mandible of a sheep, and charcoal from a hearth which gave a date of 800 BP. This site had decorated pottery and only one formal stone tool, indicating its place within the Doornfontein industry.

If this dichotomy of sites is correct, then we are probably able to separate out the different economies of hunters and herders on the basis of their material culture. Neither of these entities need have been static, however, as there are indications that Doornfontein materials later may have ended up on Swartkop-type sites, such as larger ostrich eggshell beads. This may indicate greater social overlap, with hunters either working for herders, or herders being pushed out of their territories by incoming European colonists, and entering the hunter societies as refugees.

15.4.5. Cape Fold Belt (west) and Tanqua Karoo

This region would include the Oliphants River Valley and its tributaries in the Cedarberg/Koubokkeveld. Due to its proximity to Cape Town and a considerable corpus of rock paintings, many sites are known and a number have been excavated. The soils and vegetation of this area are basically low-nutrient status, as the soils derived from the Table Mountain Sandstone are leached. The high plant species diversity was favourable, however, for use by people, and, like many other parts of the Cape, the Iridaceae were utilized. Since grasses are a minor element of the vegetation and the terrain is rugged, this area was not regarded as prime pasture country by herders. This means that, as far as we can tell, all the precolonial sites in the mountains were probably occupied by hunter-gatherers.

Excavations at De Hangen (Parkington and Poggenpoel, 1971) were the first in this region to look at the pre-colonial economies and seasonal use of resources. The cave is located in the hills at an altitude of 580 m, some 8 km to the east of the Olifants River in fynbos vegetation. At the back of the cave, on the top of the sequence, was an arc formed by a grass layer, with hearths at either end and a main ash concentration within the arms of the arc. This illustrates a common feature of cave occupation where grass bedding was laid down at the back of the cave, with day-to-day activities taking place around hearths at the front of the cave. Artifacts consisted of flaked stone tools, pottery, pieces of leather, ostrich eggshell, seed and bone beads, as well as fibre and wooden implements. A date of 1850 BP came from below the grass layer, but most of the occupation took place between 350 and 500 years ago. The dry cave preserved much of the organic remains, offering us considerable amount of information about how the people lived. *Ehrharta* and *Pennisetum* grasses, as well as *Juncus*, *Helichrysum* and *Cyperus/Mariscus* were all used as bedding. *Cyperus textilis* was used for making string, as was a member of the Thymeliaceae family (*Passerina* or *Struthiola*) (Table 15.3). The reed *Phragmites australis* was used for arrow shafts, and *Hyaenanche globosa* probably used in arrow poison. The leaves of cf. *Velthaemia* were found wrapped around several black mussel shells. Buchu (*Diosma hirsuta*) was often mixed with fat rubbed on the body. Food plants included members of the Iridaceae. Woody plants including *Rhus* sp. and members of the Proteaceae would have been used as firewood (see Table 15.3).

Seasonal use of the cave could be estimated from several strands of evidence, of which the plants were an important component. Availability of the Iridaceae as a food resource occurs after flowering, so the summer months of September to March would be most probable. This was supported by the killing of immature rock hyraxes, the age of which can be estimated from their tooth eruption sequence, and because they have a restricted birthing period around the month of October (see Milton et al., this volume). Of the 89 young dassies in the excavated sample, 58 (65%) would have been killed between September and February, and 84 (94%) between September and April.

The occurrence of black mussel shells (*Choromytilus meridionalis*) in the site, along with the seasonal data given above, was the basis for a seasonal mobility hypothesis that argued for movement between the mountains and the coast (Parkington, 1972). A great deal of discussion has revolved around this, with subsequent dietary information suggesting that some people stayed at the coast all year (Sealy and Van der Merwe, 1986). No doubt through time both of these models could have come into play depending on socio-political conditions. Regardless of spatial use, there do appear to be real seasonal differences in occupation, and marine shells were ending up in mountain caves.

Two other sites containing plant remains have been excavated in the Olifants River drainage. Andriesgrond and Renbaan overlap in age with De Hangen, and add to the list of exploited flora of this part of the Cape Fold Belt (see Table 15.3). A quantitative analysis of the plant remains from Renbaan (Liengme, 1987) showed that corm remains in three levels ranged from 9.3–29.6% of weight, and seeds/fruits 1.2–7.3%, while at Andriesgrond corm fragments from four levels ranged between 6–60% and seeds/fruits 0.2–2.5%.

On the other side of the Cape Fold Belt, with a catchment towards the Tanqua Karoo, is the site of Aspoort Cave. Dates of 5100 and 6800 BP show the site to have been

Table 15.3. *Plant species from Cape Fold Mountains (west)*

Species	De Hangen	Andriesgrond	Renbaan
Grasses/reeds			
Pennisetum sp.	x		
Ehrharta cf. *calycina*	x	x	x
Phragmites australis	x		
Lasiochloa sp.			x
Cyperus/Mariscus	x		
Ficinia sp.	x		
Restio sp.	x		
Cannamois sp.	x		
Willdenowia sp.	x	x	
Juncus sp.	x		
Geophytes			
Ledebouria sp.	x		
Veltheimia glauca	x		
Cyanella sp.	x		
Dioscorea elephantipes	x		x
Romulea sp.	x		
Moraea cf. *fugax*	x	x	
Hexaglottis sp.		x	x
Homeria sp.	x		
Chasmanthe sp.	x		
Babiana sp.	x	x	x
Gladiolus sp.	x		x
Watsonia sp.	x		
Oxalis sp.	x	x	
Succulents			
Aloe sp.		x	
Carpobrotus cf. *edulis*	x		
Cotyledon (*Tylecodon*?)	x		
Shrubs			
Brabejum stellatifolium	x	x	x
Protea laurifolia	x		
Leucadendron sp.	x		
Pelargonium sp.	x	x	x
Diosma hirsuta	x	x	
Nylandtia spinosa	x	x	x
Helichrysum sp.	x		
Trees			
Hyaenanche globosa	x		
Ricinus communis		x	x
Heeria argentea	x		x
Rhus sp.	x	x	x
Passerina/Struthiola	x		
Euclea sp.	x	x	
Diospyros sp.	x		
Olea sp.	x		x

After Liengme (1987)

occupied during the coastal hiatus mentioned above, suggesting that only parts of the coastal strip were devoid of human occupation, and not the interior. Unfortunately few plant remains were found in excavation. Only the casings of *Oxalis* sp. give an indication of plant use. The faunal component included animals that would have been more comfortable on the open country of the Tanqua karoo, such as *Equus* sp. Approximately 5 km north-east of the cave a hunters hide or windbreak was located, consisting of an arc of stones roughly 3 m in diameter (Smith and Ripp, 1978). Around this windbreak were many flaked stone pieces and fragments of ostrich eggshell. The vantage point overlooking the Tanqua karoo suggests the structure may have been multifunctional: a place to sleep out of the north winds, and a hunting blind.

Historically the low-lying Tanqua karoo has been used as a winter area for farmers in the Kouebokkeveld and the Roggeveld. They returned to the higher ground in summer when the snow had melted. There is no reason to doubt that aboriginal foragers used the area in a similar fashion.

15.4.6.   The Cape Fold Belt (south) and Little Karoo

Much of this region remains to be explored archaeologically. The exception is the eastern end, which, because of its proximity to the Albany Museum in Grahamstown, has been studied by several people when they were based there. The detailed work at Melkhoutboom (Deacon, 1976) offers us a window into later Stone Age exploitation of this area. The cave is located some 75 km north of Port Elizabeth in the Suurberg, at an elevation of 600 m. The site was occupied between 15 000 and 5900 BP, with a final date of 250 BP for the surface level. The Main Bedding Unit (5900 BP) and CAF Unit (2900 BP) are the two units from which a large percentage of the well-preserved organic remains came.

The plant remains identified are listed in Table 15.4. Of these, the Iridaceae comprise the bulk of the dry weight in all the levels, followed by Hypoxidaceae. Most of the latter would have been food plants, e.g. *Hypoxis argentea*, although the plants of this family can be used for a wide range of medicinal purposes. String was made from *Cyperus textilis*, and several knotted pieces were found, at least one suggesting a net had been made.

In addition to the plant remains, several other organic fragments were found, including several large pieces of worked leather, possibly Grysbok skin, showing stitching with sinew, and two pieces of leguaan skin.

Wooden artifacts included fire-drills, points and pegs, and, from the shavings found, one can assume that the wood was worked at the site. Bone tools included points, a polished 'wedge', and the polished plastron of a tortoise (possibly a scoop). The sense of the aesthetic can be seen in the perforated caracal atlas vertebra with ochre staining (possibly suspended around the neck), decorated ostrich eggshell, and shell pendants. Fragments of both red and yellow ochre were also found in the excavations.

Further west in the Swartberg, north of Mossel Bay, is Boomplaas, a site occupied in middle Stone Age times, with the later part of the sequence overlapping with that of Melkhoutboom (Deacon et al., 1976). Although the plant preservation was not as good as at Melkhoutboom, *Hypoxis*, *Watsonia* and *Pappea* were identified. Many of the artifact types were also duplicated here, including decorated ostrich eggshell. Four painted stones, three with mammals, and the fourth with an ostrich on it, were found in the upper part of the deposit. Many young lamb

Table 15.4. *Plant remains identified from Melkhoutboom Cave*

Species
Grasses/reeds
Merxmuellera sp.
Themeda triandra
Koeleria sp.
Cyperus textilis
C. usitatus
Geophytes
Oxalis sp.
Boophane disticha
Hypoxis sp.
Moraea sp.
Watsonia sp.
Tritonia/Freesia sp.
Schotia sp.
cf. *Mariscus tabularis*
cf. *Dioscorea* sp.
Trees
Podocarpus falcatus
P. latifolius
Calodendrum capense
Pappea capensis
Oldenburgia grandis
Syncarpha milleflora
cf. *Scolopia zeheri*
cf. *Clerodendrum* sp.
Shrubs
cf. *Solanum tomentosum*
cf. *Pimpinella caffra*
cf. *Thesium* sp.
Todea barbara
Succulents
Aloe sp.
Carpobrotus sp.

After Deacon (1976)

skeletal parts were found at the top of the sequence, leading the excavators to the conclusion that this was a kraaling area used by herders (Deacon et al., 1978).

15.4.7.   The Great Karoo

This is a large area, and tends to be under-studied archaeologically. However, one major project, the Zeekoe Valley Archaeological Project (ZVAP) has been under way for more than a decade under the direction of Garth Sampson. This intense study of part of one valley can be used as a general model for much of the area.

The Zeekoe Valley runs north/south, draining the north side of the Sneeuberg, with its confluence at the Orange River north of Colesberg. The middle and upper reaches of the valley have been surveyed, and every archaeological site plotted. Over 1000 later open-air and cave sites have been located, and the artifacts subjected to spatial analysis to see what changes occurred through time, particularly with the impact of colonial pressures on Bushmen society (Sampson, 1988, 1992; Sampson et al., 1989; Sampson and Plug, 1993; Sampson et al., 1994; Sampson and Vogel, 1995; Bollong and Sampson, 1996; Voigt et al., 1992; Moir and Sampson, 1993).

The aboriginal food procurement strategy of the valley's Bushmen would have been the exploitation of both plant and animal species. A survey of travellers through the area (Neville, 1996) indicates that underground bulbs and insects (ants and locusts) were what the travellers noted being used by the Bushmen. This is in contrast to the statement made by Sampson (1988: 35) from his archaeological observations that 'Nothing we have encountered thus far would suggest ... that the Seacow Valley Bushmen were heavily dependent on plant foods as a staple.' This is supported by the lack of plant remains in the numerous caves excavated (G. Sampson, personal communication), although problems of preservation come into play here, and no pits were seen in the caves which might have indicated storage. Thus we have to fall back on historical observations for records of plant use by hunters. The spatial distribution and density of geophytes would have been important factors in their exploitation, but there seems to be little doubt that these patchy, albeit seasonal, resources would have been attractive to foragers.

Animal exploitation, by contrast, is well documented from ZVAP excavations. As can be seen from the faunal spectrum in Table 15.5, by far the preferred target was springbok, which suggests relative ease of capture. This is not surprising, since these animals roamed the karoo in their thousands. As Dunn (1873: 31) noted in Bushmanland, halfway between Kenhart and Van Wyk's Vlei: 'One may read about the vast numbers of springboks that migrate across this continent ... we have driven through them for six hours (35 miles); while from reliable information they extend for one hundred miles in length. Imagine flocks of from 2,000 to 6,000 of these animals scattered over the plain at intervals of two miles apart over the whole of this area.'

There were also a number of small mammals which could be trapped around camp. Hares and ground squirrels could be snared by children practising to become hunters, and provided an addition to their diet.

An analysis of bone density from nine rock shelters in the Zeekoe River Valley, dating between 1800 and 400 BP, measured changes in bulk faunal mass per unit volume of deposit. It appears that at least two densely packed layers of mammal remains occurred at all the shelters in levels that are considered contemporaneous around 1100 and 800 BP (Sampson and Plug, 1993). These episodes are interpreted as resulting from warm–wet periods, and leads to the suggestion that carrying capacity was driven by climatic fluctuations. Once prey species numbers increased, these were exploited by hunters. Other clues to pre-colonial environmental shifts come from an analysis of ostrich eggshell fragments from the Zeekoe Valley (Sampson, 1994). Two dense 'sheets' of ostrich eggshell were found in several small rock shelters, dating between 1600 and 1750

Table 15.5. *Larger wild animals from Abbot's Cave*

Species	NISP/MNI[a]
Primates	
Homo sapiens sapiens (human)	4/4
Papio ursinus (chacma baboon)	8/6
Leporids	
Lepus saxatilis (scrub hare)	83/8
Pronolagus rupestris (Smith's red rock hare)	17/5
cf. Lepus/Pronolagus	401/-
Rodents	
Cryptomys hottentotus (common molerat)	8/2
Hystrix africaeaustralis (porcupine)	14/7
Pedetes capensis (springhare)	113/18
Xerus inauris (ground squirrel)	467/49
Carnivores	
Hyaena brunnea (brown hyaena)	2/1
Panthera pardus (leopard)	5/3
Felis caracal (lynx)	17/10
Felis lybica (African wild cat)	1/1
cf. Otocyon megalotis (bat-eared fox)	3/3
Lycaon pictus (wild dog)	1/1
Vulpes chama (Cape fox)	11/6
Canis mesomelas (black-backed jackal)	83/25
Ictonyx striatus (zorilla)	17/6
Suricata suricatta (suricate)	40/11
Cynictis penicillata (yellow mongoose)	60/12
Galerella sp. (mongoose)	57/5
cf. Ichneumia albicauda (white-tailed mongoose)	1/1
cf. Atilax paludinosus (water mongoose)	10/5
Hyrax	
Procavia capensis (dassie)	132/23
Equids	
Equus cf. zebra (Cape mountain zebra)	12/6
E. burchelli/quagga (Burchell's zebra/quagga)	52/31
Suids	
Phacochoerus aethiopicus (warthog)	82/21
Antelope	
Connochaetes gnou (black wildebeest)	43/16
Alcelaphus buselaphus (red hartebeest)	9/5
Damaliscus dorcas (bontebok)	57/13
Cephalophus monticola (blue duiker)	3/3
Sylvicapra grimmia (common duiker)	15/6
Antidorcas marsupialis (springbok)	4582/213
A. bondi (Bond's springbok)	4/1
Oreotragus oreotragus (klipspringer)	10/7
Raphicerus campestris (steenbok)	13/7
Pelea capreolus (vaalribbok)	34/11
Tragelaphus strepsiceros (kudu)	15/5
Taurotragus oryx (eland)	19/8
Redunca arundinum (reedbuck)	5/2
Redunca fulvorufula (mountain reedbuck)	3/2
Aves	
Struthio camelus (ostrich)	4/3

[a] number of individual specimens/minimum number of individuals
After Plug (1993)

which correlate with a marked decrease in grass pollen. This is interpreted as Bushmen having to fall back on ostrich eggs as a food resource at a time when the traditional large mammal prey was in decline.

Another attempt to interpret environmental conditions over the past 1000 years is based on the incidence of micromammals excavated from one of the ZVAP caves (Avery, 1991). Up to the seventeenth century, the micromammals were dominated by *Cryptomys hottentotus*, which is almost exclusively dependent on underground plant foods. By the seventeenth century, *Otomys irroratus* and *Mystomys albicaudatus* increased in numbers. The former is an animal which prefers waterside vegetation, and the latter would be an indicator of more mesic conditions. These data tend to contradict the bone density analysis above, by indicating generally harsh conditions at least till the sixteenth century AD, followed by amelioration in the seventeenth century just before European stock farmers moved into the area.

Because the micromammals would have been taken by owls close to the caves where their bones have been found, this probably means a fairly restricted territory. Sampson (1986) has noted that vegetation indicators of overgrazing and damaged rangeland are found around abandoned Bushmen campsites, suggesting that the rangeland had been significantly hammered by indigenous people prior to European settlement of the area, so micromammals may be reflecting this degradation around sites.

15.5.   Rock art

This general description of the karoo's aboriginal people would be incomplete if we left out the wonderful rock art found there. Rock art of the karoo comes in two forms: paintings and engraving, both of which form part of the wider Bushmen art of Southern Africa. There is very little overlap in the distribution of the two genres. Engravings are found in Bushmanland and the northern Cape, while paintings are found in the Cape Fold Belt and karoo mountains.

Engravings or Petroglyphs: are found on exposed boulders where the surface patina or wind polish was cut through with a sharp stylus or pecked with a rock hammer. The figures were produced using three techniques: (a) finelines, where an animal or human is outlined by a thin incised line; (b) scrapings, in which a fineline was filled with scratched or scraped lines; (c) peckings using percussion to create an outline (Beaumont et al., 1995).

The quality of the depiction varies, but some are truly superb, showing both skill in artistry as well as knowledge of subject. Animal depictions are of the larger mammals, with equids being the most common (24%), followed by eland (12%), kudu (8%), rhino and ostrich (each 7%) and gemsbok (6%). Often these are single animals, although a large rock could have several animals on it.

Paintings: These are usually found on cave walls which, by their size, create much larger 'canvases' than boulders, so one gets the impression more of the composition of several elements. These can include long lines of humans or animals, as well as features which indicate that they were part of what has become known as 'trance symbolism', following Lewis-Williams (1981).

It is generally accepted that many, if not most, of the paintings, and possibly the engravings as well (although this is less certain), are metaphors for trance experiences of shamans or medicine healers of Bushman society. The trance is a powerful mechanism for social and spiritual healing, and a state of altered consciousness is attained. In trance, the healer visits the land of the spirits and becomes the go-between for his people. The sweat of the trancer has potency and is used to heal. Depictions of trance, as well as animals important in the religious beliefs of the Bushmen, can be seen. In the paintings, the most common animal is the eland which is a rain animal, and is a symbol for many of the rituals (Lewis-Williams 1981).

15.6.   **Khoisan/settler interactions**

In the western Cape relations between colonists and aboriginal people were such that there was constant resistance to European expansion. The initial Khoi/Dutch wars of 1659–60 and 1673–7 set the tone for the next 100 years. The Dutch were too powerful, and not only used intergroup feuding among the Khoi to their advantage (Elphick 1985), but also interfered in raiding between groups, and in so doing took important political decision-making out of Khoi hands (Smith, 1983). Attempts by the Dutch East India Company (Verenigde Oost-Indische Compagnie, or VOC) to deal evenly with both Khoikhoi and colonist were always tempered by placing the Company's needs above all. Even *placaaten* that forbade any dealing between burger and Khoi were easily put aside when meat shortages at the Cape became critical. In 1739, arose the infamous case where certain colonists instigated raids on Khoi cattle. These individuals caused a serious backlash among their coloured servants when they did not keep their word to share the spoils of the raid with them. The result was virtual war in the countryside which was brought to a close by the Company giving in on the colonists' side, in spite of them having caused the problem in the first place. The war brought untold hardship among the Namaqua Khoikhoi, as well as other people trying to survive independently of the colony (Penn, 1987), and made it clear that the authorities were not to be trusted.

So strict were the Company's methods of maintaining control that many men deserted from VOC employment and became 'vagabonds' in the interior. These men joined with disaffected Khoi and farm workers to raid both colonist and Khoi, and, according to Penn (1987: 471), more commandos were sent out against gangs of deserters than against the Khoisan during the period 1700–38. These marauding deserters just increased the general sense of insecurity in the hinterland of the early eighteenth century Cape.

Further into the interior, the Bushmen of the later period were so well adapted to their surroundings that they were able to withstand the colonial onslaught in areas like the Sneeuberg for many years before the weight of merchant capital wore them down. A study of the history of European impact on the aboriginal people (Neville, 1996) shows that this was a slow, almost inevitable process. Between 1770 and 1798 the first trekboers to move into the Sneeuberg area were met with fierce opposition. Bushmen killed livestock and murdered stock-keepers. Commandos were sent out in punitive raids to kill Bushmen and capture women and children who could be used as virtual slaves on farms.

From 1798 to 1824, the governor at the Cape, Lord Macartney, instituted a subscription system whereby farmers shared the cost of buying stock to give to the Bushmen to stop them raiding. This brought peace, and allowed more and more trekboers to expand into the area to take over the waterholes that the Bushmen had so jealously guarded. During this period, Bushmen began to associate with the farmers who were feeding them, and more or less settled on the farms as workers. As the game was increasingly shot out by firearms, and so became scarcer, the Bushmen gradually ended up as dependents on farms. Some Bushmen went to mission stations to get away from the rough treatment of farmers, but this was only for a few years. As towns grew up in the karoo, more and more of these Bushmen became indigents on the edge of settlements with little work, or prospect of any. They could either become servants, or just hang around living off handouts. The latter succumbed to alcohol abuse, and had little or no immunity to disease episodes that occasionally struck the townships.

15.7.   Conclusions

Human use of the karoo is as old as humanity itself. Archaeological sites have produced the data that inform us of past adaptations, but much more remains to be done.

In some ways this may require the dedication to long-term projects like the Zeekoe Valley Archaeological Project, since sites tend to be mostly ephemeral and have been occupied for a very short period of time. This reflects the kind of strategies that would have been necessary for exploiting highly mobile game animals, such as springbok, as well as taking advantage of patchy plant resources that were seasonally available, like the Iridaceae.

The gradual degradation of the Bushmen was partly the function of the inability of the society to accept leadership and have spokespersons to negotiate even simple conditions of service. They were never organized as a coherent group, even when they were most successful in fighting off settler advances into the area.

The pages of manuscript collected from /Xam Bushmen who were convicts at the Breakwater Prison in Cape Town, and who became informants of Wilhelm Bleek and Lucy Lloyd (1911) at the end of the nineteenth century, are housed in the Manuscripts Division of Jagger Library, University of Cape Town. They give us clues to what has been lost with the disappearance of this cultural group. We have a small glimpse into the intricacies of stories and beliefs, but the many skills are gone forever.

Endnote
In this chapter, the reader will note the consistent use of the term 'Bushman', as opposed to 'San' which is preferred by other writers. San (= Sonqua) was the name given by the Khoikhoi or 'Hottentot' herders to people without stock. Since these hunting people stole stock, their name became synonymous with 'stock thief'. To the Khoikhoi, anyone without stock was considered inferior, so Sonqua was a pejorative name. Among the aboriginal people of the Kalahari today, such as the Ju/'hoansi, they will use their own group name, but if a collective name is needed they prefer to be called 'Bushmen'.

Khoisan is a general word used to refer to the genetics or biology of Southern African aboriginal people. It is also used as a linguistic term to describe the various click languages of southern Africa. Occasionally it might have validity as a cultural term when the people being referred to are difficult to distinguish, as became the case in the eighteenth century when refugees who lost their land and livelihood to the expanding colony at the Cape began collecting together to avoid being subjugated by the encroaching trekboers.

16 Historical and contemporary land use and the desertification of the karoo

M. T. Hoffman, B. Cousins, T. Meyer, A. Petersen and H. Hendricks

16.1. Introduction

During the last 100 years, concerns over the degradation of the karoo have had a major impact on the development of agricultural policy for South Africa. Government commissions of inquiry into stock disease (Anon, 1877), drought (Anon, 1923) and desert encroachment (Anon, 1951) and other state agricultural initiatives such as the 1960s Stock Reduction Scheme (Pringle et al., 1982), the privatization of karoo communal lands (Boonzaier et al., 1990) and the National Grazing Strategy (Du Toit et al., 1991) have all either highlighted the desertification issue or have sought to limit the rate and extent of karoo degradation. Unlike the Sahel, for example, where there has been a significant decrease in annual rainfall since 1966 (UNEP, 1992), the perceived negative changes in the karoo landscape and declining agricultural productivity have been attributed directly to agricultural impacts, and in particular to overgrazing (Acocks, 1953; Dean and Macdonald, 1994). It is generally believed that it has been the mismatch of agricultural land use practices with the production potential of the land which has led to widespread desertification in arid and semi-arid South Africa (Milton et al., 1994b; Badenhorst, 1995). Although this is a powerful indictment against the stewardship of the many generations of South Africans who have lived off the karoo, it is important to assess its veracity. To do this we need to understand how karoo landscapes have been utilized in the past and to assess the ecological impact of contemporary agricultural practices on the region.

There are two main agricultural production systems in the karoo with widely differing histories, forms of land tenure, institutional arrangements, land use practices and production objectives. These are, first, the communal systems which are generally (but not exclusively) found in a number of separate 'Reserves' in the western succulent karoo biome of the Northern Cape Province (Fig. 16.1) and, secondly, the commercial sector which dominates karoo agricultural production in terms of area utilized and in its contribution to the national economy. The two production systems are contrasted in Table 16.1.

In this chapter, we first synthesize available historical and ecological information for the main communal lands in the karoo. We describe key land use practices and production coefficients (e.g. lambing percentages, mortality) within some of the regions and discuss the important social and biophysical driving forces which affect land use practices and natural resources at various levels of scale. For the commercial sector, we first briefly outline the historical settlement of the karoo by European farmers in the eighteenth and nineteenth centuries and secondly, describe its development in the nineteenth and twentieth centuries. We discuss the major range management systems and describe the spatial distribution of current commercial land use practices.

Finally, we use the descriptive account of communal and commercial agricultural practices as core material for addressing the rich and hotly contested desertification debate in the karoo.

16.2. Communal lands

16.2.1 Historical background

With the exception of Riemvasmaak, the areas under communal tenure within the karoo are found within the 'Rural Coloured Reserves' (Fig. 16.1). These were created during the early part of the last century, when churches established mission stations to protect the indigenous

Table 16.1. *Comparison of important criteria which describe communal and commercial agricultural production systems in the arid and semi-arid rangelands of the karoo, South Africa*

Comparison	Communal areas	Commercial farms
History	Rooted in dispossession and confinement. Linked historically to indigenous Khoi subsistence economies characterized by large cycles of transhumance	Arise from European colonization in 18th and 19th centuries with a focus on market economies, private ownership and sedentarianism
Location and size	Predominantly in the west comprising 2.3% of karroid rangelands	Throughout the karoo covering probably >80% of the land area
Biophysical environment	Generally in low (<250 mm/yr) and hyper-arid (<50 mm/yr) rainfall areas of succulent karoo biome; dominated by leaf-succulent shrublands	In <50 mm/yr to >500 mm/yr rainfall environments of Nama- and succulent karoo biomes, dominated by grasses (in the east), and dwarf karroid shrubs
Land tenure	'Citizenship' grants individual rights to shared grazing areas and water sources and leased cropland areas	Ownership grants individual rights to exclusive, responsible use of rangeland resources
Institutional authority	Elected Village and Reserve council representatives linked to district, provincial and state authorities	Elected municipal, district, provincial and state authorities
State and business involvement	Community and individual benefits via limited direct infrastructural (e.g. boreholes, fencing); and remittances, pensions etc.	Individual benefits via direct subsidization of infrastructural improvements, drought aid, loan subsidies, extension services, etc.
Human density	Relatively high population (>5 people/km2) and farmer density	Relatively low population (<5 people/km2) and farmer density
Production objectives	Multiple use of the range (firewood, construction materials etc.) but primarily for increasing small stock animal numbers for meat, investment and to a lesser extent hides and milk	Primarily for cash sales focussed on the quality and quantity of wool, meat and pelts
Livestock density	Generally twice the recommended carrying capacity estimates, often sustained over decades	Generally at or below (according to 1995 data) the recommended carrying capacity estimates, also sustained over decades
Herd size	Usually between 10 and 500 animals	Usually greater than 500 animals
Dominant breeds	Boer goats and Dorper sheep, donkeys	Boer & Angora goats, Merino, Afrino, Dorper & Karakul sheep, beef cattle, ostriches
Management strategy	Animals herded daily around grazing area–water source–night kraal orbit	Free-ranging herds in paddocks with associated water point, often under rotational grazing & resting systems
Production coefficients	Relatively poor lambing and weaning percentages, slow growth rates, high mortalities	Relatively high lambing and weaning percentages, higher growth rates, lower mortalities
Marketing strategy	Poor access to markets and reliance on speculators	Markets form integral part of production system
Impacts on natural resources	Transformation from perennials to ephemeral and poisonous plant mix in heavily grazed areas	Shift from palatable to unpalatable shrubs; increase in short-lived grasses especially in the east

Khoikhoi populations, together with people of mixed descent, from further dispossession of their lands by encroaching white settlers. Riemvasmaak is different, however, in that it was settled only in the late nineteenth and early twentieth centuries by ethnically diverse semi-nomadic pastoralists comprised predominantly of Damara- and Xhosa-speaking herdsmen and their families (Hoffman et al., 1995b). In all cases, pockets of communal tenure were created within a larger framework of private land rights in the karoo (Hendricks, 1995). By the early twentieth century, the reserves had come under the control of the secular administration of the central state.

Communal tenure in the 'Coloured Reserves' has its origin in the pre-colonial period (Boonzaier et al., 1996). Indigenous Khoikhoi pastoralists were located mostly in the north-western and western regions of the Cape, where they kept mainly sheep and goats, and the southern Cape, where cattle were more important. Production involved

hunting, gathering and, in some cases, fishing as well as herding. Khoikhoi engaged in transhumance movement in order to exploit seasonal differences in the availability of grazing and water resources, and controlled territories in which a 'cycle of transhumance' could be completed (Smith, 1987). Control, however, did not necessarily imply complete exclusion of other groups, who might have paid symbolic tribute to the 'owners' when they entered their territory (Penn, 1986). Khoikhoi social organization reflected the importance of domestic stock as inheritable assets which could be accumulated, giving rise to 'ever increasing social and wealth disparities' and a patronage system (Elphick, 1985; Smith, 1986).

European colonization and settlement of South Africa disrupted indigenous social formations in three ways: through direct competition for land and grazing (leading to violent conflicts in the frontier zones (Penn, 1986)), through the penetration of mercantile capital (leading to

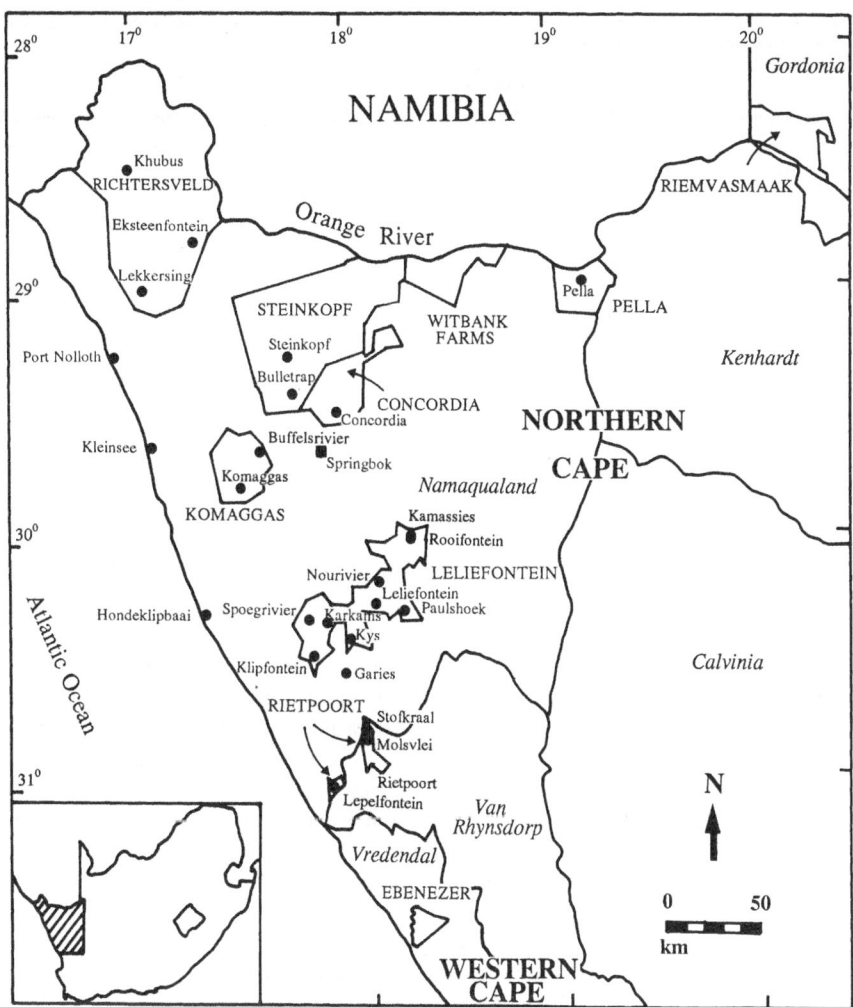

Figure 16.1 **Location of communal lands and main settlements in the western part of the karoo in relation to international, provincial and magisterial district boundaries**

a demand for labour in some regions, and the introduction of new production technologies, markets and consumption goods), and through the imposition of new forms of state power and administration.

The pastoralist Khoikhoi of the Cape suffered these changes first and most profoundly. As the early colonists expanded their zone of occupation along the east and west coasts and into the interior, they both robbed the Khoikhoi pastoralists of their herds and occupied their grazing lands (Boonzaier et al., 1996). In this process, many Khoikhoi pastoralists were forced to enter into relations of slavery or serfdom. At the van of colonial expansion were the trekboers, themselves semi-nomadic pastoralists who depended on their herds for subsistence (Penn, 1986). The trekboers also practised transhumance movement across seasonally variable rangeland, but had the advantage of superior military strength (based on guns, horses and the commando system), and, by the end of the nineteenth cen-

tury, had expropriated the means of production of most Khoikhoi peoples in the semi-arid areas. Private property and commercial livestock production slowly became dominant, and the remnants of the pastoralist economy still practised by the descendants of the Khoikhoi (in the form of semi-nomadic herding of sheep and goats) were increasingly confined to the mission stations (Carstens, 1966; Boonzaier et al., 1996).

Confinement to the mission stations and reserves severely constrained people's ability to move with their herds or to engage in transhumance. In tandem, the missionaries promoted crop cultivation and permanent village settlements (Khrone and Steyn, 1991). Reduced opportunities for local production led to movement outside the reserves to secure wage income, on settler farms and then on the copper and diamond mines which were opened in the region from the mid-nineteenth century. Over time, work opportunities opened in the west-coast fishing

Table 16.2. *Important descriptors of communal areas in the succulent karoo and Nama-karoo biomes of South Africa. All are in the magisterial district of Namaqualand, except Riemvasmaak which is in Gordonia (see Fig. 16.1)*

| Name[a] | Area (ha) | Population[b] | Farmers | No. of livestock | | | Total LSUs[d] | SR[e] | Actual SR |
				Small stock[c]	Cattle	Donkeys			
Concordia	63 383	4300	200	25 000	500	250	4 963	42–72	12.8
Komaggas	62 603	4500	160	11 000	20	2 600	3 582	44–72	17.5
Leliefontein	192 719	8000	800	26 500	450	4 000	7 600	42–60	25.4
Pella	48 276	1591	100	7 324	179	67	1 486	60–108	32.5
Richtersveld	513 919	2300	209	37 191	1 004	845	7 975	90–120	64.4
Riemvasmaak	74 563	891	52	3 517	332	250	1 126	60	66.2
Steinkopf	329 301	7500	272	54 592	837	764	10 699	54–72	30.8
Witbank farmsf	67 498	211		1 765	38	66	385	60–108	47.1

[a] The current number of farmers and stock numbers for Ebenezer (area = 18 286 ha; population 1731) in the Vredendal magisterial district and Rietpoort (area = 15 092 ha; population 2560) in Van Rhynsdorp are not available

[b] Population estimates according to First National Bank, Springbok for 1995 as well as from unpublished surveys in Surplus People Project (SPP) records

[c] Includes goats and sheep. Stock number estimates according to Department of Agriculture Extension Service numbers and SPP records

[d] Goat and sheep conversion factors to Large Stock Units = 0.17, cattle = 1.1 and donkeys = 0.65 (Anon, 1984)

[e] Recommended stocking rate (ha/LSU) (Anon, 1991); for Riemvasmaak from Hoffman et al. (1995b)

[f] Includes 9 farms, only 3 of which, totalling 18 116 ha, are farmed by Witbank community itself. Others are shared between Pella, Concordia and Steinkopf

industry, and people also began to move even further afield in search of work. As in the African reserves in other parts of the country, migrant labour became essential for household reproduction. For many residents, local agricultural production gradually assumed the character of a supplement to wages and remittances (Boonzaier, 1987; Sharp, 1984).

These generalizations must be tempered by an awareness of the specificity of each of the communal areas. These display some core similarities, but also distinct differences in climate, soils and vegetation, socio-economic structure, ethnic composition and political and institutional dynamics.

The settlement and tenure history is so different for Riemvasmaak that it warrants brief mention. Initial settlement between 1870 and 1930 was centred around the Riemvasmaak Mission Station and four main outlying villages. From the 1930s onwards, goats, sheep and, to a lesser extent, cattle, were herded within a centrally managed communal land tenure system which included rangeland resting systems and a variety of strategies developed to cope with the frequent droughts characterizing this arid region (Hoffman et al., 1995b). By the 1960s, however, only 8 of the 315 censused household heads described themselves as full-time stock farmers while the remaining stock owners worked as seasonal wage labourers or share croppers on neighbouring commercial farms. In 1974, all the inhabitants of Riemvasmaak were forcibly removed from the area and the land was used for military training and nature conservation purposes. However, in 1993, the land was returned to the Riemvasmaak community, but a formal communal land management system remains to be developed.

16.2.2.   Current land use practices and their determinants

We focus on the eight largest communal areas in the karoo (Fig. 16.1, Table 16.2). Although their combined area (together with Rietpoort and Ebenezer) is insignificant within South Africa as a whole (1.1% of the land area) and small relative to the karoo region (2.3%), in the Namaqualand magisterial district they comprise 26.3% of the land area and are home to 45.4% of the total population (Anon, 1991a). We do not include a number of communal lands from the Northern Cape Province in our analysis (e.g. the Mier area; parts of the former Bophuthatswana homeland) since these are situated more within the arid savanna biome than within the succulent or Nama-karoo biomes (*sensu* Rutherford and Westfall, 1986).

A variety of agents within a wide range of processes operating at different levels of scale influence land use practices and the natural resources of the communal lands of the karoo (Fig. 16.2). Both biophysical and social driving forces affect what occurs in these areas. Each communal area is connected in some way to the broader socio-economic environment through agents of business (especially mining), government (at district, provincial and national level), non-governmental organizations (NGOs) and research institutions. Decisions made at this level and higher, concerning policy and infrastructure development, job creation and retrenchment, have important effects on land use practices at village and household level. These decisions are in turn influenced by the production potential of the region, the conservation and aesthetic value as well as the mineral wealth of the surrounding area, amongst other biophysical factors. For example, although the Richtersveld communal area represents one of the most arid regions in South Africa, with annual rain-

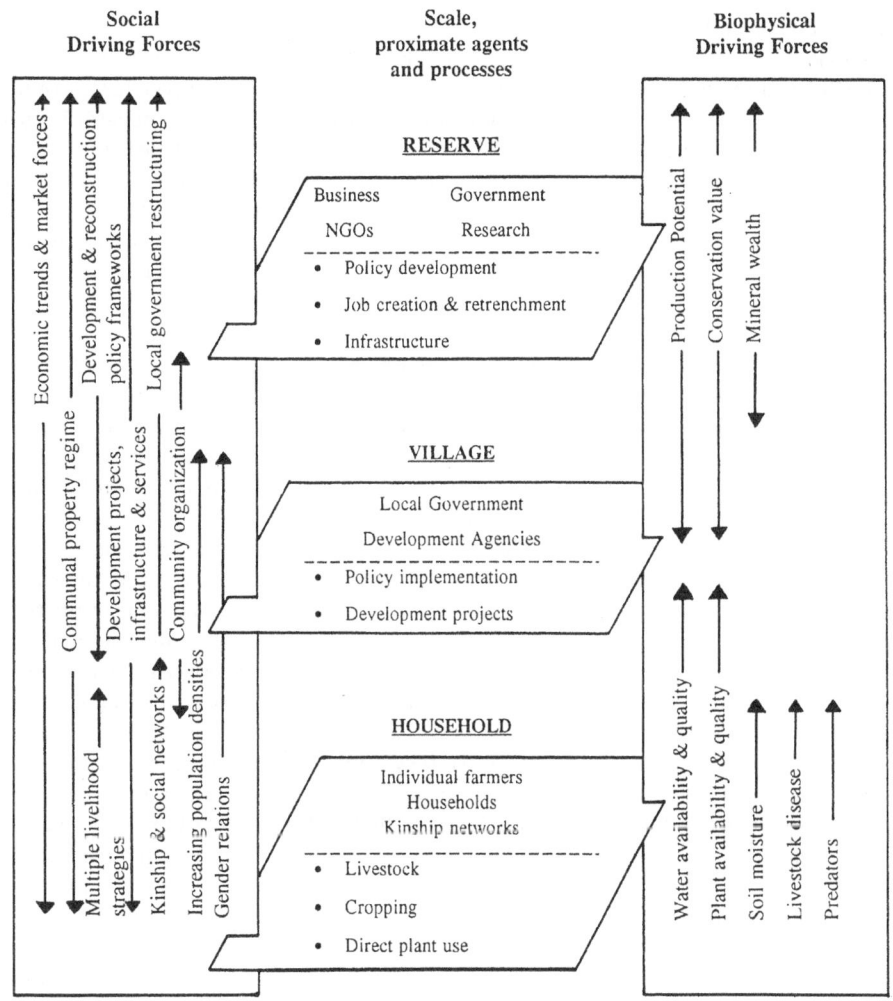

Figure 16.2 **A description of agents, processes and multiscale social and biophysical driving forces which affect the land use practices and natural resources of the karoo's communal lands**

fall below 100 mm, it contains a wealth of endemic succulents within an unusually picturesque landscape. Because of these biophysical factors, policy and infrastructural development decisions are centred not, for example, on the creation of a crop industry, which would be doomed to failure, but instead on a potentially economically and ecologically sustainable ecotourism industry.

Similarly, a range of key social driving forces, affecting decisions made at reserve level, influence directly the land use practices of individual farmers (Fig. 16.2). For example, the major mines in Namaqualand could close within the next 10–15 years largely because of dwindling reserves and increasingly uneconomical extraction procedures. It is predicted that 28 000 people (or 44% of Namaqualand's population) will be affected, either directly or indirectly, by the closure. Retrenched workers, resident within the communal lands, are likely to invest in agricultural production practices which many have been doing on a part-time basis as part of broader kinship networks.

The communal areas are usually characterized by a

number of village settlements with distinct territories, located within a broader reserve boundary (Fig. 16.1). Village councils, in association with established development forums usually manage the villages. In some cases, NGOs, state departments and parastatals work in conjunction with local structures to facilitate local development (e.g. Operation Hunger, Independent Development Trust). A wide range of social and biophysical factors influence local governance within villages. Some of these are shown in Fig. 16.2 and are explored in more detail under the discussion on household-level impacts on the natural resources of communal lands.

It is at the household level that much of the direct impact on the land occurs. This is because natural resources make important contributions to household livelihoods within the communal areas by providing water for household consumption, crop irrigation and livestock, fuelwood for household energy, building materials for housing and other structures, medicinal plants and 'veld' foods, arable land for rain-fed crops and irrigated vegetable gardens, and grazing land for domestic livestock (May and Marinus, 1995).

In what follows, the focus will be on the influence of social and biophysical factors on the livestock and cropping practices within the communal lands. Firstly, who in the Reserve is entitled to use the region's natural resources? The communal property regime suggests that residents of communal lands obtain rights to the use of natural resources if they are 'registered occupiers'. Currently, most households are located in village settlements, and share access to the designated territories within which the village's common pool resources (e.g. grazing areas) are located, as well as limited social services and infrastructure. Thus membership of 'communities' is defined partly in terms of resource use, and anyone who is registered is entitled to use the natural resources of a region.

Secondly, what proportion of people from the communal areas is engaged in agricultural practices, especially livestock and cropping? It is clear from Table 16.1 that only a small proportion (c. 3.5–10%) of the inhabitants of communal lands consider themselves farmers, although many more individuals probably own a few animals. The socioeconomic structure of these areas reveals a great deal of internal heterogeneity, in terms of sources and magnitude of income, gender of household head, livestock holdings, access to cropping land and dependence on kinship networks. The dependence of households on a variety of natural resources, the full complement of their livelihood strategies and, thus, their degree of impact on those resources differ considerably.

This is well illustrated in De Swardt's (1993) analysis of a village in Leliefontein Reserve in which he reported that income from remittances, welfare payments from the state and other non-agricultural sources together formed a much larger share of total income than agriculture. Only 57% of the 112 households kept livestock, the majority of which comprised small herds for their own use. However, when asked, 63% of households indicated that they would like to either begin keeping livestock or to increase their herd size so that more income could be earned from stock farming.

Third, what is the nature of the livestock and crop industry in communal lands? The livestock industry is based on small stock, especially Boer goats and Dorper sheep. With the exception of Riemvasmaak, where cattle numbers comprise around 10% of the total number of animals on the range, cattle generally form a minor component in the karoo communal lands (<2% of the total number of animals (Table 16.1)). The exact mix of goats and sheep in a herd is determined largely by individual farmer preference and suitability of the communal area for the different breeds. For example, a recent census shows that

in Paulshoek, Rooifontein/Kamassies and the Richtersveld National Park Boer goats comprised on average 63%, 82% and 88% respectively of the herds. However, in Concordia, which is relatively flat and more suitable for hardy Dorper sheep, Boer goats comprised only 37% of the herd.

Herd sizes also differ between individual farmers and regions. In Rooifontein/Kamassies in 1995, the 35 small stock herds ranged in size from 5 to 300 animals with a mean of 57 animals, while, in the Richtersveld National Park, the 17 sheep and goat herds ranged from 72 to 721 animals with a mean of 369. Also, different stock owners may group their animals into a single herd and kinship networks fulfil an important function in karoo communal areas. Wage labourers and city dwellers often retain links to the communal areas through stock ownership where their animals are herded by relatives or friends in collective herds.

The number of donkeys present in communal areas is difficult to ascertain since they are no longer herded or kraaled with domestic livestock. While some of the free-ranging donkeys are used for transport and seasonally for ploughing purposes, many do not service resident's needs and a large feral donkey problem exists. Some local inhabitants suggest that donkeys are highly destructive herbivores, since they remove palatable, shallow-rooted leaf-succulent shrubs with ease, especially after soaking rains. Others, however, find evidence to reject this view and contest that the benefits of owning donkeys far outweigh the disadvantages as they are an essential part of the local economy. Besides Vetter (1996), no other research to date has addressed the 'donkey problem' in Namaqualand.

Stock density is generally far higher in communal areas than on neighbouring commercial farms. On average, they exceed carrying capacity recommendations set out by the Department of Agriculture (Anon, 1991a) for commercial farms by 1.5–2.5 times (Table 16.1). A long-term dataset for Paulshoek (Fig. 16.3) indicates further that high stock numbers have been maintained on the communal lands for extended periods. In some high rainfall periods (e.g. 1975), stock numbers have exceeded the recommended values by a factor of three. In the last 26 years, stock numbers in Paulshoek have averaged nearly twice the recommended number of animals.

Farmers and their herds are usually located at stockposts (veeposte), away from village centres, yet close to water sources which may be boreholes, dug wells or fountains. Informal consensus between farmers probably determines stockpost location and associated grazing rights (Archer et al., 1996). A recent trend in some of the communal areas has been for stockposts to coalesce around villages where greater access to schools, medical assistance and other civic amenities is achieved. Water quality and availability plays an important role in deter-

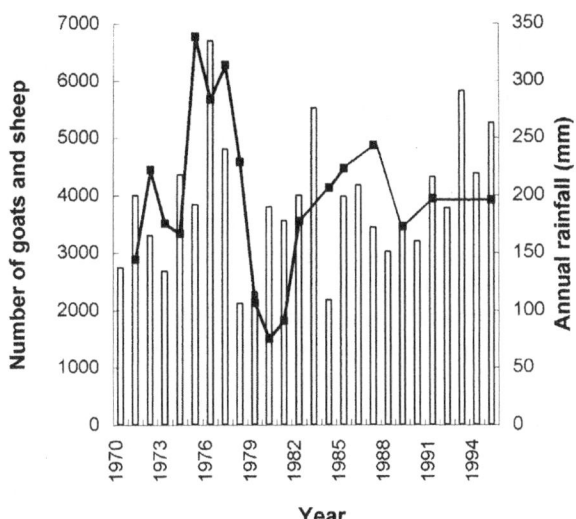

Figure 16.3 **The number of goats and sheep in Paulshoek, a 22 000 ha area in the Leliefontein 'Coloured Reserve' from 1971 to 1995 (line) and annual rainfall totals (mm) (histograms) for Springbok (mean annual rainfall of 194 mm) for the same period. The mean number of goats and sheep for the 26-year dataset is 3 906 animals which is 178% more than the 2200 goats and sheep recommended by the Department of Agriculture for the region**

mining the period of occupation at a stockpost. This varies considerably between individual farmers and between regions. In Paulshoek, for example, one farmer has occupied his post continuously since 1944, while others move every few months depending on water and grazing quality amongst a host of other biophysical and social factors.

Stockpost locations have been tracked in the Richtersveld and indicate regular movements (Archer et al., 1996) within a seasonally determined management framework. After the autumn rainfall peak, farmers disperse away from the Orange River riparian zone to the lowlands and uplands in early winter. They exploit a variety of habitats until late July when they move to the lowlands. These generally are disturbed habitats with a high ephemeral plant component. The excellent nutritional value of ephemerals is important for livestock production and reproduction. As spring advances, farmers generally move away from the lowlands and utilize a wide range of upland and lowland habitats once more. With the onset of the dry, hot summer, the Richtersveld herds begin to congregate along the riparian zones, especially those along the Orange River.

Animal mortality varies considerable between herds and regions. For example, total livestock losses in Rooifontein/Kamassies in 1995 were about 15%, in Paulshoek 7% and in the Richtersveld National Park 10%. For Pella, however, the drought conditions in 1994 meant that 63% of the new-borns died.

Reasons for losses differ between regions but are generally as a result of predators (especially jackals), diseases (e.g. 'krimpsiekte' caused by Crassulaceae), drought and theft.

The communal farmers are generally not well connected to outside commercial livestock markets. Management objectives are also very different to neighbouring commercial enterprises and data from three communal areas suggest that in 1995 between 2 and 10% of the herd was sold for cash. Slaughter for domestic consumption was also low at about 5%.

Cropping is not practised in all communal areas being confined generally to the better-watered regions south of Steinkopf, such as Leliefontein. Here, although it forms an integral part of the overall management system, it is generally practised by a minority of households. The cultivated lands, when crops are successful, provide important grains (oats, wheat, barley and rye) for both human and domestic livestock consumption. During the dry summers, when natural forage becomes increasingly scarce, supplementary feed is mandatory if abortion and lamb fatality is to be prevented.

The impact of both the livestock and cropping industries in the communal lands has generally been greatest on the lowlands. Stockposts, waterpoints and cultivated lands are generally located in the low-lying, flatter landscapes, where a common trend has been to change a perennial, leaf-succulent shrubland (*Ruschia* spp.) to an annual-dominated vegetation often with an associated mix of unpalatable (e.g. *Chrysocoma ciliata*) or poisonous plants (*Galenia africana, Tylecodon wallichii*). The effect of this transformation on the biodiversity and ecosystem functioning of the communal areas is not known. Nor do we understand the full impact of the shift in perenniality to ephemerality on the livestock industry. The long-term dataset from Paulshoek (Fig. 16.3) suggests an element of sustainability that deserves further investigation.

16.3.   Commercial agriculture

16.3.1.   Settlement and historical land use practices

Although a victualling station was established at Cape Town by European settlers in 1652, colonial expansion into the arid hinterland proceeded very slowly at first. All necessary provisions could be either grown locally or acquired through trade with Khoikhoi pastoralists frequenting the western Cape during their seasonal transhumance cycles (Smith, 1986). By the beginning of the eighteenth century, colonial influence did not extend much beyond the Berg River (Penn, 1986). As the European population expanded and the supply of suitable agricultural land and Khoikhoi cattle decreased, so colonists moved further afield (Penn, 1996). After 1700, expansion proceeded rapidly and by 1830 most of the Cape Colony was occupied by colonial farmers. Many of the original Khoikhoi and San inhabitants were exterminated in the conflict which ensued over the control of land and ecological resources. A few pastoralist groups were confined to discrete regions such as the Khamiesberg and Richtersveld, to form the nucleus of the 'Coloured Reserve' population of today, while others became incorporated into frontier colonial homesteads as labourers or slaves (Penn, 1996).

The initial movement of European farmers into the karoo was to the north-western parts of the Colony, particularly to the better-watered sites within the succulent karoo biome. By 1740, most of the area south of Namaqualand was occupied by European colonists and Namaqualand itself began supporting colonial pastoralists from 1750 (Penn, 1996). It was only after 1760 that these pastoralists began herding their animals within the north-eastern regions of the Nama-karoo biome establishing towns such as Graaff-Reinet at this time (Botha, 1926; Penn, 1996).

Colonial occupation of the karoo occurred in stages. Key, resource-rich ecological regions with suitable pasturage and water, and capable of supporting transhumance herding cycles, were rapidly settled by 'trekboers' during the eighteenth century (Penn, 1986). Once the resources became limiting, and probably exacerbated by the frequent severe droughts of the region, surplus colonial pastoralists moved off in search of better pastures within other ecologically feasible transhumance herding areas. Ensuing conflict over the ecological resources with displaced Khoikhoi pastoralists and San hunter-gatherers meant that colonial occupation was often initially quite tenuous. In fact, between 1770 and the beginning of the nineteenth century the colonial frontier was 'rolled back' (Penn, 1996) as aboriginal claims to the ecological resources of, particularly the escarpment region of the interior plateau, resulted in widespread armed conflict in the karoo (Raper and Boucher, 1988; Penn, 1996). However, the organized commandos of the colonial government, comprised usually of local farmers, crushed this resistance and, by the start of the nineteenth century, serious threat to frontier homesteads had diminished dramatically. Widespread expansion into the north-eastern karoo mixed grasslands and especially the southern Free State grasslands began in earnest at this time. By the 1830s, many eastern karoo farmers were already grazing their herds across the Orange River, and soon after this the floodgates to the interior plateau were opened as more and more colonial herders sought greener pastures, unoccupied land and escape from the restrictive colonial authorities.

Initially, land tenure was granted on the basis of a short-term grazing licence which 'conferred freely the right to as much grazing as the licensee needed in a vaguely defined area of unalienated land' (Talbot, 1961). In 1708, this was replaced by the loan farm system which provided some measure of security of tenure over key ecological resources but also enabled farmers to herd their animals within large transhumance cycles utilizing common (unalienated) grazing lands in the process. By 1813, this was replaced by 'perpetual-quitrent tenure' which gave property holders all agricultural rights and security of freehold tenure on a surveyed area in exchange for an annual rental (Talbot, 1961). Changes to the relevant act in 1878 enabled farmers to purchase their property (Talbot, 1961) thus entrenching their freehold tenurial rights. Although other key pieces of legislation, relating to land tenure and resource conservation, have been added to the statute books since 1878, today by far the majority of land in the karoo remains under private ownership and is managed on a commercial basis.

There has been an interesting evolution of grazing practices in the karoo during the historical period. Early colonial pastoralists adopted similar herding strategies to Khoikhoi herders (Van der Merwe, 1938) both for ecological as well as security reasons (Penn, 1996). Large transhumance cycles requiring great mobility were practised. These were often centred on perennial springs and rivers but also exploited the distinct seasonal differences of the many ecological zones of the karoo. However, as land tenure changed and competition for key resources between colonial pastoralists increased so too did grazing practices change. Settlement around privately owned water sources and rangeland meant that grazing orbits shrunk dramatically. Livestock was herded from rangeland to water source to kraal on a daily basis, partly as protection from the large populations of predators such as jackals, present on the range. It was this 'kraaling system' which dominated management practices of the eighteenth and nineteenth centuries and was blamed in 1877 for the degradation of the karoo rangelands amongst a host of other ills, especially those relating to stock disease (Anon, 1877). It was only with the erection of large numbers of imported windmills and the advent of fencing in the late nineteenth century that new management systems were initiated. At first, farm perimeter fences only were constructed. As farms became subdivided into camps, agricultural research began to investigate some of the potential benefits of this approach; the development and rationale for current karoo management systems will be discussed later.

The agricultural industry in the karoo has always been based on small stock, especially woolled and non-woolled sheep (Vorster and Roux, 1983). Data from published agricultural census records for the period 1838–1995 (Fig. 16.4) show that the density of cattle, sheep and goats in a core region of the karoo was initially comparatively low but fluctuated considerably between 1838 and 1904. By 1904, the density of animals had increased to the highest levels on record. Densities of more than 5 LSU per 100 ha were sustained on karoo rangelands from this time until the late 1960s when extended droughts together with state intervention schemes such as the Stock Reduction Scheme (which ran from 1969 to 1978) reduced animal numbers on the range. However, not all regions in the karoo show this general pattern. For example, stock numbers (LSUs) in the arid savanna magisterial districts have increased more than threefold since 1918, while those in the arid central karoo have decreased (Dean and Macdonald, 1994). A complex array of factors, including water-point provision, droughts, vegetation degradation, macro-economic developments and state subsidies, have all influenced stock numbers to some degree in the past. Their respective role in the desertification debate has been addressed by Dean and Macdonald (1994) and will be discussed later.

16.3.2. Current land use practices

More than 80% of land in the karoo currently belongs to private owners where extensive sheep and goat, and to a lesser extent cattle, farming on natural rangeland remains the major agricultural practice. Approximately 50% of the commercial farming enterprises are found on properties smaller than 3000 ha and less than 25% on properties larger than 6000 ha (Anon, 1986).

Although indigenous sheep and goat breeds sustained pre-colonial southern African pastoralists, they have been largely replaced by imported European and several successful composite breeds (Maree and Plug, 1993). Dual-purpose wool and mutton breeds have also been developed to exploit both the wool and meat markets and to suit production systems and variable ecological conditions of the region (Van Niekerk and Schoeman, 1993). Recently, however, the productive merits of many of the indigenous breeds have been recognized and cross-breeding experiments are underway (Van Niekerk and Schoeman, 1993).

The distribution of different livestock breeds in the karoo is determined largely by climate, especially rainfall, and vegetation. For example, the greatest current density of cattle are found in the grass-dominated, summer rainfall arid savanna districts of the east and north-east (Fig. 16.5). Sheep densities are highest in the eastern karoo where dual-purpose and pure-wool breeds, like the Merino, dominate. Sheep, bred for their mutton and pelts, such as the dorper and karakul, are found mainly in the hot, arid north-western areas of the karoo (Joubert, 1980; Roux et al., 1981; Van Niekerk and Schoeman, 1993). The angora goat and boer goat – an improved indigenous breed – occur predominantly in the eastern and south-eastern parts of the karoo. Although there exists considerable overlap in the distribution patterns (Roux et al., 1981) when calculated as Large Stock Units (Anon, 1984), the greatest density of animals in the karoo is found in the higher rainfall, more productive magisterial districts in the east, bordering the grassland biome (Rutherford and Westfall, 1986).

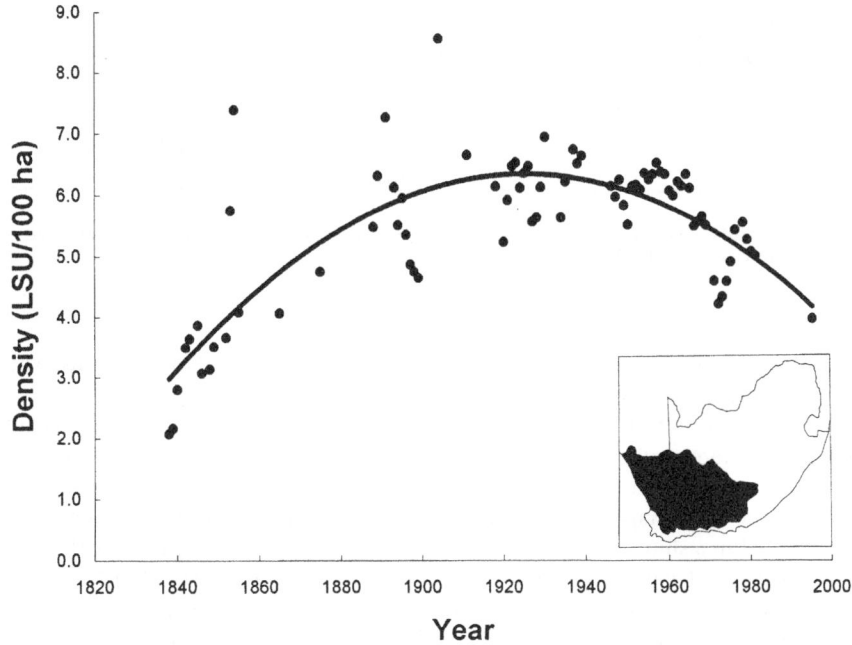

Figure 16.4 **The density (Large Stock Units per 100 ha) of cattle, sheep and goats from 1838 to 1995 for the core region of the karoo (see inset), an area comprising 407 018 km^2 and 65 magisterial districts in 1995. Stock numbers are from published agricultural census records except for 1995 which are from unpublished data from the Department of Agriculture, Directorate of Animal Health. The equation is of the form: $y = -0.0004x^2 + 1.7035x - 1633.4$; $n = 65$; $r = 0.79$; $p < 0.001$**

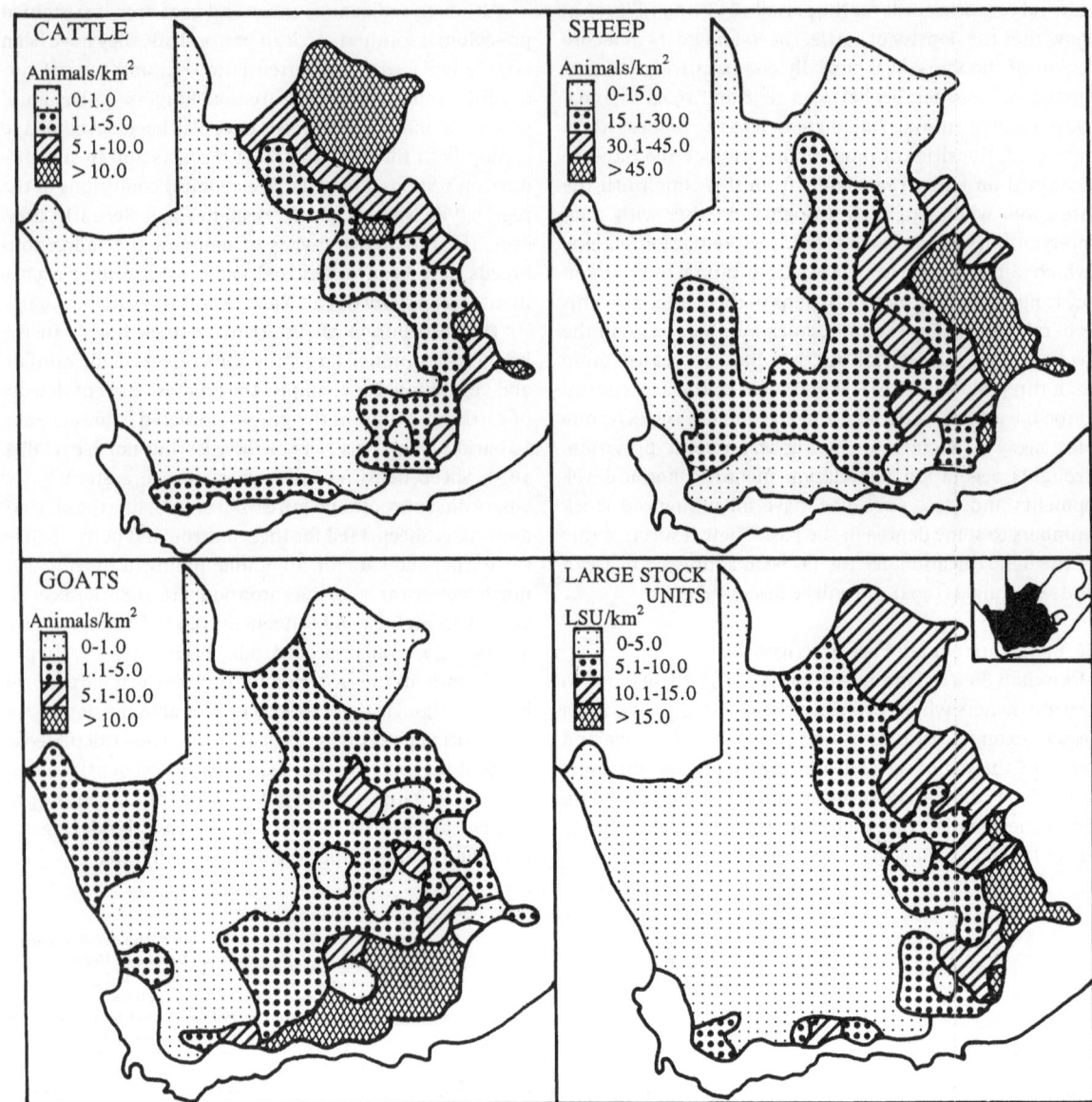

Figure 16.5 **The density (animals or units/km²) of cattle, sheep, goats and Large Stock Units in the karoo in 1995 (data are from unpublished records of Department of Agriculture, Directorate of Animal Health)**

Other factors besides climate and vegetation affect the distribution of small stock breeds in the karoo (Roux et al., 1981; Vorster, 1991). For example, in a bid to maximize their income, farmers tend to switch to more profitable breeds as the economy dictates. Also, a general decrease in unskilled and skilled labour (such as wool shearers) as a result of rural depopulation has resulted in a shift from woolled to mutton breeds in recent years. Finally, an increase in certain toxic and thorny invasive plants has prevented the expansion of angora goats and other fibre-

producing breeds into parts of the karoo, especially when goat-proof fences are absent on a farm.

The close relationship between rainfall and stocking rate (Fig. 16.5) has relevance for the grazing capacity recommendations suggested by the Department of Agriculture for the arid and semi-arid rangelands of the karoo which are significantly correlated with mean annual rainfall (Van den Berg, 1983) (Fig. 16.6). The grazing capacity is nearly ten times higher for areas in the higher rainfall environments in the east than those in the

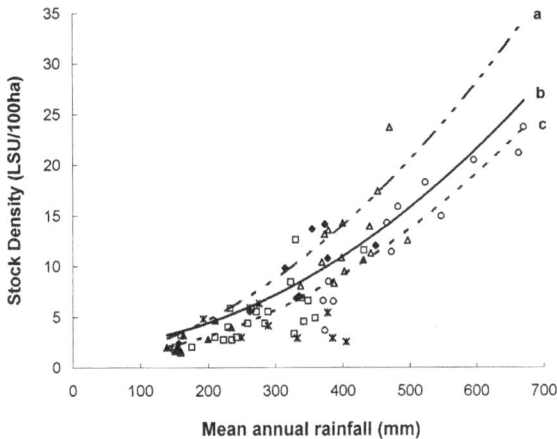

Figure 16.6 **Relationship between mean annual rainfall (mm) and the density of cattle, sheep and goats (expressed as Large Stock Units/100 ha) for 79 magisterial districts in the karoo. Different symbols reflect the relationship between MAR and stocking density in 1995 for magisterial districts within the broadly defined karoo region defined by Dean and Macdonald (1994): arid savanna = filled diamonds; eastern karoo = open circles; central karoo = open squares; succulent karoo = filled triangles; Tanqua and Little Karoo = stars; southeastern mixed grasslands of the Free State = open triangles. Line (a) is for the 1911 census where $y = 0.00006x^2 + 0.0105x + 0.2689$ ($R^2 = 0.6884$; $p < 0.001$); Line (b) reflects Van den Berg's (1983) relationship but for LSU/100 ha where $y = 0.00005x^2 + 0.0024x + 1.9331$ ($R^2 = 1$); Line (c), together with the symbols reflects the 1995 census where $y = 0.00005x^2 + 0.0009x + 1.0001$ ($R^2 = 0.7964$; $p < 0.001$)**

extreme deserts of the west with less than 100 mm of annual rainfall.

There has been a decrease in karoo stock numbers this century (Dean and Macdonald, 1994; Fig. 16.4) which can be analyzed in the light of the grazing capacity/rainfall relationship shown in Fig. 16.6. In 1911 the density (LSUS/100 ha) of animals in the karoo was generally higher than that currently recommended by the Department of Agriculture and as represented in Van den Berg's (1983) analysis. However, in 1995, stock census data from the Directorate of Animal Health suggest that, on average, the magisterial districts of the karoo had a lower density of animals than the recommended value as suggested by Van den Berg's (1983) analysis. Therefore, earlier claims that 'present stock numbers in the Karoo exceed the long term grazing capacity by approximately 30%' (Roux, 1981 in Vorster et al., 1983; see also Dean and Macdonald, 1994) should be re-examined.

Commercial livestock productions systems in the karoo, incorporating the production, reproduction, marketing and range management aspects of the livestock industry revolve mostly around extensive wool and mutton production. Livestock are primarily dependent on the range for grazing and profitability is determined by the adaptability of livestock to seasonal climatic conditions and the ability of the farmer to keep production costs to a

minimum (Roux et al., 1981; Van Niekerk and Schoeman, 1993).

One problem, however, is that the forage and nutritional supply of karoo vegetation fluctuates considerably, especially during the frequent seasonal and prolonged droughts. Many commercial farmers, therefore, find it essential to rely on supplementary feeding for maintaining stock condition, especially that of reproductive animals.

Although less than 1% of the region is cultivated under dryland or irrigated conditions, the stud breeding industry in the karoo is almost entirely dependent on production from irrigated cultivated pastures. Such pastures are also used for commercial mohair production, lambing, fattening of slaughter stock, and, in the eastern karoo, for the production of fresh milk (Steynberg and De Kock, 1987). Dual purpose breeds and animals with multiple offspring also thrive in such intensive production systems (Nel, 1980; Van Niekerk and Schoeman, 1993). Where irrigation water is either not available or limited, fodder production practices are limited to the cultivation of perennial drought-tolerant or drought-evasive fodder crops like old man saltbush (*Atriplex nummularia*), prickly pear (*Opuntia ficus-indica*) and Mexican aloe (*Agave mexicana*). Such additions are then included within the extensive production system (Joubert, 1980; Jordaan, 1993).

Modern karoo rangeland management research dates from 1934 with the establishment of a rangeland research programme at Middleburg (Roux and Vorster, 1983). By the late 1960s, the key elements of karoo management systems had been developed. All advocated the erection of fences to create separate paddocks. Animals would then be assigned to a paddock and moved or 'rotated' from one paddock to another depending on the season, grazing intensity, forage availability, prevailing environmental conditions and so on. These rotational systems replaced the traditional continuous grazing practice of karoo farmers where at least some animals were kept on all parts of the range throughout the year. However, recent reassessments which compared the effect of continuous and rotational grazing on animal production and karoo range condition are contradictory. Some studies show few stock gain benefits from rotational grazing systems (Tidmarsh, 1951; Donaldson, 1986), while others indicate clear advantages to both animal production and range condition under rotational grazing (Roberts, 1970).

The basic principles of the rotational grazing systems developed by the Department of Agriculture evolved from grazing trials at various research stations in the karoo. They were initiated to increase the plant cover on natural rangeland and so to minimize soil erosion. They seek to satisfy both the growth requirements of plants and the feeding requirements of animals. Rotational grazing systems were developed in relatively small camps and later tested on a farm scale. They have demonstrated that, if grazing capacity is not above the recommended values, then these rotational grazing systems are able to maintain plant biomass and species composition within acceptable limits (Roux, 1968b).

The main management systems in the karoo have been surrounded by much controversy (Hoffman, 1988b). Here we compare the rationale and efficacy of pauci-camp or 'conventional' management systems, where 2 to 5 paddocks are used per flock, with multicamp management systems, which utilize five or more paddocks per flock.

Conventional rangeland management systems which are generally recommended for the karoo provide for 3 to 4 camps for each group of animals (Donaldson and Vorster, 1989). They are characterized by relatively long grazing periods (periods of occupation) which may range from 2 to 6 months, and long resting periods (periods of absence) which normally exceed 6 months. They were developed from ecological principles which assessed the timing, length and duration of grazing and resting periods in terms of animal daily gains and range condition requirements. Although paddocks are grazed systematically and in a regular sequence, these conventional systems provide for a flexible management approach wherein, for example, the length of the grazing period can be adapted to forage availability. A large number of farmers in the karoo prefer to apply conventional rotational grazing systems because it simplifies their range and stock management practices. Furthermore, the high cost of fencing material is avoided as a smaller number of camps is required (Donaldson and Vorster, 1989).

The 'group camp approach' (Roux, 1968b; Roux and Skinner, 1970) to range management is a framework within which conventional rotational systems can be applied. It developed out of the numerous grazing trials conducted in the karoo, but especially those conducted in Middelburg, and remains the most widely advocated system for range management in the karoo. Although the group camp approach is flexible in design, incorporating a wide range of specific management techniques (such as intensive, non-selective grazing for short periods), it rests on two basic principles. The first is that different landscape units (e.g. mountains, pediments, rivers) differ in terms of their plant composition and production potentials and therefore possess different carrying capacities. If these units are not fenced and grazed separately within the grazing cycle then patch- and area-selective grazing will occur, especially within more productive landscape units. Detrimental changes in species composition and lowered production potentials will result. The group camp approach therefore advocates that different landscape units within a farm's boundaries be fenced, subdivided

and 'grouped' into management units. Different flocks are then assigned to the different groups of camps or paddocks. So a single flock may move from a paddock along the river course to one on a pediment, before ending the rotation within a paddock fenced along a mountain slope.

The second important principle within the group camp approach is that the grazing period within a paddock should be rotated from year to year. It is believed that different species and growth forms (e.g. grasses and shrubs) within the karoo grow, flower and set seed at slightly different times during the year (Roux, 1968b, but see Hoffman, 1989b). By staggering the interannual defoliation regime, no one species or growth form is repeatedly selected. Different species are therefore provided the opportunity to flower, set seed and establish. Diverse mixes of palatable forage grass and shrub species are predicted within this system.

The group camp approach claims to exhaust the possibilities of the practical application of rotational and special grazing systems on an ecologically justifiable and tested basis (Roux and Vorster, 1983a). It remains widely advocated and used in the karoo.

When more than five paddocks are available per flock, the management system is referred to as a multi-camp system. The primary reason for increasing the number of paddocks is to decrease the length of the grazing period while at the same time to increase the length of the resting period. Two basic utilization philosophies have arisen in the karoo. The first, non-selective grazing (NSG) (Acocks, 1966), has undergone a long and controversial development. NSG prescribes relatively high stocking rates but a short (2 weeks maximum) grazing period followed by a variable but generally long rest period (6 weeks minimum to a year or longer). This is thought to simulate the 'natural', high-intensity defoliation regime of pre-colonial migratory ungulates. NSG is an attempt to force domestic livestock to graze all species on the range, even unpalatable elements, thereby reducing the competitive advantage such species enjoy under a selective defoliation regime. In prescribing long rest periods, NSG practitioners believe that the 'root reserves' of important forage species are replenished sufficiently to withstand further defoliation (Acocks, 1966; Hoffman, 1988b). Despite direct and indirect experimental evidence which has rejected NSG as a viable management system (Roux, 1967; O'Reagain and Turner, 1992), a core group of ardent NSG practitioners remains in the karoo (McCabe, 1987).

Short duration grazing (SDG) is the second utilization philosophy and developed out of non-selective grazing. This multi-camp system prescribes a relatively fast rotation of animals between camps with short periods of stay (1–7 days) and relatively short periods of rest (40–60 days) (Savory and Parsons, 1980). Although largely untested, the physical effect (trampling, urine and dung deposits) of the prescribed, relatively large numbers of animals on the range are thought to have important consequences for the hydrological properties and nutrient-cycling processes of the soil. Also, an increase in the number of germination sites and thus enhanced seedling establishment are predicted (Savory and Parsons, 1980). Finally, an assumption inherent in the short rotation is that annual production of palatable forage species is maximized providing them with a superior competitive advantage over unpalatable elements within the sward (Hoffman, 1988b). Despite an initial enthusiasm for the use of SDG as a management system for the karoo in the 1970s and 1980s, the development of Holistic Resource Management (HRM) (Savory, 1988) as a management philosophy has eclipsed much of this earlier interest.

Despite the availability of a large number of grazing systems, only a small number of farmers in the karoo apply judicious systems of management. Ignorance, complacency, tradition, economic pressures and uncertainty of the advantages of prescribed management systems all mitigate against their adoption. However, the particular grazing system employed is probably of minor importance as it is the stocking rate, under direct control of the grazier, which has the greatest impact on range condition and animal production (O'Reagain and Turner, 1992).

The sustainability of commercial small stock farming in the karoo will depend largely on the use of adapted stock breeds and in the application of more extensive production systems in line with the arid and semi-arid environmental conditions of the region. Uneconomical external supplementation cannot be sustained in the long-term (Steynberg, 1989; Vorster, 1991). Rather than beginning with high-potential animals and attempting to adjust the environment or meet the animal's requirements, the environmental potential should be considered at the outset. For example, the emphasis of most small stock breeders in the karoo in the past has been to increase the amount of fibre produced per animal. However, an increase in fibre potential is accompanied by a reduction in the 'constitution' or fitness of animals (Herselman et al., 1993). Thus, the majority of fibre-producing breeds are unable to produce optimally under the extensive conditions of the karoo, unless they receive supplementary feeding. When the fibre and meat product prices were relatively high, external supplementation could be justified. However, these practices are no longer economically viable because input costs are high. Low risk and low input small stock farming systems should include the development of farming systems and practices which take cognizance of risk management, including the impact of periodic drought on livestock production, better adapted animal genotypes and economically viable strategies for the control of invasive plants (Vorster, 1993). Strategies which add value to agricultural products should also be developed (Vorster, 1993).

Two other animal husbandry practices, namely ostrich and game farming, are of interest in the karoo in that they have an effect on range condition. Ostriches have been farmed in parts of the southern karoo, especially around Oudtshoorn, since 1865. Early markets were centred on the fashionable trade in ostrich feathers but recent growth in the sale of other products such as ostrich skin and meat has ensured that the industry has grown exponentially in South Africa and elsewhere since the mid 1980s. The general practice is to rear slaughter stock on lucerne pastures while breeding birds are kept on natural rangeland. When free-ranging, ostriches are highly mobile covering large distances in search of the herbaceous material which comprises their preferred diet (Williams et al., 1993; Dean et al., 1994b; Milton et al., 1994a). In captivity, however, ostriches are usually confined to relatively small paddocks on the low-lying areas of a farm. Their selection of soft ephemerals, seeds, seedlings and new growth as well as the impact of their relatively large mass and footprint has caused concern over their potential effect on the production and species composition of natural rangeland (Milton et al., 1994a). While their impact may not be great in the short-term (6 months or less) (Gatimu, 1996) there is little doubt that they are able to transform natural rangeland. The development of small, intensive paddocks for the rearing of ostriches coupled with long rest periods (>5 years) between occupation has recently been recommended (Gatimu, 1996).

Although game ranching has a relatively short history in the karoo, by 1983 there were 646 registered game farms with a combined area totalling more than 3 million ha (Jooste, 1983). Springbok (*Antidorcas marsupialis*) are by far the most important game species in the karoo and in 1983 were found on 95.8% of all farms in the region (Jooste, 1983). Three years later it was estimated that there were more than 1.5 million springbok on karoo farms (Anon, 1986).

Hunting and cropping, largely for venison production, are the main consumptive uses of game in the karoo areas, and trophy hunting, sales, game viewing and general aesthetics are of less importance (Jooste, 1983). Because domestic livestock have been selected for a variety of production and behavioural traits, indigenous ungulates are unable to compete, in terms of productivity, under intensive farming conditions (Skinner et al., 1986). None the less, they have an important contribution to make especially if markets for their products were more stable. For example, prior to 1989, venison production contributed significantly to the farming income of a number of commercial farmers (Anon, 1991b) amounting to approximately 50% of that produced by merino sheep (Conroy and Gaigher, 1982). However, the subsequent dramatic drop in venison prices has ensured that it no longer plays a significant role in agriculture production (Anon, 1991b).

Because only one or two breeds are used, optimum utilization of vegetation in the karoo is not achieved with domestic livestock alone and it is more practical to maximize sustained productivity with a mix of domestic livestock and wild ungulates (Skinner, 1989). The most widespread combination in the karoo is that of springbok and sheep which differ in their feeding habits and selection of plants (Davies et al., 1986). Springbok prefer to browse karroid shrubs, spend more time foraging and feed on a wider variety of plant species than merino sheep which prefer to graze, spend more time ruminating and select a narrow range of plant species (Davies et al., 1986). Although the combination of sheep and springbok results in an increase in the number of species used, there remains considerable dietary overlap, particularly of palatable plants (Fairall et al., 1990).

All prescribed rangeland management systems in South Africa propose the periodic withdrawal of one or more paddocks from grazing to allow plants to complete essential growth and reproductive processes. This seldom happens within mixed game and domestic livestock enterprises, especially when the game species (e.g. springbok) is not dependent on the artificial provision of water (Jooste, 1983). Therefore, on most game farms in the karoo, continuous grazing occurs. Provided that paddocks are large enough and overall stocking rate estimates are kept within the recommended values this is not necessarily detrimental to the range. However, because stocking rates of game are difficult to adjust in the short term, slight understocking is advised so as to hedge against seasonal pasture shortages (Skinner, 1989).

16.4.   Desertification of the karoo

16.4.1.   A brief history of the karoo desertification debate

The desertification issue in South Africa is synonymous with the karoo. In particular, it is a relatively small region in the eastern karoo, representing a broad ecotone between grassland and shrubland vegetation, that has been at the heart of the debate (Tidmarsh, 1948; Acocks, 1953; Roux and Theron, 1987; Dean et al., 1995; Hoffman, 1995). Ecological changes in the communal areas of the succulent karoo biome, discussed earlier, and in the extensive communal areas of the grassland and savanna biomes have generally not formed part of South Africa's

desertification literature. Thus, losses in agricultural productivity due to bush encroachment, deforestation and erosion have only rarely been integrated (Roux, 1981; Roux and Opperman, 1986) within a broader dryland degradation/desertification view for the country. Although this is changing, what follows is an account of the rather narrower desertification debate as it has unfolded for the eastern karoo. This debate has its origins in the late nineteenth century when, alarmed by the increase in stock deaths due to disease in the eastern karoo, the colonial government dispatched a commission of inquiry to investigate (Anon, 1877). Together with the annual reports of the Colonial Veterinary Surgeon at the time (Branford, 1877, 1879) the commission led evidence from many farmers in the region to determine the reason for the decline in the sheep industry. Although they concluded that 'in very few instances had any change [in range composition] been observed, and in none was that change of such a serious nature as to excite any particular notice', the commission maintained that overstocking, in combination with droughts and locusts, had resulted in a shift from palatable to predominantly unpalatable forage species. It was this shift in plant species composition, they suggested, and in particular a loss of halophytes, which had resulted in a dramatic increase in stock death due to disease. A plea was made to fence and subdivide farms into paddocks and to replace the kraaling and herding system which they considered 'most injurious'.

It would take another 40 years before the next state commission of inquiry focussed on the karoo degradation issue, albeit in the context of a series of droughts in the karoo, especially that of 1919. The message contained within the Final Report of the Drought Investigation Commission (Anon, 1923) was clear. Although droughts had played a role, it was the colonial farmers who were to blame for the partial denudation of the original vegetation 'with the result that rivers, vleis, and water-holes described by old travelers [had] dried up or disappeared'. Again it was the kraaling system, overgrazing and erosion which were highlighted as the main factors responsible for 'a serious diminishing efficiency of the rainfall'. They concluded 'that the severe losses of the 1919 drought were caused principally by faulty veld and stock management'.

These themes formed the basis for much ecological research over the next three decades as the detailed changes that had taken place in the vegetation of the karoo were described (Kanthack, 1930; Hall, 1934; De Klerk, 1947; Tidmarsh, 1948). It seems, however, that state commissions and research articles had little influence on karoo farmers, as stock numbers in the 1930s were sustained at relatively high levels throughout this period (Fig. 16.4), despite the dire warnings of the Drought Investigation Commission (Anon, 1923). Even after the emergence of John Acocks' comprehensive desertification theory (Acocks, 1953), which was widely publicized in the popular press (Robertson, 1952), it took many years before a national action plan to deal with the desertification problem took effect.

In his account, Acocks (1953) mapped the vegetation of South Africa as he understood it in 1953, related it to an earlier (i.e. pre-colonial) 'pristine' era, arbitrarily set at AD 1400, and predicted changes that were likely to take place over the next 100 years. These three maps of past, present and future vegetation conditions were also included in the Report of the Desert Encroachment Committee (Anon, 1951). Their influence on the South African desertification debate cannot be overstated and it continues to the present (Huntley et al., 1989; Moll and Gubb, 1989; Hoffman, 1995; Dean et al., 1995).

Very little original research into desertification occurred in the three decades following Acocks account. Many authors were content to cite Acocks (1953) when the desertification problem came under review, and used his conclusions to support their various agricultural and conservation agendas, especially when rates of change were stated (Downing, 1978; Hilton-Taylor and Moll, 1986; Aucamp and Danckwerts, 1989; Huntley et al., 1989). The last decade, however, has seen a resurgence of interest in the field (Roux and Theron, 1987; Hoffman and Cowling, 1990b; Dean et al., 1995) and much original research, aimed at testing two central hypotheses, has emerged.

16.4.2. Hypothesis 1: The precolonial eastern karoo was a stable and extensive 'sweet' grassland

The first hypothesis was stated in many different guises prior to Acocks' (1953) treatise (e.g. De Klerk, 1947; Tidmarsh, 1948). However, his explicit spatial treatment has made it easier to use site-based analyses to test his conclusions. Simply stated, the hypothesis suggests that, prior to European colonization of the eastern karoo, a perennial, 'sweet grassveld' existed over the area (i.e. a rangeland with palatable grasses available as forage throughout the year). This was regarded as the 'climax' vegetation which Acocks (1953) considers to have been relatively stable in time. He placed the blame for its decline primarily on colonial, and subsequently on commercial farming practices. A number of archaeological, historical and ecological lines of evidence have been used to test this view (Hoffman, 1995).

Karoo micromammals have distinct habitat requirements particularly with regard to their preference for shrub- or grass-dominated landscapes (Avery, 1991). By measuring their abundance at known times in the past, it is possible to reconstruct the vegetation composition of earlier environments. Micromammal abundances in owl pellets at an archaeological site in the eastern karoo sug-

gest that grasses and shrubs have fluctuated through time and no long-term stability in grass cover is evident (Avery, 1991). In fact, eastern karoo environments in the first millennium AD are characterized, not as perennial grasslands, but as extreme deserts with 'little above ground vegetation of any kind' except around the seepage area (vlei) close to the site (Avery, 1991). Similar evidence, but from archaeological deposits of large mammal remains in the eastern karoo, also indicates fluctuations in grass and shrub conditions in pre-colonial environments (Sampson, 1994).

Since the hypothesis concerns the relative abundance of grasses and shrubs in the landscape, a study of fossil pollen in hyrax middens located in the eastern karoo has been helpful in assigning dates to major fluctuations in these growth forms (Bousman and Scott, 1994). These data confirm that fluctuations have indeed occurred in the relative abundance of grasses and shrubs. They suggest, however, that a decline in the relative proportion of grasses in the landscape began around AD 1550 (with a major decline in the early to mid 1700s), some 200 years before the arrival of the first colonial farmers in the region during the latter part of the eighteenth century (Bousman and Scott, 1994).

An additional insight into pre-colonial, eastern karoo conditions may be found in the descriptions that early travellers had of the region (e.g. Barrow, 1801; Raper and Boucher, 1988). They suggest both a more grassy environment in the late eighteenth century and one which fluctuated considerably from year to year dependent on seasonal climatic patterns (Roux, 1966; Hoffman and Cowling, 1990b).

Finally, a recent study using carbon isotopes from soils (Bond et al., 1994) replaces Acocks' (1953) idyllic view of an extensive pre-colonial grassland by describing the eastern karoo as an ecotone characterized by three distinct vegetation zones: a region in the south-west dominated by shrubs, a broad transition zone of shrubs and grasses and a grassland north and east of the Orange River. Unlike the archaeological and palynological evidence, however, this analysis of the 13C values of soil organic matter (SOM) suggests a relatively stable matrix of either grass or shrub-dominated vegetation within this ecotone, with far less fluctuation in grass and shrub conditions in pre-colonial times.

Acocks' (1953) hypothesis of an extensive pre-colonial eastern karoo grassland is rejected by most recent lines of evidence. However, the stability of the grassland in time requires further scrutiny. Some studies (Hoffman and Cowling, 1990b; Avery, 1991; Bousman and Scott, 1994) suggest considerable fluctuations in grass and shrub abundance, while evidence from stable isotopes (Bond et al., 1994) suggests greater stability. What is clear is that the influence of climate, and particularly of rainfall, on the dynamics of grasses and shrubs was never included within Acocks (1953) main thesis. It has emerged recently as an important element of the desertification debate and is discussed below.

16.4.3.   Hypothesis 2: An altered and less-productive karoo vegetation is expanding into more productive grassland environments

A second important hypothesis formulated by Acocks (1953) suggests that, for a variety of reasons, but especially as a result of selective grazing pressure of single species, domestic flocks (Acocks, 1966), the pre-colonial productive grassland has been changed to a less palatable rangeland dominated in the latter half of the twentieth century by a mixture of short-lived grasses and unpalatable shrubs (Acocks, 1953). An explicit assumption is that this new mix is less agriculturally productive than the climax grassland which once dominated the eastern karoo. An alarming addition to the hypothesis states that if management practices continued unchecked then the karoo vegetation would expand into the agriculturally more productive north-eastern grassland areas, and reach the outskirts of Pretoria by AD 2053 (Acocks 1953). Based on these and other sources (e.g. Jarman and Bosch, 1973), annual rates of expansion, ranging from 0.44 to 3.5 km yr^{-1} (see Hoffman and Cowling, 1990b), have been calculated.

Two analyses of long-term ecological survey data combined with a study of matched photography in the eastern karoo found little evidence to support these views (Hoffman and Cowling, 1990b; Palmer et al., 1990). Instead of an increase in the level of degradation at each of the study sites and a reduction in grass cover with time, as predicted by the expanding karoo hypothesis, the opposite was measured. At nearly all sites a significant increase in cover, especially of grasses, was observed. These studies concluded that the desertification debate, and the expanding karoo hypothesis in particular, had to accommodate the seasonal fluctuations in grass and shrub cover that characterize these eastern karoo environments (Roux, 1966).

The idea of an annually expanding karroid shrubland has also been rejected by the continuous pollen sequence record in the eastern karoo (Bousman and Scott, 1994). The data indicate large fluctuations in the relative abundance of Poaceae and Asteraceae pollen from 1820 to 1980. The analysis downplays the effect of stocking rate and also emphasizes the important influence that rainfall has in influencing the relative dominance of grasses and shrubs in natural karroid vegetation.

While the aforementioned studies reject the karoo degradation hypothesis, a comprehensive analysis of historical stocking rates in the karoo finds evidence to support the view (Dean and Macdonald 1994). The declining trend in stock numbers measured for most magisterial districts over the last three decades suggests that karoo rangelands can no longer support the same large number of animals which characterized these areas in the first half of the twentieth century (Fig. 16.4). It has been pointed out, however, that the role of state intervention schemes, such as the Stock Reduction Scheme, in facilitating a reduction in livestock numbers should not be underestimated (Pringle et al., 1982; Hoffman et al., 1995b). Also, the eastern karoo experienced its wettest period on record during the years 1974–6 and the rangeland response was well documented (O'Connor and Roux, 1995). Hoffman et al., (1995b) maintain, therefore, that stock numbers in the eastern karoo, at least, have been kept artificially low through government subsidy incentives, and that the reduction in stock numbers is therefore not simply an index of rangeland degradation. However, the role that state intervention has played is complex, as drought aid schemes may also maintain artificially high stock numbers on karoo rangeland through drought feeding programmes.

Finally, the analysis of stable carbon isotopes in the soil organic matter mentioned above (Bond et al., 1994) finds evidence for a substantial recent increase in shrub cover at eastern karoo sites. Together with the decline in stock numbers (Dean and Macdonald, 1994), it remains the strongest support for the expanding karoo hypothesis. Unresolved problems with the dating of the material, however, had rendered an understanding of the timing of rangeland change and consequent role of rainfall and grazing difficult (Hoffman et al., 1995b).

Any analysis of both the above hypotheses needs to consider the influence of climate, especially seasonal rainfall, and grazing on grass and shrub population dynamics (Milton and Hoffman, 1994). Long-term datasets indicate that both are important (O'Connor and Roux, 1995). While summer rainfall has a marked effect especially on the abundance of grass cover (Roux, 1966; Hoffman et al., 1990), the long-term effect of grazing on perennial species is also important (O'Connor and Roux, 1995). Uncoupling the two has proved difficult in the past, but is essential if we are to develop suitable predictive models of vegetation change in the karoo.

16.5.   Conclusions

Current trends in agricultural and biological productivity need to be assessed in terms of historical patterns and processes. A wide range of tools are needed, including those from palynology, archaeology, history, anthropology, ecology and sociology if the full impact of humans on South Africa's karroid landscapes is to be determined accurately. Although we have tried to bring an historical perspective into our understanding of land use practices and desertification in the karoo, it is clear that our information is patchy. This is especially so for the communal areas. While much archaeological, anthropological and historical research has investigated the nature of pre-colonial pastoralist economies in space and time and their relationship to early settler expansion, very little is known about current land use practices and their impacts on landscapes within the succulent karoo biome. This is unfortunate, as recent land reform policy in South Africa (Anon, 1996) has adopted rather uncritically the 'new thinking' on range ecology (e.g. Behnke et al., 1993; Scoones, 1994) as applied to communally managed farming areas. The relationship between stocking rate and land degradation, suggested by South African agricultural departments and many local range ecologists, who have worked for decades within commercial farming systems, is now questioned. Ideas developed from communal tenure systems in neighbouring African states, with often very different ecological, economic, social and cultural conditions, have been adopted as more helpful guides in our understanding of local problems of communal tenure. However, landscapes within the succulent karoo biome are unique and may not respond to the heavy utilization apparent under communal tenure in the same way that the arid savannas do in other parts of Africa. In the karoo, human populations are confined to relatively small areas with only limited possibilities for migration. Such differences are important, yet even basic descriptive information for communal areas in the karoo is lacking.

Much more is known about commercial agriculture and its impacts on the karoo. Government intervention via agricultural research, extension services, census records and subsidies has kept track of the changing ecological and economic conditions of the region. The control which has been exerted by the state over private landowners, especially through legislation, infrastructural improvement subsidies and drought subsidies has had an important influence on stocking density and grazing impacts. However, given the changing priorities of the recently elected democratic government, it is unlikely that these costly interventions will continue.

South Africa is a signatory to the United Nations Convention to Combat Desertification and aims to develop a National Action Programme (NAP) to address the problem of dryland degradation. This will require a national focus investigating the loss of secondary agricultural and biological productivity as a result of deforestation, afforestation, bush encroachment, soil erosion and overgrazing throughout South Africa. Hopefully, the desertification issue will now extend beyond the narrow confines of the expanding eastern karoo debate where it has been so firmly lodged these last several decades. More importantly, in seeking solutions to dryland degradation problems we will also need to incorporate a wider spectrum of social issues in our deliberations, as it is human agricultural production systems, whether communal or commercial, more than anything else which have so profoundly transformed the South African arid and semi-arid landscape.

17 Alien plant invaders of the karoo: attributes, impacts and control

S. J. Milton, H. G. Zimmermann and J. H. Hoffmann

17.1. Introduction

Some alien plants, including *Ricinus communis* and *Medicago polymorpha*, have been established in the karoo since the late Stone Age period (Deacon, 1986b), but none of these early colonizers has ever become particularly dominant or problematic. Apart from a prickly pear, *Opuntia ficus-indica*, which was probably introduced into South Africa before 1656 (Wells et al., 1986), the invasion of the karoo really commenced when many livestock farmers migrated northwards from the western Cape during the 1800s (Hoffman et al., this volume), inadvertently carrying with them seeds of several alien plant species. Disturbances caused by the newly settled farmers and their livestock created conditions that favoured the establishment of alien plants, predominantly annual, non-woody shrubs and herbaceous species such as *Argemone ochroleuca* (s.n. *A. subfusiformis*) and *Salsola kali* (Table 17.1).

The situation worsened with time because the landowners tried to recreate the home environment they had left and they began to introduce trees and shrubs in an otherwise almost treeless environment (Table 17.2). They propagated species that would provide fodder, shade, fuel and protection. Many of the species they selected exhibit features that characterize aggressive invaders, namely, copious seed production, efficient dispersal mechanisms, xerophytic tolerances and the ability to coppice or reproduce vegetatively. Some of the introduced species thrived in the new environment and soon 'escaped' into the natural vegetation.

In this chapter, we consider the attributes and effects of alien plant species that have become naturalized in the Great Karoo. We concentrate on species that have formed monospecific stands over large areas. Such transformer species constitute the most serious threat to the karoo and have therefore received most attention in terms of research and control efforts. Alien pioneer species and herbaceous plants sometimes form dense patches on severely disturbed sites but they are usually displaced as natural vegetation re-establishes. Projected land use and climatic scenarios indicate that the impacts of these inconspicuous plants on the karoo ecosystem are likely to increase.

17.2. Invasibility of karoo vegetation types

Information on the distribution of alien plants in the karoo has been derived from herbarium records, surveys and checklists. However, the central part of the region has been poorly covered by plant collectors, and surveys of the relative abundance and distribution of alien plants in karoo regions have been coarse-grained and generally confined to the proximity of major roads. Wells et al. (1986) listed 789 exotic plant taxa that have become naturalized in southern Africa and the list has expanded considerably during the past 10 years (Henderson, 1989, 1991a and b, 1992; Henderson and Wells, 1986). A disproportionately low number of these species have invaded the karoo and the region has apparently been less susceptible to invasion by exotic alien plants than the adjacent fynbos and sub-tropical biomes (Cronk and Fuller, 1995).

Based on roadside surveys that used 'quarter-degree' (15' latitude × 15' longitude) squares as sampling units and excluded intentionally established plants, Henderson (1991a) reported that *Prosopis* sp. and *Opuntia ficus-indica* were the most frequent species encountered in the northern Cape, followed by *Nicotiana glauca* and *Melia azedarach*. In similar surveys between 1987 and 1993, L. Henderson

Table 17.1. *The most common alien invasive grasses and forbs in the karoo–Namib area, their regions of origin and invasive potential in natural habitats (L = low, M = moderate and H = high, after Brown and Gubb, 1986; Species marked + based on information in Henderson et al., 1987)*

Species	Region of origin	Invasive potential
Grasses		
Arundo donax	Europe	L
Avena spp.	Europe	L
Bromus pectinatus	Middle East	L
Bromus diandrus agg.	Europe	L
Pennisetum setaceum	North Africa	L
Polypogon monspeliensis	Europe and Asia	H
Sorghum halepense	Circa-Mediterranean	L
Stipa tenuissima	S. America	+
S. trichotoma	S. America	+
Forbs		
Altenanthera pungens	South America	M
Amaranthus hybridus	North America	M
Argemone ochroleuca	C. America	H
Aster subulatus	Australia	M
Atriplex lindleyi	Australia	M
Atriplex semibaccata	Australia	M
Bidens biternata	N. America and Asia	H
Boerhavia diffusa		M
Chenopodium murale	Europe	L
Conyza bonariensis	Europe and Asia	M
Datura ferox	Europe and Asia	H
Datura inoxia	C. and S. America	H
Datura stramonium	C. and S. America	H
Gnaphalium luteo-album	Europe	M
Lactuca serriola	Europe	M
Malva parviflora	Europe	L
Melilotus indica	Europe and Asia	M
Oenothera rosea	Central America	M
Physalis angulata	N. and S. America	L
Plantago lanceolata	Europe	M
Polygonum aviculare	N. and S. America	M
Salsola kali	Asia	H
Sonchus oleraceus	Europe	M
Tagetes minuta	S. America	M
Taraxicum dens-leonis	Europe	M
Urtica urens	Europe	L
Xanthium spinosum	S. America	M
Zinnia peruviana	S. America	M

Table 17.2. *The most common alien invasive shrub and trees in the karoo–Namib area, their regions of origin, percentage frequency (in quarter-degree squares, 15' latitude × 15' longitude, unpublished data, L. Henderson, 1996), and invasive potential in natural habitats (L = low, M = moderate and H = high, after Brown and Gubb, 1986); n/a – data not available in Henderson surveys*

Species	Region of origin	Percentage frequency	Invasive potential
Agave americana	C. America	23	L
Arundo donax	Mediterranean	18	M
Atriplex nummularia	Australia	28	M
Eucalyptus spp.	Australia	12	M
Nicotiana glauca	S. America	38	M
Opuntia aurantiaca	S. America	n/a	H
Opuntia ficus-indica	C. America	47	M
Opuntia imbricata	N. and C. America	n/a	M
Opuntia robusta cvs	Americas	25	L
Populus × canescens	Europe and Asia	14	M
Prosopis spp.	N. and C. America	52	H
Salix babylonica	Asia	11	M
Schinus molle	S. America	19	M
Tamarix spp.	Europe and Asia	12	M

susceptibility of karoo regions to plant invasions (Milton, and Dean, in press), alien species (excluding roadside plantings), land use and soil type were recorded at 5 km intervals along 2630 km of primary and secondary roads (526 sites) in spring 1996. Patterns of alien species-richness were then analysed in relation to land use and soils, as well as vegetation type, rainfall quantity and seasonality as given in Low and Rebelo (1996). Species-richness of alien plants was lowest in the Kalahari and Bushmanland areas and highest in Renosterveld on the succulent karoo–fynbos interface in the south-western part of the region (Fig. 17.1). These patterns were explained by co-varying factors, particularly rainfall quantity, soil type and land-use. Renosterveld receives dependable winter rainfall and has fine-textured arable soil and is consequently highly disturbed (88% of sample sites used as villages, fields and orchards). Moreover, the area has been settled by Europeans for longer than other parts of the karoo. In contrast, the Kalahari and Bushmanland have low and variable rainfall, sandy soils, are used for extensive ranching and are sparsely populated. Within vegetation types, there were significantly fewer species of alien plants adjacent to rangeland than to villages, fields and orchards.

Although disturbance and relatively mesic conditions evidently facilitate invasion of karroid vegetation types by woody and herbaceous plants, they are poor predictors of the composition of the alien flora. Alien assemblages differed between dependable winter-rainfall areas in the west, and the rest of the karoo that receives either unpredictable spring and autumn rainfall, or summer rainfall (Fig. 17.2). Temperate and Mediterranean-climate species (*Acacia saligna*, *Avena sativa*, *Bromus pectinatus* and *Pennisetum setaceum*) were largely confined to the winter-rainfall region, *Atriplex lindleyi* from arid Australia

(unpublished data, Agricultural Research Council) rated the abundance of alien plant invaders of the Nama- and succulent karoo. *Prosopis* spp. and *Opuntia ficus-indica* were the most frequently recorded invaders in the Nama-karoo and accounted for more than 34% of the sightings (Table 17.3). In the succulent karoo, which lies in the winter-rainfall region, *N. glauca*, *O. ficus-indica*, *A. nummularia* and *Prosopis* spp. were identified as the most abundant woody invaders (Table 17.4). The survey results indicate a bias towards the more conspicuous, alien shrubs and trees, many of which have spread from planted trees at roadside pull-offs or from farmstead gardens. The less obvious, but often equally aggressive, alien invaders, such as *Opuntia aurantiaca*, *O. imbricata*, *O. rosea*, *Atriplex lindleyi*, *A. semibaccata* and *Argemone ochroleuca* were undoubtedly underrepresented in the surveys.

In a preliminary investigation of factors increasing the

Table 17.3. *Presence and abundance of alien plant species in 10 or more quarter-degree squares (QDS; 15' latitude × 15' longitude) in the Nama-karoo biome during roadside surveys from 1987 to 1993. The Nama-karoo biome occupies approximately 600 quarter-degree squares, of which 256 were sampled*

Species	No. QDS present	No. QDS abundant	Percentage frequency
Prosopis spp.	151	33	59
Opuntia ficus-indica	126	3	49
Opuntia robusta cvs.	78		30
Nicotiana glauca	72	1	28
Agave americana	63		25
Atriplex nummularia	61	3	24
Schinus molle	43		17
Salix babylonica	39		15
Arundo donax	35	1	14
Populus × canescens	31	2	12
Tamarix spp.	31		12
Eucalyptus spp.	25		8
Melia azedarach	17		7
Populus nigra v. italica	15		6
Gleditsia triacanthos	11		4
Ricinus communis	10		4
Echinopsis spachianus	9		4

L. Henderson (unpublished data).

Table 17.4. *Presence and abundance of alien plant species in 10 or more quarter-degree squares (15' latitude × 15' longitude) in the succulent karoo biome during roadside surveys from 1987 to 1993. The succulent karoo occupies approximately 180 quarter-degree squares, of which 102 were sampled*

Species	No. QDS present	No. QDS abundant	Percentage frequency
Nicotiana glauca	65	1	64
Opuntia ficus-indica	44		43
Atriplex nummularia	38	3	37
Prosopis spp.	37	3	36
Acacia cyclops	34	7	33
Arundo donax	28	4	27
Schinus molle	26		25
Acacia saligna	25	9	25
Ricinus communis	23		23
Agave americana	21		21
Populus canescens	19	2	19
Eucalyptus camaldulensis	17	17	4
Tamarix spp.	13		13
Opuntia robusta cvs.	12		12
Sesbania punicea	9		9
Nerium oleander	3		3

L. Henderson (unpublished data).

was equally frequent under winter and between-seasons rainfall, whereas plants of the inland deserts of Asia and the Americas (*Argemone ochroleuca, Opuntia* spp., *Prosopis* spp., *Salsola kali*) were uncommon or absent in the winter rainfall region of the karoo.

17.3. Attributes of plants invasive in the karoo

17.3.1. General trends

Demographic parameters, including low seed mass, short minimum juvenile period and a short interval between large seed crops, can predict the invasive potential of woody plants (Rejmánek and Richardson, 1996). Potentially invasive trees and shrubs thus apparently compromise between adult persistence and investment in dispersal and population growth. As in other organisms (Bruton, 1986), the life-history attributes that make some plants more invasive than others are likely to differ with environmental predictability. Species that invest in growth and survival are likely to compete successfully in predictable environments, but stress-intolerant plants that invest in reproduction would be more likely to invade unpredictable, abiotically harsh environments

Alien plants that have successfully invaded other South African biomes are generally those well adapted to the disturbance regimes and fauna characteristic of those ecosystems. Invaders of the nutrient-poor, fire-driven fynbos biome are mostly serotinous, evergreen trees and shrubs such as *Hakea*, phyllodinous *Acacia* spp. and *Pinus*

spp. that establish in high densities after hot fires (Richardson et al., 1992). Alien species naturalized in South African grassland and savanna are generally fast-growing bird- or mammal-dispersed trees, shrubs and vines such as *Cereus peruvianus, Lantana camara, Melia azedarach, Psidium guajava, Pyracantha angustifolia, Rubus* spp., *Solanum mauritianum* and Australian pinnate-leaved *Acacia* spp. (Henderson and Wells, 1986) that can reach and establish in natural bush clumps and in riverine woodland where they are protected from annual grass fires.

Drought regimes determine the regional structure and composition of karoo vegetation (Cowling, this volume) and episodic droughts may reorganize local plant communities (Milton et al., 1995). Water is a limiting factor for the establishment of survival of alien plants in the karoo. Alien plants can consequently be separated into moist habitat specialists, restricted to the region's few perennial rivers, impoundments and irrigated fields, and dryland specialists that can invade arid hills and plains. In dry karoo habitats, plants must either tolerate drought (*Prosopis* spp., *Opuntia* spp.) or avoid it through short life spans and reliance on soil-stored seedbanks (*Atriplex lindleyi, Argemone ochroleuca, Bromus* sp.). Drought avoidance appears to be more frequent among introduced species than tolerance. The prevalence of drought avoiders among arid zone aliens explains why 75% have small, soil-stored seeds (Dean et al., 1986), and employ germination cues that minimize failed germination events. Only 20% of the 160 alien species recorded in the karoo–Namib region were trees or large shrubs (Brown and Gubb, 1986). The contribution of woody plants to the alien floras in other southern African biomes has not been investigated.

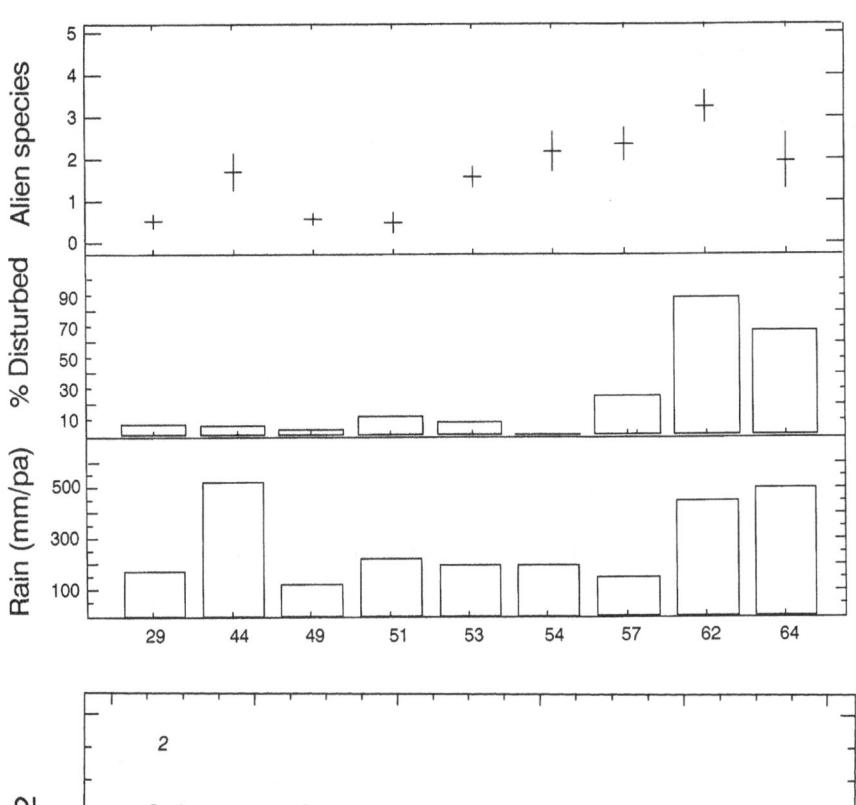

Figure 17.1 **Mean numbers of alien plant species adjacent to roads in relation to mean annual rainfall and the percentage of sites disturbed (by crop production or settlement) at 526 sites in 10 vegetation types. Vegetation type codes follow Low and Rebelo (1996): 29 Kalahari bushveld (n; 72), 44 south-eastern mountain grassveld (n = 17), 49 Bushmanland Nama-karoo (n = 121), 51 Orange River Nama-karoo (n = 51), 53 Great Nama-karoo (n = 98), 54 central lower Nama-karoo (n = 42), 57 succulent karoo (n = 73), 62 west coast Renosterveld (n = 34), 64 fynbos types (n = 18). Bars indicate 95% confidence intervals for mean number of alien species**

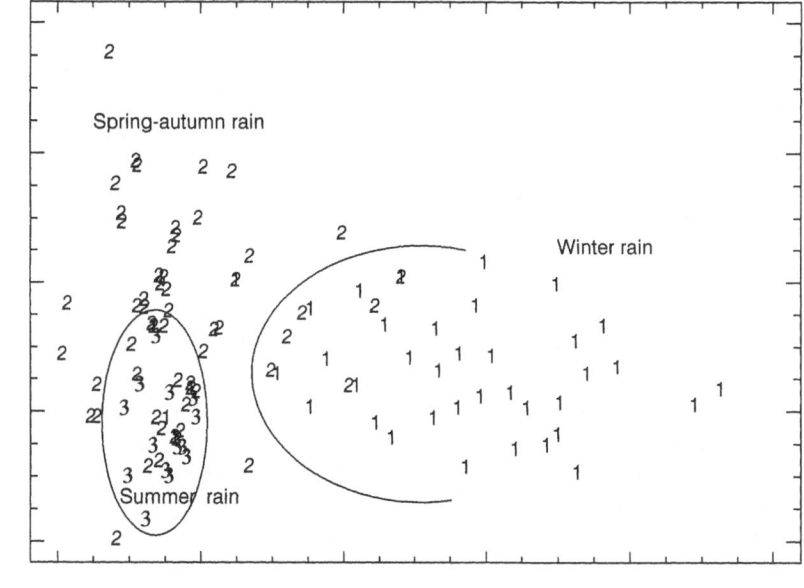

Figure 17.2 **Alien plant assemblages (based on presence–absence data 16 frequent species) as discriminated by rainfall seasonality. (1) winter rain in fynbos, Renosterveld and succulent karoo, (2) spring–autumn in all types of Nama-karoo and (3) summer rainfall in Kalahari bushveld and south-eastern mountain grassveld**

Sheep and goat ranching, the major types of land use in the karoo over the past 200 years, have selected an alien flora with burrs (9%), spines (15%) and toxins (10%), attributes that are relatively infrequent (4–8% burrs, 5–11% spines) in the indigenous flora (Milton et al., 1990). Herbaceous plants with barbs and burrs (*Lapula echinata*, *Medicago lacinata*, *M. polymorpha*, *Xanthium spinosum*, *X. strumarium*) are dispersed by sheep and angora goats. These, and endozoochorous Chenopodiaceae and Fabaceae, are associated with livestock pens, shearing sheds and watering points (Shaw, 1875; Henderson and Anderson, 1966; Shearing and Van Heerden, 1994). Some successful aliens combine spinescence with toxicity to herbivores (*Argemone ochroleuca*, *Datura* spp., *Solanum elaegnifolium*, *Xanthium* spp.). *Nicotiana glauca* and *Nerium oleander* contain toxins (Kellerman et al. 1988) but many others

(*Alhagi camelorum* (*maurorum*), *Cirsium vulgare*, *Opuntia* spp., *Prosopis* spp., *Salsola kali*, *Urtica urens*) use only physical defences to deter herbivores.

The discussion that follows focusses on morphological, physiological and reproductive attributes that have led to the success of certain alien plant species in the karoo's rare moist habitats, and in dryland habitats, across a rainfall seasonality gradient from the south-western succulent karoo, through the northern Kalahari and central Bushmanland areas, to the eastern Nama-karoo.

17.3.2.   Moist habitats

Karoo rivers, whether perennial or ephemeral, are subject to flash floods. Alien plants of ephemeral karoo watercourses are typically short-lived, being uprooted by flood water and regenerating rapidly from seed on wet sediment. The annual *Argemone ochroleuca* and short-lived *Nicotiana glauca* are common in ephemeral karoo river beds, and both produce numerous, innately dormant seeds. The shrub *Nerium oleander*, the tree *Populus canescens* (apparently a natural hybrid between *P. tremula* and *P. alba* var. *nives*; Immelman et al., 1973) and the reed *Arundo donax* have colonized the beds of perennial rivers, and survive flash floods by bending under the water and coppicing from buried and broken stems (Stirton, 1980; Wells et al., 1986). Flood waters and wind disperse the small, plumed seeds of *N. oleander* to other moist rocky sites where the drought-sensitive seedlings can establish (Henderson, 1992). Only male *P. canescens* have been introduced to South Africa, but the species suckers, forming dense clones in springs and rivers. In temporary lakes and impoundments in the karoo, salinity may increase as water levels fall. High salinity and fluctuating water levels appear to have facilitated the establishment of the Ukrainian *Tamarix ramosissima* in man-made impoundments such as Gamkaskloof Dam in the southern karoo (Brooke, 1995). The small seeds of *Tamarix* spp. are short-lived and persistence and spread depend on frequent production of flood-dispersed seeds by mature trees (Loope et al., 1988).

Moist habitats in the karoo are generally exposed to more use by mammalian herbivores, particularly during droughts, than surrounding drylands (Milton, 1991). Many of the alien species that have successfully invaded moist karoo habitats are therefore either toxic to herbivorous mammals, toxic and spinescent or simply distasteful (*Tamarix* spp.).

17.3.3.   Dryland habitats

South-western areas: succulent karoo and Little Karoo

Alien succulents have been remarkably unsuccessful as invaders of the succulent karoo, which is dominated by a species-rich, structurally poor, community of leaf succulent shrubs (Cowling and Hilton-Taylor, this volume; Esler et al., this volume). Alien succulents have probably failed to naturalize here because most of the species introduced to the karoo as forage plants or ornamentals originated from summer-rainfall regions of the Americas and Madagascar.

Winter-growing annual grasses (*Avena*, *Bromus*, *Hordeum*, *Stipa capensis*) and European forbs (*Erodium*, *Medicago*, *Plantago*), with soil-stored seedbanks, cold-cued germination and C_3 photosynthetic pathways, become abundant where indigenous shrubs have been reduced by grazing or ploughing (Vlok, 1988; Steinschen et al., 1996). The origins of *Stipa capensis*, and certain other grasses with temperate and circa-Mediterranean affiliations and small distributions in the karoo, are uncertain (Steinschen et al., 1996), but since their seeds are well adapted for dispersal in wool, they may possibly have been early introductions with European sheep.

Localized soil disturbance and amelioration by ants or termites (Dean, 1992b; Dean and Yeaton, 1993a), and selective grazing of such sites by livestock (Armstrong and Siegfried, 1990) probably generate the patchy distributions of alien grasses observed in Namaqualand and the Little Karoo (Vlok, 1994). *Stipa capensis* is particularly abundant on the circular, nutrient-enriched and highly disturbed soil patches that overlie the nests of the small harvester termite *Microhodotermes* near Vanrhynsdorp (Acocks, 1953; Steinschen et al., 1996), and *Avena* spp. show similar patterns in the Little Karoo (S. J. Milton, personal observation). *Bromus pectinatus*, which has a clumped distribution in relatively moist and sheltered microsites in Namaqualand (Steinschen et al., 1996), is locally abundant around nests of the harvester ant *Messor capensis* which collects and stores seeds for later consumption.

Australian desert saltbushes (Chenopodiaceae), including the herbaceous *Atriplex lindleyi*, *A. muelleri*, *A. semibaccata* and shrubby *A. nummularia*, were introduced in the mid 1800s to provide pasture for small stock and ostrich on saline soil (Hutchinson, 1946; Smith, 1966). All have naturalized and are widespread on fine-textured and saline soils in all regions of the karoo. They are particularly abundant in the succulent-dominated vegetation of Namaqualand, the Little Karoo and the Steytlerville karoo. Saltbushes produce large quantities of seeds, which in South Africa appear to be free of endophagous insects. Birds and mammals feed on the red berries of *A. semibac-*

cata dispersing seeds below canopies of trees and shrubs. Seeds of *A. lindleyi* (s.n., *Blackiella inflata*) are individually enclosed in a spherical spongy fruit that can float on water or roll in the wind, leading to widespread dispersal in flash floods and over bare ground. Seeds are also dispersed endo-zoochorically by tortoises (Milton, 1992a), antelope and other herbivores (Milton et al., 1995). The production of both dormant black and non-dormant brown seeds reduces the probability of seedbank depletion following recruitment failure. High concentrations of sodium and potassium salts inhibit germination of both seed types, whereas nitrogen salts break dormancy of the black seeds (Gwala et al., 1991). Efficient dispersal, large seedbanks, water- and nitrogen-broken seed dormancy and rapid growth thus enable the herbaceous saltbushes, particularly *A. lindleyi*, to colonize natural disturbances on riverbanks and termitaria, as well as such anthropogenic disturbances as abandoned fields, dry impoundments, road verges and livestock watering points.

Northern and central areas: Kalahari and Bushmanland

Very few alien plant species have invaded the arid summer-rainfall Kalahari and autumn-rainfall Bushmanland areas. High temperatures, low and variable rainfall, sand dunes and saline pans limit natural vegetation to a relatively low diversity of drought-tolerant shrubs, trees and ephemeral grasses. The most successful aliens here are the ubiquitous desert invaders, mesquite trees (*Prosopis* spp.) on sand where the water table is shallow, and Russian tumble weed (*Salsola kali*) on calcrete in endorheic pans.

Prosopis glandulosa, *P. pubescens*, *P. velutina* and numerous hybrids originate in summer-rainfall deserts of North and South America. The success of these leguminous trees in the karoo may be attributed to drought tolerance, the ability of the deep-rooted trees to use ground water, the large crops of seed pods (until recently, insect-free in South Africa), a durable soil-stored seedbank and efficient dispersal (Felker, 1979; Harding and Bate, 1991; Zimmermann, 1991). Seeds, enclosed in sugary pods are adapted for endozoochoric dispersal by large mammals. In the Kalahari and Bushmanland, dispersal has been facilitated by livestock and game. Up to 25% of the ingested seeds pass undamaged through the gut of sheep (Harding, 1991) within 1–10 days and retain 30–40% viability (Hoon et al., 1995). Livestock disperse seeds to disturbed, nutrient-enriched and moist sites, where, on heavily stocked ranches there is generally little competition from indigenous grasses or trees. Massive seedling establishment follows floods that carry seeds onto damp alluvium (Poynton, 1990).

Salsola kali, known as Russian thistle or tumble weed, is a spiny C_4 ephemeral that disperses its seeds across poorly vegetated expanses as the dead plants roll in the wind. Attributes that have led to its wide distribution in the karoo, and in Bushmanland in particular, are its tolerance of alkaline soils, abundant seed production, rapid germination, biomass accumulation and maturation after rain and effective dispersal by rolling and by adhesion of spiny seed heads to livestock, sacks and clothing. Seeds germinate at a wide range of temperature regimes (0/15 °C to 20/30 °C) and under low osmotic potentials (Young and Evans, 1972). *Salsola kali* reduced the biomass of herbaceous competitors under a high temperature regime (21/31 °C), but at lower temperatures it was out-competed by C_3 forbs (Allen, 1982a). Rapid growth may enable *S. kali* to deplete water to the detriment of grasses where water is limited, but it has no competitive advantage in moist conditions (Allen 1982b). A transient seedbank (Young and Evans, 1972) and endophagous insects indigenous to southern Africa (H. G. Zimmermann, personal observation) probably limit the abundance and persistence of *S. kali* in undisturbed karoo vegetation.

Eastern areas: Great, Lower and Eastern Nama-karoo

As summer rainfall increases and variability decreases eastwards, so the proportion of woody plants and perennial grasses from summer-rainfall regions increases in the alien flora. Moderate tolerance of drought, combined with competitive growth, defences against herbivory and adaptations for dispersal by birds and mammals characterize the most common and widespread alien plants of this region.

Cactaceae have all these characteristics, being large, succulent perennials capable of both CAM and C_3 photosynthetic pathways and well defended from herbivorous mammals by barbed spines. *Opuntia ficus-indica*, a tall, tree-cactus (previously known as *Opuntia megacantha*) originates from Mexico and may have entered South Africa as early as 1656 through Cape Town (Wells et al., 1986) from where it was distributed inland. There is a convincing evidence that a spineless variety was first introduced, and that the plants gradually reverted to the thorny 'wild-type' which, being less favoured as fodder, could survive and spread more effectively. This would explain why *O. ficus-indica* took close to 250 years to develop into a major problem (Annecke and Moran, 1978). All introduced Cactaceae bear conspicuous fleshy fruits with numerous small hard seeds that are dispersed to sheltered sites in shrubs by birds, as well as by humans and other mammals. Seeds dropped by nesting crows lead to an association of *Opuntia ficus-indica* plants with telephone poles in the karoo (Milton and Dean, 1987). Clonal spread and local dispersal in *Opuntia* spp. is achieved through shedding of cladodes which then take root near parent plants or adhere to animals that transport them further. No other alien plants have been as invasive as Cactaceae in the eastern Nama-karoo (Stirton, 1980), but a number of other toxic, distaste-ful or spiny alien shrubs, many of which are zoochorous, can establish where the natural vegetation of this region has been disturbed (*Alhagi camelorum (maurorum), Datura ferox, D. stramonium, Gleditsia triacanthos, Lantana camara, Melia azedarach, Schinus molle, Solanum elaegnifolium*).

High altitude, summer-rainfall grasslands in the east of the region are relatively mesic (400–500 mm yr^{-1}), and, unlike karoo shrublands, they occasionally burn. Narrow-leaved, fibrous and unpalatable perennial C$_3$ grasses, *Stipa tenuissima* and *S. trichotoma*, imported inadvertently with fodder from South America in 1900, have spread in cool, mountainous areas (Henderson et al., 1987). Success of *S. trichotoma* in these areas has been attributed to its avoid-ance by domestic livestock, longevity and the production of abundant (100 000 seeds/plant/annum), small seeds, 5% of which can remain dormant for up to 20 years in the soil (Wells, 1977). High-density stands of the grass establish from seed after fire (Viljoen, 1987a). A light requirement for germination and intolerance of seedlings to shading confines these *Stipa* species to disturbances in natural veg-etation. Alien C$_4$ grasses (*Digitaria sanguinalis, Pennisetum setaceum, Sorghum halepense*) are apparently confined to road verges and other disturbed, moist sites by poor toler-ance of drought or inability to compete with indigenous species.

17.4.   Impacts on the karoo ecosystem and economy

Some alien plant species were intentionally introduced to the karoo to provide shade, forage, barrier plants or as gar-den ornamentals. Many others were inadvertent imports with forage and animals. The economy of the karoo rests on three natural resources: (1) runoff and ground water essential for human habitation, ranching and small indus-try, (2) grazing land that supports sheep, goat and game ranching industries, and (3) wildflower displays, endemic succulents and game that attract tourists. All aspects of the economy consequently depend on minimizing anthro-pogenic impacts on flora, fauna and hydrology. For this reason, even economically valuable alien species can have negative impacts when they become invasive.

17.4.1.   Woody plants

The major impact of alien trees and shrubs on the almost treeless karoo ecosystem has probably been on the hydrol-ogy. Although water tables in the karoo are known to fluctuate and levels are falling in some areas (Hodgson,

1986), this cannot be directly linked to alien woody plants because ground water extraction for irrigation has influenced the hydrology. *Populus canescens* clones that fill river beds and engulf natural springs are likely to tran-spire water, reducing moisture availability in down-stream habitats (Versveld and Van Wilgen, 1986). Although there is no empirical evidence to support this assumption in the karoo, poplar clones planted to supply fuel and construction wood in the 1800s are now being removed.

If introduced *Tamarix* sp. and other phreatophytes fol-low the same trends in the karoo as they have in the Northern Territory of Australia (Cronk and Fuller, 1995) and in the southern USA (Dean and Milton, 1994; DeLoach et al., 1996), they may reduce diversity of plants and birds, as well as increasing soil salinity and rates of water loss through evapotranspiration. *Tamarix ramosissima* intro-duced from the Ukraine as an ornamental, has become naturalized in the karoo, especially along saline rivers and floodplains with fluctuating water levels. Dense stands have developed around the Gamkaspoort dam in the southern karoo (Brooke, 1995), along the banks of the Orange River, and near Lake Mentz in the south-eastern corner of the karoo.

Nicotiana glauca, although abundant in the succulent karoo, is confined to waste lands, floodplains, embank-ments and other disturbed areas. It seldom forms persis-tent dense thickets and usually gives way to secondary suc-cession after a few years. Its only known impact on the economy is the occasional poisoning of domestic live-stock. Similarly, there are no reports of drinking water contamination in rivers invaded by *Nerium oleander*. The major impacts of these species thus far have been to impede water movement and to alter the appearance of riverine vegetation. Attempts to clear *N. oleander* from rivers in nature reserves, manually and with herbicides, have been costly.

The planting of *Atriplex nummularia* as drought fodder has long been promoted by the Department of Agriculture and about 5000 km^2 (1.2%) of the karoo are covered by plan-tations of this species (Le Houérou, 1994). Although condi-tions in the karoo are generally too dry for its unaided establishment from seed, it is spreading naturally along the edges of roads and in other moist sites. Of most con-cern is its spread along water-courses. Although rain-fed *A. nummularia* plantations in areas receiving less than 300 mm yr^{-1} rainfall have been said to have 5–10 times the dry matter production of natural rangeland (Steynberg and De Kock, 1987; Le Houérou, 1994), a diet of this species does not fulfil the nutritional requirements of even the hardi-est sheep breeds (Hoon and King, 1993).

Since 1897, several species of *Prosopis* have been propa-gated on ranches, along roads and near river courses and livestock pens throughout the arid parts of South Africa to provide shade, firewood and pods (up to 140 kg tree^{-1} year^{-1}) which are a nutritious fodder for livestock (Felker, 1979; Harding, 1978, 1991; Zimmermann, 1991). However, *Prosopis* is spreading and appears to increase its range episodically in response to periods of higher rainfall. For example, a 400% increase in the range of these species occurred between 1974 and 1991 in the districts of De Aar and Britstown (Hoffman et al., 1995b). As a result of negative impacts on grazing, and possibly on ground water and biodiversity, the desirability and usefulness of *Prosopis* spp. in the karoo has become contentious.

Intraspecific competition at high densities causes *Prosopis* to grow as multi-stemmed bushes which set very few pods and provide no accessible shade. Dense thickets now cover over 180 400 ha of the karoo (Harding and Bate, 1991). Other plants in the understory of the thickets are smothered so that all potential grazing is lost in the vicinity of thickets. The problem is acute because the invasion process has been greatest in low-lying areas where ground water is close to the surface and these are the areas that otherwise would provide the best grazing (Harding and Bate, 1991). The use of such thickets by indigenous birds and mammals has apparently not been investigated, but large dung beetles avoid dense *Prosopis* thickets that impede their flight (Steenkamp, 1993). The development of monospecific stands of *Prosopis* in sandy river beds could possibly reduce water flow and carrying capacity for livestock, with potentially disastrous impacts on the natural-resource based, largely subsistence economies of the Northern Cape (Hoffman et al., 1995b). In the short-term, however, the clearing of *Prosopis* sp. in the USA yielded little additional water to off-site users and thus could not be justified on hydrological grounds alone (Douglas and Mayeux, 1991).

The distribution and abundance of some vertebrates has changed in response to the planting and spread of alien trees in the formally treeless expanses of the karoo. Red-breasted sparrowhawk *Accipiter rufiventris* (Macdonald, 1986a) and other tree-nesting raptors have expanded their ranges and now breed in stands of alien *Eucalyptus* and *Populus*, with unknown impacts on prey populations. Such fleshy-fruited, soft-wooded trees as *Schinus molle* and *Melia azedarach* have increased the distribution and abundance of the frugivorous, cavity nesting pied barbet *Tricholaema leucomelas* (Macdonald, 1986b). The development of European habitat (fruit trees, lawns) around farmhouses facilitated the spread of the alien European starling *Sturnus vulgaris* into the karoo (Maclean, 1993). Scattered thorny *Prosopis* trees, which provide ideal nest sites for masked weaver *Ploceus velatus*, appear to have led to the south-westward spread of this savanna bird into the karoo between 1930 and 1980 (Macdonald, 1990) as well as local-ized increases in the densities of porcupine (*Hystrix africaeaustralis*) that feed on the pods (H. G. Zimmermann, personal observation). The arboreal vervet monkey *Cercopithecus pygerythrus* also moved into the karoo between 1900 and 1970, probably in response to availability of alien trees as well as food crops and water (Macdonald, 1992).

17.4.2.   Cactaceae

Prickly pear *Opuntia ficus-indica* was already well established in the karoo as far east as the Sneeuberg, north of Graaff-Reinet in 1776. MacDonald (1891a) mentioned that 'for a long time (after 1750) it did not make much progress except as a cultivated (probably spineless) plant', in an area which was to become densely infested. By about 1890, the plant had become a troublesome weed in various parts of the karoo and eastern Cape Province. MacDonald (1891b) estimated that an area of 314 000 ha was already infested with prickly pear in the districts of Graaff-Reinet, Cradock, Aberdeen, Jansenville, Somerset East and Willowmore. Localized dense infestations were recorded from adjacent districts such as Albany, Oudtshoorn, Bedford, Uitenhage, Humansdorp and Fort Beaufort. A more detailed survey by Du Toit (1942) estimated that the infested area covered about 900 000 ha. In the drier parts of the karoo, prickly pear was confined mainly to the riverine vegetation and seldom spread into the shallow soils of the dry open karoo.

The thorny type was responsible for the expansive invasion of the karoo and, to a lesser extent, adjacent areas. At the peak of invasion, prickly pear had a major effect on farming and on entire communities. Land owners abandoned their properties and extensive areas became depopulated. Severe overgrazing occurred on the remaining usable veld. Feral stock, mainly pigs and donkeys, took refuge in the infested areas, making hunting for their control extremely difficult. The prickly pear invasion of the 1900s was one of the most devastating natural catastrophes in the history of South Africa. A phycitid moth, *Cactoblastis cactorum*, introduced from Australia in 1932 and a cochineal bug, *Dactylopius opuntiae*, together proved to be effective biological control agents for the weed (Brutsch and Zimmermann, 1993; Zimmermann and Malan, 1980). The residual prickly pear infestations are now a valuable source of fresh fruit, stock fodder during periods of severe drought, and the young cladodes can be used to make an appetizing vegetable dish known as 'nopalitos'. The plants also have considerable potential for the commercial production of cochineal dye (carmine) which is extracted from another cochineal species, *Dactylopius coccus*. Well-managed plantations of the spineless varieties can maintain 5–6 sheep ha^{-1} as emergency grazing during drought (Steynberg and De Kock, 1987) and plantations of these now cover about 2000 km² (Le Houérou, 1994).

Several other Cactaceae, planted in karoo gardens as ornamentals, later became invasive and troublesome in rangeland. Jointed cactus *Opuntia aurantiaca* has been a major problem in natural rangelands of eastern parts of the karoo for nearly a century. This low-growing, many-jointed, spiny cactus is most common among the denser vegetation of ephemeral rivers, and propagates vegetatively when the easily dislodged joints (cladodes) are spread by flood waters and animals. The sharp, barbed thorns readily penetrate and adhere to the hides of livestock and wild animals that brush against the plants, and joints so dislodged may be carried over considerable distances before being dropped (Moran and Annecke, 1979).

Among the many ornamental Cactaceae that have invaded the karoo are: *Opuntia imbricata*, which now has a wide distribution in the karoo; *Opuntia rosea*, which forms localized, dense infestations near Douglas in the north-eastern karoo and elsewhere; Australian pest pear *Opuntia stricta* which is common in the northern Cape Kalahari and is rapidly spreading; the pencil cactus *O. leptocaulis* limited to the Oudtshoorn district; *Harrisia martinii* in the Kimberley area; and *Tephrocactus articulatus* near Prince Albert and Rietbron (Henderson et al., 1987; Milton and Dean, 1987; Moran and Zimmermann, 1991). All have potential to reduce the value of rangeland for small stock and game, and all are costly and time-consuming to control.

17.4.3.   Herbaceous plants

Proclamation of even such spiny and noxious herbaceous plants as *Argemone ochroleuca* has been opposed because pioneer species are perceived to protect and retain denuded soil until the natural vegetation becomes re-established (Henderson and Anderson, 1966). However there is no empirical data with which to test the assumption that alien herbaceous species facilitate succession to natural vegetation in the karoo. Some have the potential to transform the karoo landscape. Among these are as *Atriplex lindleyi* and *Bromus* spp. which appear to have emerged from the incubation phase and are in the early exponential phase of invasion.

Both herbaceous *Atriplex semibaccata* and the shrub *A. nummularia* are used to cover soils disturbed by strip mining in Namaqualand (De Villiers, 1993), but there is little evidence that these species facilitate the re-establishment of indigenous vegetation (S. J. Milton, personal observation). Saltbushes store salt in bladder cells on their leaf surfaces (De Villiers, 1993), and falling leaves may increase surface soil salinity around the plants, apparently enabling saltbushes to compete successfully for space with less salt-tolerant plants. In this way they may prevent recolonization by other plant species.

Atriplex lindleyi (s.n., *A. halimioides*, *Blackiella inflata*) is an aggressive transformer species which invades overgrazed rangeland on silty soils. The plant is generally grazed by stock during the young stages only, or during drought when the plants are dead. Soil- and dung-stored dormant seeds germinate at high densities following rain that reduces salt concentrations. In this way *A. lindleyi* colonizes bare areas rapidly after drought. Its dominance during the drought recovery period appears to favour halophytes and it may therefore have a long-term influence on vegetation composition (Milton et al., 1995). The abundant seeds of *A. lindleyi* sometimes dominate the diets of indigenous granivorous birds such as canaries (W. R. J. Dean, unpublished data), with unknown effects on their demography and movements.

Although *Salsola kali* is a spiny and unpleasant weed on cleared and overgrazed land, it is browsed when soft and green, and has been used to maintain livestock on overgrazed ranches during drought in the USA (Young, 1991) and in South Africa. During succession of degraded arid rangeland in the USA, *S. kali* reduced shrub but facilitated grass establishment (Allen, 1995). Its impact on vegetation re-establishment has not been investigated in the karoo.

The presence of *Bromus pectinatus* and other winter-growing alien grasses in Namaqualand has only recently been mooted as a potential threat to the biodiversity of indigenous geophytes and annual forbs, and to the sheep-ranching (Vlok, 1988, 1994; Steinschen et al., 1996). Preliminary investigations in Namaqualand show that annual grasses are associated with disturbance and overgrazing. Species-richness of annual dicots and geophytes were reduced in dense stands of annual grass (Vlok, 1988), but grazing rather than competition from grasses appeared to be the cause (Steinschen et al., 1996). Areas dominated by annual grasses supported higher densities of granivorous ants (*Messor capensis*) but had lower ant diversity because of the absence of nectarivorous *Camponotus* and *Crematogaster* species (Görne, 1995). Species-richness of the avifauna was lower on annual grassland than in natural shrubland, and was strongly dominated by larklike bunting and grey-backed finchlark, both nomadic granivores (Görne, 1995).

The unpalatable perennial grass *Stipa trichotoma* invaded mountainous areas around Barkly East, Cradock, Graaff-Reinet and Pearston. Dense stands covered 20 km² and *S. trichotoma* occurred patchily in another 370–600 km² in 1983. Its ecological impacts are unknown, but it reduces the carrying capacity of rangeland for domestic livestock. Control measures involving burning followed by application of herbicides and fertilizers and oversowing of pasture grasses, have been effective but costly (Du Preez, 1983; Viljoen, 1987a and b).

17.5.   Control

Some invasive species have been devastating in the karoo, and the problems they have caused in the semi-desert areas are compounded because the yield from the land is generally too low to justify the costs of conventional chemical or mechanical control methods. This situation favours biological control of aliens using introduced natural enemies, mainly insects. However, biological control has limited potential where conflicts of interests exist, as in the case of *Prosopis* species. For plants such as these, it has thus far been impossible to implement a uniform management policy.

17.5.1.   *Prosopis*

The spiny nature of *Prosopis* spp. scrub and the hard wood make herbicidal and mechanical clearing operations laborious and expensive. Economic considerations dominate the planning of control operations because most of the infested land is marginal for agricultural purposes and is of low value. Although there is considerable potential for the use of biological control to alleviate the problem, its use as a management strategy has been thwarted by the fact that some landowners continue to exploit the desirable attributes of the plants (Zimmermann, 1991; Moran et al., 1993). Most owners of properties that are heavily infested with *Prosopis* spp. favour biological control and would be satisfied with a reduction in the density of the weed to levels where the plants could be utilized as originally intended. Owners of land that has not yet become densely infested are wary of biological control because they perceive that the usefulness of the plants may be diminished by the agents. A compromise has been reached whereby organisms that destroy the seeds of the plants, but not the pods or vegetative components, are considered to be acceptable for introduction for biological control (Zimmermann, 1991; Moran et al., 1993).

There are several species of insects that feed on the mature seeds of *Prosopis* spp. in their native habitat (Johnson, 1983), so the biological control option could be pursued and a programme that relies entirely on seed-feeding insects has been launched. Besides resolving the conflict of interests, there is a twofold rationale for the use of seed-feeding insects against *Prosopis* spp. in South Africa. First, a significant reduction in seed production of the plants will slow the rate of spread of the weed and curb its invasiveness. Secondly, with fewer seeds in the soil, mechanical and chemical control operations are more likely to produce lasting results because there will be less recruitment from seedlings in areas that have been cleared of the weed.

To date, three bruchid species have been released in South Africa for biological control of *Prosopis* spp.

Algarobius prosopis was first released in 1987 and has become widespread and abundant on the weed. *Algarobius bottimeri* was released in 1990 but has either failed to establish or is very scarce (Hoffmann et al., 1993). Most recently, in 1993, *Neltumius arizonensis* was released and has become established. It is spreading from the original release sites but, so far, has not become abundant and seemingly contributes very little, if at all, to control of the weed (Coetzer and Hoffmann, 1997).

One of the major constraints on the effectiveness of *A. prosopis* as a biological control agent of *Prosopis* spp. is interference by vertebrates browsing on the pods. At the start of each fruiting season (in January) the overwintering populations of *A. prosopis* are at low levels and 2 to 3 generations, each lasting approximately 6 weeks, are required before populations of the beetles reach levels that are able to destroy most of the seeds of the pods that are lying on the ground. If other herbivores have access to the areas where the pods have fallen around the trees, almost all of the pods are devoured within a few days of abscission and *A. prosopis* is unable to destroy more than a small proportion of the seeds before they are removed and dispersed. To prevent this happening, landowners are advised to fence-off infestations of the weed, to exclude livestock from the vicinity of the trees for at least eight months (i.e. until September) when the bruchids will have destroyed most of the seeds. The seed pods remain intact on the ground during this period and are still a valuable source of fodder to livestock.

Much more could probably be achieved with the biological control of *Prosopis* spp. in South Africa if there was consensus that the undesirable attributes of the plant outweigh its benefits and approval was granted for the introduction and release of additional agents that damage the vegetative parts of the plants. In the interim, landowners need to maximize the potential of *A. prosopis*.

17.5.2.   Cactaceae

Mechanical and chemical control operations against prickly pear *Opuntia ficus-indica* commenced towards the end of the 1800s (Annecke and Moran, 1978). Although the arsenic-based inorganic poison (arsenite of soda) was effective in controlling prickly pear, it was hazardous to use, being extremely poisonous, and did not prevent the plants from regrowing and from spreading. There was a demand for more favourable, alternative control methods and biological control was mooted as a possibility (Pettey, 1948). The prospects for biological control seemed promising after the successful Australian campaign against *Opuntia stricta* and *O. inermis* during the 1930s (Dodd, 1940). In 1932, a phycitid moth, *Cactoblastis cactorum*, was introduced into South Africa from Australia, where it had been a spectacularly successful control agent. This was followed by the establishment of three more insect species, namely, a cochineal, *Dactylopius opuntiae*, a cerambycid beetle, *Archlagocheirus funestus*, and a weevil, *Metamasius spinolae* (Zimmermann and Moran, 1991).

The cochineal *Dactylopius opuntiae* has been the predominant control agent against *O. ficus indica* in South Africa, in contrast with Australia where *Cactoblastis cactorum* was responsible for most of the control of *O. stricta* and *O. inermis*. Damage inflicted by the cochineal was enhanced and better control was achieved when prickly pear plants were felled manually once they had become heavily infected with the insects. Within a few years, about 75% of the originally infested area of 900 000 ha was reclaimed for pastoral use. Control was most effective in the warmer and dryer parts of the karoo which included most parts of the Little Karoo. Residual prickly pear populations persisted along the coastal areas between East London and Port Elizabeth and on the cooler and higher lying plateaus of the karoo mountains (Annecke and Moran, 1978). Although reinvasion by small plants occurs from time to time and creates the impression that biological control is ineffective, permanent establishment of prickly pear is prevented mainly by *C. cactorum* which is most effective on small plants (Zimmermann and Malan, 1981). Although biocontrol agents effectively control most prickly pear populations, some localized infestations remain troublesome and are treated chemically by stem-injecting the plants with a herbicide, monosodium methanearsonate (MSMA).

The cochineal *Dactylopius opuntiae* has been unsuccessful in the control of *Opuntia stricta* in the karoo, in spite of repeated releases of this cochineal and its effectiveness in controlling prickly pear in the karoo (Moran and Zimmermann, 1991). There is strong evidence to suggest that the *D. opuntiae* material introduced to South Africa for the biocontrol of *O. ficus-indica* is a host-adapted biotype from the related Mexican tree-cactus *O. streptacantha*. This would explain why this cochineal would not establish on *O. stricta*. New introductions of *D. opuntiae* on *O. stricta* introduced to South Africa from Australia are showing more vigour, and efforts to establish this biotype are currently underway. *Cactoblastis cactorum* has become established on *O. stricta*, but its overall effectiveness is disappointing, although it sometimes destroys isolated plants and occasionally whole clumps of plants.

The history of chemical, mechanical and biological control of jointed cactus, *Opuntia aurantiaca* has been reviewed by Moran and Annecke (1979) and by Moran and Zimmermann (1991). Of the five natural insect enemies released since 1932, only the jointed cactus cochineal *Dactylopius austrinus* and a phycitid moth, *Cactoblastis cactorum*, have become established and are exerting significant pressure on the weed. Cochineal is particularly effective in

the drier and warmer parts of the karoo, and has reduced the status of the weed in these areas. Although these biocontrol agents are able to suppress weed populations to acceptable levels if properly managed, selective chemical control with MSMA is still practised under some circumstances.

Another cochineal insect, *Dactylopius tomentosus*, released on *Opuntia imbricata* in the eastern karoo in 1970, kills many of the young plants and seedlings and debilitates the large plants which readily succumb if the plants are felled manually. This practice ensures that satisfactory control is achieved and chemical control is only necessary on isolated plants. This cochineal effectively controls localized infestations of the pencil cactus *O. leptocaulis* in the Oudtshoorn district and also kills small *O. rosea* plants, particularly during dry spells (Moran and Zimmermann, 1991). The Department of Agriculture has embarked on an intensive chemical control campaign in an effort to eradicate *O. rosea*. This programme has not yet realized expectations because the small plants and isolated cladodes are difficult to detect and serve as a reservoir for replenishment of the weed in herbicidally treated areas.

Extensive infestations of *Harrisia martinii* have been recorded in other provinces of South Africa and, in North Queensland, Australia, approximately 680 km² are infested by *H. martinii* (McFadyen and Tomley, 1981). Two insects have been introduced from South America, via Australia, to control *H. martinii* cactus near Kimberley. These are a mealybug *Hypogeococcus festerianus*, in 1983, and a cerambycid *Alcidion cereicola*, in 1990. The former was released on harrisia cactus in the Kimberley area and satisfactory control was achieved several years later. Manual redistribution of the mealybug to uninfected clumps of the weed, especially by landowners, is needed to enhance the effectiveness of this biological control programme. *Alcidion cereicola* has not yet been released in the karoo.

17.6.   Comparison with other arid and semi-arid systems

The most reliable predictor of the potential for a plant species to become invasive in Australia is evidence of weedy or invasive behaviour of that species (or congenerics) in its place of origin and of invasive behaviour elsewhere (Scott and Panetta, 1993). Repeated introductions over many years further increase the probability that a species will become invasive (Scott and Panetta, 1993). Taxa that have naturalized in other arid regions of the world are likely to invade suitable karoo habitats under similar disturbance or land use regimes. The same logic could be used to identify karoo taxa that are likely to invade other arid regions where they have been introduced.

17.6.1.   Invasive taxa and life forms

As is the case in the karoo, plant invasions into arid Australia and USA have been most severe along watercourses (Loope et al., 1988; Cronk and Fuller, 1995). Invasions of *Tamarix aphylla*, *T. chinensis* or *T. ramosissima* have followed the Finke River in the Northern Territory of Australia (Griffin et al., 1989) and the Colorado River and Rio Grande in the southern United States of America (DeLoach et al., 1996). These are major river systems and are consequently of exceptional importance to people and wildlife in these regions. All *Tamarix* species have been reported to increase soil salinity, lower water tables and to reduce diversity of birds and reptiles (Griffin et al., 1989; Ellis, 1995). For these reasons, they are now subject to biological control in America (DeLoach et al., 1996). The incidence of alien *Tamarix* species in the karoo may be underestimated because the alien species resemble the native *T. usneoides*. Other phreatophytes invasive in the karoo, that have also naturalized in rivers of arid zones elsewhere in the world, are *Nerium oleander* (USA, Mexico, Australia and several oceanic islands) and *Nicotiana glauca* (USA).

In dryland arid regions worldwide, four groups of plants have shown high potential to transform natural plant communities. These are shrubby Cactaceae (mainly *Opuntia*), Fabaceae (particularly *Prosopis*), Chenopodiaceae (*Atriplex*, *Salsola*) and Poaceae (perennial African C_4s and annual European C_3s). Of these, the Chenopodiaceae appear, thus far, to have had more positive effects on economies and relatively fewer negative impacts on ecosystems. Cactaceae become invasive in overgrazed rangelands in the Americas where they are native, as well as in Australia, where they have been introduced. Several species have had extremely negative effects on ecosystems and rangeland economies. Although most of the alien Cactaceae introduced to Australia and South Africa have responded to biological control (Dodd, 1940; Moran and Zimmermann, 1991), they remain popular garden plants and other species may yet achieve weed status in the summer-rainfall areas of the karoo.

Settled ranching with cattle and other livestock caused *Prosopis* species to become invasive in their places of origin in North and South America, even where indigenous bruchid beetles reduce production of viable seed. It was therefore predictable that these thorny leguminous shrubs would spread when introduced for agroforestry purposes to suitable habitats in Australia (Panetta and Carstairs, 1989) and the karoo (Moran et al., 1993) where they are free of their natural enemies. Alien *Parkinsonia aculeata* has reached alarming densities in Australia, and it may be in the early stages of invasion in the karoo where it

is widely planted as an ornamental tree. *Alhagi camelorum* (*maurorum*), a spiny shrub of Asian origin, is an important invader in California. It was accidentally introduced into the karoo as a seed-contaminant. Its present distribution is limited to irrigation fields of the Groot River between Willowmore and Steytlerville south-east of the Beervlei Dam, but it has potential to spread because a deep root system gives it drought resilience and makes it difficult to control.

Other leguminous trees in the genera *Cercidium*, *Ceratonia*, *Gleditsia* and *Robinia* also have a tendency to become invasive in their places of origin and on continents where they have been introduced. *Gleditsia triacanthos* remains one of the most popular agroforestry trees in South Africa where its cultivation is encouraged. Yet, in 1993, it was declared a noxious weed in Queensland, Australia, where an expensive eradication campaign against this invasive species has been launched (Csurhes and Kriticos, 1994). *Robinia pseuoacacia* has become invasive in New Zealand, the USA and Chile (Holm et al., 1991) as well in France (H. G. Zimmermann, personal observation).

The invasive potential and negative impacts of introduced grasses have been long overlooked. African lovegrasses (*Eragrostis curvula*, *E. lehmanniana*) were introduced to the USA in the 1930s in an attempt to reclaim natural grasslands damaged by drought and livestock grazing in the 1890s (Bock et al., 1986), but now appear to be spreading into undisturbed rangeland (McClaran and Anable, 1992). Mediterranean annual grasses, particularly *Bromus tectorum* (Cheatgrass), invaded arid rangelands of North America as a result of overgrazing and cultivation (Mack, 1981), and are considered to be the greatest problem facing rangeland restoration in the south-western area (Allen, 1995). Grass invasions increase abundance of certain granivorous birds and rodents, but reduce total diversity of plants, grasshoppers, birds and rodents (Bock et al., 1986). Grass invasions are widespread and, by preventing succession to natural communities, threaten biological diversity at genetic, population and species levels (D'Antonio and Vitousek, 1992). The succulent karoo holds the world's richest succulent flora (Cowling and Hilton-Taylor, this volume) and also supports an extraordinary diversity of insects, arachnids and reptiles (Vernon, this volume). This ecosystem is threatened by the incipient invasion of *Bromus* spp. and other C_3 grasses.

17.6.2.   Karoo species as invaders elsewhere

Several plants that are native to the karoo have become important invaders in parts of Australia and the United States of America that have similar climatic and edaphic conditions. These include such species as toxic *Asclepias fruticosa* and *Galenia africana* and spinescent *Acacia karroo* and *Lycium ferocissimum*, the latter a major invader in

Victoria and New South Wales, Australia (Parson and Cuthbertson, 1992). Halophytic *Mesembryanthemum crystallinum* and other species that maintain large soil-stored seedbanks, such as the ephemeral *Dimorphotheca pluvialis* and the grass *Eragrostis lehmanniana*, have also become invasive outside southern Africa (Table 17.5).

Some karoo plant species (*A. karroo, G. africana, Lycium* spp.) that are invasive on other continents are known to increase in density in overgrazed karoo rangelands (Wells et al., 1986; Steinschen et al., 1996; Milton and Dean, 1996). Others, such as the annuals *D. pluvialis* and *M. crystallinum*, are pioneers on ploughed lands (Van Rooyen, this volume). Given their evolutionary history in the karoo and response to soil and rangeland disturbance, such species should be seen as a potential threat to warm desert and semi-desert ecosystems that are used as rangelands. Arid-adapted, non-African floras that evolved with relatively little selection for tolerance of mammalian herbivores are likely to be particularly susceptible to invasions by karoo plant species.

17.7.   Future scenarios and research needs

The introduction of ornamental species has been a major source of invaders in the karoo. The worst of these are the spiny Cactaceae, tolerant of dry conditions, able to reproduce both sexually and vegetatively, efficiently dispersed by vertebrates and unimpeded by specialized phytophages. Forage plants, with relatively low forage value but high population growth rates and durable seedbanks (*Atriplex, Bromus, Opuntia, Prosopis*), were usually introduced to degraded karoo rangelands during periods of drought and grazing stress. El Niño events and the wet cycles that alternate with drought in this region may assist their spread. These, together with accidentally introduced, noxious ephemerals (*Argemone ochroleuca, Salsola kali*), have potential to inhibit recovery of natural vegetation on old fields and overgrazed rangeland. However, these statements remain speculative, because the ecological and economic impacts of alien plants on the hydrology and local plant and animal assemblages of the karoo have barely been investigated.

Climate change scenarios for the impacts of global warming on the world's arid rangelands indicate that rainfall variability may increase (Allen-Diaz et al., 1995; Bullock et al., 1995). Climatologists have not reached a consensus on the effects of anthropogenic climate change on rainfall seasonality, but, in the karoo, an eastward extension of the winter-rainfall region is likely to favour the invasion of C_3 grasses, whereas an increase in summer rainfall will favour the spread of Cactaceae and woody

Table 17.5. ***Species indigenous in the karoo which have been recorded as invasive on other continents***

Species	Country where invasive	
	Australia	Other
Trees and shrubs		
Acacia karroo	*	USA
Asclepias (Gomphocarpus) fruticosa	*	
Galenia africana	*	
Lycium ferocissimum	*	
Melianthus comosus	*	
M. major	*	
Osteospermum muricatum	*	
Sutherlandia frutescens	*	Mexico
Forbs		
Arctotis venusta	*	
Arctotheca calendula	*	
Berkheya rigida	*	
Citrillus lanatus	*	
Cucumis myriocarpus	*	
Dimorphotheca pluvialis	*	USA (Sonoran)
D. sinuata	*	
Emex australis	*	NZ, Taiwan, USA, Trinidad
Galenia pubescens	*	
Gorteria personata	*	
Lepidium africanum	*	
Osteospermum calendulaceum	*	
O. clandestinum	*	
Oxalis pes-caprae	*	
Oxalis spp.	*	
Papaver aculeatum	*	
Oncosiphon (Pentzia) suffruticosum	*	
Oncosiphon (Pentzia) piluliferum	*	
Solanum linnaeanum	*	
Ursinia anthemoides	*	
Succulents		
Aloe maculata	*	
Carpobrotus edulis	*	USA (California)
Mesembryanthemum crystallinum	*	USA, Chile, Circa-Mediterranean
M. nodiflorum	*	
Tetragonia fruticosa	*	
Stapelia grandiflora	*	
Grasses and geophytes		
Cenchrus ciliaris	*	USA
Cyanella hyacinthoides	*	
Cynodon incompletus	*	
Ehrharta calycina	*	
E. erecta	*	
Eragrostis lehmanniana		USA (Arizona)
Homeria miniata	*	
Pentaschistis airoides	*	
Pennisetum macrourum	*	
Romulea rosea	*	
Sporobolus africanus	*	

Source: Fearn (1978); Bock et al. (1986); D'Antonio and Vitousek (1992); Parsons and Cuthbertson (1992); Scott and Delfosse (1992); Unpublished data, Neser (1996)

invaders. Experts agree that human impacts on arid rangelands and on water sources in these areas will increase if the world population continues to grow (Allen-Diaz et al., 1995). Such impacts include increased stocking of domestic animals, cultivation of marginal lands and salinization of surface water, all of which are likely to facilitate the spread of *Atriplex* and *Tamarix*. There is therefore a need for research aimed at mitigating the impacts of alien plants on the diverse flora of the karoo under a scenario of increasing human utilization of the area.

Campaigns to promote and propagate agroforestry trees on farms have undoubtedly accelerated invasions by such species. Potentially invasive agroforestry species can be utilized with a very much diminished threat of them becoming problematic, if their specialized seed-feeding insects are used to neutralize their invasiveness at an early stage. There is therefore a need for pre-emptive biocontrol research that identifies host-specific seed- or flower-feeding insects that have the potential to limit the seed production of alien plants which have been recently introduced into South Africa. The Department of Water Affairs and Forestry has recently adopted this approach to agroforestry projects with the planned introduction and establishment of *Prosopis tamarugo* in salt-pans in the most arid parts of the karoo. Before these trees are propagated on a large scale, the seed-feeding bruchid *Scutobruchus gastoi* will be cleared for release. Such measures are particularly urgent in the case of introduced species that are invasive in other arid lands.

The effects of alien phreatophytes on the hydrology of karoo rivers and riverine biota need to be quantified in order to compare their possible negative impacts with the costs involved in removing and preventing their spread into rivers and impoundments in rangeland and nature reserves. Integrated control of alien plants in the karoo should include investigation of range management techniques that encourage the establishment of indigenous plants, research to develop methods for establishing and promoting indigenous shade, firewood and garden subjects, and for the consumptive utilization of problem plants.

The International Panel for Climate Change (Allen-Diaz et al., 1995) recommends that long-term monitoring of fixed sites across rainfall seasonality gradients be established in rangelands in order to understand impacts of climate change on C_3/C_4 and grassland/shrubland boundaries. In the karoo, such sites would be particularly valuable in developing a predictive understanding of the responses of alien and indigenous plants to climate change and landuse.

17.8.   Acknowledgements

The authors are most grateful to Dave Richardson for his comments on an earlier draft of this chapter, and to Lesley Henderson and Stefan Neser for permission to use their unpublished data.

Comparisons

Part five

Part five

Comparisons

18 Comparison of ecosystem processes in the Nama-karoo and other deserts

W. G. Whitford

18.1. Introduction

Most comparisons of deserts have focussed on convergent evolution of plants and animals and their adaptations to life in an extreme, hot, dry environment (Louw and Seely, 1982). There has been less attention focussed on the structural and functional comparisons of ecosystems. The synthesis volumes produced as a result of the US–IBP Desert Biome Program were primarily data summaries with limited comparisons upon which to generalize the research findings in one desert to other deserts of the world. This chapter concentrates on comparisons of ecosystem processes and some of the structural features that affect these processes. The ecosystem processes compared here are primarily those at the patch scale but, where possible, comparisons will be made at the watershed or landscape unit scale.

At the smallest scale are patches formed by individual perennial plants or groups of plants and the unvegetated spaces between these patches. Subpatterns within these patches are developed by the responses of ephemeral plants to seasonal rainfall and to soil heterogeneity resulting from perennial plant–soil interactions. Soil surface characteristics that affect erosion potential are also a product of the plant–soil interactions, such as spacing of vegetated patches, plus other interactions, such as distribution and type of disturbance by animals. Variables that affect pattern at this scale include (1) efficacy and characteristics of water redistribution, (2) redistribution of organic debris, (3) abundance and spatial–temporal distribution of animal-produced soil disturbance patches, (4) soil surface characteristics (cryptogamic crusts, rocky pavement, litter cover etc.)

Water and nutrients are the essential resources for all ecosystems and, in terrestrial ecosystems, the availability of these essential resources is a function of the characteristics of the soil. Soil stores and releases water for plant growth and is the habitat for the organisms that cycle nutrients. All organisms in terrestrial ecosystems are dependent, either directly or indirectly, on the soil. The emphasis of this chapter is on soil processes, the organisms that are involved in these processes and some of the effects of these processes on the structure and function of the karoo and other desert ecosystems.

18.2. Physical environment and primary productivity

18.2.1. Precipitation patterns

The range of physical environments of the karoo are equivalent to those of the hot deserts of North America, but, the complete range occurs within a much smaller area. Because of the scale differences, the gradients between low-rainfall, winter rainfall areas in the west and the high-rainfall, summer precipitation areas in the east are considerably steeper in the karoo than in the North American deserts. The range of average rainfall amounts and the seasonality patterns on an east–west transect parallel the North American hot deserts very closely (Cowling et al., 1986). The topography of the karoo also has some similarities with the North American deserts with ranges of semi-arid to arid mountains separated by broad valleys. This topography and the relatively young mountain ranges produce similar soils along the topographic gradients. In the North American deserts and the karoo, soils are predominately sandy, or sandy loams (Cowling et al., 1986). In North America and South Africa, mountain ranges modify moist air movement thereby affecting aridity patterns. In

North American deserts, winter precipitation is primarily from frontal storms originating in the Pacific and that move from west to east at approximately 90° to the long axis of the mountain ranges. In the summer, convectional moisture originates in the Gulf of Mexico or Sea of Cortez and moves to the north-west or north-east respectively. In the karoo, winter moisture originates in the south-west and moves inland in a north-easterly direction, while summer rainfall results from moist air moving from east to west.

18.2.2.   Topographic similarities

The topographic similarities of the karoo and North American deserts are reflected in similarities in vegetation distribution. In the karoo and North American deserts, trees and dense large shrubs form thickets along the edges of drainages. The ephemeral drainages and patterns of rills forming the dendritic drainage patterns on slopes terminate in drainage channels that are structurally similar in these deserts. The desert riparian vegetation provides the greatest structural diversity of any of the vegetation types of both deserts. This structural diversity appears to be the most important variable affecting the diversity of animals that use desert riparian corridors as habitat. Ephemeral drainages are sinks for water and possibly for nutrients. A considerable fraction of the water volume that enters an ephemeral stream enters the stream sediments as transmission loss and is stored in those sediments. The coarse materials that make up the bed-load of an ephemeral stream during the brief periods of flow, provide a large reservoir for storage of transmission loss water. Despite the importance of ephemeral stream corridors as essential features that support the biodiversity of deserts, and the potential importance of these features as sites where groundwater recharge may occur, there are few studies of ephemeral stream processes such as organic matter redistribution, water storage–plant growth relationships, nutrient relationships etc. In a study of nutrient resorption by arroyo (ephemeral stream) plants, we found that several species had highly efficient nutrient resorption (K. Killingbeck and W. G. Whitford, unpublished data). This may be an important attribute of plants rooted in coarse sediments which do not retain nutrients.

The upland slopes and flats of the North American deserts and the karoo are mostly vegetated by shrubs. The intershrub spaces are generally bare, although there are herbaceous succulents in intershrub spaces in many parts of the karoo. Although the species diversity of shrubs is higher in the karoo than in North American deserts, the structural similarities suggest that some processes such as water redistribution and the spatial patterns of nutrient pools should be similar in these deserts. In most North

American hot deserts and in the karoo, bunch grasses are a minor component of the vegetation, and plant communities are dominated by shrubs. One difference that may be important is the abundance of succulents in some areas of the karoo. There are no comparable abundant succulents or perennial herbaceous plants in North American desert ecosystems. It is not possible to assess the ecological implications of this difference without data from studies designed specifically to examine this question.

18.2.3.   Water redistribution

In order to understand and predict productivity patterns in the rangelands of the karoo or North America, we must understand the degree to which upland ecosystems retain water and how rainfall is redistributed among patches, and among landscape units. A number of variables affect infiltration and runoff. Some of the most important variables are slope length, slope angle, soil texture, soil surface stability and vegetative cover. Vegetative cover and slope are the most important of these variables. Anything that has an effect on vegetative cover and composition and/or on soil texture and soil surface stability affects water redistribution. Patterns of water redistribution are affected by the activities of domestic and native large herbivores. Reduction in plant cover, changes in the species composition (effect on morphology of the vegetation) and spatial patterns of plant distributions change runoff–run-on patterns and the quantity of water that runs off or remains within the system (Milton et al., 1994b; Tongway, 1994). Sheet flow water carries small organic debris into the rills and eventually into the large drainage channels. Redistribution of water affects productivity patterns on different parts of a watershed. The extent and patterns of redistribution depend more on the characteristics of rain events than on average or total rainfall (Ludwig, 1987) (Fig. 18.1). Much of the water from high-intensity storms runs off the slopes, whereas a larger fraction of rainfall in low-intensity storms infiltrates and is stored in the soil where it falls.

18.2.4.   Primary productivity

Generalizations about productivity in desert ecosystems are plagued by the lack of comparable data. In their synthesis of vegetation production in semi-arid and arid rangeland, Le Houérou et al. (1988) concluded that production was directly proportional to rainfall. An examination of the published studies upon which Le Houérou et al. (1988) based this conclusion showed that productivity was measured by harvesting forage species (grasses in most cases, grasses and forbs in some cases). None of the studies summarized in that analysis included productivity of shrubs in the estimation of primary production. In most desert rangelands such as the karoo and North American

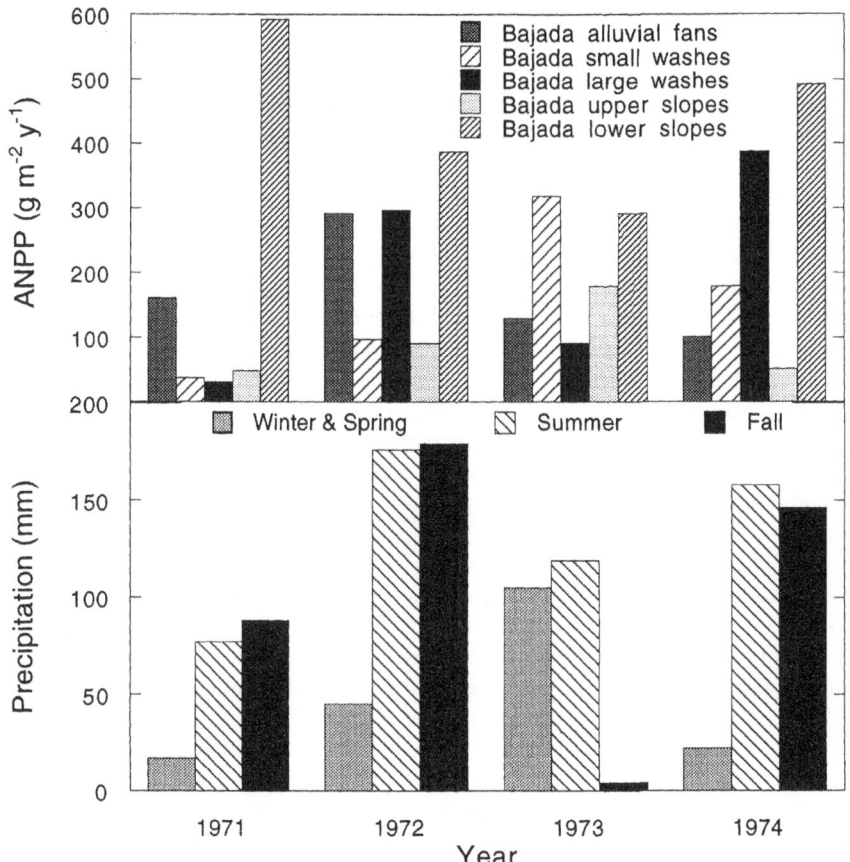

Figure 18.1 **The relationship between above-ground net primary production (AGNPP) and seasonal rainfall in different parts of a watershed in the northern Chihuahuan Desert (data from Ludwig, 1987)**

deserts, most of the standing biomass is comprised of woody shrubs or small trees. When these are included in measures of primary production, the relationship between rainfall and primary production is not linear, and watershed scale processes must be included in order to identify the variables affecting primary production (Ludwig, 1987).

Primary production in the karoo and in the Chihuahuan Desert are probably affected by the same variables and therefore are likely to exhibit the same patterns. The most important consideration is that above-ground net primary production (AGNPP) is not tightly coupled to rainfall, at least in those parts of the deserts receiving more than 100 mm long-term average annual precipitation. The data summarized by Ludwig (1987) combined with data from other studies (Gutierrez and Whitford, 1987; Fisher et al., 1988; Whitford et al., 1995) provide a basis for predicting patterns of productivity in the karoo. First, season of rainfall and the allocation of AGNPP of the previous year among annuals and perennials are important variables. Season of precipitation affects storm characteristics (intensity–duration) and determines runoff quantities and effectiveness of the runoff. Examples from

Ludwig (1987) illustrate these points. In two successive years with above-average winter and summer rainfall, AGNPP on alluvial fans was 44% of that of the first year during the second year. There was lower productivity of both shrubs and annuals during the second year because of the large residual pulse of organic matter from the first year's production. This organic matter probably resulted in nutrient immobilization by the microbial biomass using that organic matter as an energy source. This pattern of productivity is the result of decoupling nutrient availability from water availability as is discussed in detail in a later section. Considering the similarities in topography, rainfall seasonality and life form of the vegetation, the distribution of AGNPP with respect to topographic position should exhibit similar patterns in the karoo in response to variations in precipitation.

18.2.5.   Shrub morphology
In the Chihuahuan Desert of North America, the dominant shrubs redistribute stemflow water along root channels to deep storage (Martinez-Meza and Whitford, 1996). Variables affecting this water redistribution include: stem angle, canopy density and canopy area. Many karoo shrubs

are multistemmed plants with large (>45°) stem angles (Fig. 18.2). These attributes suggest that water redistribution to deep soil by stemflow and root channelization may contribute to the success of karoo shrubs.

In North American deserts, morphological variation among shrubs of the same species has a significant effect upon the density, species composition and biomass of annual plants that grow under the canopies of shrubs. Shrubs affect the nutrient status of soils because stemflow water is enriched (Whitford et al. 1997) and because of the accumulation of dead plant materials under the canopies of shrubs (Charley and West, 1975; Garcia-Moya and McKell, 1970; Parker et al., 1982). Morphological variability in shrubs therefore affects the heterogeneous spatial distribution of essential resources (water and nutrients) (De Soyza et al., 1997). The morphological characteristics of shrubs are important determinants of the relative resistance and resilience of shrub species to environmental stress such as drought (De Soyza et al., 1997). In the karoo, where the domestic livestock are primarily browsers, herbivory not only can change the composition of the plant community but can also affect the morphology of the shrubs, thereby affecting basic ecosystem properties such as soil water and nutrient distribution patterns. The effects of browsers on the morphologies of palatable species should be examined to ascertain the relative importance of morphological changes in dominant species on ecosystem processes in the karoo.

18.3.   Decomposition and nutrient cycling

Most of the nutrients required for plant productivity are made available by the process of decomposition–mineralization. While this is a continuous process from successive fragmentation of dead leaves etc. to the breakdown of the molecular structure of the fragments to carbon dioxide, water and mineral constituents, decomposition rates are generally measured only as the fragmentation process. Data on decomposition rates available from deserts clearly demonstrate that the fragmentation process occurs at high rates. The rates of decomposition are equal to or greater than those measured in more mesic environments, and are not rainfall-dependent (Whitford et al., 1981, 1986). Rates of mass loss from litter are the same for surface litter exposed to double the growing season rainfall and litter exposed to the complete exclusion of rainfall during the growing season (Whitford, 1995) (Fig. 18.3). Plant litter fragmentation appears to be a physical process due to ultraviolet light and high temperatures that break the bonds of complex structural molecules (the lignins) (Moorhead and Reynolds, 1989). Mineralization of litter on the soil surface does not occur until the small litter fragments are mixed into the soil or are buried. Because much of the decomposition of surface litter is abiotic and the complex precursors of humic compounds are lysed in the process, this process probably contributes little to soil nutrient pools except in those rare situations where litter accumulates in patches under shrubs or in debris chains deposited by overland flow.

18.3.1.   Soil organisms and soil processes

In terrestrial ecosystems, the soil is the most essential component of the ecosystem. Soil properties determine water infiltration and storage, nutrient availability patterns and the patterns of primary production. Soil also serves as a habitat for animals which by their behaviors affect soil properties and the structure and productivity of the vegetation. Soil is a living entity, and the soil biota are directly or indirectly responsible for the essential ecosystem processes that take place in the soil. The soil biota of deserts are very similar (Wallwork, 1976; Noble et al. 1995) suggesting that results of studies of soil processes in one desert may be directly applicable to other deserts especially if the climatic constraints are similar.

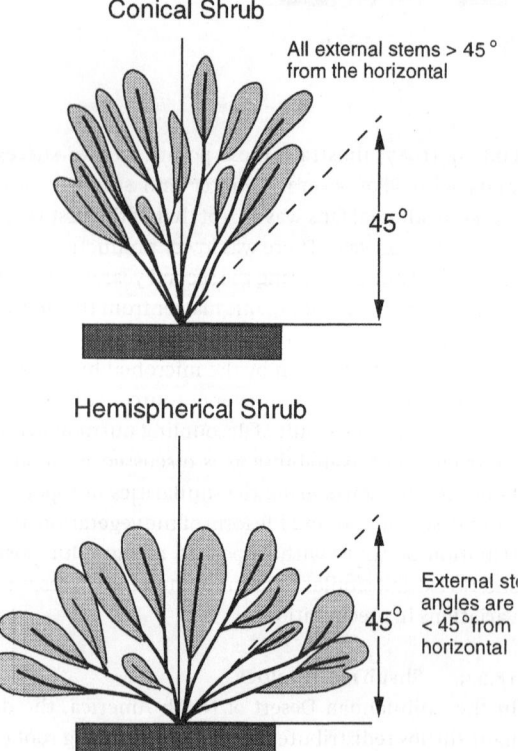

Figure 18.2 **The relationship between exterior branch–stem angles and the morphology of desert shrubs**

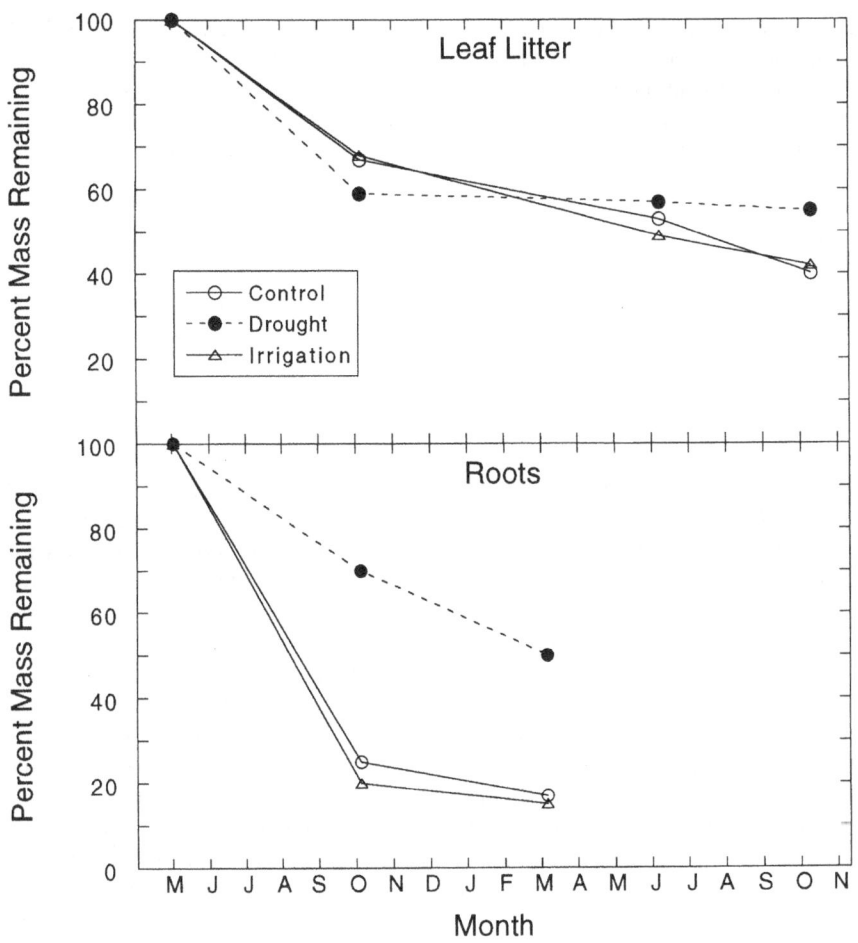

Figure 18.3 **The effects of drought and irrigation sufficient to double long-term average rainfall on the decomposition of creosotebush, *Larrea tridentata*, leaf litter on the surface of the soil (upper panel) and on the decomposition of the roots of a perennial forb (lower panel) (data from Whitford et al. (1995)**

Termites

Although both the karoo and North American deserts (Chihuahuan and part of the Sonoran) have an abundance of termites, the abundance and activity of termites in the karoo appears to be significantly higher than in the North American deserts. The harvester termites (*Hodotermes mossambicus* and *Microhodotermes viator*) of the karoo appear to be more efficient at removing dead plant material than the corresponding North American species (*Gnathamitermes* and *Amitermes* spp.). There is a virtual absence of litter accumulations under shrubs or behind obstructions in the upland areas of the karoo that I have visited. Observations of *Hodotermes* spp. harvesting dead grasses and shrub leaves suggests that these termites take a wider range of dead plant materials than North American desert termites. In the Chihuahuan Desert, subterranean termites harvest a large fraction of the herbivore dung, dead above-ground and below-ground plant parts – especially grasses and forbs (Whitford, 1991). The abundance and activity of Chihuahuan Desert termites are the most important variables affecting patterns of soil organic matter (Nash and

Whitford, 1995). Studies in semi-arid Australian rangeland showed that termites in those systems processed a large fraction of the dead plant material and some of the animal dung (Whitford et al., 1992). The apparent abundance of subterranean termites in the karoo and the breadth of materials harvested by these animals suggests: (1) karoo soils should have lower organic matter contents than Chihuahuan Desert soils and (2) soil microflora and microfauna in the karoo soils should be depauparate and of lower biomass than the soil biota of the Chihuahuan Desert (Silva et al., 1985). The ecosystem function implications of the high abundance of subterranean termites in the karoo are profound. The gut microflora of termites process dead plant materials more completely and over a shorter time span than do free-living microbes in the soil. There is evidence that the gut microflora of some termites even break down lignins (Butler and Buckerfield, 1979). The result of this is that very little carbon is returned to the soil as fecal material. This accounts for the strong negative correlation between termites and soil organic matter (Nash and Whitford, 1995). Since many termites use fecal pellets as cementing materials in their gallery tunnels or in building nests, organic matter and nutrients tend to be concentrated in subterranean or above-ground nests. Concentrations of nutrients tend to be correlated with concentration of soil organic matter in desert soils. Thus, termites may be primarily responsible for the heterogeneous patterns of nutrient concentrations and thereby indirectly responsible for the diversity and heterogeneity of the plant community.

The formation of *heuweltjies* (mima-like mounds) in the karoo have been attributed to the action of termites or molerats or both (Lovegrove and Siegfried, 1989). The soils of these mounds hold more water and are enriched in nutrients (Midgley and Musil, 1990) especially phosphorus and divalent cations such as calcium, magnesium, manganese and iron when compared to soils of the surrounding area. In the absence of data on soil organic matter and population size or activity of termites in the *heuweltjie* soils and surrounding soils, the question of soil nutrient–termite relationships in the karoo remains unresolved. Obviously there is a real need for studies of termites, soil organic matter, and nutrients in the karoo in order to confirm or reject the generalizations based on studies in other deserts.

Chihuahuan Desert termites also affect rainfall infiltration patterns because their foraging tunnels act as macropores which allow bulk flow of precipitation into the soil (Elkins et al., 1986). In the karoo, where subterranean termites are widely distributed and where there are mound-nesting species, their cumulative effects on karoo ecosystems is difficult to access. Soils eroded from termite mounds frequently have higher nutrient content but lower infiltration rates than surrounding soils (Lobry de Bruyn and Conacher, 1990). The densities of foraging ports of *Hodotermes mossambicus* may not be sufficiently high to have a significant effect on infiltration. The removal of litter from areas under shrubs may also have an effect on the infiltration of the throughfall fraction of the precipitation. If termites are responsible for the formation and maintenance of soil properties of heuweltjies, they have a direct role in maintaining the diversity and heterogeneity of karoo plant communities (Midgley and Musil, 1990; Yeaton and Esler, 1990; Knight et al., 1989). Yeaton and Esler hypothesize that the more basic calcium-rich soils of heuweltjies, which animal activity has created, erode downslope creating small environmental gradients. They conclude that the driving forces in this plant community are the mechanisms by which calcium and other elements are concentrated in the heuweltjies. Considering the potential role of termites in the formation and maintenance of heuweltjies, it is tempting to speculate that termites play even more of a keystone role in karoo ecosystems than has been documented for sub-

terranean termites in the Chihuahuan Desert (Whitford, 1991).

Buried litter and roots

Decomposition–mineralization of buried litter and roots is primarily biological and the rates are tied to soil moisture (Whitford et al., 1986; Santos et al., 1984; Whitford, 1995) (Fig. 18.3). Most of the soil nutrients available for plant growth are from decomposition–mineralization of buried litter and roots. Activities of animals that produce depressions or pits in the soil where litter can be buried are important variables affecting the temporal and spatial distribution of soil nutrients. Spatial and temporal patterns of nutrient availability in a desert is also a function of the distribution of annual plants and their below-ground biomass. Large quantities of organic matter input as dead annual plant roots can result in nutrient immobilization and consequently reduction in primary productivity. Considering the similarities in climate and soils of the karoo and North American deserts, the process of decomposition–mineralization should be very similar in these deserts. However, in order to understand the spatio-temporal variations in nutrient availability and how these variations affect plant production, it will be necessary to determine what fraction of the plant litter is buried, the spatial and temporal patchiness of annual plants and annual plant root production. The examination of annual plant roots and their contribution to nutrient pools should be a priority. However the fate of buried litter and dead roots may also be affected by subterranean termites which can change the whole picture of nutrient cycling in karoo.

Role of soil biota

Decomposition and nutrient cycling are important processes in any ecosystem. In desert ecosystems these processes are indirectly influenced by the extremes of climate because of the effects of soil water on the biota (Whitford, 1989). The soil micro- and mesofauna of the world's desert rangelands appear to be taxonomically similar at the family level based on the limited data available. Protozoans (amoebae, ciliates and flagellates) and many families of nematodes make up the microfauna. Acari and various insects (Collembola, Psocoptera and larvae of Diptera and Coleoptera) comprise the soil mesofauna (Wallwork, 1976). The micro- and mesofauna serve primarily as regulators of the rates of decomposition and mineralization. Desert rangeland soil acari are dominated by prostigmatid mites (Loots and Ryke, 1967; Cepeda-Pizarro and Whitford, 1989; Noble et al. 1995) and desert soil nematodes are dominated by bactivorous and fungivorous forms (Freckman, 1988). The prostigmatid mite genera that are most abundant in arid and semi-arid region soils feed on fungi, soil algae or nematodes.

The fraction of the soil biota which is physiologically active at any point in time is determined by soil temperature and soil water potential (Whitford, 1989). Abiotic constraints on the soil biota are especially important in semiarid and arid regions. Many taxa of the soil microflora and microfauna remain active in the film of water associated with surfaces of soil particles. As water films dry and disappear, the protozoans encyst and the nematodes enter an anhydrobiotic state. The average soil water potential at which more than half of the soil protozoans are encysted is −0.4 MPa and more than half of the nematodes are anhydrobiotic at that soil water potential. Thus, while bacteria, nematodes and protozoans require water films to remain active, fungi and soil micro-arthropods may continue to be active at soil water potentials of −6 MPa. Several studies have shown that mineralization of decomposing organic materials is primarily the result of activity of the soil micro-arthropods feeding on the heterotrophic fungi in the soil (Parker et al., 1984; Fisher et al., 1990) (Fig. 18.4). The bacteria, protozoans, and nematodes in desert soils require water films on the surfaces of soil particles and aggregates in order to be active. The micro-arthropod fauna of the karoo is dominated by the same groups of mites as the fauna of the Chihuahuan Desert (Loots and Ryke, 1967; Santos et al., 1981). Given the similarity of the micro-arthropod fauna and similarities in climates, it is reasonable to predict that mineralization processes in karoo desert soils will be similar to those in the Chihuahuan Desert.

Mineralization and immobilization

The timing of availability and the quantities of available nutrients is another critical feature of arid ecosystems. These ecosystem attributes are established by the interactions of the climate (temperature and rainfall), the soil physical properties, the soil biota and the vegetation. The timing and characteristics of inputs of dead plant materials (both above-ground and below-ground) determine the activity of the soil microbiota and the relative rates of immobilization and mineralization (Parker et al., 1984). In desert ecosystems with very seasonal rainfall and/or average rainfall of less than 100 mm, the periods favourable for plant growth are so separated in time that productivity is nearly always water limited and nutrient limitations are rare. However, in most desert rangelands, the temporal coincidence of rainfall and nutrient availability is critical and can have a large effect on the primary productivity of the system. The growth of annual plants is dependent upon seasonality, quantity of rainfall and soil characteristics. At the end of a growth cycle, the large input of carbon provides a substrate for the heterotrophic bacteria and fungi which will rapidly increase in biomass, and nutrients will be immobilized in that biomass. This results in

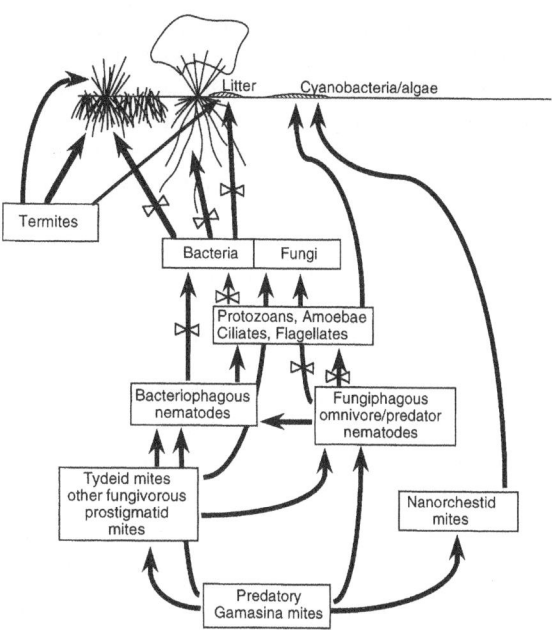

Figure 18.4 **The trophic relationships of the soil microflora and meso- and microfauna. Trophic flows affected by soil water potential are distinguished by an x on the transfer arrow**

nutrient limitation of plant production. In ecosystems where perennial grasses account for most of the vegetative cover, the relatively constant input of carbon from root turnover results in rates of mineralization that are higher than rates of immobilization most of the time (Whitford et al., 1987). Since perennial grasses account for a very small percentage of the total vegetative cover of the karoo, in the higher-rainfall areas, pulses of annual plant growth should be followed by a pulse of nutrient immobilization resulting in productivity that is decoupled from rainfall and lower than expected because of nutrient limitations. Pulsed inputs of organic matter and nutrient immobilization are probably of importance in the winter-rainfall region of Namaqualand and in parts of the karoo where drought and overgrazing has caused replacement of small shrubs with annual grasses and forbs (Steinschen et al., 1996).

The rate of mineralization of immobilized nitrogen is a function of the soil fauna. The importance of the soil fauna as grazers on bacteria and fungi, and the subsequent release of mineral nutrients in the excreta of soil animals, have been documented in a number of field studies using biocides and in microcosm studies (Freckman, 1988; Ingham et al., 1985). In deserts where litter accumulations are sparse or non-existent, soil biota are concentrated in the rhizospheres of plants. Organic inputs occur in the rhizosphere, and microflora are concentrated in this area of the soil. As a consequence, the microfauna that graze on the microflora and release the nutrients immobilized in the biomass of the microflora are also physically located in close proximity to the absorbing surfaces of the plant roots. In moist soils, dead roots and root exudates result in rapid growth of the microflora plus rapid growth in numbers of protozoans and nematodes which have the potential to overgraze the microflora and cause a reduction in the rate of decomposition (Santos et al., 1981). However, in deserts, soils are generally dry, and mineralization in the rhizosphere is primarily due to the feeding of fungivorous mites on fungal hyphae with their excretory products providing the mineral nutrients for plant growth (Parker et al., 1984). Given the similarities of climate and soil micro-arthropods, there is no reason to expect that the contributions of soil microflora and micro- and mesofauna to decomposition-mineralization are any different in the karoo.

Rhizosphere processes

As has been discussed previously, most of the decomposition-mineralization occurs in the soil and decomposition of roots is the most important source of carbon and nutrients. Turnover of roots, (live roots to dead roots) and exudates from live roots fuel a complex of biological–chemical processes known as rhizosphere processes, that are essential for the normal functioning of the ecosystem. The rhizosphere is the layer of soil in close proximity to the surface of a root and that layer is directly affected by the activity of the root. Root materials added to the inorganic matrix include sloughed cells and exuded molecules. The quantities and kinds of materials added to the rhizosphere matrix by roots vary with plant species and physiological state, and with the characteristics of the soil matrix (Van Veen et al., 1989). In some arid zone grasses, there is sufficient root exudation of complex carbohydrates to form rhizosheaths around the roots (Wullstein et al., 1979). The exudates provide the energy source for a complex of bacteria that include free-living nitrogen fixers and species (e.g. *Azospirillum* spp.) that promote mineral and water uptake and dry matter accumulation in plants. These organisms may also promote increases in root diameter, root density and length of root hairs (Okon et al., 1988). In a study of the desert grasses, *Bouteloua eriopoda* and *Sporobolus flexosus*, Herman et al. (1993) provided evidence that exudates from the roots of these grasses supported increased numbers of free-living nitrogen fixers in the rhizosphere. In a greenhouse study, *Sporobolus flexosus* showed significant increases in above-ground and below-ground biomass when grown in quartz sand culture inoculated with a native *Azospirillum* spp. isolate, and developed the largest increase in biomass when grown with a combination of *Azospirillum* and *Azotobacter* isolates (El Shahaby, 1988).

Mycorrhizal fungi are also important components of the rhizosphere microflora. Most forage plants of arid and semi-arid rangelands host mycorrhizal fungi (Trappe, 1981). Mycorrhizal fungi absorb nutrients from soil and translocate them to host plants while the host provides carbon compounds to the mycorrhizal fungi. Mycorrhizal-dependent plants cannot succeed without these fungal symbionts (Trappe, 1981). Recent studies have demonstrated rapid transfer of phosphorus from dying roots to roots of a living plant if the two root systems are connected by mycorrhizal links (Newman and Eason, 1989). Mycorrhizae are patchily distributed in some ecosystems creating a patchy environment overlying that of fertile islands created by trees in the landscape (Allsopp et al., 1994). When soil stability is lost due to overgrazing and loss of plant cover, mycorrhizal spore levels may be low or absent (Allsopp, 1994). The loss of mycorrhizal spores from mobile sands or eroding soils may compromise plant establishment and revegetation efforts.

These few examples of rhizosphere processes are indicative of the paucity of data available on these processes in arid and semi-arid ecosystems. In terrestrial ecosystems where the coincidence in timing of water and nutrient availability is critical, understanding rhizosphere processes becomes essential. The paucity of studies on these processes calls attention to a marked deficiency in research on desert ecosystems. In addition to the direct effects of rhizosheath microflora and mycorrhizae, there are many interactions between rhizosphere microfauna and mesofauna and the rhizosphere microflora that remain to be investigated.

Soil aggregates

Soil aggregates are formed when microbial products such as humic acids and hyphal fragments or bacterial cells are surrounded by, and cemented together by, clay particles. Soil biological processes are largely responsible for both the formation and stabilization of aggregates. The stability of aggregates is affected by plant roots, fungal hyphae, microbial polysaccharides and root exudates (Tisdall and Oades, 1979; Chaney and Swift, 1986; Pojasak and Kay, 1990). The composition and structure of soil aggregates affect the capacity of the soil to store and release water and nutrients, and the capacity of plant roots to access these resources. Soil available water capacity is a function of organic matter content, clay content and pore size distribution of individual aggregates and of the bulk soil. Pore size distribution also determines aeration status of the soil, controlling processes such as denitrification which affect soil nutrient availability. The capacity of the soil to store and release nutrients, and particularly nitrogen, is also a function of the quantity and quality of organic matter which is protected within soil aggregates. Finally, well-developed soil aggregates enhance the availability of water and nutrients to plant roots by providing inter- and intra-aggregate channels for roots to follow.

Ants, and particularly termites, modify soil aggregate structure. Few data exist on the role of these organisms in aggregate formation (Lee and Foster, 1991). Relatively little is known, also, about the structure of soil microbial crusts which stabilize the soil surface of many rangeland soils. These crusts, and the organisms associated with them, may play a role in nitrogen fixation and water infiltration (Eldridge and Greene, 1994; Belnap and Gardner, 1993).

The similarities in climate, topography and general soil types, together with the apparent similarities in soil biota, suggest that soil aggregate formation and stabilization will be similar in North American deserts and in the karoo. The lack of information on soil aggregates and their probable importance in desert ecosystems demonstrates the need for research in this area. Since soil is the basic resource for all terrestrial ecosystems, it is obvious that considerably more attention should be focussed on soil biota, the processes that they mediate and how the properties of soils are affected by management.

Soil perturbation

In North American and karoo desert ecosystems, soil disturbance by the activities of animals is an important feature that affects the heterogeneity and species diversity of the effected systems. In the karoo, one of the most important features is the *heuweltjie* that apparently develops from the interaction of a termite, (*Microhodotermes viator*) which produces the mound in the process of nest building and maintenance, and other animals such as the aardvark *Orycteropus afer*, common molerat *Cryptomys hottentotus* and many other burrowing animals such as foxes, rats, mongooses and porcupines (Milton and Dean, 1990a). The activities of termites and/or molerats transporting organic materials into the mounds and the turnover of soil enriches the mounds, resulting in fertile sites on which vegetation is different from the surrounding area (Midgley and Musil, 1990; Knight et al., 1989). Banner-tail kangaroo rat (*Dipodomys spectabilis*) mounds in the Chihuahuan Desert provide the same type of nutrient-enriched site at a somewhat smaller scale (Mun and Whitford, 1990; Chew and Whitford, 1992). Nutrient-enriched patches are also produced by harvester ants in the Chihuahuan Desert (Whitford and DiMarco, 1995). In the karoo, the soils of *Messor capensis* mounds are enriched with nutrients which increase leaf nitrogen of plants growing on the mounds (Dean and Yeaton, 1993b). The contribution of animal-produced soil disturbance patches in deserts appears to be scale-dependent. Quantities of subsoil moved to the surface by ant species that develop relatively short-lived colonies (less than a decade) con-

tribute to soil turnover and represent readily erodable material that may make a large contribution to soil characteristics of run-on areas (Whitford, 1995) The quantities of soil moved by the activities of ants can be quite large, and can be very important to long-term ecosystem processes such as soil genesis (Whitford, 1995). The activities of soil-nesting ants in the karoo should have a similar effect on soil turnover and soil genesis in that desert. Pits produced by animals vary in size, abundance and spatial distribution. Even small seed-cache pits serve as litter traps and eventually become small fertile microsites (Steinberger and Whitford, 1983a). Disturbance of *Messor capensis* nests by aardvarks creates pits in which seedling germination is enhanced (Dean and Yeaton, 1992). In the karoo, there are numerous digs produced by animals such as porcupines and aardvarks that serve as litter and seed accumulation sites where litter is buried by wind or water-borne soil. These sites become nutrient-enriched by the decaying litter and are essential germination sites for many kinds of plants (Dean and Milton, 1991a). Animal produced nutrient-enhanced patches may be essential in maintaining the diversity of the vegetation of deserts by providing a gradient of patches of different nutrient concentrations.

18.3.2.   Granivory and herbivory

Some of the most important and influential studies done in desert ecosystems have addressed questions about seed consumers and their effects on ecosystems (Brown et al., 1979; Brown and Heske, 1990; Davidson and Samson, 1985; Davidson et al., 1984; Morton, 1985; Kerley, 1993). In addition to the obvious importance of seeds as the reproductive products of plants, in arid ecosystems, seeds are an important adaptation to unpredictable drought. Consumption of seeds by granivores is an important process that can shape the structure of the plant community (Inouye, 1991). In North America, the primary seed consumers are heteromyid rodents and seed-harvesting ants (*Pogonomyrmex* spp. and *Pheidole* spp.), but the heteromyid rodents harvest a much larger fraction of the seed crop than the ants in these deserts (Brown et al., 1979). In the karoo, the small mammal fauna is dominated by omnivores, insectivores or herbivores and no species are predominantly granivores (Kerley, 1992c). Data from seed-tray experiments show that ants remove a larger fraction of the seed crop than rodents in the karoo (Kerley, 1991). The importance of rodents as granivores in North American deserts appears to be the exception. In Australian, South American and Eurasian deserts, granivorous rodents constitute a small fraction of the rodent community and the available data show that seed-harvesting ants remove the largest fraction of the seed crop in these deserts (Kerley and Whitford, 1994). The high rates of seed removal by heteromyid rodents in North American deserts does not necessarily mean loss of those seeds from the seedbank. Heteromyid rodents cache seeds and many of the caches are never recovered, thus contributing to the dispersal of selected seeds and maintenance of the seedbank. In the karoo, seed production, seedbanks and levels of seed predation by birds and ants were similar to those in North America (Kerley, 1993). Lower seed consumption by rodents in the karoo does not appear to have been compensated by higher rates of seed consumption by ants and birds. Obviously there are many aspects of granivory and seed dynamics that have yet to be explored.

In shrub-dominated deserts like the karoo, herbivores can shape plant communities and the morphological and chemical attributes of the plant species. Browsers select plants on the basis of the chemical, physical and physiological attributes of the plant. In the Chihuahuan Desert of North America, black-tail jackrabbit (*Lepus californicus*) is the most abundant mammalian herbivore that browses on shrubs and is the only mammal that consumes creosotebush. This animal consumes parts of the stem and drop most of the stem, leaves and twigs to the ground (Steinberger and Whitford, 1983b). Creosote bushes that have a history of being browsed have a high probability of being browsed again (Ernest, 1994). Creosote bushes that are browsed have higher stem water contents than shrubs that are not browsed. All of the shrubs that were experimentally irrigated were browsed by jackrabbits (Steinberger and Whitford, 1983b). In a study where creosote bushes were pruned by the investigators, jackrabbits continued to prune those with a history of pruning by jackrabbits and did not feed on the plants clipped by the investigators. These data were interpreted as enhanced resistance to pruning by jackrabbits by an induced response to herbivory called constitutive resistance (Ernest, 1994). Pruning of creosotebushes by jackrabbits is highest during the driest periods and higher in dry years than in wet years (Ernest, 1994, Steinberger and Whitford, 1983b) suggesting that water status of the plant is more likely to be the basis of selection than constitutive resistance. Milton (1991) has argued that the origin of spinescence in the karoo is based on the selectivity by mammalian herbivores of plants with higher water status. The selectivity of plants with higher water status or nutrient status may override the importance of toxins as feeding deterrents. These are likely to be the primary factors determining selectivity by mammalian herbivores in the karoo and in North American deserts.

Variation in herbivore loads on desert shrubs may be related to morphological differences among plants that affect the water and nutrient status of the plants as well as the chemical feeding deterrents (Lightfoot and Whitford, 1989). Most of the herbivores on desert shrubs are sucking insects with a feeding strategy that avoids most of the chemical deterrents in shrub leaves (Whitford, 1974; Schultz et al., 1977; Milton, 1993). The impact of insect herbivores on desert plants is difficult to assess. There are occasional outbreaks of insects that defoliate a species of shrub, but accounts of these are anecdotal and the effects of such outbreaks on ecosystem processes are unknown. One effect of herbivory that has been documented is enhancement of nitrogen flux by phytophagous insects on creosotebush in the Chihuahuan Desert (Lightfoot and Whitford, 1990). Herbivory can produce other feedbacks on ecosystem properties, but our understanding of these processes in desert ecosystems is inadequate for developing generalizations. Considerably more attention to the process of herbivory by non-domestic animals in desert ecosystems is obviously needed.

Domestic livestock are the most important herbivores in semi-arid and arid ecosystems, since these ecosystems are used primarily as rangelands. The effects of grazing–browsing by domestic herbivores include changing species composition of the vegetation by removal of preferred species, modifying soil properties such as bulk density, and indirectly modifying texture and nutrient stocks in soil by increasing wind and water erosion. There are parallel studies on the effects of grazing on flowering of plants in the karoo and in the Chihuahuan Desert (Milton, 1992b; Kerley et al., 1993). In the karoo, grazing by sheep, at the recommended densities during flowering and seed set of a shrub (*Osteospermum sinuatum*), reduced the potential seed set by as much as 90% (Milton, 1992b). In the Chihuahuan Desert, cattle consumed 98% of the infloresences of *Yucca elata*. The only *Y. elata* that escaped being eaten by cows were on caudices greater than 170 cm in height, higher than the cattle could reach. Cattle also affected the branching patterns of the yucca and increased the frequency of procumbent caudices. In addition to the obvious implications of grazing–browsing during flowering and seed set on the reproductive success of the affected species, removal of infloresences can affect a host of other species in the system because of the loss of food and habitat structure (Kerley et al., 1993).

18.4.   **Historical changes and desertification**

The karoo and North American deserts have a similar history with respect to the development of commercial ranching operations. Prior to the 1880s, grazing by domestic livestock in these deserts was limited to areas around permanent surface water or to short periods of use in areas around ephemeral waters. The development of deep-drilling technology which allowed for the production of borehole water from deep ground-water stores, allowed grazing on virtually all of the desert lands (Dean and Macdonald, 1994). The intensive use of these lands by domestic livestock resulted in vegetation composition change in the deserts of both continents. In the Chihuahuan Desert, the palatable grasses were replaced by unpalatable (to cattle) shrubs (Fig. 18.5) (Buffington and Herbel, 1965). This resulted in major structural change in the ecosystem as well as change in the species composition. Patterns of soil movement by wind and water were effected by the structural changes, as were the linkages between rainfall and productivity. On many ranches in the karoo, where sheep and goats are the most common livestock, palatable shrubs have been replaced by unpalatable shrubs and spinescent plants, and grass cover has been reduced (Milton et al., 1994b) (Fig. 18.5).

In addition to reduction in livestock production, desertification affects numerous ecosystem processes (Schlesinger et al., 1990). Reduction in plant cover which increases the size of unvegetated patches leads to wind and water erosion. The consequences of loss of a soil fraction (silt fraction by wind erosion) or of an entire soil horizon as a result of sheet flow and rill erosion during intense rainfall events are yet to be completely appreciated. Materials transported by wind can have global implications (Schlesinger et al., 1990). Degradation and loss of ecosystem services has prompted numerous attempts to return desertified systems to something approaching the predisturbed condition (as assessed by cover and composition of vegetation) (Herbel, 1983). Most attempts at restoring productivity of desertified rangelands have met with limited success (Whitford, 1995).

Partial success at restoring ecosystem function and some ecosystem services has been obtained by the introduction of alien plants. The introduction and spread of Lehmann's lovegrass *Eragrostis lehmanniana* in south-western New Mexico and south-eastern Arizona is one example of partial success. Lehmann's lovegrass, endemic to the southern African arid savannas, has been established and is thriving on a number of watersheds that were essentially barren with scattered small trees or shrubs. These watersheds had been desert grasslands in the nineteenth century. The high cover of this alien grass has restored the hydrologic stability of the watersheds (Whitford, 1995). While cattle utilize this grass, it does not have high forage value and is grazed most heavily in small patches adjacent to mesquite where the grass has access to higher nitrogen-content soil. Hydrological stability and forage production has been obtained at the expense of many native species of plants, vertebrates and invertebrates (Bock et al., 1986). The long-term implications of replacing native grasses with an alien grass that becomes dominant in many ecosystems are yet to be explored.

Alien plants are not limited to North American deserts. The plantings of *Agave* spp. and *Opuntia* spp. in the karoo as drought fodder represent nuclear populations that have the potential to spread rapidly to many areas of the karoo (see Milton, Zimmerman and Hoffman, this volume).

In North American deserts, structural properties of the ecosystem appear to be more important than species composition for species of birds, small mammals and lizards (Raitt and Pimm, 1976; W. G. Whitford, unpublished data). Desertification that has increased the shrub component has led to greater diversities of these animals in desertified systems and ecosystems in a transitional state between grassland and shrubland. Despite the concern about biodiversity and the need to develop sustainable management systems that allow a diverse flora and fauna to flourish, we have little quantitative data on the effects of desertification and other management consequences on the biota.

This brief comparison of the karoo with other deserts, especially the Chihuahuan Desert in North America, hopefully has demonstrated that there are many similarities in structure and function in the ecosystems of these deserts. The gaps in our knowledge, especially concerning below-ground processes, need to be addressed in southern African and North American deserts. The temporal–spatial variations in water and nutrient availability and the processes leading to that variability require considerably more attention. Understanding this variability at scales ranging from patch to watershed to region is necessary to develop management systems that support sustainable productivity. This understanding must be based on mechanisms, and is essential to design effective restoration technologies. The similarities in functional and structural properties of the karoo and North American deserts will allow direct transfer of the results of many studies that can contribute to that understanding.

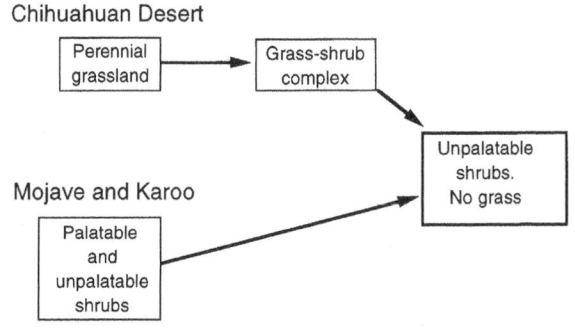

Figure 18.5 **Similarities and differences in the structural properties of desertified ecosystems in the karoo and North America**

302   W. G. Whitford

18.5.   **Acknowledgements**

The preparation of this chapter was supported by the US Environmental Protection Agency through its Office of Research and Development, National Exposure Research Laboratory. It has been subjected to the Agency's administrative review. I thank Amrita De Soyza, Kris Havstad, Jeff Herrick and Graham Kerley for their reviews and suggestions.

19 The succulent karoo in a global context: plant structural and functional comparison with North American winter-rainfall deserts

K. J. Esler, P. W. Rundel and R. M. Cowling

19.1. Introduction

In the past decade, research in the succulent karoo biome of southern Africa has highlighted the unique aspects of this winter-rainfall desert compared to other deserts with similar climates (Cowling et al., 1989; Cowling et al., 1994). In addition to long-lived succulents (e.g. *Aloe, Pachypodium*), the succulent karoo has a remarkable dominance and unique diversity of short to medium-lived leaf-succulent shrubs as well as a rich geophyte flora. In contrast, this biome supports comparatively few drought-deciduous shrubs and other long-lived perennials (Esler and Rundel in press). Despite a lack of structural diversity, due partly to growth form uniformity as discussed below, plant species diversity at both the local and regional scales in the succulent karoo is undoubtedly the highest recorded for any arid region in the world (Cowling et al., 1989; Cowling and Hilton-Taylor, this volume). The floral diversity of the biome is unparalleled for any winter-rainfall desert – 4849 species occurring in approximately 100 000 km² (Hilton-Taylor, 1996). Approximately 30% of the world's succulent species are located in this small area (Van Jaarsveld, 1987; Smith et al., 1993). Notable families with leaf- or stem-succulent species are the Aizoaceae (including Mesembryanthema) (861 taxa, 102 genera, Hilton-Taylor 1996), Asteraceae, Liliaceae, Geraniaceae, Euphorbiaceae, Asclepiadaceae and the Crassulaceae. Many of these species are rare and endangered, and approximately 40% are endemic (Hilton-Taylor, 1996). These features make the succulent karoo a unique biome of global importance.

Using a comparative approach, we propose an empirical model to explain how the special climatic conditions of the succulent karoo have provided a unique selective regime which has led to the evolution of such a diverse and unusual desert flora. Our ideas presented here are largely focussed on the Namaqualand–Namib domain of the succulent karoo; this is the strongly winter-rainfall region of the biome and excludes the southern karoo (*sensu* Jürgens, 1991) (Fig. 19.1). As discussed by Cowling and Hilton-Taylor (this volume), the Little Karoo receives a considerable amount of summer rainfall and is floristically and functionally a transition between the Namaqualand–Namib Domain and the Nama-karoo. Hereafter, we refer to the strongly winter-rainfall area of the succulent karoo as the N–N Domain.

There is a strong temptation to apply ecological principles and paradigms gained from detailed studies of the winter-rainfall deserts of North America in an attempt to understand community structure and dynamics in the N–N domain. Studies of convergent evolution in ecosystems with similar climatic regimes have been a powerful tool in understanding the evolution of community structure and biological diversity (Cody and Mooney, 1978; Ricklefs and Schluter, 1993) and this approach has been applied successfully in using functional ecosystem traits measured in one ecosystem to predict traits in another system with a similar physical environment (e.g. Mooney, 1977; Orians and Solbrig, 1977; Miller, 1981; Cowling and Witkowski, 1994). While comparative research on Mediterranean-type ecosystems over the past several decades has largely focussed on evergreen, sclerophyllous vegetation, little attention has been given to comparative research on ecological structure and community diversity in the winter-rainfall deserts (but see Shmida and Whittaker, 1974 for a comparison of New and Old World deserts).

The assumption implicit in studies of convergent evolution in similar ecosystems is that analogous environmental conditions produce a combination of climatic and edaphic factors which lead to specific physical stresses for

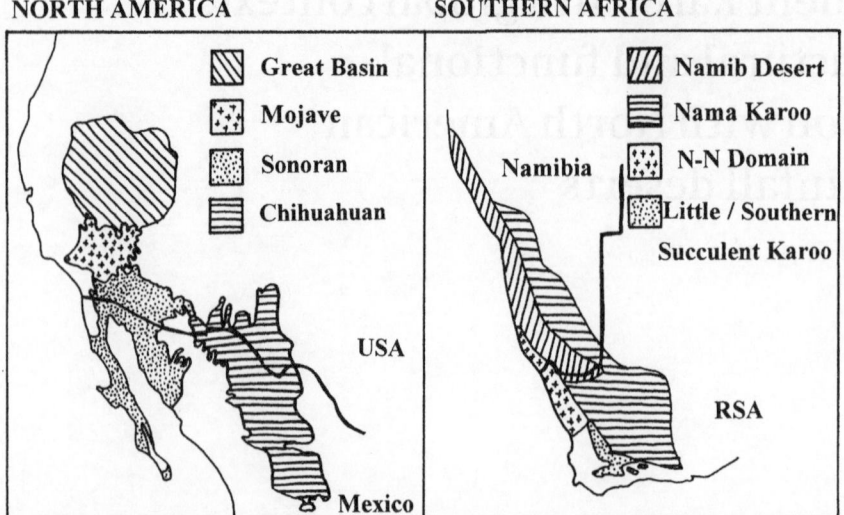

Figure 19.1 **The desert zones of South Africa and North America**

which there exist only a limited number of successful evolutionary responses. Natural selection should therefore result in convergent patterns of resource utilization in biotic communities occurring under similar selective regimes. There exists a wealth of ecological literature on the winter-rainfall Mojave and Sonoran Deserts of California and Baja California (Rundel and Gibson, 1996; Rowlands et al., 1982; MacMahon and Wagner, 1985) (Fig. 19.1) where climatic regimes appear superficially to closely approximate those of the N–N Domain. Should we therefore expect convergence in evolutionary patterns and adaptations of plant species? Furthermore, can we use the models developed in the deserts of North America to understand the N–N domain of the succulent karoo?

At the level of growth form dominance and species diversity, it is clear from simple observations that the South African and North American desert regions are dramatically different. Our studies of comparative community structure and growth dynamics along aridity gradients (Mediterranean–desert transition zones) in each continent suggest that predictably divergent seasonal patterns of resource utilization and resultant growth form diversity of perennial plants do exist in regions with seemingly similar climatic regimes (Esler and Rundel, in press). Relatively small differences in seasonal distribution and extremes of temperature and precipitation appear to be sufficient to produce this divergence, as we describe below, and these differences have provided the N–N Domain with a unique selective environment for evolution.

19.2.   Selective regime

Despite similarities in total annual rainfall along Mediterranean–desert transition zones in supposedly convergent climatic regions (Fig. 19.2), there are marked differences with respect to intra- and inter-annual patterns of rainfall events (Rundel, Esler and Cowling, unpublished data).

The Mediterranean–desert transition zone in South Africa is distinctly different from that of North America in that it has low monthly and yearly coefficients of variation in rainfall (Fig. 19.3 and 19.4). The low monthly coefficient of variation in rainfall in South Africa reflects the fact that this region receives more rain throughout the year (i.e.

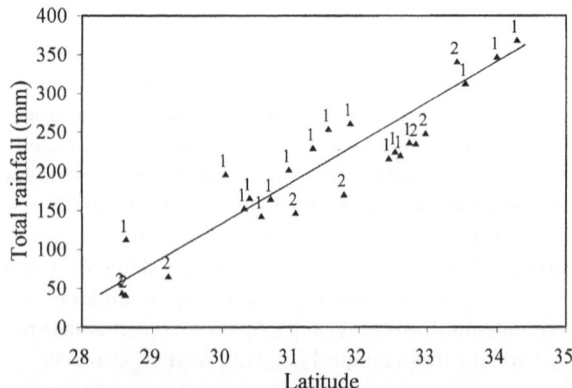

Figure 19.2 **Total annual rainfall (mm) according to latitude for sites along the Mediterranean–desert transition zones of North America (1) and South Africa (2) (data were obtained from Weather Bureau (1986) (RSA); S. H. Bullock, Ecologia, CICESE, POB 434844, San Diego, CA 92143, (MEXICO); National Climatic Data Center, Asheville, North Carolina (USA))**

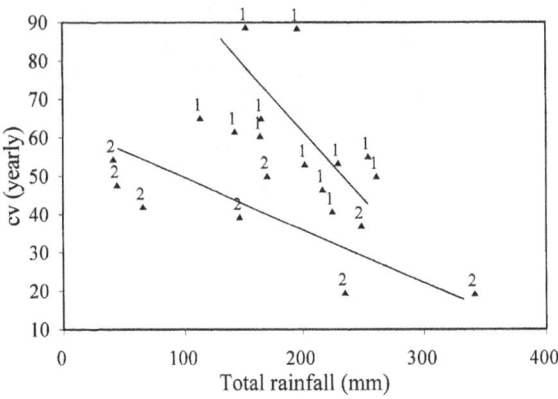

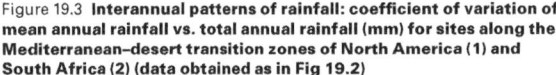

Figure 19.3 **Interannual patterns of rainfall: coefficient of variation of mean annual rainfall vs. total annual rainfall (mm) for sites along the Mediterranean–desert transition zones of North America (1) and South Africa (2) (data obtained as in Fig 19.2)**

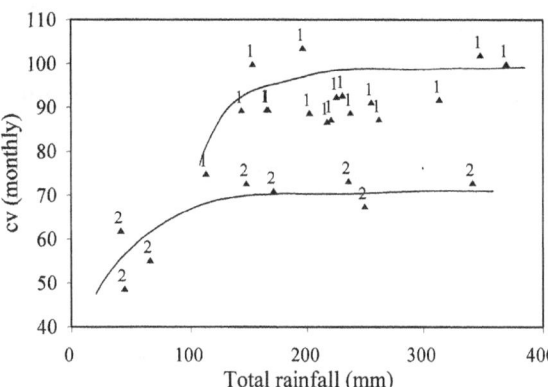

Figure 19.4 **Intra-annual patterns of rainfall: coefficient of variation in monthly rainfall (calculated from long-term monthly means) vs. total annual rainfall (mm) for sites along the Mediterranean–desert transition zones of North America (1) and South Africa (2) (data obtained as in Fig 19.2)**

although predictable, rainfall is comparatively less seasonal, despite the region being a predominantly winter-rainfall zone) than California/Baja California (Fig. 19.4).

The latter region is strongly dominated by winter rains with summer precipitation rare or absent except at the eastern margins of the California regions of the Mojave and Sonoran Deserts and in the transition to summer-rainfall regions of coastal Baja California. In addition, the N–N Domain has additional sources of precipitation in summer in the form of frequent, highly predictable fog and high nightly humidities (Jürgens, 1991; Desmet, 1996). The low coefficients of interannual variation for the N–N Domain reflects the fact that total precipitation is relatively predictable between years, even at these low annual totals (Fig. 19.3). Interannual variation in rainfall is higher in the coastal desert transition of California and Baja California. The net effect of these patterns of reduced rainfall seasonality, fog moisture input, and low interannual variation in the N–N Domain is that moisture is a relatively predictable, albeit low-level resource for plant growth.

Another unique characteristic of the South African Mediterranean desert is that temperature regimes are distinctly moderate. This is due to the proximity of the N–N domain to the west coast of South Africa and to the high incidence of fog and the cooling effect of the Benguela current in summer, when maximum upwelling of cold water occurs (Desmet, 1996). Winter-rainfall deserts of North America are largely inland and at higher latitudes than is the N–N Domain, and thus the former are more continental and temperate in temperature regime. In a comparison between mean maximum temperatures in January (summer) and mean minimum temperatures in July (winter),

stations in the N-N Domain are clearly distinguished from those in the Nama-karoo by having moderate temperatures which drop no lower than 5 °C on typical winter nights and reach 15 °C or more on most winter days (Fig. 19.5). Climatic stations in the Little Karoo are intermediate in these temperature patterns. Subzero temperatures, a characteristic of extreme winter conditions in the Nama-karoo and most winter-rainfall deserts, are absent from much of the N-N Domain (Cowling et al., 1994).

Winter temperatures are typically far lower throughout the Mojave Desert of North America, and irregular hard frosts may reach southward at infrequent intervals throughout the winter rainfall regions of the Sonoran Desert. This pattern can be seen by comparing climate stations in the eastern Mojave Desert (Rock Valley) and in the N-N Domain (Goegap/Okiep) which occur at similar elevations and have comparable mean annual rainfall (Fig. 19.6). Monthly mean maximum temperatures in the N-N Domain are 2–6 °C cooler in summer and 4–6 °C warmer in winter than those in the more continental Mojave Desert (Esler and Rundel, in press).

Monthly mean minimum temperatures at Goegap never drop below 7 °C in winter in comparison to frequent subzero temperatures at Rock Valley that are too low for growth of Mojave Desert perennials (Turner and Randall, 1989). These data for the two comparative study sites are relatively typical for their respective desert regions (Esler and Rundel, in press).

The small but significant differences in the seasonality and predictability of rainfall and seasonal patterns of temperature extremes can be extrapolated more broadly (K. J. Esler, unpublished data). This observation, along with the observation that rainfall is very predictable, is central to

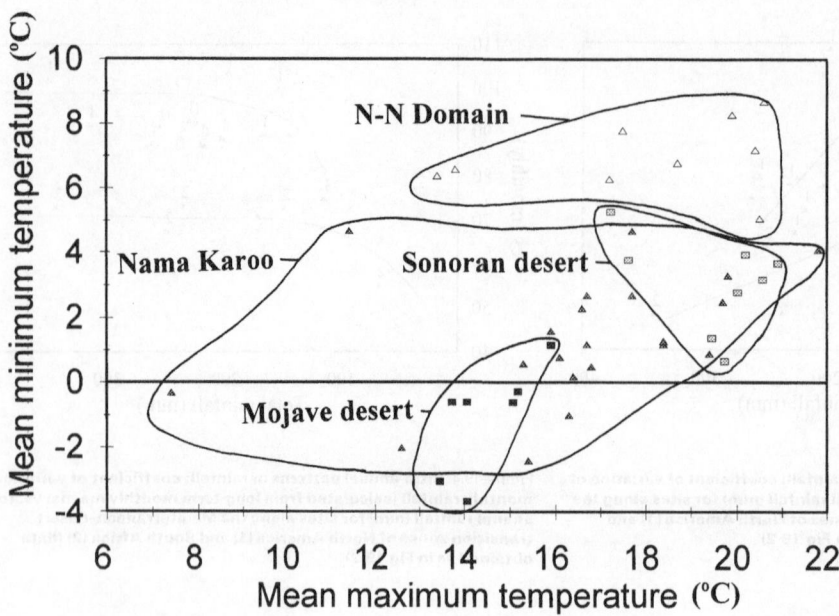

Figure 19.5 **Relationship between mean minimum and mean maximum temperatures in winter (July for South Africa; January for North America) for sites in the N–N domain and the Nama-karoo of South Africa, and the Mojave Desert and winter-rainfall regions of the Sonoran Desert in North America (data obtained as in Fig. 19.2)**

our working model to explain plant form and function in the N–N Domain. In the following sections, we outline some observations which led us to develop this working model.

19.3.   Plant form and community structure

A comparison of the life form spectra of species occurring at selected sites in the N–N Domain and Mojave Desert can be used as an indication of the differences in community structure and plant form between the regions (Table 19.1). Woody phanerophytes reaching to 0.5 m or more in height comprised less than 6% of the flora at Goegap in the N–N Domain, but twice this amount at a site in the eastern Mojave Desert. Moreover, many of the phanerophytes of the Mojave Desert are drought-deciduous shrubs, a relatively rare growth form in the N–N Domain. Small chamaephytes with perennial tissues reaching less than 0.5 m in height comprised 31% of the flora at Goegap, with the great majority of these being leaf-succulent, evergreen members of the Mesembryanthema. By comparison, such chamaephytes made up only 6% of the Mojave Desert flora. Another striking difference between the floras of each zone lies with the diversity of petaloid monocot geophytes (not included in the category: hemicryptophyte). One third of the flora of Goegap comprised geophytes, and half of this total were petaloid monocots. Hemicryptophytes were relatively more important in the Mojave Desert Region where they made up half of the flora, but geophytes

were virtually absent. Therophytes or annuals made up a similar proportion of the flora with 28% at Goegap and 32% in the Mojave Desert site.

These differences in life forms of the floristic elements of each flora tell only part of the story. The ecologically dominant species in these two communities represent an even more dramatic difference in aspects of canopy structure and leaf form. The N–N Domain presents a widespread dominance of low, leaf-succulent shrubs reaching 20–40 cm in height with scattered low-trailing leaf succulents or rosette succulents. With taller shrubs relatively uncommon, except in particular edaphic or hydrological situations, structural diversity is thus relatively limited in the N–N Domain. In contrast, Mojave Desert communities present considerably more structural diversity. Shrub height is highly varied and presents mixed dominance of a single evergreen species (*Larrea tridentata*) and drought-deciduous smaller shrubs. Stem-succulent cacti and leaf-succulent *Yucca* are commonly present, with an intermixture of low herbaceous perennials. In sharp contrast to the relatively short-lived leaf succulents of the N–N Domain, the larger shrub species in the Mojave Desert are relatively long-lived, and thus present relatively stable components of these landscapes (Vasek, 1980).

The notable above-ground structural diversity of the Californian/Baja Californian Desert is also present below-ground, where competition for soil water resources has led to the evolution of shrubs with divergent patterns of rooting depth (Cody, 1989, 1991; Rundel and Nobel, 1991) (Fig. 19.7). Leaf-succulent shrubs in the N–N domain have rooting depths which commonly reach to only 0.1–0.2 m

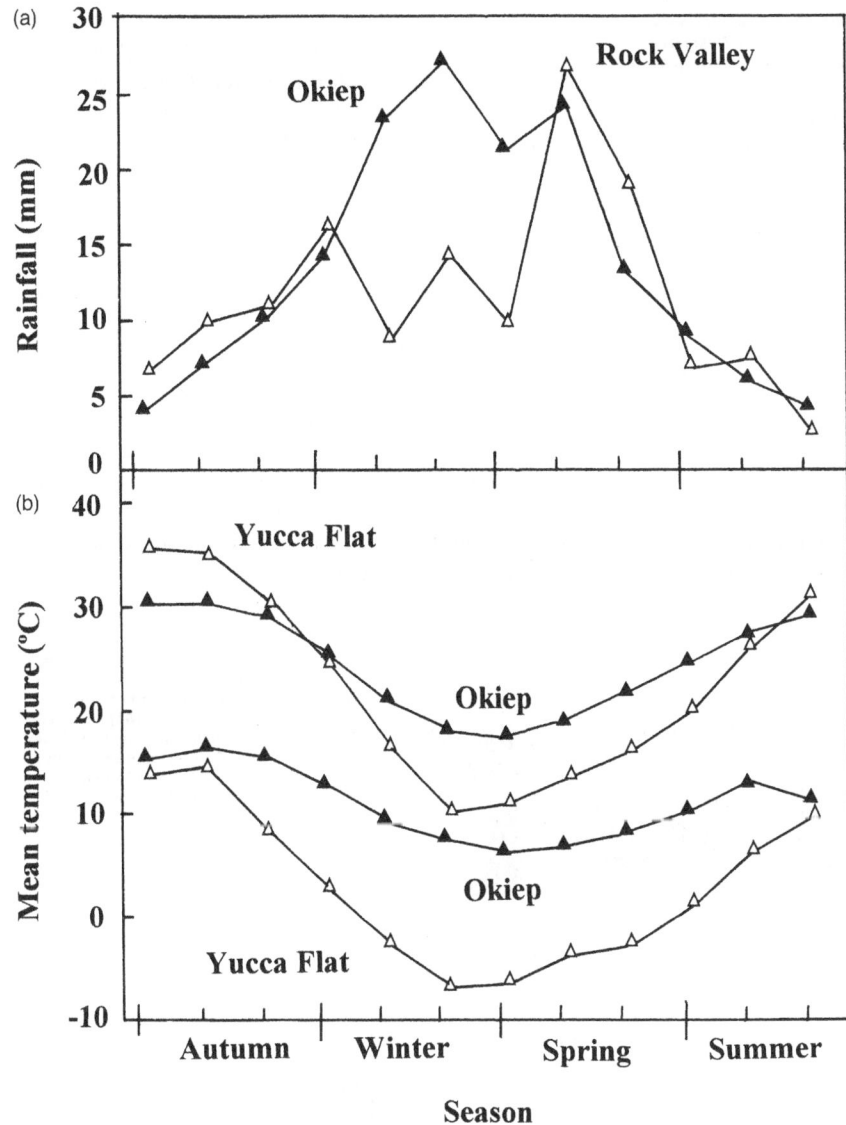

Figure 19.6 **Monthly rainfall (a) and mean monthly minimum and maximum temperatures (b) for comparable climate stations adjacent to Goegap (N–N Domain) and Rock Valley (Mojave). Goegap Nature Reserve lies in the Namaqualand region of the N–N Domain of north-western South Africa (884–1346 m elevation, 29°37′ S lat., 18°01′E long.) and has a mean annual rainfall of 162 mm. Rock Valley in the Mojave Desert area of the Nevada Test Site in south-central Nevada (1020 m elevation, 36°41′N lat., 116°12′W long.) experiences a mean annual rainfall of 145 mm. Climate data for Okiep (29°56′S, 18°24′E, 943m) in the N–N Domain were taken from a 64-year record (1920–84) (Weather Bureau 1986). Rainfall data from Rock Valley (Mojave) were taken from a six-year record collected during 1971–6 when the area served as a primary site for the International Biological Program (Rundel and Gibson 1996; Turner and Randall 1989). Temperature data from Yucca Flat (36°57′N, 116°03′W, 1186m) in the Mojave Desert was taken from a 10-year climatological summary (1962–72) (O'Farrell and Emery 1976). This site receives cold-air drainage at night, thus minimum temperatures are approximately 2–5 °C lower than typical Rock Valley temperatures (from Esler and Rundel (in press), with permission from Kluwer Academic Publishers)**

(maximally 30 cm) below the soil surface, even in sandy soils that extend far deeper.

In contrast, shrub species in the Mojave Desert present complex patterns of below-ground architecture, with some species having roots reaching to depths of 1.4 m or more (Cody, 1986b, 1989, 1991).

Shrubs with relatively small overlap in rooting patterns (shallow and deep) are more likely to coexist than those with similar rooting architectures as they are less likely to compete for the same water resources. Thus, the distinct contrast in patterns of rooting architecture found in the N–N Domain and Mojave Desert are highly significant. This low diversity of below-ground rooting architecture may well present important evolutionary feedbacks with the low above-ground structural diversity

previously discussed. Clearly, some very different rules of organization for community structure and plant form exist in the N–N Domain (see Cowling and Hilton-Taylor, this volume). These are discussed further in section 19.6.

19.4.   Community phenology

Seasonal patterns of resource allocation may well be a primary factor in the marked differences between the structural diversity of dominant plant species in the N–N Domain compared to other winter-rainfall deserts. Surprisingly, large differences exist in the vegetative and reproductive phenology between the N–N Domain and

Table 19.1. *Life form spectra of study areas at Goegap within the N–N Domain (582 spp.) and south-central Nevada (Mojave Desert) (750 spp.) (from Esler and Rundel (in press), with permission from Kluwer Academic Publishers)*

	Goegap (N–N Domain)	South-central (Mojave Desert)
Phanerophytes	5.8	11.3
Mesophanerophytes	0	7.1
Microphanerophytes	22.4	17.6
Nanophanerophytes	77.6	75.3
Chamaephytes	31.4	6.0
Hemicryptophytes	16.8	50.0
Geophytes	16.3	0.7
Therophytes	28.3	32.0

Data are given as a percentage of the total species in each flora. The phanerophyte subcategories are given as a percentage of the total phanerophyte component. Succulent species are classified as chamaephytes or phanerophytes depending on their height
Data extracted from Beatley (1976) and Van Rooyen et al. (1990)

Mojave Desert as can be illustrated with data again from sites receiving approximately the same mean annual rainfall but slightly different temperature regimes at Goegap

(N–N Domain) and Rock Valley (Mojave desert). At Goegap, moderately high daily minimum (>0 °C) and maximum (>18 °C) temperatures during winter allow growth to proceed slowly until the beginning of spring when mean monthly precipitation drops below 10 mm (Fig. 19.8a).

In contrast, vegetative growth of woody perennials in the Mojave Desert only begins in early spring when daily maximum temperatures warm to about 15 °C. This is the same season as when the N–N Domain species become dormant. At this stage, 74% of the hydrologic year (July–June) precipitation has already occurred. The frequency of vegetative growth in the Mojave Desert declines sharply in late spring as soil water availability becomes limiting, and the active growing season is confined to a much shorter period of 7 weeks compared to 30 weeks in the N–N Domain (Fig. 19.8a). Summer rains, absent most years at Rock Valley, promote a second season of vegetative growth in perennials, showing that high summer temperatures alone are not limiting when water is available.

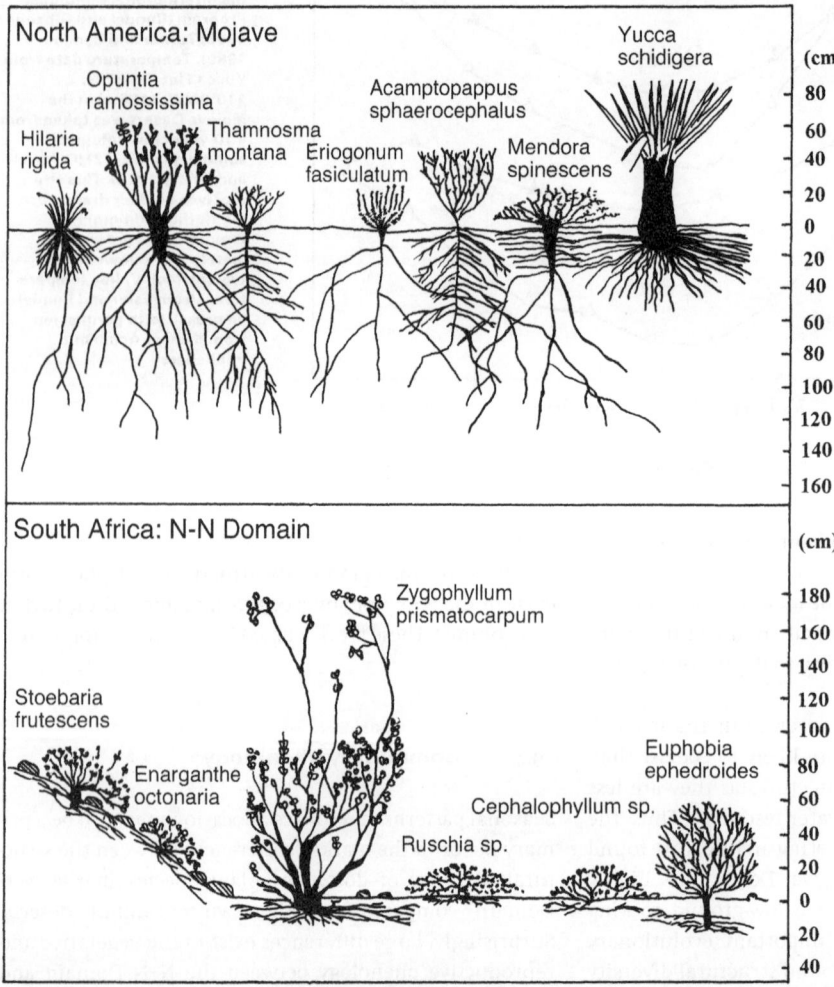

Figure 19.7 **Rooting profiles of typical species occurring in the N–N domain of South Africa and the Mojave Desert of North America (from Esler and Rundel (in press), with permission from Kluwer Academic Publishers)**

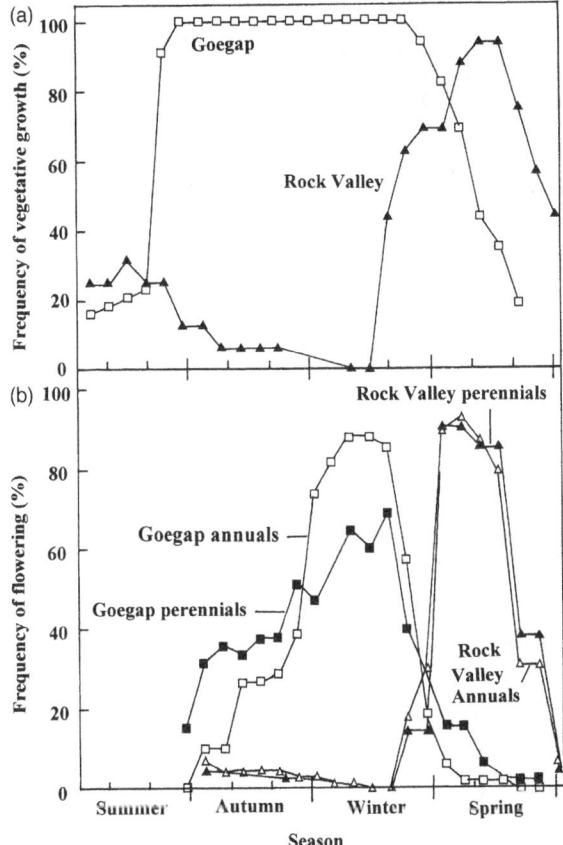

Figure 19.8 **Seasonal frequencies of vegetative growth (a) and flowering (b) of plant species at or adjacent to Goegap (N–N Domain) and Rock Valley (Mojave). Phenological data at Goegap were taken from studies by Van Rooyen et al. (1979). A species was considered in active growth or flowering when at least 10% of the individuals present were in that condition. Data were extracted at two-week intervals for 44 species of woody or semi-woody perennials and for 49 species of annuals. Vegetative growth data for 21 species of Rock Valley perennials were taken as the extremes of active growth periods over three years (1971–3) at Rock Valley and comparable stations at the Nevada Test Site (Ackerman et al. 1980). Data on flowering phenology from these 21 species of perennials and for 68 species of annuals at Rock Valley were taken from Beatley (1976). Since her observations on flowering phenology cover a broader elevational and geographic range than Rock Valley alone, our data tend to expand flowering times both earlier and later than actually present at the site itself (from Esler and Rundel (in press), with permission from Kluwer Academic Publishers)**

A very similar pattern of phenological divergence between these two systems is present for flowering. Peak flowering in the Mojave Desert occurs in spring, 12 weeks later than the midwinter peak of flowering in the N–N Domain (Fig. 19.8b). As with vegetative growth, flowering is also sharply peaked in the Mojave Desert, but occurs over more than six months in the N–N Domain.

19.5.   **A Namaqualand–Namib domain (succulent karoo) model**

The divergent patterns in climate, growth form diversity and growth phenology in seemingly convergent ecosystems allow the construction of an empirical model to explain aspects of plant form and function in the N–N Domain (Fig.19.9). This model, based on the significance of soil moisture availability for plant growth, suggests a feedback between evolutionary development of below- and above-ground architectures in perennial species (Cowling et al., in press).

The unique characteristics of the N–N Domain result, in large measure, from the selective regime of environmental temperatures and rainfall. Each of these environmental factors combine to select for shallow-rooted dominants of the N–N Domain which compete for relatively shallow soil water pools. Since mild winter-temperature conditions allow growth of the N–N Domain community through these cooler months when rainfall is present, these species harvest water resources from shallow soil horizons as rain falls (at low intensity) from the onset of the wet period in mid-autumn to the onset of drought in spring. No significant soil storage pools accumulate to allow growth into late-spring or summer. Selection is therefore towards shallow-rooted species that can successfully compete for water, while deeper-rooted species are confined to run-on habitats.

In the Mojave Desert, and most other desert regions, winter rains fall at a time when mean maximum temperatures drop to about 15 °C and mean minimum temperatures drop below freezing, too cold for growth. Initiation of new growth in perennials is delayed until late winter or early spring when daily maximum temperatures warm to 15 °C or above and mean minimum temperatures reach above freezing (Fig 19.6b; 19.8). Without significant plant growth during winter, rainfall largely enters soil storage or runoff pools. Mojave soils tend to be well-drained sandy loams or sands, but parent material varies greatly because of complex geology (physical and chemical details are in Rundel and Gibson, 1996). Species in this structurally diverse system must rely on deeper stores of soil water for growth in spring and early summer. Selection is therefore towards a variety of rooting strategies (some shallow, some deep) which can exploit these stores and at the same time minimize interspecific competition (Fig 19.7).

Classical niche theory also explains the lack of growth form diversity in the N–N Domain as a function of the low variability in rainfall regime, both at an intra- and an inter-year level. Here again, the interaction is with the seasonal availability and its variation in water resources for competing shrub species. Although both regions are characterized by an arid Mediterranean-type climatic regime

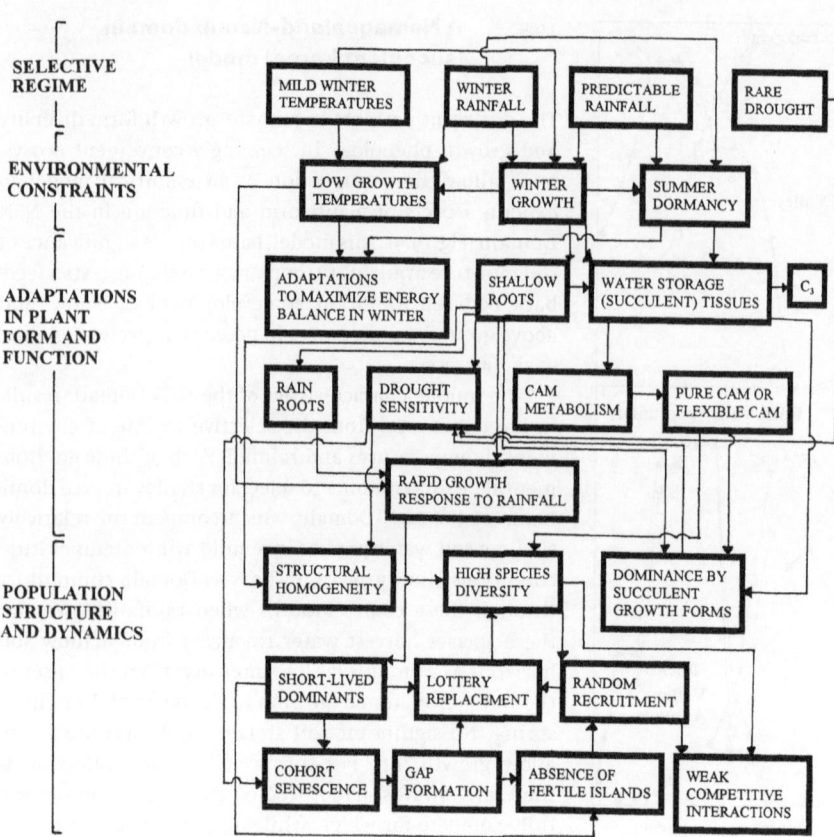

Figure 19.9 **An empirical model to explain aspects of plant form and function as well as population structure and turnover in the N–N domain (redrawn from Cowling et al. (in press), with permission from Kluwer Academic Publishers)**

with winter rainfall, inter- and intra-annual variation in rainfall is significantly greater in the Mojave Desert (discussed in section 19.2). In arid and semi-arid environments where soil moisture is the major limiting primary resource, plant growth form reflects a particular strategy for resource utilization which can be envisaged as a structural niche (Grubb, 1977; Cody, 1986b). Climate, in as much as it reflects the range of soil moisture and temperature conditions suitable for plant growth, reflects the number of growth forms that may coexist in a community. Cody (1989, 1991) found that growth form diversity in North American deserts peaked in climates which provided the widest variety of opportunities throughout the year for growth (i.e. more variable rainfall) and thus enabled the coexistence of the widest range of growth forms. We extend these ideas to suggest that the climate in the N–N Domain, being less variable, provides fewer opportunities for growth form diversification. Paradoxically, species diversity is high, which suggests that there might be other rules governing community structure and dynamics (discussed in 19.6).

The role of edaphic conditions cannot be ignored, and an alternative hypothesis to explain the divergent patterns in growth form diversity and growth phenology in seemingly convergent ecosystems might simply be that soil structure constrains the depth to which roots can grow. Soils in the N–N Domain are shallower (due to weathering and the presence of hardpans) (Desmet, 1996) than the younger and deeper apedal soils of the Mojave Desert (Rowlands et al., 1982; Rundel and Gibson, 1996), however, species in the former region do not exploit the workable soil depth which is an average of c. 0.5 to 1 m in most N–N Domain landscapes.

19.6.   Plant function

A product of the model presented is a series of predictions relating to plant function. These have been incorporated into the model, and are discussed in more detail below.

The implications of winter growth phenology are far-reaching and, apart from potentially explaining the low structural diversity of plant form in the N–N Domain, also may partially explain plant function as well as demographic and community structural features. A prediction emerging from our model is that, if plant growth does occur during winter, then we should find that these species will have relatively low optimal growth temperatures compared to species from other desert regions characterized by spring and summer growth. Although few data exist, available temperature response curves do fit this prediction, with the temperature optima of N–N Domain species ranging from 20 to 23 °C (Rutherford, 1991; P. W. Rundel, unpublished data) while those from the Nama-karoo and from the Sonoran and Mojave Deserts in North America are higher (e.g. Ehleringer and Mooney, 1983).

A further prediction emerging from the model is that there should be a variety of adaptations in the N–N Domain flora to maximize the absorption of solar irradiance and increase plant tissue temperatures to maximize rates of net photosynthesis under winter growth conditions. One such adaptation has been described by Rundel et al. (1995) for *Pachypodium namaquanum*

(Apocynaceae). By the fixed orientation of its terminal whorl of leaves at approximately 55° and a northern exposure, this unusual arborescent stem succulent with summer deciduous leaves is able to maintain nearly twice the midwinter radiation absorbance that it would have with horizontal orientation (Fig. 19.10). Leaves in *P. namaquanum* develop with the first autumn rains any time from February to April, and remain active until September or October when dry summer conditions begin. The unusual orientation in this species furthermore leads to meristem temperatures for developing reproductive tissues that are 3 to 4 °C warmer than those of ambient air temperatures. Such increased temperatures certainly speed the development of these tissues (P. W. Rundel, K. J. Esler and R. M. Cowling, unpublished data).

A further example of tissue orientation to exploit winter growth conditions may be found with the species of

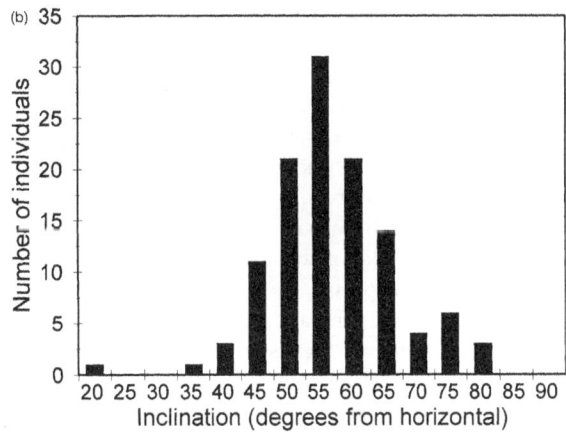

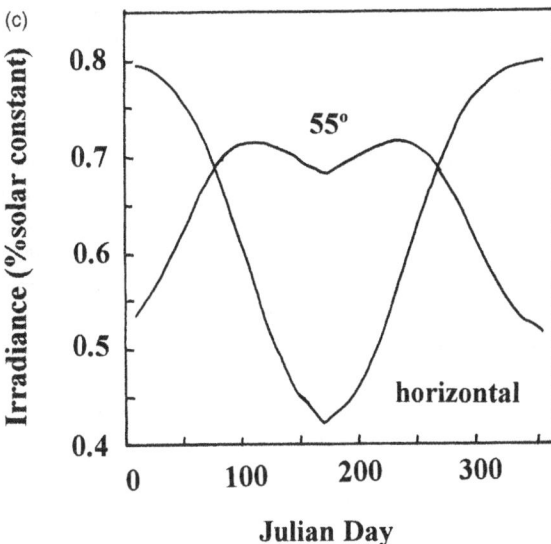

Figure 19.10 **(a)** *Pachypodium namaquanum* **(Apocynaceae), an unusual arborescent succulent from the arid Richtersveld of the N–N Domain, is characterized by a striking curvature of the terminal 20–60 cm of the trunk towards the north. (b) Frequency of angles of inclination of the apical region in 121 individuals of** *P. namaquanum.* **(c) Simulation model of solar irradiance over an annual cycle at 28° S latitude for a horizontal surface and for a surface inclined from horizontal toward north (Figures redrawn from Rundel et al., 1995, with permission of Springer-Verlag)**

geophytes with broad, flattened leaves which lie prostrate on the ground surface. At least 9 families (Amaryllidaceae, Iridaceae, Hyacynthaceae, Orchidaceae, Eriospermaceae, Colchidaceae, Asphodelaceae, Crassulaceae, Geraneaceae) and approximately 44 genera exhibit this characteristic (K. J. Esler, unpublished data). While this growth form is relatively common in many geophyte groups in the N–N Domain (particularly Namaqualand), it is absent or relatively rare in other regions worldwide. Broad and prostrate leaves which lie on the ground surface track soil temperatures which are the warmest microsites during winter days, and thereby increase leaf temperatures several degrees above ambient. These same leaves are cooler than ambient air temperatures at night, and such temperatures may be sufficiently low to allow moisture condensation in the early predawn hours when dew-point temperatures would not otherwise be reached (Esler et al., in press).

The striking dominance of leaf-succulent perennials in the N–N Domain leads to interesting hypotheses as to the possible ecophysiological significance of this trait. No other desert area in the world even remotely approaches the N–N Domain in the abundance and dominance of this growth form. Moreover, much of the diversity of the N–N Domain lies with these species (see Cowling and Hilton-Taylor, this volume). There has clearly been strong selection toward water storage in leaf and stem tissues, and there may well be a strong selective advantage of opportunistic uptake of soil moisture following winter rains (Esler et al., in press).

Researchers have generally accepted the fact that virtually all of these leaf-succulent species utilize Crassulacean acid metabolism (CAM) in their gas exchange (Von Willert et al., 1992), but the significance of CAM in these species as a strategy of drought tolerance has been questioned (Von Willert et al., 1985). Relatively little attention, however, has been given to the relative degree to which succulent-leaved species of Mesembryanthema, Asteraceae, and Portulacaceae utilize CAM in their daily metabolism. Our recent analyses of leaf carbon isotope rations of N–N Domain species and populations of such leaf succulents suggest that there is a previously unappreciated diversity of utilization of CAM present in these species. Succulents relying entirely on CAM fixation of carbon, as in *Aloe* and large species of Crassulaceae in the N–N Domain, would be expected to have $\delta^{13}C$ ratios of −14 to −16 ‰. Species having ratios of -22 ‰ or below would largely fix carbon, outside of possible CAM idling, using typical C_3 metabolism. In our surveys at several sites, we have found putative CAM leaf-succulent species with carbon isotope ratios varying from −14 ‰ to −26 ‰ (Fig. 19.11), strongly suggesting a wide range of reliance on CAM in their carbon metabolism. Intraspecific variation at a single site was relatively small, suggesting that these are species-specific patterns.

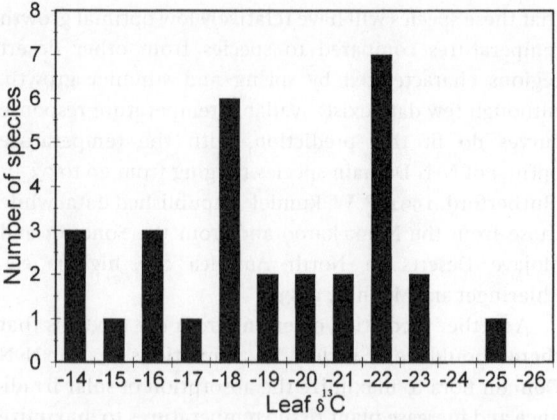

Figure 19.11 **Leaf $\delta^{13}C$ for a range of N–N Domain species (Redrawn from Rundel et al. (in press), with permission from Kluwer Academic Publishers)**

We have hypothesized that the degree of reliance on CAM is significantly related to longevity in these leaf-succulent species (P. W. Rundel, K. J. Esler and R. M. Cowling, unpublished data). Long-lived leaf succulents, as best illustrated by shrubby species of *Stoeberia* (Mesembryanthema) in the N–N Domain, appear to utilize CAM almost entirely, reflecting a conservative growth strategy and high water-use efficiency. At the other end of the spectrum, short-lived Mesembryanthema such as *Mitrophyllum (spp)* appear largely to utilize normal C_3 metabolism in their growth as indicated by highly negative values of $\delta^{13}C$, and thus have much lower efficiencies in their use of water. Other Mesembryanthema with intermediate lifespans in the N–N Domain have intermediate values of $\delta^{13}C$. These data suggest that competing leaf-succulent species with differing reliance on CAM may well compartmentalize their use of water resources temporally as well as spatially, thereby adding another potential dimension to their niche structure. This suggests that niche partitioning in the N–N Domain is extremely subtle with species reflecting different levels of water-use efficiency compared to the Mojave where niche partitioning is evident in different rooting and leaf longevity patterns.

This pattern of relationship between $\delta^{13}C$ and associated water-use efficiency with the longevity of desert plants has been previously demonstrated for non-succulent species in North American deserts (Schuster et al., 1992). We have found this same general pattern, with interesting exceptions, in the non-succulent flora of the N–N Domain as well (Rundel et al., in press). Long-lived shrubs that we have studied exhibit relatively high levels of water-use efficiency (less negative $\delta^{13}C$) compared to short-lived perennials. Annual species of the N–N Domain typically have low water-use efficiency (more negative

δ^{13}C) consistent with their short and opportunistic growth strategy. However, there are notable exceptions in this group with some annuals having high water-use efficiency.

19.7.   **Population structure and turnover**

The N–N Domain is unusual among the Mediterranean deserts of the world in that the bulk of its perennial species are relatively short-lived shrubs, most of which are succulent. This has important implications for population structure and turnover and, ultimately, the determinants of community structure. Unlike perennial species in other arid regions, where recruitment events are punctuated and populations are even-aged (Jordan and Nobel, 1982), these species exhibit uneven age structures; there is high species turnover and significant compositional change through time (see Cowling and Hilton-Taylor, this volume). Part of the explanation for this may lie with fact that these species are shallow-rooted and are therefore susceptible to even moderate drought stress. Occasional droughts do result in mass mortality (Von Willert et al., 1985; Jürgens et al., 1997). As suggested by Cowling and Hilton-Taylor (this volume), it appears that community dynamics in the N–N Domain are determined by a lottery process whereby functionally similar shrubs coexist in communities which are highly dynamic. The dynamic nature of these communities, whereby newly created sites are continuously colonized by species in the pool of available propagules from relatively short-lived seedbanks (Esler, 1993), is a unique phenomenon for desert vegetation. An obvious and testable implication of this is that competition is not a major determinant of perennial plant community structure. A further prediction is that fertile

islands which build up under long-lived shrubs in most desert systems (Rundel and Gibson, 1996) are not a feature in the N–N Domain.

19.8.   **Conclusions**

A major advantage of global comparisons is that they highlight that which is unique about the regions under comparison. The contrast between the North American deserts and the N–N Domain has highlighted some unusual differences. Climatically, the N–N Domain is unusual in its moderate winter temperatures and predictable rainfall. The mild climate has significant consequences for plant growth, and this has resulted in communities dominated by functionally similar dwarf succulent shrubs. Clearly, these features are unique compared to the North American deserts which, although less diverse in species, are more diverse structurally. The model that we have presented in this chapter attempts to explain some of the features unusual to the N–N Domain, although there are still many gaps in our understanding of this system. We see these as potentially interesting areas for future research. Succulent species in the Mojave are known to produce roots rapidly soon after rain ('rain-roots', see Nobel and Sanderson, 1984). Do species in the N–N Domain have similar adaptations? Do most species in the N–N Domain have low optimal growth temperatures to exploit the mild winter growing conditions? To what degree do N–N Domain species exhibit CAM-flexible metabolism? Are N–N seedbanks stochastic? To what extent have short generation times influenced speciation patterns in the N–N Domain of the succulent karoo? We see all of these questions as an exciting challenge to current and future researchers in the karoo.

20 The karoo: past and future

S. J. Milton and W. R. J. Dean

20.1. Current perceptions

20.1.1. Patterns

The succulent karoo is unique, with high levels of endemism and species-richness in succulent and geophytic plants. It is also extraordinarily species-rich in certain groups of invertebrates and reptiles, although not in mammals and birds (Vernon, this volume). Species-richness in succulent karoo Mesembryanthema is currently attributed to the dependable winter rainfall in a hot arid climate (Cowling and Hilton-Taylor, this volume; Esler et al., this volume). The Nama-karoo, on the other hand, may be seen as a vast ecotonal area in which the affiliation of the fauna and flora with a particular biome appears to be determined as much by local edaphic factors as it does by geographical position. Within the Nama-karoo, the structural heterogeneity of the vegetation is correlated with climatic variability (Cowling et al., 1994).

Rainfall seasonality appears to have been a powerful force in the evolution of karoo plants. Dependable, but light, winter rainfall in combination with hot arid summers have favoured CAM photosynthesis, leaf-stored water and shallow roots – doubtless accounting for the dominance of leaf-succulent Mesembryanthema in the succulent karoo (Esler et al., this volume). Highly variable and non-seasonal rainfall in the Nama-karoo enables a greater variety of structural forms to coexist in this comparatively poor vegetation. The relative abundance of deep- and shallow-rooted plants, ephemerals and perennials, succulent shrubs and grasses is spatially and temporally variable, being mediated by edaphic factors and the distribution of moisture in the landscape.

The folding of the Cape mountains along the inner edge of the succulent karoo, the subsequent isolation of patches of habitat and the high species-richness of plants

in this biome is thought to be one of the main factors influencing the high species-richness of arachnids, hopliniid beetles, aculeate hymenoptera, reptiles and tortoises (Vernon, this volume). The high barriers separating the succulent karoo from the Nama-karoo have, perhaps, discouraged movement between the two biomes even by some of the most mobile of organisms. In general, the succulent karoo tends to have resident animals, whereas the fauna of the Nama-karoo is characterized by numerous species that are irruptive and nomadic (Milton et al., this volume).

Zoogenic patterns in vegetation at all scales are a feature of the both the Succulent and Nama-karoos. The most remarkable of these are the *heuweltjies* ('little hills') apparently generated by the activities of the small harvester termite *Microhodotermes viator*. The presence of unoccupied *heuweltjies* in some areas leads some ecologists to argue that this relatively modern termite merely reuses existing termitaria generated thousands of years ago by isoptera now extinct. *Heuweltjies* influence local soil processes, decomposition rates, drainage, disturbance regimes, vegetation structure, composition and palatability, and invertebrate communities. The palatable grazing, loose soils and insect prey on and around *heuweltjies* attract large herbivores, burrowing rodents, tortoises, aardvark and burrow-dwelling carnivores to these hotspots in the landscape.

On a smaller scale, the vegetation is patterned by ants that accumulate nutrients and increase local rates of disturbance, and on a large scale by patchy grazing. The question of which factors underlie grazing patterns of wild and domestic herbivores remains controversial in karoo areas, but there is evidence to suggest that edaphic factors, including water distribution patterns, salt accumulation and positive feedback from previous intensive land use by people and animals maintain grazing patches in the karoo. Whereas some feel that these are the first stages of land degradation, others contend that grazing lawns have always been a feature of the karoo and their expansion will be limited by the very interactions of physical and biotic factors that lead to their formation (Novellie, this volume).

Surprisingly few perennial karoo plants maintain durable, soil-stored seedbanks, probably because the plants are long-lived or because seeds of the dominant taxon (Mesembryanthema) are retained in capsules on the plants (Esler, this volume). In contrast with the adjacent fire-prone fynbos where conflagration, indicating reduced competition and mineralization, releases canopy-stored seeds, capsules of many karoo plant genera are hygroscopic, shedding thin-walled seeds only onto moist soil (Esler, this volume). Seed release synchronized with rain occurs in some Acanthaceae, Asteraceae and Scrophulariaceae as well as in almost all genera of the Mesembryanthma. Subsequent short-distance, splash dispersal of seeds by raindrops leads to patchy distributions and is a possible mechanism for both pattern generation and speciation in succulent karoo vegetation (Cowling and Hilton-Taylor, this volume).

Adaptations for dispersing seeds over time are largely confined to the annual and ephemeral plants so prevalent in the succulent karoo and on dunes, sandy plains and drainage lines in the arid savanna and Nama-karoo (Van Rooyen, this volume). It may be argued that the regular winter rains in this region favour annual plants. On the other hand, the very shallow soils that occur over large parts of the karoo may reduce benefits associated with dormant seeds by providing inadequate protection from predation and temperature extremes. Short-lived plants that specialize in disturbed soils occur throughout the karoo region, and many are bet-hedgers, producing polymorphic seeds that have anatomical and morphological differences that prevent clumped dispersal and synchronized germination.

Movements of birds appear to be related to patches of resources (both food and nest material) and to the timing and spatial distribution of rainfall. Widespread rains may lead to the dispersal of certain species, so that at times the birds may occur at low densities over a wide area. Large localized rainfall events result in temporary aggregations of seed-eating and raptorial birds (Dean and Milton, this volume; Milton et al., this volume). Artificial resource concentrations such as water points, croplands and farmstead gardens have influenced both movement patterns and population sizes in small mammals and birds (Vernon, this volume; Milton et al., this volume; Hoffman et al., this volume). Similarly, major highways through the karoo *sensu lato* function as corridors for birds dependent upon the weedy plants that flourish on moist, disturbed road verges (W. R. J. Dean, unpublished data).

Fluctuations in plant and animal population sizes can, in some cases, be related to major rainfall events, or to short-term shifts in rainfall seasonality (Milton et al., this volume; Palmer et al., this volume). For example, increases in the abundance of grasses relative to shrubs, are clearly linked to periods of above-average summer rainfall in the Nama-karoo, and rodent populations may closely follow. On the other hand, sudden increases in the abundance of locusts and lepidoptera larvae cannot be explained by rainfall patterns alone, probably because of the influence of predators and parasites on their demography.

Past accounts by travellers indicated that large herds of ungulates at times moved through the karoo region (Skead, 1980, 1987). It is not clear whether their numbers fluctuated with resource availability or were contained within limits by predators. Populations of small resident vertebrates such as lagomorphs, hyraxes, rodents and birds, however, are likely to have been regulated by raptorial birds and mammalian predators (Milton et al., this volume). The drastic reduction in these predators by the settled agricultural community (Hoffman et al., this volume) is thought to explain reported changes in the abundance of rock hyraxes and caracal (lynx), and the distribution of bat-eared foxes and vultures (Macdonald, 1992; Boshoff and Vernon, 1980; Milton et al., this volume).

Past and present use of karoo landscapes by people are closely linked with availability of perennial water. The implication is that until relatively recently permanent settlements were confined to the courses of major rivers or to the proximity of endorheic lakes. All other forms of land use by people depended on mobility and a knowledge of where to find water, plants and game animals at various seasons. For this reason, indigenous karoo peoples did not practise cultivation or aggregate in villages (Smith, this volume). Once borehole technology guaranteed perennial water for people and livestock throughout the karoo, land use patterns changed (Hoffman et al. this volume), and the notion of land ownership replaced the philosophy that people belonged to the land (Troskie, 1996). Nevertheless, despite the ubiquitous availability of drinking water, the general aridity, variable rainfall and shallow aridozols place limits on the numbers of herbivorous animals and people that the karoo can support on a continuous basis.

20.1.2.   Processes

Above-ground net primary production is not tightly coupled to the quantity of rain that falls because it is also affected by rainfall season, storm characteristics, patterns of runoff and cycles in nutrient mobilization (Whitford, this volume). Pastoralists sometimes comment that the response of karoo rangelands to rain is better after a prolonged drought than in years that receive many rainfall events. The mechanism thought to explain this pattern is that nutrients become immobilized either in plant bio-mass or in micro-organisms during bursts of productivity and are thus not available to plants during repeated rains. Post-drought nutrient pulses may also explain why locust outbreaks do not occur whenever weather conditions are suitable. Annual and ephemeral plants being fast growing and productive are likely to rapidly deplete available nutrients, the release of which will depend on the rates of mineralization of the litter. Grazing management that leads to conversion of perennial plant communities to assemblages of short-lived plants has therefore implica-tions for the dynamics of nutrient availability.

Mesembryanthema, the dominant taxon of the succu-lent karoo are generally small-seeded and establish on bare ground rather than beneath other plants. Many func-tion as pioneers on disturbed and bare soils. As yet, none of the Mesembryanthema have been reported to be associ-ated with mycorrhizal fungi, and Esler (this volume) pro-poses that the capacity to establish without fungal sym-bionts gives these plants an advantage in sparsely vegetated areas. Many Mesembryanthema make use of crassulacean acid metabolic pathways during dry periods, and their leaves tend to contain acids, including citric, malic and oxalic acid (Von Willert et al., 1992). High con-centrations of calcium oxalate crystals occur in the leaves and stems of some Mesembryanthema (Milton et al., 1994a). The oxalates leached from *Salsola tragus* (syn., *S. kali*), an introduced ephemeral weed in the USA (Cannon et al., 1995), have been shown to release soil phosphorus bound to calcium, aluminium and iron, thereby making it locally available to plants. It may consequently be hypoth-esized that organic acids in some Mesembryanthema increase the availability of phosphorus to wind-dispersed shrubs that tend to establish among their decumbent stems, and in this way facilitate the development of 'fer-tile islands'.

Termites are the major mineralizers in deserts (Whitford, this volume). They generally reduce the soil organic matter. However, Midgley and Van der Heyden (this volume) report that *Microhodotermes viator* is an excep-tion to this rule because the frass piles overlying the nests contain more carbon and organic matter than surround-ing soils. This observation supports the notion that these termites have a keystone role in processing and patterning karoo soils and biota (Bond, 1992).

It has been hypothesized that grazing-induced changes to the soil surface, including salinization, accelerated runoff and loss of cryptobiotic soil crusts, lead to a more arid environment and xerophytic vegetation (Roux and Opperman, 1986; Stokes, 1994). This hypothesis could be used to explain the failure of some overgrazed rangelands to recover after prolonged rests (Palmer et al., this vol-ume). Alternative hypotheses evoke competition from established plants, rarity of suitable rainfall and tempera-ture conditions, and scarcity of seeds or establishment sites to explain the failure of certain plant species to re-establish on degraded rangeland (Milton et al., 1994b).

In a region where water resources are finite and fossil waters are already being tapped, there has been surpris-ingly little documentation of below-ground water tables and aquifers. Reports that tree deaths in the Kuiseb and Nossob rivers are a response to water extraction for agri-cultural and industrial use remain controversial (Ward and Breen, 1983; Van Rooyen et al., 1994b). Apart from agriculture, it is likely that the extensive development of alien *Prosopis* spp. in the central karoo are reducing water-table levels. Other land use practises likely to influence ground water include the extraction of saline water to pro-duce salt, and the removal of water from mines.

20.2.   Uncertainties and contradictions

The cyclicity in animal populations in the karoo is not really understood. Is it driven by isolated events, or by quasi-cyclic fluctuations in pressure systems over the southern oceans (Mason, 1990; Tennant, 1996), or by sunspot activity (Alexander, 1995)? The role of competi-tion, disease and predation in populations of many karoo animals is too poorly known to be able to test alternative hypotheses. Clearly, weather patterns influence processes such as mineralization, plant recruitment and reproduc-tion in animals, but so does grazing management. For this reason, the debate about whether vegetation change is dri-ven by weather patterns, by land use patterns or by inter-actions between the two rages on. The most controversial issue surrounding land use in the karoo concerns the browser/grazer ratio, stocking density, resting period and grazing period most likely to maintain vegetation produc-tivity, diversity and resilience over the long term.

Currently, much controversy also surrounds the con-cept of resilience in karoo rangelands. Some argue that loss of certain perennial plant species from communities reduces the resilience of the community following drought and lowers the capacity of the vegetation to sus-tain animal production. Others suggest that increased stocking rates favour resilient plants, thereby making the vegetation better able to recover rapidly after drought and disturbance. The transformation of perennial to annual plant communities arguably reduces the temporal stabil-ity of the forage flow, leading to increased yields in wet periods and reduced yields during droughts. Nevertheless, Hoffman et al. (this volume) present data to show that the numbers of domestic livestock supported annually by ephemeral plant communities on 'degraded' rangelands has no tendency to decrease over time. Thus, although ephemeral communities are clearly unstable in the short term (months), they appear to be resilient and stable over the long term (decades).

Schlesinger et al. (1990) proposed that an increase in the heterogeneity in nutrient distribution in an arid American savanna (*Prosopis*-grass mixture) signalled desertification, in that nutrients that were formerly fairly evenly distributed across grassy plains became concentrated beneath shrub clumps once grasses were removed by grazers. We suggest that this model may be applicable to some, but not all, types of karoo shrublands. In some karoo shrublands plants are characteristically arranged in multi-species clumps on fertile islands (Hutchinson, 1946, Beukman, 1991). This pattern appears to dissolve under intense and prolonged grazing, and it may be hypothesized that nutrients from fertile islands are either dispersed or lost in runoff water, as is the case when the spatially structured Chenopod shrublands of arid Australia are damaged by grazing (Tongway, 1994).

Cues for the nomadic movements of indigenous animals in the karoo are poorly known. Historical accounts suggest that the movements of massed springbok herds were unpredictable and aseasonal, and were precipitated by rapid population growth following favourable periods. Current hypotheses are that springbok movements probably coincided with poor summer rangeland conditions in the Nama-karoo caused by large population numbers and subsequent overgrazing, or that they were caused by prolonged droughts (Skinner, 1993; Lovegrove, this volume). The question is now academic; there have been no large-scale movements of springbok in the karoo since the early 1900s.

In the case of the brown locust *Locustana pardalina*, it has been suggested that the switch from the resident '*solitaria*' phase to the nomadic '*gregaria*' phase is stimulated by density increases that make the insects susceptible to predators, but when aggregations of *gregaria* locusts become very large, selection by parasites begins to favour the *solitaria* phase (Lea, 1968). An alternative hypothesis is that the nutritional status of the plants or high stress levels associated with crowding cause the switch from one phase to another. As yet, there is inadequate information to reject any of these hypotheses.

20.3. Directions for future research

While some progress has been made towards addressing many of the research questions posed by Cowling (1986) in his initial description of the Karoo Biome Project, many more questions remain unanswered. The consequences of interactions between plants, micro-organisms, large mammals and soils for productivity and diversity is a research field that has barely been touched on in the karoo. Yet this knowledge should form the basis for future conceptual models of management effects on karoo ecosystems. We still do not know how to manage the karoo, and we cannot provide guidelines for solving some of the commonest management problems. We are no closer to achieving a predictive understanding of the karoo ecosystem than we were 10 years ago. Manipulative management of plant and animal communities, karoo rangelands and other resources without any clear ideas of the outcome is likely to change plant and animal communities through species losses and replacement. This, in turn, may be reflected in reduced carrying capacity for domestic livestock, the inability of the vegetation to serve as habitat for indigenous animals, a reduction in scientific value and a loss of aesthetic appeal. These issues, and the paucity of protected areas in the karoo *sensu lato* have been extensively addressed elsewhere (Hilton-Taylor and Le Roux, 1989; Macdonald, 1989; Siegfried, 1989; Dean, 1995).

Climatic events isolated in time and space characterize the Nama-karoo, and it is difficult to devise programmes to research the responses of the biota to such events. The only long-term studies that have been set up in the karoo during the last 50 years, i.e. those that have been set up by the Department of Agriculture and various universities supported by the Foundation for Research Development, have recently been terminated or scaled down. This is a result of reallocation of public funds for supporting students or trained personnel to carry out this type of research. Commitments, in terms of funding and manpower that are needed to document responses of plant and animal communities to weather patterns, explain the scarcity of long-term research worldwide. Such research programmes have particularly low priorities in such developing countries as South Africa, where there are more urgent demands upon public funding. Modelling, using available data collected over the short-term, may be an alternative to drawing conclusions directly from a few isolated events that occur during a long-term study. Nevertheless, modelling requires substantial funding because it needs highly trained personnel, sophisticated equipment and a good empirical database.

Cowling (1986) showed that past research projects in the karoo *sensu lato* were dominated by ecosystem descriptions and autecological studies of indigenous mammals, birds and domestic livestock. Since then, many new ideas have been developed in the previously neglected disciplines of historical biogeography, Quaternary vegetation change, plant population and community dynamics, and functional aspects of plant and animal ecophysiology. Fields that show little advance in knowledge since

Cowling's (1986) review are the relationships between the soil and living organisms, including mycorrhizae, the relationship between physiology and behaviour in plants and insects, the vertical distribution of productivity and biomass, including responses of roots to nutrients and rainfall, biomass, function and diversity of underground invertebrates, mineralization processes, and practical guidelines for rehabilitation of areas damaged by mining, ploughing and grazing.

The Southern African Bird Atlas Project (SABAP) (Harrison, 1989, 1992) together with the perceived need to rationalize and justify the selection of protected areas have stimulated research into avian taxonomy, habitat discrimination, patterns of movements of various bird species and factors influencing flocking behaviour and breeding, particularly among the endemic granivorous birds. The pressures to justify conservation through consumptive utilization of game animals, birds and medicinal and firewood plants have stimulated both empirical and modelling investigations of the tolerance and resilience of selected species under a range of climatic scenarios. However, there is an urgent need for more research in order to provide guidelines for legislation to protect natural resources.

Major succulent karoo plant taxa are still undergoing rapid speciation, and, under a scenario of changing environments and anthropogenic influences, these offer researchers a golden opportunity to observe the selective forces, the very processes that have shaped the diversity of this unique region. For the imaginative researcher, with broad enough vision to link applied and academic research, opportunities for the future are almost unlimited.

References

Abbott, D. H., Barrett, J., Faulkes, C. G. and George, L. M. (1989). Social contraception in naked mole-rats and marmoset monkeys. *Journal of Zoology, London*, 219, 703–10.

Abel, N. (1993). Carrying capacity, rangeland degradation and livestock development policy for the communal rangelands of Botswana. *Pastoral Development Network*, 35c, 1–9. London: Overseas Development Institute.

Abel, N. O. J. and Blaikie, P. (1989). Land degradation, stocking rates and conservation policies for the communal rangelands of Botswana and Zimbabwe. *Land Degradation and Rehabilitation*, 1, 101–23.

Abramsky, Z. (1983). Experiments on seed predation by rodents and ants in the Israeli desert. *Oecologia*, 57, 328–32.

Ackerman, T. L., Romney, E. M., Wallace, A. and Kinnear J. E. (1980). Phenology of desert shrubs in southern Nye County, Nevada. *Great Basin Naturalist Memoirs*, 4, 4–23.

Acocks, J. P. H. (1953). Veld Types of South Africa. *Memoirs of the Botanical Survey of South Africa*, 28, 1–192.

Acocks, J. P. H. (1955). Agriculture in relation to a changing vegetation. *South African Journal of Science*, 52, 101–8.

Acocks, J. P. H. (1966). Non-selective grazing as a means of veld reclamation. *Proceedings of the Grassland Society of Southern Africa*, 1, 33–9.

Acocks, J. P. H. (1976). Riverine vegetation of the semi-arid and arid regions of South Africa. *Journal of the South African Botanical Society*, 17, 21–35.

Acocks, J. P. H. (1988). Veld Types of South Africa. 3rd edn. *Memoirs of the Botanical Survey of South Africa*, 57.

Adams, K. M. and Tainton, N. M. (1990). The function of the hygroscopic awn of *Themeda triandra*. *Journal of the Grassland Society of South Africa*, 7, 271–3.

Adamson, R. S. (1931). Notes on some petrified wood from Banke, Namaqualand. *Transactions of the Royal Society of South Africa*, 19, 255–8.

Adamson, R. S. (1938). *The vegetation of South Africa*. London: British Empire Vegetation Committee.

Ahearn, G. A. (1970). The control of water loss in desert tenebrionid beetles. *Journal of Experimental Biology*, 53, 573–95.

Alexander, W. J. R. (1995). Floods, droughts and climate change. *South African Journal of Science*, 91, 403–8.

Allan, D. (1990). More on the poison people: an indictment of pharmacists. *African Wildlife*, 44, 129–43. *South African Journal of Wildlife Research*, 19, 102–6.

Allan, D. G. (1994). The abundance and movements of Ludwig's Bustard *Neotis ludwigii*. *Ostrich*, 65, 95–105.

Allan, D. G. (1995). Habitat selection by blue cranes in the Western Cape Province and the Karoo. *South African Journal of Wildlife Research*, 25, 90–7.

Allen, E. B. (1982a). Germination and competition of *Salsola kali* with native C_3 and C_4 species under three temperature regimes. *Bulletin of the Torrey Botanical Club*, 109, 39–46.

Allen, E. B. (1982b). Water and nutrient competition between *Salsola kali* and two native grass species (*Agropyron smithii* and *Bouteloua gracilis*). *Ecology*, 63, 732–41.

Allen, E. B. (1995). Restoration ecology: limits and possibilities in arid and semiarid lands. In *Proceedings: Wildland Shrub and Arid Land Restoration Symposium*, comps. B. A. Roundy, E. D. McArthur, J. S. Haley, D. K. Mann, pp. 7–15. Ogden, Utah: US Department of Agriculture, Forest Service.

Allen-Diaz, B., Chapin, F. S., Diaz, S., Howden, M., Puigdefábregas, J. and Stafford Smith, M. (1995). Rangelands in a changing climate. In *Climate change 1995 – Impacts, adaptations and mitigation of climate change: scientific-technical analyses*, ed. R. T. Watson, M. C. Zinyowera, R. H. Moss and D. J. Dokken, pp. 131–58. Intergovernmental Panel on Climate Change, Cambridge University Press.

Allsopp, N. (1994). Restoration of Kalahari dune vegetation: a mycorrhizal perspective. *Bulletin of the Grassland Society of Southern Africa*, 5, 39.

Allsopp, N. and Stock, W. D. (1993). Mycorrhizal status of plants growing in the Cape Floristic Region, South Africa. *Bothalia*, 23, 91–104.

Allsopp, N. and Stock, W. D. (1994). VA mycorrhizal infection in relation to edaphic characteristics and disturbance regime in three lowland plant communities in the south-western Cape, South Africa. *Journal of Ecology*, 82, 271–9.

Allsopp, N., Fortuin, S., Netto, J. and Roos, P. (1994). Fertile islands in savanna ecosystems – nutrients and mycorrhizas. *Grassland Society of Southern Africa 1996 Congress Abstracts*, 96.

Andersson, M. (1984). The evolution of eusociality. *Annual Review of Ecology and Systematics*, 15, 165–89.

Andreoli, M. A. G., Smith, C. B., Watkeys, M. K., Hart, R. J., Brynard, H. J., Moore, J. M. and Ashwal, L. D. (1994). The Steenskampskraal Monazite Mine, Namaqualand Metamorphic Complex, S. Africa: a Rare Earth mineralization related to anorthosite and charnockite. *Economic Geology*, 89, 994–1016.

Annecke, D. P. and Moran, V. C. (1977). Critical reviews of biological pest control in South Africa. 1. The Karoo caterpillar, *Loxostege frustalis* Zeller (Lepidoptera: Pyralidae). *Journal of the Entomological Society of Southern Africa*, 40, 127–45.

Annecke, D. P. and Moran, V. C. (1978). Critical reviews of biological pest control in South Africa. 2. The prickly pear, *Opuntia ficus-indica* (L) Miller. *Journal of the Entomological Society of Southern Africa*, 41, 161–88.

Anon (1877). Report of the commission appointed to inquire into and report upon diseases in cattle and sheep. Cape Town: Government Printer. (G. 3–'77).

Anon (1923). Final report of the drought investigation commission. Cape Town: Government Printer. (U.G. 49–'23).

Anon (1951). Report of the desert encroachment committee. Cape Town: Government Printer. (U.G. 59/51).

Anon (1984). Conservation of Agricultural Resources Act, 1983 (Act 43 of 1983). *Government Gazette*, 227(9238), 1–56.

Anon (1986). Agricultural development programme: Karoo region. Unpublished report. Pretoria: Department of Agriculture and Water Supply.

Anon (1991a). Agricultural development programme: Northwest subregion. Unpublished report. Pretoria: Department of Agricultural Development.

Anon (1991b). Agricultural development programme: Karoo region. Unpublished report. Pretoria:

Department of Agricultural Development.

Anon (1996). Our land. Green Paper on South African Land Policy. Pretoria: Department of Land Affairs.

Applin, D. G., Cloudsley-Thompson, J. L. and Constantinou, C. (1993). Molecular and physiological mechanisms in chronobiology – their manifestations in the desert ecosystem. *Journal of Arid Environments*, 13, 187–97.

Archer, F. M. (1994). Ethnobotany of Namaqualand: the Richtersveld. MA thesis, University of Cape Town.

Archer, F., Turner, S. and Venter, F. (1996). Range management, livestock production and nature conservation: the Richtersveld National Park. In *Successful natural resource management in southern Africa*, ed. W. Critchley and S. Turner, pp. 165–95. Centre for Development Cooperation Services, Vrije Universiteit, Amsterdam.

Armesto, J. J. and Vidiella, P. E. (1993). Plant life-forms and biogeographic relations in the flora of Lagunillas (30 °S) in the fog-free Pacific coastal desert. *Annals of the Missouri Botanical Garden*, 80, 499–511.

Armitage, K. B. (1987). Do female yellow-bellied marmots adjust the sex ratios of their offspring? *American Naturalist*, 129, 501–19.

Armstrong, A. J. and Siegfried, W. R. (1990). Selective use of *heuweltjie* earth mounds by sheep in the Karoo. *South African Journal of Ecology*, 1, 77–80.

Arnold, G. (1915–1926). A monograph of the Formicidae of South Africa. *Annals of the South African Museum* 14, 1–766.

Aronson, J. and Shmida, A. (1992). Plant species diversity along a Mediterranean –desert gradient and its correlations with interannual rainfall fluctuations. *Journal of Arid Environments*, 23, 235–47.

Aronson, J., Kigel, J., Shmida, A. and Klein, J. (1992). Adaptive phenology of desert and Mediterranean populations of annual plants grown with and without water stress. *Oecologia*, 89, 17–26.

Ascaray, C. M. (1986). An ecological study of the hairy-footed gerbil, *Gerbillurus paeba*, in an eastern Cape dunefield. M. Sc. thesis, Zoology, University of Port Elizabeth.

Aschoff, J. (1964). Survival value of diurnal rhythms. *Symposia of the Zoological Society, London*, 13, 79–98.

Aschoff, J. (1984). Circadian timing. *Annals of the New York Academy of Sciences*, 423, 442–68.

Aucamp, A. and Danckwerts, J. E. (1989). Grazing management: a strategy for the future. Introduction. Pretoria: Department of Agriculture and Water Supply.

Auerbach, M. and Shmida, A. (1985). Harmony among endemic littoral plants and adjacent floras in Israel. *Journal of Biogeography*, 12, 175–87.

Austin, M. P., Cunningham, R. B. and Good, R. B. (1983). Altitudinal distribution of several eucalypt species in relation to other environmental factors in southern New South Wales. *Australian Journal of Ecology*, 8, 169–80.

Avery, D. M. (1988). The Holocene environment of central South Africa: micromammalian evidence. *Palaeoecology of Africa*, 19, 335–45.

Avery, D. M. (1990a). Holocene climatic change in southern Africa: the contribution of micromammals to its study. *South African Journal of Science*, 86, 407–12.

Avery, D. M. (1990b). Late Quaternary environmental change in southern Africa based on micromammalian evidence: a synopsis. *Palaeoecology of Africa*, 21, 131–42.

Avery, D. M. (1991). Micromammals, owls and vegetation change in the Eastern Cape Midlands, South Africa, during the last millenium. *Journal of Arid Environments* 20, 357–69.

Avery, D. M. (1992). Man and/or climate? Environmental degradation and micromammalian community structure in South Africa during the last millenium. *South African Journal of Science* 88, 483–9.

Avery, D. M. (1993). Last Interglacial and Holocene altithermal environments in South Africa and Namibia: micromammalian evidence. *Palaeogeography, Palaeoclimatology, Palaeoecology*, 101, 221–8.

Axelrod, D. I. (1972). Edaphic aridity as a factor in angiosperm evolution. *American Naturalist*, 106, 311–20.

Axelrod, D. I. (1979). Age and origin of Sonoran Desert vegetation. *Occasional Papers of the California Academy of Sciences*, 132.

Axelrod, D. I. and Raven, P. H. (1978). Late Cretaceous and Tertiary vegetation history of Africa. In *Biogeography and ecology of southern Africa*, ed. M. J. A. Werger, pp. 77–130. The Hague: W. Junk.

Baas, W. J. (1989). Secondary plant compounds, their ecological significance and consequences for the carbon budget. In *Causes and consequences of variation in growth rate and productivity of higher plants*, ed. H. Lambers, M. L. Cambridge, H. Konings and T. L. Pons, pp. 313–40. The Hague: SPB Academic Publishing.

Badenhorst, N. (1995). Proceedings of the national research and development workshop on the assessment and monitoring of desertification in South Africa, 10–12 October 1995. Pretoria: Institute for Soil, Climate and Water, Agricultural Research Council.

Balinsky, B. I. (1962). Patterns of animal distribution on the African continent. *Annals of the Cape Provincial Museums 2*, 299–310.

Bamford, M. K. and De Wit, M. C. J. (1993). Taxonomic description of fossil wood from Cainozoic Sak River terraces, near Brandvlei, Bushmanland, South Africa. *Palaeontologia Africana*, 30, 71–80.

Barbour, M. G. and Diaz, D. V. (1973). *Larrea* plant communities on bajada moisture gradients in the United States and Argentina. *Vegetatio*, 28, 335–52.

Barnard, S. A. (1987). Influence of sowing depth on the germination of Karoo shrubs and grasses. *Journal of the Grassland Society of South Africa*, 4, 123–6.

Barnes, R. D., Filer, D. L. and Milton, S. J. (1996). *Acacia karroo* monograph and annotated bibliography. *Tropical Papers*, 32. Oxford: Oxford Forestry Institute.

Barrow, J. (1801). *An account of travels into the interior of southern Africa in the years 1797 and 1798*. London: Cadell and Davies.

Bartholomew, G. A., Lighton, J. R. B. and Louw, G. N. (1985). Energetics of locomotion and patterns of respiration in tenebrionid beetles from the Namib Desert. *Journal of Comparative Physiology B*, 155, 155–62.

Bartholomew, G. A., White, F. N. and Howell, T. R. (1976). The thermal significance of the nest of the Sociable Weaver *Philetairus socius*: summer observations. *Ibis*, 118, 402–10.

Bath, G. F. (1978). The river midge plague. *Karoo Agric* 1, 37–9.

Bauer, A. M., Russell, A. P. and Edgar, B. D. (1989). Utilization of the termite *Hodotermes mossambicus* (Hagen) by gekkonid lizards near Keetmanshoop, South West Africa. *South African Journal of Zoology*, 24, 239–43.

Baxter, A. J. (1996). Late Quaternary palaeoenvironments of the sandveld, Western Cape Province. Ph. D. thesis, University of Cape Town.

Baxter, A. J. and Davies, B. R. (1994). Palaeoecological insights for the conservation of aquatic ecosystems in dryland environments: a case study of the Verlorenvlei system, South Africa. *Journal of Aquatic Conservation: Marine and Freshwater Ecosystems*, 4, 255–71.

Baxter, A. J. and Meadows, M. E. (1994). Palynological evidence for the impact of colonial settlement within lowland fynbos. *Historical Biology*, 9, 61–70.

Bayer, M. B. (1978). Pollination in Asclepiads. *Veld & Flora*, 64, 21–3.

Beatley, J. C. (1967). Survival of annuals in the northern Mojave Desert. *Ecology*, 48, 745–50.

Beatley, J. C. (1974). Phenological events and their environmental triggers in Mojave Desert ecosystems. *Ecology*, 55, 856–63.

Beatley, J. C. (1976). *Vascular plants of the Nevada Test Site and central and southern Nevada: Ecologic and geographic distributions*. Washington, DC: Energy Research and Development Administration.

Beaumont, P. B. (1990a). Pniel 1. In *Guide to archaeological sites in the northern Cape*, ed. P. B. Beaumont and D. Morris, pp. 7–9. Cape Town: Southern African Association of Archaeologists.

Beaumont, P. B. (1990b). Wonderwerk Cave. In *Guide to archaeological sites in the northern Cape*, ed. P. B. Beaumont and D. Morris, pp. 101–34. Cape Town: Southern African Association of Archaeologists.

Beaumont, P. B. (1990c). Kathu Pan. In *Guide to archaeological sites in the northern Cape*, ed. P. B. Beaumont and D. Morris, pp. 75–100. Cape Town: Southern African Association of Archaeologists.

Beaumont, P. B., Smith, A. B. and Vogel, J. C. (1995). Before the Einiqua: the archaeology of the frontier zone. In *Einiqualand: studies of the Orange River frontier*, ed. A. B. Smith, pp. 236–64. Cape Town: University of Cape Town Press.

Beaumont, P. B., van Zinderen Bakker, E. M. and Vogel, J. C. (1984). Environmental changes since 32,000 BP at Kathu Pan, Northern Cape. In *Late Cainozoic Palaeoclimates of the southern hemisphere*, ed. J. C. Vogel, pp. 329–38. Rotterdam: A. A. Balkema.

Behnke, R. H. and Scoones, I. (1993). Rethinking range ecology: Implications for rangeland management in Africa. In *Range ecology at disequilibrium: New models of natural variability and pastoral adaptation in African savannas*, ed. R. H. Behnke, I. Scoones and C. Kerven, pp. 1–30. London: Overseas Development Institute.

Behnke, R. H., Scoones, I. and Kerven, C. (1993). *Range ecology at disequilibrium. New models of natural variability and pastoral adaptation in African savannas*. London: Overseas Development Institute.

Belk, D. and Cole, G. A. (1975). Adaptational biology of desert temporary pond inhabitants. In *Environmental physiology of desert organisms*, ed. N. F. Hadley, pp. 207–66. Stroudsburg, PA: Dowden, Hutchinson and Ross, Inc.

Bell, K. L., Hiatt, H. D. and Niles, W. E. (1979). Seasonal changes in biomass allocation in eight winter annuals of the Mojave Desert. *Journal of Ecology*, 67, 781–8.

Bell, R. H. V. (1982a). Decision-making in wildlife management with reference to problems of overpopulation. In *Management of large mammals in African conservation areas*, ed. R. N. Owen-Smith, pp. 145–72. Pretoria: Haum.

Bell, R. H. V. (1982b). The effect of soil nutrient availability on community structure in African ecosystems. In *The ecology of tropical savannas*, ed. B. J. Huntley and B. H. Walker, pp. 193–216. Berlin: Springer-Verlag.

Belnap, J. and Gardner, J. S. (1993). Soil microstructure in soils of the Colorado Plateau: the role of the cyanobacterium *Microcoleus vaginatus*. *Great Basin Naturalist*, 53, 40–7.

Beneke, K., Van Rooyen, M. W. and Theron, G. K. (1992a). Fruit polymorphism in ephemeral species of Namaqualand. 5. Intramorphic competition among plants cultivated from dimorphic diaspores. *South African Journal of Botany*, 58, 461–8.

Beneke, K., Van Rooyen, M. W. and Theron, G. K. (1992b). Fruit polymorphism in ephemeral species of Namaqualand. 6. Intermorphic competition among plants cultivated from dimorphic diaspores. *South African Journal of Botany*, 58, 469–77.

Beneke, K., Van Rooyen, M. W. and Theron, G. K. (1993a). Fruit polymorphism in ephemeral species of Namaqualand. 4. Growth analysis of plants cultivated from the dimorphic diaspores. *Journal of Arid Environments*, 24, 345–60.

Beneke, K., Van Rooyen, M. W. and Theron, G. K. (1993b). Fruit polymorphism in ephemeral species of Namaqualand. 7. Diaspore production in plants cultivated from dimorphic diaspores. *Journal of Arid Environments*, 25, 233–47.

Beneke, K., Van Rooyen, M. W., Theron, G. K. and Van de Venter, H. A. (1993c). Fruit polymorphism in ephemeral species of Namaqualand. 3. Germination differences between the polymorphic diaspores. *Journal of Arid Environments*, 24, 333–44.

Beneke, K., Von Teichman, I., Van Rooyen, M. W. and Theron, G. K. (1992a). Fruit polymorphism in ephemeral species of Namaqualand. 1. Anatomical differences between polymorphic diaspores of two *Dimorphotheca* species. *South African Journal of Botany*, 58, 448–55.

Beneke, K., Von Teichman, I., Van Rooyen, M. W. and Theron, G. K. (1992b). Fruit polymorphism in ephemeral species of Namaqualand. 2. Anatomical differences between polymorphic diaspores of *Arctotis fastuosa* and *Ursinia cakilefolia*. *South African Journal of Botany*, 58, 456–60.

Bennett, N. C. (1990). Behaviour and social organization in a colony of the Damaraland mole-rat *Cryptomys damarensis*. *Journal of Zoology, London*, 220, 225–48.

Bennett, N. C. and Jarvis, J. U. M. (1988). The social structure and reproductive biology of colonies of the mole-rat, *Cryptomys damarensis* (Rodentia, Bathyergidae). *Journal of Mammalogy*, 69, 293–302.

Bennett, N. C. and Spinks, A. C. (1995). Thermoregulation and metabolism in the Cape golden mole (Insectivora:

Chrysochloris asiatica). *Journal of Zoology, London*, 236, 521–9.

Benton, M. J. (1990). *Vertebrate Palaeontology*. London: Harper Collins.

Berger, R. J. (1993). Cooling down to hibernate: Sleep and hibernation constitute a physiological continuum of energy conservation. *Neuroscience Letters*, 154, 213–16.

Berliner, R., Mitchell, D. T. and Allsopp, N. (1989). The vesicular-arbuscular mycorrhizal infectivity of sandy soils in the south-western Cape, South Africa. *South African Journal of Botany*, 55, 310–13.

Bernard, R. T. F. and Stuart, C. T. (1992). Correlates of diet and reproduction in the black-backed jackal. *South African Journal of Science*, 88, 292–4.

Bernard, R. T. F., Kerley, G. I. H., Doubell, T. and Davison, A. (1996). Aseasonal reproduction in a seasonal environment, the round-eared elephant shrew in the southern Karoo, South Africa. *Journal of Zoology (London)*, 240, 233–430.

Berry, H. H. (1980). Behavioural and ecophysiological studies on blue wildebeest (*Connochaetes taurinus*) at Etosha National Park. Ph.D. thesis, University of Cape Town.

Berry, H. H. and Siegfried, W. R. (1991). Mosaic-like events in arid and semi-arid Namibia. In *The mosaic-cycle concept of ecosystems*, ed. H. Remmert, pp. 147–59. Berlin: Springer-Verlag.

Bertram, B. C. R. (1980). Vigilance and group size in Ostriches. *Animal Behavior*, 28, 278–86.

Bertram, B. C. R. (1992). *The Ostrich Communal Nesting System*. Princeton, NJ: Princeton University Press.

Beukes, N. J. (1977). Transition from siliciclastic to carbonate sedimentation near the base of the Transvaal Supergroup at Bothithong in the northern Cape Province, South Africa. *Sedimentary Geology*, 18, 201–22.

Beukes, P. C., Cowling, R. M. and Ellis, F. (1994). Vegetation and soil changes across a succulent Karoo grazing gradient. *Abstracts of the Arid Zone Ecology Forum*, Beaufort West, September 1994.

Beukman, R. P. (1991). The role of nurse plants in the vegetation dynamics of the succulent karoo. M.Sc. thesis. University of Cape Town.

Bews, J. W. (1916). An account of the chief types of vegetation in southern Africa,

with notes on plant succession. *Journal of Ecology*, 4, 129–59.

Bittrich, V. (1987). Untersuchungen zu Merkmalsbestand, Gliederung und Abgrenzung der Unterfamilie Mesembryanthemoideae (Mesembryanthemaceae Fenzl). *Mitteilungen aus dem Institut für Allgemeine Botanik Hamburg*, 21, 5–116.

Bittrich, V. and Hartmann, H. E. K. (1988). The Aizoaceae – a new approach. *Botanical Journal of the Linnean Society*, 97, 239–54.

Bleek, W. H. I. and Lloyd, L. C. (1911). *Specimens of Bushman folklore*. London: George Allen.

Blume, H-P., Vahrson, W-G., and Meshref, H. (1985). Dynamics of water, temperature, and salts in typical aridic soils after artifical rainstorms. *Catena*, 12, 343–62.

Blumenschine, R. J. and Cavallo, J. A. (1992). Scavenging and human evolution. *Scientific American*, 267, 70–6.

Bock, C. E., Bock, J. H., Jepson, K. L. and Ortega, J. C. (1986). Ecological effects of planting Lovegrasses in Arizona. *National Geographic Research*, 2, 456–63.

Bock, C. E., Bock, J. H., Kenney, W. R. and Hawthorne, V. M. (1984). Responses of birds, rodents, and vegetation to livestock exclosure in a semi-desert grassland site. *Journal of Range Management*, 37, 239–42.

Bollong, C. A. and Sampson, C. G. (1996). Later Stone Age ceramic stratigraphy and direct dates on pottery: a comparison. *Southern African Field Archaeology*, 5(1), 3–16.

Bolus, H. (1875). Letter from Mr. Bolus to Dr J.B. Hooker. *Journal of the Linnean Society*, 14, 482–4.

Bond, W. J. (1988). Proteas as 'tumble-seeds': wind dispersal through air and over soil. *South African Journal of Botany*, 54, 455–60.

Bond, W. J. (1992). Keystone species. In *Biodiversity and ecosystem function*, ed. E.-D. Schulze and H. A. L. Mooney, pp. 237–53. Berlin: Springer-Verlag.

Bond, W. J. (1996). Towards a protocol for setting sustainable harvesting limits of wild plant populations. South African Association of Botanists 22nd Annual Congress Abstracts. Department of Botany, University of Stellenbosch.

Bond, W. J. and Richardson, D. M. (1990). What can we learn from extinctions and invasions about the effects of climatic change? *South African Journal of Science*, 86, 429–33.

Bond, W. J. and Slingsby, P. (1983). Seed dispersal by ants in shrublands of the Cape Province and its evolutionary implications. *South African Journal of Science*, 79, 231–3.

Bond, W. J. and Van Wilgen, B. W. (1995). *Fire and plants*. London: Chapman and Hall.

Bond, W. J., Stock, W. D. and Hoffman, M. T. (1994). Has the Karoo spread? A test for desertification using carbon isotopes from soils. *South African Journal of Science*, 90, 391–7.

Boobyer, M. G. (1989). The eco-ethology of the Karoo Korhaan *Eupodotis vigorsii*. M.Sc. thesis, University of Cape Town.

Boobyer, M. G. and Hockey, P. A. R. (1994). Dietary opportunism in the Karoo Korhaan: consequence of a sedentary lifestyle in an unpredictable environment. *Ostrich* 65, 32–8.

Boonzaier, E. (1987). From communal grazing to 'economic' units: Changing access to land in a Namaqualand reserve. *Development Southern Africa*, 4, 479–91.

Boonzaier, E. A., Hoffman, M. T., Archer, F. M. and Smith, A. B. (1990). Communal land use and the 'Tragedy of the Commons': Some problems and development perspectives with specific reference to semi-arid regions of southern Africa. *Journal of the Grassland Society of Southern Africa*, 7, 77–80.

Boonzaier, E. A., Malherbe, C., Berens, P. and Smith, A. B. (1996). *The Cape herders. A history of the Khoikhoi of southern Africa*. Cape Town: David Philip.

Bosch, O. J. H. (1988). Vegetation deterioration in southern Africa: A research and researcher problem. *Journal of the Grassland Society of Southern Africa*, 5, 3–4.

Bosch, O. J. H. (1989). Degradation of the semi-arid Grasslands of southern Africa. *Journal of Arid Environments*, 16, 143–7.

Bosch, O. J. H. and Booysen, J. (1992). An integrative approach to rangeland condition and capability assessment. *Journal of Range Management*, 45, 116–22.

Bosch, O. J. H. and Gauch, H. G. (1991). The use of degradation gradients for the assessment and ecological interpretation of range condition. *Proceedings of the Grassland Society of Southern Africa*, 8, 138–45.

Bosch, O. J. H. and Kellner, K. (1991). The use of a degradation gradient for the ecological interpretation of condition assessments in the western grassland biome of southern Africa. *Journal of Arid Environments*, 21, 21–9.

Bosch, O. J. H., Gauch, H. G., Booysen, J., Gouws, G. A., Nel, M. E., Stols, S. H. E. and Van Zyl, E. (1992). *Users guide: Integrated system for plant dynamics (ISPD)*. Potchefstroom University for CHE, Potchefstroom.

Boshoff, A. F. (1986). Stability of a population of Martial eagles in a sheep farming area in the Great Karoo, South Africa, project outline and early results. *Gabar*, 1, 9–13.

Boshoff, A. F. and De Kock, A. C. (1988). Further evidence of organo-chlorine contamination in Cape Vultures. *Ostrich*, 59, 40–1.

Boshoff, A. F. and Robertson, A. S. (1985). A conservation plan for the Cape vulture colony at Potberg, De Hoop Nature Reserve, southwestern Cape Province. *Bontebok*, 4, 25–31.

Boshoff, A. F. and Vernon, C. J. (1980). The past and present distribution and status of the Cape Vulture in the Cape Province. *Ostrich* 51, 230–50.

Boshoff, A. F., Palmer, N. G. and Avery, G. (1990). Regional variation in the diet of martial eagles in the Cape Province, South Africa. *South African Journal of Wildlife Research*, 20, 57–68.

Boshoff, A. F., Palmer, N. G., Avery, G., Davies, R. A. G. and Jarvis, M. J. F. (1991). Biogeographical and topographical variation in the prey of the Black Eagle in the Cape Province, South Africa. *Ostrich*, 62, 59–72.

Boshoff, C. (1987). Report on the preliminary investigation into phytomass consumed and plant species selected for by hoppers of the brown locust (*Locustana pardalina*) in the Kimberley district. Unpublished report, Botany Dept, University of Cape Town.

Botha, B. J. V., Grobler, N. J., Linström, W. and Smit, C. A. (1976). Stratigraphic correlation between the Kheis and Matsap Formations and their relation to the Namaqualand Metamorphic Complex. *Transactions of the Geological Society of South Africa*, 79, 304–11.

Botha, C. G. (1926). *Place names in the Cape Province*. Cape Town: Juta.

Botha, D. H. (1967). The viability of brown locust eggs, *Locustana pardalina* (Walker). *South African Journal of Science*, 10, 445–60.

Botha, D. H., Ross, W. F., Van Ark, H. and Pick, F. E. (1974). Residues in sheep exposed to BHC treated Karoo veld in the outbreak region of the brown locust (*Locustana pardalina* Walker). *Phytophylactica*, 6, 235–48.

Bouaichi, A., Roessingh, P. and Simpson, S. J. (1995). An analysis of the behavioural effects of crowding and re-isolation on solitary-reared adult desert locusts (*Schistocerca gregaria*) and their offspring. *Physiological Entomology*, 20, 199–208.

Boucher, C. and Le Roux, A. (1993). Dry coastal ecosystems of the South African west coast. In *Ecosystems of the World*, 2B, *Dry Coastal Ecosystems. Africa, America, Asia and Oceania*, ed. E. van der Maarel, pp. 75–88. Amsterdam: Elsevier.

Boucot, A. J., Johnson, J. G. and Talent, J. A. (1969). Early Devonian brachiopod zoogeography. *Geological Society of America Special Paper*, 199, 1–113.

Bousman, B. and Scott, L. (1994). Climate or overgrazing?: The palynological evidence for vegetation change in the eastern Karoo. *South African Journal of Science*, 90, 575–8.

Bousman, B., Partridge, T. C., Scott, L., Metcalf, S. C., Vogel, J. C., Seaman, M. and Brink, J. S. (1988). Palaeoenvironmental implications of late Pleistocene and Holocene valley fills in Blydefontein basin, Noupoort, CP, South Africa. *Palaeoecology of Africa*, 19, 43–68.

Bowers, M. A. (1987). Precipitation and the relative abundances of desert winter annuals: a 6-year study in the northern Mojave desert. *Journal of Arid Environments*, 12, 141–9.

Boycott, R. C. and Bourquin, O. (1988). *The South African tortoise book*. Johannesburg: Southern Books.

Boyer, H. J. (1988). Breeding biology of the Dune Lark. *Ostrich*, 59, 30–7.

Bozinovic, F. and Marquet, P. A. (1991). Energetics and torpor in the Atacama desert-dwelling rodent *Phyllotis darwini rupestris*. *Journal of Mammalogy*, 72, 734–8.

Bozinovic, F. and Rosenmann, M. (1988).

Daily torpor in *Calomys musculinus*, a South American rodent. *Journal of Mammalogy*, 69, 150–2.

Bradley, W. G., Miller, J. S. and Yousef, M. K. (1974). Thermoregulatory patterns in pocket gophers: desert and mountain. *Physiological Zoology*, 47, 172–9.

Brain, C. (1988). Water gathering by baboons in the Namib desert. *South African Journal of Science*, 84, 590–1.

Branch, W. R. (Bill) (1988a). *Field guide to the snakes and other reptiles of southern Africa*. Cape Town: Struik.

Branch, W. R. (ed.) (1988b). South African Red Data Book – reptiles and amphibians. *South African National Scientific Programmes Report*, 151.

Branch, W. R., Benn, G. A. and Lombard, A. T. (1995). The tortoises (Testudinidae) and terrapins (Pelomedusidae) of southern Africa, their diversity, distribution and conservation. *South African Journal of Zoology*. 30, 91–102.

Brand, D. J. (1992). Probleemdiere. *Arid Zone Ecology Forum Abstracts*, p. 18. Foundation for Research Development, Pretoria.

Brand, D. J. (1995). Baboon capture and factors influencing capture success in the Western Cape Province, South Africa. *South African Journal of Wildlife Research*, 25, 141–3.

Brand, D. J., Fairall, N. and Scott, W. M. (1995). The influence of the regular removal of black-backed jackals on the efficiency of coyote getters. *South African Journal of Wildlife Research*, 25, 44–8.

Brandon, R. N. (1990). *Adaptation and environment*. Princeton, New Jersey: Princeton University Press.

Branford, W. C. (1877). Report of the Colonial Veterinary Surgeon on sheep and cattle diseases in the colony of the Cape of Good Hope. Cape Town: Government Printer. (G. 8–'77).

Branford, W. C. (1879). Report of Professor Branford, Colonial Veterinary Surgeon for the years 1878–'79. Cape Town: Government Printer. (G. 54–'79).

Breytenbach, G. J. (1988). Why are myrmecochorous plants limited to fynbos (Macchia) vegetation types? *South African Forestry Journal*, 144, 3–5.

Brigham, R. M. (1992). Daily torpor in a free-ranging goatsucker, the Common Poorwill (*Phalaenoptilus nuttallii*).

Physiological Zoology, 65, 457–72.

Brink, J. S. (1988). The taphonomy and palaeoecology of the Florisbad spring fauna. *Palaeoecology of Africa*, 19, 169–79.

Broekhuysen, G. J. (1963). The Thick-billed Lark (*Calendula magnirostris*) feeding on bulbs. *Ostrich*, 34, 251.

Brooke, R. K. (1995). Alien *Tamarix ramosissima* alert. *Veld and Flora*, 81, 26.

Brooke, R. K. and Dean, W. R. J. (1990). On the biology and taxonomic position of *Drymoica substriata* Smith, the so-called Namaqua Prinia. *Ostrich* 61, 50–5.

Brooke, R. K. and Vernon, C. J. (1988). Historical records of the Wattled Crane *Bugeranus carunculatus* (Gmelin) in the Cape Province and the Orange Free State, South Africa. *Annals of the Cape Provincial Museums (Natural History)*, 16, 363–71.

Brown, C. J. and Gubb, A. A. (1986). Invasive alien organisms in the Namib Desert, Upper Karoo and the arid and semi-arid savannas of western southern Africa. In *The ecology and management of biological invasions in southern Africa*, ed. I. A. W. Macdonald, F. J. Kruger and A. A. Ferrar, pp. 93–108. Cape Town: Oxford University Press.

Brown, E. and Willan, K. (1991). Microhabitat selection and use by the bush Karoo rat *Otomys unisulcatus* in the Eastern Cape Province. *South African Journal of Wildlife Research*, 21, 69–75.

Brown, H. D. (1989). A remarkable new genus of Acrididae from the upper Karoo, South Africa, with North American affinities (Orthoptera: Acridoidea). *Annals of the Transvaal Museum* 35(4), 67–74.

Brown, J. H. and Heske, E. J. (1990). Control of a desert–grassland transition by a keystone rodent guild. *Science*, 250, 1705–7.

Brown, J. H., Reichman, O. J. and Davidson, D. W. (1979). Granivory in desert ecosystems. *Annual Review of Ecology and Systematics*, 10, 201–227.

Brundrett, M. (1991). Mycorrhizas in natural ecosystems. *Advances in Ecological Research*, 21, 171–313.

Bruton, M. N. (1986). Life history styles of invasive fishes in southern Africa. In *The ecology and management of biological invasions in southern Africa*, ed. I. A. W. Macdonald, F. J. Kruger and A. A. Ferrar, pp. 201–8. Cape Town: Oxford

University Press.

Brutsch, M. O. and Zimmermann, H. G. (1993). The prickly pear (*Opuntia ficus-indica* (Cactaceae)) in South Africa, utilization of the naturalized weed and of the cultivated plant. *Economic Botany*, 47, 154–62.

Bruyns, P. and Linder, H. P. (1991). A revision of *Microloma* R. Br. (Asclepidaceae–Asclepiadeae). *Botanische Jahrbücher für Systematik*, 112, 453–527.

Bryant, J. P., Chapin, F. S. and Klein, D. R. (1983). Carbon/nutrient balance of boreal plants in relation to vertebrate herbivory. *Oikos*, 40, 357–68.

Bryant, J. P., Tuomi, J. and Niemala, P. (1988). Environmental constraints of constitutive and long-term inducible defenses in woody plants. In *Chemical mediation of coevolution*, ed. K. Spencer, pp. 367–89. New York: American Institute of Biological Sciences.

Bucher, T. L. and Chappell, M. A. (1992). Ventilatory and metabolic dynamics during entry into and arousal from torpor in *Selasphorus* hummingbirds. *Physiological Zoology*, 65, 978–93.

Budack, K. F. R. (1983). A harvesting people on the south Atlantic coast. *South African Journal of Ethnology* 6(2), 1–7.

Buffenstein, R. (1984). The importance of microhabitat in thermoregulation and thermal conductance in two Namib rodents – a crevice dweller, *Aethomys namaquensis*, and a burrow dweller, *Gerbillurus paeba*. *Journal of Thermal Biology*, 9, 235–41.

Buffenstein, R., Cambell, W. E. and Jarvis, J. U. M. (1985). Identification of crystalline allantoin in the urine of African Cricetidae (Rodentia) and its role in their water economy. *Journal of Comparative Physiology B*, 155, 493–9.

Buffington, L. C. and Herbel, C. H. (1965). Vegetational changes on a semidesert grassland range from 1858 to 1963. *Ecological Monographs*, 35, 139–64.

Bullock, P., Le Houérou, H., Hoffman, M. T., Rounsevell, M., Sehgal, J. and Várallyay, G. (1995). Land degradation and desertification. In *Climate change 1995 – Impacts, adaptations and mitigation of climate change: scientific-technical analyses*, ed. R. T. Watson, M. C. Zinyowera, R. H. Moss and D. J. Dokken, pp. 171–89. Intergovernmental Panel on Climate Change. Cambridge

University Press.

Buol, S. W. (1965). Present soil forming factors and processes in arid and semi arid regions. *Soil Science*, 99, 45–9.

Burger, M. and Branch, W.R. (1994). Tortoise mortality caused by electrified fences in the Thomas Baines Nature Reserve. *South African Journal of Wildlife Research*, 24, 32–7.

Burns, M. (1994). *ALEXKOR Environmental Management Programme Report*, vol. 3. *Report Number* EMAS-C 94. Stellenbosch: EMATEK.

Burrage, B. R. (1973). Comparative ecology and behaviour of *Chamaeleo pumilus pumilus* (Gmelin) and *C. namaquensis* A. Smith (Sauria: Chamaeleonidae). *Annals of the South African Museum*, 61, 1–158.

Butler, J. H. A. and Buckerfield, J. C. (1979). Digestion of lignin by termites. *Soil Biology and Biochemistry*, 11, 507–13.

Butzer, K. W., Fock, G. L., Stuckenrath, R. and Zilch, A. (1973). Paleohydrology of late Pleistocene lake Alexandersfontein, Kimberley, South Africa. *Nature*, 243, 328–30.

Bykov, B. A. (1974). Fluctuations in the desert and semi-desert vegetation of the Turanian Plain. In *Handbook of Vegetation Science*, vol. 8, *Vegetation dynamics*, ed. R. Knapp, pp. 243–51. The Hague: W. Junk.

Cade, T. J. and Maclean, G. L. (1967). Transport of water by adult sandgrouse to their young. *Condor*, 69, 323–43.

Cade, T. J., Willoughby, E. J. and Maclean, G. L. (1986). Drinking behaviour of sandgrouse in the Namib and Kalahari Deserts, Africa. *Auk*, 83, 124–6.

Cain, S. A. (1950). Life-forms and phytoclimate. *Botanical Review*, 16, 1–32.

Cairncross, B., Anderson, J. M. and Anderson, H. M. (1995). Palaeoecology of the Triassic Molteno formation, Karoo basin, South Africa: sedimentological and palaeoentological evidence. *South African Journal of Geology*, 98, 452–78.

Cannon, J. P., Allen, E. B., Allen, M. F., Dudley, L. M. and Jurinak, J. J. (1995). The effects of oxalates produced by *Salsola tragus* on the phosphorous nutrition of *Stipa pulchra*. *Oecologia*, 102, 265–72.

Cannon, W. A. (1911). The root habits of desert plants. *Carnegie Institution of Washington, Publication*, 131.

Washington, DC: Gibson Bros.

Cannon, W. A. (1924). General and physiological features on the vegetation of the more arid portions of South Africa with notes on the climatic environment. *Yearbook of the Carnegie Institution of Washington*, 8 (354), 159.

Cannon, W. A. (1949). A tentative classification of root systems. *Ecology*, 30, 542–8.

Carroll, R. L. (1977). The origin of lizards. In *Problems in Vertebrate Evolution*, ed. S. M. Andrews, R. S. Miles and A. D. Walker, pp. 359–96. London: Academic Press.

Carstens, W. P. (1966). *The social structure of a Cape Coloured reserve*. Cape Town: Oxford University Press.

Caughley, G. and Gunn, A. (1993). Dynamics of large herbivores in deserts: kangaroos and caribou. *Oikos*, 67, 47–55.

Caviiedes-Vidal, E., Codelia, E. C., Roig, V. and Dona, R. (1990). Facultative torpor in the South American rodent *Calomys venustus* (Rodentia: Cricetidae). *Journal of Mammalogy*, 71, 72–5.

Cepeda-Pizarro, J. G. and Whitford, W. G. (1989). Species abundance distribution patterns of microarthropods in surface decomposing leaf-litter and mineral soil on a desert watershed. *Pedobiologia*, 33, 254–68.

Chaloner, W. G., Forey, P. L., Gardiner, B. G., Hill, A. J. and Young, V. T. (1980). Devonian fish and plants from the Bokkeveld Series of South Africa. *Annals of the South African Museum*, 81, 127–57.

Chaney, K and Swift, R. S. (1986). Studies on aggregate stability. I. Re-formation of soil aggregates. *Journal of Soil Science*, 37, 329–35.

Charig, A. J. and Crompton, A. W. (1974). The alleged synonymy of *Lycorhinus* and *Heterodontosaurus*. *Annals of the South African Museum*, 64, 167–89.

Charley, J. L. and Cowling, S. L. (1968). Changes in soil nutrient status resulting from overgrazing and their consequences in plant communities of semi-arid areas. *Proceedings of the Ecological Society of Australia*, 3, 28–38.

Charley, J. L. and West, N. E. (1975). Plant-induced soil chemical patterns in some shrub-dominated semi-desert ecosystems of Utah. *Journal of Ecology*, 63, 945–64.

Chew, R. M. and Whitford, W. G. (1992). A

long-term positive effect of kangaroo rats, (*Dipodomys spectabilis*) on creosotebushes (*Larrea tridentata*). *Journal of Arid Environments*, 22, 375–86.

Chhotani, O. B. (1988). The termites of Oman. *Journal of Oman Studies*, Special Report 3, 363–71.

Christian, D. P. (1978). Physiological correlates of demographic patterns in three sympatric Namib Desert rodents. *Physiological Zoology*, 52, 329–39.

Christian, D. P. (1979). Comparative demography of three desert rodents: responses to the provision of supplementary water. *Journal of Mammalogy*, 60, 679–90.

Christian, D. P. (1980). Patterns of recovery from low numbers in the Namib Desert rodents. *Acta Theriologica*, 25, 431–50.

Christian, D. P., Enders, J. E. and Shump, K. A. Jr. (1977). A laboratory study of caching in *Desmodillus auricularis*. *Zoologica Africana*, 12, 505–7.

Clancey, P. A. (1986). Endemicity in the Southern African avifauna. *Durban Museum Novitates*, 13, 245–54.

Clancey, P. A. (1994). Combined biogeographic role of river valleys and aridity in southern African bird distribution. *Durban Museum Novitates*, 19, 13–29.

Clark, A. B. (1979). Sex ratio and local resource competition in a prosimian primate. *Science*, 201, 163–5.

Clark, G. C. and Dickson, C. G. C. (1971). *Life histories of the South African Lycaenid butterflies*. London: Purnell.

Clarke, B. C. and Nicolson, S. W. (1994). Water, energy, and electrolyte balance in captive Namib sand-dune lizards (*Angolosaurus skoogi*). *Copeia*, 4, 962–74.

Clarkson, E. N. K. (1993). *Invertebrate palaeontology and evolution*, 3rd edn. London: Chapman and Hall.

Cloudsley-Thompson, J. L. (1979). Adaptive functions of the colours of desert animals. *Journal of Arid Environments*, 2, 95–104.

Cloudsley-Thompson, J. L. (1991). *Ecophysiology of desert arthropods and reptiles*. Berlin: Springer-Verlag.

Coaton, W.G.H. (1948). *Trinervitermes* species – the Snouted Harvester Termites. *Department of Agriculture Bulletin*, 261, 1–19.

Coaton, W. G. H. (1958). The hodotermitid harvester termites of South Africa.

Department of Agriculture Science Bulletin, 375, *Entomology Series*, 43, 1–112.

Coaton, W. G. H. (1963). Survey of the termites (Isoptera) of the Kalahari thornveld and shrub bushveld of R.S.A. *Koedoe*, 6, 38–50.

Coaton, W. G. H. and Sheasby, J. L. (1973a). National survey of the Isoptera of southern Africa. 1. The genus *Baucaliotermes* Sands (Termitidae: Nasutitermitinae). *Cimbebasia*, **Ser. (A)** 3, 1–7.

Coaton, W. G. H. and Sheasby, J. L. (1973b). National survey of the Isoptera of southern Africa. 2. The genus *Schedorhinotermes* Silvestri (Rhinotermitidae). *Cimbebasia*, **Ser. (A)** 3, 9–17.

Coaton, W. G. H. and Sheasby, J. L. (1973). National survey of the Isoptera of southern Africa. 3. The genus *Psammotermes* Desneux (Rhinotermitidae). *Cimbebasia*, **Ser. (A)** 3, 19–28.

Coaton, W. G. H. and Sheasby, J. L. (1973d). National survey of the Isoptera of southern Africa. 4. The genus *Fulleritermes* Coaton (Termitidae: Nasutitermitinae). *Cimbebasia*, **Ser. (A)** 3, 29–38.

Coaton, W. G. H. and Sheasby, J. L. (1974a). National survey of the Isoptera of southern Africa. 5. The genus *Rhadinotermes* Sands (Termitidae: Nasutitermitinae). *Cimbebasia*, **Ser. (A)** 3, 39–46

Coaton, W. G. H. and Sheasby, J. L. (1974b). National survey of the Isoptera of southern Africa. 6. The genus *Microhodotermes* Sjöstedt (Hodotermitidae). *Cimbebasia*, **Ser. (A)** 3, 47–59.

Coaton, W. G. H. and Sheasby, J. L. (1974c). National survey of the Isoptera of southern Africa. 7. The genus *Apicotermes* Holmgren (Termitidae: Apicotermitinae). *Cimbebasia*, **Ser. (A)** 3, 61–72.

Coaton, W. G. H. and Sheasby, J. L. (1975a). National survey of the Isoptera of southern Africa. 9. The genus *Ancistrotermes* Silvestri (Termitidae: Macrotermitinae) *Cimbebasia*, **Ser. (A)** 3, 95–104.

Coaton, W. G. H. and Sheasby, J. L. (1975b). National survey of the Isoptera of southern Africa. 10. The genus *Hodotermes* Hagen (Hodotermitidae) *Cimbebasia*, **Ser. (A)** 3, 105–38.

Coaton, W. G. H. and Sheasby, J. L. (1976a). National survey of the Isoptera of southern Africa. 11. The genus *Coptotermes* Wasman (Rhinotermitidae: Coptotermitinae) *Cimbebasia*, **Ser. (A)** 3, 139–72.

Coaton, W. G. H. and Sheasby, J. L. (1976b). National survey of the Isoptera of southern Africa. 12. The genus *Porotermes* Hagen (Termopsidae: Porotermitinae) *Cimbebasia*, **Ser. (A)** 3, 173–82.

Coaton, W. G. H. and Sheasby, J. L. (1977). National survey of the Isoptera of southern Africa. 13. The genus *Pseaudacanthotermes* Sjöstedt (Termitidae: Macrotermitinae) *Cimbebasia*, **Ser. (A)** 3, 183–206.

Coaton, W. G. H. and Sheasby, J. L. (1978). National survey of the Isoptera of southern Africa. 14. The genus *Stolotermes* Hagen (Termopsidae: Stolotermitinae) *Cimbebasia*, **Ser. (A)** 3, 207–13.

Cody, M. L. (1986a). Diversity, rarity and conservation in Mediterranean-climate regions. In *Conservation biology. The science of scarcity and diversity*, ed. M. E. Soulé, pp. 122–52. Sunderland, MA: Sinauer Associates.

Cody, M. L. (1986b). Structural niches in plant communities. In *Community ecology*, ed. J. Diamond and T. J. Case, pp. 381–405. New York: Harper and Row.

Cody, M. L. (1989). Growth form diversity and community structure in desert plants. *Journal of Arid Environments*, 17, 199–209.

Cody, M. L. (1991). Niche theory and plant growth form. *Vegetatio* 97, 39–55.

Cody, M. L. and Mooney, H. A. (1978). Convergence versus nonconvergence in mediterranean-climate ecosystems. *Annual Review of Ecology and Systematics*, 9, 265–321.

Coe, M. J. and Skinner, J. D. (1992). Connections, disjunctions and endemism in the eastern and southern African mammal faunas. *Transactions of the Royal Society of Southern Africa*, 48, 233–55.

Coetzee, J. A. (1967). Pollen analytical studies in East and Southern Africa. *Palaeoecology of Africa*, 3, 1–146.

Coetzer, W. R. and Hoffmann, J. H. (1997).

Establishment of *Neltumius arizonensis* (Coleoptera: Bruchidae) on mesquite (*Prosopis* species: Mimosaceae) in South Africa. *Biological Control*, 10, 187–92.

Cohen, D. (1966). Optimising reproduction in a randomly varying environment. *Journal of Theoretical Biology*, 12, 119–29.

Cohen, D. (1968). A general model of reproduction in variable environments. *Journal of Ecology*, 56, 219–28.

Cole, D. T. (1988). *Lithops – Flowering Stones*. Johannesburg: Acorn Books.

Cole, M. M. (1980). Geobotanical expression of ore bodies. *Transactions of the Institution of Mining and Metallurgy*, B89, 73–91.

Cole, M. M. and Le Roex, H. D. (1978). The role of geobotany, biochemistry and geochemistry in mineral exploration in South West Africa and Botswana – a case history. *Transactions of the Geological Society of South Africa*, 81, 277–317.

Coley, P. D., Bryant, J. P. and Chapin, F. S., III. (1985). Resource availability and plant antiherbivore defense. *Science*, 230, 895–9.

Conroy, A. M. and Gaigher, I. G. (1982). Venison, aquaculture and ostrich meat production: Action 2003. *South African Journal of Animal Science*, 12, 219–33.

Cooper, R. L. and Skinner, J. D. (1979). Importance of termites in the diet of the aardwolf *Proteles cristatus* in southern Africa. *South African Journal of Zoology*, 14, 5–8.

Cottrell, C. B. (1985). The absence of coevolutionary associations with Capensis floral element plants in the larval/plant relationships of Southwestern Cape butterflies. In *Species and speciation*, ed. E. S. Vrba. *Transvaal Museum Monograph*, 4, 115–24.

Coughenour, M. B., McNaughton, S. J. and Wallace, L. L. (1985). Shoot growth and morphometric analyses of Serengeti graminoids. *African Journal of Ecology*, 23, 179–94.

Coutchie, P. A. and Crowe, J. H. (1979). Transport of water vapor by tenebrionid beetles. I. Kinetics. *Physiological Zoology*, 52, 67–87.

Cowling, R. M. (1985). The formulation of hypotheses on Quaternary vegetation history: general approach and an example from the south-eastern Cape. *Palaeoecology of Africa*, 17, 155–72.

Cowling, R. M. (1986). A description of the

Karoo Biome Project. *South African National Scientific Programmes Report*, 122, 1–43.

Cowling, R. M. and Campbell, B. M. (1983). A comparison of fynbos and non-fynbos coenoclines in the lower Gamtoos river valley, south eastern Cape, South Africa. *Vegetatio*, 53, 161–78.

Cowling, R. M. and Hilton-Taylor, C. (1994). Patterns of plant diversity and endemism in southern Africa: an overview. In *Botanical Diversity in Southern Africa*, ed. B. J. Huntley. *Strelitzia*, 1, 31–52.

Cowling, R. M. and Holmes, P. M. (1991). Subtropical thicket in the southeastern Cape: a biogeographical perspective. In *Proceedings of the First Valley Bushveld/Subtropical Thicket Symposium*, ed. P. J. K. Zacharias, G. C. Stuart-Hill and J. J. Midgley, pp. 3–4. Howick, KwaZulu-Natal: Grassland Society of Southern Africa (Special Publication).

Cowling, R. M. and Holmes, P. M. (1992). Flora and vegetation. In *The ecology of fynbos. Nutrients, fire and diversity*, ed. R. M. Cowling, pp. 23–61. Cape Town: Oxford University Press.

Cowling, R. M. and McDonald, D. J. (1997). Local endemism and plant conservation in the Cape Floristic Region. In *Landscape degradation in mediterranean-type ecosystems*, ed. P. W. Rundel, G. Montenegro and F. Jaksic, in press. Heidelberg: Springer-Verlag.

Cowling, R. M. and Witkowski, E. (1994). Convergence and non-convergence of plant traits in climatically and edaphically matched sites in mediterranean Australia and South Africa. *Australian Journal of Ecology*, 19, 220–32.

Cowling, R. M., Esler, K. J. and Rundel, P. W. (in press). Namaqualand, South Africa – a unique winter-rainfall desert ecosystem. *Plant Ecology*.

Cowling, R. M., Esler, K. J., Midgley, G. F. and Honig, M. A. (1994). Plant functional diversity, species diversity and climate in arid and semi-arid southern Africa. *Journal of Arid Environments*, 27, 141–58.

Cowling, R. M., Gibbs Russell, G. E., Hoffman, M. T. and Hilton-Taylor, C. (1989). Patterns of plant species diversity in southern Africa. In *Biotic diversity in southern Africa. Concepts and conservation*, ed. B. J. Huntley, pp. 19–50. Cape Town: Oxford University Press.

Cowling, R. M., Holmes, P. M. and Rebelo, A. G. (1992). Plant diversity and endemism. In *The ecology of fynbos. Nutrients, fire and diversity*, ed. R. M. Cowling, pp. 62–112. Cape Town: Oxford University Press.

Cowling, R. M., Richardson, D. M., Schulze, R. E., Hoffman, M. T., Midgley, J. J. and Hilton-Taylor, C. (1997). Species diversity at the regional scale. In *Vegetation of southern Africa*, ed. R. M. Cowling, D. M. Richardson and S. M. Pierce, pp. 447–73. Cambridge University Press.

Cowling, R. M., Roux, P. W. and Pieterse, A. J. H. (eds.) (1986). *The Karoo biome: a preliminary synthesis. Part 1 – physical environment. South African National Scientific Programmes Report*, 124.

Cowling, R. M., Rundel, P. W. and Desmet, P. G. (1998). Regional-scale plant diversity in southern African arid lands: subcontinetal and global comparisons. *Diversity and Distributions*, 4, 27–36.

Cowling, R. M., Rundel, P. W., Lamont, B. B., Arroyo, M. K. and Arianoutsou, M. (1996). Plant diversity in Mediterranean-climate regions. *Trends in Ecology and Evolution*, 11, 362–6.

Cramer, F. H., Rust, I. C. and Diez de Cramer, M. de C. R. (1974). Upper Ordovician chitinozoans from the Cedarberg Formation of South Africa. Preliminary note. *Geologische Runschau*, 64, 340–5.

Cronk, Q. C. B. and Fuller, J. L. (1995). *Plant invaders: the threat to natural ecosystems*. London: Chapman and Hall.

Cronwright-Schreiner, S. C. (1925). *The migratory springbucks of South Africa (the trekbokke). Also an essay on the ostrich and a letter descriptive of the Zambezi falls*. London: Hamilton, Adams.

Crowe, T. M. (1978). Limitation of population in the helmeted guineafowl. *South African Journal of Wildlife Research*, 8, 117–26.

Csurhes, S. M. and Kriticos, D. (1994). *Gleditsia triacanthos* L. (Caesalpinaceae), another thorny, exotic fodder tree gone wild. *Plant Protection Quarterly*, 9, 101–5.

Cunliffe, R. N., Jarman, M. L., Moll, E. J. and Yeaton, R. I. (1990). Competitive interactions between the perennial shrub *Leipoldtia constricta* and an annual forb, *Gorteria diffusa*. *South African Journal of Botany*, 56, 34–8.

Currie, D. J. (1991). Energy and large-scale patterns of animal- and plant-species richness. *American Naturalist*, 137, 27–49.

Curtis, B. (1991). Notes on the behaviour and potential food plants of the snail *Trigonephrus haughtoni* Connolly (Mollusca: Dorcasiidae), in the southern Namib desert. *Cimbebasia*, 13, 99–104.

Curtis, B. A. (1985a). Observations on the natural history and behaviour of the dune ant, *Camponotus detritus* Emery, in the central Namib Desert. *Madoqua*, 14, 279–89.

Curtis, B. A. (1985b). Activity of the Namib Desert dune ant, *Camponotus detritus*. *South African Journal of Zoology*, 20, 41–8.

Curtis, B. A. (1985c). Temperature tolerances in the Namib Desert dune ant, *Camponotus detritus*. *Journal of Insect Physiology*, 31, 463–66.

Curtis, B. A. (1985d). The dietary spectrum of the Namib Desert dune ant *Camponotus detritus*. *Insectes Sociaux*, 32, 78–85.

Curtis, B. A. (1988). Do ant-mimicking *Cosmophasis* spiders prey on their *Camponotus* models? *Cimbebasia*, 10, 67–70.

Curtis, B. A. and Seely, M. K. (1987). Effect of an environmental gradient upon the distribution and abundance of the dune ant *Camponotus detritus*, in the central Namib Desert. *Journal of Arid Environments*, 13, 259–66.

D'Antonio, C. M. (1990). Seed production and dispersal in the non-native, invasive succulent *Carpobrotus edulis* L. (Aizoaceae), in coastal strand communities of central California. *Journal of Applied Ecology*, 27, 693–702.

D'Antonio, C. M. and Mahall, B. E. (1991). Root profiles and competition between the invasive, exotic perennial, *Carpobrotus edulis*, and two native shrub species in California coastal scrub. *American Journal of Botany*, 78, 885–94.

D'Antonio, C. M. and Vitousek, P. M. (1992). Biological invasions by exotic grasses, the grass/fire cycle, and global change. *Annual Reviews of Ecology and Systematics*, 23, 63–87.

D'Antonio, C. M., Odion, D. C. and Tyler, C. M. (1993). Invasion of maritime chaparral by the introduced succulent *Carpobrotus edulus*. *Oecologia*, 95, 14–21.

Daan, S. and Aschoff, J. (1982). Circadian

contributions to survival. In *Vertebrate circadian systems: structure and physiology*, ed. J. Aschoff, S. Daan and G. A. Groos, pp. 305–21. Berlin: Springer-Verlag.

Dan, J. (1973). Arid-zone soils. In *Arid zone irrigation*. ed. B. Yaron, E. Danfors and Y. Vaadi, *Analysis and synthesis*, 5, pp. 61–77. Berlin: Springer-Verlag.

Danckwerts, J. C. and Stuart-Hill, G. C. (1988). The effect of severe drought and management after drought on the mortality and recovery of semi-arid grassveld. *Journal of the Grassland Society of Southern Africa*, 5, 218–22.

Danin, A. (1983). *Desert vegetation of Israel and Sinai*. Jerusalem: Cana Publishing House.

Danin, A. (1991). Plant adaptations in desert dunes. *Journal of Arid Environments*, 21, 193–212.

David, J. H. M. (1975). Fidelity to a fixed territory in some male bontebok in the Bontebok National Park, Swellendam, Cape Province. *Journal of the southern African Wildlife Management Association*, 5, 111–14.

Davidson, D. W. and Samson, D. A. (1985). Granivory in the Chihuahuan Desert: interactions within and between trophic levels. *Ecology*, 66, 486–502.

Davidson, D. W., Inouye, R. S. and Brown, J. H. (1984). Granivory in a desert ecosystem: experimental evidence for indirect facilitation of ants by rodents. *Ecology*, 65, 1780–6.

Davidson, J. L. and Milthorpe, F. L. (1966). The effect of defoliation on the carbon balance in *Dactylis glomerata*. *Annals of Botany*, 30, 185–98.

Davies, R. A. G. (1985). A comparison of Springbok *Antidorcas marsupialis* and Merino Sheep *Ovis aries* on Karoo veld. M.Sc. thesis, University of Pretoria.

Davies, R. A. G. (1994). Black Eagle *Aquila verreauxii* predation on rock hyrax *Provavia capensis* and other prey in the Karoo. Ph.D. thesis, University of Pretoria. Pretoria.

Davies, R. A. G., Botha, P. and Skinner, J. D. (1986). Diet selected by springbok *Antidorcas marsupialis* and Merino sheep *Ovis aries* during Karoo drought. *Transactions of the Royal Society of South Africa*, 46, 165–76.

Davies, S. J. J. F. (1984). Nomadism as a response to desert conditions in Australia. *Journal of Arid Environments*, 7, 183–95.

Davis, D. H. S. (1953). Plague in South Africa, a study of the epizootic cycle in gerbils (*Tatera brantsi*) in the northern Orange Free State. *The Journal of Hygiene*, 51, 427–49.

Dawson, T. J. (1989). Responses to cold of monotremes and marsupials. In *Advances in comparative and environmental physiology 4: Animal adaptation to cold*, ed. C. H. Wang, pp. 255–88. Berlin: Springer-Verlag.

Day, J. A. (1990). Environmental correlates of aquatic faunal distribution in the Namib Desert. In *Namib ecology: 25 years of Namib research*, ed. M. K. Seely. *Transvaal Museum Monograph*, 7, 99–107.

Day, T. A. and Detling, J. K. (1990). Grassland patch dynamic sand herbivore grazing preference following urine deposition. *Ecology*, 71, 180–8.

De Bie, G. and Hewitt, P. H. (1990). Thermal responses of the semi-arid zone ants *Ocymyrmex weitzeckeri* (Emery) and *Anoplolepis custodiens* (Smith). *Journal of the Entomological Society of Southern Africa*, 53, 65–74.

De Jager, L. A. and Joubert, D. M. (1968). Benutting van natuurlike weiding in die Transvaalse Bankenveld deur merinohamels. *Tydskrif vir Natuurwetenskappe*, 8, 1–26.

De Klerk, J.C. (1947). Pastures of the southern O.F.S., a century ago and today. *Farming in South Africa*, 22, 347–54.

De Soyza, A., Whitford, W. G., Martinez-Meza, E. and Van Zee, J. W. (1997). Variation in creosotebush (*Larrea tridentata*) canopy morphology in relation to habitat, soil fertility and associated annual plant communities. *American Midland Naturalist*, 137, 13–26.

De Swardt, C. (1993). Report on a survey at Spoegrivier, Leliefontein Rural Reserve, Namaqualand. Unpublished report, University of the Western Cape, Cape Town.

De Villiers, A. J. (1993). Ecophysiological studies on several Namaqualand pioneer species, with special reference to the revegetation of saline mined soil. M.Sc. thesis, University of Pretoria.

De Villiers, A. J., Van Rooyen, M. W. and Theron, G. K. (1994a). Comparison of two methods for estimating the size of the viable seed bank of two plant communities in the strandveld of the West Coast, South Africa. *South African Journal of Botany*, 60, 81–4.

De Villiers, A. J., Van Rooyen, M. W. and Theron, G. K. (1995a). Removal of sodium and chloride from a saline soil by *Mesembryanthemum barklyi*. *Journal of Arid Environments*, 29, 325–30.

De Villiers, A. J., Van Rooyen, M. W., Theron, G. K. and Van de Venter, H. A. (1994b). Germination of three Namaqualand pioneer species as influenced by salinity, temperature and light. *Seed Science and Technology*, 22, 427–33.

De Villiers, A. J., Van Rooyen, M. W., Theron, G. K. and Claassens, A. S. (1995b). The effect of leaching and irrigation on the growth of *Atriplex semibaccata*. *Land Degradation and Rehabilitation*, 6, 125–31.

De Villiers, W. M. (1988). On the plague dynamics of the brown locust, *Locustana pardalina* (Walk). In *Proceedings of the Locust Symposium*, ed. B. McKenzie and M. Longridge, pp. 41–9. *South African Institute of Ecologists Bulletin* (Special Issue).

De Wit, M. C. J. and Bamford, M. K. (1993). Fossil wood from the Brandvlei area, Bushmanland as an indication of palaeoenvironmental changes during the Cainozoic. *Palaeontologia Africana*, 30, 81–9.

De Wit, M. J., Roering, C., Hart, R. J., Armstrong, R. A., De Ronde, C. E. J., Green, R. W. E., Tredoux, M., Peberdy, E. and Hart, R. A. (1992). Formation of an Archaean continent. *Nature*, 357, 553–62.

Deacon, H. J. (1976). *Where Hunters Gathered*. South African Archaeological Society Monograph, 1.

Deacon, H. J., Deacon, J. and Brooker, M. (1976). Four painted stones from Boomplaas Cave, Oudtshoorn District. *South African Archaeological Bulletin*, 31, 141–5.

Deacon, H. J., Deacon, J., Brooker, M. and Wilson, M. L. (1978). The evidence for herding at Boomplaas Cave in the southern Cape, South Africa. *South African Archaeological Bulletin*, 33, 39–65.

Deacon, H. J., Hendey, Q. B. and Lamprechts, J. J. N. (eds.) (1983). *Fynbos palaeoecology: A preliminary synthesis*. *South African National Scientific Programmes Report*, 75.

Deacon, H. J., Jury, M. R. and Ellis, F. (1992).

Selective regime and time. In *The ecology of fynbos*, ed. R. M. Cowling, pp. 6–22. Cape Town: Oxford University Press.

Deacon, J. (1984). Later Stone Age people and their descendants in southern Africa. In *Southern African prehistory and paleoenvironments*, ed. R. G. Klein, pp. 221–328. Rotterdam: A. A. Balkema.

Deacon, J. (1986a). 'My place is the Bitterpits': the home territory of Bleek and Lloyd's /Xam informants. *African Studies*, 45(2), 135–55.

Deacon, J. (1986b). Human settlement in South Africa and archaeological evidence for alien plants and animals. In *The ecology and management of biological invasions in southern Africa*, ed. I. A. W. Macdonald, F. J. Kruger and A. A. Ferrar, pp. 3–20. Cape Town: Oxford University Press.

Deacon, J. and Lancaster, N. (1988). *Late Quaternary palaeoenvironments of southern Africa*. Oxford: Clarendon Press.

Dean, S. J., Holmes, P. M. and Weiss, P. W. (1986). Seed biology of invasive alien plants in South Africa and South West Africa/Namibia. In *The ecology and management of biological invasions in southern Africa*, ed. I. A. W. Macdonald, F. J. Kruger and A. A. Ferrar, pp. 157–70. Cape Town: Oxford University Press.

Dean, W. R. J. (1975). Dry season roadside raptor counts in the northern Cape, S.W.A. and Angola. *Journal of the Southern African Wildlife Management Association*, 5, 99–102.

Dean, W. R. J. (1988). Spider predation on termites. *Journal of the Entomological Society of Southern Africa*, 51, 147–8.

Dean, W. R. J. (1989). Foraging and forager-recruitment in *Ophthalmopone hottentota* Emery (Hymenoptera: Formicidae). *Psyche*, 96, 123–30.

Dean, W. R. J. (1991). Ecological effects of mound-building by the harvester ant *Messor capensis* on Karoo plants. M.Sc. thesis, University of Natal.

Dean, W. R. J. (1992a). Temperatures determining activity patterns of some ant species in the southern Karoo, South Africa. *Journal of the Entomological Society of Southern Africa*, 55, 149–56.

Dean, W. R. J. (1992b). Effects of animal activity on the absorption rate of soils in the southern Karoo, South Africa. *Journal of the Grassland Society of southern Africa*, 9, 178–80.

Dean, W. R. J. (1993). Unpredictable foraging behaviour in *Microhodotermes viator* (Isoptera: Hodotermitidae): an antipredator tactic? *Journal of African Zoology*, 107, 281–5.

Dean, W. R. J. (1995). Where birds are rare or fill the air: the protection of the endemic and nomadic avifaunas of the Karoo. Ph.D. thesis, University of Cape Town.

Dean, W. R. J. and Bond, W. J. (1990). Evidence for rapid faunal changes on islands in a man-made lake. I. Ants. *Oecologia*, 83, 388–91.

Dean, W. R. J. and Griffin, E. (1993). Seasonal activity patterns and habitats in Solifugae (Arachnida) in the southern Karoo, South Africa. *South African Journal of Zoology*, 28, 91–4.

Dean, W. R. J. and Hockey, P. A. R. (1989). An ecological perspective of lark (Alaudidae) distribution and diversity in the southwest-arid zone of Africa. *Ostrich*, 60, 27–34.

Dean, W. R. J. and Macdonald, I. A. W. (1994). Historical changes in stocking rates of domestic livestock as a measure of semi-arid and arid rangeland degradation in the Cape Province, South Africa. *Journal of Arid Environments*, 26, 281–98.

Dean, W. R. J. and Milton, S. J. (1991a). Patch disturbances in arid grassy dunes: antelope, rodents and annual plants. *Journal of Arid Environments*, 20, 231–7.

Dean, W. R. J. and Milton, S. J. (1991b). Disturbances in semi-arid shrubland and arid grassland in the Karoo, South Africa: mammal diggings as germination sites. *African Journal of Ecology*, 29, 11–16.

Dean, W. R. J. and Milton, S. J. (1991c). Emergence and oviposition of *Quintillia* cf. *conspersa* Karsch (Homoptera, Cicadidae) in the southern Karoo, South Africa. *Journal of the Entomological Society of Southern Africa*, 54, 111–19.

Dean, W. R. J. and Milton, S. J. (1993). The use of *Galium tomentosum* (Rubiaceae) as nest material by birds in the southern Karoo. *Ostrich*, 64, 187–9.

Dean, W. R. J. and Milton, S. J. (1994). Semi-arid rangelands in the south-western United States: used, abused and in need of rescue. *South African Journal of Science*, 90, 261–2.

Dean, W. R. J. and Milton, S. J. (1995). Plant

and invertebrate assemblages on old fields in the arid southern Karoo, South Africa. *African Journal of Ecology*, 33, 1–13.

Dean, W. R. J. and Siegfried, W. R. (1991). Orientation of diggings of the aardvark. *Journal of Mammalogy*, 72, 823–4.

Dean, W. R. J. and Turner, J. S. (1991). Ants nesting under stones in the semi-arid Karoo, South Africa: predator avoidance or temperature benefits? *Journal of Arid Environments*, 21, 59–69.

Dean, W. R. J. and Yeaton, R. I. (1992). The importance of harvester ant *Messor capensis* nest mounds as germination sites in the southern Karoo, South Africa. *African Journal of Ecology*, 30, 335–45.

Dean, W. R. J. and Yeaton, R. I. (1993a). The effects of harvester ant *Messor capensis* nest-mounds on the physical and chemical properties of soils in the southern Karoo, South Africa. *Journal of Arid Environments*, 25, 249–60.

Dean, W. R. J. and Yeaton, R. I. (1993b). The influence of harvester ant *Messor capensis* nest mounds on the productivity and distribution of some plant species in the southern Karoo, South Africa. *Vegetatio*, 106, 21–35.

Dean, W. R. J., Hoffman, M. T., Meadows, M. E. and Milton, S. J. (1995). Desertification in the semi-arid Karoo, South Africa: review and reassessment. *Journal of Arid Environments*, 30, 247–64.

Dean, W. R. J., Midgley, J. J. and Stock, W. D. (1994a). The distribution of mistletoes in South Africa: patterns of species richness and host choice. *Journal of Biogeography*, 21, 503–10.

Dean, W. R. J., Milton, S. J. and Siegfried, W. R. (1990). Dispersal of seeds as nest material by birds in semi-arid Karoo shrubland. *Ecology*, 71, 1299–306.

Dean, W. R. J., Milton, S. J., Siegfried, W. R. and Jarvis, M. J. F. (1994b). Diet, mobility and reproductive potential of ostriches: successful tactics for life in arid regions. In *Wildlife ranching: A celebration of biodiversity*, ed. W. van Hoven, H. Edebes, and A. Conroy, pp. 8–16. Pretoria: Promedia.

Dean, W. R. J., Milton, S. J., Watkeys, M. K. and Hockey, P. A. R. (1991). Distribution, habitat preference and conservation status of the red lark *Certhilauda burra* in Cape Province, South Africa. *Biological Conservation* 58, 257–74.

Dean, W. R. J., Williams, J. B. and Milton, S. J. (1993). Breeding of the White-backed Mousebird *Colius colius* in relation to rainfall and the phenology of fruiting plants in the southern Karoo, South Africa. *Journal of African Zoology*, 107, 105–11.

Delany, M. J. (1986). Ecology of small rodents in Africa. *Mammal Review*, 16, 1–41.

DeLoach, C. J., Gerling, D., Fornasari, L., Sobhian, R., Myartseva, S., Mityaev, I. D., Lu, Q. G., Tracy, J. L., Wang, R., Wang, J. F., Kirk, A., Pemberton, R. W., Chikatunov, V., Jashenko, R. V., Johnson, J. E., Zheng, H., Jiang, S. L., Liu, M. T., Liu, A. P. and Cisneroz, J. (1996). Biological control programme against saltceder (*Tamarix* spp.) in the United States of America: progress and problems. In *Proceedings of the IX International Symposium on Biological Control of Weeds*, ed. V. C. Moran and J. H. Hoffmann, pp. 253–260. 19–26 January 1996, Stellenbosch, South Africa.

Dent, M. C., Lynch, S. D. and Schulze, R. E. (1989). Mapping mean annual and other rainfall statistics in southern Africa. *ACRU Report*, 27, Department of Agricultural Engineering, University of Natal.

DePuit, E. J. (1979). Photosynthesis, respiration and biochemical transformations. In *Arid-land ecosystems: structure, functioning and management*, vol. 1, IBP 16, ed. D. W. Goodall and P. A. Perry, pp. 509–36. Cambridge University Press.

Desmet, P. G. (1996). Vegetation and restoration potential of the arid coastal belt between Port Nolloth and Alexander Bay, Namaqualand, South Africa. M.Sc. thesis, University of Cape Town.

Diamond, J. M. (1988). Factors controlling species diversity: overview and synthesis. *Annals of the Missouri Botanical Garden*, 75, 117–29.

Diazuriarte, R. and Garland, T. (1996). Testing hypotheses of correlated evolution using phylogenetically independent contrasts: Sensitivity to deviations from Brownian motion. *Systematic Biology*, 45, 27–47.

Dickson, C. G. C. and Kroon, D. M. (eds.) (1978). *Pennington's butterflies*. Johannesburg: A. D. Donker.

Dingle, R. V. and Hendey, Q. B. (1984). Late Mesozoic and Tertiary sediment supply to the eastern Cape Basin (S. E. Atlantic) and palaeo-drainage systems in southwestern Africa. *Marine Geology*, 56, 13–26.

Dingle, R. V., Siesser, W. G. and Newton, A. R. (1983). *Mesozoic and Tertiary geology of southern Africa*. Rotterdam: A. A. Balkema.

Dippenaar, S. M. and Ferguson, J. W. H. (1994). Towards a captive breeding programme for the Riverine Rabbit *Bunolagus monticularis*. *South African Journal of Science*, 90, 381–5.

Distefano, F. V. (1990). The effect of temperature and chemicals on the germination of some Conophytums. *Cactus and Succulent Journal (U.S)*, 62, 60–4.

Dixon, J. and Louw, G. (1978). Seasonal effects on nutrition, reproduction and aspects of thermoregulation in the Namaqua Sandgrouse (*Pterocles namaqua*). *Madoqua*, 11, 19–29.

Dixon, J. B. and Weed, S. B. (eds.) (1989). *Minerals in soil environments*, 2nd edn. Madison, Wisconsin: Soil Science Society of America.

Dobzhansky, T. (1950). Evolution in the tropics. *American Scientist*, 38, 209–21.

Dodd, A. P. (1940). *The biological campaign against prickly pear*. Brisbane: Commonwealth Prickly Pear Board Bulletin.

Donaldson, C. H. (1986a). The camp No. 6 veld grazing trial: An important milestone in the development of pasture research at the Grootfontein College of Agriculture. *Karoo Agric*, 3(8), 1–6.

Donaldson, C. H. (1986b). Observations on the Karoo caterpillar. *Karoo Agric*, 3(8), 17–18.

Donaldson, C. H. (1989). Seedling survival and dryland production potential of grasses and shrubs during the establishment period. *Karoo Agric*, 4, 3–10.

Donaldson, C. H. and Vorster, M. (1989). Veld management in the Karoo. Unpublished report. Pretoria: Department of Agriculture and Water Supply.

Douglas, W. A. and Mayeux, H. S. (1991). Evaporation from rangeland with and without honey mesquite. *Journal of Range Management*, 44, 161–70.

Downing, B. H. (1978). Environmental consequences of agricultural expansion in South Africa since 1850. *South African Journal of Science*, 74, 420–2.

Downs, C. T. and Perrin, M. R. (1991). Urinary concentrating ability of four *Gerbillurus* species of southern African arid regions. *Journal of Arid Environments*, 20, 71–81.

Dreisig, H. (1990). Thermoregulatory stilting in tiger beetles, *Cicindela hybrida* L. *Journal of Arid Environments*, 19, 297–302.

Drever, J. I. and Smith, C. L. (1978). Cyclic wetting and drying of the soil zone as an influence on the chemistry of groundwater in arid terrains. *American Journal of Science*, 278, 1448–52.

Drinkrow, D. R. and Cherry, M. I. (1995). Anuran distribution, diversity and conservation in South Africa, Lesotho and Swaziland. *South African Journal of Zoology*, 30, 82–90.

Dropkin, V. H. (1989). *Introduction to plant nematology*. New York: John Wiley.

Du Plessis, A. (1989). Ecophysiology of the bush Karoo rat (*Otomys unisulcatus*) and the whistling rat (*Parotomys brantsii*). M.Sc. thesis, University of Port Elizabeth, Port Elizabeth.

Du Plessis, A. and Kerley, G. I. H. (1991). Refuge strategies and habitat segregation in two sympatric rodents *Otomys unisulcatus* and *Parotomys brantsii*. *Journal of Zoology, London*, 224, 1–10.

Du Plessis, A., Kerley, G. I. H. and Winter, P. E. D. (1991). Dietary patterns of two herbivorous rodents, *Otomys unisulcatus* and *Parotomys brantsii*, in the Karoo. *South African Journal of Zoology*, 26, 51–4.

Du Plessis, A., Kerley, G. I. H. and Winter, P. E. D. (1992). Refuge microclimates of rodents: a surface nesting *Otomys unisulcatus* and a burrowing *Parotomys brantsii*. *Acta Theriologica*, 37, 351–8.

Du Plessis, H. M. and Shainberg, I. (1985). Effect of exchangeable sodium and phosphogypsum on the hydraulic properties of several South African soils. *South African Journal of Plant & Soil*, 2, 179–86.

Du Plessis, N. and Duncan, G. (1989). *Bulbous plants of Southern Africa – a guide to their cultivation and propagation*. Cape Town: Tafelberg.

Du Plessis, S. S. (1972). Ecology of blesbok with special reference to productivity. *Wildlife Monographs*, 30.

Du Plooy, F. (1993). Germination of *Lithops*

seed and what to expect. *Aloe*, 30, 14–15.

Du Preez, C. M. R. (1964). Die transpirasie-tempo van 'n aantal karrobossoorte. M.Sc. thesis, University of the Orange Free State.

Du Preez, K. (1983). Nassella situation worse than statistics show. *Karoo Regional Newsletter*, Autumn 1983, 8–9. Pretoria, Department of Agriculture.

Du Toit, E. (1942). The spread of prickly pear in the Union. *Farming in South Africa*, 17, 300–4.

Du Toit, P. C. V. and Blom, C. D. (1995). Diet selection by sheep and goats in the Noorsveld. *African Journal of Range and Forage Science*, 12, 27–37.

Du Toit, P. C. V., Blom, C. D. and Immelman, W. F. (1995). Diet selection by sheep and goats in the Arid Karoo. *African Journal of Range and Forage Science*, 12, 16–26.

Du Toit, P. F., Aucamp, A. J. and Bruwer, J. J. (1991). The national grazing strategy of the Republic of South Africa. Objectives, achievements and future challenges. *Journal of the Grassland Society of Southern Africa*, 8, 126–30.

Du Toit, P. J., Louw, J. G. and Malan, A. I. (1940). A study of the mineral content and feeding value of natural pastures in the Union of South Africa (Final Report). *Onderstepoort Journal of Veterinary Science and Animal Industry*, 14, 123–327.

Du Toit, S. (1993). Ecophysiology and host status of the rock elephant shrew, *Elephantulus myurus* (Thomas and Schwann, 1906). M.Sc. thesis, University of the Orange Free State, Bloemfontein.

Duncan, A. R., Joubert, P., Reid, A. M., Watkeys, M. K., Betton, P. J., Reid, D. L., Erlank, A. J., and Cleverly, R. W. (1985). Geochemical studies on the Floor rocks of Namaqualand. Final Report, National Geoscience Programme, Pretoria: CSIR.

Dunlevey, J. N. and Hiller, N. (1979). The Witteberg-Dwyka contact in the south-western Cape. *Transactions of the Geological Society of South Africa*, 82, 251–6.

Dunn, E. J. (1872). Through Bushmanland, Part 1. *Cape Monthly Magazine*, 5(30), 374–84.

Dunn, E. J. (1873). Through Bushmanland, Part 2. *Cape Monthly Magazine*, 6(31), 31–42.

Dunne, T., Dietrich, W. E. and Brunengo,

M. J. (1978). Recent and past erosion rates in semi-arid Kenya. *Zeitschrift für Geomorphologie, Supplement*, 29, 130–40.

Duthie, A. G. (1989). The ecology of the Riverine Rabbit *Bunolagus monticularis*. M.Sc. thesis, University of Pretoria.

Dyer, R. A. (1937). The vegetation of the divisions of Albany and Bathurst. *Memoirs of the Botanical Survey of South Africa*, 17, 11–38.

Dyson-Hudson, N. (1984). Adaptive resource use strategies of African pastoralists. In *Ecology in practice. Part 1: Ecosystem management*, ed. F. di Castri, F. W. G. Baker and M. Hadley, pp. 262–73. Paris: UNESCO.

Eardley, C. D. (1983). A taxonomic revision of the genus *Xylocopa* Latreille (Hymenoptera: Anthophoridae) in southern Africa. Republic of South Africa, Department of Agriculture. *Entomology Memoir*, 58, 1–67.

Eardley, C. D. (1988). A revision of the genus *Lithurge* Latreille (Hymenoptera: Megachilidae) of sub-Saharan Africa. *Journal of the Entomological Society of Southern Africa* 51, 251–63.

Eardley, C. D. (1989). The Afrotropical species of *Eucara* Friese, *Tetralonia* Spinola and *Tetraloniella* Ashmead (Hymenoptera: Anthophoridae). Republic of South Africa, Department of Agriculture and Water Supply. *Entomology Memoir*, 75, 1–62.

Eardley, C. D. (1991a). The genus *Epeolus* Latreille from subsaharan Africa (Hymenoptera: Anthophoridae). *Journal of Natural History*, 25, 711–31.

Eardley, C. D. (1991b). The Melectini in Subsaharan Africa (Hymenoptera: Anthophoridae). Republic of South Africa, Department of Agricultural Development. *Entomology Memoir*, 82, 1–49.

Eardley, C. D. (1991c). The southern African Panurginae (Andrenidae: Hymenoptera). *Phytophylactica*, 23, 115–36.

Eardley, C. D. (1993). The African species of *Pachymelus* Smith (Hymenoptera: Anthophoridae). *Phytophylactica*, 25, 217–29.

Eardley, C. D. (1994). The genus *Amegilla* Friese (Hymenoptera: Anthophoridae) in southern Africa. Republic of South Africa, Department of Agriculture. *Entomology Memoir*, 91, 1–68.

Eardley, C. D. (1996). The genus *Scrapter*

Lepeletier & Serville (Hymenoptera: Colletidae). *African Entomology*, 4, 37–92.

Eardley, C. D. and Brooks, R. W. (1989). The genus *Anthophora* Latreille in southern Africa (Hymenoptera: Anthophoridae). Republic of South Africa, Department of Agriculture and Water Supply. *Entomology Memoir*, 76, 1–55.

Eardley, C. D. and Schwarz, M. (1991). The Afrotropical species of *Nomada* Scopoli (Hymenoptera: Anthophoridae). *Phytophylactica*, 23, 17–27.

Earlé, R. A. and Herholdt, J. J. (1988). Breeding and moult of the Anteating Chat *Myrmecocichla formicivora*. *Ostrich*, 59, 155–61.

Edney, E. B. (1971). Some aspects of water balance in Tenebrionid beetles and a thysanuran from the Namib Desert of southern Africa. *Physiological Zoology*, 44, 61–76.

Edwards, D. (1970). Vegetation map of South Africa. Botanical Research Institute, Pretoria. (Unpublished).

Ehleringer, J. R. and Monson, R. K. (1993). Evolutionary and ecological aspects of photosynthetic pathway variation. *Annual Review of Ecology and Systematics*, 24, 411–39.

Ehleringer, J. R. and Mooney, H. A. (1983). Productivity of desert and mediter-ranean-climate plants. In *Encyclopaedia of plant physiology*, vol. 12B, *Physiological plant ecology*, IV, ed. O. L. Lange, P. S. Nobel, C. B. Osmond and H. Ziegler, pp. 205–31. New York: Springer-Verlag.

Eisenberg, J. F. and Kleiman, D. G. (1977). Communication in lagomorphs and rodents. In *How animals communicate*, ed. T. A. Sebeok, pp. 634–54. Bloomington, In: Indiana University Press.

El Shahaby, A. F. (1988). Associative nitrogen fixation with C_4 grasses of the northern Chihuahuan Desert. Ph.D. dissertation, New Mexico State University, Las Cruces, New Mexico.

Eldridge, D. J. and Greene, R. S. B. (1994). Microbiotic soil crusts: a review of their roles in soil and ecological processes in the rangelands of Australia. *Australian Journal of Soil Research*, 32, 389–415.

Elfatih, A. and Eltahir, B. (1996). El Niño and the natural variability in the flow of the Nile river. *Water Resource Research*, 32, 131–7.

Elkins, N. Z., Sabol, G. V., Ward, T. J. and Whitford, W. G. (1986). The influence of subterranean termites on the hydrological characteristics of a Chihuahuan Desert ecosystem. *Oecologia*, 68, 521–8.

Eller, B. M., Ruess, B. R. and Ferrari, S. (1991). Re-establishment of water uptake by succulents after drought: potometric field determinations in the Richtersveld (Cp., Rep. South Africa). *Botanica Helvetica*, 101, 259–65.

Ellery, W. N., Scholes, R. J. and Scholes, M. C. (1995). The distribution of sweetveld and sourveld in South Africa's grassland biome in relation to environmental factors. *African Journal of Range and Forage Science*, 12, 38–45.

Ellis, F. (1987). *Soil map of the Karoo*. Pretoria: Institute for Soil Climate and Water.

Ellis, F. (1988). Die Gronde van die Karoo. Ph.D. thesis, University of Stellenbosch.

Ellis, F. and Lambrechts, J. J. N. (1986). Soils. In *The Karoo biome: a preliminary synthesis. Part 1. Physical environment*, ed. Cowling, R. M., Roux, P. W. and Pieterse, A. J. H. *South African National Scientific Programmes Report*, 124, 18–38.

Ellis, F. and Schloms, B. H. A. (1982). A note on the dorbanks (duripans) of South Africa. *Palaeoecology of Africa and Surrounding Islands*, 15, 149–57.

Ellis, J. and Galvin, K. A. (1994). Climate patterns and land-use practices in the dry zones of Africa. *BioScience*, 44, 340–9.

Ellis, L. M. (1995). Bird use of salt cedar and cottonwood vegetation in the middle Rio Grande Valley of New Mexico, USA. *Journal of Arid Environments*, 30, 339–49.

Ellis, R. P., Vogel, J. C. and Fuls, A. (1980). Photosynthetic pathways and the geographical distribution of grasses in South West Africa/Namibia. *South African Journal of Science*, 76, 307–12.

Ellison, G. T. H. (1988). *Dorylus helvolus* L. (Formicidae: Dorylinae) preying on *Saccostomus campestris* (Rodentia: Cricetidae). *Journal of the Entomological Society of Southern Africa*, 51, 296.

Ellison, G. T. H. (1993). Evidence of climatic adaptation in spontaneous torpor among Pouched Mice *Saccostomus campestris* from southern Africa. *Acta Theriologica*, 38, 49–59.

Ellison, G. T. H. and Skinner, J. D. (1991). Thermoregulation and torpor in African woodland dormice, *Graphiurus murinus*, following cold acclimation. *Zeitschrift für Saügertierkunde*, 56, 41–7.

Ellison, G. T. H. and Skinner, J. D. (1992). The influence of ambient temperature on spontaneous daily torpor in Pouched Mice (*Saccostomus campestris*: Rodentia-Cricetidae) from South Africa. *Journal of Thermal Biology*, 17, 25–31.

Ellner, S. P. and Shmida, A. (1981). Why are adaptations for long-range seed dispersal rare in desert plants? *Oecologia*, 51, 133–44.

Eloff, F. C. (1959a). Observations on the migration and habits of the antelopes of the Kalahari Gemsbok Park. Part 1. *Koedoe*, 2, 1–29.

Eloff, F. C. (1959b). Observations on the migration and habits of the antelopes of the Kalahari Gemsbok Park. Part 2. *Koedoe*, 2, 30–51.

Eloff, F. C. (1961). Observations on the migration and habits of the antelopes of the Kalahari Gemsbok Park. Part 3. *Koedoe*, 4, 18–30.

Eloff, F. C. (1962). Observations on the migration and habits of the antelopes of the Kalahari Gemsbok Park. Part 4. *Koedoe*, 5, 128–36.

Elphick, R. (1985). *Khoikhoi, and the founding of white South Africa*. Johannesburg: Ravan Press.

Endrödy-Younga, S. (1986). The *Cardiosus* evolutionary lineages of the genus *Zophosis* Latreille (Coleoptera, Tenebrionidae, Zophosini). *Cimbebasia*, 7, 209–33.

Endrödy-Younga, S. (1988). Evidence for the low-altitude origin of the Cape Mountain biome derived from the systematic revision of the genus *Colophon* Gray (Coleoptera, Lucanidae). *Annals of the South African Museum*, 96, 359–424.

Endrödy-Younga, S. and Peck, S. B. (1983). Onychophora from mesic grassveld in South Africa (Onychophora: Peripatopsidae). *Annals of the Transvaal Museum*, 33, 347–52.

Engel, A. E. J., Nagy, B., Nagy, L. A., Engel, C. G., Kremp, G. O. W. and Drew, C. M. (1968). Algal-like forms in the Onverwacht Series, South Africa: oldest recognized life-like forms on Earth. *Science*, 161, 1005–8.

Ennos, A. R. (1997). Wind as an ecological factor. *Trends in Ecology and Evolution*, 12, 108–11.

Enright, J. T. (1970). Ecological aspects of endogenous rhythmicity. *Annual Review of Ecology and Systematics*, 1, 221–38.

Erasmus, B. H. (1988). Natural predators and parasites of the brown locust – an ecological appraisal. In *Proceedings of the Locust Symposium*, ed. B. McKenzie and M. Longridge, pp. 118–37. *South African Institute of Ecologists Bulletin* (Special Issue).

Erlank, A. J. (ed.) (1984). Petrogenesis of the volcanic rocks of the Karoo Province. *Special Publications of the Geological Society of South Africa*, 13, 1–395.

Ernest, K. A. (1994). Resistance of creosotebush to mammalian herbivory: temporal consistency and browsing-induced changes. *Ecology*, 75, 1684–92.

Esler, K. J. (1993). Vegetation patterns and plant reproductive processes in the Succulent Karoo. Ph.D. thesis. University of Cape Town, Cape Town.

Esler, K. J. and Cowling, R. M. (1993). Edaphic factors and competition as determinants of pattern in South African Karoo vegetation. *South African Journal of Botany*, 59, 287–95.

Esler, K. J. and Cowling, R. M. (1995). The comparison of selected life history characteristics of *Mesembryanthema* species occurring on and off Mima-like mounds (*heuweltjies*) in semi-arid southern Africa. *Vegetatio*, 116, 41–50.

Esler, K. J. and Phillips, N. (1994). Experimental effects of water stress on semi-arid Karoo seedlings: implications for field seedling survivorship. *Journal of Arid Environments*, 26, 325–37.

Esler, K. J. and Rundel, P. W. (in press) Comparative patterns of phenology and growth form diversity in two winter rainfall deserts: the Succulent Karoo and Mojave desert ecosystem. *Plant Ecology*.

Esler, K. J., Cowling, R. M. and Ivey, P. (1992). Seed biology of three species of Mesembryanthema in the southern succulent Karoo. *South African Journal of Botany*, 58, 343–8.

Esler, K. J., Rundel, P. W. and Cowling, R. M. (1994). Intercontinental comparisons along mediterranean/desert transition zones: climate and vegetation structure. *Noticieros de Biologia*, 2, 22.

Esler, K. J., Rundel, P. W. and Vorster, P. (in press). Biogeography of prostrate-leaved geophytes in semi-arid South

Africa: hypotheses on functionality. *Plant Ecology*.

Esterhuizen, W. C. N. and Norton, P. M. (1985). The leopard as a problem animal in the Cape Province, as determined by the permit system. *Bontebok*, 4, 9–16.

Estes, R. (1969). Territorial behaviour of the wildebeest *Connochaetes taurinus*. *Zeitschrift fur Tierpsychologie*, 26, 284–370.

Estes, R. (1977). Relationships of the South African fossil frog *Eoxenopoides reuningi* (Anura, Pipidae). *Annals of the South African Museum*, 73, 49–80.

Estes, R. D. (1966). Behaviour and life history of the wildebeest (*Connochaetes taurinus* Burchell). *Nature*, 212, 999–1000.

Estes, R. D. (1976). The significance of breeding synchrony in the wildebeest. *East African Wildlife Journal*, 14, 135–42.

Evenari, M. and Gutterman, Y. (1966). The photoperiodic response of some desert plants. *Zeitschrift für Pflanzenphysiologie*, 54, 7–27.

Evenari, M. and Gutterman, Y. (1985). Desert plants. In *CRC handbook of flowering*, vol. 1, ed. A. H. Halevy, pp. 41–59. Florida: CRC Press.

Ewer, R. F. (1965). The anatomy of the thecodont reptile *Euparkeria capensis* Broom. *Philosophical Transactions of the Royal Society of London*, B248, 379–435.

Fabricius, C. (1994). The relation between herbivore density and relative resource density at the landscape level: kudu in semi-arid savanna. *African Journal of Range & Forage Science*, 11, 7–10.

Fabricius, C. and Mentis, M. T. (1992). Modelling the habitat relations of kudu in arid savanna. *South African Journal of Science*, 88, 280–4.

Fahse, L. (1994). Konsequenzen stochastischer abiotischer Faktoren für die Populationsdynamik von Lerchen in der Nama Karoo. M.Sc. thesis, Philips Universität, Marburg.

Fairall, N. and Le Roux, A. (1991). Game management in arid areas, non-equilibrium alternative. In *Wildlife Production, Conservation and Sustainable Development*, ed. L. A. Renecker and R. J. Hudson, pp. 251–4. Fairbanks, Alaska: University of Alaska.

Fairall, N., Jooste, J. F. and Conroy, A. M. (1990). Biological evaluation of a springbok-farming enterprise. *South*

African Journal of Wildlife Research, 20, 73–80.

Fairall, N., Vermeulen, P. and Van der Merwe, M. (1986). A general model of population growth in the hyrax *Procavia capensis* (Pallas, 1766). *Ecological Modelling*, 34, 115–32.

Faulkes, C. G., Abbott, D. H. and Jarvis, J. U. M. (1990). Social suppression of ovarian cyclicity in captive and wild colonies of naked mole-rats, *Heterocephalus glaber*. *Journal of Reproduction and Fertility*, 88, 559–68.

Faure, J. C. (1932). The phase of locusts in South Africa. *Bulletin of Entomological Research*. London, 23, 293–405.

Faure, J. C. (1937). Some recent advances in research on locust problems. *South African Journal of Science*, 33, 797–811.

Fearn, B. (1978). The world and southern African distribution of the Mesembryanthemaceae. *Excelsa*, 8, 75–88.

Felker, P. (1979). Mesquite. An all purpose leguminous arid land tree. In *New Agricultural Crops*, ed. G. A. Ritchie, pp. 89–132. Boulder, Colorado: Westview Press.

Felstenstein, J. (1985). Phylogenies and the comparative method. *The American Naturalist*, 125, 1–15.

Fenner, M. (1991). The effects of the parent environment on seed germinability. *Seed Science Research*, 1, 75–84.

Fielden, L. J., Waggoner, J. P., Perrin, M. R. and Hickmann, G. C. (1990). Thermoregulation in the Namib Desert Golden Mole, *Eremitalpa granti namibensis* (Chrysochloridae). *Journal of Arid Environments*, 18, 221–37.

Fisher, F. M., Freckman, D. W. and Whitford, W. G. (1990). Decomposition and soil nitrogen availability in Chihuahuan Desert field microcosms. *Soil Biology and Biochemistry*, 22, 241–9.

Fisher, F. M., Zak, J. C., Cunningham, G. L. and Whitford, W. G. (1988). Water and nitrogen effects on growth and allocation patterns of creosotebush in the northern Chihuahuan Desert. *Journal of Range Management*, 41, 387–91.

Fisher, M. (1994). Another look at the variability of desert climates using examples from Oman. *Global Ecology and Biogeography Letters*, 4, 79–87.

FitzPatrick, R. W. (1987). Iron compounds as indicators of pedogenic processes: examples from the Southern

Hemisphere. In *Iron in soils and clay minerals*, ed. J. W. Stucki, B. A. Goodman and U. Schwertman, pp. 351–89. Dordrecht: D. Reidel.

Flach, K. W., Nettleton, W. D., Gile, L. H. and Cady, J. G. (1969). Pedocementation: induration by silicate, carbonates and sequioxides in the Quaternary. *Soil Science*, 107, 442–53.

Fock, G. J. (1968). Rooidam, a sealed site of the First Intermediate. *South African Journal of Science*, 64, 153–9.

Forrester, J. (1988). Fire in the Karoo N.B.G. Floral Reserve. *Veld and Flora*, 74, 5.

Fouche, H. J. (1984). Ondersoek na die gebruik van die Putu II simulasiemodel en Palmer-indeks vir die karakterisering van droogtetoestande. M.Sc. thesis, University of the Orange Free State, Bloemfontein.

Fouche, H. J. (1992). Simulering van die produksiepotensiaal van veld en die kwantifisering van droogte in die sentrale Oranje-Vrystaat. Ph.D. thesis, University of the Orange Free State, Bloemfontein.

Fourie, J. H., De Wet, N. J. and Page, J. J. (1987). Veld condition and trend in Kalahari duneveld under an extensive stock production system. *Journal of the Grassland Society of Southern Africa*, 4, 48–54.

Fourie, L. J. (1983). The population dynamics of the rock hyrax *Procavia capensis* (Pallas, 1766) in the mountain Zebra National Park. Ph.D. thesis, Rhodes University, Grahamstown.

Fourie, L. J. and Perrin, M. R. (1987). Social behaviour and spatial relationships in the rock hyrax. *South African Journal of Wildlife Research*, 17, 91–9.

Fox, G. A. (1989). Consequences of flowering time variation in a desert annual: adaptation and history. *Ecology*, 70, 1294–306.

Fox, G. A. (1990a). Components of flowering time variation in a desert annual. *Evolution*, 44, 1404–23.

Fox, G. A. (1990b). Drought and the evolution of flowering time in desert annuals. *American Journal of Botany*, 77, 1508–18.

Fox, G. A. (1992). The evolution of life history traits in desert annuals: adaptation and constraint. *Evolutionary Trends in Plants*, 6, 25–31.

Franco, A. C. and Nobel, P. S. (1990).

Influences of root distribution and growth on predicted water uptake and interspecific competition. *Oecologia*, 82, 151–7.

Frank, C. L. (1992). The influence of dietary fatty acids on hibernation by golden-mantled ground squirrels (*Spermophilus lateralis*). *Physiological Zoology*, 65, 906–20.

Frank, C. L. (1994). Polyunsaturate content and diet selection by ground squirrels (*Spermophilus lateralis*). *Ecology*, 75, 458–63.

Frean, J. L., Downs, C. T. and Lovegrove, B. G. (1998). Osmoregulatory capacity of an arboreal rodent *Thallomys nigricauda*. *Journal of Arid Environments*, 39 (in press).

Freas, K. E. and Kemp, P. R. (1983). Some relationships between environmental variability and seed dormancy in desert plants. *Journal of Ecology*, 71, 211–17.

Freckman, D. W. (1988). Bacterivorous nematodes and organic matter decomposition. *Agriculture, Ecosystems, and Environment*, 24, 195–217.

Friedel, M. H. (1991). Range condition assessment and the concept of thresholds: A view point. *Journal of Range Management*, 44, 422–6.

Friedländer, M. and Scholtz, C. H. (1993). Two different patterns of interrupted spermatogenesis in winter diapause and summer quiescence in the desert beetle, *Omorgus freyi*. *Journal of Morphology*, 218, 347–58.

Fryxell, J. M. and Sinclair, A. R. E. (1988). Causes and consequences of migration by large herbivores. *Ecology and Evolution*, 3, 237–41.

Fuls, E. R. (1992). Semi-arid and arid rangelands: a resource under siege due to patch selective grazing. *Journal of Arid Environments*, 22, 191–3.

Gaff, D. F. (1977). Desiccation tolerant vascular plants of southern Africa. *Oecologia*, 31, 95–109.

Gandar, M. V. (1988). Feeding ecology and food selection of grasshoppers (Acridoidea): some general principles and experimental techniques. In *Proceedings of the Locust Symposium*, ed. B. McKenzie and M. Longridge, pp. 60–69. *South African Institute of Ecologists Bulletin* (Special Issue).

Garcia-Moya, E. and McKell, C. M. (1970). Contribution of shrubs to the nitrogen economy of a desert-wash plant community. *Ecology*, 51, 81–8.

Gardiner, B. G. (1969). New palaeoniscoid fish from the Witteberg Series of South Africa. *Zoological Journal of the Linnean Society*, 48, 423–52.

Garland, T., Harvey, P. H. and Ives, A. R. (1992). Procedures for the analysis of comparative data using phylogenetically independent contrasts. *Systematic Biology*, 41, 18–32.

Garside, S. and Lockyer, S. (1930). Seed dispersal from the hygroscopic fruits of *Carpanthea* (*Mesembryanthemum*), *pomeridiana* N.E. Br. *Annals of Botany*, 44, 639–55.

Gaston, K. G. (1994). *Rarity*. London: Chapman and Hall.

Gatimu, L. N. (1996). The effects of free-ranging ostrich farming on arid rangeland in the Prince Albert district, South Africa. M.Sc. thesis (Conservation Biology), University of Cape Town.

Geiser, F. and Broome, L. S. (1993). The effect of temperature on the pattern of torpor in a marsupial hibernator. *Journal of Comparitive Physiology. B. Biochemical, Systematic and Environmental Physiology*, 163, 133–7.

Geiser, F. and Kenagy, G. J. (1993). Dietary fats and torpor patterns in hibernating ground squirrels. *Canadian Journal of Zoology*, 71, 1182–5.

Geiser, F., Stahl, B. and Learmonth, R. P. (1992). The effect of dietry fatty acids on the pattern of torpor in a marsupial. *Physiological Zoology*, 65, 1236–45.

George, T. N. (1976). Charles Lyell: the present is the key to the past. *Philosophical Journal*, 13, 3–24.

Germs, G. J. B. (1974). The Nama Group in South West Africa and its relationship to the Pan-African geosyncline. *Journal of Geology*, 82, 301–17.

Gess, F. W. (1980a). Some aspects of the ethology of *Dasyproctus westermanni* (Dahlbom) (Hymenoptera: Sphecidae: Crabroninae) in the Eastern Cape Province of South Africa. *Annals of the Cape Provincial Museums (Natural History)*, 13, 95–106.

Gess, F. W. (1980b). Prey and nesting sites of some sympatric species of *Cerceris* (Hymenoptera: Sphecidae) with a review and discussion of the prey diversity of the genus. *Annals of the Cape Provincial Museums (Natural History)*, 13, 85–93.

Gess, F. W. (1981). Some aspects of an ethological study of the aculeate wasps and the bees of a karroid area in the vicinity of Grahamstown, South Africa. *Annals of the Cape Provincial Museums (Natural History)*, 14, 1–80.

Gess, F. W. (1986). Three new species of southern African *Bembix*, a new synonymy and biological notes on other species of the genus (Hymenoptera: Sphecidae: Nyssoninae). *Annals of the Cape Provincial Museums (Natural History)*, 16, 137–60.

Gess, F. W. and Gess, S. K. (1974). An ethological study of *Dichragenia pulchichroma* (Arnold) (Hymenoptera: Pompilidae), a southern African spider-hunting wasp which builds a turreted, subterranean nest. *Annals of the Cape Provincial Museums (Natural History)*, 9, 21–46.

Gess, F. W. and Gess, S. K. (1975). Ethological studies of *Bembecinus cinguliger* (Smith) and *B. oxydorcus* (Handl.) (Hymenoptera: Sphecidae), two South African turret-building wasps. *Annals of the Cape Provincial Museums (Natural History)* 11, 21–46.

Gess, F. W. and Gess, S. K. (1976). An ethological study of *Parachilus insignis* (Saussure) (Hymenoptera: Eumenidae). *Annals of the Cape Provincial Museums (Natural History)*, 11, 83–102.

Gess, F. W. and Gess, S. K. (1980a). 'Mesemb bread' – the nesting of the wasp *Ceramius lichtensteinii* (Klug) (Masaridae). *The Eastern Cape Naturalist*, 24, 7–9.

Gess, F. W. and Gess, S. K. (1980b). Ethological studies of *Jugurtia confusa* Richards, *Ceramius capicola* Brauns, *C. linearis* Klug and *C. lichtensteinii* (Klug) (Hymenoptera: Masaridae) in the Eastern Cape Province of South Africa. *Annals of the Cape Provincial Museums (Natural History)*, 13, 63–83.

Gess, F. W. and Gess, S. K. (1982). Ethological studies of *Isodontia simoni* (du Buysson), *I. pelopoeiformis* (Dahlbom) and *I. stanleyi* (Kohl) (Hymenoptera: Sphecidae: Sphecinae) in the Eastern Cape Province of South Africa. *Annals of the Cape Provincial Museums (Natural History)*, 14, 151–71.

Gess, F. W. and Gess, S. K. (1988). A further contribution to the knowledge of the ethology of the genus *Ceramius* Latreille (Hymenoptera: Masaridae) in the southern and western Cape Province of South Africa. *Annals of the Cape Provincial*

Museums (Natural History) 18, 1–30.

Gess, F. W. and Gess, S. K. (1991a). Some aspects of the ethology of five species of Eumenidae (Hymenoptera) in southern Africa. *Annals of the Cape Provincial Museums (Natural History)*, 18, 245–70.

Gess, F. W. and Gess, S. K. (1991b). A preliminary survey of the aculeate wasps and the bees of the lower reaches of the Nossob River Valley, Kalahari Gemsbok National Park, South Africa. *Koedoe*, 34(2), 77–88.

Gess, F. W. and Gess, S. K. (1993). Effects of increasing land utilization on species representation and diversity of aculeate wasps and bees in the semi-arid areas of southern Africa. In *Hymenoptera and biodiversity*, ed. J. La Salle and I. D. Gauld, pp. 83–113. Wallingford, UK: CAB International.

Gess, S. K. (1992). Biogeography of the masarine wasps (Hymenoptera: Vespidae: Masarinae), with particular emphasis on the southern African taxa and on correlations between masarine and forage plant distributions. *Journal of Biogeography*, 19, 491–503.

Gess, S. K. (1993). The Karoo violet and a Masarid wasp – a mutually beneficial association. *The Naturalist*, 37, 32–6.

Gess, S. K. (1996). *The pollen wasps. Ecology and natural history of the Masarinae.* Cambridge, MA: Harvard University Press.

Gess, S. K. and Gess, F. W. (1989). Flower visiting by masarid wasps in southern Africa (Hymenoptera: Vespoidea: Masaridae). *Annals of the Cape Provincial Museums* 18(5), 95–134.

Gess, S. K. and Gess, F. W. (1994). Potential pollinators of the Cape Group of Crotalarieae (*sensu* Polhill) (Fabales: Papilionaceae), with implications for seed production in cultivated rooibos tea. *African Entomology* 2(2), 97–106.

Gibbs Russell, G. E. (1987). Preliminary floristic analysis of the major biomes in southern Africa. *Bothalia*, 17, 213–27.

Gibbs Russell, G. E., Watson, L., Koekemoer, M., Smook, L., Barker, N. P., Anderson, H. M. and Dallwitz, M. J. (1990). Grasses of southern Africa. *Memoirs of the Botanical Survey of South Africa*, 58, 1–437.

Gibson, D. A. (1995). Modelling the distribution of *Portulacaria afra* in the Eastern and Western Cape Provinces, South Africa, in relation to environ-

mental variables and the normalised difference vegetation index. B.Sc. Hons. thesis (Botany), Rhodes University.

Gilchrist, A. R., Kooi, H. and Beaumont, C. (1994). Post-Gondwana geomorphic evolution of southwestern Africa: implications for the controls on landscape development from observations and numerical experiments. *Journal of Geophysical Research*, 99, 12211–28.

Giliomee, J. H. (1986). Hunting for insect pollinators on fynbos flowers. *Veld and Flora*, 72, 6–7.

Ginzburg, C. (1973). Some anatomic features of splitting of desert shrubs. *Phytomorphology*, 13, 92–7.

Givnish, T. J. (1986). *On the economy of plant form and function.* Cambridge University Press.

Goldberg, S. R. and Robinson, M. D. (1979). Reproduction in two Namib Desert lacertid lizards (*Aporosaura anchietae* and *Meroles cuneirostris*). *Herpetologica*, 35, 169–75.

Goldblatt, P. (1978). An analysis of the flora of southern Africa: its characteristics, relationships and origins. *Annals of the Missouri Botanical Garden*, 65, 369–436.

Goldblatt, P. (1984). A revision of *Hesperantha* (Iridaceae) in the winter rainfall area of southern Africa. *Journal of South African Botany*, 50, 15–14.

Goldblatt, P. (1991). An overview of the systematics, phylogeny and biology of the African Iridaceae. *Contributions from the Bolus Herbarium*, 13, 1–74.

Goldblatt, P. and Manning, J. C. (1996). Phylogeny and speciation in *Lapeirousia* subgenus *Lapeirousia* (Iridaceae: Ixioideae). *Annals of the Missouri Botanical Garden*, 83, 346–61.

Goldblatt, P., Manning, J. C. and Bernhardt, P. (1995). Pollination biology of *Lapeirousia* subgenus *Lapeirousia* (Iridaceae) in southern Africa; floral divergence and adaptation for long-tong fly pollination. *Annals of the Missouri Botanical Garden*, 82, 517–34.

Gonzales, B., Boucaud, J., Salette, J., Langlois, J. and Duyme, M. (1989). Changes in stubble carbohydrate content during regrowth of defoliated perennial ryegrass (*Lolium perenne* L.) on two nitrogen levels. *Grass and Forage Science*, 44, 411–15.

Goodman, P. S. (1990). Soil, vegetation and large herbivore relations in Mkuzi Game Reserve, Natal. Ph.D. thesis, University of the Witwatersrand.

Gordon, C. J. (1993). *Temperature regulation in laboratory rodents.* Cambridge University Press.

Görne, A. (1995). Auswirkung von Beweidung auf ausgewählte Tiergruppen (Rodentes, Aves, Formicidae) in einem Halbwüstenökosystem Südafrikas. M.Sc. thesis, Biology Department, University of Marburg, Germany.

Goudie, A. S. (1991). *Environmental change*, 3rd edition Oxford: Clarendon Press.

Gould, S. J. (1979). An allometric interpretation of species-area curves: the meaning of the coefficient. *American Naturalist*, 114, 335–43.

Gould, S. J. (1989). *Wonderful Life. Burgess shale and the nature of history.* London: Penguin Books.

Gould, S. J. and Vrba, E. S. (1982). Exaptation – a missing term in the science of form. *Paleobiology*, 8, 4–15.

Grace, J. (1977). *Plant responses to wind.* London: Academic Press.

Gray, D. A. and Brown, C. R. (1995). Saline-infusion-induced increases in plasma osmolality do not stimulate nasal gland secretion in the Ostrich (*Struthio camelus*). *Physiological Zoology*, 68, 164–75.

Gray, F. J. (1970). Isoptera. In *The insects of Australia*, ed. I. M. Mackerras, pp. 275–93. Melbourne University Press.

Green, D. G. (1989). Simulated effects of fire, dispersal and spatial pattern in competition within forest mosaics. *Vegetatio*, 82, 139–53.

Greig, J. C. and P. D. Burdett (1976). Patterns in the distribution of southern African terrestrial tortoises (Cryptodira: Testudinidae). *Zoologica Africana*, 11, 249–73.

Griffin, G. F., Stafford Smith, D. M., Morton, S. R., Allan, G. E. and Masters, K. A. (1989). Status and implications of the invasion of Tamarisk (*Tamarix aphylla*) on the Finke River, Northern Territory, Australia. *Journal of Environmental Management*, 29, 297–315.

Griffin, M. (1990). A review of taxonomy and ecology of gerbilline rodents of the central Namib Desert, with keys to the species (Rodentia: Muridae). In *Namib ecology: 25 years of Namib research*, ed.

M. K. Seely. *Transvaal Museum Monograph*, 7, 83–98.

Grindley, J. R. (1987). Gondwana mollusc relicts in the Langeberg mountains. *South African Journal of Science*, 83, 67.

Grobler, H. (1987). Rooikatte word meer. *Landbouweekblad*, 469, 44–5.

Grobler, N. J. and Emslie, D. P. (1986). Stromatolitic limestone and chert in the Ventersdorp Supergroup at the T'Kuip Hills area and surroundings, Britstown district, South Africa. *Transactions of the Geological Society of South Africa*, 79, 49–52.

Grubb, P. J. (1977). The maintenance of species-richness in plant communities: the importance of the regeneration niche. *Biological Review of the Cambridge Philosophical Society*, 52, 107–45.

Günster, A. (1994). Seed bank dynamics, longevity, viability and predation of seeds of serotinous plants in the central Namib desert. *Journal of Arid Environments*, 28, 195–205.

Günster, A. (1995). Grass cover distribution in the central Namib – a rapid method to assess regional and local rainfall patterns of arid regions? *Journal of Arid Environments*, 29, 107–14.

Gussmann, S. M. V. (1995). New species and subspecies of *Julodis* from southern Africa. *African Entomology*, 3, 111–29.

Gutierrez, J. R. and Whitford, W. G. (1987). Chihuahuan Desert annuals: importance of water and nitrogen. *Ecology*, 68, 2032–45.

Gutterman, Y. (1980/1981). Annual rhythm and position effect in the germinability of Mesembryanthemum nodiflorum. *Israel Journal of Botany*, 29, 93–7.

Gutterman, Y. (1982). Survival mechanisms of desert winter annual plants in the Negev highlands of Israel. *Scientific Review of Arid Zone Research*, 1, 249–83.

Gutterman, Y. (1983). Mass germination of plants under desert conditions. Effects of environmental factors during seed maturation, dispersal, germination and establishment of desert annual and perennial plants in the Negev Highlands, Israel. In *Development in ecology and environmental quality*, ed. H. I. Shuval, pp. 1–10. Rehovot: Balaban.

Gutterman, Y. (1994). *Seed germination in desert plants*. Berlin: Springer-Verlag.

Gwala, P. P., Drennan, P. M. and Van Staden, J. (1991). Factors affecting germination of seeds of *Atriplex lindleyi*. *South African Association of Botanists Congress Abstracts*, 17, 122. Pietermaritzburg: University of Natal.

Haarhoff, P. J. (1982). Karoo harvester termite alates found in the stomach of a Kelp Gull *Parus dominicanus*. *Cormorant*, 10, 119.

Hadley, N. F. (1972). Desert species and adaptation. *American Scientist*, 60, 338–47.

Hadley, N. F. (1974). Adaptational biology of desert scorpions. *Journal of Arachnology*, 2, 11–23.

Hadley, N. F. (1975). *Environmental phsyiology of desert inhabitants*. Stroudsburg, PA: Dowden, Hutchinson and Ross, Inc.

Hadley, N. F. (1981). Cuticular lipids of terrestrial plants and arthropods: a comparison of their structure, composition, and waterproofing function. *Biological Reviews*, 56, 23–47.

Hadley, N. F. (1994). *Water relations of terrestrial arthropods*. San Diego, California: Academic Press.

Haim, A., Skinner, J. D. and Robinson, T. J. (1987). Bioenergetics, thermoregulation and urine analysis of the genus *Xerus* from an arid environment. *South African Journal of Zoology*, 22, 45–9.

Hall, B. P. and R. E. Moreau (1970). *An Atlas of Speciation in African Passerine Birds*. London: British Museum (Natural History).

Hall, C. A. S. and Day, J. W. (ed.) (1977). *Ecosystem modelling in theory and practice: An introduction with case histories*. New York: John Wiley.

Hall, T. D. (1934). South African pastures: Retrospective and prospective. *South African Journal of Science*, 21, 59–97.

Hamilton, W. J., III, and Heppner, F. (1967). Radiant solar energy and the function of black homeotherm pigmentation: an hypothesis. *Science*, 155, 196–7.

Hamilton, W. J., III. (1973). *Life's color code*. New York: McGraw-Hill.

Hamilton, W. J., III. (1976). Fog basking by the Namib Desert beetle, *Onymacris unguicularis*. *Nature*, 262, 284–5.

Hamilton, W. J., III. (1985). Demographic consequences of a food and water shortage to desert Chacma baboons, *Papio ursinus*. *Journal of Primatology*, 6, 451–62.

Hamilton, W. J., III. (1986). Namib desert Chacma baboon (*Papio ursinus*) use of food and water resources during a food shortage. *Madoqua*, 14, 397–408.

Hammer, S. A. (1991). Scents and sensibility. *Veld & Fora*, 77, 70–1.

Hammer, S. A. (1993). *The Genus Conophytum*. Pretoria: Succulent Plant Publications.

Hammer, S. A. and Liede, S. (1990). Natural and artificial hybrids in Mesembryanthemaceae. *South African Journal of Botany*, 56, 356–62.

Hanrahan, S. A. (1988). The locust problem and intergrated pest management. *South African Journal of Science*, 84, 5–8.

Hanrahan, S. A. and Seely, M. K. (1990). Food and habitat use by three tenebrionid beetles (Coleoptera) in a riparian desert environment. In *Namib ecology: 25 years of research*, ed. M. K. Seely. *Transvaal Museum Monograph*, 7, 143–8.

Hanrahan, S. A., McClain, E. and Gernecke, D. (1984). Dermal glands concerned with production of wax blooms in desert tenebrionid beetles. *South African Journal of Science*, 80, 176–81.

Haq, B. U., Hardenbol, J. and Vail, P. R. (1987). Chronology of fluctuating sea levels since the Triassic. *Science*, 235, 1156–67.

Hara, T. (1993). Mode of competition and size-structure dynamics in plant communities. *Plant Species Biology*, 8, 75–84.

Harding, G. B. (1978). Mesquite. In *Plant invaders, beautiful but dangerous*, ed. C. H. Stirton, pp. 128–31. Cape Town: Department of Nature and Environmental Conservation.

Harding, G. B. (1991). Sheep can reduce seed recruitment of invasive *Prosopis* species. *Applied Plant Science*, 5, 25–7.

Harding, G. B. and Bate, G. C. (1991). The occurrence of invasive *Prosopis* species in the north-western Cape, South Africa. *South African Journal of Science*, 87, 188–92.

Harper, J. L. (1977). *Population biology of plants*. New York: Academic Press.

Harrison, J. A. (1989). Atlassing as a tool in conservation, with special reference to the Southern African Bird Atlas Project. In *Biotic diversity in southern Africa. Concepts and conservation*, ed. B. J. Huntley, pp. 157–69. Cape Town:

Oxford University Press.

Harrison, J. A. (1992). The Southern African Bird Atlas Project databank: five years of growth. *South African Journal of Science*, 88, 410–13.

Hartmann, H. (1982). Monographie der Subtribus Leipoldtiinae. III. Monographie der Gattung *Fenestraria* (Mesembrynathemacea). *Botanische Jahrbücher*, 103, 145–83.

Hartmann, H. E. K. (1983a). Interaction of ecology, taxonomy and distribution in some Mesembryanthemaceae. *Bothalia*, 14, 653–9.

Hartmann, H. E. K. (1983b). Monographien der Subtribus Leipoldtiinae. IV. Monographie der Gattung *Vanzijlia* (Mesembryanthemaceae). *Botanische Jahrbücher für Systematik*, 103, 499–538.

Hartmann, H. E. K. (1988). Fruit types in Mesembryanthema. *Beiträge zur Biologie der Pflanzen*, 63, 313–49.

Hartmann, H. E. K. (1989). Biological adaptation in Mesembryanthema. *Excelsa*, 14, 39–45.

Hartmann, H. E. K. (1991). Mesembryanthema. *Contributions from the Bolus Herbarium*, 13, 75–157.

Hartmann, H. E. K. and Dehn, M. (1987). Monographien der Leipoldtiinae V11. Monographie der Gettung *Cheiridopsis* (Mesembryanthemaceae). *Botanische Jahrbücher*, 108, 567–663.

Hartmann, H. E. K. and Gölling, H. (1993). A monograph of the genus *Glottiphyllum* (Mesembryanthema, Aizoaceae). *Bradleya*, 11, 1–49.

Hartmann, H. E. K. and Stüber, D. (1993). On the spiny Mesembryanthema and the genus Eberlanzia (Aizoaceae). *Contributions from the Bolus Herbarium*, 15, 1–75.

Harvey, P. H. and Pagel, M. D. (1991). *The comparative method in evolutionary biology*. Oxford University Press.

Hattingh, J. (1984). Body fluid composition in dehydrated and hydrated fog baskers and trench diggers (Coleoptera: Tenebrionidae). *South African Journal of Science*, 80, 191.

Haughton, S. (1915). On some dinosaur remains from Bushmanland. *Transactions of the Geological Society of South Africa*, 5, 259–64.

Haughton, S. (1931). On a collection of fossil frogs from the clays at Banke. *Transactions of the Royal Society of South Africa*, 19, 233–49.

Hayssen, V. and Lacy, R. C. (1985). Basal metabolic rates in mammals: taxonomic differences in the allometry of BMR and body mass. *Comparative Biochemistry and Physiology*, 81A, 741–54.

Heard, H. W. and Stephenson, A. (1987). Electrification of a fence to control the movements of black-backed jackals. *South African Journal of Wildlife Research*, 17, 20–4.

Helldén, U. (1991). Desertification – time for an assessment? *Ambio*, 20, 372–83.

Henderson, L. (1989). Invasive alien woody plants of Natal and the north-eastern Orange Free State. *Bothalia*, 19, 237–61.

Henderson, L. (1991a). Invasive alien woody plants of the northern Cape. *Bothalia*, 21, 177–89.

Henderson, L. (1991b). Invasive alien woody plants of the Orange Free State. *Bothalia*, 21, 73–89.

Henderson, L. (1992). Oleander: an invasive riverside shrub from the Mediterranean. *Veld & Flora*, 78, 84–5.

Henderson, L. and Wells, M. J. (1986). *Alien plant invasions in the grassland and savanna biomes*. In *The ecology and management of biological invasions in southern Africa*. ed. I. A. W. Macdonald, F. J. Kruger and A. A. Ferrar, pp. 109–18. Cape Town: Oxford University Press.

Henderson, M. and Anderson, J. G. (1966). Common weeds in South Africa. *Botanical Survey of South Africa Memoir*, 37. Pretoria: Department of Agriculture.

Henderson, M., Fourie, D. M. C., Wells, M. J. and Henderson, L. (1987). *Declared weeds and alien invader plants in South Africa*. Bulletin 413. Pretoria: Botanical Research Institute.

Hendey, Q. B. (1982). *Langebaanweg: a record of past life*. Cape Town: South African Museum.

Hendricks, F. T. (1995). Antinomies of access: Social differentiation and communal tenure in a Namaqualand reserve, South Africa. Paper presented at the fifth annual conference of the International Association for the Study of Common Property, Bodö, Norway, 24–28 May 1995.

Hendry, G. A. F., Thompson, K., Moss, C. J., Edwards, E. and Thorpe, P. C. (1994). Seed persistence: a correlation between seed longevity in the soil and ortho-dihydroxyphenol concentration.

Functional Ecology, 8, 658–64.

Henning, S. F. (1983). Chemical communication between lycaenid larvae (Lepidoptera: Lycaenidae) and ants (Hymenoptera: Formicidae). *Journal of the Entomological Society of Southern Africa*, 46, 341–66.

Henrici, M. (1935). Germination of Karoo bush seeds. Part 1. *South African Journal of Science*, 32, 223–34.

Henrici, M. (1939). Germination of Karoo bush seeds. Part 2. *South African Journal of Science*, 36, 212–19.

Henrici, M. (1940). Fodder plants of the Broken Veld (Fauresmith District), Part 2. *South African Department of Agriculture, Science Bulletin*, 213. Pretoria: Government Printer.

Henschel, J. R. (1990). Spiders wheel to escape. *South African Journal of Science*, 86, 151–2.

Henschel, J. R. (1995). Tool use by spiders: Stone selection and placement by corolla spiders Ariadna (Segestriidae) of the Namib Desert. *Ethology*, 101, 187–99.

Henschel, J. R. and Jocqué, R. (1994). Bauble spiders: a new species of *Achaearanea* (Araneae: Theridiidae) with ingenious spiral retreats. *Journal of Natural History*, 28, 1287–95.

Henwood, K. (1975). A field-tested thermoregulation model for two diurnal Namib desert tenebrionid beetles. *Ecology*, 56, 1329–42.

Herbel, C. H. (1983). Principles of intensive range improvements. *Journal of Range Management*, 36, 140–4.

Herman, R. P., Provencio, K. R., Torrez, R. J. and Seager, G. W. (1993). Effect of water and nitrogen additions on free-living nitrogen fixer populations in desert grass root zone. *Applied and Environmental Microbiology*, 59, 3021–6.

Herppich, W. (1989). CAM-Ausprung in *Plectranthus marruboides* Benth. (fam. Lamiaceae). Einfluss der Faktoren Licht, Blattentemperatur, Luchtfeuchtigkeit, Bodenwasserverfugbarkeit und Blattwasserzustand. Inaugural dissertation, Münster University.

Herppich, W., Midgley, G. F., Herppich, M., Tuffers, A., Veste, M. and Von Willert, D. J. (1998). Interactive effects of photon fluence rates and drought on CAM-cycling in *Delosperma tradescantioides* (Mesembryanthemaceae). *Physiologia Plantarum*, 102, 148–54.

Herppich, W., Midgley, G. F., Von Willert,

D. J. and Veste, M. (1996). CAM variations in the leaf succulent *Delosperma tradescantioides* (Mesembryanthemaceae), native to southern Africa. *Physiologia Plantarum*, 98, 485–92.

Herselman, M. J. (1989). Model for management decisions. *Karoo Region Newsletter*, 1, 6–7.

Herselman, M. J., Olivier, J. J. and Wentzel, D. (1993). Varying fibre production potentials under veld conditions. *Karoo Agric*, 5(1), 8–10.

Heske, E. J., Brown, J. H. and Mistry, S. (1994). Long-term experimental study of a Chihuahuan Desert rodent community: 13 years of competition. *Ecology*, 75, 438–45.

Hesse, A. J. (1938). A revision of the Bombyliidae (Diptera) of southern Africa. Part 1. *Annals of the South African Museum*, 34, 1–1053.

Hewitt, P. H. and Nel, J. J. C. (1969). Toxicity and repellency of *Chrysocoma tenuifolia* Berg (Compositae to the harvester termite *Hodotermes mossambicus* (Hagen) (Hodotermitidae). *Journal of the Entomological Society of Southern Africa*, 32, 133–6.

Heydorn, A. E. F. and Tinley, K. L. (1980). Estuaries of the Cape. Part 1. Synopsis of the Cape coast. Natural features, dynamics and utilization. *Council for Scientific and Industrial Research, Research Report*, 380.

Hiebert, S. M. (1990). Energy costs and temporal organization of torpor in the Rufus Hummingbird (*Selasphorus rufus*). *Physiological Zoology*, 63, 1082–97.

Higuchi, H. (1994). Photoperiodic induction of diapause, hibernation and voltinism in *Piezodorus hybneri* (Heteroptera: Pentatomidae). *Applied Entomological Zoology*, 29, 585–92.

Hiller, N. and Dunlevey, J. N. (1978). The Bokkeveld–Witteberg boundary in the Montagu-Touws River area, Cape Province. *Transactions of the Geological Society of South Africa*, 81, 101–4.

Hilton-Taylor, C. (1987). Phytogeography and origins of the Karoo flora. In *The Karoo biome: a preliminary synthesis. Part 2 – vegetation and history*, ed. R. M. Cowling and P. W. Roux. *South African National Scientific Programme Reports*, 142, 70–95.

Hilton-Taylor, C. (1994a). The Kaokoveld, Namibia and Angola. In *Centres of plant diversity. A guide and strategy for their conservation*, Vol. 1, *Europe, Africa, South West Asia and the Middle East*, ed. S. D. Davis, V. H. Heywood and A. C. Hamilton, pp. 201–3. Cambridge: IUCN Publications Unit.

Hilton-Taylor, C. (1994b). Western Cape Domain (Succulent Karoo). Republic of South Africa and Namibia. In *Centres of plant diversity. A guide and strategy for their conservation*, Vol. 1, *Europe, Africa, South West Asia and the Middle East*, ed. S. D. Davis, V. H. Heywood and A. C. Hamilton, pp. 204–17. Cambridge: IUCN Publications Unit.

Hilton-Taylor, C. (1996). Patterns and characteristics of the flora of the Succulent Karoo Biome, southern Africa. In *The biodiversity of African plants*, ed. L. J. E. Van der Maesen, X. M. Van der Burgt and J. M. Van Medenbach de Rooy, pp. 58–72. Dordrecht: Kluwer Academic Publishers.

Hilton-Taylor, C. and Le Roux, A. (1989). Conservation status of the fynbos and karoo biomes. In *Biotic diversity in southern Africa. Concepts and conservation*, ed. B. J. Huntley, pp. 202–23. Cape Town: Oxford University Press.

Hilton-Taylor, C. and Moll, E. J. (1986). The Karoo – a neglected biome. *Veld & Flora*, 72, 33–6.

Hincks, W. D. (1955). Dermaptera. In *South African animal life*, vol. 4, ed. B. Hanström, P. Brinck and G. Rudebeck, pp. 33–94. Stockholm: Almqvist Wiksell.

Hobbs, N. T. and Swift, D. M. (1988). Grazing in herds: when are nutritional benefits realized? *American Naturalist*, 131, 760–4.

Hobson, N. K., Jessop, J. P. Van Der Ginn, M. C. (1970). *Karoo plant wealth*. Pearston: Pearston Publications.

Hockey, P. A. R. (1988). The brown locust – war or peace? *South African Journal of Science*, 84, 8–10.

Hockey, P. A. R. and Boobyer, M. G. (1994). Territoriality and determinants of group size in the Karoo Korhaan *Eupodotis vigorsii* (Otididae). *Journal of Arid Environments*, 28, 325–32.

Hodgson, F. D. I. (1986). Geohydrology. In *The Karoo Biome: a preliminary synthesis. Part 1. Physical environment*, ed. R. M. Cowling, P. W. Roux and A. J. H. Pieterse. *South African National Scientific Programmes Report*, 124, 84–91.

Hoeck, H. N. (1982). Population dynamics, dispersal and genetic isolation in two species of hyrax (*Heterohyrax brucei* and *Procavia johnstoni*) on habitat islands in the Serengeti. *Zeitschrift für Tierpsycologie*, 59, 177–210.

Hoeck, H. N. (1989). Demography and competition in hyrax, a 17 year study. *Oecologia*, 79, 353–60.

Hoeck, H. N., Klein, H. and Hoeck, P. (1982). Flexible social organization in hyrax. *Zeitschrift für Tierpsycologie*, 59, 265–98.

Hoffman, M. T. (1988a). The pollination of *Aloe ferox* Mill. *South African Journal of Botany*, 54, 345–50.

Hoffman, M. T. (1988b). The rationale for Karoo grazing systems: criticisms and research implications. *South African Journal of Science*, 84, 556–9.

Hoffman, M. T. (1989a). *Vegetation studies and the impact of grazing in the semi-arid Eastern Cape*. Ph.D. thesis, University of Cape Town.

Hoffman, M. T. (1989b). A preliminary investigation of the phenology of subtropical thicket and karroid shrubland in the lower Sundays River Valley, SE Cape. *South African Journal of Botany*, 55, 586–98.

Hoffman, M. T. (1995). Environmental history and desertification of the Karoo, South Africa. *Gionale Botanico Italiano*, 129(1), 261–73.

Hoffman, M. T. (1996). Nama Karoo Biome. In *Vegetation of South Africa, Lesotho and Swaziland*, ed. A. B. Low and A. G. Rebelo, pp. 52–61. Pretoria: Department of Environmental Affairs and Tourism.

Hoffman, M. T. and Cowling, R. M. (1987). Plant physiognomy, phenology and demography. In *The Karoo biome: a preliminary synthesis. Part 2 – vegetation and history*, ed. R. M. Cowling and P. W. Roux. *South African National Scientific Programmes Report*, 142, 1–34.

Hoffman, M. T. and Cowling, R. M. (1990a). Desertification in the lower Sundays River Valley, South Africa. *Journal of Arid Environments*, 19, 105–17.

Hoffman, M. T. and Cowling, R. M. (1990b). Vegetation change in the semi-arid eastern Karoo over the last two hundred years: an expanding Karoo – fact or fiction. *South African Journal of Science*, 86, 286–94.

Hoffman, M. T. and Cowling, R. M. (1991).

Phytochorology and endemism along aridity and grazing gradients in the Sundays River Valley, South Africa. *Journal of Biogeography*, 18, 189–201.

Hoffman, M. T., Barr, G. D. and Cowling, R. M. (1990). Vegetation dynamics in the semi-arid eastern Karoo, South Africa: the effect of seasonal rainfall and competition on grass and shrub basal cover. *South African Journal of Science*, 86, 462–3.

Hoffman, M. T., Bond, W. J. and Stock, W. D. (1995a). Desertification of the eastern Karoo, South Africa, conflicting palaeoecological, historical, and soil isotopic evidence. *Environmental Monitoring and Assessment*, 37, 159–77.

Hoffman, M. T., Cowling, R. M., Douie, C. and Pierce, S. M. (1989). Seed predation and germination of *Acacia erioloba* in the Kuiseb River Valley, Namib Desert. *South African Journal of Botany*, 55, 103–6.

Hoffman, M. T., Midgley, G. F. and Cowling, R. M. (1994). Plant richness is negatively related to energy availability in semi-arid southern Africa. *Biodiversity Letters*, 2, 35–8.

Hoffman, M. T., Sonnenberg, D., Hurford, J. L. and Jagger, B. W. (1995b). *The Ecology and Management of Riemvasmaak's Natural Resources*. Claremont: National Botanical Institute.

Hoffmann, J. H., Impson, F. A. C. and Moran, V. C. (1993). Competitive interactions between two bruchid species (*Algarobius* spp.) introduced into South Africa for biological control of mesquite weeds (*Prosopis* spp.). *Biological Control*, 3, 215–20.

Hofmeyr, M. D. and Louw, G. N. (1987). Thermoregulation, pelage conductance and renal function in the desert-adapted springbok, *Antidorcas marsupialis*. *Journal of Arid Environments*, 13, 137–51.

Hollmann, J., Myburgh, S., Van der Schijff, M. and Van Wyk, B. (1995). Aardvark and cucumber: a remarkable relationship. *Veld and Flora*, 81, 108–9.

Holm, E. (1990). Notes on faunas bordering on the Namib Desert. In *Namib ecology: 25 years of Namib research*, ed. M. K. Seely. *Transvaal Museum Monograph*, 7, 55–60.

Holm, E. and Edney, E. B. (1973). Daily activity of Namib Desert arthropods in relation to climate. *Ecology*, 54, 45–56.

Holm, E. and Kirsten, J. F. (1979).

Pre-adaptation and speed mimicry among Namib Desert scarabaeids with orange elytra. *Journal of Arid Environments*, 2, 263–71.

Holm, L. G., Pancho, J. V., Herberger, J. P. and Plucknett, D. L. (1991). *A geographical atlas of world weeds*. Florida: Krieger Publishing Company.

Holmes, P. J. and Marker, M. E. (1995). Evidence for environmental change from Holocene valley fills from three central Great Karoo upland sites. *South African Journal of Science*, 91, 617–20.

Hoon, J. H. and King, B. R. (1993). Garingboomaanvulling vir dorpers op oumansoutbos. *Karoo Agric*, 5(1), 1–2.

Hoon, J. H., King, B. R. and Jordaan, G. (1995). Die ontrekkinsperiode van vee vanaf veld nadat prosopispeule ingeneem is. *Grootfontein Agric*, 1(1), 7–12.

Horowitz, A. (1992). *Palynology of Arid Lands*. Amsterdam: Elsevier.

Horowitz A., Sampson, C. G., Scott, L. and Vogel, J. C. (1978). Analysis of the Voightspost site, O.F.S., South Africa. *South African Archaeological Bulletin*, 33, 152–9.

Howe, H. F. and Smallwood, J. (1982). Ecology of seed dispersal. *Annual Review of Ecology and Systematics*, 13, 201–28.

Huey, R. B. and Pianka, E. R. (1977). Natural selection for juvenile lizards mimicking noxious beetles. *Science*, 195, 201–3.

Huey, R. B. and Pianka, E. R. (1981). Ecological consequences of foraging mode. *Ecology*, 62, 991–9.

Huey, R. B., Bennett, A. F., John-Alder, H. and Nagy, K. A. (1984). Locomotor capacity and foraging behaviour of Kalahari lacertid lizards. *Animal Behavior*, 32, 41–50.

Huey, R. B., Pianka, E. R., Egan, M. E. and Coons, L. W. (1974). Ecological shifts in sympatry: Kalahari fossorial lizards (*Typhlosaurus*). *Ecology*, 55, 304–16.

Hughes, J. J., Ward, D. and Perrin, M. R. (1994). Predation risk and competition affect habitat selection and activity of Namib Desert gerbils. *Ecology*, 75, 1397–405.

Hunter, R. B., Romney, E. M. and Wallace, A. (1982). Nitrate distribution in the Mojave Desert soils. *Soil Science*, 134, 22–30.

Huntley, B. J. and Matos, E. M. (1994). Botanical diversity and its conservation

in Angola. In *Botanical Diversity in southern Africa*, ed. B. J. Huntley. *Strelitzia*, 1, 53–74.

Huntley, B. J., Siegfried, R. and Sunter, C. (1989). *South African environments into the 21st Century*. Cape Town: Human and Rousseau and Tafelberg.

Hurt, C. R. and Bosch, O. J. H. (1991). A comparison of some range condition assessment techniques used in southern African grasslands. *Journal of the Grasslands Society of South Africa*, 8, 131–7.

Hutchinson, J. (1946). *A botanist in South Africa*. London: Crawthorn.

Ihlenfeldt, H.-D. (1971). Some aspects of the biology of dissemination of the Mesembryanthemaceae. In *The genera of the Mesembryanthemaceae*, ed. H. Herre, pp. 28–34. Cape Town: Tafelberg Publications.

Ihlenfeldt, H.-D. (1983). Dispersal of Mesembryanthemaceae in arid habitats. *Sonerbände Naturwissenschaftlichen Vereins in Hamburg*, 7, 381–90.

Ihlenfeldt, H.-D. (1994). Diversification in an arid world: the Mesembryanthemaceae. *Annual Review of Ecology and Systematics*, 25, 521–46.

Immelman, W. F. E., Dicht, C. L. and Ackerman, D. P. (ed.) (1973). *Our Green Heritage*. Cape Town: Tafelberg.

Immelmann, K. (1971). Ecological aspects of periodic reproduction. In *Avian biology*, ed. D. S. Farner, J. R. King and K. C. Parkes, pp. 341–89. New York: Academic Press.

Immelmann, K. (1972). Erörterungen zur Definition und Anwendbarkeit der Begriffe 'Ultimate Factor', 'Proximate Factor' und 'Zeitgeber'. *Oecologia*, 9, 259–64.

Ingham, R. E., Trofymow, J. A., Ingham, E. R. and Coleman, D. C. (1985). Interactions of bacteria and fungi, and their nematode grazers: effects on nutrient cycling and plant growth. *Ecological Monographs*, 55, 119–40.

Ingrouille, M. (1992). *Diversity and evolution of land plants*. London: Chapman and Hall.

Inouye, R. S. (1980). Density-dependent germination response by seeds of desert annuals. *Oecologia*, 46, 235–8.

Inouye, R. S. (1991). Population biology of desert annual plants. In *The ecology of desert communities*, ed. G. Polis, pp.

27–54. Tucson, AZ: The University of Arizona Press.

Irish, J. (1990). Namib biogeography, as exemplified mainly by the Lepismatidae (Thysanura:Insecta). In *Namib ecology: 25 years of Namib research*, ed. M. K. Seely. *Transvaal Museum Monograph*, 7, 163–8.

Jackson, S. P. (1951) Climates of southern Africa. *South Africa Geographical Journal*, 33, 17–37.

Jacobs, D. S., Bennett, N. C., Jarvis, J. U. M. and Crowe, T. M. (1991). The colony structure and dominance hierarchy of the Damaraland mole-rat, *Cryptomys damarensis* (Rodentia: Bathyergidae) from Namibia. *Journal of Zoology, London*, 224, 553–76.

Jacobsen, H. (1960). *A handbook of succulent plants*, vols. 1–3. Dorset, UK: Blandford Press.

Jacobsen, W. B. G. (1967a). The influence of copper content of the soil on trees and shrubs of Molly South Hill, Mangula. *Kirkia*, 6, 63–84.

Jacobsen, W. B. G. (1967b). The influence of copper content of the soil on trees and shrubs of Silverside North, Mangula. *Kirkia*, 6, 259–77.

Jago, N. (1987). The return of the eighth plague. *New Scientist*, 1565, 47–51.

Janzen, D. H. (1976). Why bamboos take so long to flower. *Annual Review of Ecology and Systematics*, 7, 347–91.

Jarman, N. and Bosch, O. (1973). The identification and mapping of extensive secondary invasive and degraded ecological types (test site D). In *To assess the value of satellite imagery in resource evaluation on a national scale*, ed. O. G. Malan, pp. 77–80. Pretoria: Council for Scientific and Industrial Research.

Jarvis, J. U. M. (1978). Energetics of survival in *Heterocephalus glaber* (Rüppell), the naked mole-rat (Rodentia: Bathyergidae). *Bulletin of the Carnegie Museum of Natural History*, 6, 81–7.

Jarvis, J. U. M. (1981). Eusociality in a mammal: cooperative breeding in naked mole-rat colonies. *Science*, 212, 571–3.

Jarvis, J. U. M. and Bennett, N. C. (1990). The evolutionary history, population biology and social structure of African mole-rats: Family Bathyergidae. In *Evolution of subterranean mammals at the organismal and molecular levels*, ed. E. Nevo and O. A. Reig, pp. 97–128. New York: Wiley-Liss.

Jarvis, J. U. M. and Bennett, N. C. (1993). Eusociality has evolved independently in two genera of bathyergid mole-rats – but occurs in no other subterranean mammal. *Behavioural Ecology and Sociobiology*, 33, 253–60.

Jeltsch, F., and Wissel, C. (1994). Modelling dieback phenomena in natural forests. *Ecological Modelling*, 75/76, 111–21.

Jeltsch, F., Milton, S. J., Dean, W. R. J. and Van Rooyen, N. (1996). Tree spacing and coexistence in semi-arid savannas. *Journal of Applied Ecology*, 84, 583-95.

Jeltsch, F., Milton, S. J., Dean, W. R. J. and Van Rooyen, N. (1997a). Simulated pattern formation around artificial waterholes in the semi-arid Kalahari. *Journal of Vegetation Science*, 8, 177–88.

Jeltsch, F., Milton, S. J., Dean, W. R. J. and Van Rooyen, N. (1997b). Analysing shrub encroachment in the southern Kalahari: a grid-based modelling approach. *Journal of Applied Ecology*, 34, 1497–509.

Jessen, C., Laburn, H. P., Knight, M. H., Kuhnen, G., Goelst, K. and Mitchell, D. (1994). Blood and brain temperatures of free-ranging black wildebeest in their natural environment. *American Journal of Physiology: Regulation, Integration and Comparative Physiology*, 267, R1528–R1536.

Joffe, P. (1993). *The gardener's guide to South African plants*. Cape Town: Tafelberg Publishers Ltd.

Johnson, A. W. (1968). The evolution of desert vegetation in western North America. In *Desert biology*, vol. 1, ed. G. W. Brown Jr., pp. 101–4. New York: Academic Press.

Johnson, C. D. (1983). Handbook of seed insects of *Prosopis* species. Ecology, control and identification of seed-infesting insects of New World *Prosopis* (Leguminosae), *FAO Handbook*.

Johnson, S. D. (1992). Climatic and phylogenetic determinants of flowering seasonality in the Cape flora. *Journal of Ecology*, 81, 567–72.

Johnson, S. D. (1992). Plant–animal interactions. In *The ecology of fynbos. Nutrients, fire and diversity*, ed. R. M. Cowling, pp. 175–205. Cape Town: Oxford University Press.

Johnson, S. D and Midgley, J. J. (1996a). The adaptive significance of floral patterns in monkey beetle and bee-fly pollinated plants. *South African Association of Botanists AGM abstracts*.

Johnson, S. D and Midgley, J. J. (1996b). Fly pollination of *Gorteria diffusa* (Asteraceae), and a possible mimetic function for dark spots on the capitulum. *American Journal of Botany*, 84, 429–36.

Jooste, J. F. (1980). *A study of the phytosociology and small mammals of the Rolfontein Nature Reserve, Cape Province*. M.Sc. thesis, University of Stellenbosch.

Jooste, J. F. (1983). Game farming as a supplementary farming activity in the Karoo. *Proceedings of the Grassland Society of Southern Africa*, 18, 46–9.

Jooste, J. F. and Palmer, N. G. (1982). The distribution and habitat preference of some small mammals in the Rolfontein Nature Reserve. *South African Journal of Wildlife Research*, 12, 26–35.

Jordaan, G. (1993). 'n Ondersoek na die voervloeisituasie op plase in die Karoostreek en die bepaling van die rol wat *Triticum aestivum* en *Lolium multiflorum/Trifolium resupinatum* weidings in kleinveeproduksie kan speel. M.Sc. Agriculture thesis, University of Fort Hare, Alice.

Jordan, P. W. and Nobel, P. S. (1982). Height distribution of two species of cacti in relation to rainfall, seedling establishment and growth. *Botanical Gazette*, 143, 511–17.

Joubert, J. G. V. (1980). Veld and pastures in animal production systems in the Western Cape and South West Africa/Namibia. *South African Journal of Animal Science*, 10, 299–303.

Joubert, J. G. V. and Van Breda, P. A. B. (1976). Vestigingsmetodes vir *Osteospermum sinuatum* in die veld van die Klein Karoo. *Proceedings of the Grassland Society of Southern Africa*, 11, 123–4.

Jubb, R. A. (1976). Freshwater mussels, Unionidae, what is their distribution in South African inland waters today? *Piscator*, 97, 73–5.

Juhren, M., Went, F. W. and Phillips, E. (1956). Ecology of desert plants. 4. Combined field and laboratory work on germination of annuals in the Joshua Tree National Monument, California.

Ecology, 37, 318–30.

Jump, J. A. (1988). Phytogeographic and evolutionary trends in Lithops. Aloe, 25, 33.

Jürgens, N. (1986). Untersuchungen zur ökologie sukkulenter pflanzen des südlichen Afrika. Mitteilungen aus dem Institut für Allgemeine Botanik, Hamburg, 21, 139–365.

Jürgens, N. (1991a) Psammophorous plants and other adaptations to desert ecosystems with high incidence of sand storms. Feddes Repertorium, 107, 345–59.

Jürgens, N. (1991b). A new approach to the Namib Region. I: Phytogeographic subdivision. Vegetatio, 97, 21–38.

Jürgens, N., Gunster, A., Seely, M. K. and Jacobsen, K. M. (1997). Desert. In Vegetation of southern Africa, ed. R. M. Cowling, D. M. Richardson and S. M. Pierce, pp. 189–215. Cambridge University Press.

Kalisz, S. (1991). Experimental determination of seed bank age structure in the winter annual Collinsia verna. Ecology, 72, 575–85.

Kanthack, F. E. (1930). The alleged desiccation of South Africa. Geographical Journal, 76, 516–21.

Keating, B. A., McCown, R. L. and Cresswell, H. P. (1995). Paddock-scale models and catchment-scale problems: the role of APSim in the Liverpool plains. In Proceedings of the International Congress on Modelling and Simulation, Vol. I, Agriculture, Catchment Hydrology and Industry, ed. P. Binnings, H. Bridgman and B. Williams, pp. 158–65. University of Newcastle Press.

Keeley, L. H. (1988). Hunter-gatherer economic compexity and 'population pressure': a cross-cultural analysis. Journal of Anthropological Archaeology, 7, 373–411.

Keightley, A. I., Struthers, J. K., Johnson, S. and Barnard, B. J. H. (1987). Rabies in South Africa, 1980–1984. South African Journal of Science, 83, 466–72.

Keller, C. M. (1973). Montagu Cave in prehistory. University of California Anthropological Records, 28, 1–150.

Kellerman, T. S., Coetzer, J. A. W. and Naudé, T. W. (1988). Plant poisonings and mycotoxicoses of livestock in South Africa. Cape Town: Oxford University Press.

Kellner, K. (1995). Vegetation dynamics during the processes of recovery and degradation in parts of the grassveld and Karoo biomes of Southern Africa. Ph.D. thesis. Potchefstroom University for CHE, Potchefstroom.

Kellner, K. and Bosch, O. J. H. (1992). Influence of patch formation in determining stocking rate for southern African grasslands. Journal of Arid Environments, 22, 99–105.

Kemp, P. R. (1983). Phenological pattern of Chihuahuan desert plants in relation to the timing of water availability. Journal of Ecology, 71, 427–36.

Kemp, P. R. (1989). Seed banks and vegetation processes in deserts. In Ecology of soil seed banks, ed. M. A. Leck, V. T. Parker and R. L. Simpson, pp. 257–81. San Diego: Academic Press.

Kemp, T. S. (1982). Mammal-like reptiles and the origin of mammals. London: Academic Press.

Kenagy, G. J. (1976). Field observations of male fighting, drumming, and copulation in the Great Basin kangaroo rat, Dipodomys microps. Journal of Mammalogy, 57, 781–5.

Keogh, H. J. (1973). Behaviour and breeding in captivity of the Namaqua gerbil Desmodillus auricularis (Cricetidae, Gerbillinae). Zoologica Africana, 8, 231–40.

Kerley, G. I. H. (1989). Diet of small mammals from the Karoo, South Africa. South African Journal of Wildlife Research, 19, 67–72.

Kerley, G. I. H. (1990). Browsing by Lepus capensis in the Karoo. South African Journal of Zoology, 25, 199–200.

Kerley, G. I. H. (1991). Seed removal by rodents, birds and ants in the semi-arid Karoo, South Africa. Journal of Arid Environments, 20, 63–9.

Kerley, G. I. H. (1992a). Ecological correlates of small mammal community structure in the semi-arid Karoo, South Africa. Journal of Zoology, London, 227, 17–27.

Kerley, G. I. H. (1992b). Small mammal seed consumption in the Karoo, South Africa, further evidence for divergence in desert biotic process. Oecologia, 89, 471–5.

Kerley, G. I. H. (1992c). Trophic status of small mammals in the semi-arid Karoo, South Africa. Journal of Zoology, London, 226, 563–72.

Kerley, G. I. H. (1993). Small mammal seed consumption in the semiarid Karoo, South Africa – further evidence for divergence in desert biotic processes. Oecologia, 89, 471–5.

Kerley, G. I. H. (1995). The round-eared elephant-shrew Macroscelides proboscideus (Macroscelidea) as an omnivore. Mammal Review, 25, 39–44.

Kerley, G. I. H. and Erasmus, T. (1992a). Fire and the range limits of the bush Karoo rat Otomys unisulcatus. Global Ecology and Biogeography Letters 2, 11–15.

Kerley, G. I. H. and Erasmus, T. (1992b). Small mammals in the semi-arid Karoo, South Africa, biomass and energy requirements. Journal of Arid Environments, 22, 251–60.

Kerley, G. I. H. and Whitford, W. G. (1994). Desert-dwelling small mammals as granivores: intercontinental variations. Australian Journal of Zoology, 42, 543–55.

Kerley, G. I. H., Tiver, F. and Whitford, W. G. (1993). Herbivory of clonal populations: cattle browsing affects reproduction and population structure of Yucca elata. Oecologia, 93, 12–17.

Kevan P. G. and Baker, H. G. (1983). Insects as flower visitors and pollinators. Annual Review of Entomology, 28, 407–53.

Keyser, A. W. and Smith, R. M. H. (1979). Vertebrate biozonation of the Beaufort Group with special reference to the western Karoo basin. Annals of the South African Geological Survey, 12, 1–35.

Khrone, H. and Steyn, L. (1991). Land use in Namaqualand. Cape Town: Surplus People Project.

Kilduff, T. S., Krilowicz, B., Milsom, W. K., Trachsel, L. and Wang, L. C. H. (1993). Sleep and mammalian hibernation: Homologous adaptations and homologous processes. Sleep, 16, 372–86.

Kinahan, J. (1991). Pastoral nomads of the Central Namib Desert. Windhoek: New Namibia Books.

Kinahan, J. and Kinahan, J. H. A. (1984). Holocene subsistence and settlement on the Namib coast: the example of the Ugab River mouth. Cimbebasia, Ser (B), 4(6), 59–72.

King, G. (1990). The Dicynodonts. A study in palaeobiology. London: Chapman and Hall.

King, L. C. (1955). Pediplanation and isotasy: an example from South Africa. Quarterly Journal of the Geological Society of London, 111, 353–9.

King, L. C. (1967). The morphology of the

Earth. Edinburgh: Oliver and Boyd.

Kingdon, J. (1971). *East African mammals*, vol. 1. London: Academic Press.

Kingdon, J. (1974). *East African mammals*. vol. 2A. London: Academic Press.

Kirchheimer, F. (1934). On pollen from the upper Cretaceous dysodil of Banke, Namaqualand (South Africa). *Transactions of the Royal Society of South Africa*, 19, 41–50.

Kitching, J. W. (1977). Distribution of the Karoo vertebrate fauna. *Memoirs of the Bernard Price Institute of Palaeontological Research*, 1, 1–131.

Klein, R. G. (1988). The archaeological significance of animal bones from Acheulean sites in southern Africa. *The African Archaeological Review*, 6, 3–25.

Klein, R. G., Cruz-Uribe, K. and Beaumont, P. B. (1991). Environmental, ecological and palaeoanthropological implications of the late Pleistocene mammalian fauna from Equus Cave, northern Cape Province, South Africa. *Quaternary Research*, 36, 94–119.

Knight, M. H. (1989). The importance of water to doves and sandgrouse in the semi-arid southern Kalahari. *South African Journal of Wildlife Research*, 19, 42–6.

Knight, M. H. (1991). Ecology of the gemsbok *Oryx gazella gazella* (Linnaeus) and blue wildebeest *Connochaetes taurinus* (Burchell) in the southern Kalahari, Ph.D. thesis, University of Pretoria.

Knight, M. H. (1995a). Drought-related mortality of wildlife in the southern Kalahari and the role of man. *African Journal of Ecology*, 33, 377–94.

Knight, M. H. (1995b). Tsama melons, *Citrullus lanatus*, a supplementary water supply for wildlife in the southern Kalahari. *African Journal of Ecology*, 33, 71–80.

Knight, M. H. and Skinner, J. D. (1981). Thermoregulatory, reproductive and behavioural adaptations of the big eared desert mouse, *Malacothrix typica* to its arid environment. *Journal of Arid Environments*, 4, 137–45.

Knight, R. S. (1988). Aspects of plant dispersal in the southwestern Cape with particular reference to the roles of birds as dispersal agents. Ph.D. thesis, University of Cape Town.

Knight, R. S., Rebelo, A. G. and Siegfried, W. R. (1989). Plant assemblages on

Mima-like earth mounds in the Clanwilian district, South Africa. *South African Journal of Botany*, 55, 465–72.

Kok, O. B. (1980). Voedseliname van volstruise in die Namib-Naukluft park, Suidwes-Afrika. *Madoqua*, 12, 155–61.

Kok, O. B. (1996). Dieëtsamestelling van enkele karnivoorsoorte in die Vrystaat, Suid Afrika. *South African Journal of Science*, 92, 393–8.

Kok, O. B. and Hewitt, P. H. (1990). Bird and mammal predators of the harvester termite *Hodotermes mossambicus* (Hagen) in semi-arid regions of South Africa. *South African Journal of Science*, 86, 34–7.

Kok, O. B. and Louw, S. (1994). Bird and mammal predators of curculionid and tenebrionid beetles in semi-arid regions of South Africa. *Journal of African Zoology*, 108, 555–63.

Kokot, D. F. (1948). An investigation into the evidence bearing on recent climatic change over southern Africa. *South African Department of Irrigation Memoirs*, 160 pp.

Kolata, G. B. (1974). !Kung hunter-gatherers: feminism, diet and birth control. *Science*, 185, 932–4.

Krause, G. H., Grafflage, S., Rumich-Bayer, S. and Somersalo, S. (1988). Effects of freezing on plant mesophyll cells. In *Plants and temperature*, ed. S. P. Long and F. I. Woodward, pp. 311–28. Cambridge: Society for Experimental Biology.

Kruuk, H. (1972). *The spotted hyaena: a study of predation and social behaviour*. University of Chicago Press.

Kuhnen, G. and Jessen, C. (1991). Threshold and slope of selective brain cooling. *Pflügers Archiv*, 418, 176–83.

Kuhnen, G. and Jessen, C. (1994). Thermal signals in control of selective brain cooling. *American Journal of Physiology: Regulation, Integration and Comparative Physiology*, 267, R355–R359.

Kuman, K. and Clarke, R. J. (1986). Florisbad: new excavations at a Middle Stone Age site in South Africa. *Geoarchaeology*, 1, 103–25.

Kuntzsch, V. and Nel, J. A. J. (1992). Diet of bat-eared foxes in the Karoo. *Koedoe*, 35(2), 37–48.

Kuyper, M. A. (1979). A biological study of the golden mole *Amblysomus hottentotus*. M.Sc. thesis, University of Natal.

Laburn, H. P., Mitchell, D., Mitchell, G. and

Saffy, K. (1988). Effects of tracheostomy breathing on brain and body temperatures in hyperthermic sheep. *Journal of Physiology*, 406, 331–44.

Lamoral, B. H. (1979). The scorpions of Namibia (Arachnida: Scorpionida). *Annals of the Natal Museum*, 23, 497–783.

Lancaster, J., Lancaster, N. and Seely, M. K. (1984). Climate of the central Namib Desert. *Madoqua*, 14, 5–61.

Lancaster, N. (1989). *The Namib Sand Sea*. Rotterdam: A. A. Balkema.

Land Type Survey Staff (1984). Land types of the maps 2522 Bray, 2622 Morokweng, 2524 Mafeking, 2624 Vryburg. *Memoirs of the Agricultural National Resource South Africa*, 1.

Langman, V. A., Roberts, T. J., Black, J., Maloiy, G. M. O., Heglund, N. C., Webers, J. M., Kram, R. and Taylor, C. R. (1995). Moving cheaply: Energetics of walking in the African elephant. *Journal of Experimental Biology*, 198, 629–32.

Lanza, B. (1981). A check-list of the Somali amphibians. *Monitore Zoologica Italia (N.S.) Suppl.* 15, 151–86.

Lavorel, S., O'Neill, R. V. and Gardner, R. H. (1994). Spatio-temporal dispersal strategies and annual plant species coexistence in a structured landscape. *Oikos*, 71, 75–88.

Lawrence, R. F. (1955). Solifugae, Scorpions and Pedipalpi, with checklists and keys to South African families, genera and species. In *South African animal life*, vol. 1, ed. B. Hanström, P. Brinck and G. Rudebeck, pp. 151–262. Stockholm: Almqvist Wiksell.

Lawrence, R. F. (1963). The Solifugae of South West Africa. *Cimbebasia*, 8, 1–28.

Lawrence, R. F. (1964). A conspectus of South African spiders. *Bulletin of the Department of Agricultural Technical Services*, 369.

Lawrence, R. F. (1975). The chilopoda of south west Africa. *Cimbebasia* (A) 4(2), 35–45.

Lay, D. M. (1972). The anatomy, physiology, functional significance, and evolution of specialized hearing organs of gerbilline rodents. *Journal of Morphology*, 138, 41–93.

Le Bars, H. (1967). The endogenous urea cycle in ruminants. In *Urea as a protein supplement*, ed. M. H. Briggs, pp. 155–72.

Oxford: Pergamon Press.

Le Houérou, H. N. (1994). *Drought tolerant and water efficient fodder shrubs (DTFS), their role as a 'drought insurance' in the agricultural development of arid and semi-arid zones in southern Africa.* Report to the Water Research Commission of South Africa, 130 pp.

Le Houérou, H. N., Bingham, R. L. and Skerbek, W. (1988). Relationship between the variability of primary production and the variability of annual precipitation in world arid lands. *Journal of Arid Environments*, 15, 1–18.

Le Houérou, H. N., Popov, G. F. and See, L. (1993). Agro-bioclimatic classification of Africa. *Agrometeorology Series Working Paper*, 6. Rome: FAO.

Le Roux, A. (1984). 'n Fitososiologiese studie van die Hester Malan-Natuurreservaat. M.Sc. thesis, University of Pretoria.

Le Roux, A., Perry, P. and Kyriacou, X. (1989). South Africa. In *Plant phenomorphological studies in Mediterranean type ecosystems*, ed. G. Orshan, pp. 159–346. Dordrecht: Kluwer Academic Publishers.

Lea, A. (1964). Some major factors in the population dynamics of the brown locust *Locustana pardalina* (Walker). In *Ecological studies in southern Africa*, ed. D. H. S. Davis, pp. 269–83. The Hague: W. Junk.

Lea, A. (1968). Natural regulation and artificial control of brown locust numbers. *Journal of the Entomological Society of Southern Africa*, 31, 97–112.

Lea, A. (1969). The distribution and abundance of brown locusts, *Locustana pardalina* (Walk), between 1954 and 1965. *Journal of the Entomological Society of Southern Africa*, 32, 367–98.

Lebedev, A. N. (1970). *The climate of Africa. Part I: Air, temperature, precipitation.* Jerusalem: Israel Program for Scientific Translation.

Ledger, J. A. (1969). Notes on the tropical nest fly. *Bokmakierie*, 21, 28–30.

Lee, K. E. and Foster, R. C. (1991). Soil fauna and soil structure. *Australian Journal of Soil Research*, 29, 745–75.

Lee, K. E. and Wood, T. G. (1971). *Termites and soils.* London: Academic Press.

Lee, R. B. (1969). !Kung Bushman subsistence: an input–output analysis. In *Environment & cultural behavior*, ed.

A. P. Vayda, pp. 47–79. New York: Natural History Press.

Lee, R. B. (1979). *The !Kung San.* Cambridge University Press.

Lee, R. B. and Devore, I. (1976). *Kalahari Hunter-Gatherers.* Cambridge, MA: Harvard University Press.

Lee-Thorpe, J. A and Beaumont, P. B. (1990). Environmental shifts in the last 20,000 years: isotopic evidence from Equus Cave. *South African Journal of Science*, 86, 452–3.

Leistner, O. A. (1961). On dispersal of *Acacia giraffae* by game. *Koedoe*, 4, 101–4.

Leistner, O. A. (1967). The plant ecology of the southern Kalahari. *Botanical Survey of South Africa Memoirs*, 38.

Leistner, O. A. (1991). Annuals of the arid transition zone between winter and summer rainfall regions in southern Africa. Proceedings of the AETFAT Congress.

Lensing, J. E. (1983). Feeding strategy of the rock hyrax and its relation to the rock hyrax problem in southern South West Africa. *Madoqua*, 13, 177–96.

Leonard, S. G., Miles, R. L. and Tueller, P. T. (1988). Vegetation–soil relationships on arid and semi-arid rangelands. In *Vegetation science applications for rangeland analysis and management*, ed. P. T. Tueller, pp. 226–51. Dordrecht: Kluwer Academic Publishers.

Leroi, A. M., Rose, M. R. and Lauder, G. V. (1994). What does the comparative method reveal about adaptation? *The American Naturalist*, 143, 381–402.

Levin, M., Nieman, N. and Le Roux, J. P. (1986). The development of Tertiary formations on the Bushmanland Plateau, pp. 1035–9. Extended Abstracts of the 21st Biennial Congress, Geological Society of South Africa, Johannesburg.

Levin, S. A., Cohen, D. and Hastings, A. (1984). Dispersal strategies in patchy environments. *Theoretical Population Biology*, 26, 165–91.

Lewis-Williams, J. D. (1981). *Believing and seeing: Symbolic meanings in southern San Roque paintings.* New York: Academic Press.

Liede, S., Hammer, S. A. and Whitehead, V. (1991). Observations on pollination and hybridization in the genus *Conophytum* (Mesembryanthemaceae). *Bradleya*, 9, 93–9.

Liengme, C. (1987). Botanical remains

from archaeological sites in the Western Cape. In *Papers in the prehistory of the Western Cape, South Africa*, ed. J. Parkington and M. Hall, pp. 237–61. Oxford: BAR International Series, 332.

Lightfoot, D. C. and Whitford, W. G. (1989). Interplant variation in creosotebush foliage characteristics and canopy arthropods. *Oecologia*, 81, 166–75.

Lightfoot, D. C. and Whitford, W. G. (1990). Phytophagous insects enhance nitrogen flux in a desert creosotebush community. *Oecologia*, 82, 18–25.

Lighton, J. B. (1991). Ventilation in Namib Desert Tenebrionid beetles: mass scaling and evidence of a novel quantized flutter-phase. *Journal of Experimental Biology*, 159, 249–68.

Lighton, J. R. B. (1996). Discontinuous gas exchange in insects. *Annual Review of Entomology*, 41, 309–24.

Liljelund, L. E., Årgren, G. I. and Fagerström, T. (1988). Succession in stationary environments generated by interspecific differences in life history parameters. *Annales Zoologici Fennici*, 25, 17–22.

Linder, H. P., Meadows, M. E. and Cowling, R. M. (1992). The history of the Cape flora. In *The ecology of fynbos*, ed. R. M. Cowling, pp. 113–34. Cape Town: Oxford University Press.

Linn, I. J. (1991). Influence of 6-methoxy-benzoxazolonone and green vegetation on reproduction of the multimammate rat *Mastomys coucha*. *South African Journal of Wildlife Research*, 21, 33–7.

Little, R. M., Crowe, T. M. and Villacastin-Herrero, C. A. (1996). Conservation implications of long-term population trends, environmental correlates and predictive models for Namaqua Sandgrouse *Pterocles namaqua*. *Biological Conservation*, 75, 93–101.

Liversidge, R. (1961). The Wattled Starling (*Creatophora cinerea*). *Annals of the Cape Provincial Museums*, 1, 71–80.

Liversidge, R. (1980). Seasonal changes in the use of avian habitat in southern Africa. In *Proceedings of the 17th International Ornithological Congress: Symposium on evolution of habitat utilization*, ed. A. Keast, pp. 1019–24. Berlin: Deutsch Ornithologen Gesellschaft.

Liversidge, R. (1984). The importance of national parks for raptor survival.

Proceedings of the Fifth Pan-African Ornithological Congress, pp. 589–600.

Lloyd, J. W. (1985). *A plant ecological study of the farm 'Vaalputs', Bushmanland, with special reference to edaphic factors.* M.Sc. thesis, University of Cape Town.

Lloyd, J. W. (1989a). Discriminant analysis and ordination of vegetation and soils on the Vaalputs radioactive waste disposal site, Bushmanland, South Africa. *South African Journal of Botany*, 55, 127–36.

Lloyd, J. W. (1989b). Phytosociology of the Vaalputs radioactive waste disposal site, Bushmanland, South Africa. *South African Journal of Botany*, 55, 372–82.

Lloyd, P. H. and Millar, J. C. G. (1983). A questionnaire survey (1969–1974) of some of the larger mammals of the Cape Province. *Bontebok*, 3, 1–49.

Lloyd, P., Little, R. M. and Crowe, T. M. (1994). The Namaqua Sandgrouse – supreme patchy resource specialist of the arid zone. Paper presented at the Arid Zone Ecology Forum, Beaufort West, September 1994.

Lobry de Bruyn, L. A. and Conacher, A. J. (1990). The role of termites and ants in soil modification: a review. *Australian Journal of Soil Research*, 28, 55–93.

Lock, B. E. (1980). Flat-plate subduction and the Cape Fold Belt of South Africa. *Geology*, 8, 35–9.

Lockyer, S. (1932). Seed dispersal from hygroscopic Mesembryanthemum fruits; *Bergeranthus scapigerus* Schw., and *Dorotheanthus bellidiformis* N.E. Br., with a note on *Carpanthea pomeridiana* N.E. Br. *Annals of Botany*, 46, 323–42.

Loope, L. L., Sanchez, P. G., Tarr, P. W., Loope, W. L. and Anderson, R. L. (1988). Biological invasions of arid land nature reserves. *Biological Conservation*, 44, 95–118.

Loots, G. C. and Ryke, P. A. J. (1967). The ratio Oribatei : Trombidiformes with reference to organic matter content in soils. *Pedobiologia*, 7, 121–4.

Lounsbury, C. P. (1915). Some phases of the locust problem. *South African Journal of Science*, 11, 33–45.

Lourens, S. and Nel, J. A. J. (1990). Winter activity of bat-eared foxes *Otocyon megalotis* on the Cape west coast. *South African Journal of Zoology*, 25, 124–32.

Louw, G. N. (1972). The role of advective fog in the water economy of certain Namib Desert animals. *Symposia of the Zoological Society, London*, 31, 297–314.

Louw, G. N. (1993). *Physiological animal ecology.* London: Longman.

Louw, G. N. and Seely, M. K. (1982). *Ecology of desert organisms.* London and New York: Longman.

Louw, G. N., Belonje, P. C. and Coetzee, H. J. (1969). Renal function, respiration, heart rate and thermoregulation in the Ostrich (*Struthio camelus*). *Scientific papers of the Namib Desert Research Station*, 42, 43–54.

Louw, G. N., Nicolson, S. W. and Seely, M. K. (1986). Respiration beneath desert sand: carbon dioxide diffusion and respiratory patterns in a tenebrionid beetle. *Journal of Experimental Biology*, 120, 443–7.

Lovegrove, B. G. (1986). The metabolism of social subterranean rodents: adaptation to aridity. *Oecologia*, 69, 551–5.

Lovegrove, B. G. (1991a). Mimalike mounds (heuweltjies) of South Africa: the topographical, ecological and economic impact of burrowing animals. *Symposia of the Zoological Society of London*, 63, 183–98.

Lovegrove, B. G. (1991b). The evolution of eusociality in molerats (Bathyergidae): a question of risks, numbers, and costs. *Behavioural Ecology and Sociobiology*, 28, 37–45.

Lovegrove, B. G. (1993). *The Living Deserts of Southern Africa.* Cape Town: Fernwood Press.

Lovegrove, B. G. (1996). The low basal metabolic rates of marsupials: the influence of torpor and zoogeography. In *Adaptations to the cold: Tenth International Hibernation Symposium*, ed. F. Geiser, A. J. Hulbert and S. C. Nicol, pp. 141–51. Armidale: University of New England Press.

Lovegrove, B. G. and Heldmaier, G. (1994). The amplitude of circadian body temperature rythms in three rodents (*Aethomys namaquensis, Thallomys paedulcus* and *Cryptomys damarensis*) along an arboreal-subterranean gradient. *Australian Journal of Zoology*, 42, 65–78.

Lovegrove, B. G and Jarvis, J. U. M. (1986). Coevolution between mole-rats (Bathyergidae) and a geophyte, *Micranthus* (Iridaceae). *Cimbebasia*, 8, 79–85.

Lovegrove, B. G. and Knight-Eloff, A. (1988). Soil and burrow temperatures, and the resource characteristics of the social mole-rat *Cryptomys damarensis* (Bathyergidae) in the Kalahari Desert. *Journal of Zoology, London*, 216, 403–16.

Lovegrove, B. G. and Painting, S. (1987). Variations in the foraging behaviour and burrow structures of the Damara Molerat *Cryptomys damarensis* in the Kalahari Gemsbok National Park. *Koedoe*, 30, 149–63.

Lovegrove, B. G. and Raman, J. (1998). Torpor patterns in the pouched mouse (*Saccostomus campestris, Rodentia*): a model animal for unpredictable environments. *Journal of Comparative Physiology*, 168, 303–12.

Lovegrove, B. G. and Siegfried, W. R. (1989). Spacing and origin(s) of Mima-like earth mounds in the western Cape Province of South Africa. *South African Journal of Science* 85, 108–12.

Lovegrove, B. G. and Wissel, C. (1988). Sociality in molerats: metabolic scaling and the role of risk sensitivity. *Oecologia*, 74, 600–6.

Lovegrove, B. G., Heldmaier, G. and Knight, M. H. (1991a). Seasonal and circadian energetic patterns in an arboreal rodent, *Thallomys paedulcus*, and a burrow-dwelling rodent, *Aethomys namaquensis*, from the Kalahari Desert. *Journal of Thermal Biology*, 16, 199–209.

Lovegrove, B. G., Heldmaier, G. and Ruf, T. (1991b). Perspectives of endothermy revisited: the endothermic temperature range. *Journal of Thermal Biology*, 16, 185–97.

Low, A. B. and Rebelo, A. G. (eds.) (1996). *Vegetation of South Africa, Lesotho and Swaziland.* Pretoria: Department of Environmental Affairs and Tourism.

Ludwig, J. A. (1987). Primary productivity in arid lands: myths and realities. *Journal of Arid Environments*, 13, 1–7.

Lyman, C. P. and O'Brien, R. C. (1974). A comparison of temperature regulation in hibernating rodents. *American Journal of Physiology*, 227, 218–23.

Lyman, C. P., Willis, J. S., Malan, A. and Wang, L. C. H. (1982). *Hibernation and torpor in mammals and birds.* New York: Academic Press.

Lynch, C. D. (1974). A behavioural study of the blesbok with special reference to territoriality. *Memoirs of the National Museum, Bloemfontein*, 8.

Lynch, J. (1995). Root architecture and

plant productivity. *Plant Physiology*, 109, 7–13.

MacDonald, A. C. (1891a). Prickly pear in South Africa. Report laid before a Select Committee of the House of Assembly. *Agricultural Journal of the Cape of Good Hope*, 4, 21–5.

MacDonald, A. C. (1891b). Does Kaalblad change into Doornblad? *Agricultural Journal of the Cape of Good Hope*, 4, 106–7.

Macdonald, I. A. W. (1982). The influence of short term climatic fluctuations on the distribution of savanna organisms in southern Africa. M.Sc. thesis, University of Natal.

Macdonald, I. A. W. (1986a). Do Redbreasted Sparrowhawks belong in the Karoo? *Bokmakierie* 38, 3–4.

Macdonald, I. A. W. (1986b). Range expansion in the Pied Barbet and the spread of alien tree species in southern Africa. *Ostrich* 57, 75–94.

Macdonald, I. A. W. (1989). Man's role in changing the face of southern Africa. In *Biotic diversity in southern Africa: concepts and conservation*, ed. B. J. Huntley, pp. 51–78. Cape Town: Oxford University Press.

Macdonald, I. A. W. (1990). Range expansion of the Masked Weaver *Ploceus velatus* in the Karoo facilitated by the spread of alien mesquite trees. *Ostrich*, 61, 85–6.

Macdonald, I. A. W. (1992). Vertebrate populations as indicators of environmental change in southern Africa. *Transactions of the Royal Society of South Africa* 48, 87–122.

Macdonald, I. A. W. and Macdonald, S. A. (1985). The demise of the solitary scavengers in southern Africa – the early rising crow hypothesis. In *Proceedings of the Birds and Man Symposium*, ed. L. J. Bunning, pp. 321–35. Johannesburg: Witwatersrand Bird Club.

Macdonald, I. A. W., Richardson, D. M. and Powrie, F. J. (1986). Range expansion of the hadeda ibis *Bostrychia hagedash* in southern Africa. *South African Journal of Zoology* 21, 331–42.

Macdonald, J. T. and Nel, J. A. J. (1986). Comparative diets of sympatric small carnivores. *South African Journal of Wildlife Research*, 16, 115–21.

Mack, R. N. (1981). Invasion of *Bromus tectorum* L. into western North America: an ecological chronicle. *Agro-Ecosystems*, 7, 145–65.

Mackerras, I. M. (1970). The composition and distribution of the fauna. In *The insects of Australia*, ed. I. M. Mackerras, pp. 187–204. Melbourne University Press.

Mackie, A. J. and Nel, J. A. J. (1989). Habitat selection, home range use, and group size of bat-eared foxes in the Orange Free State. *South African Journal of Wildlife Research*, 19, 135–9.

Maclean, G. L. (1967). The breeding biology and behaviour of the Double-banded Courser *Rhinoptilus africanus* (Temminck). *Ibis*, 109, 556–69.

Maclean, G. L. (1970a). The biology of the larks (Alaudidae) in the Kalahari Sandveld. *Zoologica Africana*, 5, 7–39.

Maclean, G. L. (1970b). Breeding behaviour of the larks in the Kalahari Sandveld. *Annals of the Natal Museum*, 20, 381–401.

Maclean, G. L. (1970c). The breeding season of birds in the south-western Kalahari. *Ostrich Supplement*, 8, 179–92.

Maclean, G. L. (1970d). An analysis of the avifauna of the southern Kalahari Gemsbok National Park. *Zoologica Africana*, 5, 249–65.

Maclean, G. L. (1971). Larks in the Kalahari. *African Wildlife*, 25, 143–7.

Maclean, G. L. (1973). The Sociable Weaver, Parts 1–5. *Ostrich*, 44, 176–261.

Maclean, G. L. (1974). Arid-zone adaptations in southern African birds. *Cimbebasia*, (A) 2, 163–76.

Maclean, G. L. (1983). Water transport by sandgrouse. *BioScience*, 33, 365–9.

Maclean, G. L. (1984). Avian adaptations in the Kalahari environment: a typical continental semidesert. *Koedoe, Supplement*, 27, 187–93.

Maclean, G. L. (1993). *Roberts' birds of southern Africa*. Cape Town: John Voelcker Bird Book Fund.

Maclean, G. L. (1996). *Ecophysiology of desert birds*, Berlin: Springer-Verlag.

MacMahon, J. A. and Wagner, F. H. (1985). The Mojave, Sonoran and Chihuahuan Deserts of North America. In *Hot deserts and arid shrublands. Ecosystems of the world*, ed. M. Evenari, I. Noy-Meir and D. W. Goodall, vol. 12A, pp. 105–202. Amsterdam: Elsevier.

Major, J. (1988). Endemism: a botanical perspective. In *Analytical biogeography*, ed. A. A. Myers and P. S. Giller, pp. 117–46. New York: Chapman and Hall.

Malan, G. (1992). Nest-lining used by Pale Chanting Goshawks in the Little Karoo, South Africa. *Gabar*, 7, 56–61.

Malan, G. (1995). Cooperative breeding and delayed dispersal in the Pale Chanting Goshawk *Melierax canorus*. Ph.D. thesis, University of Cape Town.

Malan, G. and Branch, W. R. (1992). Predation on tent tortoise and leopard tortoise hatchlings by the pale chanting goshawk in the Little Karoo. *South African Journal of Zoology*, 27, 33–5.

Malan, G. and Crowe, T.M. (1996). The diet and conservation of monogamous and polyandrous pale chanting goshawks in the Little Karoo, South Africa. *South African Journal of Wildlife Research*, 26, 1–10.

Malan, G., Little, R. M. and Crowe, T. M. (1993). The effects of hunting effort and weather on hunting success and population dynamics of Namaqua sandgrouse. *South African Journal of Wildlife Research*, 23, 107–11.

Malan, G., Little, R. M. and Crowe, T. M. (1994). Temporal and spatial patterns of abundance and breeding activity of Namaqua sandgrous in South Africa. *South African Journal of Zoology*, 29, 162–7.

Maloiy, G. M. O. (1973). The water metabolism of a small East African antelope: the dik-dik. *Proceedings of the Royal Society of London. B. Biological Sciences*, 184, 167–78.

Maloiy, G. M. O. and Kamau, J. M. Z. (1982). Thermoregulation and metabolism in a small desert carnivore: the Fennec fox (*Fennecus zerda*) (Mammalia). *Journal of Zoology, London*, 198, 279–91.

Manning, J. C. and Goldblatt, P. (1996). The *Proseca peringuyei* (Diptera: Nemenstrinidae) pollination guild in southern Africa: long-tongued flies and their tubular flowers. *Annals of the Missouri Botanical Garden*, 83, 67–86.

Manry, D. E. (1985). Distribution, abundance and conservation of the Bald Ibis *Geronticus calvus* in southern Africa. *Biological Conservation*, 33, 351–62.

Maree, C. and Plug, I. (1993). Origin of southern African livestock and their potential role in the industry. In *Livestock production systems: Principles and practices*, ed. C. Maree and N. H. Casey, pp. 5–13. Pretoria: Agri Development Foundation.

Marincowitz, C. P. (1992). Tekort aan 'jakkals-sielkunde'. *Goue Vag*, August, p. 2.

Marker, M. E. (1995). Further data for a Pleistocene periglacial gradient in southern Africa. *Transactions of the Royal Society of South Africa* 50, 49–58.

Marker, M. E. and Holmes, P. J. (1993). A Pleistocene sand deposit in the northeastern Cape, South Africa: palaeoenvironmental implications. *Journal of African Earth Sciences*, 17, 479–85.

Markotter, E. I. (1936). Die lewens-geskiedenis van sekere geslagte van die Amaryllidaceae. *Annals of the University of Stellenbosch*, 14.

Marloth, R. (1894). On the means of the distribution of seeds in the South African flora. *Transactions of the South African Philosophical Society*, 8, 74–88.

Marloth, R. (1908). *Das Kapland*. Jena: Gustav Fischer.

Marloth, R. (1925). The flora of Southern Africa. vol. 2, Sect 1. London: Weldon and Wesley.

Marsh, A. C. (1985a). *Aspects of the ecology of Namib Desert ants*. Ph.D. thesis, University of Cape Town.

Marsh, A. C. (1985b). Thermal responses and temperature tolerance in a diurnal desert ant, *Ocymyrmex barbiger*. *Physiological Zoology*, 58, 629–36.

Marsh, A. C. (1985c). Microclimatic factors influencing foraging patterns and success of the thermophilic desert ant, *Ocymyrmex barbiger*. *Insectes Sociaux*, 32, 286–96.

Marsh, A. C. (1987a). Thermal responses and temperature tolerance of a desert ant-lion larva. *Journal of Thermal Biology*, 12, 295–300.

Marsh, A. C. (1987b). The foraging ecology of two Namib Desert harvester ant species. *South African Journal of Zoology*, 22, 130–6.

Marsh, A. C. (1988). Activity patterns of some Namib Desert ants. *Journal of Arid Environments*, 14, 61–73.

Marsh, A. C. (1990). The biology and ecology of Namib Desert ants. In *Namib ecology: 25 years of Namib research*, ed. M. K. Seely. *Transvaal Museum Monograph*, 7, 109–14.

Marsh, A. C., Louw, G. and Berry, H. H. (1978). Aspects of renal physiology, nutrition and thermoregulation in the ground squirrel, *Xerus inauris*. *Madoqua*, 2, 129–35.

Marshall, L. (1976). *The !Kung of Nyae Nyae*. Cambridge, MA: Harvard University Press.

Martens, J. C., Danckwerts, J. E., Stuart-Hill, G. C. and Aucamp, A. J. (1990). Use of multivariate techniques to identify vegetation units and monitor change on a livestock production system in a semi arid savanna of the eastern Cape. *Journal of the Grassland Society of Southern Africa*, 7, 184–9.

Martin, A. R. H. and Noel, A. R. A. (1960). *The flora of Albany and Bathurst*. Grahamstown: Rhodes University.

Martinez-Meza, E. and Whitford, W. G. (1996). Stemflow, throughfall, and channelization of stemflow by roots in three Chihuahuan Desert shrubs. *Journal of Arid Environments* 32, 271–87.

Mason, R. J. (1966). The excavation of Doornlaagte Earlier Stone Age Camp, Kimberley District. In *Actas del V Congreso Panafricano de Prehistoria y de Estudio del Cuaternario*, pp. 187–8. Tenerife: Museo Arquelogico Santa Cruz de Tenerife.

Mason, S. J. (1990). Temporal variability of sea surface temperatures around southern Africa: a possible forcing mechanism for the 18-year rainfall oscillation? *South African Journal of Science*, 86, 243–52.

Matthee, J. J. (1950). Egg of the brown locust. *Farming in South Africa*, 25, 255–7.

Matthee, J. J. (1978). Induction of diapause in eggs of *Locustana pardalina* (Walker) (Acrididae) by high temperatures. *Journal of the Entomological Society of Southern Africa*, 41, 25–30.

Matthews, S. (1995). The Riverine Rabbit, a case study of what not to do? *African Wildlife*, 49, 15.

Maxwell, W. D. (1989). The end-Permian mass extinction. In *Mass extinctions, processes and evidence*, ed. S. K. Donovan, pp. 152–73. London: Belhaven Press.

May, H. and Marinus, T. (1995). Natural resource management and rural livelihoods in Leliefontein, Namaqualand. Interim report. Cape Town: Surplus People Project.

Maynard, L. A., Loosli, J. K., Hintz, H. F. and Warner, R. G. (1979). *Animal nutrition*. New York: McGraw-Hill.

McAuliffe, J. R. (1988). Markovian dynamics of simple and complex desert plant communities. *American Naturalist*, 131, 459–90.

McCabe, K. (1987). Veld management in the Karoo. *The Naturalist*, 31(1), 8–15.

McCarthy, T. S., Moon, B. P. and Levin, M. (1985). Geomorphology of the western Bushmanland plateau. *South African Geographical Journal*, 67, 160–78.

McClain, E. (1984). Wax blooms on Namib Desert tenebrionids: Indicators of aridity. *South African Journal of Science*, 80, 191.

McClain, E., Pretorius, R. L., Gernecke, D. and Hanrahan, S. A. (1984). Dynamics of wax bloom production in a seasonal Namib desert beetle, *Cauricara phalangium* (Coleoptera: Tenebrionidae). *South African Journal of Science*, 80, 191.

McClain, E., Seely, M. K., Hadley, N. F. and Gray, V. (1985). Wax blooms in tenebrionid beetles of the Namib Desert: correlations with environment. *Ecology*, 66, 112–18.

McClaran, M. and Anable, M. E. (1992). Spread of introduced Lehman lovegrass along a grazing intensity gradient. *Journal of Applied Ecology*, 29, 92–8.

McCook, L. J. (1994). Understanding ecological community succession: causal models and theories, a review. *Vegetatio*, 110, 115–47.

McFadyen, R. E. and Tomley, A. J. (1981). Biological control of harrisia cactus, *Eriocereus martinii*, in Queensland by the mealybug *Hypogeococcus festerianus*. In *Proceedings of the Fifth International Symposium on the Biological Control of Weeds*, ed. E. S. Delfosse, pp. 589–94. Brisbane, Australia: CSIRO, Canberra.

McGinnis, W. G. (1979). General description of desert areas. In *Arid-land ecosystems: structure, functioning and management*, ed. D. W. Goodall and R. A. Perry, vol. 1, pp. 5–20. Cambridge University Press.

McKenzie, B. and Longridge, M. (ed.) (1988). *Proceedings of the Locust Symposium*, May 1987. *South African Institute of Ecologists Bulletin* (Special Issue).

McLachlan, I. R. and Anderson, A. (1973). A review of the evidence for marine conditions in southern Africa during Dwyka times. *Palaeontologica Africana*, 15, 37–64.

McLatern, F. (1987). Ticks, tortoises . . . trouble? *Rostrum*, 17.

McNab, B. K. (1970). Body weight and the

energetics of temperature regulation. *Journal of Experimental Biology*, 53, 329–48.

McNab, B. K. (1974). The energetics of endotherms. *The Ohio Journal of Science*, 74, 370–9.

McNab, B. K. (1979). Climatic adaptation in the energetics of heteromyid rodents. *Comparative Biochemistry and Physiology*, 62A, 813–20.

McNab, B. K. (1983). Energetics, body size, and the limits to endothermy. *Journal of Zoology, London*, 199, 1–29.

McNab, B. K. and Bonaccorso, F. J. (1995). The energetics of Australasian swifts, frogmouths, and nightjars. *Physiological Zoology*, 68, 245–61.

McNamara, J. M. (1994). Timing of entry into diapause: Optimal allocation to 'growth' and 'reproduction' in a stochastic environment. *Journal of Theoretical Biology*, 168, 201–9.

McNaughton, S. J. (1979). Grassland herbivore dynamics. In *Serengeti: Dynamics of an ecosystem*, ed. A. R. E. Sinclair and M. Norton Griffiths, pp. 46–81. University of Chicago Press.

McNaughton, S. J. (1983a). Compensatory plant growth as a response to herbivory. *Oikos*, 40, 329–36.

McNaughton, S. J. (1983b). Serengeti grassland ecology: the role of composite environmental factors and contingency incommunity organization. *Ecological Monographs*, 53, 291–320.

McNaughton, S. J. (1985). Ecology of a grazing ecosystem: the Serengeti. *Ecological Monographs*, 53, 259–94.

McNaughton, S. J. (1988). Mineral nutrition and spatial concentrations of African ungulates. *Nature*, 334, 343–5.

Meadows, M. E. (1988). Vlei sediments and sedimentology: a tool in the reconstruction of palaeoenvironments of southern Africa. *Palaeoecology of Africa*, 19, 249–60.

Meadows, M. E. and Meadows, K. F. (1988). Late Quaternary vegetation history of the Winterberg, Eastern Cape. *South African Journal of Science*, 84, 253–9.

Meadows, M. E. and Sugden, J. M. (1988). Late Quaternary environmental changes in the Karoo, South Africa. In *Geomorphological Studies in Southern Africa*, ed. G. F. Dardis and B. P. Moon, pp. 337–53. Rotterdam: A. A. Balkema.

Meadows, M. E. and Sugden, J. M. (1991). The last 14500 years on the Cederberg,

Western Cape. *South African Journal of Science*, 87, 34–43.

Meadows, M. E., Baxter, A. J. and Adams, T. (1994). Late Holocene vegetation history of lowland fynbos. *Historical Biology*, 9, 47–59.

Meeuse, A. D. J. (1958). A possible case of interdependence between a mammal and a higher plant. *Archives Neérlandaises de Zoologie*, 13, 314–18.

Melton, D. A. (1975). Environmental heterogeneity produced by termitaria in Western Uganda, with special reference to mound usage by vertebrates. M.Sc. thesis, University of British Columbia.

Menaut, J. C., Gignoux, J., Prado, C. and Clobert, J. (1990). Tree community dynamics in a humid savanna of the Cote-d'Ivoire: modelling the effects of fire and competition with grass and neighbours. *Journal of Biogeography*, 17, 471–81.

Mentis, M. T. (1980). The effects of animal size and adaptation on defoliation, selective defoliation, animal production and veld condition. *Proceedings of the Grassland Society of Southern Africa*, 15, 147 53.

Mentis, M. T. and Duke, R. R. (1976). Carrying capacities of natural veld in Natal for large wild herbivores. *South African Journal of Wildlife Research*, 6, 65–74.

Meve, U. (1994). The genus *Piaranthus* R. Br. (Asclepiadaceae). *Bradleya*, 12, 57–102.

Meve, U. and Liede, S. (1994). Floral biology and pollination in stapeliads – new results and a literature review. *Plant Systematics and Evolution*, 192, 99–116.

Meyer, M. K. P. and G. C. Loots. (1978). Acari. In *Biogeography and ecology of southern Afric*, vol. 1, ed. M. J. A. Werger, pp. 703–18. The Hague: W. Junk.

Meyer, T. C. (1992). Weidingskapasiteitstudies op veld in die dorre karoo. M.Sc. thesis, University of the Orange Free State.

Midgley, G. F. and Bosenberg, J. de W. (1990). Seasonal and diurnal plant water potential changes in relation to water availability in the mediterranean climate western Karoo. *South African Journal of Ecology*, 1, 45–52.

Midgley, G. F. and Moll, E. J. (1993). Gas exchange in arid adapted shrubs: When is efficient water use a disadvantage?

South African Journal of Botany, 59, 491–5.

Midgley, G. F. and Musil, C. F. (1990). Substrate effects of zoogenic soil mounds on vegetation composition in the Worcester – Robertson valley, Cape Province. *South African Journal of Botany* 56, 158–66.

Midgley, G. F., Midgley, J. J., Bond, W. J. and Linder, H. P. (1994). C_3 mistletoes on CAM hosts: An ecophysiological perspective on an unusual combination. *South African Journal of Science*, 90, 482–5.

Midgley, J. J. (1991). Beetle daisies and daisy beetles. *African Wildlife*, 45, 318–319.

Midgley, J. J. (1993). An evaluation of Hutchinson's 'beetle-daisy' hypothesis. *Bothalia*, 23, 70–2.

Midgley, J. J. and Cowling, R. M. (1993). Regeneration patterns in Cape subtropical transitional thickets, where are all the seedlings? *South African Journal of Botany*, 59, 496–9.

Midgley, J. J, Cowling, R. M., Hendricks, H., Desmet, P. G., Esler, K. J. and Rundel, P. W. (1997). Population ecology of tree succulents (*Aloe* and *Pachypodium*) in the arid western Cape: decline of keystone species. *Biodiversity and Conservation*, 6, 869–76.

Migdoll, I. (1987). *Field guide to the butterflies of Southern Africa*. Cape Town: Struik.

Milewski, A. V. (1978). Diet of *Serinus* species in the southwestern Cape, with special reference to the Protea Seedeater. *Ostrich* 49, 174–84.

Milewski, A. V., Abensperg-Traun, M. and Dickman, C. R. (1994). Why are termite- and ant-eating mammals smaller in Australia than in southern Africa: history or ecology? *Journal of Biogeography*, 21, 529–43.

Millar, A. R. (1971). Reproduction in the rock hyrax (*Procavia capensis*). *Zoologica Africana*, 6, 243–61.

Miller, D. E., Yates, R., Parkington, J. E. and Vogel, J. C. (1993). Radiocarbon-dated evidence relating to a mid-Holocene relative high sea-level on the southwestern Cape coast, South Africa. *South African Journal of Science*, 89, 35–44.

Miller, P. C. (1981). *Resource use by Chaparral and Matorral: A comparison of vegetation function in two mediterranean type ecosystems*. New York: Springer-Verlag.

Mills, M. G. L. (1978). Foraging behaviour

of the brown hyaena (*Hyaena brunnea* Thunberg, 1820) in the southern Kalahari. *Zeitschrift für Tierphychologie*, 48, 113–41.

Mills, M. G. L. (1983). Behavioural mechanisms in territory and group maintenance of the browm hyaena, *Hyaena brunnea* in the southern Kalahari. *Animal Behavior*, 31, 503–10.

Mills, M. G. L. (1985). Related spotted hyaenas forage together but do not cooperate in rearing young. *Nature*, 316, 61–2.

Mills, M. G. L. (1990). *Kalahari hyaenas: comparative behavioural ecology of two species*. London: Unwin Hyman.

Milton, S. J. (1990). Life styles of plants in four habitats in an arid Karoo shrubland. *South African Journal of Ecology*, 1, 63–72.

Milton, S. J. (1991). Plant spinescence in arid southern Africa: does moisture mediate selection by mammals? *Oecologia*, 87, 279–87.

Milton, S. J. (1992a). Plants eaten and dispersed by *Geochelone pardalis* (Reptilia: Chelonii) in the southern Karoo. *South African Journal of Zoology*, 27, 45–9.

Milton, S. J. (1992b). Effects of rainfall, competition and grazing on flowering of *Osteospermum sinuatum* (Asteraceae) in arid Karoo rangeland. *Journal of the Grassland Society of southern Africa*, 9, 158–64.

Milton, S. J. (1992c). Studies of herbivory and vegetation change in Karoo shrublands. Ph.D. thesis, University of Cape Town.

Milton, S. J. (1993). Insects from the shrubs *Osteospermum sinuatum* and *Pteronia pallens* (Asteraceae) in the southern Karoo. *African Entomology*, 1, 257–61.

Milton, S. J. (1994a). Growth, flowering and recruitment of shrubs in grazed and in protected rangeland in the arid Karoo, South Africa. *Vegetatio*, 111, 17–27.

Milton, S. J. (1994b). Small-scale reseeding trials in arid rangeland: effects of rainfall, clearing and grazing on seedling survival. *African Journal of Range and Forage Science*, 11, 54–8.

Milton, S. J. (1995a). Spatial and temporal patterns in the emergence and survival of seedlings in arid Karoo shrubland. *Journal of Applied Ecology*, 32, 145–56.

Milton, S. J. (1995b). Effects of rain, sheep and tephritid flies on seed production of two arid Karoo shrubs in South Africa. *Journal of Applied Ecology*, 32, 137–44.

Milton, S. J. and Collins, H. (1989). Hail in the southern Karoo. *Veld and Flora*, 75, 69–73.

Milton, S. J. and Dean, W. R. J. (1987). Cactus in the Karoo, a thorny problem. *Veld and Flora*, 73, 128–31.

Milton, S. J. and Dean, W. R. J. (1988). Flower and fruit production of *Rhigozum obovatum* (Bignoniaceae) in road reserves and grazing land. *South African Journal of Science*, 84, 798–9.

Milton, S. J. and Dean, W. R. J. (1990a). Mima-like mounds in the southern and western Cape: are the origins so mysterious? *South African Journal of Science* 86, 207–8.

Milton, S. J. and Dean, W. R. J. (1990b). Seed production in rangelands of the southern Karoo. *South African Journal of Science*, 86, 231–3.

Milton, S. J. and Dean, W. R. J. (1991). Disturbances in dune grassland: colourful consequences of clearing. *African Wildlife*, 45, 199–203.

Milton, S. J. and Dean, W. R. J. (1992). An underground index of rangeland degradation: cicadas in arid southern Africa. *Oecologia*, 91, 288–91.

Milton, S. J. and Dean, W. R. J. (1993). Selection of seeds by harvester ants (*Messor capensis*) in relation to condition of arid rangeland. *Journal of Arid Environments*, 24, 63–74.

Milton, S. J. and Dean, W. R. J. (1995a). Factors influencing recruitment of forage plants in arid Karoo shrublands, South Africa. In: *Proceedings: Wildland Shrub and Arid Land Restoration Symposium*, ed. B. A. Roundy, D. E. McArthur, J. S. Haley and D. K. Mann. Ogden, Utah: US Department of Agriculture.

Milton, S. J. and Dean, W. R. J. (1995b). How useful is the keystone species concept, and can it be applied to *Acacia erioloba* in the Kalahari Desert. *Zeitschrift für Ökologie und Naturschutz*, 4, 147–56.

Milton, S. J. and Dean, W. R. J. (1996). *Karoo Veld – Ecology and Management*. Pretoria, Forage Research Institute, Agricultural Research Council.

Milton, S. J. and Hoffman, M. T. (1994). The application of state-and-transition models to rangeland research and management in arid succulent and semi-arid grassy Karoo, South Africa. *African Journal of Range Forage Science*, 11, 18–26.

Milton, S. J., Dean, W. R. J. and Kerley, G. I. H. (1992a). Tierberg Karoo Research centre: history, physical environment, flora and fauna. *Transactions of the Royal Society of South Africa* 48, 15–46.

Milton, S. J., Dean, W. R. J. and Linton, A. (1993). Consumption of termites by captive Ostrich chicks. *South African Journal of Wildlife Research*, 23, 58–61.

Milton, S. J., Dean, W. R. J. and Marincowitz, C. P. (1992b). Preferential utilization of pans by springbok *Antidorcas marsupialis*. *Journal of the Grassland Society of Southern Africa*, 9, 114–18.

Milton, S. J., Dean, W. R. J. and Siegfried, W. R. (1994a). Food selection by Ostrich in Southern Africa. *Journal of Wildlife Management*, 58, 234–48.

Milton, S. J., Dean, W. R. J., Du Plessis, M. A. and Siegfried, W. R. (1994b). A conceptual model of arid rangeland degradation, the escalating cost of declining productivity. *BioScience*, 44, 70–6.

Milton, S. J., Dean, W. R. J., Marincowitz, C. P. and Kerley, G. I. H. (1995). Effects of the 1990/91 drought on rangeland in the Steytlerville Karoo. *South African Journal of Science*, 91, 78–84.

Milton, S. J., Gourlay, I. D. and Dean, W. R. J. (1997) Shrub growth and demography in arid Karoo, South Africa: inference from wood rings. *Journal of Arid Environments*, 37, 487–96.

Milton, S. J., Siegfried, W. R. and Dean, W. R. J. (1990). The distribution of epizoochoric plant species: a clue to the prehistoric use of arid karoo range-lands by large herbivores. *Journal of Biogeography*, 17, 25–34.

Milton, S. J., Yeaton, R. I., Dean, W. R. J. and Vlok, J. H. J. (1997). Succulent Karoo. In *The vegetation of southern Africa*, ed. R. M. Cowling, D. M. Richardson and S. M. Pierce, pp. 131–66. Cambridge University Press.

Mitchell, B. L. (1965). An unexpected association between a plant and an insectivorous mammal. *Puku*, 3, 178.

Mitchell, B. L. (1980). Report on a survey of the termites of Zimbabwe. *Occasional Papers of the National Museums of Rhodesia*,

B., *Natural Science*, 6, 187–323.

Möhr, J. D. (1982). The Karoo caterpillar *Loxostege frustalis* Zeller in relation to its host plants and natural enemies. Ph.D. thesis, Rhodes University, Grahamstown.

Moir, R. W. and Sampson, C. G. (1993). European and Oriental ceramics from rock shelters in the upper Seacow valley. *Southern African Field Archaeology*, 2, 35–43.

Moll, E. J. and Gubb, A. A. (1989). Southern African shrublands. In *The biology and utilization of shrubs*, ed. C. M. Mackell, pp. 145–75. New York: Academic Press.

Montana, C. (1990). A floristic-structural gradient related to land forms in the southern Chihuahan Desert. *Journal of Vegetation Science*, 1, 669–74.

Moon, B. P. and Dardis, G. F. (eds.) (1988) *The Geomorphology of Southern Africa*. Johannesburg: Southern Books.

Mooney, H. A. (1977). *Convergent evolution in Chile and California*. Stroudsburg, PA: Dowden, Hutchinson and Ross.

Mooney, H. A., Ehleringer, J. R. and Berry, J. A. (1976). High photosynthetic capacity of a winter annual in Death Valley. *Science*, 194, 322 4.

Mooney, H. A., Troughton, J. H. and Berry, J. A. (1977). Carbon isotope measurements of succulent plants in southern Africa. *Oecologia*, 30, 295–305.

Moore, A. (1989). Die ekologie en ekofisiologie van *Rhigozum trichotomum* (driedoring). Ph.D. thesis, University of Port Elizabeth.

Moore, J. M. and Picker, M. D. (1991). Heuweltjies (earth mounds) in the Clanwilliam district, Cape Province, South Africa: 4000 year-old termite nests. *Oecologia*, 86, 424–32.

Moore, J. M., Watkeys, M. K. and Reid, D. L. (1990). The regional setting of the Aggeneys/Gamsberg base metal deposits, Namaqualand, South Africa. In *Regional Metamorphism of Ore Deposits*, ed. P. Spry and T. Bryndzia, pp. 77–95. The Netherlands: VNU Science Publishers.

Moorhead, D. L. and Reynolds, J. F. (1989). Mechanisms of surface litter mass loss in the northern Chihuahuan Desert: A reinterpretation. *Journal of Arid Environments*, 16, 157–63.

Moran, V. C. and Annecke, D. P. (1979). Critical reviews of biological pest control in South Africa. 3. The jointed cactus *Opuntia aurantiaca* Lindley. *Journal of the Entomological Society of Southern Africa*, 42, 299–329.

Moran, V. C. and Zimmermann, H. G. (1991). Biological control of cactus weeds of minor importance in South Africa. *Agriculture, Ecosystems and Environment*, 37, 37–55.

Moran, V. C., Hoffmann, J. H. and Zimmermann, H. G. (1993). Objectives, constraints, and tactics in the biological control of mesquite weeds (*Prosopis*) in South Africa. *Biological Control*, 3, 80–93.

Morton, S. R. (1985) Granivory in arid regions: comparisons of Australia with North and South America. *Ecology*, 66, 1859–66.

Morton, S. R. and Davies, P. H. (1983). Food of the zebra finch (*Poephila guttata*), and an examination of granivory in birds of the Australian arid zone. *Australian Journal of Ecology*, 8, 235–43.

Mott, J. J. and Chouard, P. (1979). Flowering, seed formation and dispersal. In *Arid-land ecosystems: structure, functioning and management*, ed. P. A. Perry and D. W. Goodall, pp. 627–645. Cambridge University Press.

Mott, J. J. and McComb, A. J. (1975a). The role of photoperiod and temperature in controlling the phenology of three annual species from an arid region in Western Australia. *Journal of Ecology*, 63, 633–41.

Mott, J. J. and McComb, A. J. (1975b). Effects of moisture stress on the growth and reproduction of three annual species from an arid region of Western Australia. *Journal of Ecology*, 63, 825–34.

Mueller-Dombois, D. and Ellenberg, H. (1974). *Aims and methods of vegetation ecology*. New York: John Wiley.

Muirhead, W. A. (1962). The effect of cultivation techniques on the establishment of *Ehrhata calycina* Sm. in a semi-arid environment. *The Journal of the Australian Institute of Agricultural Science*, 28, 148–9.

Müller, E. F. and Lojewski, U. (1986). Thermoregulation in the meerkat (*Suricata suricata* Schreber, 1776). *Comparative Biochemistry and Physiology*, 83A, 217–24.

Mulroy, T. W. and Rundel, P. W. (1977). Annual plants: adaptations to desert environments. *BioScience*, 27, 109–14.

Mun, H. T. and Whitford, W. G. (1990).

Factors affecting annual plant assemblages on banner-tailed kangaroo rat mounds. *Journal of Arid Environments*, 18, 165–73.

Murbeck, S. (1919). Beiträge zur Biologie der Wüstenpflanzen. Vorkommen und Bedeutung von Schleimabsonderung aus Samenhullen. *Lunds Universitets Arsskrift N. F. avd. 2*, 15, 1–36.

Murbeck, S. (1920). Beiträge zur Biologie der Wüstenpflanzen. Die Synaptospermie. *Lunds Universitets Arsskrif N. F. avd. 2*, 17, 1–52.

Murray, G. W. (1984). Food of the wedge-snouted sand lizard, *Meroles cuneirostris* (Strauch) (Lacertidae), and comparison with that of the sand-diving lizard, *Aporosaura anchietae* (Bocage) (Lacertidae) in the Namib desert. *South African Journal of Science*, 80, 189.

Murray, G. W. and Schramm, D. (1987). A comparative study of the diet of the wedge-snouted sand lizard, *Meroles cuneirostris* (Strauch) and the sand diving lizard, *Aposaura achietae* (Bocage), (Lacertidae), in the Namib Desert. *Madoqua*, 15, 55–61.

Nailand, P. (1992). The analysis of vegetation change in the Karoo using remote sensing and image processing techniques. M.Sc. thesis, University of the Witwatersrand, Johannesburg.

Nailand, P. (1993). The feasibility of using remote sensing to predict and monitor irruptions of the brown locust, *Locusta pardalina* (Walker). *South African Journal of Science*, 89, 425–6.

Nailand P. and Hanrahan, S. A. (1993). Modelling brown locust, *Locustana pardalina* (Walker), outbreaks in the Karoo. *South African Journal of Science*, 89, 420–4.

Narins, P. M., Reichman, O. J., Jarvis, J. U. M. and Lewis, E. R. (1992). Seismic signal transmission between burrows of the Cape mole-rat, *Georychus capensis*. *Journal of Comparitive Physiology. A. Sensory, neutral and Behavioral Physiology*, 170, 13–21.

Nash, M. H. and Whitford, W. G. (1995). Subterranean termites: regulators of soil organic matter in the Chihuahuan Desert. *Biology and Fertility of Soils*, 19, 15–18.

Neff, J. L., Simpson, B. B. and Moldenke, A. R. (1977). Flowers – flower visitor systems. In *Plant phenomorphological studies in Mediterranean type ecosystems*, ed. G. H. Orians and O. T. Solbrig, pp. 389–99. Dordrecht: Kluwer Academic Publishers.

Nel, J. A. (1980). Ekstensiewe skaappro-duksie. *South African Journal of Animal Science*, 10, 305–9.

Nel, J. A. J. (1967). Burrow systems of *Desmodillus auricularis* in the Kalahari Gemsbok National Park. *Koedoe*, 10, 118–21.

Nel, J. A. J. (1984). Behavioural ecology of canids in the south-western Kalahari. *Koedoe Suppl.*, 27, 229–35.

Nel, J. A. J. (1990). Foraging and feeding by bat-eared foxes *Otocyon megalotis* in the south-western Kalahari. *Koedoe*, 33, 9–16.

Nel, J. A. J. and Mackie, A. J. (1990). Food and foraging behaviour of bat-eared foxes in the south-eastern Orange Free State. *South African Journal of Wildlife Research*, 20, 162–6.

Nel, J. A. J. and Rautenbach, I. L. (1975). Habitat use and community structure of rodents in the southern Kalahari. *Mammalia*, 39, 9–29.

Nel, J. A. J., Mills, M. G. L. and Van Aarde, R. J. (1984). Fluctuating group size in bat-eared foxes (*Otocyon m. megalotis*) in the Kalahari. *Journal of Zoology, London*, 203, 294–8.

Nel, J. J. C. and Hewitt, P. H. (1969). A study of the food eaten by a field population of the Harvester Termite *Hodotermes mossambicus* (Hagen) and its relation to population density. *Journal of the Entomological Society of Southern Africa*, 32, 123–31.

Nel, J. J. C., Hewitt, P. H. and Joubert, L. (1970). The collection and utilization of redgrass (*Themeda triandra* Forsk) by the harvester termite, *Hodotermes mossambicus* (Hagen) and its relation to population density. *Journal of the Entomological Society of Southern Africa*, 33, 331–41.

Netterberg, F. (1980). Geology of southern African calcretes. I. Terminology, description, macrofeatures and classification. *Transactions of the Geological Society of South Africa*, 83, 255–83.

Neville, D. E. (1996). European impacts on the Seacow River valley and its hunter-gatherer inhabitants, AD 1770–1900. MA thesis, University of Cape Town.

Newlands, G. (1978). Arachnida. In *Biogeography and ecology of southern*

Africa, vol. 1, ed. M. J. A. Werger, pp. 685–702. The Hague: W. Junk.

Newlands, G. and Ruhberg, H. (1978). Onychophora. In *Biogeography and ecology of southern Africa*, vol. 2, ed. M. J. A. Werger, pp. 123–56. The Hague: W. Junk.

Newman, E. I. and Eason, W. R. (1989). Cycling of nutrients from dying roots to living plants, including the role of mycorrhizas. *Plant and Soil*, 115, 133–7.

Nicolson, S. W., Bartholomew, G. A. and Seely, M. K. (1984a). Ecological correlates of locomotion speed, morphometrics and body temperature in three Namib Desert tenebrionid beetles. *South African Journal of Zoology*, 19, 131–4.

Nicolson, S. W., Louw, G. N. and Edney, E. B. (1984b). Use of a ventilation capsule and tritiated water to measure evaporative water losses in a tenebri-onid beetle. *Journal of Experimental Biology*, 108, 477–81.

Nilssen, P-J. (1989). Refitting pottery and eland body parts as a way of recon-structing hunter-gatherer behaviour: An example from the Later Stone Age at Verlorenvlei. Archaeology Hons. thesis, University of Cape Town.

Noble, J. (1886). *Official handbook of the Cape and South Africa*. Cape Town: Juta.

Noble, J. C., Whitford, W. G. and Kaliszweski, M. (1995). Surface-soil and litter microarthropod populations in contrasting rangeland ecosystems of semi-arid Eastern Australia. *Journal of Arid Environments*, 32, 329–346.

Nobel, P. S. and Sanderson, J. (1984). Rectifier-like activities of roots of two desert succulents. *Journal of Experimental Botany* 35, 727–37.

Nordenstam, B. (1967). Phytogeography of the genus *Euryops* (Compositae). A contribution to the phytogeography of southern Africa. *Opera Botanica*, 23, 1–77.

Nordenstam, B. (1974). The flora of the Brandberg. *Dinteria*, 11, 3–67.

Norton, P. (1986). Historical changes in the distribution of leopards in the Cape Province, South Africa. *Bontebok*, 5, 1–9.

Norton, P. (1990). How many leopards? A criticism of Martin de Meulenaer's population estimates for Africa. *South African Journal of Science*, 86, 218–20.

Norton, P. M. (1984). Food selection by klipspringers in two areas of the Cape

Province. *South African Journal of Wildlife Research*, 14, 33–41.

Norton, P. M. (1989). Population dynamics of mountain reedbuck on three Karoo nature reserves. Ph.D. thesis, University of Stellenbosch.

Norton, P. M. (1994). Simple spreadsheet models to study population dynamics, as illustrated by a mountain reedbuck model. *South African Journal of Wildlife Research* 24(4), 73–81.

Novellie, P. A. (1988). The Karoo region. In *Long-term data series relating to southern Africa's renewable natural resources*, ed. I. A. W. Macdonald and R. J. M. Crawford, pp. 280–9. *South African National Scientific Programmes Report*, 157. Pretoria: Foundation for Research Development.

Novellie, P. A. (1990a). Habitat use by indigenous grazing ungulates in relation to sward structure and veld condition. *Journal of the Grassland Society of Southern Africa*, 7, 16–23.

Novellie, P. A. (1990b). The impact of a controlled burn on Karroid *Merxmuellera* mountain veld in the Mountain Zebra National Park. *South African Journal of Ecology*, 1, 33–7.

Novellie, P. A. (1991). Seasonal movements and habitat use by African grazing ungulates in a small conservation area. In *Proceedings of the IVth International Rangeland Congress*, ed. A. Gaston, M. Kernick and H. N. Le Houerou, pp. 709–13. Montpelier: CIRAD.

Novellie, P. A. and Bezuidenhout, H. (1994). The influence of rainfall and grazing on vegetation changes in the Mountain Zebra National Park. *South African Journal of Wildlife Research*, 24, 60–71.

Novellie, P. A. and Strydom, G. (1987). Monitoring the response of vegetation to use by large herbivores: an assessment of some techniques. *South African Journal of Wildlife Research*, 17, 109–17.

Noy-Meir, I. (1973). Desert ecosystems: environment and producers. *Annual Review of Ecology and Systematics*, 4, 25–51.

Noy-Meir, I. (1974). Desert ecosystems: higher trophic levels. *Annual Review of Ecology and Systematics*, 5, 195–214.

Noy-Meir, I. (1985). Desert ecosystem structure and function. In *Hot deserts and arid shrublands*, A, ed. M. Evenari, I.

Noy-Meir and D. W. Goodall, pp. 93–104. Amsterdam: Elsevier.

O'Connor, T. G. (1991). Local extinction in perennial grasslands, a life history approach. *American Naturalist*, 137, 753–73.

O'Connor, T. G. (1993). The influence of rainfall and grazing on the demography of some African savanna grasses: a matrix modelling approach. *Journal of Applied Ecology*, 30, 119–32.

O'Connor, T. G. (1995). Transformation of a savanna grassland by drought and grazing. *African Journal of Range and Forage Science*, 12, 53–60.

O'Connor, T. G. and Pickett, G. A. (1992). The influence of grazing on seed production and seed banks of some African savanna grasslands. *Journal of Applied Ecology*, 29, 247–60.

O'Connor, T. G. and Roux, P. W. (1995). Vegetation changes (1949–71) in a semi-arid, grassy dwarf shrubland in the Karoo, South Africa, influence of rainfall and grazing by sheep. *Journal of Applied Ecology*, 32, 612–26.

O'Farrell, T. P. and Emery, L. A. (1976). Ecology of the Nevada Test Site: a narrative summary and annotated bibliography. Report No. NVO-167. US Dept. of Energy, Operations Office, Nevada.

O'Reagain, P. J. and Turner, J. R. (1992). An evaluation of the empirical basis for grazing management recommendations for rangeland in southern Africa. *Journal of the Grassland Society of Southern Africa*, 9, 38–49.

Oatley, T. B., Earlé, R. A. and Prins, A. J. (1989). The diet and foraging behaviour of the Ground Woodpecker. *Ostrich*, 60, 75–84.

Odendaal, F. J. (1979). Notes on the adaptive ecology and behaviour of four species of *Rhoptropus* (Gekkonidae) from the Namib Desert with special reference to a thermoregulatory mechanism empoyed by *Rhoptropus afer*. *Madoqua*, 2, 255–60.

Oelofsen, B. W. and Loock, J. C. (1987). Palaeontology. In *The Karoo biome: a preliminary synthesis. Part 2 – vegetation and history*, ed. R. M. Cowling and P. W. Roux. *South African National Scientific Programme Report*, 142, 102–16.

Okon, Y., Fallik, E., Sarig, S., Yahalom, E. and Tal, S. (1988). Plant growth promoting effects of *Azospirillum*. In *Nitrogen fixation: Hundred years after*, ed. H. Bothe, F. De Bruijn and W. E. Newton, pp. 741–6. Stuttgart: Gustav Fischer.

Olivier, J. (1995). Spatial distribution of fog in the Namib. *Journal of Arid Environments*, 29, 129–38.

Ollier, C. D. and Marker, M. E. (1985). The great escarpment of southern Africa. *Zeitschrift für Geomorphologie, NF, Supplement*, 54, 37–46.

Olsvig-Whittaker, L., Shachak, M. and Yair, Y. (1983). Vegetation patterns related to environmental factors in a Negev Desert watershed. *Vegetatio*, 54, 153–65.

Onderstall, D. (1984). Descriptions of two new subspecies of *Afroedura pondolia* (Hewitt) and a discussion of species groups within the genus (Reptilia: Gekkonidae). *Annals of the Transvaal Museum* 33, 497–509.

Oosthuizen, M. A. (1994). The effect of competition on three Namaqualand ephemeral plant species. M.Sc. thesis, University of Pretoria, Pretoria.

Oosthuizen, M. A., Van Rooyen, M. W. and Theron, G. K. (1996a). A replacement series evaluation of the effect of competition on three ephemeral plant species. *South African Journal of Botany*, 62, 342–5.

Oosthuizen, M. A., Van Rooyen, M. W. and Theron, G. K. (1996b). Neighbourhood analysis of competition between Namaqualand ephemeral plant species. *South African Journal of Botany*, 62, 231–5.

Orians, G. H. and Solbrig, O. T. (1977a). A cost-income model of leaves and roots with special reference to arid and semi-arid areas. *The American Naturalist*, 111, 677–90.

Orians, G. H. and Solbrig, O. T. (eds.) (1977b). *Convergent evolution in warm deserts*. Stroudsburg, PA: Dowden, Hutchinson and Ross, Inc.

Orshan, G. (1964). Seasonal dimorphism of desert and mediterranean chamae-phytes and its significance as a factor in their water economy. In *Water in relation to plants*, ed. A. J. Rutter and F. H. Whitehead, pp. 206–22. Oxford: Blackwell.

Orshan, G. (ed.) (1989). *Plant pheno-morphological studies in mediterranean type ecosystems*. Dordrecht: Kluwer Academic Publishers.

Orshan, G., Floret, C., Le Floc'h, E., Le Roux, A., Montenegro, G. and Romane, F. (1989). General synthesis. In *Plant phenomorphological studies in mediter-ranean type ecosystems*, ed. G. Orshan, pp. 389–99. Dordrecht: Kluwer Academic Publishers.

Orshan G., Le Roux, A. and Montenegro, G. (1984). Distribution of monocharacter growth form types in mediterranean plant communities of Chile, South Africa and Israel. *Bulletin de la societe botanique de France*, 131, 427–39.

Osmond, C. B., Winter, K. and Ziegler, H. (1982). Functional significance of different pathways of CO_2 fixation in photosynthesis. In *Physiological plant ecology II. Encyclopaedia of plant physiology*, vol. 12B, ed. O. L. Lange, P. S. Nobel, C. B. Osmond and H. Ziegler. pp. 479–548. New York: Springer-Verlag.

Owen-Smith, R. N. (1985). Niche separation among African ungulates. In *Species and speciation*, ed. E. Vrba. *Transvaal Monographs*, 4, 167–71.

Owen-Smith, R. N. (1988). Feeding ecology of the brown locust. In *Proceedings of the Locust Symposium*, ed. B. McKenzie and M. Longridge, pp. 50–9. Cape Town: South African Institute of Ecologists.

Palmer, A. R. (1988). Vegetation ecology of the Camdebo and Sneeuberg regions of the karoo biome, South Africa. Ph.D. thesis, Rhodes University.

Palmer, A. R. (1989). The vegetation of the Karoo Nature Reserve, Cape Province. I. A phytosociological reconnaissance. *South African Journal of Botany*, 55, 215–30.

Palmer, A. R. (1991a). A syntaxonomic and synecological account of the vegetation of the eastern Cape midlands. *South African Journal of Botany*, 57, 76–94.

Palmer, A. R. (1991b). The potential vegetation of the upper Orange River, South Africa: Concentration analysis and its application to rangeland assessment. *Coenoses*, 6, 131–8.

Palmer, A. R. (1991c). Vegetation/environ-ment relationships in the central area of the Cape midlands, South Africa. *Coenoses*, 6, 29–38.

Palmer, A. R. and Cowling, R. M. (1994). An investigation of topo-moisture gradients in the eastern Karoo, South Africa, and the identification of factors responsible for species turnover. *Journal of Arid Environments*, 26, 135–47.

Palmer, A. R and Hoffman, M. T. (1997).

Nama-karoo. In *The vegetation of southern Africa*, ed. R. M. Cowling, D. M. Richardson and S. M. Pierce, pp. 167–89. Cambridge University Press.

Palmer, A. R. and Van der Heyden, F. (1997). Patchiness in semi-arid dwarf shrublands – evidence from satellite-derived indices of elevated photosynthesis. *African Journal of Range & Forage Science*, 14, 75–80.

Palmer, A. R. and Van Rooyen, A. (1995). Detecting change in the southern Kalahari. ARC–ISCW Desertifcation Workshop, Pretoria. 10–11 October 1995.

Palmer, A. R., Crook, B. J. S. and Lubke, R. A. (1988). Aspects of the vegetation and soil relationships in the Andries Vosloo Kudu Reserve, Cape Province. *South African Journal of Botany*, 54, 309–14.

Palmer, A. R., Hobson, C. G. and Hoffman, M. T. (1990). Vegetation change in a semi-arid succulent dwarf shrubland in the eastern Cape, South Africa. *South African Journal of Science*, 86, 392–6.

Palmer, R. and Fairall, N. (1988). Caracal and African wild cat diet in the Karoo National Park and the implications thereof for hyrax. *South African Journal of Wildlife Research*, 18, 30–4.

Palmer, W. C. (1965). Meteorogical drought. *Research Paper* 45, US Weather Bureau, Washington, DC.

Panetta, F. D. and Carstairs, S. A. (1989). Isozymic discrimination of tropical Australian populations of mesquite (*Prosopis* spp.): implications for biological control. *Weed Research*, 29, 157–65.

Parker, L. W., Fowler, H. G., Ettershank, G. and Whitford, W. G. (1982). The effects of subterranean termite removal on desert soil nitrogen and ephemeral flora. *Journal of Arid Environments*, 5, 53–9.

Parker, L. W., Santos, P. F., Phillips, J. and Whitford, W. G. (1984). Carbon and nitrogen dynamics during decomposition of litter and roots of a Chihuahuan Desert annual, *Lepidium lasiocarpum*. *Ecological Monographs*, 54, 339–60.

Parkington, J. and Poggenpoel, C. (1987). Diepkloof rock shelter. In *Papers in the prehistory of the western Cape, South Africa*, ed. J. Parkington and M. Hall, pp. 269–93. Oxford: BAR International Series, 332.

Parkington, J. E. (1972). Seasonal mobility in the Late Stone Age. *African Studies*, 31, 223–43.

Parkington, J. E. (1976). Coastal settlement between the mouths of the Berg and Olifants Rivers, Cape Province. *South African Archaeological Bulletin* 31, 127–40.

Parkington, J. E. (1981). The effects of environmental changes on the scheduling of visits to the Elands Bay Cave, Cape Province, South Africa. In *Pattern of the past*, ed. I. Hodder, G. Isaac and N. Hammond, pp. 341–59. Cambridge University Press.

Parkington, J. E. (1992). Making sense of sequence at the Elands Bay Cave, western Cape, South Africa. In *Guide to archaeological sites in the south-western Cape*, ed. A. B. Smith and B. J. Mütti, pp. 6–12. Cape Town: Southern African Association of Archaeologists.

Parkington, J. E. (in press). *Elands Bay Cave: a view on the past*. University of Cape Town Press.

Parkington, J. E. and Poggenpoel, C. (1971). Excavations at De Hangen, 1968. *South African Archaeological Bulletin* 26, 3–36.

Parkington, J. E., Poggenpoel, C. Buchanan, B., Robey, T., Manhire, T. and Sealy, J. (1988). Holocene coastal settlement patterns in the western Cape. In *The archaeology of prehistoric coastlines*, ed. G. Bailey and J. Parkington, pp. 22–41. Cambridge University Press.

Parrish, J. M., Parrish, J. T. and Ziegler, A. M. (1986). Permian-Triassic paleogeography and paleoclimatology and implications for therapsid distribution. In *Ecology and biology of mammal-like reptiles*, ed. N. Hotton III, P. D. MacLean, J. J. Roth and E. C. Roth, pp. 109–32. Washington, DC: Smithsonian Institution Press.

Parsons, W. T. and Cuthbertson, E. G. (1992). *Noxious weeds of Australia*. Melbourne: Inkata Press.

Partridge, T. C. (1993). Warming phases in southern Africa during the last 150,000 years: an overview. *Palaeogeography, Palaeoclimatology, Palaeoecology*, 101, 237–44.

Partridge, T. C. and Dalbey, T. S. (1986). Geoarchaeology of the Haaskraal Pan: a preliminary palaeoenvironmental model. *Palaeoecology of Africa*, 17,

69–78.

Partridge, T. C. and Maud, R. R. (1987). Geomorphic evolution of southern Africa since the Mesozoic. *South African Journal of Geology*, 90, 179–208.

Partridge, T. C., Avery, D. M., Botha, G. A., Brink, J. S., Deacon, J., Herbert, R. S., Maud, R. R., Scholtz, A., Scott, L., Talma, A. S. and Vogel, J. C. (1990). Late Pleistocene and Holocene climatic change in southern Africa. *South African Journal of Science*, 86, 302–6.

Passmore, N. I. and Carruthers, V. C. (1979). *South African frogs*. Johannesburg: Witwatersrand University Press.

Pearce, M. J. (1997). *Termites. Biology and Pest Management*. Wallingford: CAB International.

Penn, N. G. (1986). Pastoralists and pastoralism in the northern Cape frontier zone during the eighteenth century. *South African Archaeological Society, Goodwin Series*, 5, 62–8.

Penn, N. G. (1987). The frontier in the western Cape, 1700–1740. In *Papers in the prehistory of the western Cape, South Africa*, ed. J. Parkington and M. Hall, pp. 462–503. Oxford: BAR International Series, 332.

Penn, N. G. (1996). The Orange River frontier zone, *c.* 1700–1805. In *Einiqualand: Studies of the Orange River frontier*, ed. A. B. Smith, pp. 21–109. University of Cape Town Press.

Penrith, M-L. (1984a). New taxa of *Onymacris* Allard, and relationships within the genus (Coleoptera: Tenebrionidae). *Annals of the Transvaal Museum* 33, 511–13.

Penrith, M-L. (1984b). Origin of sand adapted tenebrionid beetles of the Kalahari. *Koedoe suppl.*, 159–65

Penrith, M-L. (1984c). Two new species of *Stizopina* (Coleoptera: Tenebrionidae: Opatrini) from Namaqualand, southern Africa, and the relationships between psammophilous genera. *Annals of the Transvaal Museum*, 33, 353–63.

Penrith, M-L. and Endrödy-Younga, S. (1994). Revision of the subtribe Cryptochelina (Coleoptera: Tenebrionidae: Cryptochilini). *Transvaal Museum Monograph*, 9.

Peringuey, L. (1902). Descriptive catalogue of the Coleoptera of South Africa. *Transactions of the South African*

Philosophical Society, 12, 1–627.

Perkins, J. S. and Thomas, D. S. G. (1993). Spreading deserts or spatially confined environmental impacts? Land degradation and cattle ranching in the Kalahari desert of Botswana. *Land Degradation and Rehabilitation*, 4, 179–94.

Pettey, F. W. (1948). The biological control of prickly pear in South Africa. *Department of Agriculture, Union of South Africa, Science Bulletin*, 271.

Pettifer, H. L. and Nel, J. A. J. (1977). Hoarding in four southern African rodent species. *Zoologica Africana*, 12, 409–18.

Petty, G. L. (1973a). Value of the egg-pod material to the sustained viability of eggs of the brown locust, *L pardalina* (Walk). *Phytophylactica*, 5, 155–8.

Petty, G. L. (1973b). Effect on sand particle size on egg viability of the brown locust, *L pardalina* (Walk). *Phytophylactica*, 5, 159–61.

Petty, G. L. (1974). Effects on humidity on the hatching of eggs of the brown locust *L pardalina* (Walk). *Phytophylactica*, 6, 305–6.

Philander, S. G. (1990). *El Niño, La Niña, and the Southern Oscillation*. San Diego, CA: Academic Press.

Philander, S. G. (1992). El Niño. *Oceanos* (Summer), 56–61.

Phillips, D. L. and MacMahon, J. A. (1981). Competition and spacing patterns of desert shrubs. *Journal of Ecology*, 69, 97–115.

Pianka, E. R. and Huey, R. B. (1978). Comparative ecology, resource utilization and niche segregation among gekkonid lizards in the southern Kalahari. *Copeia*, 4, 691–701.

Picker, M. D. and Midgley, J. J. (1996). Pollination by monkey beetles (Coleoptera: Scarabaeidae: Hopliini): flower and colour preferences. *African Entomology*, 4, 7–14.

Pierce, S. M., Esler, K. J. and Cowling, R. M. (1995). Smoke-induced germination of succulents (Mesembreanthemaceae) from fire-prone and fire-free habitats in South Africa. *Oecologia*, 102, 520–2.

Pietruszka, R. D. and Seely, M. K. (1985) Predictability of two moisture sources in the Namib Desert. *South African Journal of Science*, 81, 682–5.

Pietruszka, R. D., Hanrahan, S. A., Mitchell, D. and Seely, M. K. (1986).

Lizard herbivory in a sand dune environment: the diet of *Angolosaurus skoogi*. *Oecologia*, 70, 587–91.

Plug, I. (1993). The macrofaunal remains of wild animals from Abbot's cave and Lame Sheep Shelter, Seacow Valley, Cape. *Koedoe*, 36, 15–26.

Plumstead, E. P. (1967). A general review of the Devonian fossil plants found in the Cape System of South Africa. *Palaeontologia Africana*, 10, 1–83.

Plumstead, E. P. (1969). Three thousand million years of plant life in Africa. *Geological Society of South Africa*, 72 (annexure), 1–72.

Poggenpoel, C. A. (1996). The exploitation of fish during the Holocene in the southwestern Cape, South Africa. MA thesis, University of Cape Town.

Pojasak, T. and Kay, B. D. (1990). Effect of root exudates from corn and bromegrass on soil structural stability. *Canadian Journal of Soil Science*, 70, 351–62.

Pole-Evans, I. B. (1936). A vegetation map of South Africa. *Memoirs of the Botanical Survey of South Africa*, 15.

Polis, G. A. (ed.) (1991a). *The ecology of desert communities*. Tucson, AZ: University of Arizona Press.

Polis, G. A. (1991b). Complex trophic interactions in deserts: an empirical critique of food-web theory. *American Naturalist* 138, 123–55.

Polis, G. A. (1991c). Food webs in desert communities: complexity via diversity and omnivory. In *The ecology of desert communities*, ed. G. A. Polis, pp. 383–438. Tucson, AZ: University of Arizona Press.

Powrie, L. W. (1993). Responses of Karoo plants to hail damage near Williston, Cape Province. *South African Journal of Botany*, 59, 65–8.

Poynton, J. C. (1962). Patterns in the distribution of the southern African amphibia. *Annals of the Cape Provincial Museums*, 2, 252–72.

Poynton, J. C. (1983). The dispersal versus vicariance debate in biogeography. *Bothalia*, 14, 455–60.

Poynton, J. C. (1995). The 'arid corridor' distribution in Africa: a search for instances among amphibians. *Madoqua*, 19, 45–8.

Poynton, R. J. (1990). The genus *Prosopis* in southern Africa. *South African Forestry Journal*, 152, 62–6.

Prendini, L. (1995). Patterns of scorpion

distribution in southern Africa: a GIS approach. Hons. thesis (Zoology), University of Cape Town.

Preston-Whyte, R. A. (1974). Climatic classification of South Africa: A multivariate approach. *South African Geographical Journal*, 56, 79–86.

Preston-Whyte, R. A. and Tyson, P. D. (1988). *The atmosphere and weather of southern Africa*. Cape Town: Oxford University Press.

Pretorius, H. (1989). Worried about Karoo caterpillar? *Karoo Region Newsletter*, 1, 18–21.

Pringle, J. A., Bond, C. and Clark, J. (1982). *The conservationists and the killers. The story of game protection and the Wildlife Society of southern Africa*. Cape Town: T. V. Bulpin and Books of Africa.

Prinzinger, R. and Siedle, K. (1988). Ontogeny of metabolism, thermoregulation and torpor in the house martin *Delichon u. urbica* (L.) and its ecological significance. *Oecologia*, 76, 307–12.

Prinzinger, R., Lübben, I. and Jackel, S. (1986). Vergleichende Untersuchungen zum Energiestoffwechsel bei *Kolibris* und Nektarvögeln. *Journal für Ornithologie*, 127, 303–13.

Prinzinger, R., Preßmar, A. and Schleucher, E. (1991). Body temperature in birds. *Comparative Biochemistry and Physiology*, 99A, 499–506.

Pullen, R. A. (1979). Termite hills in Africa: their characteristics and evolution. *Catena*, 6, 267–91.

Raitt, R. J. and Pimm, S. L. (1976). Dynamics of bird communities in the Chihuahuan Desert, New Mexico. *Condor*, 78, 427–42.

Rall, M. and Fairall, N. (1993). Diets and food preferences of two South African tortoises *Geochelone pardalis* and *Psammobates oculifer*. *South African Journal of Wildlife Research*, 23, 63–70.

Randall, J. A. and Stevens, C. M. (1987). Footdrumming and other anti-predator responses in the bannertail kanagaroo rat (*Dipodomys spectabilis*). *Behavioural Ecology and Sociobiology*, 20, 187–94.

Raper. P. E. and Boucher, M. (1988). *Robert Jacob Gordon Cape Travels, 1777–1786*. Johannesburg: Brenthurst.

Ras, A. M. (1990). The influence of defoliation and moisture on the tannin and polyphenol contents of *Portulacaria afra*. *Proceedings of the Grassland Society of*

Southern Africa, 7, 139–43.

Rasa, O. A. E. (1994). Behavioural adaptations to moisture as an enviromental constraint in a nocturnal burrow-inhabiting Kalahari detritivore *Parastizopus armaticeps* Peringuey (Coleoptera: Tenebrionidae). *Koedoe*, 37, 57–66.

Rasa, O. A. E. (1995). Ecological factors influencing burrow location, group size and mortality in a nocturnal fossorial Kalahari detritivore, *Parastizopus armaticeps* Periguey (Coleoptera: Tenebrionidae). *Journal of Arid Environments*, 29, 353–65.

Rasa, O. A. E., Wenhold, B. A., Howard, P., Marais, A. and Pallett, J. (1992). Reproduction in the yellow mongoose revisited. *South African Journal of Zoology* 27, 192–5.

Rathbun, G. B. (1979). *The social structure and ecology of elephant shrews*. Berlin: Springer-Verlag.

Rathcke, B. and Lacey, E. P. (1985). Phenological patterns of terrestrial plants. *Annual Review of Ecology and Systematics*, 16, 179–214.

Raunkiaer, C. (1934). *The life forms of plants and statistical plant geography*. Oxford: Clarendon Press.

Raven, P. H. (1963). Amphitropical relationships in the floras of North and South America. *Quarterly Review of Biology*, 38, 151–77.

Raven, P. H. and Axelrod, D. I. (1978). Origin and relationships of the California flora. *University of Californian Publications in Botany*, 72, 1–134.

Rayner, N. A. and Bowland, A. E. (1985a). A note on *Triops granarius* (Lucas), *Lynceus truncatus* (Barnard) and *Streptocephalus cafer* (Loven) (Branchiopoda: Crustacea) from Umfolozi Game Reserve, Natal, South Africa. *Journal of the Limnological Society of Southern Africa*, 11, 11–13.

Rayner, N. A. and Bowland, A. E. (1985b). Notes on the taxonomy and ecology of *Triops granarius* (Lucas) (Notostraca: Crustacea) in South Africa. *South African Journal of Science*, 81, 500–5.

Rayner, R. J. (1995). The palaeoclimate of the Karoo: evidence from plant fossils. *Palaeogeography, Palaeoclimatology, Palaeoecology*, 119, 385–94.

Rebelo, A. G. (1987). Bird pollination in the Cape Flora. In *A preliminary synthesis of pollination biology in the Cape flora*, ed. A. G. Rebelo. *South African National Scientific Programmes Report*, 141, 83–108.

Rebelo, A. G. (1992). Red data book species in the Cape Floristic region: threats, priorities and target species. *Transactions of the Royal Society of South Africa*, 48, 55–86.

Reeve, H. K. and Sherman, P. W. (1993). Adaptation and the goals of evolutionary research. *The Quarterly Review of Biology*, 68, 1–32.

Rejmánek, M. and Richardson, D. M. (1996). What attributes make some plant species more invasive? *Ecology*, 77, 1655–61.

Rennie, J. V. L. (1931). Note on fossil leaves from the Banke clays. *Transactions of the Royal Society of South Africa*, 19, 251–3.

Retief, J. A. (1978). Kunsmatige bestuiwing by die Apocynaceae. *Aloe*, 16, 39–40.

Retief, J. A. (1988). Cultivation of *Pachypodium namaquanum*. *Aloe*, 25, 6–7.

Richards, J. H. and Caldwell, M. M. (1985). Soluble carbohydrates, concurrent photosynthesis and efficiency in regrowth following defoliation: a field study with Agropyron species. *Journal of Applied Ecology*, 22, 907–20.

Richardson, D. M., Macdonald, I. A. W., Holmes, P. M. and Cowling, R. M. (1992). Plant and animal invasions. In *The ecology of fynbos. Nutrients, fire and diversity*, ed. R. M. Cowling, pp. 271–308. Oxford University Press.

Richardson, F. D., Hahn, B. D. and Wilke, P. I. (1991). Simulation models as an aid to the study of livestock production systems in the arid zone. *Proceedings of the Arid Zone Ecology Forum*, 25–27 November 1991, Elsenburg, South Africa.

Richardson, P. R. K. (1987a). Food consumption and seasonal variation in the diet of the aardwolf *Proteles cristatus* in southern Africa. *Zietschrift für Säugtierkunde*, 52, 307–25.

Richardson, P. R. K. (1987b). Aardwolf mating system: overt cuckoldry in an apparently monogamous mammal? *South African Journal of Science*, 83, 405–10.

Ricklefs, R. E. and Schluter, D. (1993). Species diversity: regional and historical influences. In *Species diversity in ecological communities: Historical and geographical perspectives*, ed. R. E. Ricklefs and D. Schluter, pp. 350–63. Chicago, MI: University of Chicago Press.

Rilett, M. H. P. (1951). A fossil lamellibranch from the middle Ecca beds in the Klip-river coalfield near Dundee, Natal. *Transactions of the Royal Society of South Africa*, 53, 93–6.

Ringrose, S. (1987). Monitoring desertification in Botswana using Landsat MSS Data with consideration as to the infra red paradox. In *African Resources*, vol. 1, *Appraisal and monitoring*, ed. A. C. Millington, S. K. Mutiso and J. A. Binns, pp. 12–20. Reading, UK: University of Reading.

Roach, D. A. and Wulff, R. D. (1987). Maternal effects in plants. *Annual Review of Ecology and Systematics*, 18, 209–35.

Roberts, B. R. (1970). Why multicamp layouts? *Proceedings of the Grassland Society of Southern Africa*, 5, 17–22.

Robertson, T. C. (1952). Erosion: is climate or man the culprit? *Veldtrust*, 13, 16–19.

Robinson, E. R. (1978). Phytogeography of the Namib Desert of South West Africa (Namibia) and its significance to discussions of the age and uniqueness of this desert. *Palaeoecology of Africa*, 10, 67–74.

Robinson, M. D. (1987). Diet diversity and prey utilization by the omnivorous Namib dune lizard *Aporosaura anchietae* (Bocage) during two years of very different rainfall. *Journal of Arid Environments*, 13, 279–86.

Robinson, M. D. (1990). Summer field energetics of the Namib Desert dune lizard *Aporosaura anchietae* (Lacertidae) and its relation to reproduction. *Journal of Arid Environments*, 18, 207–15.

Rong, J-Y. and Harper, D. A. T. (1988). A global synthesis of the latest Ordovician Hirnantan brachiopod fauna. *Transactions of the Royal Society of Edinburgh, Earth Sciences*, 79, 383–402.

Rösch, H. (1996). Life history strategies of Namaqualand pioneer plant species. M.Sc. thesis, University of Pretoria, Pretoria.

Rösch, M. W. (1977). Enkele plantekologiese aspekte van die Hester Malan-natuurreservaat. M.Sc. thesis, University of Pretoria, Pretoria.

Rosenzweig, M. L. (1992). Species diversity gradients: we know more and less than we thought. *Journal of Mammalogy*, 73, 715–30.

Rosenzweig, M. L. (1995). *Species diversity in space and time*. Cambridge University Press.

Rosenzweig, M. L. and Abramsky, Z. (1993). How are diversity and productivity related? In *Species diversity in ecological communities: Historical and geographical perspectives*, ed. R. E. Ricklefs and D. Schluter, pp. 52–65. University of Chicago Press.

Roux, E. (1969). *Grass: a story of Frankenwald*. London and New York: Oxford University Press.

Roux, F. A. (1992). The influence of the composition of mixed karoo vegetation on the grazing habits of merino and dorper wethers. M.Sc. thesis, Rhodes University.

Roux, F. A., Du Pisani, L. G., Roux, J. A., Venter, J. C. and Verloren van Themaat, R. (1995). Bestuursbeplanning met behulp van klimaatsdata. *Grootfontein Agric*, 1, 23–8.

Roux, P. W. (1960). *Tetrachne dregei* Nees: Fruit production, germination and survival of seedlings. M.Sc. thesis, University of Natal, Pietermaritzburg.

Roux, P. W. (1966). Die uitwerking van seisoenrënval en beweiding op gemengde karooveld. *Proceedings of the Grassland Society of Southern Africa*, 1, 103 10.

Roux, P. W. (1967). Die onmiddelike uitwerking van intensiewe beweiding op gemengde karooveld. *Proceedings of the Grassland Society of Southern Africa*, 2, 83–90.

Roux, P. W. (1968a). The autecology of *Tetrachne dregei* Nees. Ph.D. thesis, University of Natal, Pietermaritzburg.

Roux, P. W. (1968b). Principles of veld management in the Karoo and adjacent dry sweet-grass veld. In *The small stock industry*, ed. W. J. Hugo, pp. 318–40. Pretoria: Government Printer.

Roux, P. W. (1980). Elements of the trampling factor in stock. *Karoo Agric*, 1, 9–12.

Roux, P. W. (1981a). Interrelationships between climate, vegetation and run-off in the Karoo. *Karoo Agric*, 2(1), 4–8.

Roux, P. W. (1981b). Interaction between climate, vegetation and run-off in the Karoo. In *Proceedings of the Workshop on the effect of rural land use and catchment management on water resources*, ed. H. Maaren, pp. 90–106. Pretoria: Department of Water Affairs, Forestry and Environmental Conservation.

Roux, P. W. (1993). Relatiewe droogtebe-

standheidsindekse vir die Karoostreek. *Karoo Agric*, 5(2), 25–8.

Roux, P. W. and Opperman, D. P. J. (1986). Soil erosion. In *The karoo biome: a preliminary synthesis. Part 1 – Physical environment*, ed. R. M. Cowling, P. W. Roux and A. J. H. Pieterse. *South African National Scientific Programmes Report*, 124, 92–111.

Roux, P. W. and Skinner, T. E. (1970). The Group-Camp system. *Farming in S.A.*, 45(10), 25–8.

Roux, P. W. and Smart, C. W. (1977). Pros and cons of veld burning in the eastern Cape. *Eastern Cape Naturalist*, 62, 4–7.

Roux, P. W. and Smart, C. W. (1979). Veld burning in the Karoo mountains. Cyclostyled report, Department of Agricultural Technical Services, Karoo Region, Middelburg, CP.

Roux, P. W. and Theron, G. K. (1987). Vegetation change in the Karoo biome. In *The Karoo biome: a preliminary synthesis. Part 2 – vegetation and history*, ed. R. M. Cowling and P. W. Roux. *South African National Scientific Programmes Report*, 142, 50–69.

Roux, P. W. and Vorster, M. (1983a). Development of veld management research in the Karoo Region. *Proceedings of the Grassland Society of Southern Africa*, 18, 30–4.

Roux, P. W. and Vorster, M. (1983b). Vegetation change in the Karoo. *Proceedings of the Grassland Society of South Africa*, 18, 25–9.

Roux, P. W., Vorster, M., Zeeman, P. J. L. and Wentzel, D. (1981). Stock production in the Karoo Region. *Proceedings of the Grassland Society of Southern Africa*, 16, 29–35.

Rowan, M. K. (1967). A study of the colies of southern Africa. *Ostrich*, 38, 63–115.

Rowell, D. (1994). *Soil science. Methods and applications*. Harlow: Longman Scientific & Technical.

Rowlands, P., Johnson, H., Ritter, E. and Endo, A. (1982). The Mojave Desert. In *Reference handbook of the deserts of North America*, ed. G. L. Bender, pp. 103–62. Westport, CT: Greenwood Press.

Rubenstein, D. I. and Wrangham, R. W. (1986). *Ecological aspects of social evolution: birds and mammals*. Princeton University Press.

Rubidge, B. S. (1987). South Africa's oldest land-living reptiles from the Ecca-Beaufort transition in the southern

Karoo. *South African Journal of Science*, 83, 165–6.

Rubin, F. and Palmer, A. R. (1996). The physical environment and major plant communities of the Karoo National Park, South Africa. *Koedoe*, 39(2), 25–52.

Ruelle, J. E., Coaton, W. G. H. and Sheasby, J. L. (1975). National survey of the Isoptera of southern Africa. 8. The genus *Macrotermes* Holmgren (Termitidae: Macrotermitinae). *Cimbebasia*, Ser. (A) 3, 73–94.

Ruess, R. W. and Seagle, S. W. (1994). Landscape patterns in soil microbial processes in the Serengeti National Park. *Ecology*, 75, 892–904.

Ruf, T. and Heldmaier, G. (1992). The impact of daily torpor on energy requirements in the Djungarian Hamster, *Phodopus sungorus*. *Physiological Zoology*, 65, 994–1010.

Rundel, P. W. and Gibson, A. C. (1996). *Ecological communities and processes in a Mohave Desert ecosystem: Rock Valley, Nevada*. Cambridge University Press.

Rundel, P. W. and Mahu, M. (1976). Community structure and diversity in a coastal fog desert in northern Chile. *Flora*, 165, 493–505.

Rundel, P. W. and Nobel, P. S. (1991). Structure and function in desert root systems. In *Plant root growth. An ecological perspective*, ed. D. Atkinson, pp. 349–78. Oxford: Blackwell Scientific Publications.

Rundel, P. W., Cowling, R. M., Esler, K. J., Mustart, P., Van Jaarsveld, E. and Bezuidenhout, H. (1995). Winter growth, phenology and leaf orientation in *Pachypodium namaquanum* (Apocynaceae) in the Succulent Karoo of the Richtersveld, South Africa. *Oecologia*, 101, 472–7.

Rundel, P. W., Esler, K. J. and Cowling, R. M. (in press). Biodiversity and conservation biology of coastal transition zones from mediterranean to desert ecosystems: an intercontinental comparison. In *Landscape degradation in mediterranean-type ecosystems*, ed. P. W. Rundel, G. Montenegro and F. Jaksic. Heidelberg: Springer-Verlag.

Rundel, P. W., Esler, K. J. and Cowling, R. M. (in press). Ecological and phylogentic patterns of carbon isotope discrimination in the winter-rainfall flora of the Richtersveld, South Africa. *Plant Ecology*.

Russel, A. P. and Bauer, A. M. (1990). Substrate excavation in the Namibian web-footed gecko, *Palmatogecko rangei* Andersson 1908, and its ecological significance. *Tropical Zoology*, 3, 197–207.

Rutherford, M. C. (1991). Diversity of photosynthetic responses in the mesic and arid mediterranean-type climate regions of southern Africa. In *Modern ecology: Basic and applied aspects*, ed. G. Esser and D. Overdieck, pp. 133–59. Amsterdam: Elsevier.

Rutherford, M. C. and Westfall, R. H. (1986). The biomes of southern Africa – an objective categorization. *Memoirs of the Botanical Survey of South Africa*, 54, 1–98.

Ryan, P. G. (1996). Barlow's Lark. A new endemic lark for southern Africa. *Africa Birds & Birding*, 1(4), 65–70.

Ryan, P. G. and Bloomer, P. (1997). Geographical variation in Red Larks *Certhilauda burra* plumage, morphology, song and mitochondrial DNA haplotypes. *Ostrich*, 68, 31-60.

Sale, J. B. (1970). Unusual external adaptations in the Rock Hyrax. *Zoologica Africana*, 5, 101–13.

Sampson, C. G. (1968). The Middle Stone Age industries of the Orange River Scheme area. *Memoir of the National Museum*, Bloemfontein, 4, 1–111.

Sampson, C. G. (1974). *The Stone Age archaeology of southern Africa*. New York: Academic Press.

Sampson, C. G. (1986). Veld damage in the Karoo caused by its pre-trekboer inhabitants: preliminary observations in the Seacow valley. *The Naturalist*, 30, 37–42.

Sampson, C. G. (1988). *Stylistic boundaries among mobile hunter-foragers*. Washington. DC: Smithsonian Institution Press.

Sampson, C. G. (1992). Acquisition of European livestock by the Seacow River Bushmen between AD 1770-1890. *Southern African Field Archaeology*, 4, 30–6.

Sampson, C. G. (1994). Ostrich eggs and Bushmen survival on the north-east frontier of the Cape Colony, South Africa. *Journal of Arid Environments*, 26, 383–99.

Sampson, C. G. and Plug, I. (1993). Late Holocene and historical bone midden density in rock shelters of the upper Seacow River valley. *Southern African Field Archaeology*, 2(2), 59–66.

Sampson, C. G. and Vogel, J. C. (1995). Radiocarbon chronology of Later Stone Age pottery decorations in the upper Seacow Valley. *Southern African Field Archaeology*, 4, 84–94.

Sampson, C. G., Hart, T., Wallsmith, D. L. and Blagg, J. D. (1989). The ceramic sequence in the upper Seacow valley: problems and implications. *South African Archaeological Bulletin*, 44, 3–16.

Sampson, C. G., Sampson, B. E. and Neville, D. (1994). An early Dutch settlement pattern on the north east frontier of the Cape Colony. *Southern African Field Archaeology*, 3, 74–81.

Sandelowsky, B. H. (1977). Mirabib: an archaeological study in the Namib. *Madoqua*, 10, 221–83.

Sanders, H. L. (1968). Benthic marine diversity: a comparative study. *American Naturalist*, 102, 2432–82.

Santos, P. F., Elkins, N. Z. Steinberger, Y. and Whitford, W. G. (1984). A comparison of surface and buried *Larrea tridentata* leaf litter decomposition in North American hot deserts. *Ecology*, 65, 278–84.

Santos, P. F., Phillips, J. and Whitford, W. G. (1981). The role of mites and nematodes in early stages of buried litter decomposition in a desert. *Ecology*, 62, 664–9.

Sauer, E. G. F. and Sauer, E. M. (1966). Social behaviour of the South African Ostrich *Struthio camelus australis*. *Ostrich*, Suppl. 6, 183–91.

Savory, A. (1988). *Holistic resource management*. Washington, DC: Island Press.

Savory, A. and Parsons, S. D. (1980). The Savory grazing method. *Rangelands*, 2, 234–7.

Scarola, S. (1995). Foraging and food selection in laboratory colonies of the harvester termite *Microhodotermes viator* (Isoptera: Hodotermitidae). Hons. thesis (Zoology), University of Cape Town.

Schaffer, W. M. and Gadgil, M. D. (1975). Selection for optimal life histories in plants. In *Ecology and evolution of communities*, ed. M. L. Cody and J. M. Diamond, pp. 142–57. Cambridge, MA: Belknap.

Schlesinger, W. H., Reynolds, J. F.,

Cunningham, G. L., Huenneke, L. F., Jarrell, W. M., Virginia, R. A. and Whitford, W. G. (1990). Biological feedbacks in global desertification. *Science*, 247, 1043–8.

Schmida, A. and Whittaker, R. H. (1984). Convergence and non-convergence of mediterranean-type communities in the Old and New World. In *Being alive on land*, ed. N. S. Margaris, M. Arianoutsou-Farragitaki and W. C. Oechel, pp. 5–11. The Hague: W. Junk.

Schmidt-Nielsen, K. (1983). *Animal physiology. Adaptation and environment*. Cambridge University Press.

Schmitt, M. B., Baur, S. and Von Maltitz, F. (1987). Observations on the Jackal Buzzard in the Karoo. *Ostrich*, 58, 97–102.

Scholes, R. J. and Walker, B. H. (1993). *An African savanna. Synthesis of the Nylsvley study*. Cambridge University Press.

Scholtz, A. (1985). The palynology of the upper lacustrine sediments of the Arnot pipe, Banke, Namaqualand. *Transactions of the Royal Society of South Africa*, 95, 1–109.

Scholtz, C. H. and Holm, E. (1985). *Insects of southern Africa*. Durban: Butterworths.

Schultz, J. C., Otte, D. and Enders, F. (1977). *Larrea* as a habitat component for desert arthropods. In *Creosotebush*, ed. T. J. Mabry, J. H. Hunziker and D. R. Difeo Jr., pp. 176–208. Stroudsburg, PA: Dowden, Hutchinson and Ross Inc.

Schulze, B. R. (1947). The climates of South Africa according to the classifications of Köppen and Thornthwaite. *South African Geographical Journal*, 29, 32–42.

Schulze, B. R. (ed.) (1965) *Climate of South Africa*, Part 8. General Survey (WB28). Pretoria: Weather Bureau.

Schulze, R. E. and McGee, O. S. (1978). Climatic indices and classifications in relation to the biogeography of southern Africa. In *Biogeography and ecology of southern Africa*, ed. M. J. A. Werger, pp. 19–52. The Hague: W. Junk.

Schumann, T. E. W. and Thompson, W. R. (1934). A study of South African rainfall secular variations and agricultural aspects. *University of Pretoria*, Series 1, (28), 46 pp.

Schuster, W. S. F., Sandquist, D. R., Phillips, S. L. and Ehleringer, J. R. (1992). Comparison of carbon isotope discrimination in populations of aridland plant species differing in life span. *Oecologia*, 91, 332–7.

Schwertmann, U. (1971). Transformation of haematite to geothite in soils. *Nature*, 232, 624–5.

Scoones, I. (1992). Land degradation and livestock production in Zimbabwe's communal areas. *Land Degradation and Rehabilitation*, 3, 99–113.

Scoones, I. (ed.) (1994). *Living with uncertainty. New directions in pastoral development in Africa*. London: Intermediate Technology Publications.

Scotese, C. R. and Sager, W. W. (eds.). (1989). *Mesozoic and Cenozoic plate reconstructions. Tectonophysics*, 155, 1–399.

Scott, J. D. (1959). Principles of pasture management. In *Grasses and pastures of South Africa*, ed. D. Meredith, pp. 601–23. Cape Town: Central News Agency, Cape Town.

Scott, J. D. and Van Breda, N. G. (1937a). Preliminary studies on the root systems of renosterbos on the Worcester Veld Reserve. *South African Journal of Science*, 33, 560–9.

Scott, J. D. and Van Breda, N. G. (1937b). Preliminary studies on the root systems of *Galenia africana* on the Worcester Veld Reserve. *South African Journal of Science*, 34, 268–74.

Scott, J. D. and Van Breda, N. G. (1938). Preliminary studies on the root systems of *Pentzia incana-forma* on the Worcester Veld Reserve. *South African Journal of Science*, 35, 280–7.

Scott, J. D. and Van Breda, N. G. (1939). Preliminary studies on the root systems of *Euphorbia mauretanica*, *Euphorbia burmannii* and *Ruschia multiflora* on the Worcester Veld Reserve. *South African Journal of Science*, 36, 227–35.

Scott, J. K. and Delfosse, E. S. (1992). Southern African plants naturalized in Australia: a review of weed status and biological control potential. *Plant Protection Quarterly*, 7, 70–80.

Scott, J. K. and Panetta, F. D. (1993). Predicting the Australian weed status of southern African plants. *Journal of Biogeography*, 20, 87–93.

Scott, L. (1987). Pollen analysis of hyena coprolites and sediments from Equus Cave, Taung, Southern Kalahari (South Africa). *Quaternary Research*, 28, 144–56.

Scott, L. (1988). Holocene environmental change at western Orange Free State

pans, South Africa, inferred from pollen analysis. *Palaeoecology of Africa*, 19, 109–18.

Scott, L. (1990). Hyrax (Procaviidae) and dassie rat (Petromuridae) middens in palaeoenvironmental studies in Africa. In *Packrat middens: the last 40,000 years of biotic change*, ed. J. Betancourt, T. R. Van Devender and P. S. Martin, pp. 398–407. Tucson, AZ: University of Arizona Press.

Scott, L. (1994). Palynology of late Pleistocene hyrax middens, southwestern Cape Province, South Africa: a preliminary report. *Historical Biology*, 9, 71–81.

Scott, L. and Bousman, C. B. (1990). Palynological analysis of hyrax middens from southern Africa. *Palaeogeography, Palaeoclimatology, Palaeoecology*, 76, 367–79.

Scott, L. and Brink, J. S. (1992). Quaternary palaeoenvironments of pans in central South Africa: palynological and palaeontological evidence. *South African Geographer*, 19, 22–34.

Scott, L. and Cooremans, B. (1990). Late Quaternary pollen from a hot spring in the upper Orange River basin, South Africa. *South African Journal of Science*, 86, 154–6.

Scott, L. and Cooremans, B. (1992). Pollen in recent *Procavia* (hyrax), *Petromus* (dassie rat) and bird dung in South Africa. *Journal of Biogeography*, 19, 205–15.

Scott, L. and Vogel, J. C. (1992). Short-term changes of climate and vegetation revealed by pollen analysis of hyrax dung in South Africa. *Review of Palaeobotany and Palynology*, 74, 283–91.

Scott, L., Anderson, H. M. and Anderson, J. M. (1997). Vegetation history. In *Vegetation of southern Africa*, ed. R. M. Cowling, D. M. Richardson and S. M. Pierce, pp. 62–90. Cambridge University Press.

Sealy, J. and Van der Merwe, N. J. (1986). Isotope assessment and the seasonal mobility hypothesis in the southwestern Cape, South Africa. *Current Anthropology*, 27, 135–50.

Sealy, J. and Yates, R. J. (1994). The chronology of the introduction of pastoralism to the Cape, South Africa. *Antiquity*, 68, 58–67.

Seely, M. K. (1989). Desert invertebrate physiological ecology: is anything

special? *South African Journal of Science*, 85, 266–70.

Seely, M. K. (1990). Patterns of establishment on a linear desert dune. *Israel Journal of Botany*, 39, 443–51.

Seely, M. K. and Hamilton, W. J., III. (1976). Fog catchment sand trenches constructed by tenebrionid beetles, *Lepidochora*, from the Namib Desert. *Science*, 193, 484–6.

Seemann, J. R., Downton, W. J. S. and Berry, J. A. (1986). Temperature and leaf osmotic potential as factors in the acclimation of photosynthesis to high temperature in desert plants. *Plant Physiology*, 80, 926–30.

Seleshi, Y. and Demaree, G. R. (1995). Rainfall variability in the Ethiopian and Eritrean highlands and its links with the Southern Oscillation Index. *Journal of Biogeography*, 22, 945–52.

Shaanker, R. U., Ganeshaiah, K. N. and Radhamani, T. R. (1990). Associations among the modes of pollination and seed dispersal – ecological factors and phylogenetic constraints. *Evolutionary Trends in Plants*, 4, 107–11.

Shackleton, N. J. (1986). Paleogene stable isotope events. *Palaeogeography, Palaeclimatology, Palaeoecology*, 57, 91–102.

Shannon, L. V., Boyd, A. J., Brundrit, G. B. and Taunton-Clark, J. (1986). On the existence of an El Niño-type phenomenon in the Benguela system. *Journal of Marine Research*, 44, 495–520.

Sharon, D. (1981). The distribution in space of local rainfall in the Namib Desert. *Journal of Climatology*, 1, 69–75.

Sharp, J. S. (1984). Rural development schemes and the struggle against impoverishment in the Namaqualand reserves. Paper presented to the Carnegie Conference on Poverty and Development, April 1984. University of Cape Town, Cape Town.

Shaw, J. (1875). On the changes going on in the vegetation of S.A. through the introduction of the Merino sheep. *Journal of the Linnean Society*, 14, 202–8.

Shearing, D. and Van Heerden, K. (1994). *Karoo: South African Wild FLower Guide* 6. Kirstenbosch, Botanical Society of South Africa.

Sherman, P. W., Jarvis, J. U. M. and Braude, S. H. (1992). Naked mole rats. *Scientific American*, 243, 42–8.

Shewry, P. R. and Petersen, P. J. (1976). Distribution of chromium and nickel in plants from soil from serpentine and other sites. *Journal of Ecology*, 64, 195–212.

Shipman, P. (1983). Early hominid lifestyle: hunting and gathering or foraging and scavenging? In *Animals and archaeology, vol. 1: Hunters and their prey*, ed. J. Clutton Brock and C. Grigson, pp. 31–49. Oxford: BAR International Series, 163.

Shipman, P. and Phillips-Conroy, J. (1977). Hominid tool-making versus carnivore scavenging. *American Journal of Physical Anthropology*, 46, 77–86.

Shmida, A. (1985). Biogeography of the desert flora. In *Hot deserts and arid shrublands, A*, ed. M. Evenari, I. Noy-Meir and D. W. Goodall, pp. 23–78. Amsterdam: Elsevier.

Shmida, A. and Werger, M. J. A. (1992). Growth form diversity on the Canary Islands. *Vegetatio*, 102, 183–99.

Shmida, A. and Whittaker, R. H. (1979). Convergent evolution of deserts in the Old and New Worlds. In *Werden und Vergehen von Pflanzengesellschaften*, ed. O. Wilmanns and R. Tuxen, pp. 437–50. Vaduz: J. Kramer.

Siegfried, W. R. (1963a). A preliminary evaluation of the economic status of Corvidae and their control on sheep farms in the Great Karoo. *Cape Nature Conservation Investigational Report*, 4, 1–16.

Siegfried, W. R. (1963b). A preliminary report on Black and Martial Eagles in the Laingsburg and Philipstown Divisions. *Cape Nature Conservation Investigational Report*, 5, 1–16.

Siegfried, W. R. (1965a). A survey of wildlife mortality on roads in the Cape Province. *CDNEC Investigational Report*, 6, 1–20.

Siegfried, W. R. (1965b). Rock Kestrels and road casualties. *Ostrich*, 36, 146.

Siegfried, W. R. (1966). The past and present distribution of the Bald Ibis in the province of the Cape of Good Hope. *Ostrich* 37, 216–18.

Siegfried, W. R. (1980). Vigilance and group size in springbok. *Madoqua*, 12, 151–4.

Siegfried, W. R. (1989). Preservation of species in southern African nature reserves. In *Biotic diversity in southern Africa: concepts and conservation*, ed. B. J.

Huntley, pp. 186–201. Cape Town: Oxford University Press.

Siegfried, W. R. and Brooke, R. K. (1989). Alternative life-history styles of South African birds. In *Alternative life-history styles of animals*, ed. M. N. Bruton, pp. 385–420. Dordrecht: Kluwer Academic Publishers.

Siegfried, W. R. and Underhill, L. G. (1975). Flocking as an anti-predator strategy in doves. *Animal Behavior*, 23, 504–8.

Silanikove, N. (1994). The struggle to maintain hydration and osmoregulation in animals experiencing severe dehydration and rapid rehydration: The story of ruminants. *Experimental Physiology*, 79, 281–300.

Silberbauer, G. B. (1981). *Hunter & habitat in the Central Kalahari Desert*. Cambridge University Press.

Silva, S. I., MacKay, W. P. and Whitford, W. G. (1985). The relative contributions of termites and microarthropods to fluff grass litter disappearance in the Chihuahuan Desert. *Oecologia*, 67, 31–4.

Sinclair, A. R. E. (1977). Lunar and timing of the breeding season in Serengeti wildebeest. *Nature*, 267, 832–3.

Sinclair, A. R. E. and Fryxell, J. M. (1985). The Sahel of Africa: ecology of a disaster. *Canadian Journal of Zoology*, 63, 987–94.

Skaife, S. H. (1979). *African insect life*, revised edn., ed. J. Ledger. Cape Town and Johannesburg: C. Struik Publishers.

Skead, C. J. (1960). *The canaries, seedeaters and buntings of southern Africa*. Cape Town: The South African Bird Book Fund.

Skead, C. J. (1962). A study of the Crowned Guinea Fowl *Numida meleagris coronata* Gurney. *Ostrich*, 33, 51–65.

Skead, C. J. (1967). *The sunbirds of southern Africa*. Cape Town: A. A. Balkema.

Skead, C. J. (1980). *Historical mammal incidence in the Cape Province, vol. 1. The western and northern Cape*. Cape Town. Chief Directorate Nature and Environmental Conservation.

Skead, C. J. (1987). *Historical mammal incidence in the Cape Province, vol. 2. The eastern half of the Cape Province, including the Ciskei, Transkei and East Griqualand*. Cape Town. Chief Directorate Nature and Environmental Conservation.

Skelton, P. (1993). *A complete guide to the freshwater fishes of southern Africa*.

Halfway House, Gauteng: Southern Books.

Skelton, P. H. (1986). Distribution patterns and biogeography of non-tropical southern African freshwater fishes. *Palaeoecology of Africa*, 17, 211–30.

Skinner, J. D. (1989). Game ranching in southern Africa. In *Wildlife production systems – economic utilisation of wild ungulates*, ed. R. J. Hudson, pp. 286–306. Cambridge University Press.

Skinner, J. D. (1993). Springbok (*Antidorcas marsupialis*) treks. *Transactions of the Royal Society of South Africa*, 48, 291–305.

Skinner, J. D. and Louw, G. N. (1996). The springbok *Antidorcas marsupialis* (Zimmermann, 1780). *Transvaal Museum Monograph*, 10.

Skinner, J. D. and Smithers, R. H. N. (1990). *The mammals of the southern African subregion*. 2nd edn. Pretoria: University of Pretoria.

Skinner, J. D. and Van Jaarsveld, A. S. (1987). Adaptive significance of restricted breeding in southern African ruminants. *South African Journal of Science*, 83, 657–63.

Skinner, J. D., Davies, R. A. G., Conroy, A. M. and Dott, H. M. (1986). Productivity of springbok *Antidorcas marsupialis* and Merino sheep *Ovis aries* during Karoo drought. *Transactions of the Royal Society of South Africa*, 46, 149–64.

Skinner, J. D., Dott, H. M., Van Aarde, R. A., Davies, R. A. G. and Conroy, A. M. (1987). Observations on a population of springbok *Antidorcas marsupialis* prior to, and during a severe drought. *Transactions of the Royal Society of South Africa*, 46, 191–7.

Skinner, J. D., Nel, J. A. J. and Millar, R. P. (1977). Evolution of time of parturition and differing litter sizes as an adaptation to changes in environmental conditions. *Proceedings of the Fourth International Symposium on Reproduction, Australia*, 39–44.

Skinner, J. D., Van Aarde, R. J. and Van Jaarsveld, A. S. (1984). Adaptations in three species of large mammals (*Antidorcas marsupialis*, *Hystrix africaeaustralis*, *Hyaena brunnea*) to arid environments. *South African Journal of Zoology*, 19, 82–6.

Skinner, T. E. (1964). 'n Fisiologiese-ekologiese studie van *Stipagrostis ciliata* en *Stipagrostis obtusa*. M.Sc. thesis, University of Pretoria.

Skinner, T. E. (1965). Save the bushman grasses. *Farming in South Africa*, 41, 9–12.

Smale, D. (1973). Silcretes and associated silica diagenesis in southern Africa and Australia. *Journal of Sedimentary Petrology*, 43, 1077–89.

Smit, C. J. B. (1941). Forecasts of incipient outbreaks of the brown locust. *Journal of the Entomological Society of Southern Africa*, 4, 206–21.

Smith, A. B. (1983). The disruption of Khoi society in the 17th century. *African Seminar: Collected Papers*, 3, 257–71. Cape Town: Centre for African Studies, University of Cape Town.

Smith, A. B. (1986). Competition, conflict and clientship: Khoikhoi and San relationships in the western Cape. *South African Archaeological Society, Goodwin Series*, 5, 36–41.

Smith, A. B. (1987). Seasonal exploitation of resources on the Vredenburg Peninsula after 2 000 BP. In *Papers in the prehistory of the western Cape, South Africa*, ed. J. Parkington and M. Hall, pp. 393–402. Oxford: British Archaeological Reports, International Series, 332(ii).

Smith, A. B. (1995). Archaeological observations along the Orange River and its hinterland. In *Einiqualand: Studies of the Orange River Frontier*, ed. A. B. Smith, pp. 265–300. Cape Town: University of Cape Town Press.

Smith, A. B. and Kinahan, J. (1984). The invisible whale. *World Archaeology*, 16, 89–97.

Smith, A. B. and Pheiffer, R. H. (1992). Col. Robert Jacob Gordon's notes on the Khoikhoi 1779–80. *Annals of the South African Cultural History Museum*, 5, 1–56.

Smith, A. B. and Ripp, M. R. (1978). An archaeological reconnaissance of the Doorn/Tanqua Karoo. *South African Archaeological Bulletin*, 33, 118–33.

Smith, A. G., Briden, J. C. and Drewry, G. E. (1973). Phanerozoic world maps. In *Organisms and continents through time*, ed. N. F. Hughes. *Special Papers in Palaeontology*, 12, 1–42.

Smith, C. A. (1966). Common names of South African plants. *Memoirs of the Botanical Survey of South Africa*, 35, 1–642. Pretoria: Government Printer.

Smith, G. F., Hobson, S. R., Meyer, N. L., Chesselet, P., Archer, R. H., Burgonye, P. M., Glen, H. F., Herman, P. P. J., Retief, E., Smithies, S. J., Van Jaarsveld, E. J. and

Welman, W. G. (1993). Southern African succulent plants – an updated synopsis. *Aloe*, 30, 32–74.

Smith, R. M. H. (1987). Helical burrow casts of therapsid origin from the Beaufort Group (Permian) of South Africa. *Palaeogeography, Palaeoclimatology, Palaeoecology*, 60, 155–70.

Smith, R. M. H. (1990). A review of the stratigraphy and sedimentary environments of the Karoo basin of South Africa. *Journal of African Earth Sciences*, 10, 117–37.

Smitheman, J. and Perry, P. (1987). A vegetation survey of the Karoo National Botanic Garden Reserve, Worcester. *South African Journal of Botany*, 56, 525–41.

Smithers, R. H. N. (1971). The mammals of Botswana. *Museum Memoir*, 4, 1–340. Salisbury: National Museums and Monuments of Rhodesia.

Smithers, R. H. N. (1983). *The mammals of the southern African subregion*. University of Pretoria.

Smithers, R. H. N. (1986). South African Red Data Book – terrestrial mammals. *South African National Scientific Programmes Scientific Report* 125.

Smuts, R. and Bond, W. (1995). Namaqualand's carpets of colour conceal an insect desert. *Veld & Flora*, 81, 70–1.

Snijman, D. (1984). The genus *Haemanthus*. *Journal of South African Botany*. Supplementary volume 12.

Snow, D. W. (1978). Relationships between the European and African avifaunas. *Bird Study* 25, 134–48.

Snyder, G. K. and Nestler, J. R. (1990). Relationships between body temperature, thermal conductance, Q_{10} and energy metabolism during daily torpor and hibernation in rodents. *Journal of Comparative Physiology B*, 159, 667–75.

Snyder, T. E. (1961). Supplement to the annotated, subject-heading bibliography of termites 1955–1960. *Smithsonian Institution Miscellaneous Collection* 143(3), 1–136.

Snyder, T. E. (1968). Second supplement to the annotated, subject-heading bibliography of termites 1961–1965. *Smithsonian Institution Miscellaneous Collection* 152(3), 1–188.

Snyman, H. A. and Fouché, H. J. (1991). Production and water use efficiency of

semi-arid grasslands of South Africa as affected by veld condition and rainfall. *Water SA*, 17, 263–8.

Snyman, H. A. and Van Rensburg, W. L. J. (1986). Effect of slope and plant cover on run off, soil loss and water use efficiency of natural veld. *Journal of the Grassland Society of Southern Africa*, 3, 153–8.

Soane, B. D. and Saunder, H. (1959). Nickel and chromium toxicity of serpentine soils in southern Rhodesia. *Soil Science*, 88, 322–9.

Sober, E. (1984). *The nature of selection*. Cambridge, MA: MIT Press.

Soil Classification Working Group. (1991). *Soil classification. A taxonomic system for South Africa*. Agricultural Natural Resources of South Africa, Memoir 15, 257 pp. Department of Agricultural Development, Pretoria.

Solbrig, O. T., Barbour, M. A., Cross, J., Goldstein, G., Lowe, C. H., Morello, J. and Yang, T. W. (1977). The strategies and community patterns of desert plants. In *Convergent evolution in warm deserts*, ed. G. H. Orians and O. T. Solbrig, pp. 68–106. Stroudsburg, PA: Dowden, Hutchinson and Ross, Inc.

Spinage, C. A. (1992). The decline of the Kalahari wildebeest. *Oryx*, 26, 147–9.

Squeo, F. A., Polastri, A. and Ehleringer, J. R. (1994). Water sources and water-use efficiency in desert plants from the mediterranean-climate regions of Chile. *Noticiero de Biologia*, 2, 53.

Stafford Smith, M. and Pickup, G. (1993). Out of Africa, looking in. In *Range ecology at disequilibrium: New models of natural variability and pastoral adaptation in African savannas*, ed. R. H. Behnke, I. Scoones and C. Kerven, pp. 196–226. London: Overseas Development Institute.

Stafford-Smith, D. M. (1996). Management of rangelands: paradigms at their limits. In *The ecology and management of grazing*, ed. J. Hodgson and A. W. Ilius, pp. 325–57. Wallingford, UK: CAB International.

Starfield, A. M. and Bleloch, A. L. (1986). *Building models for conservation and wildlife management*. New York: Macmillan.

Stearns, S. C. (1986). Natural selection and fitness, adaptation and constraint. In *Patterns and processes in the history of life,* ed. D. M. Raup and D. Jablonski, pp. 23–46. Berlin: Springer-Verlag.

Stearns, S. C. (1992). *The evolution of life histories*, Oxford University Press.

Stebbins, G. L. (1952). Aridity as a stimulus to plant evolution. *American Naturalist*, 86, 33–44.

Steenkamp, H. E. (1993). The influence of an invasive plant (*Prosopis* sp.) on dung beetle assemblages in the northern Cape. *Entomological Society of Southern Africa Congress Abstracts*, 9, 96.

Steinberger, Y. and Whitford, W. G. (1983a). The contribution of rodents to decomposition processes in a desert ecosystem. *Journal of Arid Environments*, 6, 177–81.

Steinberger, Y. and Whitford, W. G. (1983b). The contribution of shrub pruning by jackrabbits to litter input in a Chihuahuan Desert ecosystem. *Journal of Arid Environments*, 6, 183–7.

Steinschen, A. (1995). Vergleich unterschiedlich beweideter Flächen in einem semiariden Ökosystem (Succulent Karoo/Südafrika) unter besonderer Berücksichtigung der annuellen Grasart *Stipa capensis*. M.Sc. thesis, Biology Department, University of Marburg, Germany.

Steinschen, A. K., Görne, A. and Milton, S. J. (1996). Threats to the Namaqualand flowers: outcompeted by grass or exterminated by grazing? *South African Journal of Science*, 91, 237–42.

Stephan, T., Jeltsch, F., Wiegand, T., Wissel, C. and Breiting, H. A. (1998). Sustainable farming at the edge of the Namib: analysis of land use strategies by a computer simulation model. In *Advances in Ecological Sciences 1, Ecosystems and Sustainable Management*, ed. J. L. Usó, C. A. Brebbia and H. Power. pp 41–51. Southampton: Computational Mechanics.

Steyn, D. (1980). Reproduction in the hyrax, *Procavia capensis*, in relation to environmental factors. M.Sc. thesis, University of Natal, Pietermaritzburg.

Steyn, D. G. (1934). *The toxicology of plants in South Africa*. Johannesburg: Central News Agency.

Steyn, H. M. (1994). Namaqualand ephemerals: flowering time prediction. *South African Association of Botanists Congress Abstracts*, 29, 93.

Steyn, H. M., Van Rooyen, N., Van Rooyen, M. W. and Theron, G. K. (1996a). The

phenology of Namaqualand ephemeral species: the effect of sowing date. *Journal of Arid Environments*, 32, 407–20.

Steyn, H. M., Van Rooyen, N., Van Rooyen, M. W. and Theron, G. K. (1996b). The phenology of Namaqualand ephemeral species. The effect of water stress. *Journal of Arid Environments*, 33, 49–62.

Steyn, H. M., Van Rooyen, N., Van Rooyen, M. W. and Theron, G. K. (1996c). The prediction of phenological stages in four Namaqualand ephemeral species. *Israel Journal of Plant Sciences*, 44, 147–60.

Steyn, P. (1982). *Birds of prey of southern Africa*. Cape Town: David Phillip.

Steyn, P. and Myburgh, N. (1991). Further notes on Sclater's Lark. *Birding in South Africa*, 43, 73–4.

Steynberg, H. (1989). The historic development of the Afrino. *Afrino Manual*, 4, 4–6.

Steynberg, H. and De Kock, G. C. (1987). Aangeplante weidings in die veeproduksiestelsels van die Karoo en ariede gebiede. *Karoo Agric*, 3(10), 4–13.

Stirton, C. H. (1980). *Plant invaders, beautiful but dangerous*. Cape Town: Cape Nature Conservation.

Stock, W. D. and Allsopp, N. (1992). Functional perspective of ecosystems. In *The ecology of fynbos. Nutrients, fire and diversity*, ed. R. M. Cowling, pp. 241–59. Cape Town: Oxford University Press .

Stock, W. D., Le Roux, D. and Van der Heyden, F. (1993). Regrowth and tannin production in woody and succulent Karoo shrubs in response to simulated browsing. *Oecologia*, 96, 562–8.

Stock, W. D., Allsopp, N., Van der Heyden, F. and Witkowski, E. T. F. (1997). Plant form and function. In *Vegetation of southern Africa*, ed. R. M. Cowling, D. M. Richardson and S. M. Pierce, pp. 376–97. Cambridge University Press.

Stokes, C. J. (1994). Degradation and dynamics of succulent Karoo vegetation. M.Sc. thesis, Botany, University of Natal, Pietermartizburg, South Africa.

Stokes, C. J. and Yeaton, R. I. (1995). Population dynamic pollination ecology and the significance of plant heights in *Aloe candelabrum*. *African Journal of Ecology*, 33, 101–13.

Stopp, K. (1958). Die verbreitungshemmenden Einrichtungen der Südafrikanischen Flora. *Botanische Studien*, 8, 1–103.

Story, R. (1952). A botanical survey of the

Keiskammahoek District. *Botanical Survey of South Africa Memoirs*, 27, 1–184. Pretoria: Government Printer.

Stowe, C. W., Hartnady, C. J. H. and Joubert, P. (1984). Proterozoic tectonic provinces of southern Africa. *Precambrian Research*, 25, 229–31.

Straka, H. (1955). Anatomische und entwicklungsgeschichtliche Untersuchungen an Früchten paraspermer Mesembryanthemen. *Nova Acta Leop Neue Folge*, 17, 127–90.

Struck, M. (1989). The biology of the fruit capsules in *Dorotheanthus* Schwantes (Mesemb.) *Veld & Flora*, 75, 41–3.

Struck, M. (1992). Pollination ecology in the arid winter rainfall region of southern Africa: a case study. *Mitteilungen aus dem Institut für Allgemeine Botanik Hamburg*, 24, 61–90.

Struck, M. (1994a). Flowers and their insect visitors in the arid winter rainfall region of southern Africa: observations on permanent plots. Composition of the anthophilous insect fauna. *Journal of Arid Environments*, 28, 45–50.

Struck, M. (1994b). Flowers and their insect visitors in the arid winter rainfall region of southern Africa: observations on permanent plots. Insect visitation behaviour. *Journal of Arid Environments*, 28, 51–74.

Struck, M. (1994c). Phenology of flowering of permanent plots in the arid winter rainfall region of southern Africa. *Bothalia*, 24, 77–90.

Struck, M. (1995). Land of blooming pebbles: flowers and their pollinators in the Knersvlakte. *Aloe*, 32, 56–64.

Stuart, C. T., Macdonald, I. A. W. and Mills, M. G. L. (1985). History, current status and conservation of large mammalian predators in Cape Province, Republic of South Africa. *Biological Conservation*, 31, 9–17.

Stuart, C. T. (1981). Notes on the mammalian carnivores of the Cape Province, South Africa. *Bontebok*, 1, 1–58.

Stuart-Hill, G. C. and Mentis, M. T. (1982). Co-evolution of African grasses and large herbivores. *Proceedings of the Grassland Society of Southern Africa*, 17, 122–8.

Stuckenberg, B. R. (1978). Two new species of *Nemopalpus* (Diptera: Psychodidae) found in rock hyrax abodes in South-

West Africa. *Annals of the Natal Museum*, 23, 367–74.

Sugden, J. M. and Meadows, M. E. (1989). The use of multiple discriminant analysis in reconstructing recent vegetation changes on the Nuweveldberg, South Africa. *Review of Palaeobotany and Palynology*, 60, 131–47.

Swart, J., Perrin, M. R., Hearne, J. W. and Fourie, L. J. (1986). Mathematical model of the interaction between rock hyrax and caracal lynx, based on demo-graphic data from populations in the Mountain Zebra National Park, South Africa. *South African Journal of Science*, 82, 289–94.

Tainton, N. M. (1981). The ecology of the main grazing lands of South Africa. In *Veld and pasture Management in South Africa*, ed. N. M. Tainton, pp. 25–56, Pietermaritzburg: Shuter & Shooter.

Talbot, W. J. (1961). Land utilization in the arid regions of southern Africa. Part I. South Africa. In *A history of land use in arid regions. Arid Zones Research*, 17, ed. L. D. Stamp, pp. 299–338. Paris: UNESCO.

Tallis, J. H. (1991). *Plant community history*. London: Chapman and Hall.

Tankard, A. J, Jackson, M. P. A., Eriksson, K. A., Hobday, D. K., Hunter, D. R. and Minter, W. E. L. (1982). *Crustal evolution of southern Africa. 3.8 Billion years of Earth history*. New York: Springer-Verlag.

Tanser, F. (1997). The development of landscape diversity indices using remote sensing and GIS to identify degradation patterns in the Great Fish River valley, Eastern Cape Province. M.Sc. thesis (Geography), Rhodes University.

Taylor, C. R. (1969). The eland and the oryx. *Scientific American*, 220, 88–95.

Taylor, C. R., Schmidt-Nielsen, K. and Raab, J. L. (1970). Scaling of energetic cost of running to body size in mammals. *American Journal of Physiology*, 219, 1104–7.

Taylor, J. S. (1949). Dipterous parasites of nestling birds. *Ostrich*, 20, 171.

Teague, W. R. and Smit, G. N. (1992). Relations between woody and herbaceous components and the effect of bush-clearing in southern African savannas. *Journal of the Grassland Society of South Africa*, 9, 60–71.

Tennant, W. J. (1996). Influence of Indian Ocean sea-surface temperature anomalies on the general circulation of southern Africa. *South African Journal of Science*, 92, 289–95.

Terborgh, J. (1973). On the notion of favorableness in plant ecology. *American Naturalist*, 107, 481–501.

Tevis, L. Jr. (1958a). Germination and growth of ephemerals induced by sprinkling a sandy desert. *Ecology*, 39, 681–8.

Tevis, L. Jr. (1958b). A population of desert ephemerals germinated by less than one inch of rain. *Ecology*, 39, 688–95.

Thackeray, A. I. (1992). The Middle Stone Age south of the Limpopo. *Journal of World Prehistory*, 6, 385–440.

Theiler, G. (1959). African ticks and birds. *Ostrich Supplement*, 3, 353–78.

Theiler, G. (1965). Ecogeographical aspects of tick distribution. In *Ecological studies in southern Africa*, ed. D. H. S. Davis, pp. 284–300. The Hague: W. Junk.

Theron, G. K. (1964). 'n Outelologiese studie van *Plinthus karooicus* Verdoorn. M.Sc. thesis, University of Pretoria.

Theron, G. K., Van Rooyen, N., Van Rooyen, M. W. and Jankowitz, W. J. (1985). Vegetation structure and vitality in the lower Kuiseb. In *The Kuiseb environment*, ed. B. J. Huntley, pp. 81–91. *South African National Scientific Programmes Report*, 106.

Theron, G. K., Van Rooyen, M. W. and Van Rooyen, N. (1993). Skilpad Wildflower Reserve: Botanical Survey. Unpublished report to WWF South Africa.

Thiéry, J. M., D'Herbés, J. M. and Valentin, C. (1995). A model simulating the genesis of banded vegetation patterns in Niger. *Journal of Ecology*, 83, 497–507.

Thomas, D. H. and Maclean, G. L. (1981). Comparison of physiological and behavioural thermoregulation and osmoregulation in two sympatric sandgrouse species (Aves: Pteroclididae). *Journal of Arid Environments*, 4, 335–58.

Thompson, K., Band, S. R. and Hodgson, J. G. (1993). Seed size and shape predict persistence in soil. *Functional Ecology*, 7, 236–41.

Thompson, W. A., Vertinsky, I. and Krebs, J. R. (1974). The survival value of flocking in birds: a simulation model. *Journal of Applied Ecology*, 43, 785–820.

Tidmarsh, C. E. (1936). Ecology of a karroid area in the Bloemfontein district. M.Sc. thesis, University of the Orange Free State.

Tidmarsh, C. E. (1948). Conservation problems of the Karoo. *Farming in South Africa*, 23, 19–530.

Tidmarsh, C. E. (1951). Veld management studies 1934–1950. Pasture Research in South Africa, Unpublished Progress Report III. Pretoria: Department of Agriculture.

Tisdall, J. M. and Oades, J. M. (1979). Stabilization of soil aggregates by the root systems of ryegrass. *Australian Journal of Soil Research*, 17, 429–41.

Tölken, H. R. (1977). A revision of the genus *Crassula* in southern Africa. *Contributions from the Bolus Herbarium*, 8, 1–560.

Tongway, D. (1994). *Rangeland soil condition assessment manual*. Canberra: CSIRO Division of Wildlife and Ecology.

Tongway, D. and Hindley, N. (1995). *Manual assessment of soil condition of tropical grasslands*. Canberra: Division of Wildlife and Ecology, CSIRO.

Toolson, E. C. (1987). Water profligacy as an adaptation to hot deserts: water loss rates and evaporative cooling in the Sonoran Desert cicada, *Dicerprocta apache* (Homoptera: Cicadidae). *Physiological Zoology*, 60, 130–2.

Toolson, E. C. and Hadley, N. F. (1987). Energy-dependent facilitation of transcuticular water flux contributes to evaporative cooling in the Sonoran Desert cicada, *Dicerprocta apache* (Homoptera: Cicadidae). *Journal of Experimental Biology*, 131, 439–44.

Trappe, J. M. (1981). Mycorrhizae and productivity of arid and semi-arid rangelands. In *Advances in food producing systems for arid and semiarid lands*, ed. J. T. Manassah and E. J. Briskey, pp. 582–99. New York: Academic Press.

Troskie, D. (1996). Sustainable utilization of agricultural resources: the need for a new theoretical framework. *Elsenburg Journal*, 1, 39–44.

Tuffers, A. V., Martin, C. E. and Von Willert, D. J. (1995). Possible water movement from older to younger leaves and photosynthesis during drought stress in two succulent species from South Africa, *Delosperma tradescantioides* Bgr. and *Prenia sladeniana* L. Bol. (Mesembryanthemaceae). *Journal of Plant Physiology*, 146, 177–82.

Turner, F. B. and Randall, D. C. (1989). Net production by shrubs and winter annuals in southern Nevada. *Journal of Arid Environments*, 17, 23–6.

Turner, J. S. and Lombard, A. T. (1990). Body color and body temperature in white and black Namib Desert beetles. *Journal of Arid Environments*, 19, 303–15.

Turner, J. S. and Picker, M. D. (1993). Thermal ecology of an embedded dwarf succulent from southern Africa (*Lithops* spp: Mesembryanthemaceae). *Journal of Arid Environments*, 24, 361–85.

Tyson, J. J. (1984). Evolution of eusociality in diploid species. *Theoretical Population Biology*, 26, 283–95.

Tyson, P. D. (1987). *Climatic change and variability in southern Africa.* Cape Town: Oxford University Press.

UNEP (1992). *World atlas of desertification.* London: Edward Arnold.

Usher, M. B. (1981). Modelling ecological succession, with particular reference to Markovian models. *Vegetatio*, 46, 11–18.

Uvarov, B. (1921). A revision of the genus *Locusta*, with a new theory as to the periodicity and migrations of locusts. *Bulletin of Entomological Research*, 12, 135.

Van Aarde, R. J., Willis, C. K., Skinner, J. D. and Haupt, M. A. (1992). Range utilisation by the aardvark, *Orycteropus afer* (Pallas, 1766) in the karoo, South Africa. *Journal of Arid Environments*, 22, 387–94.

Van Ark, H. (1967). Some bio-ecological observations on *Microhodotermes viator* (Latreille) colonies (Hodotermitidae, Isoptera). *Phytophylactica*, 1, 107–10).

Van Blerk, J. F. (1987). Die verwerking van Acocks se ongepubliseerde velddata van die Sukkulente Karoo met behulp van moderne rekenaarmetodes – is dit moontlik? B.Sc. (Hons.) thesis (Botany), University of Pretoria.

Van Breda, N. G. (1939). An improved method of sowing grass and Karoo shrub seed. *South African Journal of Science*, 36, 328–35.

Van Breda, P. A. B. and Barnard, S. A. (1991). *100 Veld plants of the winter-rain-fall region.* Pretoria: Department of Agriculture, *Bulletin* 422, 1–210.

Van Bruggen, A. C. (1978). Land molluscs. In *Biogeography and ecology of southern Africa*, vol. 2, ed. M. J. A. Werger, pp. 877–923. The Hague: W. Junk.

Van Coller, A. and Stock, W. D. (1994). Cold

tolerance of the southern African succulent, *Cotyledon orbiculata* L. across its geographical range. *Flora*, 198, 89–94.

Van den Berg, J. A. (1983). The relationship between the long term average rainfall and the grazing capacity of natural veld in the dry areas of South Africa. *Proceedings of the Grassland Society of Southern Africa*, 18, 165–7.

Van der Heyden, F. (1992). Effects of defoliation on regrowth and carbon budgets of three semi-arid karoo shrubs. Ph.D. thesis, University of Cape Town.

Van der Heyden, F. and Stock, W. D. (1995). Nonstructural carbohydrate allocation following different frequencies of simulated browsing in three semi-arid shrubs. *Oecologia*, 102, 238–45.

Van der Heyden, F. and Stock, W. D. (1996). Regrowth of a semi-arid shrub following simulated browsing: the role of reserve carbon. *Functional Ecology*, 10, 647–53.

Van der Heyden, F., Roux, F., Cupido, C. N., Leeuw, M. W. P. and Malo, N. (in press). Responses to sheep browsing at different stocking rates: water relations, photosynthesis and resource allocation in two semi-arid shrubs. *African Journal of Range and Forage Science.*

Van der Hulst, R. (1980). Vegetation dynamics or ecosystem dynamics: Dynamic sufficiency in succession theory. *Vegetatio*, 43, 147–51.

Van der Merwe, C. A. (1987). Die saadbiologie van *Augea capensis.* *Abstracts of the Karoo Annual Research Meeting*, 1987, p. 20.

Van Der Merwe, C. A. (1991). 'n ondersoek na aspekte rakende die ontkiemingsfisiologie van *Augea capensis.* M.Sc. thesis, University of Stellenbosch.

Van der Merwe, C. R. (1955). Clay minerals of soil of the desert and adjoining semi-arid regions. *Soil Science*, 80, 479–94.

Van der Merwe, C. R. (1962). Soil groups and subgroups of South Africa. *Science Bulletin*, 356. Pretoria: Department of Agriculture and Technical Services.

Van der Merwe, M. and Skinner, J. D. (1982). Annual reproductive pattern in the dassie *Procavia capensis. South African Journal of Zoology*, 17, 130–5.

Van der Merwe, P. (1966). Die flora van Swartbosklooof, Stellenbosch en die

herstel van die soorte na brand. *Annale Universiteit van Stellenbosch*, 41, 691–736.

Van der Merwe, P. J. (1938). *Die noordwaartse begweging van die boere voor die Groot Trek (1770–1842).* The Hague: Van Stockum and Zoon.

Van der Pijl, L. (1982). *Principles of dispersal in higher plants.* 3rd edn. Berlin: Springer-Verlag.

Van der Schijff, H. P. (1948). 'n Fisiologies-Ekologiese studie van *Geigeria passerinoides* (L. Her) Harv. M.Sc. thesis, Potchefstroom University for C. H. E.

Van der Walt, P. T. (1980). A phytosociological reconnaissance of the Mountain Zebra National park. *Koedoe*, 23, 1–132.

Van der Walt, P. T. and Le Riche, E. A. N. (1984). The influence of veld fire on an *Acacia erioloba* community in the Kalahari Gemsbok National Park. *Koedoe Supplement*, 27, 103–6.

Van der Walt, S. J. (1949). Oorlog in die Karoo (Jakkalsjag). *Huisgenoot*, 34(1245), 24–25.

Van Heerden, J. and Dauth, J. (1987). Aspects on adaptations to an arid environment in free-living ground squirrels *Xerus inauris. Journal of Arid Environments*, 13, 83–9.

Van Jaarsveld, A. S., Richardson, P. R. K. and Anderson, M. D. (1995). Post-natal growth and sustained lactational effort in the Aardwolf: life-history implications. *Functional Ecology*, 9, 492–7.

Van Jaarsveld, E. (1987). The succulent riches of South Africa and Namibia. *Aloe*, 24, 45–92.

Van Jaarsveld, E. (1994). *Gasterias of South Africa – A new revision of a major succulent group.* Cape Town: Fernwood Press.

Van Niekerk, W. A. and Schoeman, S. J. (1993). Sheep and goat production. In *Livestock production systems: Principles and practices*, ed. C. Maree and N. H. Casey, pp. 124–49. Pretoria: Agri Development Foundation.

Van Rensburg, D. J. (1978). Fenologiese aspekte van enkele Namakwalandse spesies. B.Sc. (Hons.) mini-thesis, University of Pretoria, Pretoria.

Van Rooyen, M. W. (1978). Anemochorie by enkele plantspesies van die Hester Malan-natuurreservaat, Namaqualand. *Journal of South African Biological Society*, 19, 26–37.

Van Rooyen, M. W. (1988). Ekofisiologiese studies van die efemere van

Namakwaland. Ph.D. thesis, University of Pretoria.

Van Rooyen, M. W. and Grobbelaar, N. (1982). Saadbevolkings in die grond van die Hester Malan-natuurreservaat in die Namakwalandse Gebroke veld. *South African Journal of Botany*, 1, 41–50.

Van Rooyen, M. W., Barkhuizen, A. and Myburg, S. E. (1980). Saadverspreiding van enkele verteenwoordigers van die Mesembryanthemaceae. *Journal of South African Botany*, 46, 173–92.

Van Rooyen, M. W., Grobbelaar, N. and Theron, G. K. (1979b). Phenology of the vegetation in the Hester Malan Nature Reserve in the Namaqualand Broken Veld. 2. The therophyte population. *Journal of South African Botany*, 45, 433–52.

Van Rooyen, M. W., Grobbelaar, N., Theron, G. K. and Van Rooyen, N. (1991). The ephemerals of Namaqualand: effects of photoperiod, temperature and moisture stress on development and flowering of three species. *Journal of Arid Environments*, 20, 15–29.

Van Rooyen, M. W., Grobbelaar, N., Theron, G. K. and Van Rooyen, N. (1992a). The ephemerals of Namaqualand: effects of germination date on parameters of growth analysis of three species. *Journal of Arid Environments*, 22, 117–36.

Van Rooyen, M. W., Grobbelaar, N., Theron, G. K. and Van Rooyen, N. (1992b). The ephemerals of Namaqualand: effects of germination date on development of three species. *Journal of Arid Environments*, 22, 51–66.

Van Rooyen, M. W., Theron, G. K. and Grobbelaar, N. (1979a). Phenology of the vegetation in the Hester Malan Nature Reserve in the Namaqualand Broken Veld. 1. General observations. *Journal of South African Botany*, 45, 279–93.

Van Rooyen, M. W., Theron, G. K. and Grobbelaar, N. (1990). Life form and dispersal spectra of the flora of Namaqualand. *Journal of Arid Environments*, 19, 133–45.

Van Rooyen, M. W., Theron, G. K. and Van Rooyen, N. (1992c). The ephemerals of Namaqualand: effects of density on yield and biomass allocation. *Journal of Arid Environments*, 23, 249–62.

Van Rooyen, N., Bredenkamp, G. J.,

Theron, G. K., Bothma, J. Du P. and Le Riche, E. A. N. (1994a). Vegetational gradients around artificial watering points in the Kalahari Gemsbok National Park. *Journal of Arid Environments*, 26, 349–62.

Van Rooyen, N., Theron, G. K. and Bredenkamp, G. J. (1994b). Population trends of woody species in the Kalahari National Park from 1978–1994. *Abstracts of the Arid Zone Ecology Forum*, 1994, 30–1.

Van Rooyen, N., Theron, G. K. and Schmidt, A. (1994c). Age structure of woody species in the Kalahari Gemsbok National Park. Poster presented to the *Arid Zone Ecology Forum*, Beaufort West, 6–8 September 1994.

Van Rooyen, N., Van Rensburg, D. J., Theron, G. K. and Bothma, J. Du P. (1984). A preliminary report on the dynamics of the vegetation of the Kalahari Gemsbok National Park. *Koedoe Supplement*, 27, 83–102.

Van Rooyen, T. H. and Burger, R. T. (1974). Plant ecological significance of the soils of the Central Orange River Basin. *South African Geographical Journal*, 56, 60–6.

Van Schilfgaarde, J. (1974). *Drainage for agriculture*. Agronomy No. 17. Madison, Wisconsin: American Society of Agronomy.

Van Veen, J. A., Merckx, R. and Van De Geijn, S. C. (1989). Plant and soil related controls of the flow of carbon from roots through the soil microbial biomass. *Plant and Soil*, 115, 179–88.

Van Zinderen Bakker, E. M. (1978). Quaternary vegetation changes in southern Africa. In *Biogeography and ecology of southern Africa*, vol. 1, ed. M. J. A. Werger, pp. 131–43. The Hague: W. Junk.

Van Zinderen Bakker, E. M. (1982). African palaeoenviroments 18 000 yrs BP. *Palaeoecology of Africa* 15, 77–99.

Van Zinderen Bakker, E. M. (1989). Middle Stone Age palaeoenvironments at Florisbad (South Africa). *Palaeoecology of Africa*, 20, 133–54.

Vari, I. (1971). Revision of the genus *Tarsocera* Butler, 1898 and descriptions of new Saturninae (Lepidoptera: Nymphalidae). *Annals of the Transvaal Museum*, 27, 193–224.

Vasek, F. C. (1980). Creosote bush: long-lived clones in the Mojave Desert. *American Journal of Botany*, 67, 246–55.

Venable, D. L. and Lawlor, L. (1980). Delayed germination and escape in desert annuals: escape in space and time. *Oecologia*, 46, 272–82.

Venter, I. G. and Potgieter, W. J. (1967). Systematic discontinuities between eggs and hoppers from individual egg-pods of the brown locust. *South African Journal of Agricultural Science*, 10, 293–8.

Venter, J. C. (1962). Die produksie van groen materiaal deur vyf Karoobos soorte gedurende vier sesoene van die jaar: *P. incana, A. muricatus, E. glaber, C. tenuifolia, P. tricephala*. Taak van Tegnikus: Karoo Streek, Middelberg.

Venter, J. C. (1992). Drought characterization based on Karoo shrubland productivity. *South African Journal of Science*, 88, 154–7.

Venter, J. M., Mocke, C. and de Jager, J. M. (1986). Climate. In *The karoo biome: a preliminary synthesis. Part 1. Physical Environment*, ed. R. M. Cowling, P. W. Roux and A. J. H. Pieterse. *South African National Scientific Programmes Report*, 124, 39–52.

Venter, P. S. and Rethman, N. F. G. (1992). Germination of fresh seed of thirty *Cenchrus ciliaris* ecotypes as influenced by seed treatment. *Journal of the Grassland Society of South Africa*, 9, 181–2.

Vermeulen, H. C. and Nel, J. A. J. (1988). The bush Karoo rat *Otomys unisulcatus* on the Cape West coast. *South African Journal of Zoology*, 23, 103–11.

Vernon, C. J., Boshoff, A. F. and Stretton, W. S. (1992). The status and conservation of cranes in the eastern Cape Province. In *Proceedings of the First Southern African Crane Conference*, ed. D. J. Porter, H. S. Craven, D. N. Johnson and M. J. Porter, pp. 47–72. Durban: Southern African Crane Foundation.

Versveld, D. B. and Van Wilgen, B. W. (1986). Impact of woody aliens on ecosystem properties. In *The ecology and management of biological invasions in southern Africa*. ed. I. A. W. Macdonald, F. J. Kruger and A. A. Ferrar, pp. 239–46. Cape Town: Oxford University Press.

Vetter, S. (1996). Investigating the impacts of donkeys on a communal range in Namaqualand: How much does a donkey 'cost' in goat units? Unpublished B.Sc. (Hons.) thesis, University of Cape Town.

Viljoen, B. D. (1987a). Effect of rate and time of application on nassella tussock (*Stipa trichotoma* Nees) in South Africa. *South African Journal of Plant and Soil*, 4, 79–81.

Viljoen, B. D. (1987b). Pasture recovery after nassella tussock control with tetrapion. *Applied Plant Science*, 1, 18–22.

Viljoen, P. J. (1983). Distribution, numbers and group size of the Karoo Korhaan in Koakoland, South West Africa. *Ostrich*, 54, 50–1.

Vincent, A. S. (1985). Plant foods in savanna environments: a preliminary report of tubers eaten by the Hadza of northern Tanzania. *World Archaeology*, 17, 131–48.

Visser, J. H., De La Harpe, A. C. and Sauer, N. (1976). Germination of the seed of *Dinteranthus wilmotianus* L.Bol. *Aloe*, 14, 65–8.

Visser, J. N. J. and Joubert, P. (1991). Cyclicity in the late Pleistocene to Holocene spring and lacustrine deposits at Florisbad, Orange Free State. *South African Journal of Geology*, 94, 123–31.

Visser, L. (1993). Saadontkiemingstudies van geselekteerde efemeerspesies van Namakwaland. M.Sc. thesis, University of Pretoria, Pretoria.

Vlok, J. H. J. (1988). Alpha diversity of lowland fynbos herbs at various levels of infestation by alien annuals. *South African Journal of Botany*, 54, 623–7.

Vlok, J. H. J. (1994). How green is your valley? *South African Association of Botanists, Congress Abstracts*, p. 108.

Vogel, J. C., Fuls, A. and Danin, A. (1986). Geographical and environmental distribution of C3 and C4 grasses in the Sinai, Negev, and Judean deserts. *Oecologia*, 70, 258–65.

Vogel, J. C., Fuls, A. and Ellis, R. P. (1978). The geographical distribution of Kranz grasses in South Africa. *South African Journal of Science*, 74, 209–15.

Voigt, E. A., Plug, I. and Sampson, C. G. (1992). European livestock from rock shelters in the upper Seacow River valley. *Southern African Field Archaeology*, 4, 37–49.

Volk, O. H. (1960). Flowers and fruits of the Mesembryanthemums. In *A handbook of succulent plants, vol. 3, Mesembryanthemums*, ed. H. Jacobsen, pp. 896–901. Dorset, UK: Blandford Press.

Von Richter, W. (1971). The black wildebeest. Orange Free State Provincial Administration, Department of Nature Conservation. *Miscellaneous Publication*, 2.

Von Willert, D. J. and Wagner-Douglas, U. (1994). Water relations, CO_2 exchange, water-use efficiency and growth of *Welwitschia mirabilis* Hook. *fil.* in three contrasting habitats of the Namib Desert. *Botanica Acta*, 107, 291–9.

Von Willert, D. J., Brinckmann, E., Scheitler, B. and Eller, B. M. (1985). Availability of water controls crassulacean acid metabolism in succulents of the Richtersveld (Namib Desert, South Africa). *Planta*, 164, 44–55.

Von Willert, D. J., Eller, B. M., Werger, M. J. A. and Brinkmann, E. (1990). Desert succulents and their life strategies. *Vegetatio*, 30, 133–43.

Von Willert, D. J., Eller, B. M., Werger, M. J. A., Brinckmann, E. and Ihlenfeldt, H.-D. (1992). *Life strategies of succulents in deserts, with special reference to the Namib Desert.* Cambridge University Press.

Vorster, F. (1994). Physiological response of springbok to grazing. In *Wildlife ranching, a celebration of diversity*, ed. W. Van Hoven and H. Ebedes, pp. 310–17. Pretoria: Promedia.

Vorster, H., Hewitt, P. H. and Van der Westhuizen, M. C. (1992). Nest density of the granivorous ant *Messor capensis* (Mayr) (Hymenoptera: Formicidae) in semi-arid grassland of South Africa. *Journal of African Zoology*, 106, 445–50.

Vorster, M. (1982). The development of the Ecological Index for assessing veld condition in the karoo. *Proceedings of the Grasslands Society of Southern Africa*, 17, 84–9.

Vorster, M. (1985). Die ordening van landtipes in die Karoostreek in redelike homogene boerdery gebiede deur middel van plantegroei- en omgewings-faktore. D.Sc. thesis, Potchefstroom University for Christian Higher Education.

Vorster, M. (1986). Enkele plant-grondvorm-assosiasies in die Karoostreek. *Karoo Agric*, 3(7), 59–61.

Vorster, M. (1991). Livestock production systems and problems in the Karoo. Paper presented at the Arid Zone Ecology Forum Meeting, Stellenbosch.

Vorster, M. (1993). Sustainable commercial stock farming in the arid parts of the Karoo in the Republic of South Africa. Paper presented at the Arid Zone Ecology Forum Meeting, Middelburg.

Vorster, M. and Roux, P. W. (1983). Veld of the Karoo areas. *Proceedings of the Grassland Society of Southern Africa*, 18, 18–24.

Vorster, M., Becker, H. R., Greyling, J. S. and Bosch, O. J. H. (1987). Ordination of landtypes in the Karoo region in reasonably homogeneous farming areas based on vegetation and environmental factors. *Journal of the Grassland Society of Southern Africa*, 4, 13–17.

Vorster, M., Botha, P, and Hobson, F. O. (1983). The utilisation of Karoo veld by livestock. *Proceedings of the Grassland Society of Southern Africa*, 18, 35–39.

Vrba, E. S. (1980). Evolution, species and fossils: how does life evolve. *South African Journal of Science*, 76, 61–84.

Vrba, E. S. (1985). Environment and evolution: alternative causes of the temporal distribution of evolutionary events. *South African Journal of Science* 81, 229–236.

Wagner, J. C. and Bokkenheuser, V. (1961). The mycobacterium isolated from the dassie *Procavia capensis* (Pallas). *Tubercle*, 42, 47–56.

Wagner, J. C., Buchanan, G., Bokkenheuser, V. and Leviseur, S. (1958). An acid-fast bacillus isolated from lungs of the Cape hyrax *Procavia capensis* (Pallas). *Nature*, 181, 284–5.

Walker, B. H. (1993). Rangeland ecology: Understanding and managing change. *Ambio*, 22, 80–7.

Walker, B. H. and Noy-Meir, I. (1982). Aspects of the stability and resilience of savanna ecosystems. In *Ecology of tropical savannas*, ed. B. J. Huntley and B. H. Walker, pp. 577–90. Berlin: Springer-Verlag.

Walker, B. H., Matthews, D. A. and Dye, P. J. (1986). Management of grazing systems: existing versus an event-oriented approach. *South African Journal of Science*, 82, 172.

Wallis, A. H. (1935). Is our rainfall getting less? *The 1820 Magazine*, 6, 35–7.

Wallwork, J. A. (1976). *The distribution and diversity of soil fauna*. New York: Academic Press.

Walsberg, G. E. and Wolf, B. O. (1995). Solar heat gain in a desert rodent:

Unexpected increases with wind speed and implications for estimating the heat balance of free living animals. *Journal of Comparative Physiology. B. Biochemical, Systematic and Environmental Physiology*, 165, 306–14.

Walsberg, G. E., Campbell, G. S. and King, J. R. (1978). Animal coat colour and radiative heat gain: a re-evaluation. *Journal of Comparative Physiology*, 126, 211–22.

Walter, H. (1986). The Namib Desert. In *Hot deserts and arid shrublands*, B, ed. M. Evenari, I. Noy-Meir and D. W. Goodall, vol. 12B, pp. 245–82. Amsterdam: Elsevier.

Walters, M. M. (1951). Bare patches in the Eastern Mixed Karoo. *Farming in South Africa*, 26, 12–16.

Wannenburgh, A., Johnson, P. and Bannister, A. (1979). *The Bushmen*. Cape Town: Struik.

Ward, D. and Seely, M. K. (1996a). Adaptation and constraint in the evolution of the physiology and behaviour of the Namib Desert tenebrionid beetle genus *Onymacris*. *Evolution*, 50, 1231–40.

Ward, D. and Seely, M. K. (1996b). Behavioral thermoregulation of six Namib Desert tenebrionid beetle species (Coleoptera). *Annals of the Entomology Society of America*, 89, 442–51.

Ward, J. D. and Breen, C. M. (1983). Drought stress and the demise of *Acacia albida* along the lower Kuiseb River, Central namib Desert: preliminary findings. *South African Journal of Science*, 79, 444–7.

Ward, J. D., Seely, M. K. and Lancaster, N. (1983). On the antiquity of the Namib. *South African Journal of Science*, 79, 175–83.

Watson, R. T. (1989). Niche separation in Namib Desert dune Lepismatidae (Thysanura: Insecta): detrivores in an allochthonous detritus ecosystem. *Journal of Arid Environments*, 17, 37–48.

Watt, J. M. and Breyer-Brandwijk, M. G. (1962). *Medicinal and poisonous plants of southern and eastern Africa*. Edinburgh and London: E. & S. Livingstone Ltd.

Weather Bureau. (1986). *Climate of South Africa: climate statistics up to 1984*. Report No. WB40. Pretoria: Government Printer.

Weaving, A. J. S. (1988). Prey selection in several sympatric species of *Ammophila*

W. Kirby (Hymenoptera: Sphecidae) in southern Africa. *Annals of the Cape Provincial Museums (Natural History)* 16, 327–49.

Webb, J. W. (1977). The life history and population dynamics of *Acizzia russellae* (Homoptera, Psillidae). *Journal of the Entomological Society of Southern Africa*, 40, 37–46.

Webb, J. W. and Moran, V. C. (1978). The influence of the host plant on the population dynamics of *Acizzia russellae* (Homoptera, Psillidae). *Ecological Entomology*, 3, 313–21.

Webley, L. E. (1984). Archaeology and ethnoarchaeology in the Leliefontein Reserve and surrounds, Namaqualand. MA thesis, University of Stellenbosch.

Webley, L. E. (1986). Pastoralist ethnoar-chaeology in Namaqualand. In *Prehistoric pastoralism in southern Africa*, ed. M. Hall and A. B. Smith. *South African Archaeological Society, Goodwin Series*, 5, 57–61.

Webley, L. E. (1992a). The history and archaeology of pastoralist and hunter-gatherer settlement in the north-western Cape, South Africa. Ph.D. thesis, University of Cape Town.

Webley, L. E. (1992b). Early evidence for sheep from Spoeg River Cave, Namaqualand. *Southern African Field Archaeology*, 1, 3–13.

Weiher, E. and Keddy, P. A. (1995). Assembly rules, null models, and trait dispersion: new questions from old patterns. *Oikos*, 74, 159–64.

Wells, M. J. (1977). Progress with research on Nasella Tussock. *Proceedings of the National Weeds Conference of South Africa*, 2, 47–55. Cape Town: A. A. Balkema.

Wells, M. J., Balsinhas, A. A., Joffe, H., Engelbrecht, M. W., Harding, G. and Stirton, C. H. (1986). A catalogue of problem plants in Southern Africa. *Memoirs of the Botanical Survey of South Africa*, 53.

Wendt, W. E. (1976). 'Art mobilier' from the Apollo 11 Cave, South West Africa. *South African Archaeological Bulletin*, 31, 5–11.

Went, F. W. (1948). Ecology of desert plants. 1. Observations on germination in the Joshua Tree National Monument, California. *Ecology*, 29, 242–53.

Went, F. W. (1949). Ecology of desert plants. 2. Effect of rain and tempera-

ture on germination and growth. *Ecology*, 30, 1–13.

Went, F. W. (1955). The ecology of desert plants. *Scientific American*, 193, 68.

Went, F. W. (1979). Germination and seedling behaviour. In *Hot deserts and arid shrublands*, ed. D. W. Goodall and P. A. Perry, pp. 477–89. Amsterdam: Elsevier.

Went, F. W. and Westergaard, M. (1949). Ecology of desert plants. 3. Development of plants in the Death Valley National Monument, California. *Ecology*, 30, 26–38.

Wentzel, H. E. (1993). *Mesembryanthemum guerichianum* – Enkelspesiedominansie op ou krale. B.Sc. (Hons.) mini-thesis, University of Pretoria, Pretoria.

Wentzel, H. E., Van Rooyen, M. W., Theron, G. K. and De Villiers, A. J. (1994). *Mesembryanthemum guerichianum*: dominance on old kraals. *Abstracts of the Arid Zone Ecology Forum*, Beaufort West, September 1994.

Werger, M. J. A. (1973). An account of the plant communities of the Tussen die Riviere Game Farm, Orange Free State. *Bothalia*, 11, 165–76.

Werger, M. J. A. (1978a). The Karoo-Namib Region. In *Biogeography and ecology of southern Africa*, ed. M. J. A. Werger, pp. 231–99. The Hague: W. Junk.

Werger, M. J. A. (1978b). Biogeographical divisions of southern Africa. In *Biogeography and ecology of southern Africa*, ed. M. J. A. Werger, pp. 145–70. The Hague: W. Junk.

Werger, M. J. A. (1980). A phytosociologi-cal study of the Upper Orange River Valley. *Memoirs of the Botanical Survey of South Africa*, 46, 1–98.

Werger, M. J. A. (1983). Vegetation geographical patterns as a key to the past, with emphasis on the dry vegetation types of South Africa. *Bothalia*, 14, 405–10.

Werger, M. J. A. (1986). The Karoo and southern Kalahari. In *Hot deserts and arid shrublands*, B, ed. M. Evenari, I. Noy-Meir and D. W. Goodall, pp. 283–360. Amsterdam: Elsevier.

Werger, M. J. A. and Ellis, R. P. (1981). Photosynthetic pathways in the arid regions of South Africa. *Flora*, 171, 64–75.

Westoby, M., Cousins, J. M. and Grice, A. C. (1982). Rate of decline of some soil seed populations during drought in western

New South Wales. In *Ant–plant interactions in Australia*, ed. R. C. Buckley, pp. 7–16. The Hague: W. Junk.

Westoby, M., Walker, B. H. and Noy-Meir, I. (1989). Opportunistic management for rangelands not at equilibrium. *Journal of Range Management*, 42, 266–74.

Wharton, G. W. (1985). Water balance of insects. In *Comprehensive insect physiology, biochemistry and pharmacology*, ed. G. A. Kerkut and L. I. Gilbert, pp. 565–610. Oxford: Pergamon Press.

Wharton, R. A. (1980). Colouration and diurnal activity patterns in some Namib Desert Zophosini (Coleoptera: Tenebrionidae). *Journal of Arid Environments*, 3, 309–17.

Wharton, R. A. (1981). Namibian Solifuga (Arachnidae). *Cimbebasia Memoir*, 5, 1–87.

Wharton, R. A. and Seely, M. K. (1982). Species composition of and biological notes on Tenebrionidae of the lower Kuiseb river and adjacent gravel plain. *Madoqua*, 13, 5–25.

White, F. (1976). The vegetation map of Africa – the history of a completed project. *Boissiera*, 24, 659–66.

White, F. (1983). *The vegetation of Africa*. Paris: UNESCO.

White, F. N., Bartholomew, G. A. and Howell, T. R. (1975). The thermal significance of the nest of the Sociable Weaver *Philetairus socius*: winter observations. *Ibis*, 117, 171–9.

White, F. N., Bartholomew, G. A. and Howell, T. R. (1981). The thermal significance of the nest of the Sociable Weaver. *National Geographic Society Research Reports*, 13, 107–15.

White, R. M. and Bernard, R. T. F. (1996). Secondary plant compound and photperiod influences on the reproduction of two southern African rodent species, *Gerbillurus paeba* and *Saccostomus campestris*. *Mammalia*, 60, 639–49.

White, R. M., Kerley, G. I. H. and Bernard, R. T. F. (1996). Pattern and controls of reproduction of the southern African rodent *Gerbillurus paeba* in the semi-arid Karoo, South Africa. *Journal of Arid Environments*, 37, 529–49.

Whitehead, V. B. (1984). Distribution, biology and flower relationships of fideliid bees of southern Africa (Hymenoptera, Apoidea, Fideliidae).

South African Journal of Zoology, 19, 87–90.

Whitehead, V. B. and Steiner, K. (1985). Oil collecting bees in South Africa. *African Wildlife*, 39, 144–7.

Whitehead, V. B., Giliomee, J. H. and Rebelo, A. G. (1987). Insect pollination in the Cape flora. In *A preliminary synthesis of pollination biology in the Cape flora*, ed. A. G. Rebelo. *South African National Scientific Programmes Report*, 141, 52–82.

Whitford, W. G. (1974). Jornada Validation Site Report. *US/IBP Desert Biome Research Memorandum*, 74(4).

Whitford, W. G. (1976). Temporal fluctuations in density and diversity of desert rodent populations. *Journal of Mammalogy*, 57, 351–69.

Whitford, W. G. (1989). Abiotic controls on the functional structure of soil food webs. *Biology and Fertility of Soils*, 8, 1–6.

Whitford, W. G. (1991). Subterranean termites and long-term productivity of desert rangelands. *Sociobiology*, 19, 235–43.

Whitford, W. G. (1995). Desertification: Implications and limitations of the ecosystem health metaphor. In *Evaluating and monitoring the health of large-scale ecosystems*, ed. D. J. Rapport, C. L. Gaudet and P. Calow, pp. 273–93. Berlin: Springer-Verlag.

Whitford, W. G. and DiMarco, R. (1995). Variability in soils and vegetation associated with harvester ant (*Pogonomyrmex rugosus*) nests on a Chihuahuan Desert watershed. *Biology and Fertility of Soils*, 20, 169–73.

Whitford, W. G., Anderson, J. and Rice, P. M. (1997). Stemflow contribution to the 'fertile island' effect in creosote-bush, *Larrea tridentata*. *Journal of Arid Environments*, 35, 451–7.

Whitford, W. G., Ludwig, J. A. and Noble, J. C. (1992). The importance of subterranean termites in semi-arid ecosystems in south-eastern Australia. *Journal of Arid Environments*, 22, 87–91.

Whitford, W. G., Martinez-Turanzas, G. and Martinez-Meza, E. (1995). Persistence of desertified ecosystems: explanations and implications. *Environmental Monitoring and Assessment*, 37, 319–32.

Whitford, W. G., Meentemeyer, V., Seastedt, T. R., Cromack, J. K. Jr.,

Crossley, D. A., Santos, P. F., Todd, R. L. and Waide, J. B. (1981). Exceptions to the AET model: deserts and clear-cut forest. *Ecology*, 62, 275–7.

Whitford, W. G., Reynolds, J. F. and Cunningham, G. L. (1987). How desertification affects nitrogen limitation of primary production of Chihuahuan Desert watersheds. In *Proceedings of the Symposium on Strategies for Classification and Management of Native Vegetation for Food Production in Arid Zones*, ed. E. F. Aldon, C. E. Gonzales-Vincente, and W. H. Moir, pp. 143–53. USDA Forest Service, Rocky Mountain Forest and Range Experiment Station, Fort Collins, CO.

Whitford, W. G., Steinberger, Y., MacKay, W., Parker, L. W. Freckman, D., Wallwork, J. A. and Weems, D. (1986). Rainfall and decomposition in the Chihuahuan Desert. *Oecologia*, 68, 512–15.

Whitford, W. G., Stinnett, K. and Anderson, J. (1989). Decomposition of roots in a Chihuahuan Desert ecosystem. *Oecologia*, 75, 8–11.

Whittaker, R. H. (1977). Evolution of species diversity in land communities. *Evolutionary Biology*, 10, 1–67.

Whittaker, R. H. and Niering, W. A. (1975). Vegetation of the Santa Catalina mountains, Arizona. V. Biomass, production and diversity along the gradient. *Ecology*, 56, 432–43.

Wiegand, T. and Milton, S. J. (1996). Vegetation change in semiarid communities, simulating probabilities and time scales. *Vegetatio*, 125, 169–83.

Wiegand, T., Dean, W. R. J. and Milton, S. J. (1997). Simulated plant population responses to small scale disturbances in semi-arid shrublands. *Journal of Vegetation Science*, 8, 163–76.

Wiegand, T., Milton, S. J. and Wissel, C. (1995). A simulation model for a shrub ecosystem in the semiarid Karoo, South Africa. *Ecology*, 76, 2205–21.

Wiens, J. A. (1977). On competition in variable environments. *American Scientist*, 65, 590–7.

Wiens, J. A. (1991). The ecology of desert birds. In *The ecology of desert communities*, ed. G. A. Polis, pp. 278–310. Tucson, AZ: University of Arizona Press.

Wilcox, B. P., Wood, M. K. and Tromble, J. M. (1988). Factors influencing infiltrability of semi-arid mountain slopes. *Journal of Range Management*, 41, 197–206.

Wild, H. (1978). The vegetation of heavy metal and other toxic soils. In *Biogeography and ecology of southern Africa*, ed. M. J. A. Werger, pp. 1301–32. The Hague: W. Junk.

Williams, G. C. (1966). *Adaptation and natural selection: a critique of some current evolutionary thought*, Princeton, NJ: Princeton University Press.

Williams, J. B. (1996). A phylogenetic perspective of evaporative water loss in birds. *Auk*, 113, 457–72.

Williams, J. B. and Du Plessis, M. A. (1996). Field metabolism and water flux of Sociable Weavers *Philetairus socius* in the Kalahari Desert. *Ibis*, 138, 168–71.

Williams, J. B., Siegfried, W. R., Milton, S. J., Adams, N. J., Dean, W. R. J., du Plessis, M. A., Jackson, S. and Nagy, K. A. (1993). Field metabolism and water requirements of wild ostriches in the Namib Desert. *Ecology*, 74, 390–404.

Williams, M. A. J., Dunkerley, D. L., De Dekker, P. Kershaw, A. P. and Stokes, T. (1993). *Quaternary environments*. London: Edward Arnold.

Williamson, D. T. (1987). Plant underground storage organs as a source of moisture for Kalahari wildlife. *African Journal of Ecology*, 25, 63–4.

Williamson, M. (1988). Relationship of species number to area, distance and other variables. In *Analytical biogeography. An integrated approach to the study of animal and plant distributions*, ed. A. A. Myers and P. S. Giller, pp. 91–115. London: Chapman & Hall.

Willis, C. K., Skinner, J. D. and Robertson, H. G. (1992). Abundance of ants and termites in the False Karoo and their importance in the diet of the aardvark *Orycteropus afer*. *African Journal of Ecology* 30, 322–4.

Willoughby, E. J. (1969). Desert colouration of birds in the central Namib Desert. *Scientific Papers of the Namib Desert Research Station*, 44, 59–68.

Willoughby, E. J. (1971). Biology of larks (Aves: Alaudidae) in the central Namib Desert. *Zoologica Africana*, 6, 133–76.

Wilson, A. C., Cann, R. L., Thorne, A. G. and Wolpoff, M. H. (1992). Where did modern humans originate? *Scientific American*, 266, 20–33.

Wilson, D. S. and Clark, A. B. (1977). Above ground defense in the harvester

termite *Hodotermes mosambicus* (Hagen). *Journal of the Entomological Society of Southern Africa*, 40, 271–82.

Wilson, M. V. and Shmida, A. (1984). Measuring beta diversity with presence–absence data. *Journal of Ecology*, 72, 1055–64.

Wilson, R. T. (1989). *Ecophysiology of the Camelidae and desert ruminants*. Berlin: Springer-Verlag.

Winter, K., Luttge, U., Winter, E. and Troughton, J. H. (1978). Seasonal shift from C_3 photosynthesis to CAM in *Mesembryanthemum crystallinum* growing in its natural environment. *Oecologia*, 34, 225–37.

Winterbottom, J. M. (1973). Note on the ecology of *Serinus* spp. in the Western Cape. *Ostrich*, 44, 31–3.

Winterbottom, J. M. (1975). Notes on the South African species of *Corvus*. *Ostrich*, 46, 236–50.

Winterbottom, J. M. and Rowan, M. K. (1962). Effect of rainfall on breeding of birds in arid areas. *Ostrich*, 33, 77–8.

Withers, P. C. (1979). Ecology of a small mammal community on rocky outcrop in the Namib Desert. *Madoqua*, 11, 229–46.

Withers, P. C. (1983a). Seasonal reproduction by small mammals of the Namib Desert. *Mammalia*, 47, 195–204.

Withers, P. C. (1983b). Energy, water, and solute balance of the Ostrich *Struthio camelus*. *Physiological Zoology*, 56, 568–79.

Withers, P. C. (1992). *Comparative animal physiology*. Orlando, FL: Saunders College.

Withers, P. C., Louw, G. N. and Henschel, J. (1980). Energetics and water relations of Namib desert rodents. *South African Journal of Zoology* 15, 131–7.

Withers, P. C., Siegfried, W. R. and Louw, G. N. (1981). Desert ostrich exhales unsaturated air. *South African Journal of Science*, 77, 569–70.

Wolfe, J. A. (1985). Distribution of major vegetational types during the Tertiary. In *The carbon cycle and atmospheric CO_2: Natural variations Archaean to present*, ed. E. T. Sundquist and W. S. Broecker, pp. 357–75. Washington, DC: American Geophysical Union.

Wolfram, S. 1986. *Theory and applications of cellular automata*. Singapore: World Science Publications.

Woodburn, J. (1988). African hunter-gatherer social organisation: is it best

understood as a product of encapsulation? In *Hunters and gatherers I: History, evolution and social change*, ed. T. Ingold, D. Riches and J. Woodburn, pp. 31–64. Oxford: Berg.

Woodell, S. R. J. and King, T. J. (1991). The influence of mound-building ants on British lowland vegetation. In *Ant–plant interactions*, ed. C. R. Huxley and D. F. Cutler, pp. 521–35. Oxford University Press.

Woodward, F. I. (1987). Climate and plant distribution. *Nature*, 329, 25.

Wright, D. H. (1983). Species-energy theory an extension of species-area theory. *Oikos*, 41, 495–506.

Wulfse, A. D. and Holdsworth, R. (1994). Application of regional geochemical surveys to environmental studies; a case study from Namaqualand copper district. *Journal of African Earth Sciences*, 18, 343–6.

Wullstein, L. H., Bruening, M. L. and Bollen, W. B. (1979). Nitrogen fixation associated with sand grain root sheaths (rhizosheaths) of certain xeric grasses. *Plant Physiology*, 46, 1–4.

Yeaton, R. I. (1988). Structure and function of the Namib dune grasslands. characteristics of the environmental gradients and species distributions. *Journal of Ecology*, 76, 744–58.

Yeaton, R. I. and Cody, M. L. (1976). Competition and spacing in plant communities of the northern Mojave Desert. *Journal of Ecology*, 64, 686–96.

Yeaton, R. I. and Esler, K. J. (1990). The dynamics of a succulent Karoo vegetation: a study of species association and recruitment. *Vegetatio*, 88, 103–13.

Yeaton, R. I., Moll, E. J., Jarman, M. L. and Cunliffe, R. N. (1993). The impact of competition on the structure of early successional plant species of the Atlantic Coast of South Africa. *Journal of Arid Environments*, 25, 211–19.

Young, J. A. (1991). Tumble weed. *Scientific American*, 264, 58–63.

Young, J. A. and Evans, R. A. (1972). Germination and establishment of *Salsola* in relation to seedbed environment. 1. Temperature, afterripening, and moisture relations of *salsola* seeds as determined by laboratory studies. *Agronomy Journal*, 64, 214–18.

Zedler, P. H. (1981). Vegetation change in chaparral and desert communities in

San Diego County, California. In *Forest succession. Concepts and applications*, ed. D. C. West, H. H. Shugart and D. B. Botkin, pp. 406–30. Berlin: Springer-Verlag.

Zedler, P. H. and Scheid, G. A. (1988). Seedling establishment of *Carpobrotus edulis* and *Salix lasiolepis* in a coastal strand at Morro Bay, California. *Bulletin of the Torrey Botanical Club*, 111, 145–52.

Zimmermann, H. G. (1991). Biological control of mesquite, *Prosopis* spp. (Fabaceae), in South Africa. *Agriculture, Ecosystems and Environment*, 37, 175–86.

Zimmermann, H. G. and Malan, D. E. (1980). Present status of prickly pear (*Opuntia ficus-indica* (L) Mill) control in South Africa. In *Proceedings of the Third National Weeds Conference of South Africa*, pp. 79–85. Pretoria, South Africa, Weeds Society of South Africa.

Zimmermann, H. G. and Malan, D. E. (1981). The role of imported natural enemies in suppressing regrowth of prickly pear, *Opuntia ficus-indica*, in South Africa. In *Proceedings of the Fifth International Symposium on the Biological Control of Weeds*, ed. E. S. Delfosse, pp. 375–81. Canberra, Australia: CSIRO.

Zimmermann, H. G. and Moran, V. C. (1991). Biological control of prickly pear, *Opuntia ficus-indica* (Cactaceae), South Africa. *Agriculture, Ecosystems and Environment*, 37, 29–35.

Zohary, M. (1937). Die verbreitungsökologischen Verhältnisse der Pflanzen Palestinas. 1. *Beihefte zum Botanischen Zentralblatt*, A., 1–155.

Zucchini, W. (1992). Program GENRAIN. Dept of Statistical Sciences, University of Cape Town.

Zucchini, W., Adamson, P. and McNeill, L. (1992). A model of southern African rainfall. *South African Journal of Science*, 88, 103–9.

Index